HYDROSTATICS

HYDROSTATICS

Dr Binoy Bhattacharyya PhD (Moscow), MIE
Formerly Senior Scientific Officer
Structural Engineering Research Centre, Roorkee and Chennai
Formerly Superintending Engineer (Civil)
Damodar Valley Corporation, Kolkata

New Central Book Agency (P) Ltd
LONDON
HYDERABAD ERNAKULAM BHUBANESWAR
DELHI KOLKATA PUNE GUWAHATI

NCBA

REGD OFFICE

8/1 Chintamoni Das Lane, Kolkata 700 009, India
ncbapvtltd@eth.net

OVERSEAS

NCBA (UK) Ltd, 149 Park Avenue North, Northampton, NN3 2HY, UK
ncbauk@yahoo.com

EXPORT

NCBA Exports Pvt. Ltd, 212 Shahpur Jat, New Delhi 110 049
ncbaexports@ncbaindia.com

BRANCHES

208B Peacock Lane, Shahpur Jat
New Delhi 110 049
ncbadel@ncbapvtltd.com

Shop Nos. 3 & 4, Vinayak Towers
681/B Budhwar Peth, Appa Balwant Chowk
Pune 411 002
ncbapune@ncbapvtltd.com

Shop No. 15, Bharati Towers, Ground Floor
Forest Park, **Bhubaneswar** 751 001
ncbabub@ncbapvtltd.com

House No. 3-1-315, 1st Floor
Nimboliadda, Kachiguda, **Hyderabad** 500 027
ncbahydb@ncbapvtltd.com

GSS Shopping Complex, 1st Floor
Opposite College Post Office, Convent Road
Ernakulam 682 035
ncbaernk@ncbapvtltd.com

Radhamayee Complex (Second Floor)
Opp. SBI, Panbazar, Jaswanta Road
Guwahati 781 001
ncbaguwahati@ncbapvtltd.com

HYDROSTATICS • Dr Binoy Bhattacharyya

First Published: 2013

PUBLISHER & TYPESETTER
New Central Book Agency (P) Ltd
8/1 Chintamoni Das Lane, Kolkata 700 009

PRINTER
New Central Book Agency (P) Ltd
Web-Offset Division, Dhulagarh, Sankrail, Howrah

COVER DESIGNER
Soumen Paul

COVER PRINTER
Techno Process, Kolkata

PROOF CHECKING
Dipan Roy, and Author

PROJECT TEAM
Subir Pal, and Prabhat Jas

ISBN 978 81 7381 801 1

Price ₹ 350.00

To

Munna

My dearest sister-in-law

Contents

Preface

This book on "Hydrostatics" has been written to fulfil the requirements of Mathematics Honours students of Indian Universities. The book will also cater to the needs of those who aspire to sit for various competitive examinations in India.

Some theoretical portions in the book have been discussed in a modern style in order to update the concepts. The role of gravitating liquids has been introduced and discussed in details.

In contrast with other books available, topics on "Surface Tension" and "Hydrostatic Machines" have been discussed in detail for the benefit of those who really have interest in the subject. Every chapter in this book has been enriched with sufficient number of worked-out examples which will help the students to elevate their confidence level in solving problems.

I will feel honoured if the book becomes useful to the students and teachers of different academic institutions.

I am thankful to Sri Amitabha Sen, Director, New Central Book Agency (P) Ltd, for taking interest in speedy publication of the book. Thanks are also due to Sri Prabhat Jas and Sri Subir Pal for their constant cooperation in bringing the book to light.

Valuable comments and suggestions from any corner will be gratefully accepted.

28 July 2012
Tollygunge
Kolkata 700 040

Dr. Binoy Bhattacharyya

CHAPTER 1

Fundamentals

1.1 Introduction

Hydrostatics is that branch of mechanics which deals with fluids at rest under different forces. It is concerned with the conditions of relative equilibrium of fluids and of solid bodies wholly or partially immersed in a fluid at rest. The subject also deals with the reaction forces exerted by the fluids upon solid bodies which are in contact with them.

1.2 Definition of a Fluid

In nature, the matters exist in three states—solid, liquid and gaseous. In solids, intermolecular attraction forces are very strong due to which they are rigid. In liquids, these forces are weak and in gases, these forces are extremely small. Hence, in liquids and gases, the molecules have some mobility relative to each other.

Liquids and gases are combined to give a common name of fluids. Fluids have no definite shape like solids—they take the shape of the vessels where they are stored.

Liquids offer great resistance to compression and are not greatly affected by change of temperature. Gases can be easily compressed and are susceptible to temperature changes.

The coefficient of thermal expansion of liquids are small and can be ignored when the liquids are subjected to a small range of temperature difference. We, therefore, will assume no effect of temperature on the physical properties of liquids.

1.3 Normal and Shearing Stress

We consider a body ABC remaining at rest on a horizontal plane (FIG. 1.1). We imagine a plane DE passing through the body and separating it into two portions—upper part DBE and the lower part $ADEC$.

Let W be the weight of the upper part DBE acting vertically downwards and S be the area of the surface at DE common to both parts.

The weight W is balanced by the two forces N and T acting along directions normal and tangential to the plane DE respectively.

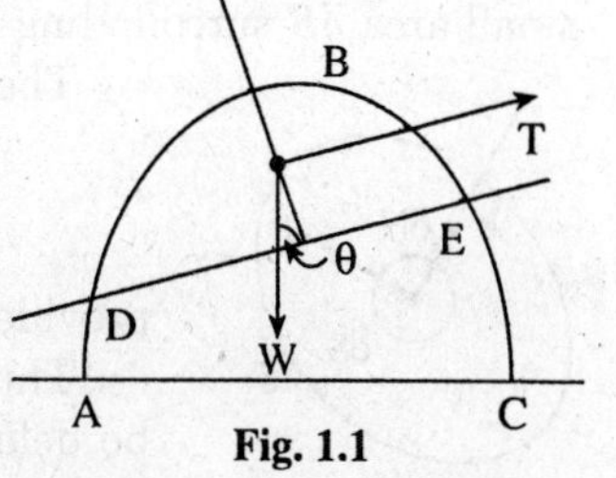

Fig. 1.1

Then

$$\left.\begin{aligned} N &= W\sin\theta \\ T &= W\cos\theta \end{aligned}\right\} \tag{1.1}$$

where θ is the angle between the plane DE and the vertical.

The ***normal stress*** per unit area is defined as

$$\frac{N}{S} = \frac{W\sin\theta}{S} \tag{1.2}$$

and the ***shearing stress*** per unit area is

$$\frac{T}{S} = \frac{W\cos\theta}{S}. \tag{1.3}$$

The shearing stress prevents the upper part DBE from sliding down the plane ED. But this is true in the case of solids only. In the case of fluids, the upper part DBE will not remain in equilibrium, but, as we experience, will flow down the plane DE. This is so, because fluids in equilibrium cannot resist shearing stress. Thus we give another definition of fluids:

A fluid is a substance which will yield to any continued shearing stress, however small it may be.

1.4 Pressure

We have seen that on any plane across a fluid mass, there acts only the normal thrust N. This normal thrust is said to be ***uniform*** when the thrust on any part of the area is proportional to the area of that part. In this case, the pressure is defined to be equal to the total normal thrust N divided by the area S on which it is acting. Thus,

$$p = \frac{N}{S}. \tag{1.4}$$

Hence the pressure is actually equivalent to the normal thrust per unit area.

If, however, the normal thrust is not uniform, the quantity in the RHS of Eqn (1.4) defines ***average*** or ***mean pressure***, and we write

$$p_{\text{av}} = \frac{N}{S}. \tag{1.5}$$

Thus, the average pressure on a given plane surface is the uniform pressure which would produce the same normal thrust as the actual pressure produces.

1.5 Pressure at a Point

We consider a point P on the surface S which is subjected to a normal thrust N. We take a small area δS surrounding the point P. Let δN be the normal thrust on the small area δS.

Then the pressure at the point P is defined as (FIG. 1.2)

$$p = \lim_{\delta S \to 0} \frac{\delta N}{\delta S}, \tag{1.6}$$

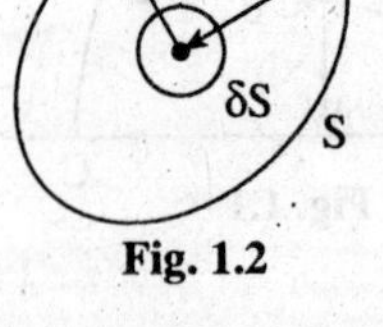

Fig. 1.2

provided that the limit exists.

Thus, the pressure at a point of a surface subjected to a fluid thrust may be defined as the limit of the mean pressure on a small area surrounding the point, as the area is diminished indefinitely.

Now, we can draw an infinite number of plane surfaces surrounding a point. Hence the value of the pressure will be unique only if the quantity in the RHS of Eqn (1.6) is independent of the orientation of the surface S. We will, however, see below that the orientation of the surface S is immaterial.

1.6 Equality of Pressure in all Directions

We consider a prismatic element in a fluid at rest as shown in the FIG. 1.3. Let the average pressures in x-, y- and z-directions be p_x, p_y and p_z respectively. Let p_n be the average pressure on the inclined surface of length δs, the angle of inclination with the vertical being θ.

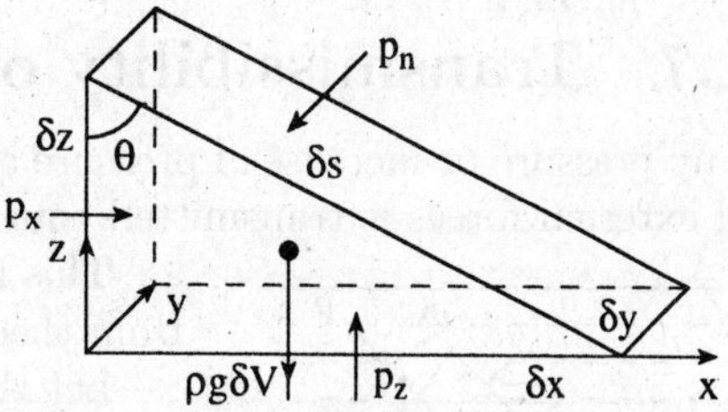

Fig. 1.3

Now we write the equations of equilibrium for the prismatic element along the x-, y- and z-coordinate axes. The forces acting on the element are:

(i) $p_x(\delta y \cdot \delta z)$ on the vertical surface

(ii) $p_z(\delta x \cdot \delta y)$ on the bottom horizontal surface

(iii) $p_n(\delta s \cdot \delta y)$ on the top inclined surface, and

(iv) $g\rho\delta V$, the weight of the prism acting through its centre of gravity along vertically downward direction. Here δV is the volume of the prism, ρ is the density of fluid and g is the acceleration due to gravity.

The forces acting on the vertical triangular surfaces on two sides will be equal in magnitude and opposite in directions and hence they balance each other.

The equations of equilibrium are:

$$\left.\begin{aligned} p_x(\delta y \cdot \delta z) - p_n(\delta s \cdot \delta y)\cos\theta = 0, &\text{ along } x\text{-direction} \\ p_z(\delta x \cdot \delta y) - p_n(\delta s \cdot \delta y)\sin\theta - \rho g \delta V = 0, &\text{ along } z\text{-direction} \end{aligned}\right\} \tag{1.7}$$

We note from the figure that

$$\delta s \cos\theta = \delta z. \tag{1.8}$$

The first of Eqns (1.7) with the use of relation (1.8) yields

$$p_x = p_n. \tag{1.9}$$

Again, the volume of the elementary prism is

$$\delta V = \frac{1}{2}\delta x \cdot \delta y \cdot \delta z. \tag{1.10}$$

Since

$$\delta x = \delta s \sin\theta, \tag{1.11}$$

we get from the second of Eqns (1.7),

$$p_z - p_n - \frac{1}{2}\rho g \delta z = 0. \tag{1.12}$$

In the limit $\delta z \to 0$ when the prismatic element is shrunk to zero surface area, Eqn (1.12) yields

$$p_z = p_n. \tag{1.13}$$

The total force on the inclined surface remains on the xz-plane. By rotating the prism in such a way that the total force on the inclined surface remains on yz-plane, we can show that

$$p_y = p_n. \tag{1.14}$$

Thus we get that at a point within a fluid at rest,

$$p_x = p_y = p_z = p_n. \tag{1.15}$$

Hence, ***the pressure at any point within a fluid at rest remains the same in all directions***.

1.7 Transmissibility of Fluid Pressure

Any pressure or increase of pressure applied to any point of a fluid at rest under a given system of external forces is transmitted equally to every other point of the fluid.

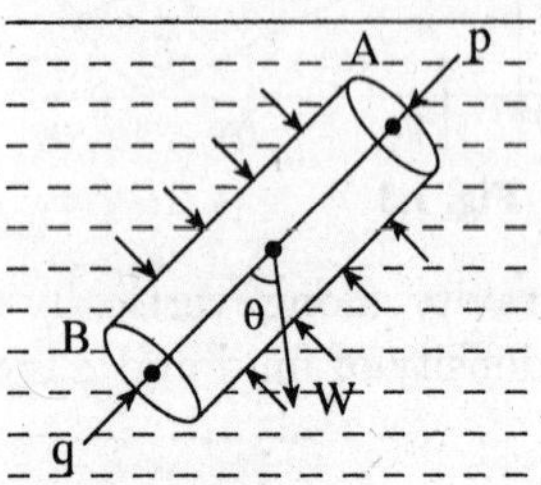

Fig. 1.4

This is known as ***Pascal's Law***. Let us prove the proposition from theoretical considerations.

Let A and B be two points within a fluid at rest under gravity. We join A and B and construct a cylinder of small cross-sectional area S with AB as its axis. The end planes at A and B are at right angles to the axis AB (FIG. 1.4).

Let p and q be the fluid pressures at the points A and B respectively. Then the cylinder of fluid remains in equilibrium under the action of following forces:

(i) Fluid thrust pS on the plane A along AB

(ii) Fluid thrust qS on the plane B along BA

(iii) Weight W of the cylinder of fluid acting vertically downwards and

(iv) Pressure on the curved surface of the cylinder, which is, at every point, perpendicular to the axis AB. Hence its component along AB is zero.

We write the equation of equilibrium of the fluid cylinder along AB in the form

$$pS - qS + W\cos\theta = 0, \tag{1.16}$$

where θ is the angle which the axis AB makes with the vertical as shown in the figure.

We now apply an additional pressure δp at the end A. Let there be an increase of pressure of an amount δq at the end B due to the additional pressure at A. Then the forces acting at the end faces of the fluid-cylinder are $(p + \delta p)S$ and $(q + \delta q)S$. The equation of equilibrium in this case assumes the form

$$(p + \delta p)S - (q + \delta q)S + W\cos\theta = 0. \tag{1.17}$$

Using Eqn (1.16), we find from Eqn (1.17),

$$\delta pS - \delta qS = 0, \text{ i.e., } \delta p = \delta q. \tag{1.18}$$

This implies that, increase of pressure at A = increase of pressure at B. Hence the proposition.

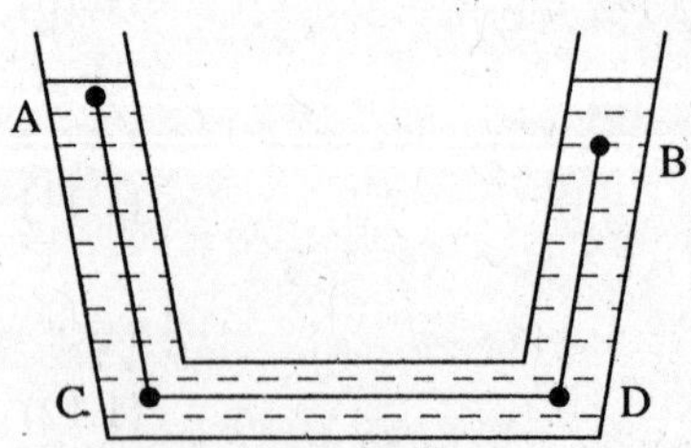

Fig. 1.5

Now, the points A and B may not be joined by a straight line lying wholly in the liquid as shown in the (FIG. 1.5). In this case, the points can be joined by a succession of lines AC, CD and DB which lie in the fluid.

Then following the above reasoning, we write, increase of pressure at A

$$= \text{increase of pressure at } C$$
$$= \text{increase of pressure at } D$$
$$= \text{increase of pressure at } B.$$

1.8 Compressibility of Fluids

Fluids subjected to external pressure undergo a change in volume. Compressibility is a measure of this volume change and the coefficient of compressibility η is the reciprocal of bulk modulus of elasticity β of the substance.

If a substance of volume V under external pressure dp undergoes a change in volume dV, then the bulk modulus of elasticity is defined as

$$\beta = -\frac{dp}{dV/V} = -V\frac{dp}{dV}. \tag{1.19}$$

Then the compressibility of the substance is

$$\eta = -\frac{1}{V}\frac{dV}{dp}. \tag{1.20}$$

The compressibility of liquids is very small. For practical purposes, ***all liquids are considered incompressible***.

For an ideal gas undergoing isothermal process, we have

$$pV = RT = \text{constant}. \tag{1.21}$$

Differentiating, we get

$$pdV + Vdp = 0,$$

which yields

$$-\frac{dV}{V} = \frac{dp}{p}. \tag{1.22}$$

From Eqns (1.19) and (1.22), we get the bulk modulus of elasticity of an ideal gas for an isothermal process as

$$\beta = p, \tag{1.23}$$

that is, equal to pressure itself.

For an ideal gas undergoing an adiabatic process, we have

$$pV^\gamma = \text{constant}, \tag{1.24}$$

where γ is the ratio of specific heat of the gas at constant pressure to the specific heat at constant volume. For any particular gas, $\gamma = c_p/c_v$ is constant. For air, $\gamma = 1 \cdot 4$.

Differentiating Eqn (1.24), we get

$$V^\gamma dp + \gamma p V^{\gamma-1} dV = 0,$$

which yields

$$\frac{dp}{p} = -\gamma\frac{dV}{V}. \tag{1.25}$$

From Eqns (1.19) and (1.25), we get the bulk modulus of an ideal gas undergoing adiabatic process as

$$\beta = \gamma p, \tag{1.26}$$

that is, γ times the pressure.

1.9 Viscosity of Fluids

Fluids undergo continuous deformation due to externally applied shearing forces. During this process, molecular frictional forces appear within all real fluids which oppose the continuous deformation. This property of fluids by virtue of which they resist the continuous deformation under external shearing forces is called ***viscosity***.

Let us consider a thin layer of fluid of thickness dy in between two plates (FIG. 1.6). The lower plate is kept fixed while the upper one is movable. A tangential shearing force F is applied to the upper plate in horizontal direction. Due to the action of the shearing force F, the upper plate moves steadily with a velocity du relative to the lower plate, so that a vertical line AC in the fluid gets displaced to BC in time dt.

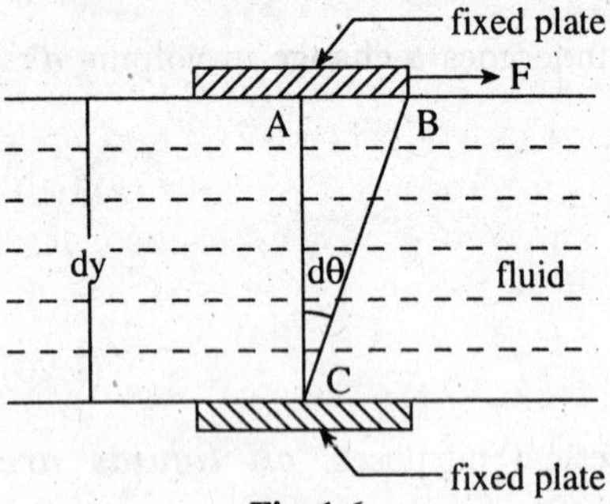

Fig. 1.6

Then

$$AB = (du)dt = (dy)d\theta, \text{ i.e., } \frac{d\theta}{dt} = \frac{du}{dy}. \tag{1.27}$$

If S be the area of the upper plate in contact with the fluid, then the applied shear stress is

$$\tau = \frac{F}{S}. \tag{1.28}$$

For a Newtonian fluid, the external shear stress is proportional to the time rate of shear strain. Thus

$$\tau \propto \frac{d\theta}{dt}, \text{ or, } \tau = \mu\frac{d\theta}{dt} = \mu\frac{du}{dy}, \tag{1.29}$$

where μ is known as ***Newtonian coefficient of viscosity*** or ***dynamic viscosity*** or simply ***viscosity***.

For ideal fluids, there is no viscosity and hence $\mu = 0$. Thus ideal fluids offer no resistance to shear deformation. But real fluids have finite viscosity and hence shearing stresses exist in the fluid during its flow and deformation.

The concept of viscosity, as we have seen, is associated with the relative motion of fluid layers. Hence ***viscosity has no influence on the equilibrium of fluid at rest under a given system of external forces.***

1.10 Mass, Weight and Density

The mass of a body is defined as the quantity of matter contained in that body.

The weight of a body of mass m is obtained by taking the product of mass by the acceleration due to gravity g.

The density of a homogeneous substance is its mass per unit volume.

Thus, if a homogeneous substance of mass m occupies a volume V, then its weight W and density ρ are given by

$$W = mg, \ \rho = \frac{m}{V}, \ W = V\rho g. \tag{1.30}$$

If, however, the substance is heterogeneous, then the density at a point within the substance is defined as

$$\rho = \lim_{\delta V \to 0} \frac{\delta m}{\delta V}, \tag{1.31}$$

where δm is the mass of a small portion of substance occupying volume δV containing that point. For a heterogeneous body, the density varies from point to point while for a homogeneous body, it is constant.

In F.P.S. system, unit of density is pounds per cubic foot and in C.G.S. system, it is grammes per cubic centimetre.

The dimensions of density in terms of fundamental units is

$$\frac{M}{V} = \frac{M}{L^3} = ML^{-3}.$$

1.11 Specific Gravity

The specific gravity of a substance is the ratio of the density of the substance to the density of a standard substance. For liquids, pure water at a temperature of 4 °C when it has maximum density, is usually taken as the standard substance. For gases, hydrogen at N.T.P. is taken as the standard substance. Thus, the specific gravity being the ratio of two densities, is a pure number independent of units.

The specific gravity may also be defined as the ratio of the weight of a given volume of a substance to the weight of an equal volume of the standard substance. If ρ and δ denote the densities of a given substance and the standard substance respectively, then the specific gravity of the substance is

$$\begin{aligned} s &= \frac{\rho}{\delta} = \frac{\text{mass of unit volume of substance}}{\text{mass of unit volume of standard substance}} \\ &= \frac{\rho V}{\delta V} = \frac{m}{m'} = \frac{\text{mass of any volume of substance}}{\text{mass of same volume of standard substance}} \\ &= \frac{g\rho V}{g\delta V} = \frac{gm}{gm'} = \frac{W}{W'} = \frac{\text{weight of any volume of substance}}{\text{weight of same volume of standard substance}}. \end{aligned}$$

This gives the relation

$$W = sW'. \tag{1.32}$$

If w denotes the weight of unit volume of the standard substance, then the weight of V volume of a substance of specific gravity s is

$$W = Vsw. \tag{1.33}$$

"The specific gravity of mercury is 13.6"—it means that mercury of some volume is 13.6 times heavier than equal volume of water at 4 °C.

1.12 Specific Gravity of Mixtures

We consider a mixture of n number of substances whose specific gravities are $s_1, s_2, \ldots, s_n$. We assume that there does not occur any chemical reaction when the substances are mixed together.

Case 1. *When Volumes are given.*

Let $V_1, V_2, \ldots, V_n$ be the volumes of different fluids which are mixed together. Let w be the weight of a unit volume of standard substance. Then the weight of the mixture is

$$W_1 = \sum_{i=1}^{n} V_i s_i w = w \sum_{i=1}^{n} V_i s_i. \tag{1.34}$$

Let s be the specific gravity of the mixture. Assuming no expansion or shrinkage due to mixing of the fluids, the total volume of the mixture will be

$$V = \sum_{i=1}^{n} V_i. \tag{1.35}$$

The weight of the mixture is

$$W_2 = Vsw = \left(\sum V_i\right) sw. \tag{1.36}$$

The two weights in Eqns (1.34) and (1.36) must be equal. Hence

$$\left(\sum V_i\right) sw = w \sum V_i s_i,$$

which yields

$$s = \frac{\sum V_i s_i}{\sum V_i}. \tag{1.37}$$

Let on mixing the different fluids, there happens expansion or shrinkage due to some reason or other. In this case, if U be the total volume of the mixture, then by the reasoning as above, we have

$$Usw = w \cdot \sum V_i s_i$$

which yields

$$s = \frac{\sum V_i s_i}{U}. \tag{1.38}$$

Here we are to note that

$$U = \xi \sum V_i = \xi V, \tag{1.39}$$

where ξ may be greater or less than unity.

If, in particular,

$$V_1 = V_2 = \cdots = V_n = V^* \text{ (say)} \tag{1.40}$$

then, from Eqn (1.37), we get

$$s = \frac{V^* \sum s_i}{nV^*} = \frac{\sum s_i}{n}. \tag{1.41}$$

Thus, the specific gravity of a mixture of equal volumes of different substances, when there happens no change in volume due to mixing, is the arithmetic mean of the specific gravities of the given substances.

If there is a change in volume, the specific gravity of a mixture of equal volumes of different substances is from Eqns (1.38) to (1.40),

$$s = \frac{\sum V_i s_i}{\xi \sum V_i} = \frac{V^* \sum s_i}{\xi n V^*} = \frac{\sum s_i}{\xi n}. \tag{1.42}$$

Case 2. *When weights are given.*

Let $W_1, W_2, \ldots, W_n$ be the weights of n different substances of specific gravities $s_1, s_2, \ldots, s_n$. Then the volumes of the different substances are respectively

$$\frac{W_1}{s_1 w}, \frac{W_2}{s_2 w}, \ldots, \frac{W_n}{s_n w}.$$

The total volume of the mixture, when there is no expansion or shrinkage due to mixing, is

$$V = \sum \frac{W_i}{s_i w} = \frac{1}{w} \sum \frac{W_i}{s_i}. \tag{1.43}$$

Hence, the total weight of the mixture is

$$W = Vsw = \left(\frac{1}{w} \sum \frac{W_i}{s_i}\right) sw = s \sum \frac{W_i}{s_i}, \tag{1.44}$$

where as before, s is the specific gravity of the mixture and w is the weight of unit volume of the standard substance.

But the total weight of the mixture, as given, is

$$W = W_1 + W_2 + \cdots + W_n = \sum W_i. \tag{1.45}$$

Equating the R.H.S. quantities in Eqns (1.44) and (1.45), we get

$$s \sum \frac{W_i}{s_i} = \sum W_i.$$

This yields

$$s = \frac{\sum W_i}{\sum (W_i/s_i)}. \tag{1.46}$$

Let on mixing the different substances, there happens a change in volume due to some reason or other. In this case, if U be the total volume of the mixture, then we can have a relation as in Eqn (1.39). In this case, the specific gravity of the mixture is

$$s = \frac{\sum W_i}{\xi V w} \quad \text{[by Eqn (1.43)]}. \tag{1.47}$$

If, in particular,

$$W_1 = W_2 = \cdots = W_n = W^* \text{ (say)}, \tag{1.48}$$

then, from Eqn (1.46), we get

$$s = \frac{nW^*}{W^* \left(\sum \frac{1}{s_i}\right)} = \frac{n}{\sum \frac{1}{s_i}}. \tag{1.49}$$

Thus, the specific gravity of a mixture of equal weights of different substances, when there happens no change in volume due to mixing, is the harmonic mean of the specific gravities of the given substances.

If the change in volume due to mixing is considered, then the specific gravity of the mixture of equal weights of different substances is given by

$$s = \frac{\sum W_i}{\xi V w} = \frac{nW^*}{\xi W^* \sum \frac{1}{s_i}} \quad \text{[by Eqn (1.43)]}$$

$$\text{i.e., } s = \frac{n}{\xi \sum \frac{1}{s_i}}, \tag{1.50}$$

where the constant quantity ξ is $>$ or $<$ 1.

EXAMPLE 1.1 *If m be the mass of a substance and ρ its density, then show that its compressibility is*

$$\eta = \frac{1}{\rho}\frac{d\rho}{dp}.$$

Solution. We know that the bulk modulus of a substance occupying a volume V and subjected to pressure p is

$$\beta = -V\frac{dp}{dV}. \tag{1}$$

Since its mass is m, we have

$$V\rho = m = \text{constant}. \tag{2}$$

Differentiating Eqn (2), we get

$$Vd\rho + \rho dV = 0,$$

which yields

$$-\frac{dV}{V} = \frac{d\rho}{\rho}. \tag{3}$$

Using Eqn (3) in Eqn (1), we get

$$\beta = \rho\frac{dp}{d\rho}. \tag{4}$$

Hence the compressibility of the substance is

$$\eta = \frac{1}{\rho}\frac{d\rho}{dp} \quad \textbf{(Proved)}.$$

EXAMPLE 1.2 *There are n different substances of densities $\rho_1, \rho_2, \ldots, \rho_n$. They are mixed by volumes and by weights. Determine the density of the mixture in the two cases, assuming no change in volume due to mixing.*

Solution. In the first case, let $V_1, V_2, \ldots, V_n$ be the volumes of the substances. Then total volume of the mixture is

$$V = \sum_{i=1}^{n} V_i. \tag{1}$$

Total mass of the substances mixed is

$$M = \sum V_i\rho_i. \tag{2}$$

If ρ be the density of the mixture, then its mass is

$$M = V\rho = \left(\sum V_i\right)\rho. \tag{3}$$

From Eqns (2) and (3), we get

$$\left(\sum V_i\right)\rho = \sum V_i\rho_i. \tag{4}$$

Hence

$$\rho = \frac{\sum V_i\rho_i}{\sum V_i}. \tag{5}$$

In the second case, let $W_1, W_2, \ldots, W_n$ be the weights of the substances. Then their volumes are

$$\frac{W_1}{g\rho_1}, \frac{W_2}{g\rho_2}, \ldots, \frac{W_n}{g\rho_n}.$$

The total volume of the mixture is

$$V = \frac{1}{g}\sum \frac{W_i}{\rho_i}. \tag{6}$$

If ρ be the density of the mixture, then its total weight is

$$W = V\rho g = \left(\sum \frac{W_i}{\rho_i}\right)\rho. \tag{7}$$

But the total weight given is

$$W = W_1 + W_2 + \cdots + W_n = \sum W_i. \tag{8}$$

From Eqns (7) and (8), we have

$$\left(\sum \frac{W_i}{\rho_i}\right)\rho = \sum W_i,$$

which yields

$$\rho = \frac{\sum W_i}{\sum \frac{W_i}{\rho_i}}.$$

EXAMPLE 1.3 *Show that the specific gravity of a mixture of n liquids is greater, when equal volumes are taken than when equal weights are taken, assuming no change in the volume as the result of mixing.*

Solution. Let s be the specific gravity of a mixture of n liquids when equal volumes are taken. Then, by Eqn (1.41)

$$s = \frac{\sum s_i}{n}. \tag{1}$$

Similarly, if s' be the specific gravity of a mixture of n liquids when equal weights are taken, then by Eqn (1.49),

$$s' = \frac{n}{\sum \frac{1}{s_i}}. \tag{2}$$

From Eqns (1) and (2), we find that s is the arithmetic mean (A.M.) and s' is the harmonic mean (H.M.) of positive quantities $s_1, s_2, \ldots, s_n$. But we know that A.M. of n distinct positive numbers is greater than their H.M. Hence $s > s'$.

EXAMPLE 1.4 *A mixture is formed of equal volumes of n liquids, the densities of which are in the ratio of the numbers* $1, 2, 3, \ldots, n$. *Find the density of the mixture.*

Also find the density of the mixture when the volumes are in the ratio (a) of the numbers $1, 2, 3, \ldots, n$ *and (b) of the numbers* $n, (n-1), (n-2), \ldots, 3, 2, 1$.

Solution. 1st part : Let the volume of each of n liquids be V and the densities of n liquids be $\rho, 2\rho, 3\rho, \ldots, n\rho$ respectively. Then the density δ of the mixture is given by

$$\delta = \frac{V\cdot\rho + V\cdot 2\rho + \cdots + V\cdot n\rho}{V + V + \cdots + V} = \frac{V(1+2+\cdots+n)\rho}{nV}$$

$$= \frac{n(n+1)\rho}{2n} = \frac{1}{2}(n+1)\rho. \tag{1}$$

Hence the density of the mixture is in the ratio of the number $\frac{1}{2}(n+1)$.

2nd part (a) : Let the volumes of n liquids be $V, 2V, 3V, \ldots, nV$. The densities of n liquids are, as before $\rho, 2\rho, 3\rho, \ldots, n\rho$. Then the density of the mixture is given by

$$\begin{aligned}\delta &= \frac{V\cdot\rho + 2V\cdot 2\rho + \cdots + nV\cdot n\rho}{V + 2V + \cdots + nV}\\ &= \frac{1^2+2^2+\cdots+n^2}{1+2+\cdots+n}\rho = \frac{\frac{1}{6}n(n+1)(2n+1)}{\frac{1}{2}n(n+1)}\rho\\ &= \frac{1}{3}(2n+1)\rho. \end{aligned} \tag{2}$$

Thus, in this case, the density of the mixture is in the ratio of the number $\frac{1}{3}(2n+1)$.

(b) Let the volumes of n liquids be $nV, (n-1)V, (n-2)V, \ldots, 2V, V$. Then the density of the mixture is given by

$$\begin{aligned}\delta &= \frac{nV\cdot\rho + (n-1)V\cdot 2\rho + \cdots + 2V\cdot(n-1)\rho + V\cdot n\rho}{nV + (n-1)V + \cdots + 2V + V}\\ &= \frac{n\cdot 1 + (n-1)\cdot 2 + (n-2)\cdot 3 + \cdots + 2\cdot(n-1) + 1\cdot n}{n + (n-1) + \cdots + 2 + 1}\rho \end{aligned} \tag{3}$$

Let S be the sum of the series in the numerator. Then,

$$S = n\cdot 1 + (n-1)\cdot 2 + (n-2)\cdot 3 + \cdots + 2\cdot(n-1) + 1\cdot n. \tag{4}$$

The r-th term of this series is given by

$$t_r = (n-r+1)\cdot r = nr - r^2 + r. \tag{5}$$

Putting $r = 1, 2, \ldots, n$, we get

$$\begin{aligned} t_1 &= 1\cdot n - 1^2 + 1\\ t_2 &= 2\cdot n - 2^2 + 2\\ &\cdots\cdots\cdots\cdots\cdots\\ t_n &= n\cdot n - n^2 + n \end{aligned}$$

Hence, adding all the terms, we get

$$\begin{aligned} S &= n(1+2+3+\cdots+n) - (1^2+2^2+3^2+\cdots+n^2) + (1+2+3+\cdots+n)\\ &= n\cdot\frac{n(n+1)}{2} - \frac{1}{6}n(n+1)(2n+1) + \frac{n(n+1)}{2} = \frac{1}{6}n(n+1)(n+2). \end{aligned} \tag{6}$$

Hence, the density of the mixture is

$$\delta = \frac{\frac{1}{6}n(n+1)(n+2)}{\frac{n(n+1)}{2}}\rho, \text{ i.e., } \delta = \frac{1}{3}(n+2)\rho. \tag{7}$$

Thus the density of the mixture is in the ratio of the number $\frac{1}{3}(n+2)$.

EXAMPLE 1.5 *Pure water is added, drop by drop, to a vessel of volume V, filled with a salt solution of specific gravity s which is allowed to overflow.*

Find the specific gravity of the solution when a volume v of the water has been poured in.

Solution. Let ρ be the specific gravity of the mixture when a volume v of water has been poured in. The specific gravity of the mixture becomes $\rho + d\rho$ when a volume of water equal to $v + dv$ is added, so that dv may be taken to denote the volume of a drop of water.

At this stage, if a drop of volume dv and specific gravity 1 is added to a volume V of specific gravity ρ, then the resulting volume becomes $V + dv$ of specific gravity $\rho + d\rho$. Consequently, a drop overflows.

Since the weight of the mixture is equal to the sum of the component parts, we have

$$(V + dv)(\rho + d\rho) = 1 \cdot dv + \rho V,$$

which, after ignoring the small quantity $dvd\rho$, yields

$$V d\rho + (\rho - 1)dv = 0, \text{ i.e., } \frac{d\rho}{\rho - 1} + \frac{dv}{V} = 0. \tag{1}$$

Integrating, we get

$$\ln(\rho - 1) + \frac{v}{V} = C, \tag{2}$$

where C is a constant.

Initially, $v = 0$ and $\rho = s$. Putting these values in Eqn (2), we find

$$C = \ln(s - 1). \tag{3}$$

Substituting this value of C into Eqn (2), $\ln(\rho - 1) = \ln(s - 1) - \dfrac{v}{V}$.

Hence $\rho = 1 + (s - 1)\exp\left(-\dfrac{v}{V}\right)$.

EXAMPLE 1.6 *A vessel contains a certain volume of a liquid of specific gravity s_1 and another contains an equal volume of a liquid of specific gravity s_2. From each vessel, $\frac{1}{n}$th of volume is taken out and put into the other and well mixed. This process is repeated m times. Show that the final specific gravities are*

$$s_1 + \frac{s_2 - s_1}{2}\left\{1 - \left(1 - \frac{2}{n}\right)^m\right\} \text{ and } s_2 + \frac{s_1 - s_2}{2}\left\{1 - \left(1 - \frac{2}{n}\right)^m\right\}.$$

Solution. Let

V = volume of each vessel

ρ_1 = sp. gr. of the mixture in the first vessel when $\frac{1}{n}$th volume of liquid of each vessel of sp. gr. s_1 and s_2 is taken out and put into the other

δ_1 = sp. gr. of the mixture in the second vessel after the above process.

Then

$$\rho_1 = \frac{\frac{n-1}{n}Vs_1 + \frac{1}{n}Vs_2}{\frac{n-1}{n}V + \frac{1}{n}V} = \frac{n-1}{n}s_1 + \frac{1}{n}s_2. \tag{1}$$

Expression (1) can be written as

$$\rho_1 = s_1 + \frac{s_2 - s_1}{2} \cdot \frac{2}{n} = s_1 + \frac{s_2 - s_1}{2}\left\{1 - \left(1 - \frac{2}{n}\right)\right\}. \tag{2}$$

Similarly,

$$\delta_1 = \frac{\frac{n-1}{n}Vs_2 + \frac{1}{n}Vs_1}{\frac{n-1}{n}V + \frac{1}{n}V} = \frac{n-1}{n}s_2 + \frac{1}{n}s_1, \tag{3}$$

which can be written as

$$\delta_1 = s_2 + \frac{s_1 - s_2}{2}\left\{1 - \left(1 - \frac{2}{n}\right)\right\}. \tag{4}$$

Again, let $\rho_2 =$ sp. gr. of the mixture in the first vessel when $\frac{1}{n}$th volume of liquid of each vessel of sp. gr. ρ_1 and δ_1 is taken out and put into the other. Let δ_2 be the sp. gr. of the mixture in the second vessel after this process. Then

$$\rho_2 = \frac{\frac{n-1}{n}V\rho_1 + \frac{1}{n}V\delta_1}{\frac{n-1}{n}V + \frac{1}{n}V} = \frac{n-1}{n}\rho_1 + \frac{1}{n}\delta_1. \tag{5}$$

Putting the values of ρ_1 and δ_1 from Eqns (1) and (3),

$$\rho_2 = \frac{n-1}{n}\left(\frac{n-1}{n}s_1 + \frac{1}{n}s_2\right) + \frac{1}{n}\left(\frac{1}{n}s_1 + \frac{n-1}{n}s_2\right)$$

$$= \left(\frac{n-1}{n}\right)^2 s_1 + \frac{1}{n^2}s_1 + \frac{2(n-1)}{n^2}s_2 = s_1 + \frac{2(n-1)}{n^2}(s_2 - s_1)$$

$$= s_1 + \frac{s_2 - s_1}{2}\left\{1 - \left(1 - \frac{2}{n}\right)^2\right\}. \tag{6}$$

Similarly,

$$\delta_2 = \frac{\frac{n-1}{n}V\delta_1 + \frac{1}{n}\rho_1 V}{\frac{n-1}{n}V + \frac{1}{n}V} = \frac{n-1}{n}\delta_1 + \frac{1}{n}\rho_1. \tag{7}$$

Putting the values of ρ_1 and δ_1 from Eqns (1) and (3),

$$\delta_2 = \frac{n-1}{n}\left(\frac{n-1}{n}s_2 + \frac{1}{n}s_1\right) + \frac{1}{n}\left(\frac{n-1}{n}s_1 + \frac{1}{n}s_2\right)$$

$$= \left(\frac{n-1}{n}\right)^2 s_2 + \frac{2(n-1)}{n}s_1 + \frac{1}{n^2}s_2$$

$$= s_2 + \frac{s_1 - s_2}{2}\left\{1 - \left(1 - \frac{2}{n}\right)^2\right\}. \tag{8}$$

Again, let ρ_3 and δ_3 be the sp. gr. of the mixtures in the first and second vessel respectively when $\frac{1}{n}$th volume of liquid of each vessel of sp. gr. ρ_2 and δ_2 is taken out and put into the other and mixed well. Then

$$\rho_3 = \frac{\frac{n-1}{n}V\rho_2 + \frac{1}{n}V\delta_2}{\frac{n-1}{n}V + \frac{1}{n}V} = \frac{n-1}{n}\rho_2 + \frac{1}{n}\delta_2. \tag{9}$$

Putting here the values of ρ_2 and δ_2 from Eqns (6) and (8), we get

$$\rho_3 = \frac{n-1}{n}\left[s_1 + \frac{s_2 - s_1}{2}\left\{1 - \left(1 - \frac{2}{n}\right)^2\right\}\right] + \frac{1}{n}\left[s_2 + \frac{s_1 - s_2}{2}\left\{1 - \left(1 - \frac{2}{n}\right)^2\right\}\right]$$

$$= \frac{n-1}{n}s_1 + \frac{1}{n}s_2 + \frac{s_1 - s_2}{2}\left\{1 - \left(1 - \frac{2}{n}\right)^2\right\}\left(\frac{1}{n} - \frac{n-1}{n}\right)$$

$$= s_1 - \frac{1}{n}(s_1 - s_2) + \frac{s_1 - s_2}{2}\left\{1 - \left(1 - \frac{2}{n}\right)^2\right\}\frac{2-n}{n}$$

$$= s_1 + \frac{s_1 - s_2}{2}\left\{-\frac{2}{n} + \left(\frac{2}{n} - 1\right) + \left(1 - \frac{2}{n}\right)^3\right\}$$

$$= s_1 + \frac{s_2 - s_1}{2}\left\{1 - \left(1 - \frac{2}{n}\right)^3\right\}. \tag{10}$$

Similarly, we have

$$\delta_3 = \frac{\frac{n-1}{n}V\delta_2 + \frac{1}{n}V\rho_2}{\frac{n-1}{n}V + \frac{1}{n}V} = \frac{n-1}{n}\delta_2 + \frac{1}{n}\rho_2. \tag{11}$$

Putting the values of ρ_2 and δ_2 from Eqns (6) and (8), we get

$$\begin{aligned}
\delta_3 &= \frac{n-1}{n}\left[s_2 + \frac{s_1 - s_2}{2}\left\{1 - \left(1 - \frac{2}{n}\right)^2\right\}\right] + \frac{1}{n}\left[s_1 + \frac{s_2 - s_1}{2}\left\{1 - \left(1 - \frac{2}{n}\right)^2\right\}\right] \\
&= \frac{n-1}{n}s_2 + \frac{1}{n}s_1 + \frac{s_1 - s_2}{2}\left\{1 - \left(1 - \frac{2}{n}\right)^2\right\}\left(\frac{n-1}{n} - \frac{1}{n}\right) \\
&= s_2 + \frac{1}{n}(s_1 - s_2) + \frac{s_1 - s_2}{2}\left\{1 - \left(1 - \frac{2}{n}\right)^2\right\}\left(1 - \frac{2}{n}\right) \\
&= s_2 + \frac{s_1 - s_2}{2}\left\{\frac{2}{n} + \left(1 - \frac{2}{n}\right) - \left(1 - \frac{2}{n}\right)^3\right\} \\
&= s_2 + \frac{s_1 - s_2}{2}\left\{1 - \left(1 - \frac{2}{n}\right)^3\right\}
\end{aligned} \tag{12}$$

Following similar procedure after m-operations, the sp. gr. of the final mixture will be obtained as

$$\rho_m = s_1 + \frac{s_2 - s_1}{2}\left\{1 - \left(1 - \frac{2}{n}\right)^m\right\}$$

$$\delta_m = s_2 + \frac{s_1 - s_2}{2}\left\{1 - \left(1 - \frac{2}{n}\right)^m\right\} \quad \textbf{(Proved)}.$$

EXAMPLE 1.7 *Equal weights of gold, silver and an alloy of gold and silver are dipped successively in a cylinder of water and caused water to rise a, b, c units respectively. Prove that the alloy contains gold and silver in the proportion of $(b - c) : (c - a)$ by weight.*

Solution. Let

S = cross-section of the cylinder

ρ_1, ρ_2, ρ_3 = sp. gravities of gold, silver and the alloy respectively

and w = weight of unit volume of standard substance.

Then the volumes of gold, silver and the alloy are Sa, Sb and Sc respectively. Hence the weights of gold, silver and the alloy are

$$W_1 = Sa\rho_1 w,\ W_2 = Sb\rho_2 w,\ W_3 = Sc\rho_3 w$$

respectively. These weights, by condition given, are equal. Hence

$$Sa\rho_1 w = Sb\rho_2 w = Sc\rho_3 w,$$

which yields

$$\rho_1 = \frac{\lambda}{a},\ \rho_2 = \frac{\lambda}{b},\ \rho_3 = \frac{\lambda}{c}, \tag{1}$$

where λ is a constant.

Now, let w_1 and w_2 be the weights of gold and silver contained in the alloy. Then

$$W_3 = w_1 + w_2. \tag{2}$$

But, volume of alloy = volume of gold + volume of silver. Hence,

$$\frac{W_3}{\rho_3 w} = \frac{w_1}{\rho_1 w} + \frac{w_2}{\rho_2 w} \text{ or, } \frac{w_1 + w_2}{\rho_3} = \frac{w_1}{\rho_1} + \frac{w_2}{\rho_2}$$

$$\text{or, } w_1\left(\frac{1}{\rho_3} - \frac{1}{\rho_1}\right) = w_2\left(\frac{1}{\rho_2} - \frac{1}{\rho_3}\right). \tag{3}$$

Using the relations in (1), we get from Eqn (3)

$$w_1\left(\frac{c}{\lambda} - \frac{a}{\lambda}\right) = w_2\left(\frac{b}{\lambda} - \frac{c}{\lambda}\right) \text{ or, } w_1(c-a) = w_2(b-c)$$

$$\therefore \quad \frac{w_1}{w_2} = \frac{b-c}{c-a} \quad \textbf{(Proved)}.$$

EXAMPLE 1.8 *The mixture of a volume of A with* w_1 lbs *of B has a specific gravity* s_1, *with* w_2 lbs *of B, a specific gravity* s_2, *with* w_3 lbs *of B, a specific gravity* s_3. *Find the specific gravity of A and B.*

Solution. Let V = volume of the mixture A

ρ = specific gravity of the mixture A

δ = specific gravity of the mixture B

and w = weight of unit volume of standard substance.

Then weight of mixture $A = V\rho w$. By the first condition, we have

$$s_1 = \frac{w_1 + V\rho w}{\dfrac{w_1}{\delta} + \dfrac{V\rho w}{\rho}}.$$

This yields

$$s_1\left(\frac{w_1}{\delta} + Vw\right) = w_1 + V\rho w. \tag{1}$$

Similarly, by the second condition, we have

$$s_2 = \frac{w_2 + V\rho w}{\dfrac{w_2}{\delta} + \dfrac{V\rho w}{\rho}}.$$

Hence

$$s_2\left(\frac{w_2}{\delta} + Vw\right) = w_2 + V\rho w. \tag{2}$$

In the same way, we get by using the third condition,

$$s_3\left(\frac{w_3}{\delta} + Vw\right) = w_3 + V\rho w. \tag{3}$$

Subtracting (2) from (1),

$$\frac{1}{\delta}(s_1 w_1 - s_2 w_2) + Vw(s_1 - s_2) = w_1 - w_2,$$

which yields

$$w = \frac{\frac{1}{\delta}(s_1 w_1 - s_2 w_2) + w_2 - w_1}{V(s_2 - s_1)} \tag{4}$$

Similarly, subtracting (3) from (2),

$$\frac{1}{\delta}(s_2 w_2 - s_3 w_3) + Vw(s_2 - s_3) = w_2 - w_3,$$

which yields

$$w = \frac{\frac{1}{\delta}(s_2 w_2 - s_3 w_3) + w_3 - w_2}{V(s_3 - s_2)}. \tag{5}$$

From (4) and (5), we get

$$\frac{\frac{1}{\delta}(s_1 w_1 - s_2 w_2) + w_2 - w_1}{V(s_2 - s_1)} = \frac{\frac{1}{\delta}(s_2 w_2 - s_3 w_3) + w_3 - w_2}{V(s_3 - s_2)}.$$

This after simplification yields

$$\delta = \frac{\sum w_1 s_1 (s_2 - s_3)}{\sum w_1 (s_2 - s_3)}. \tag{6}$$

Now, multiplying Eqn (1) by $s_2 w_2$ and Eqn (2) by $s_1 w_1$, we get

$$\frac{s_1 s_2 w_1 w_2}{\delta} = w_1 w_2 s_2 + V \rho w s_2 w_2 - V w s_1 s_2 w_2 \tag{7}$$

$$\text{and} \quad \frac{s_1 s_2 w_1 w_2}{\delta} = w_1 w_2 s_1 + V \rho w s_1 w_1 - V w s_1 s_2 w_1. \tag{8}$$

From Eqns (7) and (8), we have upon subtraction,

$$w_1 w_2 (s_2 - s_1) + V \rho w (s_2 w_2 - s_1 w_1) + V w s_1 s_2 (w_1 - w_2) = 0,$$

which yields

$$V w = \frac{w_1 w_2 (s_1 - s_2)}{\rho (s_2 w_2 - s_1 w_1) + s_1 s_2 (w_1 - w_2)}. \tag{9}$$

Similarly, multiplying Eqn (2) by $s_3 w_3$ and Eqn (3) by $s_2 w_2$, we get

$$V w = \frac{w_2 w_3 (s_2 - s_3)}{\rho (s_3 w_3 - s_2 w_2) + s_2 s_3 (w_2 - w_3)}. \tag{10}$$

From Eqns (9) and (10), we get

$$\frac{w_1 w_2 (s_1 - s_2)}{\rho (s_2 w_2 - s_1 w_1) + s_1 s_2 (w_1 - w_2)} = \frac{w_2 w_3 (s_2 - s_3)}{\rho (s_3 w_3 - s_2 w_2) + s_2 s_3 (w_2 - w_3)},$$

which, after simplification, yields

$$\rho = \frac{\sum w_2 w_3 s_1 (s_2 - s_3)}{\sum w_2 w_3 (s_2 - s_3)}.$$

EXERCISE

1. A mixture is formed of two liquids. The density ρ of the mixture, the ratio $m : 1$ of the volumes and the ratio $n : 1$ of the densities are given. Prove that the densities of the fluids are

$$\frac{mn + n}{mn + 1}\rho \text{ and } \frac{m + 1}{mn + 1}\rho.$$

2. Two liquids of sp. grs. s_1, s_2 and of volumes V_1, V_2 having been mixed, the sp. gr. of the mixture is found to be δ. Find the volume of the mixture.

3. The sp. gr. of a mixture of equal volumes of two substances is p and that of a mixture of equal weights of the same substances is q. Show that the sp. grs. of two substances are given by

$$x^2 - 2px + pq = 0.$$

4. A vessel contains a liquid of sp. gr. s_1 to start with; $\frac{1}{n}$th of it is removed and replaced by a liquid of sp. gr. s_2. The operation is repeated successively n times. Find the sp. gr. of the final mixture left in the vessel.

5. n cylinders of equal volumes are each half-filled with a liquid of sp. gr. s_1. If the first be filled with a liquid of sp. gr. s_2, the second filled from the first, the third filled from the second and so on, find the sp. gr. of the liquid in the last cylinder.

6. The density of milk in the vessel of the milkman was found to be ρ. If the density of pure milk is δ, show that the ratio of weights of water and milk in the adulterated milk is

$$\frac{\delta-\rho}{\delta(\rho-1)}.$$

Also prove that the volumes of water and milk in the mixture bear a ratio $(\delta-\rho):(\rho-1)$.

7. Three equal vessels A, B, C are half full of liquids of densities ρ_1, ρ_2 and ρ_3 respectively. If now B be filled from A and then C from B, prove that the density of the liquid now contained in C is

$$\frac{1}{4}(\rho_1+\rho_2+2\rho_3),$$

the liquid being supposed to mix completely.

8. Having given sp. gr. ρ of the mixture formed of equal volumes of two fluids and also the sp. gr. δ of a mixture formed by taking a volume of one fluid double that of the other, prove that the sp. grs. of fluids are $(3\delta-2\rho)$ and $(4\rho-3\delta)$.

9. Equal weights of gold, silver and an alloy of gold and silver are dipped successively in a cylinder of water and caused water to rise a, b, c units respectively. Prove that the alloy contains gold and silver in the proportion $(b-c):(c-a)$ by weight.

10. Two cylinders of equal volumes are filled with liquids of sp. gr. s_1 and s_2. One-third of the contents of each cylinder is taken out and put in the other and the operation is repeated n times. Find the sp. gr. of the final mixture in each cylinder.

ANSWERS

1. $\rho_A = \dfrac{\sum w_2w_3s_1(s_2-s_3)}{\sum w_2w_3(s_2-s_3)}$, $\rho_B = \dfrac{\sum w_1s_1(s_2-s_3)}{\sum w_1(s_2-s_3)}$. **2.** $\dfrac{V_1s_1+V_2s_2}{\delta}$. **4.** $\left(1-\dfrac{1}{n}\right)^n(s_1-s_2)+s_1$.

5. $s_1+\dfrac{1}{2^n}(s_2-s_1)$. **10.** $\rho_n = \dfrac{1}{3^n}s_1+\dfrac{s_1+s_2}{2}\left(1-\dfrac{1}{3^n}\right)$; $\delta_n = \dfrac{1}{3^n}s_2+\dfrac{s_1+s_2}{2}\left(1-\dfrac{1}{3^n}\right)$.

CHAPTER 2

Fluids Under Gravity

2.1 Introduction

Fluids, in general, may be subject to a variable field of force. In this chapter, we will discuss equilibrium of fluids at rest under the action of its weight due to gravity. Thus we shall assume that the fluid is heavy.

2.2 Pressure Function

We consider a point P with coordinates (x, y, z) within the mass of a fluid in equilibrium. Let p be the pressure of the fluid at this point. The pressure, in general, will be a function of coordinates x, y, z of the point P, so that we write

$$p = p(x, y, z). \tag{2.1}$$

Its total differential is

$$dp = \frac{\partial p}{\partial x}dx + \frac{\partial p}{\partial y}dy + \frac{\partial p}{\partial z}dz. \tag{2.2}$$

If we consider a point Q close to P with coordinates $(x + dx, y, z)$ so that $dy = dz = 0$, the total differential of the pressure function is

$$dp = \frac{\partial p}{\partial x}dx. \tag{2.3}$$

Hence the pressure at the point Q is

$$p + dp = p + \frac{\partial p}{\partial x}dx. \tag{2.4}$$

In the same manner, we can find the pressure at any other point close to the point P.

2.3 Equation of Equilibrium

We take a rectangular coordinate system within the fluid mass such that x- and y-axes are in a horizontal plane and the z-axis is along the vertically upward direction.

Let a point $P(x, y, z)$ be within the fluid mass and we construct an elementary parallelopiped with P at a corner and its faces parallel to the rectangular coordinate planes (FIG. 2.1).

Let the sides of the elementary parallelopiped be dx, dy and dz.

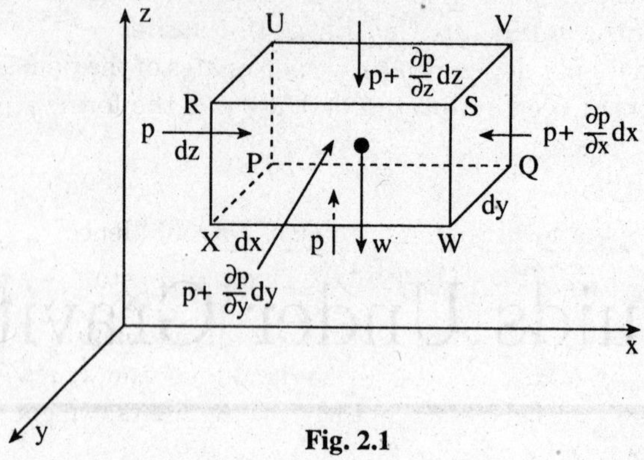

Fig. 2.1

The areas of the surfaces of the parallelopiped are very small so that the thrust on the face $PURX$ is approximately equal to $pdydz$ neglecting terms of higher order of smallness. We shall consider thrusts on other faces in the same manner. Thus the fluid element under gravity remains in equilibrium subject to the following forces:

(i) $pdydz$ along +ve x-direction and $\left(p + \dfrac{\partial p}{\partial x}dx\right) dydz$ along −ve x-direction,

(ii) $pdxdz$ along +ve y-direction and $\left(p + \dfrac{\partial p}{\partial y}dy\right) dxdz$ along −ve y-direction,

(iii) $pdxdy$ along +ve z-direction and $\left(p + \dfrac{\partial p}{\partial z}dz\right) dxdy$ along −ve z-direction,

(iv) the weight of the fluid element $w = (dxdydz)\rho g$ along −ve z-direction.

For equilibrium, we resolve the forces along x-axis and write

$$pdydz - \left(p + \frac{\partial p}{\partial x}dx\right) dydz = 0,$$

which yields

$$\frac{\partial p}{\partial x} = 0. \tag{2.5}$$

Resolving the forces along y-axis, we get

$$pdxdz - \left(p + \frac{\partial p}{\partial y}dy\right) dxdz = 0,$$

This yields

$$\frac{\partial p}{\partial y} = 0. \tag{2.6}$$

Similarly, resolving the forces along z-axis, we get

$$pdxdy - \left(p + \frac{\partial p}{\partial z}dz\right) dxdy - \rho g(dxdydz) = 0.$$

This yields

$$\frac{\partial p}{\partial z} = -\rho g. \tag{2.7}$$

Eqns (2.5) and (2.6) imply that the pressure at any point in a fluid mass at rest is independent of x- and y-coordinates. Hence, ***in a fluid at rest under gravity, the pressure at all points in the same horizontal plane is the same.***

Since the pressure p does not depend on x-, y-coordinates of the point, we can write Eqn (2.7) in terms of total derivative instead of partial derivative in the form

$$\frac{dp}{dz} = -\rho g. \tag{2.8}$$

On the free surface, there is no pressure, so that $dp = 0$. Hence,

$$\rho g dz = 0, \text{ i.e., } dz = 0, \text{ i.e., } z = \text{constant}. \tag{2.9}$$

Hence, ***the free surface of a fluid at rest under gravity is horizontal.***

Let p_1 and p_2 be the pressures at points with heights z_1 and z_2 respectively in a fluid mass at rest under gravity. Integrating Eqn (2.8) within these limits, we get

$$\int_{p_1}^{p_2} dp = -\int_{z_1}^{z_2} \rho g dz, \text{ i.e., } (p_2 - p_1) = -\rho g(z_2 - z_1). \tag{2.10}$$

Here we have considered the fluid to be homogeneous so that its density ρ is constant. From Eqn (2.10), we find that

$$(p_2 - p_1) \propto (z_2 - z_1). \tag{2.11}$$

This implies that, ***in a homogeneous fluid at rest under gravity, the pressure difference between two points is proportional to the difference of their heights.***

Let γ denote the specific weight of the fluid, i.e., weight of unit volume of the fluid. Hence

$$\gamma = \rho g. \tag{2.12}$$

Eqn (2.10) can be written in the form

$$\frac{p_1}{\gamma} + z_1 = \frac{p_2}{\gamma} + z_2 = \text{constant}. \tag{2.13}$$

The quantity

$$\Pi = \frac{p}{\gamma} + z \tag{2.14}$$

is called ***specific potential energy*** and has the dimension of length. The specific potential energy represents the energy stored in a unit weight of the fluid at rest with reference to a certain plane where energy is taken to be zero.

Thus, Eqn (2.13) implies that ***the specific potential energy of a fluid at rest under gravity is constant.*** This is shown in FIG. 2.2.

An indefinite integration of Eqn (2.8) leads to

$$p = -\gamma z + \text{constant}. \tag{2.15}$$

Assuming $p = 0$ when $z = 0$, we get

$$p = -\gamma z. \tag{2.16}$$

Thus, ***the pressure at any point in a fluid at rest under gravity is proportional to its depth below the free surface.***

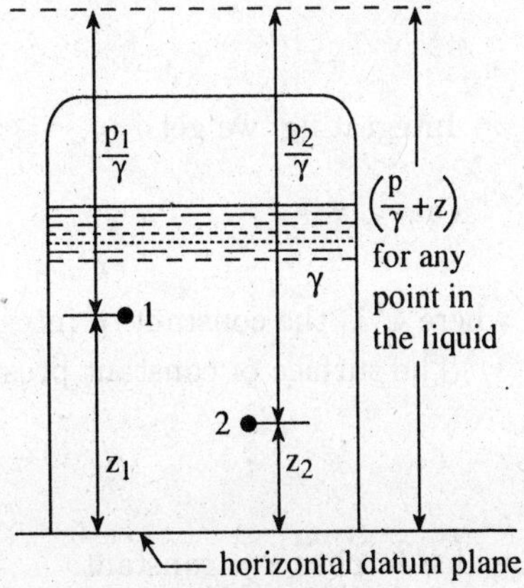

Fig. 2.2

2.4 Heterogeneous Fluids

In a heterogeneous fluid, the density at any point within the fluid is a function of z, so that

$$\rho = \rho(z). \tag{2.17}$$

Then, from Eqn (2.8), we get

$$\int dp = -g \int \rho(z)dz. \tag{2.18}$$

This can be integrated when a certain relation between the pressure and density is provided or the density function is given.

Suppose, for example, that the density of a fluid at a depth z is given by

$$\rho(z) = \lambda z, \lambda = \text{constant}. \tag{2.19}$$

Taking the x- and y-axes on the free surface and z-axis vertically downwards, we write Eqn (2.8) in the form

$$dp = g\rho dz = \lambda gz\, dz. \tag{2.20}$$

Integrating, we get

$$p = \frac{1}{2}\lambda gz^2 + \text{constant}. \tag{2.21}$$

Assuming the atmospheric pressure over the free surface to be π, we get

$$p = \pi + \frac{1}{2}\lambda gz^2. \tag{2.22}$$

In this case, the pressure at any point is proportional to the square of its depth.

For ***elastic*** fluids, $p = \xi\rho$, where ξ = constant. In this case,

$$\left.\begin{aligned} dp &= g\rho dz \\ \text{and} \quad p &= \xi\rho \end{aligned}\right\} \tag{2.23}$$

which yields

$$\frac{dp}{p} = \frac{g}{\xi}dz. \tag{2.24}$$

Integrating, we get

$$p = C\exp\left(\frac{gz}{\xi}\right), \tag{2.25}$$

where C is the constant of integration to be determined from boundary condition.

The surface of constant pressures is given by

$$C\exp\left(\frac{gz}{\xi}\right) = \text{constant}, \tag{2.26}$$

which implies z = constant.

Hence, ***for elastic fluids also, the surfaces of equal pressure are horizontal planes.***

2.5 Immiscible Liquids

We consider two heavy homogeneous liquids of densities ρ_1, $\rho_2 (\rho_2 > \rho_1)$ which do not mix (FIG. 2.3). Let P be any point in the heavier liquid. The vertical through P up to the free surface (which is horizontal) meets the surface of separation of two liquids at Q and the free surface at R.

Let $QR = h_1$ and $PQ = h_2$. Then the pressure at Q is

$$p_1 = \pi + \gamma_1 h_1, \tag{2.27}$$

where π is the atmospheric pressure.

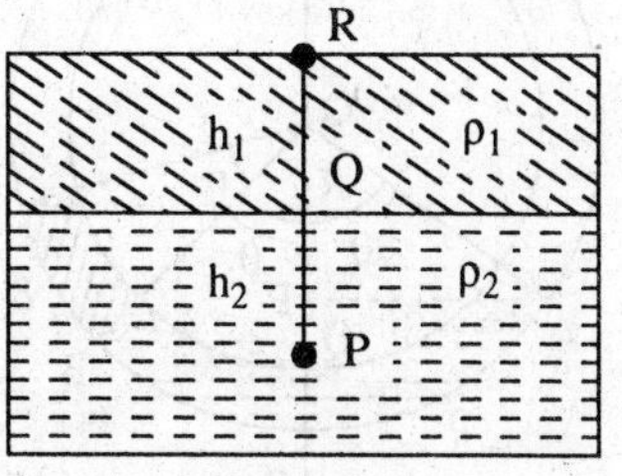

Fig. 2.3

The pressure at P

$$\begin{aligned} &= \text{pressure at } Q + \text{pressure due to} \\ &\qquad \text{liquid of density } \rho_2 \text{ and depth } h_2 \\ &= \pi + \gamma_1 h_1 + \gamma_2 h_2. \end{aligned} \tag{2.28}$$

Similarly, if there be n different liquids (which do not mix) occupying depths of $h_1, h_2, \dots, h_n$ and of densities $\rho_1, \rho_2, \dots, \rho_n$, the pressure at the bottom of the liquid is given by

$$p = \pi + \sum_{i=1}^{n} \gamma_i h_i = \pi + \sum_{i=1}^{n} g\rho_i h_i. \tag{2.29}$$

Let us now prove that ***the surface of separation of two heavy homogeneous liquids that do not mix and are at rest under gravity is a horizontal plane.***

Let P_1, P_2 be two points in the lower liquid on the same horizontal line. Let the verticals through these points meet the common surface at Q_1, Q_2 and the free surface at R_1, R_2 respectively (FIG. 2.4).

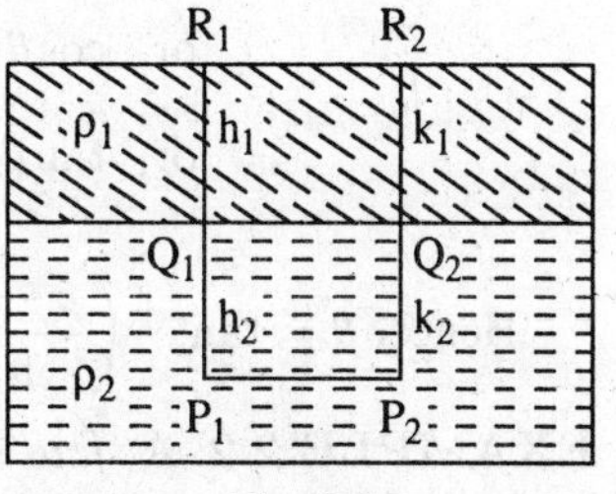

Fig. 2.4

Let

$$\left.\begin{aligned} Q_1R_1 = h_1, \; Q_2R_2 = k_1 \\ P_1Q_1 = h_2, \; P_2Q_2 = k_2 \end{aligned}\right\} \tag{2.30}$$

and ρ_1, ρ_2 are the densities of the two liquids.

Since the free surface is a horizontal plane, the line R_1R_2 is horizontal. Hence,

$$P_1R_1 = P_2R_2, \text{ i.e., } h_1 + h_2 = k_1 + k_2. \tag{2.31}$$

Multiplying both sides of Eqn (2.31) by $\gamma_2 = g\rho_2$, we get

$$(h_1 + h_2)\gamma_2 = (k_1 + k_2)\gamma_2. \tag{2.32}$$

Now, since P_1P_2 is horizontal, pressure at P_1 = pressure at P_2.
Hence,

$$\pi + h_1\gamma_1 + h_2\gamma_2 = \pi + k_1\gamma_1 + k_2\gamma_2, \text{ i.e., } h_1\gamma_1 + h_2\gamma_2 = k_1\gamma_1 + k_2\gamma_2. \tag{2.33}$$

Subtracting Eqn (2.32) from Eqn (2.33), we get $h_1(\gamma_1 - \gamma_2) = k_1(\gamma_1 - \gamma_2)$ which yields $h_1 = k_1$. Putting these into Eqn (2.31), we find $h_2 = k_2$. Hence, the line Q_1Q_2 is horizontal.

EXAMPLE 2.1 *A fine circular tube in a vertical plane contains a column of liquid of density ρ which subtends a right angle at the centre and a column of density δ, subtending an angle α. Prove that the radius through the common surface makes with the vertical an angle of*

$$\tan^{-1}\left(\frac{\rho - \delta + \delta\cos\alpha}{\rho + \delta\sin\alpha}\right).$$

Solution. Let the two liquids occupy the portions ADB and BC of the tube which remains in a vertical plane (FIG. 2.5). Let DO be the vertical radius. Then, according to the problem

$$\angle AOB = 90^\circ,\ \angle BOC = \alpha. \tag{1}$$

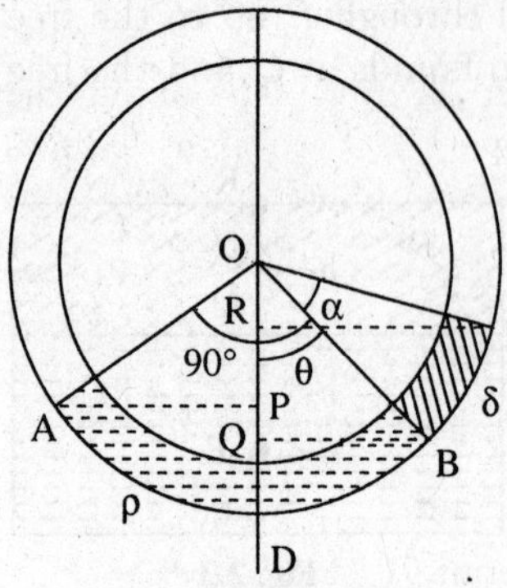

Fig. 2.5

Let the radius OB through the common surface make with the vertical an angle θ. We draw perpendiculars AP, BQ and CR on the vertical radius OD.

For equilibrium of the two columns of liquids, pressure at the lowest point D from both sides must be equal. Hence,

$$\rho g \cdot DP = \rho g \cdot DQ + \delta g \cdot QR$$

$$\text{or, } \rho(DP - DQ) = \delta \cdot QR$$

$$\text{or, } \rho(OQ - OP) = \delta \cdot (OQ - OR). \tag{2}$$

From the figure, we have when r = radius of the tube,

$$\left.\begin{aligned} OQ = r\cos\theta,\ OP = r\cos(90-\theta) = r\sin\theta \\ OR = r\cos(\theta+\alpha) = r(\cos\theta\cos\alpha - \sin\theta\sin\alpha) \end{aligned}\right\}. \tag{3}$$

Putting these values into (2), we get

$$\rho(r\cos\theta - r\sin\theta) = r\delta(\cos\theta - \cos\theta\cos\alpha + \sin\theta\sin\alpha)$$

$$\text{or, } \cos\theta(\rho - \delta + \delta\cos\alpha) = \sin\theta(\rho + \delta\sin\alpha)$$

$$\text{or, } \tan\theta = \frac{\rho - \delta + \delta\cos\alpha}{\rho + \delta\sin\alpha}.$$

Hence, $\theta = \tan^{-1}\left(\dfrac{\rho - \delta + \delta\cos\alpha}{\rho + \delta\sin\alpha}\right)$.

EXAMPLE 2.2 *A fine tube ABC of uniform bore having parts AB, BC straight and perpendicular to one another, is held in a vertical plane and contains several liquids of different densities. If the tube be turned about the point B in its own plane and the liquids be allowed to settle down without spilling, prove that the weight of the liquid which has passed from one branch to the other bears to the weight of the whole the ratio* $(\tan\alpha - \tan\beta) : 2$, *where α, β are the inclinations to the vertical of the bisector of the angle ABC in the two positions.*

Solution. In the initial position of the tube as shown in the (FIG. 2.6), let w_1, w_2 be weights of liquids in the branches AB, BC respectively. BP is the bisector of $\angle ABC$ and BQ is the vertical through B. Then,

$$\phi = \angle ABQ = \frac{\pi}{4} - \alpha$$

$$\text{and } \angle CBQ = \frac{\pi}{2} - \phi. \tag{1}$$

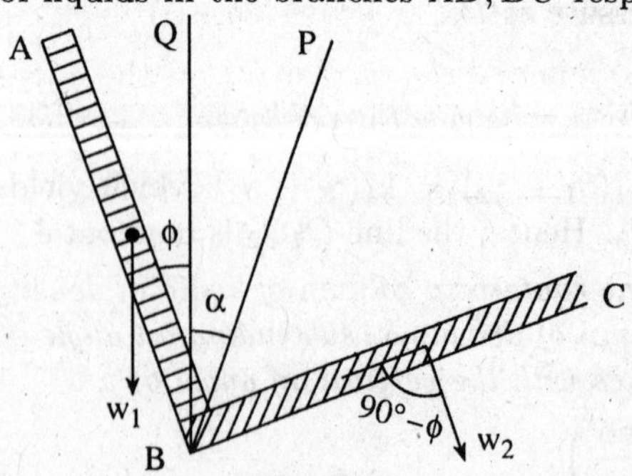

Fig. 2.6

For equilibrium, the pressure at the point B should be equal from both sides. Hence

$$w_1\cos\left(\frac{\pi}{4} - \alpha\right) = w_2\sin\left(\frac{\pi}{4} - \alpha\right)$$

$$\text{or, } \frac{w_1}{w_2} = \tan\left(\frac{\pi}{4} - \alpha\right) = \frac{1 - \tan\alpha}{1 + \tan\alpha}.$$

Hence

$$\frac{w_2 - w_1}{w_2 + w_1} = \frac{(1+\tan\alpha) - (1-\tan\alpha)}{(1+\tan\alpha) + (1-\tan\alpha)} = \tan\alpha. \tag{2}$$

In the second position of the tube when it has been turned in its own plane about the point B, angle α becomes β. Let due to this turning, liquid from part BC of weight w flows to the part AB. Then weights of liquids in two parts now are $(w + w_1)$ in AB and $(w_2 - w)$ in BC.

Equating the pressure at B from both sides

$$(w_1 + w)\cos\left(\frac{\pi}{4} - \beta\right) = (w_2 - w)\sin\left(\frac{\pi}{4} - \beta\right)$$

$$\text{or,} \quad \frac{w_1 + w}{w_2 - w} = \tan\left(\frac{\pi}{4} - \beta\right) = \frac{1 - \tan\beta}{1 + \tan\beta}$$

$$\text{or,} \quad \frac{(w_1 + w) - (w_2 - w)}{(w_1 + w) + (w_2 - w)} = \frac{(1 - \tan\beta) - (1 + \tan\beta)}{(1 - \tan\beta) + (1 + \tan\beta)}$$

$$\text{or,} \quad \frac{w_1 - w_2 + 2w}{w_1 + w_2} = \frac{-2\tan\beta}{2} = -\tan\beta. \tag{3}$$

Adding (2) and (3), we get

$$\tan\alpha - \tan\beta = \frac{(w_2 - w_1) + (w_1 - w_2 + 2w)}{w_1 + w_2}$$

$$\text{i.e.,} \quad \tan\alpha - \tan\beta = \frac{2w}{w_1 + w_2}. \tag{4}$$

If we assume that due to turning of the tube about the point B, liquid of weight w flows from part AB to BC, then we will get

$$\tan\beta - \tan\alpha = \frac{2w}{w_1 + w_2}. \tag{5}$$

Combining Eqns (4) and (5), we write

$$\left|\frac{\tan\alpha - \tan\beta}{2}\right| = \frac{w}{w_1 + w_2}. \tag{6}$$

The RHS expression is the ratio of the weight of the liquid passed from one branch to the other to that of the whole liquid. Hence proved.

EXAMPLE 2.3 *A fine uniform cycloidal tube is held with its axis vertical and vertex downward and equal weights of two liquids occupy lengths a and b. Prove that the heights of the free surfaces of the liquids above the vertex are in the ratio* $(3a + b)^2 : (3b + a)^2$.

Solution. We take the vertex O of the cycloid as the origin and the vertical OD as the y-axis (FIG. 2.7).

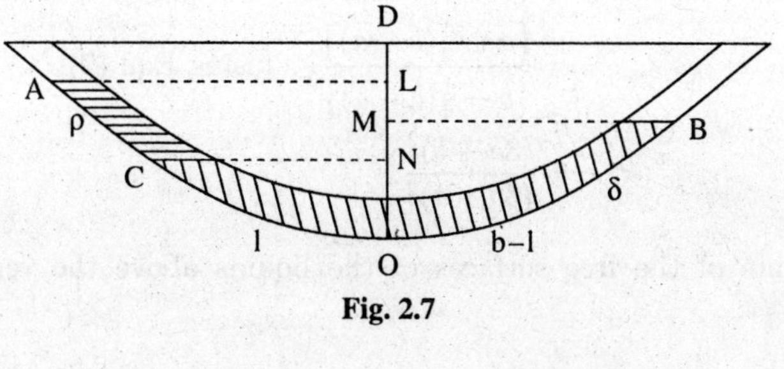

Fig. 2.7

Let r be the radius of the generating circle of the cycloid. Then the equation of the cycloid is taken in the form

$$s^2 = 8ry, \tag{1}$$

where s is the length of the arc measured from the vertex of the point whose ordinate is y. We are given that arc $AC = a$ and arc $BC = b$. Since the liquids are of equal weights, we have

$$a\rho = b\delta, \text{ i.e., } \rho = \frac{b}{a}\delta, \tag{2}$$

where ρ, δ are the densities of the liquids in lengths AC and BC respectively.

We draw AL, BM and CN perpendiculars to the vertical OD from the free surfaces A, B and the common surface C respectively.

For equilibrium, pressure at O due to the columns of liquids on two sides must be equal.

Denoting $OL = y_1, OM = y_2$ and $ON = y_3$, we write $g\rho \cdot LN + g\delta \cdot ON = g\delta \cdot OM$ or, $\rho(y_1 - y_3) + \delta y_3 = \delta y_2$ or, $\frac{b}{a}\delta(y_1 - y_3) = \delta(y_2 - y_3)$, using (2).

Hence,

$$\frac{a}{b} = \frac{y_1 - y_3}{y_2 - y_3}. \tag{3}$$

Let the arc $OC = l$. Then arc $OB = b - l$ and arc $OA = a + l$.

By Eqn (1) for the points A, B, C, we have

$$(a + l)^2 = 8ry_1, \ (b - l)^2 = 8ry_2, \ l^2 = 8ry_3 \tag{4}$$

From Eqn (4), we find

$$\left.\begin{aligned} 8r(y_1 - y_3) &= (a + l)^2 - l^2 = a(a + 2l) \\ \text{and } 8r(y_2 - y_3) &= (b - l)^2 - l^2 = b(b - 2l) \end{aligned}\right\} \tag{5}$$

Hence,

$$\frac{y_1 - y_3}{y_2 - y_3} = \frac{a(a + 2l)}{b(b - 2l)}. \tag{6}$$

From (3) and (6), we get

$$\frac{a}{b} = \frac{a(a + 2l)}{b(b - 2l)}$$

which yields

$$l = \frac{1}{4}(b - a). \tag{7}$$

Now,

$$\begin{aligned} \frac{y_1}{y_2} &= \frac{(a + l)^2}{(b - l)^2}, \text{ [by Eqn (4)]} \\ &= \frac{\left\{a + \frac{1}{4}(b - a)\right\}^2}{\left\{b - \frac{1}{4}(b - a)\right\}^2}, \text{ [using Eqn (7)]} \\ &= \frac{(3a + b)^2}{(3b + a)^2}. \end{aligned}$$

Thus, the heights of the free surfaces of the liquids above the vertex are in the ratio $(3a + b)^2 : (3b + a)^2$.

EXAMPLE 2.4 *Equal volumes of three fluids of different densities, which do not mix, together completely fill a circular tube which is kept in a vertical plane. Prove that, if the densities of the fluids are in A.P., the common surface of the lightest and the heaviest fluids is at an extremity of a horizontal diameter of the circle.*

Solution. Let the densities of the fluids be $\rho-\lambda$, ρ, $\rho+\lambda$. Since the volumes of the fluid are equal, each fluid portion makes an angle 120° at the centre.

Let O be the centre of the circle of radius r and OA, OB, OC be the radii through common surfaces of the fluids of densities $\rho-\lambda, \rho$ and $\rho+\lambda$ respectively.

OD is the vertical through the centre of the circle (FIG. 2.8).

We draw AL, BM and CN perpendiculars from the common surfaces on the vertical OD. Let $\angle DOC = \theta$. We are to prove that $\theta = \pi/2$.

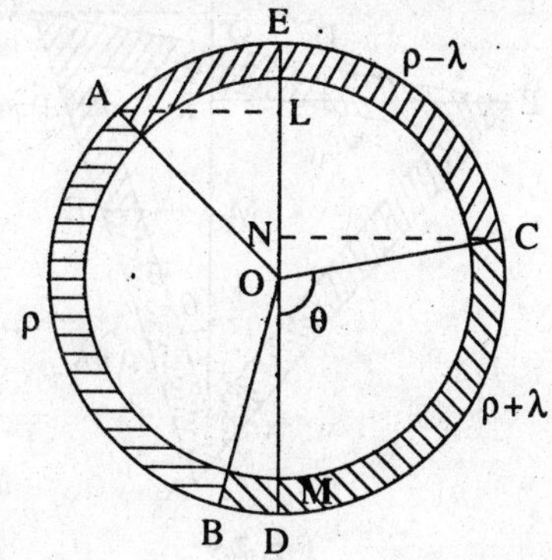

Fig. 2.8

For equilibrium, the pressure at the lowest point D due to fluids on two sides must be the same.

Hence, we write

$$g(\rho+\lambda)MD + g\rho \cdot LM + g(\rho-\lambda)EL = g(\rho+\lambda)ND + g(\rho-\lambda)EN. \tag{1}$$

From the figure, we get

$$MD = OD - OM = r - r\cos(120^\circ - \theta)$$
$$LM = OL + OM = r\cos(\theta - 60^\circ) + r\cos(120^\circ - \theta)$$
$$\text{since } \angle AOL = 120^\circ - \angle EOC = 120^\circ - (180^\circ - \theta) = \theta - 60^\circ$$
$$EL = r - r\cos(\theta - 60^\circ)$$
$$ND = ON + OD = r + r\cos(180^\circ - \theta)$$
$$EN = OE - ON = r - r\cos(180^\circ - \theta)$$

Using these relations, we write Eqn (1) as

$$(\rho+\lambda)\{r - r\cos(120^\circ-\theta)\} + \rho\{r\cos(\theta-60^\circ) + r\cos(120^\circ-\theta)\}$$
$$+(\rho-\lambda)\{r - r\cos(\theta-60^\circ)\} = (\rho+\lambda)\{r + r\cos(180^\circ-\theta)\}$$
$$+(\rho-\lambda)\{r - r\cos(180^\circ-\theta)\}$$
$$\text{or, } 2\rho + \lambda\{-\cos(120^\circ-\theta) + \cos(\theta-60^\circ)\} = 2\rho + 2\lambda\cos(180-\theta)$$
$$\text{or, } -\cos(120^\circ-\theta) + \cos(\theta-60^\circ) = 2\cos(180-\theta) = -2\cos\theta$$
$$\text{or, } \cos(60^\circ+\theta) + \cos(60^\circ-\theta) = -2\cos\theta$$
$$\text{or, } 2\cos\theta\cos 60^\circ + 2\cos\theta = 0$$
$$\text{or, } 3\cos\theta = 0, \text{ i.e. } \cos\theta = 0$$

Hence, $\theta = \pi/2$.

EXAMPLE 2.5 *A fine glass tube in the shape of an equilateral triangle is filled with equal volumes of three liquids which do not mix and whose densities are in arithmetical progression. The tube is held in a vertical plane and the side that contains portions of heaviest and lightest liquids makes an angle θ with the vertical.*

Show that the surfaces of separation divide the side in the ratio of $\cos\left(\frac{\pi}{6}-\theta\right) : \cos\left(\frac{\pi}{6}+\theta\right)$.

Solution. The glass tube filled with liquids is shown in the FIG. 2.9. Let A, B, C be the corners of the tube and D, E, F be the surfaces of separation. Since volumes of three liquids are equal, the lengths of the portions DCE, EAF, FBD are equal. These portions contain liquids of densities $(\rho - \lambda)$, $(\rho + \lambda)$ and ρ respectively.

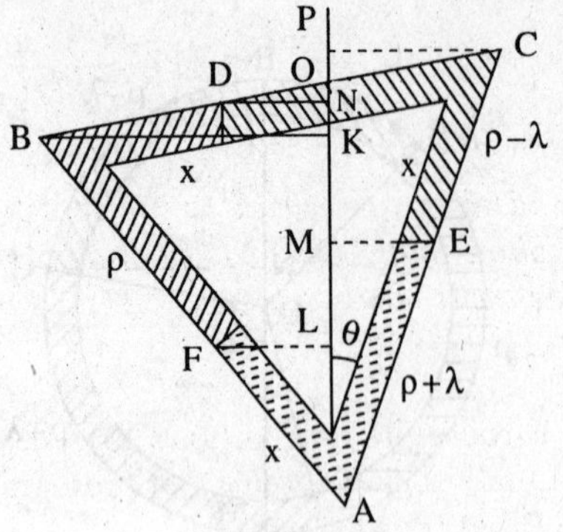

Fig. 2.9

Let a = length of each side of the triangle, and $BD = CE = AF = x$.

The vertical through A meets the side BC at O. Then by questions, $\angle CAO = \theta$.

We draw perpendiculars from D, E, F and C on the vertical shown in dotted lines. For equilibrium, pressures at A from both sides must be equal. Hence,

$$g(\rho+\lambda)\cdot AL + g\rho\cdot LN + g(\rho-\lambda)\cdot ON = g(\rho+\lambda)\cdot AM + g(\rho-\lambda)\cdot MP$$

$$\text{or,}\ (\rho+\lambda)\cdot AL + \rho\cdot LN + (\rho-\lambda)\cdot ON = (\rho+\lambda)\cdot AM + (\rho-\lambda)\cdot MP \tag{1}$$

Now, from the figure, we have

$$AL = x\cos(60^\circ - \theta), AM = (a-x)\cos\theta, MP = x\cos\theta, LN = LK + KN.$$

Also, $\angle BOK = 60^\circ + \theta$. Hence, $LN = LK + KN = (a-x)\cos(60^\circ - \theta) + x\cos(60^\circ + \theta)$.
Putting these values in Eqn (1), we get

$$(\rho+\lambda)x\cos(60^\circ-\theta) + \rho\{(a-x)\cos(60^\circ-\theta) + x\cos(60^\circ+\theta)\} + (\rho-\lambda)\cdot ON = (\rho+\lambda)(a-x)\cos\theta + (\rho-\lambda)x\cos\theta$$

$$\text{or,}\ \rho\{x\cos(60^\circ-\theta) + (a-x)\cos(60^\circ-\theta) + x\cos(60^\circ+\theta) + ON - (a-x)\cos\theta - x\cos\theta\} = \lambda\{(a-x)\cos\theta - x\cos\theta - x\cos(60^\circ-\theta) + ON\} \tag{2}$$

Since λ and ρ are arbitrary, Eqn (2) can be satisfied only if both sides are zero. Hence,

$$\left.\begin{aligned} x\cos(60^\circ-\theta) + (a-x)\cos(60^\circ-\theta) + x\cos(60^\circ+\theta) + ON \\ -(a-x)\cos\theta - x\cos\theta = 0 \\ (a-x)\cos\theta - x\cos\theta - x\cos(60^\circ-\theta) + ON = 0 \end{aligned}\right\} \tag{3}$$

Eliminating ON from Eqn (3), we get

$$(a-x)\cos(60^\circ-\theta) + x\cos(60^\circ-\theta) + x\cos(60^\circ+\theta) - (a-x)\cos\theta - x\cos\theta = (a-x)\cos\theta - x\cos\theta - x\cos(60^\circ-\theta)$$

$$\text{or,}\ (a-x)\{\cos(60^\circ-\theta) - 2\cos\theta\} = x\{-2\cos(60^\circ-\theta) - \cos(60^\circ+\theta)\}$$

$$\therefore\ \frac{a-x}{x} = \frac{2\cos(60^\circ-\theta) + \cos(60^\circ+\theta)}{2\cos\theta - \cos(60^\circ-\theta)} \tag{4}$$

Numerator of RHS of Eqn (4) is

$$= \cos(60^\circ-\theta) + 2\cos 60^\circ\cos\theta$$

$$= \cos(60^\circ-\theta) + \cos\theta = 2\cos 30^\circ\cos(30^\circ-\theta) \tag{5}$$

Denominator of RHS of Eqn (4) is

$$= \cos\theta + \cos\theta - \cos(60^\circ-\theta) = \cos\theta + 2\sin 30^\circ\sin(30^\circ-\theta)$$

$$= \cos\theta + \sin(30^\circ-\theta) = \cos\theta + \cos(60^\circ+\theta)$$

$$= 2\cos 30^\circ\cos(30^\circ+\theta) \tag{6}$$

Hence, Eqn (4) assumes the form

$$\frac{a-x}{x}=\frac{2\cos 30^\circ \cos(30^\circ-\theta)}{2\cos 30^\circ \cos(30^\circ+\theta)}=\frac{\cos(30^\circ-\theta)}{\cos(30^\circ+\theta)}. \tag{7}$$

Thus the surface of separation divides the side in the ratio

$$(a-x):x=\cos\left(\frac{\pi}{6}-\theta\right):\cos\left(\frac{\pi}{6}+\theta\right).$$

EXAMPLE 2.6 *A vertical circular cylinder of height 2h and radius r, closed at the top, is just filled by equal volumes of two liquids of densities ρ and δ. Show that, if the axis be gradually tilted to the vertical, the pressure at the lowest point of the base will never exceed*

$$g(\rho+\delta)\sqrt{r^2+h^2}.$$

Solution. Let $ABCD$ be the plane section of the cylinder through its axis (FIG. 2.10). Let the point B always touch the ground when the cylinder is gradually being tilted. The pressure at the lowest point B will be maximum when the diagonal BD will be vertical.

Since the volumes of two liquids are same, we have

$$DE=CF=h,\ AE=BF=h. \tag{1}$$

Also, the radius of the cylinder being r, we have

$$EO=OF=r. \tag{2}$$

Hence,

$$CO=DO=\sqrt{h^2+r^2}. \tag{3}$$

In this position, the axis of the cylinder makes an angle θ with the vertical given by

$$\tan\theta=\frac{r}{h},\ \text{i.e.,}\ \theta=\tan^{-1}\left(\frac{r}{h}\right). \tag{4}$$

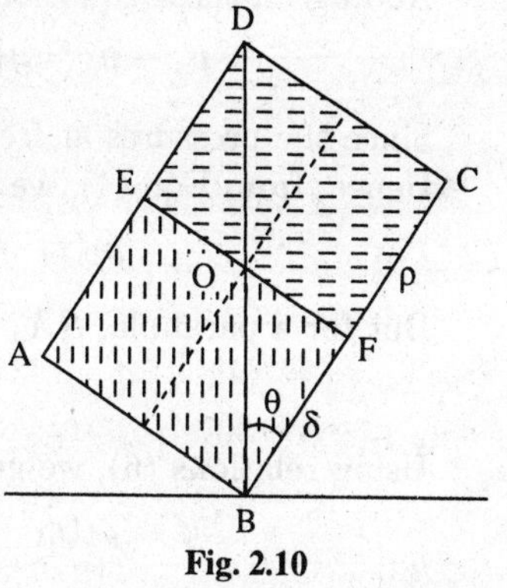

Fig. 2.10

Total pressure at the point B = pressure due to height OD of liquid of density ρ + pressure due to height OB of liquid of density δ.

Thus, $p=g\rho\cdot OD+g\delta\cdot OC=g(\rho+\delta)\sqrt{h^2+r^2}$, by using Eqn (3).

Hence proved.

EXAMPLE 2.7 *A fine parabolic tube is held with its axis vertical and vertex downward. It contains liquids of successive densities $\rho_1,\rho_2,\ldots,\rho_n$ beginning from the top of one side and ending with the last liquid on the other. If $r_0,r_1,r_2,\ldots,r_n$ be the focal distances of the boundaries, prove that $\rho_1(r_0-r_1)+\rho_2(r_1-r_2)+\cdots+\rho_n(r_{n-1}-r_n)=0$.*

Solution. We take origin O at the vertex, the vertical through O as the y-axis and the line PQ as the directrix of the parabolic tube (FIG. 2.11). Let S be the focus of the parabola and $A_0A_1, A_1A_2,\ldots,A_{n-1}A_n$ be the portions of the tube containing liquids of densities $\rho_1,\rho_2,\ldots,\rho_n$ respectively.

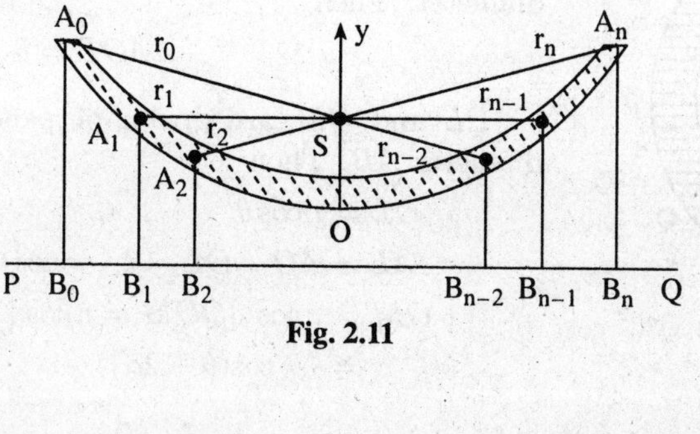

Fig. 2.11

Then, by given condition, we have

$$A_0S = r_0,\ A_1S = r_1, \dots, A_nS = r_n. \tag{1}$$

We draw perpendiculars $A_0B_0, A_1B_1, \dots, A_nB_n$ upon the directrix PQ.
Let

$$A_0B_0 = z_0, A_1B_1 = z_1, \dots, A_nB_n = z_n \tag{2}$$

and $p_0, p_1, p_2, \dots, p_n$ be the pressures at the points $A_0, A_1, \dots, A_n$ respectively. Then we have

$$\left.\begin{aligned} p_1 - p_0 &= g\rho_1(z_1 - z_0)\\ p_2 - p_1 &= g\rho_2(z_2 - z_1)\\ &\dots\dots\dots\dots\\ &\dots\dots\dots\dots\\ p_n - p_{n-1} &= g\rho_n(z_n - z_{n-1}) \end{aligned}\right\} \tag{3}$$

Adding all these equations, we get

$$p_n - p_0 = g\{\rho_1(z_1 - z_0) + \rho_2(z_2 - z_1) + \cdots + \rho_n(z_n - z_{n-1})\}. \tag{4}$$

Since the pressures at free surfaces are zero, we have $p_n = p_0 = 0$.
Hence, from Eqn (4), we get

$$\rho_1(z_1 - z_0) + \rho_2(z_2 - z_1) + \cdots + \rho_n(z_n - z_{n-1}) = 0. \tag{5}$$

But for a parabola, $SA_0 = A_0B_0, SA_1 = A_1B_1, \dots, SA_n = A_nB_n$,

$$\text{i.e., } r_0 = z_0, r_1 = z_1, \dots, r_n = z_n. \tag{6}$$

Using relations (6), we get from Eqn (5)

$$\rho_1(r_1 - r_0) + \rho_2(r_2 - r_1) + \cdots + \rho_n(r_n - r_{n-1}) = 0.$$

Hence proved.

EXAMPLE 2.8 *A fine uniform circular tube is held in a vertical plane and contains two liquids of densities ρ and δ which subtend angles 2α and 2β at the centre. If the radius passing through the surface of separation makes an angle θ with the vertical, show that*

$$\rho \sin\alpha \sin(\alpha \pm \theta) = \delta \sin\beta \sin(\beta \mp \theta)$$

Solution. Let the portion PQ of the tube contain liquid of density ρ and the portion $PRBQ$ contain liquid of density δ.

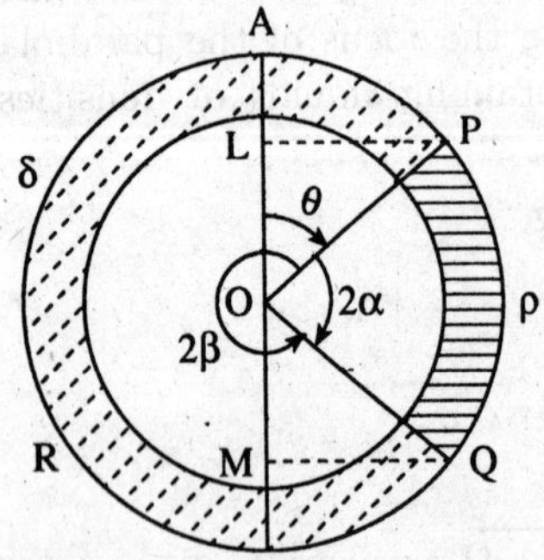

Fig. 2.12

Then by question (FIG. 2.12),

$$\angle POQ = 2\alpha \text{ and } \angle POQ \text{ (obtuse)} = 2\beta. \tag{1}$$

Let r be the radius of the circle and AB the vertical diameter. Then

$$\angle AOP = \theta \tag{2}$$

PL and QM are drawn perpendiculars on the vertical diameter AB. Then

$$\begin{aligned} OL &= r\cos\theta\\ AL &= AO - OL = r - r\cos\theta\\ OM &= r\cos\angle QOM = r\cos\{180° - (\theta + 2\alpha)\}\\ &= -r\cos(\theta + 2\alpha) \end{aligned} \tag{3}$$

Since $2\alpha + 2\beta = 360°, 2\alpha = 360° - 2\beta$, we can write

$$OM = -r\cos\{\theta + 360° - 2\beta\} = -r\cos(2\beta - \theta). \tag{4}$$

For equilibrium at the lowest point B, the pressures from two sides must be the same. Hence,

$$g\rho(OL + OM) + g\delta(AL + BM) = g\delta(AO + BO)$$
$$\text{i.e., } g\rho(OL + OM) + g\delta \cdot AL = g\delta(AO + OM). \tag{5}$$

Using Eqns (3) and (4), we write Eqn (5) as

$$g\rho\{r\cos\theta - r\cos(\theta + 2\alpha)\} + g\delta(r - r\cos\theta) = g\delta\{r - r\cos(2\beta - \theta)\}$$
$$\text{i.e., } \rho\{\cos\theta - \cos(\theta + 2\alpha)\} = \delta\{\cos\theta - \cos(2\beta - \theta)\}$$
$$\text{or, } \rho\sin\alpha\sin(\theta + \alpha) = \delta\sin\beta\sin(\beta - \theta) \tag{6}$$

Similarly, by taking the radius OP joining the surface of separation on the other side of OA when we are to write $-\theta$ for θ in Eqn (6), we get

$$\rho\sin\alpha\sin(\alpha - \theta) = \delta\sin\beta\sin(\beta + \theta). \tag{7}$$

Thus, combining Eqns (6) and (7), we get $\rho\sin\alpha\sin(\alpha \pm \theta) = \delta\sin\beta\sin(\beta \mp \theta)$.

EXAMPLE 2.9 *A fine bent tube in a vertical plane has its branches AB, BC inclined to the vertical at angles α, β on opposite sides. It contains two liquids which fill lengths a, b of the tube, the fluids meeting in the branch AB at a distance c from B. Prove that they will meet at B, if the tube be turned in its own plane through an angle whose cotangent is*

$$\frac{b}{c}\sin\frac{1}{2}(\alpha + \beta)\sec\frac{1}{2}(\alpha - \beta)\sec\alpha - \tan\alpha.$$

Solution. Let ABC be the bent tube meeting at B with branches AB, BC. Let D be the surface of separation in AB of the two liquids. Then

$$\left.\begin{aligned} ED = a, DBF = b, \\ DB = c, BF = b - c. \end{aligned}\right\} \tag{1}$$

Let the densities of liquids in column ED and DBF be ρ and δ respectively (FIG. 2.13). Considering equilibrium at the lowest point B, we write

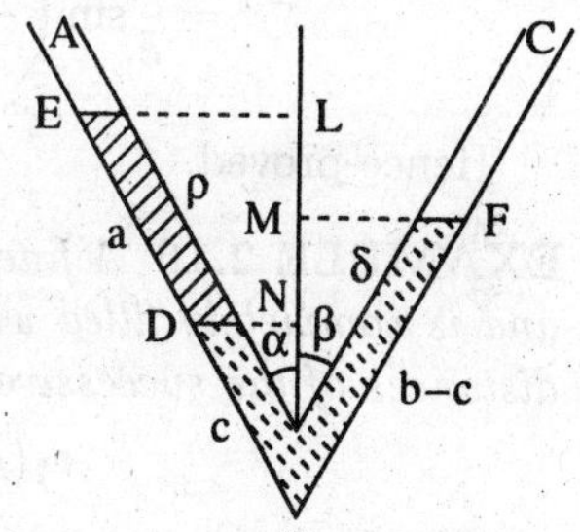

Fig. 2.13

$$g\rho \cdot LN + g\delta \cdot NB = g\delta \cdot MB. \tag{2}$$

Now, from the figure, we have

$$LN = a\cos\alpha, NB = c\cos\alpha, MB = (b - c)\cos\beta. \tag{3}$$

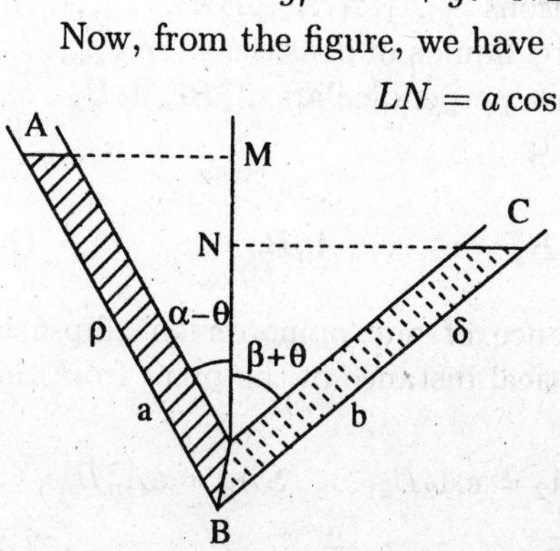

Fig. 2.14

Using the relations in Eqn (2), we get

$$\rho a\cos\alpha + \delta c\cos\alpha = \delta(b - c)\cos\beta$$
$$\text{or, } \rho a\cos\alpha = -\delta\{c\cos\alpha - (b - c)\cos\beta\}$$
$$\text{or, } \frac{\rho}{\delta} = \frac{c\cos\alpha - (b - c)\cos\beta}{a\cos\alpha} \tag{4}$$

Now, let the tube be turned in its own plane through an angle θ so that the surface of separation is at the point B. In this position, angle α becomes $(\alpha - \theta)$ and angle β becomes $(\beta + \theta)$ (FIG. 2.14).

For equilibrium at the lowest point B, we have

$$g\rho \cdot BM = g\delta \cdot BN \text{ or, } \rho a\cos(\alpha-\theta) = \delta b\cos(\beta+\theta).$$

Hence,

$$\frac{\rho}{\delta} = \frac{b\cos(\beta+\theta)}{a\cos(\alpha-\theta)}. \tag{5}$$

From Eqns (4) and (5), we get

$$\frac{b\cos(\beta+\theta)}{a\cos(\alpha-\theta)} = -\frac{c\cos\alpha-(b-c)\cos\beta}{a\cos\alpha}$$

$$\text{or, } b\cos\alpha\cos(\beta+\theta) = \cos(\alpha-\theta)\{(b-c)\cos\beta - c\cos\alpha\}$$

$$\text{or, } b\cos\alpha(\cos\beta\cos\theta-\sin\beta\sin\theta) = (\cos\alpha\cos\theta+\sin\alpha\sin\theta)\{(b-c)\cos\beta-c\cos\alpha\}$$

$$\text{or, } \cos\theta[b\cos\alpha\cos\beta - \cos\alpha\{(b-c)\cos\beta - c\cos\alpha\} = \sin\theta[b\cos\alpha\sin\beta+\sin\alpha\{(b-c)\cos\beta-c\cos\alpha\}$$

$$\text{or, } c\cos\theta\cos\alpha(\cos\alpha+\cos\beta) = \sin\theta\{b\sin(\alpha+\beta)-c\sin\alpha(\cos\alpha+\cos\beta)\}$$

$$\therefore \cot\theta = \frac{b\sin(\alpha+\beta)-c\sin\alpha(\cos\alpha+\cos\beta)}{c\cos\alpha(\cos\alpha+\cos\beta)} = \frac{b\sin(\alpha+\beta)}{c\cos\alpha(\cos\alpha+\cos\beta)} - \frac{\sin\alpha}{\cos\alpha}$$

$$= \frac{b\cdot 2\sin\left(\frac{\alpha+\beta}{2}\right)\cos\left(\frac{\alpha+\beta}{2}\right)}{c\cos\alpha\cdot 2\cos\left(\frac{\alpha+\beta}{2}\right)\cos\left(\frac{\alpha-\beta}{2}\right)} - \tan\alpha$$

$$= \frac{b}{c}\sin\left(\frac{\alpha+\beta}{2}\right)\sec\left(\frac{\alpha-\beta}{2}\right)\sec\alpha - \tan\alpha$$

Hence proved.

EXAMPLE 2.10 *A fine bent tube in the form of an ellipse is held with its major axis vertical and is completely filled with n liquids of densities $\rho_1, \rho_2, \ldots, \rho_n$. If $r_1, r_2, \ldots, r_n$ be the focal distances of the successive points of separation of liquids, prove that*

$$r_1(\rho_1-\rho_2)+r_2(\rho_2-\rho_3)+\cdots+r_n(\rho_n-\rho_1)=0.$$

Solution. Let S be the focus, PQ the directrix and e the eccentricity of the elliptic tube (FIG. 2.15). Let the portions $A_0A_1, A_1A_2, A_2A_3, \ldots, A_nA_0$ of the tube be occupied by liquids of densities $\rho_1, \rho_2, \rho_3, \ldots, \rho_n$ respectively. We draw perpendiculars $A_1B_1, A_2B_2, \ldots, A_nB_n$ on the directrix PQ.

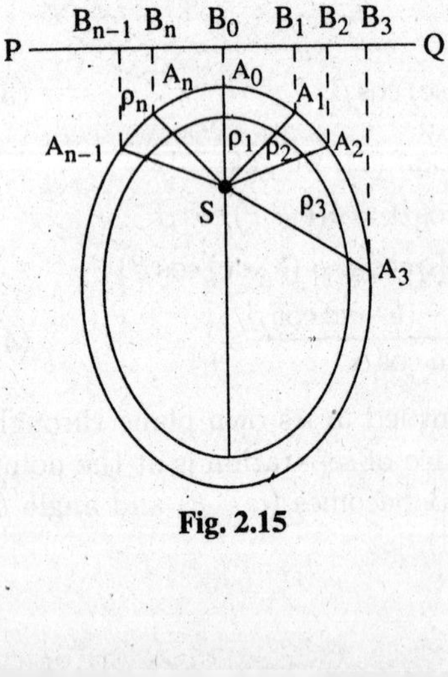

Fig. 2.15

Let

$$A_1B_1 = z_1, A_2B_2 = z_2, \ldots, A_nB_n = z_n. \tag{1}$$

Since the focal distance of any point on an ellipse is equal to e times the vertical distance of the point from the directrix, we have

$$SA_1 = eA_1B_1, SA_2 = eA_2B_2, \ldots, SA_n = eA_nB_n$$

$$\text{i.e., } z_1 = \frac{r_1}{e}, z_2 = \frac{r_2}{e}, \ldots, z_n = \frac{r_n}{e} \tag{2}$$

Now, let $p_1, p_2, \ldots, p_n$ be the pressures at points $A_1, A_2, \ldots, A_n$ respectively. Then we have

$$\begin{aligned} p_2 - p_1 &= g\rho_2(z_2 - z_1) \\ p_3 - p_2 &= g\rho_3(z_3 - z_2) \\ &\ldots\ldots\ldots\ldots \\ &\ldots\ldots\ldots\ldots \\ p_n - p_{n-1} &= g\rho_n(z_n - z_{n-1}) \\ p_1 - p_n &= g\rho_1(z_1 - z_n) \end{aligned}$$

Adding all these equations, we get

$$0 = g[\rho_2(z_2 - z_1) + \rho_3(z_3 - z_2) + \cdots + \rho_n(z_n - z_{n-1}) + \rho_1(z_1 - z_n)]$$

$$\text{or, } 0 = \rho_2\left(\frac{r_2}{e} - \frac{r_1}{e}\right) + \rho_3\left(\frac{r_3}{e} - \frac{r_2}{e}\right) + \cdots + \rho_n\left(\frac{r_n}{e} - \frac{r_{n-1}}{e}\right) + \rho_1\left(\frac{r_1}{e} - \frac{r_n}{e}\right)$$

$$\text{or, } 0 = \rho_2(r_2 - r_1) + \rho_3(r_3 - r_2) + \cdots + \rho_n(r_n - r_{n-1}) + \rho_1(r_1 - r_n).$$

Hence $r_1(\rho_1 - \rho_2) + r_2(\rho_2 - \rho_3) + \cdots + r_n(\rho_n - \rho_1) = 0$ **(Proved)**.

EXAMPLE 2.11 *A circular tube with centre O is filled with three fluids of densities ρ_1, ρ_2, ρ_3 (in descending order of magnitude) and placed in a vertical plane. If $2\alpha, 2\beta, 2\gamma$ be the angles subtended at the centre by the fluids and P be the point on the circumference midway between the ends of the lightest fluid, then angle θ which OP makes with the vertical is given by*

$$\frac{\rho_2 - \rho_3}{\rho_1 - \rho_3} = \frac{\sin\alpha}{\sin(\alpha + \beta)} \cdot \frac{\sin(\beta - \theta)}{\sin\beta}.$$

Solution. Let the fluids of densities ρ_1, ρ_2, ρ_3 occupy the portions LN, NM, ML of the tube respectively. The liquid of lowest density ρ_3 is in the portion ML where P is its middle point. The line CD is the vertical diameter of the tube (FIG. 2.16).

Then by conditions given,

$$\left.\begin{aligned} &\angle LON = 2\alpha, \angle NOM = 2\beta, \\ &\angle MOL = 2\gamma, \angle POD = \theta, \\ &\angle POM = \angle POL = \gamma. \end{aligned}\right\} \quad (1)$$

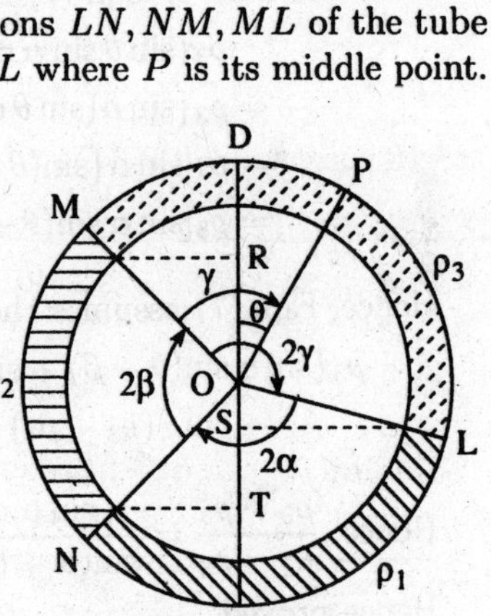

Fig. 2.16

Let the radius of the circle be r. We draw perpendiculars from L, M, N on the vertical CD. For equilibrium at C, the pressures from both sides must be equal. Hence

$$\begin{aligned} &g\rho_3 \cdot DS + g\rho_1 \cdot SC \\ &= g\rho_3 \cdot DR + g\rho_2 \cdot RT + g\rho_1 \cdot TC. \end{aligned} \quad (2)$$

Now

$$\left.\begin{aligned} &DS = DO + OS, SC = OC - OS, DR = OD - OR, \\ &RT = RO + OT, TC = CO - OT. \end{aligned}\right\} \quad (3)$$

Putting these values into Eqn (2), we get after simplification

$$\rho_3(OS + OR) = \rho_1(OS - OT) + \rho_2(OR + OT) \quad (4)$$

Again,

$$\left.\begin{aligned} OS &= r\cos\angle SOL = r\cos(\pi-\theta-\gamma) = r\cos(\alpha+\beta-\theta) \\ &\quad [\text{since } 2\alpha+2\beta+2\gamma = 360^\circ, \alpha+\beta+\gamma=\pi] \\ OR &= r\cos\angle MOR = r\cos(\gamma-\theta) \\ OT &= r\cos\angle NOT = r\cos\{2\alpha-(\alpha+\beta-\theta)\} = r\cos(\alpha-\beta+\theta) \end{aligned}\right\} \quad (5)$$

Putting these values in Eqn (4), we get

$$\rho_3\{r\cos(\alpha+\beta-\theta)+r\cos(\gamma-\theta)\} = \rho_1\{r\cos(\alpha+\beta-\theta)-r\cos(\alpha-\beta+\theta)\}$$
$$+\rho_2\{r\cos(\gamma-\theta)+r\cos(\alpha-\beta+\theta)\}$$

$$\text{or, } \rho_3\cdot 2\cos\left(\frac{\alpha+\beta+\gamma}{2}-\theta\right)\cos\left(\frac{\beta+\alpha-\gamma}{2}\right) = \rho_1\cdot 2\sin\alpha\sin(\theta-\beta)$$
$$+\rho_2\cdot 2\cos\left(\frac{\gamma+\alpha-\beta}{2}\right)\cos\left(\frac{\gamma-\alpha+\beta-2\theta}{2}\right) \quad (6)$$

But we have $\alpha+\beta+\gamma=\pi$, i.e., $\gamma=\pi-\alpha-\beta$. Using these relations in Eqn (6), we get

$$\rho_3\cos\left(\frac{\pi}{2}-\theta\right)\cos\left(\frac{2\alpha+2\beta-\pi}{2}\right)$$
$$= \rho_1\sin\alpha\sin(\theta-\beta)+\rho_2\cos\left(\frac{\pi-2\beta}{2}\right)\cos\left(\frac{\pi-2\alpha-2\theta}{2}\right)$$

$$\text{or, } \rho_3\sin\theta\sin(\alpha+\beta) = \rho_1\sin\alpha\sin(\theta-\beta)+\rho_2\sin\beta\sin(\alpha+\theta) \quad (7)$$

Now, the LHS of Eqn (7) can be written as

$$\rho_3\{\sin\theta\sin\alpha\cos\beta+\sin\theta\cos\alpha\sin\beta\}$$
$$= \rho_3\{\sin\alpha(\sin\theta\cos\beta)+\sin\beta(\sin\theta\cos\alpha)\}$$
$$= \rho_3[\sin\alpha\{\sin(\theta-\beta)+\cos\theta\sin\beta\}+\sin\beta\{\sin(\alpha+\theta)-\cos\theta\sin\alpha\}]$$
$$= \rho_3[\sin\alpha\sin(\theta-\beta)+\sin\beta\sin(\alpha+\theta)] \quad (8)$$

Hence, Eqn (7) assumes the form

$$\rho_3\{\sin\alpha\sin(\theta-\beta)+\sin\beta\sin(\alpha+\theta)\} = \rho_1\sin\alpha\sin(\theta-\beta)+\rho_2\sin\beta\sin(\alpha+\theta)$$
$$\text{or, } (\rho_3-\rho_1)\sin\alpha\sin(\theta-\beta) = (\rho_2-\rho_3)\sin\beta\sin(\alpha+\theta)$$

Hence, $\dfrac{\rho_2-\rho_3}{\rho_1-\rho_3} = \dfrac{\sin\alpha}{\sin(\alpha+\theta)}\cdot\dfrac{\sin(\beta-\theta)}{\sin\beta}$ [since $\sin(\theta-\beta) = -\sin(\beta-\theta)$].

Hence proved.

EXERCISE

1. In the lower half of a uniform circular tube, one quadrant is occupied by a liquid of density 2ρ and the other by two liquids which do not mix and are of densities 3ρ and ρ. Prove that the volume of the lower of the two latter liquids is twice that of the other.

2. If a liquid be heterogeneous and of density $\rho z/a$ at a depth z, show that the pressure is $\pi+\dfrac{g\rho z^2}{2a}$, where π is the atmospheric pressure.

3. Equal volumes of two liquids of densities ρ and 3ρ, which do not mix together just fill a cone which is held with its axis vertical and vertex uppermost. Prove that the pressure at any point of the base is $(3-\sqrt[3]{4})$ times the pressure at the same point when the cone is filled with the lighter liquid.

4. A small uniform tube is bent into the form of a circle whose plane is vertical. Equal quantities of fluids whose densities are ρ and δ, fill half of the tube. Show that the radius passing through the common surface makes with the vertical an angle $\tan^{-1}\left(\dfrac{\rho-\delta}{\rho+\delta}\right)$.

5. Three fluids, whose densities are in A.P. fill a semi-circular tube whose bounding diameter is horizontal. Prove that the depth of one of the common surfaces is double that of the other.

6. A tube in the form of a parabola held with its vertex downward and axis vertical, is filled with two different liquids of densities ρ and ρ'. If the distances of the free surfaces of the liquids from the focus be r and r' respecively, show that the distance of their common surface from the focus is

$$\frac{r\rho-r'\rho'}{\rho-\rho'}.$$

7. A closed tube in the form of an equilateral triangle contains equal volumes of three liquids which do not mix and is placed with the lower side horizontal. Prove that if the densities of the liquids are in A.P., their surfaces of separation will be at points of trisection of the sides of the triangle.

8. A fine circular tube in the vertical plane, contains a column of liquid of density ρ which subtends an angle α at the centre and a column of another liquid of density δ which subtends an angle β at the centre. If θ be the angle which the radius through the common surface makes with the vertical, prove that

$$\tan\theta=\frac{\rho-\delta-\rho\cos\alpha+\delta\cos\beta}{\delta+\rho\sin\alpha+\delta\sin\beta}.$$

9. A thin unform cycloidal tube is held with its axis vertical and vertex downward and contains liquids of densities ρ and δ in such a proportion that their surface of separation is at the vertex. Prove that the weights of the liquids are in the ratio $\sqrt{\rho}:\sqrt{\delta}$.

10. A fine tube bent into the form of an ellipse is held with its major axis vertical and is filled with three liquids whose densities are ρ_1,ρ_2 and ρ_3, taken in order, round the elliptic tube. If r_1,r_2 and r_3 be the distances of the points of separation from either focus, prove that

$$r_1(\rho_1-\rho_2)+r_2(\rho_2-\rho_3)+r_3(\rho_3-\rho_1)=0.$$

11. Liquids of densities ρ and δ occupy areas subtending angles α and β at the centre of a fine vertical circular tube. If the surface of separation be at the lowest point, prove that

$$\sin\frac{\beta}{2}=\sqrt{\frac{\rho}{\delta}}\sin\frac{\alpha}{2}.$$

12. A semi-circular tube has its bounding diameter horizontal and contains equal volumes of n fluids of densities successively equal to $\rho, 2\rho, 3\rho, \ldots$ arranged in this order.

If each fluid subtends an angle 2α at the centre and the tube just holds them all, show that $\tan n\alpha=(2n+1)\tan\alpha$.

13. Three liquids which do not mix and whose densities are ρ_1,ρ_2,ρ_3 fill a circular tube in vertical plane and α,β,γ are the angles which the radius to the common surfaces make with the vertical diameter measured in the same direction. Prove that

$$\rho_1(\cos\beta-\cos\gamma)+\rho_2(\cos\gamma-\cos\alpha)+\rho_3(\cos\alpha-\cos\beta)=0.$$

14. Three liquids whose densities $\rho_1,\rho_2,\rho_3(\rho_1>\rho_2>\rho_3)$ are in A.P., completely fill a circular tube in vertical plane and occupy lenghs of arcs subtending at the centre angles $2\alpha, 2\beta, 2\gamma$ respectively. If the radius to the point midway between the ends of the lightest liquid makes with the vertical an angle θ, prove that $\cot\theta=\cot\alpha+2\cot\beta$.

15. A fine tube of uniform bore is bent into the form of a square and filled with equal volumes of three liquids of densities $\rho_1, \rho_2, \rho_3 (\rho_1 < \rho_2 < \rho_3)$.

If the tube is placed with one side vertical, show that a side of the square will be filled with the liquid of the first, third or second kind only according as

$$\rho_2 > \frac{1}{3}(\rho_1 + 2\rho_3)$$

$$\text{or,}\quad \rho_2 < \frac{1}{3}(2\rho_1 + \rho_3)$$

or, ρ_2 lies between these values.

CHAPTER 3

Fluids in a Given Field of Force

3.1 Introduction

We now consider equilibrium of fluids under the action of arbitrary given forces. But before that we like to have some basic concepts necessary for our discussion.

A ***field of force*** is a region of space in which every particle is subjected to a force proportional to its mass. The force per unit mass is called the ***intensity of force***.

A ***line of force*** is a curve in the field of force, such that the tangent at any point on the curve coincides with the line of action of the resultant external force at that point.

For example, in the gravitational field of force, the lines of force are directed towards the centre of the earth.

3.2 Conservative Force

Work done by an external force $\boldsymbol{F}$ upon a particle to displace it from point 1 to point 2 is defined as

$$W = \int_1^2 \boldsymbol{F} \cdot d\boldsymbol{r} \tag{3.1}$$

Here we have denoted in vector notation.

If the work done by the force in moving a particle from point 1 to point 2 is the same for any possible path between the points, then the force is said to be ***conservative***. The region in which the particle is experiencing a conservative force is called ***conservative field of force***.

For example, the gravitational field of force is conservative, because the work done in moving a particle from one point to another depends on the difference in heights of the points only.

Thus, if we consider a conservative force $\boldsymbol{F}$, work done along two paths P and Q as shown in FIG. 3.1 are the same, i.e.,

$$\left(\int_1^2 \boldsymbol{F} \cdot d\boldsymbol{r}\right)_{\text{on } P} = \left(\int_1^2 \boldsymbol{F} \cdot d\boldsymbol{r}\right)_{\text{on } Q}$$

$$\text{or,} \quad \int_1^2 \boldsymbol{F} \cdot d\boldsymbol{r} + \int_2^1 \boldsymbol{F} \cdot d\boldsymbol{r} = 0, \text{ i.e., } \oint \boldsymbol{F} \cdot d\boldsymbol{r} = 0 \tag{3.2}$$

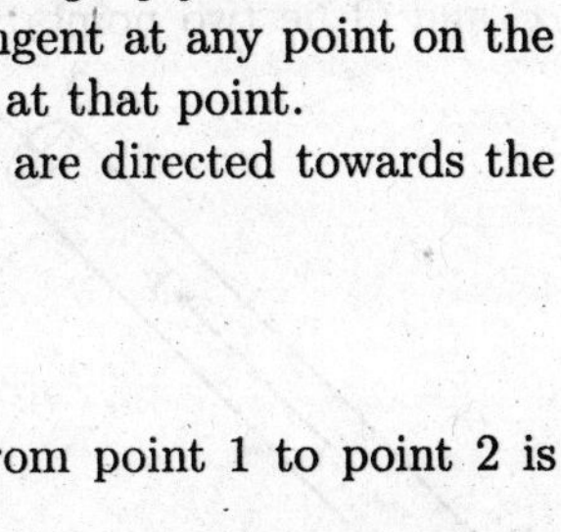

Fig. 3.1

Thus, ***in a conservative field of force, the work done on a particle around a closed path is zero***.

By Stokes' theorem in vector analysis, we write Eqn (3.1) in the form

$$\oint \boldsymbol{F} \cdot d\boldsymbol{r} = \iint (\nabla \times \boldsymbol{F}) \cdot d\boldsymbol{S} = 0, \tag{3.3}$$

where S represents the area within the closed path.

Since the work done is zero around any closed path in the conservative force field, the area dS is arbitrary and hence not equal to zero. Thus,

$$\nabla \times \boldsymbol{F} = 0. \tag{3.4}$$

Therefore, the force $\boldsymbol{F}$ can be represented in terms of a scalar function as

$$\boldsymbol{F} = -\nabla V = -\left(\boldsymbol{i}\frac{\partial V}{\partial x} + \boldsymbol{j}\frac{\partial V}{\partial y} + \boldsymbol{k}\frac{\partial V}{\partial z}\right), \tag{3.5}$$

where $\boldsymbol{i}, \boldsymbol{j}, \boldsymbol{k}$ are unit vectors along rectangular coordinate axes.

It can be verified that

$$\nabla \times \boldsymbol{F} = \nabla \times \nabla V = 0 \tag{3.6}$$

The scalar function V is called the ***potential*** or ***potential energy*** and depends on position.

3.3 Pressure Derivative in Terms of Force

We consider a fluid in equilibrium under the action of a given system of external forces. Let P and Q be two points very near to each other so that PQ is a very short length δs in any assigned direction.

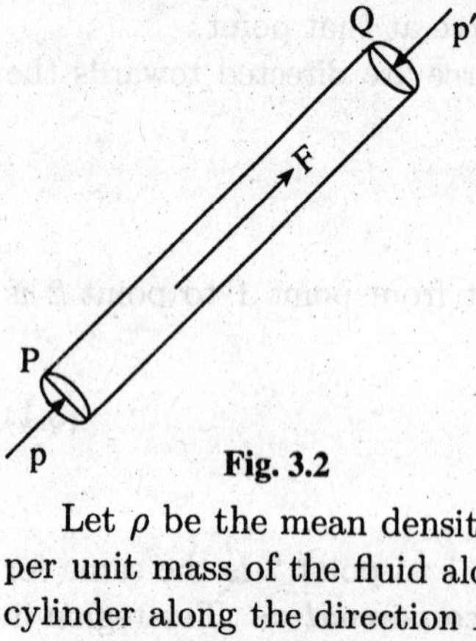

Fig. 3.2

Let p and p' be the pressures at points P and Q respectively (FIG. 3.2). Assuming that the pressure p is a function of s, i.e., $p = p(s)$, we express the pressure p' at Q in Taylor's series,

$$p'(s + \delta s) = p(s) + \frac{\delta p}{\delta s}\delta s + \cdots = p(s) + \frac{\delta p}{\delta s}\delta s \tag{3.7}$$

neglecting higher powers of δs.

Now, with PQ as axis, we construct a very thin cylinder of cross-sectional area α having the plane ends at right angles to the axis.

Let ρ be the mean density of the fluid-cylinder and F be the component of external forces per unit mass of the fluid along the diection PQ. Then the external force acting on the fluid-cylinder along the direction PQ is $= \rho\alpha F \delta s$.

For equilibrium of the fluid-cylinder, we have

$$(p' - p)\alpha = \rho\alpha F \delta s$$

$$\text{or,} \quad \left(p + \frac{\delta p}{\delta s}\delta s - p\right)\alpha = \rho\alpha F\delta s, \text{ or, } \frac{\delta p}{\delta s} = \rho F \tag{3.8}$$

In the limit $\delta s \to 0$, ρ becomes the density at P and F becomes the component of external forces per unit mass at P along the direction PQ. Hence, Eqn (3.8) yields on taking the limit

$$\frac{\partial p}{\partial s} = \rho F \tag{3.9}$$

Thus, ***the pressure derivative along any direction in a fluid at rest under a given system of external forces is equal to density times the component of external forces along that direction.***

3.4 Pressure Equation

Let $P(x, y, z)$ be a point in the fluid with reference to a rectangular coordinate system. Let $Q(x+dx, y, z)$ be another point on x-axis very near to P. We construct a small parallelopiped about PQ whose surfaces are perpendicular to coordinate axes (FIG. 3.3).

Let X, Y, Z be the components of external forces per unit mass along the direction of coordinate axes. If ρ be the mass density of the fluid, then its mass is $= \rho dx\, dy\, dz$, where dx, dy, dz are the lengths of sides of the elemental parallelopiped. Then total external forces acting along coordinate axes are:

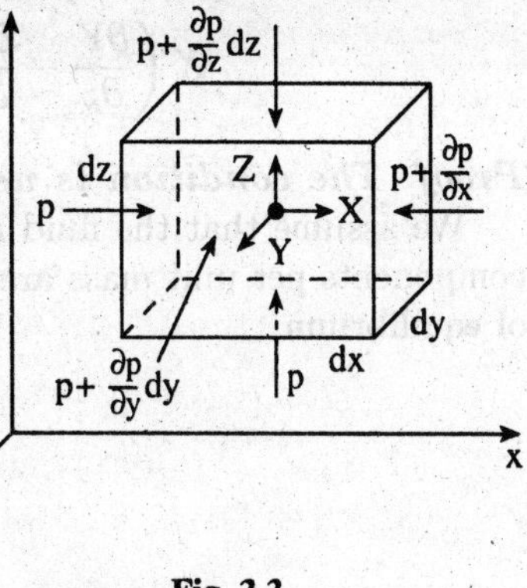

Fig. 3.3

$$\left.\begin{array}{ll} \text{(i) } \rho X(dx\, dy\, dz) & \text{along } x\text{-axis} \\ \text{(ii) } \rho Y(dx\, dy\, dz) & \text{along } y\text{-axis} \\ \text{and (iii) } \rho Z(dx\, dy\, dz) & \text{along } z\text{-axis} \end{array}\right\} \tag{3.10}$$

The fluid thrust on the faces are:

(i) $p\, dy\, dz$ along +ve x-direction and $\left(p + \dfrac{\partial p}{\partial x} dx\right) dy\, dz$ along −ve x-direction,

(ii) $p\, dx\, dz$ along +ve y-direction and $\left(p + \dfrac{\partial p}{\partial y} dy\right) dx\, dz$ along −ve y-direction,

(iii) $p\, dx\, dy$ along +ve z-direction and $\left(p + \dfrac{\partial p}{\partial z} dz\right) dx\, dy$ along −ve z-direction.

For equilibrium, we resolve the forces along x-, y- and z-axes. Thus we get

$$p\, dy\, dz - \left(p + \frac{\partial p}{\partial x} dx\right) dy\, dz + \rho X(dx\, dy\, dz) = 0 \tag{3.11}$$

$$p\, dx\, dz - \left(p + \frac{\partial p}{\partial y} dy\right) dx\, dz + \rho Y(dx\, dy\, dz) = 0 \tag{3.12}$$

$$p\, dx\, dy - \left(p + \frac{\partial p}{\partial z} dz\right) dx\, dy + \rho Z(dx\, dy\, dz) = 0. \tag{3.13}$$

These equations yields

$$\left.\begin{array}{l} \dfrac{\partial p}{\partial x} = \rho X \\ \dfrac{\partial p}{\partial y} = \rho Y \\ \dfrac{\partial p}{\partial z} = \rho Z \end{array}\right\} \tag{3.14}$$

Multiplying the first equation by dx, second by dy and third by dz, and then adding all, we get

$$\frac{\partial p}{\partial x} dx + \frac{\partial p}{\partial y} dy + \frac{\partial p}{\partial z} dz = \rho(X dx + Y dy + Z dz), \text{ i.e., } dp = \rho(X dx + Y dy + Z dz) \tag{3.15}$$

This equation is known as ***pressure equation*** and it determines pressure at any point in the fluid.

Note. Eqn (3.14) can be directly obtained by using the principle of pressure derivative as discussed in the previous article.

3.5 Condition of Equilibrium

The necessary and sufficient condition for equilibrium of a fluid mass under the action of external forces whose components per unit mass are X, Y, Z along the coordinate axes is that

$$X\left(\frac{\partial Y}{\partial z}-\frac{\partial Z}{\partial y}\right)+Y\left(\frac{\partial Z}{\partial x}-\frac{\partial X}{\partial z}\right)+Z\left(\frac{\partial X}{\partial y}-\frac{\partial Y}{\partial x}\right)=0 \tag{3.16}$$

Proof. The condition is necessary.

We assume that the fluid mass is in equilibrium under the action of external forces whose components per unit mass are X, Y, Z along the coordinate axes. Then we have the equations of equilibrium

$$\left.\begin{aligned}\frac{\partial p}{\partial x}&=\rho X\\ \frac{\partial p}{\partial y}&=\rho Y\\ \frac{\partial p}{\partial z}&=\rho Z\end{aligned}\right\}, \tag{3.17}$$

where p is the pressure at any point in the fluid and ρ is the density of the fluid.

Since the pressure p is a function of independent variables x, y, z, we have

$$\left.\begin{aligned}\frac{\partial^2 p}{\partial y\partial z}&=\frac{\partial^2 p}{\partial z\partial y}\\ \frac{\partial^2 p}{\partial z\partial x}&=\frac{\partial^2 p}{\partial x\partial z}\\ \frac{\partial^2 p}{\partial x\partial y}&=\frac{\partial^2 p}{\partial y\partial x}\end{aligned}\right\} \tag{3.18}$$

From Eqns (3.17), we get on using Eqns (3.18)

$$\left.\begin{aligned}\frac{\partial}{\partial y}(\rho Z)&=\frac{\partial}{\partial z}(\rho Y)\\ \frac{\partial}{\partial z}(\rho X)&=\frac{\partial}{\partial x}(\rho Z)\\ \frac{\partial}{\partial x}(\rho Y)&=\frac{\partial}{\partial y}(\rho X)\end{aligned}\right\} \tag{3.19}$$

These are the conditions to be satisfied in order that a given distribution of force with components X, Y, Z may maintain a mass of heterogenous fluid in equilibrium.

Let us consider the first of Eqns (3.19). That is

$$\frac{\partial}{\partial y}(\rho Z)=\frac{\partial}{\partial z}(\rho Y),\text{ i.e., }\rho\frac{\partial Z}{\partial y}+Z\frac{\partial \rho}{\partial y}=\rho\frac{\partial Y}{\partial z}+Y\frac{\partial \rho}{\partial z}$$

$$\text{i.e., }\rho\left(\frac{\partial Y}{\partial z}-\frac{\partial Z}{\partial y}\right)=Z\frac{\partial \rho}{\partial y}-Y\frac{\partial \rho}{\partial z} \tag{3.20}$$

We can get similar two equations like

$$\rho\left(\frac{\partial Z}{\partial x}-\frac{\partial X}{\partial z}\right)=X\frac{\partial \rho}{\partial z}-Z\frac{\partial \rho}{\partial x} \tag{3.21}$$

$$\text{and }\rho\left(\frac{\partial X}{\partial y}-\frac{\partial Y}{\partial x}\right)=Y\frac{\partial \rho}{\partial x}-X\frac{\partial \rho}{\partial y} \tag{3.22}$$

Multiplying Eqns (3.20), (3.21) and (3.22) by X, Y and Z respectively and then adding all these resulting equations, we get

$$X\left(\frac{\partial Y}{\partial z}-\frac{\partial Z}{\partial y}\right)+Y\left(\frac{\partial Z}{\partial x}-\frac{\partial X}{\partial z}\right)+Z\left(\frac{\partial X}{\partial y}-\frac{\partial Y}{\partial x}\right)=0. \tag{3.23}$$

This is the required necessary condition.

The condition is sufficient.

We assume that

$$X\left(\frac{\partial Y}{\partial z}-\frac{\partial Z}{\partial y}\right)+Y\left(\frac{\partial Z}{\partial x}-\frac{\partial X}{\partial z}\right)+Z\left(\frac{\partial X}{\partial y}-\frac{\partial Y}{\partial x}\right)=0 \tag{3.24}$$

We know that the necessary and sufficient condition that the expression

$$X dx+Y dy+Z dz \tag{3.25}$$

is an exact differential is that

$$\left.\begin{aligned}\frac{\partial Y}{\partial z}&=\frac{\partial Z}{\partial y}\\ \frac{\partial Z}{\partial x}&=\frac{\partial X}{\partial z}\\ \frac{\partial X}{\partial y}&=\frac{\partial Y}{\partial x}\end{aligned}\right\} \tag{3.26}$$

Thus Eqn (3.23) shows that the expression (3.25) is an exact differential. Hence the expression $\rho(X dx+Y dy+Z dz)$ is also an exact differential.

Let $\rho(X dx+Y dy+Z dz)=dp$.

Then $\rho(X dx+Y dy+Z dz)=\dfrac{\partial p}{\partial x}dx+\dfrac{\partial p}{\partial y}dy+\dfrac{\partial p}{\partial z}dz$ which yields

$$\left.\begin{aligned}\frac{\partial p}{\partial x}&=\rho X\\ \frac{\partial p}{\partial y}&=\rho Y\\ \frac{\partial p}{\partial z}&=\rho Z\end{aligned}\right\} \tag{3.27}$$

But these are the equations of equilibrium. Hence the condition is sufficient.

3.6 Specific Potential Energy

We consider that the components of external forces acting on unit mass of fluid constitute a conservative system. Then there is a scalar function $U(x, y, z)$ called potential function, such that

$$\left.\begin{aligned}X&=-\frac{\partial U}{\partial x}\\ Y&=-\frac{\partial U}{\partial y}\\ Z&=-\frac{\partial U}{\partial z}\end{aligned}\right\} \tag{3.28}$$

Now, the expression

$$X dx+Y dy+Z dz$$

represents the work done by the external forces on the fluid of unit mass due to infinitesimal displacements dx, dy, dz from the point (x, y, z). By virtue of Eqn (3.28), this work done under conservative system of forces is

$$= -\left(\frac{\partial U}{\partial x}dx + \frac{\partial U}{\partial y}dy + \frac{\partial U}{\partial z}dz\right) = -dU \tag{3.29}$$

which represents the loss of potential energy. Then the pressure Eqn (3.15) can be written in the form

$$\frac{1}{\rho}dp + dU = 0 \tag{3.30}$$

For homogeneous fluids, ρ = constant. In this case Eqn (3.30) may be written as

$$d\left(\frac{p}{\rho} + U\right) = 0,$$

which implies that

$$\frac{p}{\rho} + U = \text{constant}. \tag{3.31}$$

Writing, $\gamma = \rho g$, we get

$$\Pi \equiv \frac{p}{\gamma} + \left(\frac{U}{g}\right) = \text{constant}. \tag{3.32}$$

The quantity Π is called the ***specific potential energy*** and has the dimension of length. It represents the potential energy of unit weight of the fluid measured with reference to some initial plane. The total potential energy of a fluid mass of volume V would be equal to $\gamma V \Pi$.

In view of Eqn (3.32), Eqns (3.27) of equilibrium are reduced to the system

$$\left.\begin{aligned} \frac{\partial \Pi}{\partial x} &= 0 \\ \frac{\partial \Pi}{\partial y} &= 0 \\ \frac{\partial \Pi}{\partial z} &= 0 \end{aligned}\right\} \tag{3.33}$$

This implies that, ***for a homogeneous fluid in equilibrium, change in its specific potential energy in any direction is zero.***

3.7 Surfaces of Equi-pressure

The surface passing through all the points of a fluid at which the pressure is the same is called a surface of equi-pressure.

As we know, the pressure at any point in a fluid at rest is given by

$$dp = \rho(X dx + Y dy + Z dz). \tag{3.34}$$

We can integrate this equation to obtain

$$p = \phi(x, y, z). \tag{3.35}$$

If p = constant = c (say), then

$$\phi(x, y, z) = c, \tag{3.36}$$

which determines the equation of a surface at all points on which the pressure is constant. By assigning different values to the constant c, we obtain a series of surfaces of equi-pressure. The equation of free surface is obtained by taking p equal to the pressure on the free surface. Thus, if the external pressure is zero, the equation of the free surface will be given by

$$\phi(x, y, z) = 0. \tag{3.37}$$

Now, a surface of equal pressure is given by

$$\phi(x, y, z) = p = \text{constant}. \tag{3.38}$$

The quantities

$$\frac{\partial\phi}{\partial x}, \frac{\partial\phi}{\partial y}, \frac{\partial\phi}{\partial z}, \tag{3.39}$$

which are proportional to the direction cosines of the normal at the point (x, y, z) of the surface (3.37), are equal to

$$\frac{\partial p}{\partial x}, \frac{\partial p}{\partial y}, \frac{\partial p}{\partial z} \tag{3.40}$$

respectively.

But the quantities in (3.40) are equal to

$$\rho X, \rho Y, \rho Z \tag{3.41}$$

respectively, where ρ is the density of the fluid and X, Y, Z are components of external forces per unit mass acting at the point (x, y, z) in the direction of coordinate axes.

Hence, the direction cosines of the normal to the surface (3.38) at the point (x, y, z) are proportional to X, Y, Z.

Hence the resultant force at any point is in the direction of the normal to the surface of equal pressure passing through the point. Therefore, ***the surfaces of equal pressure are cut orthogonally by the lines of force***.

It follows from this result that a necessary condition of equilibrium of fluid under a given system of forces is the existence of a system of surfaces orthogonal to the lines of force.

3.8 Equi-potential Surfaces

Let the components of forces X, Y, Z acting in the fluid per unit mass constitute a conservative system. Then the pressure Eqn (3.15) becomes

$$dp = -\rho dU, \tag{3.42}$$

where U is the potential energy function for a unit mass of fluid in the given field.

Since dp is a perfect differential, ρdU must also be a perfect differential. Hence ρ must be a function of U. Thus, when $\rho =$ constant, $U =$ constant.

Again, writing Eqn (3.42) as

$$\frac{dp}{\rho} = -dU \tag{3.43}$$

we conclude that $\dfrac{dp}{\rho}$ must be a perfect differential since dU is so. Hence, ρ is also a function of p. Thus, when $\rho =$ constant, $p =$ constant.

Therefore, in a conservative field of forces, the surfaces of equal pressure are equi-potential surfaces, and are also surfaces of equal density. That is, ***in a conservative field of forces, the surfaces of equi-pressure, equi-potential energy and equi-density coincide.***

Now suppose that the fluid be elastic and the temperature be variable. The relation between pressure p, density ρ and temperature T in this case is given by

$$p = k\rho T. \tag{3.44}$$

Then, in a conservative field of forces, Eqn (3.44), together with Eqn (3.42) yields

$$\frac{dp}{p} = -\frac{dU}{kT}. \tag{3.45}$$

It follows by a similar process of reasoning that T is a function of p. Hence, ***surfaces of equal pressure are also surfaces of equal temperature.***

3.9 Differential Equations of Equi-pressure Surfaces

Let the fluid be heterogeneous and incompressible. Then the density ρ of the fluid is a function of (x, y, z), so that

$$\rho = \rho(x, y, z). \tag{3.46}$$

Hence,

$$d\rho = \frac{\partial \rho}{\partial x}dx + \frac{\partial \rho}{\partial y}dy + \frac{\partial \rho}{\partial z}dz. \tag{3.47}$$

The surfaces of equal density are given by $d\rho = 0$. Hence, from Eqn (3.47), we have

$$\frac{\partial \rho}{\partial x}dx + \frac{\partial \rho}{\partial y}dy + \frac{\partial \rho}{\partial z}dz = 0. \tag{3.48}$$

The surfaces of equal pressure are given by $dp = 0$. Hence, from Eqn (3.34), we have

$$\rho(X dx + Y dy + Z dz) = 0, \text{ or, } X dx + Y dy + Z dz = 0, \tag{3.49}$$

since ρ is a non-zero function of (x, y, z).

Eqns (3.48) and (3.49) are the differential equations of surfaces which by their intersections determine the curves of equal pressure and density.

Solving these equations, we have

$$\frac{dx}{Y\frac{\partial \rho}{\partial z} - Z\frac{\partial \rho}{\partial y}} = \frac{dy}{Z\frac{\partial \rho}{\partial x} - X\frac{\partial \rho}{\partial z}} = \frac{dz}{X\frac{\partial \rho}{\partial y} - Y\frac{\partial \rho}{\partial x}}. \tag{3.50}$$

But from the conditions of equilibrium, we have as shown in Eqns (3.20), (3.21) and (3.22),

$$\left.\begin{aligned} Z\frac{\partial \rho}{\partial y} - Y\frac{\partial \rho}{\partial z} = \rho\left(\frac{\partial Y}{\partial z} - \frac{\partial Z}{\partial y}\right), \quad X\frac{\partial \rho}{\partial z} - Z\frac{\partial \rho}{\partial x} = \rho\left(\frac{\partial Z}{\partial x} - \frac{\partial X}{\partial z}\right) \\ \text{and } Y\frac{\partial \rho}{\partial x} - X\frac{\partial \rho}{\partial y} = \rho\left(\frac{\partial X}{\partial y} - \frac{\partial Y}{\partial x}\right) \end{aligned}\right\} \tag{3.51}$$

In view of Eqn (3.51), Eqn (3.50) assume the form

$$\frac{dx}{\frac{\partial Y}{\partial z} - \frac{\partial Z}{\partial y}} = \frac{dy}{\frac{\partial Z}{\partial x} - \frac{\partial X}{\partial z}} = \frac{dz}{\frac{\partial X}{\partial y} - \frac{\partial Y}{\partial x}}. \tag{3.52}$$

These are the equations of the curves of equal pressure and equal density.

3.10 Fluids under Central Forces

We consider a fluid mass in which the particles are under the action of forces which tend to or from a fixed point and are functions of distances from those fixed points. Thus, the forces in this case can be written as

$$F = \Sigma F(r). \tag{3.53}$$

If a, b, c be the coordinates of the fixed point, then we have

$$r^2 = (x-a)^2 + (y-b)^2 + (z-c)^2. \tag{3.54}$$

Hence,

$$rdr = (x-a)dx + (y-b)dy + (z-c)dz \tag{3.55}$$

$$\text{and } \frac{\partial r}{\partial x} = \frac{x-a}{r}, \frac{\partial r}{\partial y} = \frac{y-b}{r}, \frac{\partial r}{\partial z} = \frac{z-c}{r} \tag{3.56}$$

The components of forces X, Y, Z along the coordinate axes are given by

$$\left.\begin{aligned} X &= \Sigma F(r)\frac{x-a}{r} \\ Y &= \Sigma F(r)\frac{y-b}{r} \\ Z &= \Sigma F(r)\frac{z-c}{r} \end{aligned}\right\} \tag{3.57}$$

Hence,

$$Xdx + Ydy + Zdz = \Sigma F(r)dr, \tag{3.58}$$

by using Eqn (3.55).

Then the pressure Eqn (3.34) in this case assumes the form

$$dp = \rho\Sigma F(r)dr. \tag{3.59}$$

Taking partial derivatives of Eqns (3.57) and using Eqn (3.56), we get

$$\left.\begin{aligned} \frac{\partial X}{\partial y} &= \Sigma\left[F'(r)\frac{x-a}{r}\cdot\frac{y-b}{r} - F(r)\frac{x-a}{r^2}\cdot\frac{y-b}{r}\right] \\ \frac{\partial Y}{\partial x} &= \Sigma\left[F'(r)\frac{y-b}{r}\cdot\frac{x-a}{r} - F(r)\frac{y-b}{r^2}\cdot\frac{x-a}{r}\right] \end{aligned}\right\} \tag{3.60}$$

$$\left.\begin{aligned} \frac{\partial X}{\partial z} &= \Sigma\left[F'(r)\frac{x-a}{r}\cdot\frac{z-c}{r} - F(r)\frac{x-a}{r^2}\cdot\frac{z-c}{r}\right] \\ \frac{\partial Z}{\partial x} &= \Sigma\left[F'(r)\frac{z-c}{r}\cdot\frac{x-a}{r} - F(r)\frac{z-c}{r^2}\cdot\frac{x-a}{r}\right] \end{aligned}\right\} \tag{3.61}$$

$$\left.\begin{aligned} \frac{\partial Y}{\partial z} &= \Sigma\left[F'(r)\frac{y-b}{r}\cdot\frac{z-c}{r} - F(r)\frac{y-b}{r^2}\cdot\frac{z-c}{r}\right] \\ \frac{\partial Z}{\partial y} &= \Sigma\left[F'(r)\frac{z-c}{r}\cdot\frac{y-b}{r} - F(r)\frac{z-c}{r^2}\cdot\frac{y-b}{r}\right] \end{aligned}\right\} \tag{3.62}$$

It is obvious that Eqn (3.23) is always satisfied. If the fluid is heterogeneous, the equilibrium of the fluid may not be always possible, since Eqn (3.19) may not be satisfied. But for a homogeneous fluid, $\rho =$ constant, we have from Eqn (3.19)

$$\left.\begin{aligned}\frac{\partial Z}{\partial y} = \frac{\partial Y}{\partial z}\\ \frac{\partial X}{\partial z} = \frac{\partial Z}{\partial x}\\ \frac{\partial Y}{\partial x} = \frac{\partial X}{\partial y}\end{aligned}\right\} \tag{3.63}$$

which are in this case always satisfied. Hence, equilibrium of ***the homogeneous fluid under the central forces is always possible.***

The situation of fluids under the action of central forces occurs when the particles of fluids attract one another according to the law of gravitation. We call such fluids as ***gravitating fluids.***

For example, let us consider a sphere of gravitating fluid of radius R. If P be any point within the fluid at a distance r from the centre O, then the pressure equation is (FIG. 3.4)

$$dp = \rho F(r)dr, \tag{3.64}$$

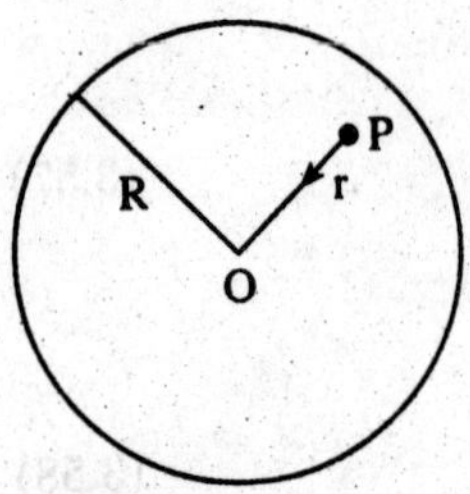

Fig. 3.4

where ρ is the density. The force of attraction $F(r)$ is given by

$$F = -G\frac{M \cdot 1}{r^2}, \tag{3.65}$$

where M = mass of the sphere of radius $r = \frac{4}{3}\pi r^3 \rho$, and G is the gravitational constant. Hence

$$F = -G\frac{\frac{4}{3}\pi r^3 \rho}{r^2} = -\frac{4}{3}\pi G\rho r. \tag{3.66}$$

Thus, if the fluid be homogeneous, equilibrium of the sphere of gravitating fluid under the central force given by Eqn (3.66) is always possible.

Hence, the pressure equation becomes

$$dp = -\rho\left(\frac{4}{3}\pi G\rho\right) r dr = -\xi r dr, \tag{3.67}$$

where we have denoted

$$\xi = -\frac{4}{3}\pi G\rho^2. \tag{3.68}$$

Integrating Eqn (3.67), we get

$$p = C - \frac{\xi}{2}r^2. \tag{3.69}$$

Considering the boundary of the sphere to be free from pressure, we have the condition

$$p = 0, \text{ for } r = R. \tag{3.70}$$

This yields,

$$p = \frac{\xi}{2}(R^2 - r^2). \tag{3.71}$$

The pressure at the centre will be maximum and is given by

$$p_{\max} = \frac{\xi}{2}R^2 = \frac{2}{3}\pi G\rho^2 R^2. \tag{3.72}$$

3.11 Ellipsoid of Gravitating Fluid

We consider the equilibrium of a homogeneous mass of gravitating fluid in the form of an ellipsoid. We take a point $P(x,y,z)$ within the fluid of constant density ρ. Then the attraction on a unit mass at the point P due to a similar ellipsoid through P are (FIG. 3.5).

$$\left.\begin{aligned} X &= -A\rho x \\ Y &= -B\rho y \\ Z &= -C\rho z \end{aligned}\right\}. \tag{3.73}$$

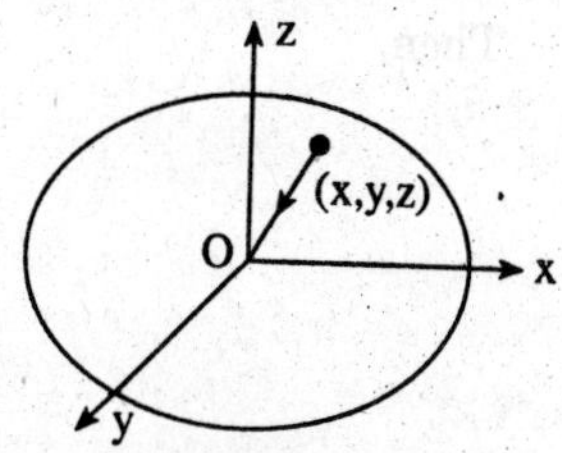

Fig. 3.5

Let the equation of ellipsoid be given by

$$\frac{x^2}{a^2}+\frac{y^2}{b^2}+\frac{z^2}{c^2}=1, \tag{3.74}$$

where a,b,c are semi-axes of the ellipsoid. Then the factors A,B,C in Eqn (3.73) are (Ref. 3)

$$\left.\begin{aligned} A &= 2\pi abc\int_0^\infty \frac{du}{(a^2+u)D} \\ B &= 2\pi abc\int_0^\infty \frac{du}{(b^2+u)D} \\ C &= 2\pi abc\int_0^\infty \frac{du}{(c^2+u)D} \end{aligned}\right\}, \tag{3.75}$$

where we have denoted

$$D=\{(a^2+u)(b^2+u)(c^2+u)\}^{\frac{1}{2}}. \tag{3.76}$$

Hence the pressure equation in this case is

$$\frac{dp}{\rho}=-\rho(Ax+By+Cz). \tag{3.77}$$

Integrating Eqn (3.77), we get

$$\frac{p}{\rho}=\lambda-(Ax^2+By^2+Cz^2)\rho. \tag{3.78}$$

The surfaces of zero pressure are, therefore,

$$\rho(Ax^2+By^2+Cz^2)=\text{constant}=\lambda. \tag{3.79}$$

If we assume the surface of the ellipsoid to be free, then comparing Eqns (3.74) and (3.79), we get

$$A\rho a^2=B\rho b^2=C\rho c^2=\lambda. \tag{3.80}$$

Thus the value of integration constant λ in Eqn (3.78) is obtained. Hence the pressure at any point can be determined.

The pressure at the centre $(0,0,0)$ is maximum and its value from Eqn (3.78) is obtained as

$$p_{\text{max}}=\lambda=A\rho a^2=B\rho b^2=C\rho c^2. \tag{3.81}$$

EXAMPLE 3.1 *The components of forces X,Y,Z acting on an element of fluid at (x,y,z) are given as*

$$\begin{aligned} X &= y^2+2\lambda yz+z^2 \\ Y &= z^2+2\mu zx+x^2 \\ Z &= x^2+2\nu xy+y^2. \end{aligned}$$

Show that, if equilibrium be possible, then $2\lambda=2\mu=2\nu=1$.

Solution. We are given,

$$\left.\begin{aligned} X &= y^2 + 2\lambda yz + z^2 \\ Y &= z^2 + 2\mu zx + x^2 \\ Z &= x^2 + 2\nu xy + y^2 \end{aligned}\right\} \tag{1}$$

Then,

$$\frac{\partial X}{\partial y} = 2y + 2\lambda z, \quad \frac{\partial X}{\partial z} = 2z + 2\lambda y \tag{2}$$

$$\frac{\partial Y}{\partial x} = 2x + 2\mu z, \quad \frac{\partial Y}{\partial z} = 2z + 2\mu x \tag{3}$$

$$\frac{\partial Z}{\partial x} = 2x + 2\nu y, \quad \frac{\partial Z}{\partial y} = 2y + 2\nu x. \tag{4}$$

For eqilibirum to be possible, we must have

$$X\left(\frac{\partial Y}{\partial z} - \frac{\partial Z}{\partial y}\right) + Y\left(\frac{\partial Z}{\partial x} - \frac{\partial X}{\partial z}\right) + Z\left(\frac{\partial X}{\partial y} - \frac{\partial Y}{\partial x}\right) = 0. \tag{5}$$

Putting the values from Eqns (1) - (5), we get

$$\begin{aligned} &(y^2 + 2\lambda yz + z^2)\{(2z + 2\mu x) - (2y + 2\nu x)\} \\ &+(z^2 + 2\mu zx + x^2)\{(2x + 2\nu y) - (2z + 2\lambda y)\} \\ &+(x^2 + 2\nu xy + y^2)\{2y + 2\lambda z) - (2x + 2\mu z)\} = 0 \\ \text{or, } &x^2y(1 - \lambda - \nu) + x^2z(\lambda + \mu - 1) + y^2z(1 - \lambda - \mu) \\ &+y^2x(\mu + \nu - 1) + z^2x(1 - \mu - \nu) + z^2y(\lambda + \nu - 1) = 0. \end{aligned} \tag{6}$$

This implies that

$$\left.\begin{aligned} 1 - \lambda - \nu = 0,\ \lambda + \mu - 1 = 0,\ 1 - \lambda - \mu = 0 \\ \mu + \nu - 1 = 0,\ 1 - \mu - \nu = 0,\ \lambda + \nu - 1 = 0 \end{aligned}\right\} \tag{7}$$

Hence, $\lambda + \mu = 1, \mu + \nu = 1, \nu + \lambda = 1$, or, $\lambda = \mu = \nu = \frac{1}{2}$

$$\therefore 2\lambda = 2\mu = 2\nu = 1.$$

EXAMPLE 3.2 *The components of forces acting in a fluid per unit mass are given as*

$$X = y(y + z),\ Y = z(z + x),\ Z = y(y - x).$$

Show that the surfaces of equal pressure are hyperbolic paraboloids

$$y(x + z) = c(y + z),\ c = \textit{constant},$$

and the curves of equal pressure and density are given by

$$y(x + z) = \textit{constant},\ y + z = \textit{constant respectively}.$$

Solution. We are given

$$X = y(y + z), Y = z(z + x), Z = y(y - x). \tag{1}$$

The pressure equation, as we know, is

$$dp = \rho(X dx + Y dy + Z dz), \tag{2}$$

where ρ is the density of the fluid.

For surfaces of equal pressure, $p =$ constant, which implies that $dp = 0$. Hence, from Eqn (2), we have

$$Xdx + Ydy + Zdz = 0, \text{ (since } \rho \neq 0). \tag{3}$$

Putting the values of X, Y, Z from Eqn (1) into Eqn (3), we get

$$y(y+z)dx + z(z+x)dy + y(y-x)dz = 0$$

$$\text{or, } \frac{dx}{z+x} + \frac{zdy}{y(y+z)} + \frac{(y-x)dz}{(y+z)(z+x)} = 0$$

$$\text{or, } \frac{dx}{z+x} + dy\left(\frac{1}{y} - \frac{1}{y+z}\right) + \frac{(y+z)-(x+z)}{(y+z)(z+x)}dz = 0$$

$$\text{or, } \frac{dx}{z+x} + dy\left(\frac{1}{y} - \frac{1}{y+z}\right) + \frac{dz}{z+x} - \frac{dz}{y+z} = 0$$

$$\text{or, } \frac{d(x+z)}{x+z} + \frac{dy}{y} - \frac{d(y+z)}{y+z} = 0.$$

Integrating,

$$\ln(x+z) + \ln y - \ln(y+z) = \ln c, \tag{4}$$

where $c > 0$ is the constant of integration.

From Eqn (4), we get

$$\ln(y(x+z)] = \ln[c(y+z)] \text{ or, } y(x+z) = c(y+z), \tag{5}$$

which are the required surfaces of equal pressure. Again, the curves of equal pressure and density are given by

$$\frac{dx}{\dfrac{\partial Y}{\partial z} - \dfrac{\partial Z}{\partial y}} = \frac{dy}{\dfrac{\partial Z}{\partial x} - \dfrac{\partial X}{\partial z}} = \frac{dz}{\dfrac{\partial X}{\partial y} - \dfrac{\partial Y}{\partial x}}. \tag{6}$$

From Eqn (1), we find

$$\left.\begin{aligned} \frac{\partial Y}{\partial z} - \frac{\partial Z}{\partial y} &= (2z+x) - (2y-x) = 2(x+z-y) \\ \frac{\partial Z}{\partial x} - \frac{\partial X}{\partial z} &= -y - y = -2y \\ \frac{\partial X}{\partial y} - \frac{\partial Y}{\partial x} &= (2y+z) - z = 2y \end{aligned}\right\} \tag{7}$$

Putting the values of (7) into Eqn (6), we get

$$\frac{dx}{2(x+z-y)} = \frac{dy}{-2y} = \frac{dz}{2y} \text{ or, } \frac{dx}{x+z-y} = \frac{dy}{-y} = \frac{dz}{y} = \lambda \text{ (say)}. \tag{8}$$

From last two fractions, we have $ydy = -ydz$, i.e., $dy + dz = 0$.
This yields

$$y + z = \text{constant}. \tag{9}$$

Again, $ydx + (x+z)dy + ydz = \lambda[y(x+z-y) - y(x+z) + y^2] = 0$, i.e., $d\{y(x+z)\} = 0$.
This yields

$$y(x+z) = \text{constant}. \tag{10}$$

Eqns (9) and (10) together represent curves of equal pressure and density.

EXAMPLE 3.3 *Prove that, if the forces per unit mass at the point (x, y, z) parallel to axes are $y(a-z), x(a-z), xy$, the surfaces of equal pressure are hyperbolic paraboloids and the curves of equal pressure and density are rectangular hyperbolas.*

Solution. We are given

$$X = y(a-z), Y = x(a-z), Z = xy. \tag{1}$$

The pressure equation is

$$dp = \rho(X dx + Y dy + Z dz). \tag{2}$$

For surfaces of equal pressure, $dp = 0$. Hence,

$$X dx + Y dy + Z dz = 0. \tag{3}$$

Putting the values of X, Y, Z from Eqn (1), we get

$$y(a-z)dx + x(a-z)dy + xydz = 0$$

$$\text{or, } \frac{dx}{x} + \frac{dy}{y} + \frac{dz}{a-z} = 0. \tag{4}$$

Integrating, we get

$$\ln x + \ln y - \ln(a-z) = \ln c$$

$$\text{or, } \ln xy = \ln\{c(a-z)\}, \text{ or, } xy = c(a-z). \tag{5}$$

Eqn (5) determines the surfaces of equal pressure and represents hyperbolic paraboloids. The curves of equal pressure and density are given by

$$\frac{dx}{\dfrac{\partial Y}{\partial z} - \dfrac{\partial Z}{\partial y}} = \frac{dy}{\dfrac{\partial Z}{\partial x} - \dfrac{\partial X}{\partial z}} = \frac{dz}{\dfrac{\partial X}{\partial y} - \dfrac{\partial Y}{\partial x}}. \tag{6}$$

From Eqn (1), we find

$$\left.\begin{aligned} \frac{\partial Y}{\partial z} - \frac{\partial Z}{\partial y} &= -x - x = -2x \\ \frac{\partial Z}{\partial x} - \frac{\partial X}{\partial z} &= y + y = 2y \\ \frac{\partial X}{\partial y} - \frac{\partial Y}{\partial x} &= a - a = 0. \end{aligned}\right\} \tag{7}$$

Putting these values into Eqn (6), we get

$$\frac{dx}{-2x} = \frac{dy}{2y} = \frac{dz}{0}, \text{ or, } \frac{dx}{x} = \frac{dy}{-y} = \frac{dz}{0}. \tag{8}$$

From the first two fractions, we have $xdy + ydx = 0$, i.e., $d(xy) = 0$. Hence,

$$xy = \text{constant} = A \text{ (say)}. \tag{9}$$

From the last fraction, we have $dz = 0$. Hence,

$$z = \text{constant} = B \text{ (say)}. \tag{10}$$

Thus, intersections of curves (9) and (10), which are rectangular hyperbolas are the curves of equal pressure and density.

EXAMPLE 3.4 *A container carrying heavy liquid of density ρ moves down a rough plane inclined at an angle α with the horizon. Determine the pressure at any point in the liquid if gravity is the only external force acting in the liquid and μ is the coefficient of friction between the wheels of the container and the plane.*

Also discuss the situation when $\mu = 0$.

Solution. The cross section of the container by a vertical plane passing through its centre of gravity is shown in the FIG. 3.6. We take the origin of coordinates at the middle point in the bottom of the container.

Let f be the resultant acceleration of the container down the plane. Since the reaction force R per unit mass is $g\cos\alpha$, the acceleration along upward direction caused due to friction is $\mu g\cos\alpha$. Hence, the resltant acceleration down the plane is

$$f = g\sin\alpha - \mu g\cos\alpha = g(\sin\alpha - \mu\cos\alpha). \quad (1)$$

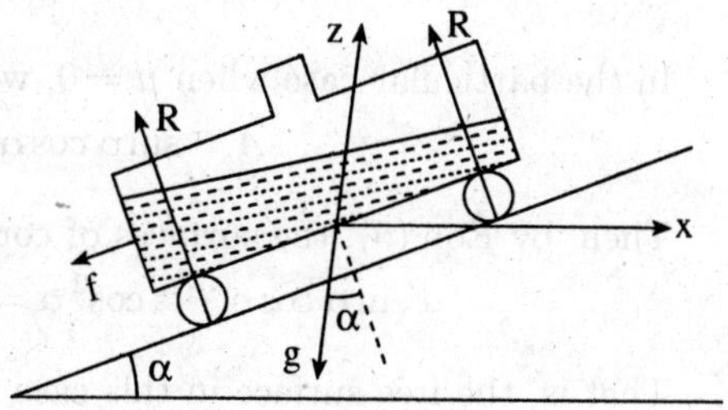

Fig. 3.6

Now we treat the problem as a statical one by applying a reversed effective force f per unit mass along the upward direction of the plane.

Hence, we have

$$\left.\begin{aligned} X &= f\cos\alpha = g(\sin\alpha - \mu\cos\alpha)\cos\alpha \\ Y &= 0 \\ Z &= -g + f\sin\alpha = g\{-1 + (\sin\alpha - \mu\cos\alpha)\sin\alpha\} \end{aligned}\right\} \quad (2)$$

Then the pressure equation $dp = \rho\{X dx + Y dy + Z dz\}$ assumes the form

$$dp = \rho g[\{(\sin\alpha - \mu\cos\alpha)\cos\alpha\}dx + \{-1 + (\sin\alpha - \mu\cos\alpha)\sin\alpha\}dz]. \quad (3)$$

Integrating, we get

$$p = \rho g[\{(\sin\alpha - \mu\cos\alpha)\cos\alpha\}x + \{-1 + (\sin\alpha - \mu\cos\alpha)\sin\alpha\}z + C].$$

This can be written as

$$\frac{p}{g\rho} = Ax + Bz + C, \quad (4)$$

where

$$A = (\sin\alpha - \mu\cos\alpha)\cos\alpha,\ B = (\sin\alpha - \mu\cos\alpha)\sin\alpha - 1 \quad (5)$$

and C is the integration constant.

Hence the system of planes

$$Ax + Bz = \text{constant} \quad (6)$$

determines the equi-pressure surfaces.

Let the surface of the liquid be open to the atmosphere. If Π be the atmospheric pressure and λ be the value of the constant in Eqn (6) at the free surface, then from Eqn (4), we have

$$\frac{\Pi}{\rho g} = \lambda + C, \text{ i.e., } C = \frac{\Pi}{\rho g} - \lambda. \quad (7)$$

Then the pressure at any point in the liquid is given by

$$\frac{p}{g\rho} = Ax + Bz + \frac{\Pi}{g\rho} - \lambda. \quad (8)$$

Let $z = z_0$ be the height of the liquid surface at the origin. Then for the point $(0, z_0)$, we have from Eqn (8),

$$\frac{\Pi}{g\rho} = Bz_0 + \frac{\Pi}{g\rho} - \lambda,$$

which yields

$$\lambda = Bz_0 = \sin\alpha(\sin\alpha - \mu\cos\alpha)z_0 - z_0. \tag{9}$$

Hence the pressure at any point in the liquid is

$$p = \Pi + g\rho\{Ax + B(z - z_0)\}. \tag{10}$$

In the particular case when $\mu = 0$, we have from Eqn (5)

$$A = \sin\alpha\cos\alpha,\ B = \sin^2\alpha - 1 = -\cos^2\alpha. \tag{11}$$

Then, by Eqn (6), the surfaces of constant pressure are given by

$$x\sin\alpha\cos\alpha - z\cos^2\alpha = \text{constant, or, } z = x\tan\alpha + \text{constant}. \tag{12}$$

That is, the free surface in this case is parallel to the inclined plane.

From Eqns (10) and (11), we get the pressure at any point in this case as

$$p = \Pi + g\rho\left\{x\sin\alpha\cos\alpha - (z - z_0)\cos^2\alpha\right\}. \tag{13}$$

EXAMPLE 3.5 *A given volume V of a liquid is acted upon by forces*

$$-\frac{\mu x}{a^2}, -\frac{\mu y}{b^2}, -\frac{\mu z}{c^2}$$

per unit mass along the x-, y-, z-directions respectively. Find the equation to the free surface.

Solution. We are given

$$X = -\frac{\mu x}{a^2}, Y = -\frac{\mu y}{b^2}, Z = -\frac{\mu z}{c^2}. \tag{1}$$

The pressure equation is

$$dp = \rho(X\,dx + Y\,dy + Z\,dz), \tag{2}$$

where ρ is the density of the liquid.

From Eqns (1) and (2), we get

$$dp = -\mu\rho\left(\frac{x\,dx}{a^2} + \frac{y\,dy}{b^2} + \frac{z\,dz}{c^2}\right). \tag{3}$$

On the free surface, $p = 0$, i.e., $dp = 0$.

Hence, Eqn (3) yields

$$\frac{x\,dx}{a^2} + \frac{y\,dy}{b^2} + \frac{z\,dz}{c^2} = 0. \tag{4}$$

Integrating this equation,

$$\frac{x^2}{a^2} + \frac{y^2}{b^2} + \frac{z^2}{c^2} = \text{constant} = \lambda^2 \text{ (say)}$$

$$\text{or, } \frac{x^2}{(\lambda a)^2} + \frac{y^2}{(\lambda b)^2} + \frac{z^2}{(\lambda c)^2} = 1, \tag{5}$$

which represents an ellipsoid with semi-axes equal to $\lambda a, \lambda b, \lambda c$.

Its volume is

$$= \frac{4}{3}\pi abc\lambda^3 \tag{6}$$

But the volume of the liquid is given to be V. Hence,

$$V = \frac{4}{3}\pi abc\lambda^3, \tag{7}$$

which yields

$$\lambda = \left(\frac{3V}{4\pi abc}\right)^{\frac{1}{3}}. \tag{8}$$

From Eqns (5) and (7), we get the required equation of the free surface as

$$\frac{x^2}{a^2} + \frac{y^2}{b^2} + \frac{z^2}{c^2} = \left(\frac{3V}{4\pi abc}\right)^{\frac{2}{3}}.$$

EXAMPLE 3.6 *Show that the pressure at a small depth z below the surface of a sphere of water attracted to the centre of the sphere with a force producing an acceleration μ/r^2 at a distance r is approximately*

$$\pi + g\rho\left(z + \frac{z^2}{R}\right),$$

where π = pressure at free surface, ρ = density of the liquid, R = radius of the sphere and g = attraction force on unit mass on the surface of the sphere.

Solution. We take a point P at a depth z below the top of the sphere with centre at C (FIG. 3.7). Since radius of the sphere is R, the distance

$$PC = R - z \tag{1}$$

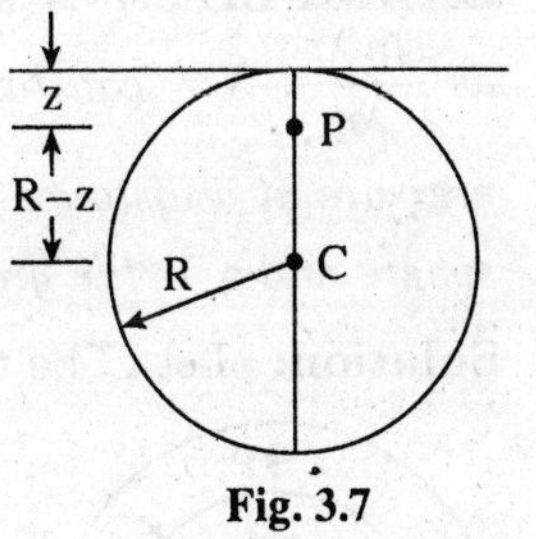

Fig. 3.7

Since the force of attraction F on unit mass at a distance r from the centre is $\frac{\mu}{r^2}$, we have

$$F = -\frac{\mu}{r^2}. \tag{2}$$

Hence the pressure equation is

$$dp = -\frac{\rho\mu}{r^2}dr \tag{3}$$

Integrating, we get

$$p = \frac{\rho\mu}{r} + C, \tag{4}$$

where C is the constant of integration.

On the free surface, the pressure $p = \pi$, where $r = R$. Using this condition in Eqn (4), we get

$$\pi = \frac{\rho\mu}{R} + C,$$

which yields

$$C = \pi - \frac{\rho\mu}{R}. \tag{5}$$

Hence, the pressure at any point at a distance r from the centre by Eqns (4) and (5) is

$$p = \pi + \rho\mu\left(\frac{1}{r} - \frac{1}{R}\right). \tag{6}$$

Then the pressure at the point P, where $r = R - z$ is given by

$$p = \pi + \rho\mu\left(\frac{1}{R-z} - \frac{1}{R}\right) \quad \text{or, } p = \pi + \rho\mu\frac{z}{R(R-z)}$$

$$\text{or, } p = \pi + \rho\mu z\frac{1}{R^2\left(1-\frac{z}{R}\right)} \quad \text{or, } p = \pi + \frac{\rho\mu z}{R^2}\left(1-\frac{z}{R}\right)^{-1}$$

$$\text{or, } p = \pi + \frac{\rho\mu z}{R^2}\left(1+\frac{z}{R}+\frac{z^2}{R^2}+\cdots\right) \tag{7}$$

Since the distance z is very small, higher powers of z/R can be neglected. In this case, the pressure at the point P is

$$p = \pi + \frac{\rho\mu z}{R^2}\left(1+\frac{z}{R}\right) = \pi + \frac{\rho\mu}{R^2}\left(z+\frac{z^2}{R}\right). \tag{8}$$

Since g is the attraction force on unit mass at the free surface, we have

$$g = \frac{\mu}{R^2}. \tag{9}$$

Hence, the pressure at P is

$$p = \pi + \rho g\left(z + \frac{z^2}{R}\right). \tag{10}$$

EXAMPLE 3.7 *A sphere of fluid, whose density ρ at a distance r from the centre varies as $\frac{\sin\lambda r}{\lambda r}$, λ = constant, is self-attracting according to the law of gravitation. Prove that the pressure at a distance r from the centre is $\frac{2\pi\gamma}{\lambda^2}(\rho^2 - \rho_1^2)$, where ρ_1 is the surface density of the sphere and γ is the gravitational constant.*

Solution. Let C be the centre of the sphere. We consider a point P at a distance r from the centre and draw a spherical shell of radius x and thickness dx (FIG. 3.8).

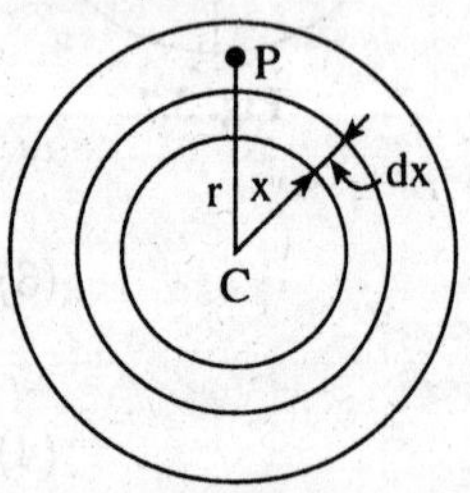

Fig. 3.8

The attraction of this shell at an internal point is zero and at an external point is

$$F = \gamma\frac{M\cdot 1}{r^2}, \tag{1}$$

where M is the mass of the sphere of radius r given by

$$M = \int_0^r 4\pi x^2\rho(x)dx. \tag{2}$$

The density $\rho(x)$ at any distance x from the centre by condition is

$$\rho(x) = K\frac{\sin\lambda x}{\lambda x}, \tag{3}$$

where K is a constant.

Hence, from Eqns (2) and (3), we get

$$M = \int_0^r \frac{4\pi K}{\lambda}x\sin\lambda x dx = \frac{4\pi K}{\lambda^3}(\sin\lambda r - \lambda r\cos\lambda r). \tag{4}$$

From Eqns (1) and (4),

$$F = \frac{4\pi K\gamma}{\lambda^3} \cdot \frac{\sin \lambda r - \lambda r \cos \lambda r}{r^2}. \tag{5}$$

The pressure equation for a point at a distance r from the centre is

$$\frac{dp}{dr} = -\rho F. \tag{6}$$

From Eqns (5) and (6), we get

$$\frac{dp}{dr} = -\left(K\frac{\sin \lambda r}{\lambda r}\right)\frac{4\pi K\gamma}{\lambda^3 r^2}(\sin \lambda r - \lambda r \cos \lambda r)$$

$$= \frac{2\pi\gamma K^2}{\lambda^4}\left[-\frac{2\sin^2 \lambda r}{r^3} + \frac{2\lambda \sin \lambda r \cos \lambda r}{r^2}\right]$$

$$= \frac{2\pi\gamma K^2}{\lambda^4}\frac{d}{dr}\left(\frac{\sin^2 \lambda r}{r^2}\right) = \frac{2\pi\gamma}{\lambda^2}\frac{d}{dr}\left(\frac{K^2 \sin^2 \lambda r}{\lambda^2 r^2}\right) \tag{7}$$

Integrating Eqn (7), we get

$$p = \frac{2\pi\gamma}{\lambda^2}\frac{K^2 \sin^2 \lambda r}{\lambda^2 r^2} + C = \frac{2\pi\gamma}{\lambda^2}\rho^2 + C, \tag{8}$$

where C is the constant of integration.

On the free surface, we have

$$p = 0 \text{ and } \rho = \rho_1. \tag{9}$$

Putting these values in Eqn (8), we get

$$0 = \frac{2\pi\gamma}{\lambda^2}\rho_1^2 + C, \text{ i.e., } C = -\frac{2\pi\gamma}{\lambda^2}\rho_1^2. \tag{10}$$

From Eqns (8) and (10), we get $p = \dfrac{2\pi\gamma}{\lambda^2}(\rho^2 - \rho_1^2)$, which is the required result.

EXAMPLE 3.8 *A mass of liquid rests upon a plane and is subject to a central force* $\dfrac{\mu}{r^2}$, *situated at a distance c from the plane on the side opposite to that on which is the liquid. Show that the pressure on the plane is* $\dfrac{\pi\rho\mu(a-c)^2}{a}$, *where a is the radius of the free spherical surface of the liquid and* ρ *is its density.*

Solution. Let O be the centre of the spherical surface, a part of which is the liquid resting on the plane AB. At any point P at a distance r from the centre, the force acting per unit mass is (FIG. 3.9).

$$F = -\frac{\mu}{r^2}. \tag{1}$$

Hence, the pressure equation is

$$dp = -\frac{\rho\mu}{r^2}. \tag{2}$$

Upon integration, we get

$$p = \frac{\rho\mu}{r} + C, \tag{3}$$

where C is the constant of integration.

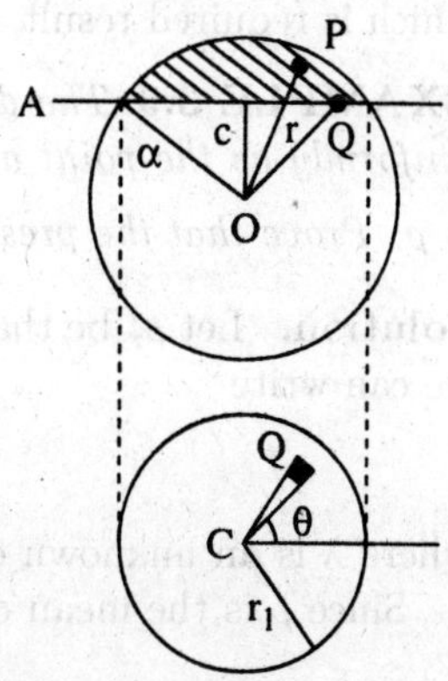

Fig. 3.9

On the free surface, $p = 0, r = a$. Putting these values in Eqn (3),

$$0 = \frac{\rho\mu}{a} + C, \text{ i.e., } C = -\frac{\rho\mu}{a}. \tag{4}$$

Hence, the pressure at any point is

$$p = \rho\mu\left(\frac{1}{r} - \frac{1}{a}\right). \tag{5}$$

Now, the liquid occupies on the plane a circular area of radius r_1, where

$$r_1 = \sqrt{a^2 - c^2}. \tag{6}$$

We take a point Q on the plane base of the liquid at a distance x from centre O, such that OQ is x. Then the pressure at Q by Eqn (5) is

$$p = \rho\mu\left(\frac{1}{x} - \frac{1}{a}\right). \tag{7}$$

We now consider an elementary area $RdRd\theta$ in polar coordinates surrounding the point Q which is at a distance R from the centre C of the circular liquid base, i.e., $CQ = R$. Then

$$x = \sqrt{c^2 + R^2}. \tag{8}$$

Hence, from Eqn (7), pressure at point Q is

$$p = \rho\mu\left(\frac{1}{\sqrt{c^2 + R^2}} - \frac{1}{a}\right). \tag{9}$$

Then the total pressure on the elementary area $RdRd\theta$ at Q is

$$p = \rho\mu\left(\frac{1}{\sqrt{c^2 + R^2}} - \frac{1}{a}\right) RdRd\theta.$$

Hence the total pressure on the circular plane is

$$= \int_0^{r_1}\int_0^{2\pi} \rho\mu\left(\frac{1}{\sqrt{c^2 + R^2}} - \frac{1}{a}\right) RdRd\theta = 2\pi\rho\mu\left[\sqrt{c^2 + R^2} - \frac{R^2}{2a}\right]_0^{\sqrt{a^2 - c^2}}$$

$$= 2\pi\rho\mu\left[a - c - \frac{a^2 - c^2}{2a}\right] = \frac{\pi\rho\mu}{a}(a - c)^2,$$

which is required result.

EXAMPLE 3.9 *The density of a gravitating liquid sphere of radius a at any point increases uniformly as the point approaches the centre. The surface density is ρ_0 and the mean density is ρ. Prove that the pressure at the centre is $\frac{2}{9}\pi a^2\left\{3\rho_0^2 + 10(\rho - \rho_0)\right\}$.*

Solution. Let ρ_r be the density at a point distant r from the centre. Then by condition given, we can write

$$\rho_r = \rho_0 + \lambda(a - r), \tag{1}$$

where λ is an unknown constant.

Since ρ is the mean density, the mass of the liquid sphere is

$$M = \frac{4}{3}\pi a^3 \rho. \tag{2}$$

We take an elementary spherical shell of thickness dx and of radius x (FIG. 3.10). Mass of this elementary shell is

$$dM = 4\pi x^2[\rho_0 + \lambda(a-x)]dx. \tag{3}$$

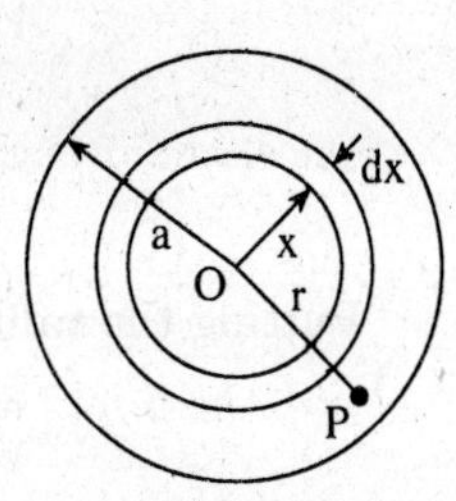

Fig. 3.10

Hence, the total mass of the sphere is

$$M = \int_0^a 4\pi x^2[\rho_0 + \lambda(a-x)]dx$$

$$= \frac{4\pi a^3}{3}\left[\rho_0 + \lambda\left(a - \frac{3a}{4}\right)\right] = \frac{4\pi a^3}{3}\left(\rho_0 + \frac{\lambda a}{4}\right) \tag{4}$$

Equating the mass of the sphere from (2) and (4),

$$\frac{4}{3}\pi a^3 \rho = \frac{4}{3}\pi a^3\left(\rho_0 + \frac{\lambda a}{4}\right), \text{ or, } \rho = \rho_0 + \frac{\lambda a}{4}, \text{ i.e., } \lambda = \frac{4(\rho - \rho_0)}{a}. \tag{5}$$

Now we take a point P at a distance r from the centre. The force of attraction due to the elementary shell will be zero when P is inside the shell. If P is outside the shell, the force of attraction is

$$dF = \frac{\gamma dM}{r^2} = \frac{4\pi\gamma}{r^2}x^2\{\rho_0 + \lambda(a-x)\}dx.$$

Then the total force of attraction at P is

$$F = \frac{4\pi\gamma}{r^2}\int_0^r x^2\{\rho_0 + \lambda(a-x)\}dx = 4\pi\gamma\left[\frac{\rho_0 r}{3} + \lambda\left(\frac{ar}{3} - \frac{r^2}{4}\right)\right]. \tag{6}$$

Here γ is the gravitational constant.
The pressure equation of the liquid is

$$dp = -\rho_r F dr$$

$$\text{i.e., } dp = -4\pi\gamma[\rho_0 + \lambda(a-r)]\left[\frac{\rho_0 r}{3} + \lambda\left(\frac{ar}{3} - \frac{r^2}{4}\right)\right] dr \tag{7}$$

$$\text{or, } dp = -4\pi\gamma\left[\rho_0^2\frac{r}{3} + \lambda\rho_0\frac{r(a-r)}{3} + \lambda\rho_0\left(\frac{ar}{3} - \frac{r^2}{4}\right) + \lambda^2\left(\frac{a^2r}{3} - \frac{ar^2}{3} - \frac{ar^2}{4} + \frac{r^3}{4}\right)\right] dr.$$

Integrating,

$$p = A - 4\pi\gamma\left[\rho_0^2\frac{r^2}{6} + \frac{\lambda\rho_0}{3}\left(\frac{ar^2}{2} - \frac{r^3}{3}\right) + \frac{\lambda\rho_0}{3}\left(\frac{ar^2}{2} - \frac{r^3}{4}\right) + \lambda^2\left(\frac{a^2r^2}{6} - \frac{ar^3}{9} - \frac{ar^3}{12} + \frac{r^4}{16}\right)\right]$$

$$\text{or, } p = A - 4\pi\gamma r^2\left[\frac{\rho_0^2}{6} + \frac{\lambda\rho_0}{3}\left(a - \frac{7}{12}r\right) + \lambda^2\left(\frac{a^2}{6} - \frac{7ar}{36} + \frac{r^2}{16}\right)\right] \tag{8}$$

Here A is the constant of integration which is determined from the condition that the pressure on the free surface is zero. That is

$$p = 0, \text{ for } r = a. \tag{9}$$

Applying this condition in Eqn (8), we get

$$A = 4\pi\gamma a^2 \left[\frac{\rho_0^2}{6} + \frac{\lambda\rho_0}{3}\left(a - \frac{7}{12}a\right) + \lambda^2\left(\frac{a^2}{6} - \frac{7a^2}{36} + \frac{a^2}{16}\right)\right]$$

$$= 4\pi\gamma a^2 \left[\frac{\rho_0^2}{6} + \frac{\lambda\rho_0}{3}\cdot\frac{5a}{12} + \lambda^2\cdot\frac{5a^2}{144}\right] \tag{10}$$

Putting the value of λ from Eqn (5) into Eqn (10), we get

$$A = 4\pi\gamma a^2 \left[\frac{\rho_0^2}{6} + \frac{4(\rho-\rho_0)}{a}\cdot\frac{5\rho_0 a}{36} + \frac{16(\rho-\rho_0)^2}{a^2}\cdot\frac{5a^2}{144}\right]$$

$$= 4\pi\gamma a^2 \left[\frac{\rho_0^2}{6} + \frac{5\rho_0(\rho-\rho_0)}{9} + \frac{5(\rho-\rho_0)^2}{9}\right]$$

$$= \frac{4\pi\gamma a^2}{6}\left[\rho_0^2 + \frac{10}{3}\left\{\rho_0(\rho-\rho_0) + (\rho-\rho_0)^2\right\}\right]$$

$$= \frac{2\pi\gamma a^2}{9}\left[3\rho_0^2 + 10(\rho_0\rho - \rho_0^2 + \rho^2 - 2\rho\rho_0 + \rho_0^2)\right]$$

$$= \frac{2}{9}\pi\gamma a^2\left[3\rho_0^2 + 10\rho(\rho-\rho_0)\right]. \tag{11}$$

The pressure at the centre of the sphere will be obtained by putting $r = 0$ in Eqn (8). Hence, the pressure at the centre is

$$p = A = \frac{2}{9}\pi\gamma a^2\left[3\rho_0^2 + 10\rho(\rho-\rho_0)\right],$$

which is the required result.

EXAMPLE 3.10 *A uniform spherical mass of liquid of density* $(\rho+\delta)$ *and radius* a *is surrounded by another incompressible liquid of density* ρ *and external radius* b. *The whole is in equilibrium under its own gravitation, but with no external pressure. Show that the pressure at the centre is*

$$\frac{2}{3}\pi\gamma(\rho+\delta)^2 a^2 + \frac{2}{3}\pi\gamma\rho\left\{\frac{2a^2}{b}\delta + \rho(a+b)\right\}(b-a)$$

where γ *is the gravitational constant.*

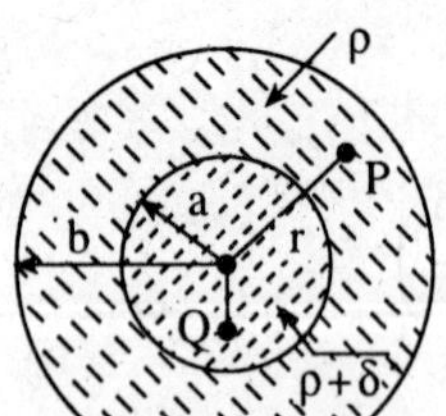

Fig. 3.11

Solution. We consider a point P at a distance r from the centre within the liquid in the annular part as shown in FIG. 3.11. The force of attraction due to the fluid of internal sphere at P is

$$F_1 = \frac{4}{3}\pi\gamma(\rho+\delta)\frac{a^3}{r^2} \tag{1}$$

and that due to the fluid in the annular part is

$$F_2 = \frac{4}{3}\pi\gamma\rho\frac{r^3 - a^3}{r^2}. \tag{2}$$

Hence the total force of attraction at P is

$$F = F_1 + F_2 = \frac{4}{3}\pi\gamma\left\{(\rho+\delta)a^3 + \rho(r^3 - a^3)\right\}\frac{1}{r^2}$$

$$\text{i.e., } F = \frac{4}{3}\pi\gamma\left(\frac{\delta a^3}{r^2} + \rho r\right). \tag{3}$$

Then the pressure equation for the point P is

$$dp = -\rho F dr, \text{ or, } dp = -\frac{4}{3}\pi\gamma\rho\left(\frac{\delta a^3}{r^2} + \rho r\right) dr. \tag{4}$$

Integrating,

$$p = A - \frac{4}{3}\pi\gamma\rho\left(\frac{\rho r^2}{2} - \frac{\delta a^3}{r}\right), \tag{5}$$

where A is the constant of integration.

The outermost surface being free, we have

$$p = 0, \text{ for } r = b. \tag{6}$$

Using this condition in Eqn (5), we get

$$0 = A - \frac{4}{3}\pi\gamma\rho\left(\frac{\rho b^2}{2} - \frac{\delta a^3}{b}\right),$$

which yields

$$A = \frac{4}{3}\pi\gamma\rho\left(\frac{\rho b^2}{2} - \frac{\delta a^3}{b}\right). \tag{7}$$

Hence pressure at any point P is

$$p = \frac{4}{3}\pi\gamma\rho\left[\left(\frac{\rho b^2}{2} - \frac{\delta a^3}{b}\right) - \left(\frac{\rho r^2}{2} - \frac{\delta a^3}{r}\right)\right]. \tag{8}$$

The pressure on the common surface is obtained by putting $r = a$ in Eqn (8). Hence,

$$\begin{aligned}(p)_{r=a} &= \frac{4}{3}\pi\gamma\rho\left[\left(\frac{\rho b^2}{2} - \frac{\delta a^3}{b}\right) - \left(\frac{\rho a^2}{2} - \frac{\delta a^3}{a}\right)\right] \\ &= \frac{4}{3}\pi\gamma\rho\left[\delta a^3\left(\frac{1}{a} - \frac{1}{b}\right) + \frac{\rho}{2}(b^2 - a^2)\right] \\ &= \frac{4}{3}\pi\gamma\rho\left[\delta a^3\frac{b-a}{ab} + \frac{\rho}{2}(b+a)(b-a)\right] \\ &= \frac{4}{3}\pi\gamma\rho(b-a)\left[\frac{\delta a^2}{b} + \frac{\rho}{2}(b+a)\right] \\ &= \frac{2}{3}\pi\gamma\rho(b-a)\left[\frac{2\delta a^2}{b} + \rho(b+a)\right]\end{aligned} \tag{9}$$

Now, we consider a point Q at a distance r from the centre within the liquid of inner sphere. The force of attraction at Q due to the liquid in annular part is zero and that due to the liquid in the inner sphere is

$$F = \frac{4}{3}\pi\gamma(\rho + \delta)r. \tag{10}$$

Hence, the pressure equation for point Q is

$$dp = -(\rho + \delta)\cdot\frac{4}{3}\pi\gamma(\rho + \delta)r\,dr = -\frac{4}{3}\pi\gamma(\rho + \delta)^2 r\,dr. \tag{11}$$

Integrating, we get

$$p = B - \frac{2}{3}\pi\gamma(\rho + \delta)^2 r^2, \tag{12}$$

where B is the constant of integration.

Pressure on the common surface is obtained by putting $r = a$ in Eqn (12). Thus

$$(p)_{r=a} = B - \frac{2}{3}\pi\gamma(\rho+\delta)^2 a^2. \qquad (13)$$

This yields,

$$B = (p)_{r=a} + \frac{2}{3}\pi\gamma(\rho+\delta)^2 a^2. \qquad (14)$$

The pressure at the centre is obtained by putting $r = 0$ in Eqn (12). Hence

$$(p)_{r=0} = B. \qquad (15)$$

From Eqns (14) and (15),

$$(p)_{r=0} = \frac{2}{3}\pi\gamma(\rho+\delta)^2 a^2 + (p)_{r=a}. \qquad (16)$$

Putting the value of $(p)_{r=a}$ from Eqn (9), we get

$$(p)_{r=0} = \frac{2}{3}\pi\gamma(\rho+\delta)^2 a^2 + \frac{2}{3}\pi\gamma\rho(b-a)\left\{\frac{2\delta a^2}{b} + \rho(b+a)\right\},$$

which is the required result.

EXERCISE

1. A given volume of liquid is at rest on a fixed plane under the action of a force which varies as the distance from a fixed point in the plane. If the free surface be hemispherical, show that the thrust on the plane is $\frac{1}{4}\pi\mu\rho a^4$, where ρ = density of the liquid, a = radius of the hemisphere and μ = a constant.

2. Show that the forces represented by $X = \mu(y^2+yz+z^2)$, $Y = \mu(z^2+zx+x^2)$, $Z = \mu(x^2+xy+y^2)$ will keep a mass of liquid at rest, if

$$\text{density} \propto \frac{1}{(\text{distance})^2}$$

from the plane $x + y + z = 0$; and the curves of equal pressure and density are circles.

3. An open tank containing heavy liquid is moved horizontally with uniform acceleration f. Show that the free surface is inclined to the vertical at an angle $\tan^{-1}(f/g)$ and that the pressure at any point is $\Pi + \rho z\sqrt{f^2+g^2}$, where Π = atmospheric pressure, ρ = density of the liquid, z = distance of the point from free surface and g = acceleration due to gravity.

4. A given volume V of a heavy liquid is acted upon by forces $-\mu x, -\mu y, -\mu z$. Show that equation of the free surface is a sphere given by $x^2 + y^2 + z^2 \pm \frac{2g}{\mu}z = \frac{2C}{\mu}$, where

$$V = \frac{4}{3}\pi\left(\frac{2C}{\mu} + \frac{g^2}{\mu^2}\right)^{\frac{3}{2}}.$$

5. A fluid is in equilibrium under a given system of forces. If $\rho_1 = \phi(x,y,z), \rho_2 = \psi(x,y,z)$ be two possible values of the density at any point, show that the equations of surfaces of equal pressure in either case are given by $\phi(x,y,z) + \lambda\psi(x,y,z) = 0$, where λ is an arbitrary parameter.

6. Prove that, if the density of a liquid at rest under gravity varies as the square root of pressure, then the density increases uniformly with the depth.

7. A uniform solid sphere of radius a and density ρ is surrounded by a concentric shell of gravitating liquid of outer radius b and density δ. Show that the pressure at a point within the liquid at a distance r from the centre is

$$p = \frac{4}{3}\pi\gamma\delta\left\{(\rho-\delta)a^3\left(\frac{1}{r}-\frac{1}{b}\right) - \frac{1}{2}\delta(r^2-b^2)\right\}.$$

8. A spherical vessel is just filled with a heavy liquid the particles of which attract one another according to the law of gravitation. If the pressure at the highest point vanishes, show that the resultant thrust across a vertical diametral plane is $\pi g\rho a^2 + \frac{1}{3}\pi^2\rho^2\gamma a^4$, where a = radius of the spherical vessel, ρ = density of the liquid and γ = the constant of gravitation.

9. A hollow sphere half-filled with water is moved with constant acceleration f in a horizontal straight line. When the contents are in relative equilibrium, show that the pressure at any point is $\rho z\sqrt{f^2+g^2}$, where z is the distance of the point from the free surface.

10. A uniform solid sphere of radius a and density ρ is surrounded by a concentric shell of liquid of outer radius b and density δ. The liquid is subjected to an attractive force

$$\frac{g}{b\delta}\left[\frac{(\rho-\delta)a^3}{r^2}+r\delta\right]$$

per unit mass towards the centre, where r is the distance from the centre. Assuming the outer boundary of the liquid to be free from pressure, show that the pressure at any point on the surface of the sphere is

$$p=\frac{g(b-a)}{2b^2}[2(\rho-\delta)a^2+\delta b(a+b)].$$

11. A uniform spherical mass of incompressible fluid of density ρ and radius a is surrounded by another incompressible fluid of density δ and external radius b. The total fluid is in equilibrium under its gravitation, but with no external pressure or forces. The two fluids are now mixed into a homogeneous fluid of the same volume, and the mass is again in equilibrium in a spherical form. Prove that the pressure at the centre in the first case exceeds the pressure at the centre in the second case by

$$\frac{8}{3}\pi\delta(\rho-\delta)a^2\left(1-\frac{a}{b}\right)\left[1+\frac{1}{4}\left(\frac{\rho}{\delta}-1\right)\left(1+\frac{a}{b}\right)\left(1+\frac{a^2}{b^2}\right)\right].$$

CHAPTER 4

Thrust on Plane Surfaces

4.1 Introduction

We know that liquid exerts pressure on any surface in contact with it. This pressure on every element of the surface is in the direction of normal to the surface at that point. Hence, in the case of a plane surface, the direction of all the thrusts at every point on the surface being same, the forces will constitute a parallel system. These parallel forces can be compounded into a single force whose magnitude is equal to the sum of all the parallel forces and whose direction is along the normal to the surface. This single force is called the ***resultant thrust***. The whole pressure of a liquid on any plane surface in contact with the liquid is equal to the resultant thrust.

4.2 Thrust of Heavy Homogeneous Liquids

We consider a plane surface of area S within a heavy homogeneous liquid of density ρ and is subjected to gravity only.

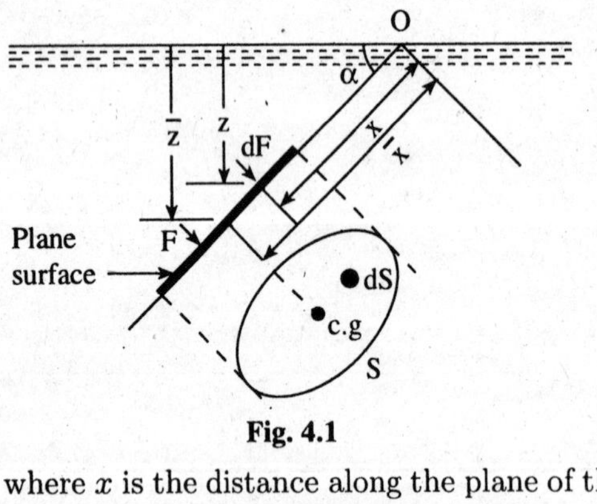

Fig. 4.1

Let dS be an elementary area on the surface S at a depth z below the effective surface (FIG. 4.1). The fluid pressure at a point at a depth z below the effective surface being $g\rho z$, the thrust on the elementary area is

$$dF = g\rho z dS. \tag{4.1}$$

If the plane surface be inclined at an angle α to the horizon, then $z = x \sin\alpha$, so that

$$dF = g\rho x \sin\alpha dS, \tag{4.2}$$

where x is the distance along the plane of the elementary area from the effective surface. Hence, the total thrust on the plane surface is

$$F = \int_S g\rho x \sin\alpha dS = g\rho \sin\alpha \int_S x dS. \tag{4.3}$$

If $\bar{x}$ be the distance of the centre of gravity of the plane area S from the point O on the liquid surface, then by definition $\bar{x} = \dfrac{\int x dS}{\int dS} = \dfrac{\int x dS}{S}$.

This yields

$$\int x dS = S\bar{x}. \tag{4.4}$$

If $\bar{z}$ be the vertical depth of the centre of gravity, then

$$\bar{z} = \bar{x} \sin \alpha \tag{4.5}$$

From Eqns (4.3) - (4.5), we get

$$F = g\rho \sin \alpha \bar{x} S = g\rho S\bar{z}. \tag{4.6}$$

Eqn (4.6) implies that the total thrust on a plane surface dipped in a heavy homogeneous liquid is equal to the area of the plane surface S multiplied by the pressure at its centre of gravity $= g\rho\bar{z}$. Hence, the following theorem:

THEOREM 4.1 *The thrust of a heavy homogeneous liquid on a plane area is equal to the product of the area and the pressure at its centre of gravity.*

Eqn (4.6) also shows that ***the total thrust on a plane area is independent of its inclination to the horizontal, provided that the position of its centre of gravity remains unaltered.***

Eqn (4.6) is equally applicable when the plane surface of area S can be divided into a finite number of discrete areas $s_1, s_2, \dots, s_n$, whose centres of gravity below the effective surface are at depths $z_1, z_2, \dots, z_n$ respectively. In this case, the total thrust on the plane area is

$$F = g\rho \sum_{i=1}^{n} z_i s_i. \tag{4.7}$$

The depth $\bar{z}$ of the plane area S from the effective surface is given by

$$\bar{z} = \frac{\Sigma z_i s_i}{\Sigma s_i} = \frac{\sum_{i=1}^{n} z_i s_i}{S},$$

which yields

$$\sum_{i=1}^{n} s_i z_i = S\bar{z}. \tag{4.8}$$

Hence, the total thrust on the plane area S is

$$F = g\rho\bar{z}S. \tag{4.9}$$

4.3 Thrust on a Horizontal Surface

On a horizontal surface, the depth of the centre of gravity $\bar{z}$ from the effective liquid surface is constant. Hence the total thrust F on the surface depends only on the area of the surface S.

It is equal to the weight of a column of liquid whose base is equal to the area of the given plane surface and whose height is equal to the depth of the centre of gravity of the plane area below the effective surface. The liquid of this imaginary cylinder is known as the ***superincumbent liquid***.

This definition of thrust on a horizontal surface leads to a paradox known as ***hydrostatical paradox***. For example, if we consider two vessels, one cylindrical and the other conical, both having same base area and both containing same liquid up to same height, then the thrust on the base of each of them is (FIG. 4.2)

$$F = g\rho h(\pi R^2) = \pi g\rho h R^2. \tag{4.10}$$

Here R is the radius of the base and h is the height of liquid surface from the base.

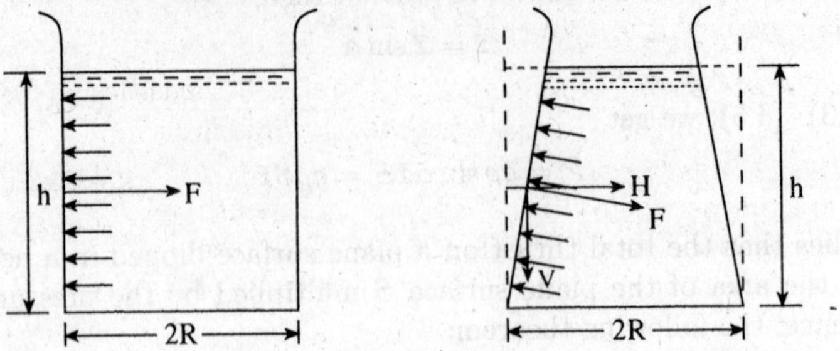

Fig. 4.0

Thus the total thrust on the base is same, although the amount of liquid in the cylindrical vessel is much more than that in the conical vessel. This is the paradox. In the first vessel, the total thrust on the base is equal to the weight of the amount of liquid contained in the vessel; but in the second case, the total thrust on the base is more than the weight of the liquid contained in the vessel. Wherefrom does this extra thrust come?

The paradox, however, can be explained very simply. The liquid in the vessels exerts pressure on the vertical or slant sides along the normals to the surfaces. There will be equal and opposite reaction force transmitted to the liquid. In the cylindrical vessel, the thrusts on vertical sides are horizontal and has no vertical component. In the conical vessel, however, the thrusts on slant surfaces are not horizontal and the reaction force F on the liquid has got two components—one along horizontal (H) and the other along vertically downward (V). This vertical component of the reaction force F after being transmitted through the liquid provides the extra thrust on the base.

4.4 Plane Surface below Layers of Different Liquids

We consider a plane surface of area S dipped into layers of different immiscible liquids. Let there be n different liquids of densities $\rho_1, \rho_2, \ldots, \rho_n$ having layers of thickness $h_1, h_2, \ldots, h_n$ respectively (FIG. 4.3). There may be two cases as discussed below:

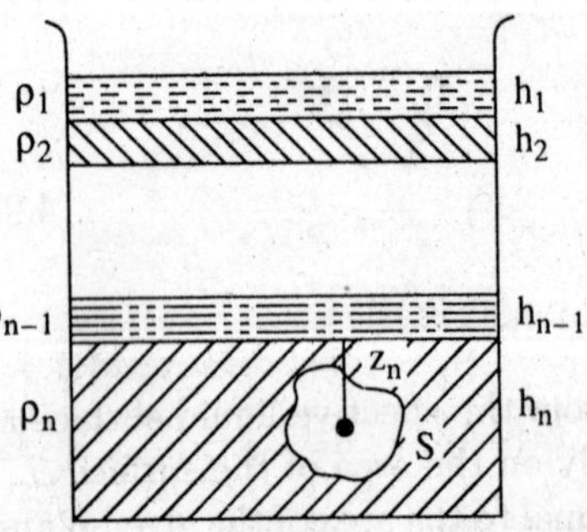

Fig. 4.3

Case 1. *The plane surface is completely immersed in the lowest liquid.*

In this case, the total thrust on the plane surface due to all the layers of liquids excepting the lowest liquid is

$$F_1 = (g\rho_1 h_1 + g\rho_2 h_2 + \cdots + g\rho_{n-1}h_{n-1})\,S$$
$$= \left(\sum_{i=1}^{n-1} \rho_i h_i\right) gS. \tag{4.11}$$

If $\bar{z}$ be the depth of centre of gravity of the plane surface below the lowest surface of separation, then by Eqn (4.6), the total thrust on the surface due to the lowest liquid is

$$F_2 = g\rho_n \bar{z} S. \tag{4.12}$$

Hence the total thrust on the plane surface of area S due to all the layers of liquids is

$$F = F_1 + F_2 = \left(\sum_{i=1}^{n-1} \rho_i h_i\right) gS + g\rho_n \bar{z} S, \text{ i.e., } F = \left(\sum_{i=1}^{n-1} \rho_i h_i + \rho_n \bar{z}\right) gS. \quad (4.13)$$

Case 2. *The plane surface in contact with all liquids.*

We consider a plane surface dipped into n liquids of densities $\rho_1, \rho_2, \dots, \rho_n$ which are immiscible. The plane surface is in contact with all the liquids (FIG. 4.4).

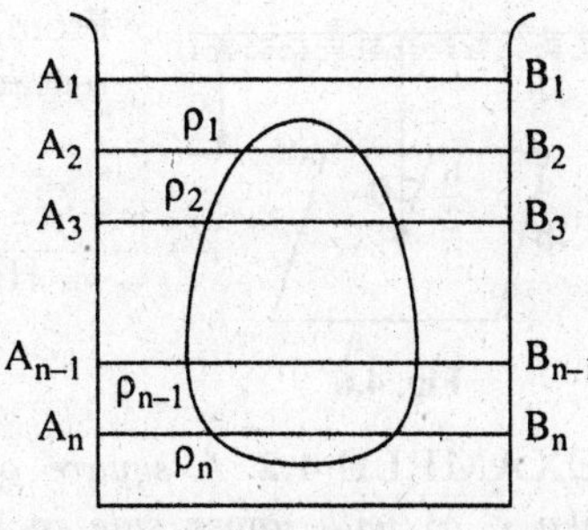

Fig. 4.4

Let S be the area of the whole surface and $\bar{z}_1$ be the depth of its centre of gravity from the free surface A_1B_1. Then the total thrust on the surface assuming the whole liquid to be of density ρ_1 is

$$F_1 = g\rho_1 S_1 \bar{z}_1. \quad (4.14)$$

Now we remove the liquid of density ρ_1 so that A_2B_2 is the free surface and assume the remaining liquid to be of density $(\rho_2 - \rho_1)$.

Let S_2 be the area of the remaining portion of surface and let the depth of its centre of gravity from free surface A_2B_2 be $\bar{z}_2$. Then the total thrust on area S_2 due to the whole liquid of density $(\rho_2 - \rho_1)$ is

$$F_2 = g(\rho_2 - \rho_1) S_2 \bar{z}_2. \quad (4.15)$$

Proceeding in the same way, we find the thrust on the remaining portion of area S_r whose depth of centre of gravity from free surface A_rB_r is $\bar{z}_r$ due to the whole liquid of density $(\rho_{r-1} - \rho_r)$ is

$$F_r = g(\rho_r - \rho_{r-1}) S_r \bar{z}_r. \quad (4.16)$$

The total thrust on the surface will be the same as the sum of all the above pressures. Thus the total thrust on the surface is

$$\begin{aligned} F &= F_1 + F_2 + \cdots + F_r + \cdots \\ &= g\rho_1 S_1 \bar{z}_1 + g(\rho_2 - \rho_1) S_2 \bar{z}_2 + \cdots + g(\rho_r - \rho_{r-1}) S_r \bar{z}_r + \cdots \\ &= g\left[\rho_1 S_1 \bar{z}_1 + \sum_{r=2}^{n} (\rho_r - \rho_{r-1}) S_r \bar{z}_r\right] \end{aligned} \quad (4.17)$$

4.5 Heterogeneous Liquids

We now consider heterogeneous liquids whose density ρ is a function of z, so that

$$\rho = \rho(z). \quad (4.18)$$

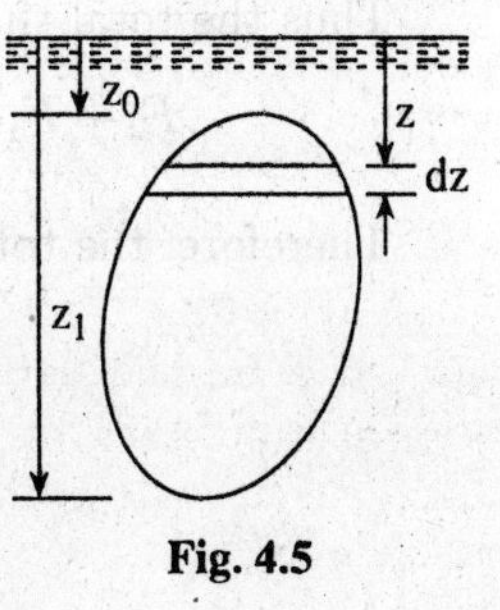

Fig. 4.5

Let there be a plane surface of area S vertically immersed in this liquid. The total thrust in this case cannot be determined by the formula (4.6).

We consider a thin strip of width dz at a depth z from free surface. Let its breadth be $B(z)$ (FIG. 4.5). The thrust on this elementary strip is $g\rho(z) z B(z) dz$.

Hence the total thrust on the surface is

$$F = \int_{z_0}^{z_1} g\rho(z)zB(z)dz = g\int_{z_0}^{z_1} \rho(z)B(z)zdz. \tag{4.19}$$

EXAMPLE 4.1 *Find the thrust on a rectangle whose sides are a and b, the side a being horizontal and at a depth c below the surface. The plane of the rectangle is inclined to the vertical at an angle θ.*

Solution. Let ρ be the density of the liquid. The surface area of the rectangle is, $S = ab$. From the FIG. 4.6, it is clear that the depth of centre of gravity of the rectangle from the free surface is

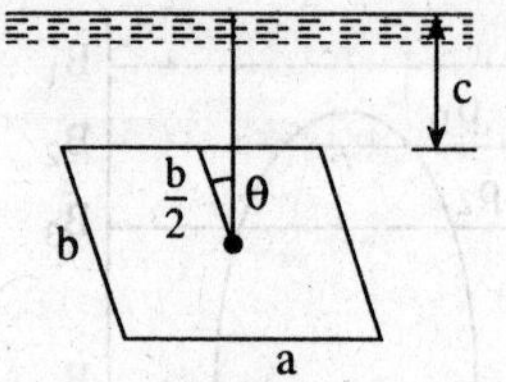

Fig. 4.6

$$\bar{z} = c + \frac{b}{2}\cos\theta. \tag{1}$$

Hence, by Eqn (4.6), the total thrust on the rectangle is

$$F = g\rho S\bar{z} = g\rho ab\left(c + \frac{b}{2}\cos\theta\right).$$

EXAMPLE 4.2 *A square of side a is dipped vertically in two liquids of densities ρ and $\delta(\rho < \delta)$ with upper side in the surface. The depth of upper liquid of density ρ is $b(< a)$. Find the thrust on the square.*

Also, if the thrusts on the two portions of the square in contact with the two liquids be equal, then show that $\rho b(3b - 2a) = \delta(a - b)^2$.

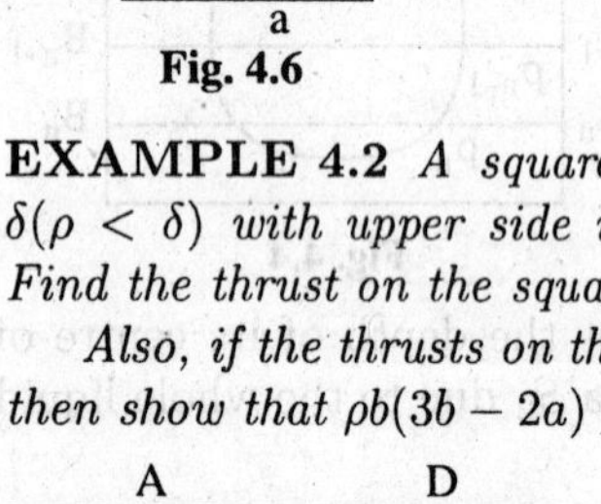

Fig. 4.7

Solution. Let $ABCD$ be the square and EF the surface of separation. Then (FIG. 4.7)

$$AB = a, DF = b.$$

Area of the portion $AEFD$ is $S_1 = ab$ and depth of its c.g. from the surface is $\bar{z}_1 = \frac{b}{2}$.

Hence the thrust on the portion $AEFD$ of area S_1 is

$$F_1 = g\rho S_1\bar{z}_1 = g\rho(ab)\frac{b}{2} = \frac{1}{2}g\rho ab^2. \tag{1}$$

Now, the pressure at EF level is $= g\rho b$, which will cause a thrust on the area $EBCF$

$$F_2 = g\rho bS_2 = g\rho ba(a - b). \tag{2}$$

The depth of c.g. of the area $EBCF$ below the line of separation EF is $\bar{z}_2 = \frac{a-b}{2}$.
Hence the thrust on the area $EBCF$ due to lower liquid is

$$F_3 = g\delta a(a - b)\frac{a - b}{2} = \frac{1}{2}g\delta a(a - b)^2. \tag{3}$$

Thus the total thrust on the area $EBCF$ is

$$F_2 + F_3 = g\rho ba(a - b) + \frac{1}{2}g\delta a(a - b)^2 = ga(a - b)\left[\rho b + \frac{1}{2}\delta(a - b)\right]. \tag{4}$$

Therefore, the total thrust on the square $ABCD$ is

$$\begin{aligned} F &= F_1 + F_2 + F_3 \\ &= \frac{1}{2}g\rho ab^2 + g\rho ab(a - b) + \frac{1}{2}g\delta a(a - b)^2 \\ &= \frac{1}{2}g\rho ab(2a - b) + \frac{1}{2}g\delta a(a - b)^2. \end{aligned} \tag{5}$$

Second part. By the condition given, we have

$$F_1 = F_2 + F_3$$

$$\text{or, } \frac{1}{2}g\rho ab^2 = g\rho ab(a-b) + \frac{1}{2}g\delta a(a-b)^2$$

$$\text{or, } \rho\{b^2 - 2b(a-b)\} = \delta(a-b)^2$$

$$\therefore\ \rho b(3b-2a) = \delta(a-b)^2 \ \textbf{(Proved)}$$

Note. The total thrust on the square $ABCD$ could also be obtained by using Eqn (4.17). Thus, we first assume that the whole liquid is of uniform density ρ. Area of the rectangle $ABCD$ is $S = a^2$ and its c.g. is at a depth of $\bar{z} = \dfrac{a}{2}$.

Hence, thrust on $ABCD$ is

$$F_1 = g\rho S\bar{z} = g\rho a^2 \cdot \frac{a}{2} = \frac{1}{2}g\rho a^3.$$

Now we remove the top layer of liquid of density ρ and assume that the remaining liquid is of density $(\delta - \rho)$.

Area of rectangle $EBCF$ is $S = a(a-b)$ and depth of its c.g. from the surface EF is

$$\bar{z} = \frac{1}{2}(a-b).$$

Hence, the thrust on $EBCF$ is

$$F_2 = g(\delta-\rho)a(a-b)\frac{1}{2}(a-b) = \frac{1}{2}g(\delta-\rho)a(a-b)^2$$

Thus the total thrust is

$$F = F_1 + F_2 = \frac{1}{2}g\rho a^3 + \frac{1}{2}g(\delta-\rho)a(a-b)^2$$

$$= \frac{1}{2}ga[\rho a^2 - \rho(a-b)^2 + \delta(a-b)^2]$$

$$= \frac{1}{2}ga[\rho a(2a-b) + \delta(a-b)^2]$$

EXAMPLE 4.3 *A rectangular area is immersed in a heavy liquid with two sides horizontal and is divided by horizontal lines into strips on which total thrusts are equal. Prove that, if a, b, c are the breadths of the consecutive strips, then $a(a+b)(b-c) = c(b+c)(a-b)$.*

Solution. Let ρ be the density of liquid, y = length of horizontal edge and x be the depth of the top edge of the rectangular area below the free surface (FIG. 4.8).

Then the thrust on portion I is

$$F_1 = g\rho ay\left(x + \frac{a}{2}\right). \qquad (1)$$

The thrust on portion II is

$$F_2 = g\rho by\left(x + a + \frac{b}{2}\right) \qquad (2)$$

and the thrust on portion III is

$$F_3 = g\rho cy\left(x + a + b + \frac{c}{2}\right). \qquad (3)$$

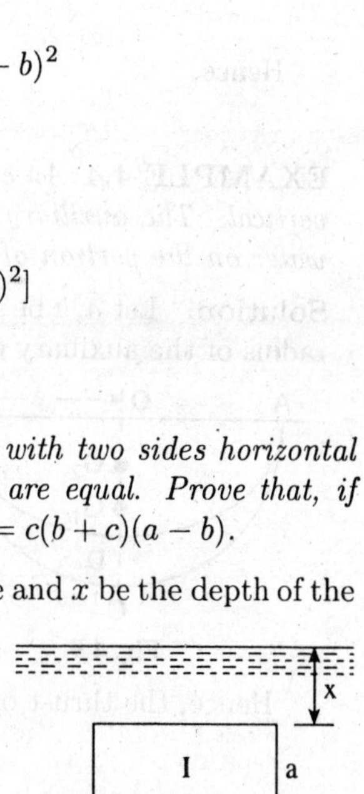

Fig. 4.8

By condition given

$$F_1 = F_2 = F_3$$

$$\text{i.e., } g\rho ay\left(x+\frac{a}{2}\right) = g\rho by\left(x+a+\frac{b}{2}\right) = g\rho cy\left(x+a+b+\frac{c}{2}\right)$$

$$\therefore \frac{x+\frac{a}{2}}{bc} = \frac{x+a+\frac{b}{2}}{ca} = \frac{x+a+b+\frac{c}{2}}{ab} = \lambda \text{ (say)} \tag{4}$$

Then

$$x+\frac{a}{2} = \lambda bc, x+a+\frac{b}{2} = \lambda ca, x+a+b+\frac{c}{2} = \lambda ab. \tag{5}$$

From Eqns (5), we find by subtracting first from second,

$$\lambda(ca-bc) = \left(x+a+\frac{b}{2}\right) - \left(x+\frac{a}{2}\right) = \frac{1}{2}(a+b), \text{ i.e., } \lambda = \frac{a+b}{2c(a-b)}. \tag{6}$$

Similarly, by subtracting the second from third of Eqns (5),

$$\lambda(ab-ca) = \left(x+a+b+\frac{c}{2}\right) - \left(x+a+\frac{b}{2}\right) = \frac{1}{2}(b+c), \text{ i.e., } \lambda = \frac{b+c}{2a(b-c)}. \tag{7}$$

From Eqns (6) and (7),

$$\frac{a+b}{2c(a-b)} = \frac{b+c}{2a(b-c)}$$

Hence,

$$a(a+b)(b-c) = c(b+c)(a-b) \textbf{ (Proved)}.$$

EXAMPLE 4.4 *An ellipse is placed with its major axis on the surface of water and its plane vertical. The auxiliary circle is described on the major axis as diameter. Find the thrust of water on the portion of the area enclosed between the auxiliary circle and the ellipse.*

Solution. Let a, b be the lengths of semi-major and semi-minor axes of the ellipse. Then the radius of the auxiliary circle will be a (FIG. 4.9).

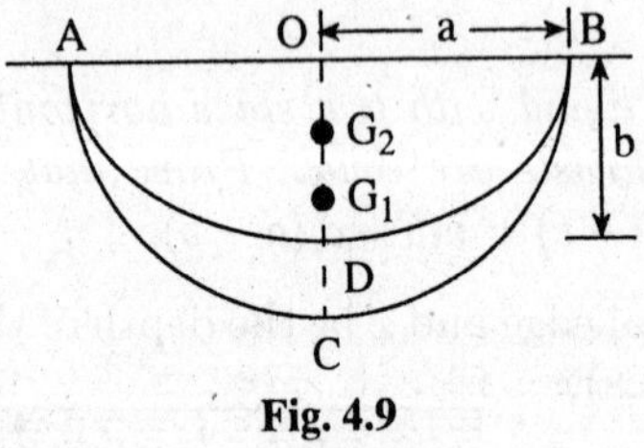

Fig. 4.9

Let G_1, G_2 be the centres of gravity of the semi-circle and semi-ellipse respectively. Now, the area of the semi-circle is

$$S_1 = \frac{1}{2}\pi a^2 \tag{1}$$

and depth of its c.g. from the surface is

$$\bar{z}_1 = OG_1 = \frac{4a}{3\pi}. \tag{2}$$

Hence, the thrust on the semi-circle is

$$F_1 = g\rho\left(\frac{1}{2}\pi a^2\right)\frac{4a}{3\pi} = \frac{2}{3}g\rho a^3. \tag{3}$$

Similarly, the area of the semi-ellipse is

$$S_2 = \frac{1}{2}\pi ab \tag{4}$$

and depth of its c.g. below the surface is

$$\bar{z}_2 = OG_2 = \frac{4b}{3\pi}. \tag{5}$$

Hence, the thrust on the semi-ellipse is

$$F_2 = g\rho\left(\frac{1}{2}\pi ab\right)\frac{4b}{3\pi} = \frac{2}{3}g\rho ab^2. \tag{6}$$

Hence, the thrust on the area $ACBDA$ enclosed between the auxiliary circle and the ellipse is

$$F = F_1 - F_2 = \frac{2}{3}g\rho a^3 - \frac{2}{3}g\rho ab^2, \text{ i.e., } F = \frac{2}{3}g\rho a(a^2 - b^2).$$

EXAMPLE 4.5 *A hollow cone whose axis is vertical and base downwards is filled with equal volumes of two liquids whose densities are in the ratio* 3 : 1. *Show that the thrust on the base is* $(3 - 4^{1/3})$ *times as much as it is when the vessel is filled with the lighter liquid.*

Solution. Let h = the height and 2α = vertical angle of the cone, x = depth of the surface of separation and 3ρ, ρ be the densities of the two liquids (FIG. 4.10).

Then volume of the whole cone is $V_1 = \frac{1}{3}\pi(h\tan\alpha)^2 h = \frac{1}{3}\pi h^3\tan^2\alpha$ and the volume of the liquid of density ρ is

$$V_2 = \frac{1}{3}\pi(x\tan\alpha)^2 x = \frac{1}{3}\pi x^3\tan^2\alpha.$$

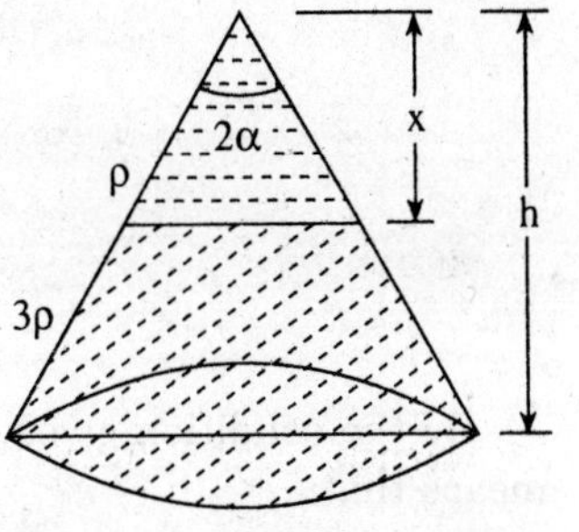

Fig. 4.10

By condition given,

$$V_2 = \frac{1}{2}V_1, \text{ i.e., } \frac{1}{3}\pi x^3\tan^2\alpha = \frac{1}{2}\cdot\frac{1}{3}\pi h^3\tan^2\alpha,$$

which yields

$$x = \left(\frac{1}{2}\right)^{1/3} h. \tag{1}$$

Pressure at any point on the base when the cone is filled with both the liquids of densities ρ and 3ρ is

$$p_1 = g\rho x + 3g\rho(h - x) = (3h - 2x)g\rho. \tag{2}$$

By virtue of Eqn (1), we get

$$p_1 = \left[3h - 2\left(\frac{1}{2}\right)^{1/3} h\right] g\rho, \text{ or, } p_1 = \left(3 - 4^{1/3}\right) g\rho h. \tag{3}$$

Hence, the thrust on the base due to the two liquids is

$$\begin{aligned} F_1 &= p_1 \times \text{area of base} \\ &= \left(3 - 4^{1/3}\right) g\rho h\left(\pi h^2\tan^2\alpha\right) = \left(3 - 4^{1/3}\right)\pi h^3 g\rho\tan^2\alpha. \end{aligned} \tag{4}$$

Again, pressure at any point on the base of the cone when filled with liquid of density ρ only is $g\rho h$.

Hence, the thrust on the base in this case is

$$F_2 = g\rho h\left(\pi h^2\tan^2\alpha\right) = \pi h^3 g\rho\tan^2\alpha. \tag{5}$$

From (4) and (5), we get

$$\frac{F_1}{F_2} = \frac{\left(3 - 4^{1/3}\right)\pi h^3 g\rho\tan^2\alpha}{\pi h^3 g\rho\tan^2\alpha} = \left(3 - 4^{1/3}\right),$$

which shows that $F_1 = (3 - 4^{1/3})$ times F_2.

EXAMPLE 4.6 *The side AB of a triangle ABC is in the surface of a liquid and points D, E are taken on AC such that the pressures on the triangles BAD, BDE and BEC are equal. Show that* $AD : DE : EC = 1 : (\sqrt{2}-1) : (\sqrt{3}-\sqrt{2})$.

Solution. The triangle ABC with its side AB in the surface of the liquid is shown in FIG. 4.11. Let $AB = x$ and z_1, z_2, z_3 be the depths of the points D, E, C below the liquid surface. Then the depths of c.g. of the triangles ABD, ABE and ABC from the liquid surface AB are:

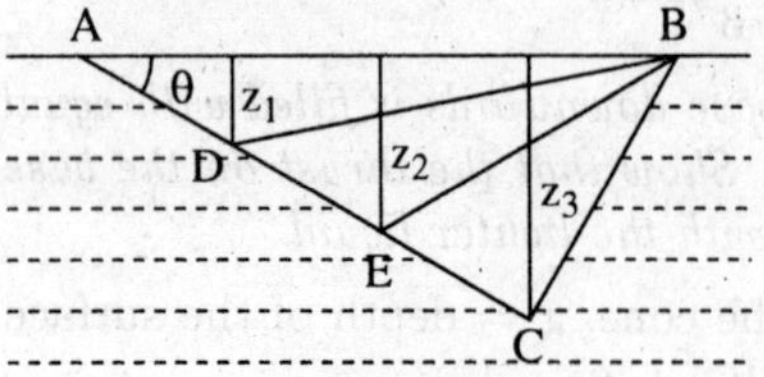

Fig. 4.11

$$\left.\begin{aligned} \bar{z}_1 &= \frac{1}{3}(0+0+z_1) = \frac{z_1}{3} \\ \bar{z}_2 &= \frac{1}{3}(0+0+z_2) = \frac{z_2}{3} \\ \bar{z}_3 &= \frac{1}{3}(0+0+z_3) = \frac{z_3}{3}. \end{aligned}\right\} \tag{1}$$

Hence, the thrusts on triangles ABD, ABE and ABC are:

$$\left.\begin{aligned} F_1 &= g\rho\left(\frac{1}{2}xz_1\right)\frac{z_1}{3} = \frac{1}{6}g\rho x z_1^2 \\ F_2 &= g\rho\left(\frac{1}{2}xz_2\right)\frac{z_2}{3} = \frac{1}{6}g\rho x z_2^2 \\ F_3 &= g\rho\left(\frac{1}{2}xz_3\right)\frac{z_3}{3} = \frac{1}{6}g\rho x z_3^2 \end{aligned}\right\} \tag{2}$$

By the condition given, the whole pressure on triangles ABD, BDE, BEC are equal. This means that:

Thrust on $\triangle ABD = \frac{1}{2} \times$ thrust on $\triangle ABE = \frac{1}{3} \times$ thrust on $\triangle ABC$.

Hence, $F_1 = \frac{1}{2}F_2 = \frac{1}{3}F_3$, which by (2) yields

$$\frac{1}{6}g\rho x z_1^2 = \frac{1}{2}\cdot\frac{1}{6}g\rho x z_2^2 = \frac{1}{3}\cdot\frac{1}{6}g\rho x z_3^2$$

$$\text{or, } z_1^2 = \frac{1}{2}z_2^2 = \frac{1}{3}z_3^2, \text{ i.e., } z_1 = \frac{1}{\sqrt{2}}z_2 = \frac{1}{\sqrt{3}}z_3 \tag{3}$$

Let $\angle BAD = \theta$. Then, from the figure, we have

$$z_1 = AD\sin\theta, z_2 = AE\sin\theta, z_3 = AC\sin\theta. \tag{4}$$

From (3) and (4), we get

$$AD = \frac{AE}{\sqrt{2}} = \frac{AC}{\sqrt{3}} = \lambda \text{ (say)}.$$

Then

$$AD = \lambda, AE = \sqrt{2}\lambda, AC = \sqrt{3}\lambda.$$

Hence, $\quad DE = AE - AD = \sqrt{2}\lambda - \lambda = (\sqrt{2}-1)\,\lambda$

and $EC = AC - AE = \sqrt{3}\lambda - \sqrt{2}\lambda = (\sqrt{3}-\sqrt{2})\,\lambda.$

Hence, $AD : DE : EC = 1 : (\sqrt{2}-1) : (\sqrt{3}-\sqrt{2})$.

EXAMPLE 4.7 *A hollow weightless hemisphere filled with a liquid is suspended freely from a point in the rim of its base. Show that the thrust on the base is to the weight of the liquid contained as* $12 : \sqrt{73}$.

Solution. Let ABC be the hemisphere with AB as a diameter on the plane base. The hemisphere filled with liquid of density ρ (say) is suspended from the point A on the rim of its base (FIG. 4.12).

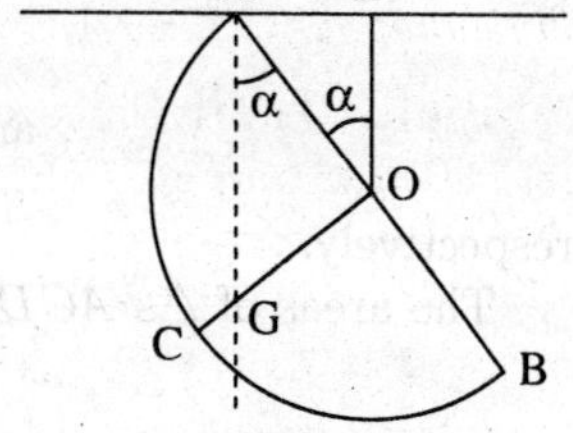

Fig. 4.12

OD is drawn perpendicular to the horizontal through A, where O is the centre of the base. Let G be the centre of gravity of the hemisphere. Then the line AG will be vertical.

Let $\angle AOD = \angle OAG = \alpha$.

If R be the radius of the hemisphere, then

$$OG = \frac{3}{8}R. \qquad (1)$$

In the $\triangle AOG$,

$$\tan\alpha = \frac{GO}{AO} = \frac{3R}{8}\cdot\frac{1}{R} = \frac{3}{8} \text{ so that } \cos\alpha = \frac{8}{\sqrt{3^2+8^2}} = \frac{8}{\sqrt{73}}.$$

Hence,

$$OD = AO\cos\alpha = \frac{8R}{\sqrt{73}}. \qquad (2)$$

Now, the thrust on the plane base is

$$F = g\rho\left(\pi R^2\right)\times OD = g\rho\pi R^2 \times \frac{8R}{\sqrt{73}} = \frac{8}{\sqrt{73}}g\rho\pi R^3. \qquad (3)$$

Weight of the liquid contained in the hemisphere is

$$W = \frac{2}{3}\pi R^3 g\rho. \qquad (4)$$

Hence, from (3) and (4)

$$\frac{F}{W} = \frac{8}{\sqrt{73}}g\rho\pi R^3\cdot\frac{3}{2\pi R^3 g\rho} = \frac{12}{\sqrt{73}} \textbf{ (Proved)}$$

EXAMPLE 4.8 *A triangle ABC is immersed vertically in a liquid with its vertex C in the surface and the sides AC, BC equally inclined to the surface. Show that the vertical through C divides the triangle into two others, the thrust on which are in the ratio*

$$(b^3+ab^2):(a^3+3a^2b).$$

Solution. Let ABC be a triangle immersed vertically with its vertex C on the surface; CD be a vertical from C meeting AB at D; G_1, G_2 be the centres of gravity of $\triangle$s ACD, BCD respectively (FIG. 4.13). Since the sides AC and BC make equal angles α (say) with the surface,

$$\angle ACD = \angle BCD = 90^\circ - \alpha.$$

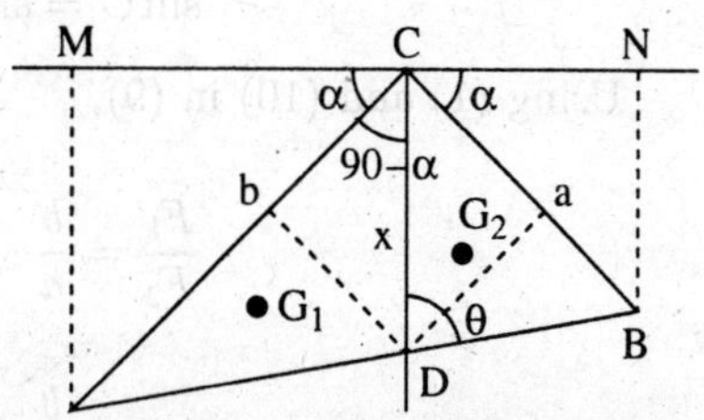

Fig. 4.13

Thus CD is the bisector of $\angle C$ of the $\triangle ABC$. Hence, from geometry

$$\frac{AD}{b} = \frac{BD}{a} = \frac{AD+BD}{a+b} = \frac{c}{a+b}$$

$$\therefore AD = \frac{bc}{a+b}, BD = \frac{ac}{a+b} \qquad (1)$$

From the figure,

$$AM = AC\sin\alpha = b\sin\alpha,\ BN = BC\sin\alpha = a\sin\alpha. \qquad (2)$$

Let $CD = x$. Then the depths of G_1 and G_2 from the surface are

$$\left.\begin{aligned} \bar{z}_1 &= \frac{1}{3}(0 + b\sin\alpha + x) = \frac{1}{3}(b\sin\alpha + x) \\ \text{and } \bar{z}_2 &= \frac{1}{3}(0 + a\sin\alpha + x) = \frac{1}{3}(a\sin\alpha + x) \end{aligned}\right\} \tag{3}$$

respectively.

The areas of Δs ACD and BCD are

$$\left.\begin{aligned} S_1 &= \frac{1}{2}AC \cdot x\cos\alpha = \frac{1}{2}bx\cos\alpha \\ \text{and } S_2 &= \frac{1}{2}BC \cdot x\cos\alpha = \frac{1}{2}ax\cos\alpha \end{aligned}\right\} \tag{4}$$

respectively.

Hence, the thrusts on the Δs ACD and BCD are

$$\left.\begin{aligned} F_1 &= g\rho\left(\frac{1}{2}bx\cos\alpha\right)\bar{z}_1 = \frac{1}{6}g\rho(b\sin\alpha + x)bx\cos\alpha \\ \text{and } F_2 &= g\rho\left(\frac{1}{2}ax\cos\alpha\right)\bar{z}_2 = \frac{1}{6}g\rho(a\sin\alpha + x)ax\cos\alpha \end{aligned}\right\} \tag{5}$$

respectively. Thus

$$\frac{F_1}{F_2} = \frac{b}{a} \cdot \frac{b\sin\alpha + x}{a\sin\alpha + x}. \tag{6}$$

Now, from Δs ACD and BCD, we have from sine formula,

$$\left.\begin{aligned} \frac{x}{\sin A} &= \frac{AD}{\sin(90° - \alpha)} = \frac{AD}{\cos\alpha} \\ \text{and } \frac{x}{\sin B} &= \frac{BD}{\sin(90° - \alpha)} = \frac{BD}{\cos\alpha} \end{aligned}\right\} \tag{7}$$

Hence,

$$x = \frac{\sin A}{\cos\alpha}AD = \frac{\sin B}{\cos\alpha}BD. \tag{8}$$

Putting the value of x in Eqn (6), we get

$$\frac{F_1}{F_2} = \frac{b}{a} \cdot \frac{b\sin\alpha\cos\alpha + AD\sin A}{a\sin\alpha\cos\alpha + BD\sin B} \tag{9}$$

Now, in ΔABC, $\angle C = 180° - 2\alpha$, so that

$$\sin C = \sin(180° - 2\alpha) = \sin 2\alpha = 2\sin\alpha\cos\alpha. \tag{10}$$

Using (1) and (10) in (9),

$$\begin{aligned} \frac{F_1}{F_2} &= \frac{b}{a} \cdot \frac{b\sin C + \dfrac{2bc}{a+b}\sin A}{a\sin C + \dfrac{2ac}{a+b}\sin B} \\ &= \frac{b}{a} \cdot \frac{b(a+b)c + 2bca}{a(a+b)c + 2acb} \quad \text{[using sine formula]} \\ &= \frac{b^2}{a^2} \cdot \frac{(a+b) + 2a}{(a+b) + 2b} = \frac{b^3 + 3ab^2}{a^3 + 3a^2b} \quad \textbf{(Proved)} \end{aligned}$$

EXAMPLE 4.9 *A hollow weightless cone of semi-vertical angle α and of height h is filled with liquid and freely hung from a point on the rim of the base. Show that the thrust of liquid on the base is*

$$\frac{4\pi g\rho h^3 \tan^3\alpha \sin\alpha}{\sqrt{1+15\sin^2\alpha}}$$

Also show that the ratio of the thrust on base to the weight of the liquid which the cone would contain is $12\sin^2\alpha : \cos\alpha\sqrt{1+15\sin^2\alpha}$.

Solution. Let ABC be the cone suspended from the point A on the rim of its base (FIG. 4.14). If G be the centre of gravity of the cone, then the line AG should be vertical. Radius of the base is

$$R = h\tan\alpha. \tag{1}$$

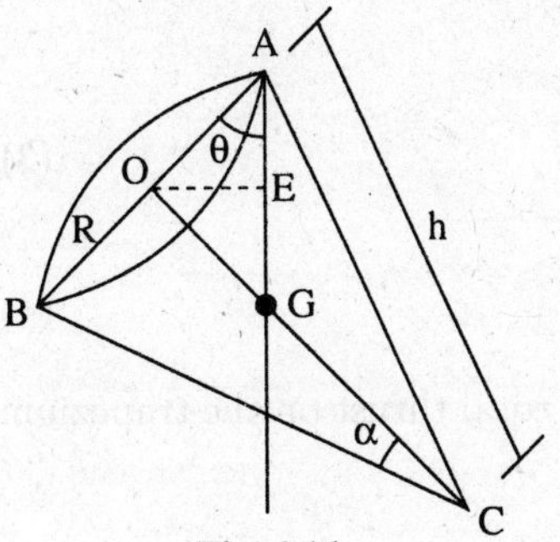

Fig. 4.14

We draw the perpendicular OE on AG. Let $\angle OAE = \theta$. The point G being the centre of gravity of the cone, we have

$$CG = \frac{3}{4}h; GO = \frac{1}{4}h.$$

Now,

$$\tan\theta = \frac{GO}{AO} = \frac{1}{4}\frac{h}{R} = \frac{1}{4}\cot\alpha \text{ [using (1)]}. \tag{2}$$

Hence,

$$\cos\theta = \frac{1}{\sqrt{1+\tan^2\theta}} = \frac{4\sin\alpha}{\sqrt{16\sin^2\alpha+\cos^2\alpha}} = \frac{4\sin\alpha}{\sqrt{1+15\sin^2\alpha}}. \tag{3}$$

The area of the base is

$$S = \pi R^2 = \pi h^2\tan^2\alpha \tag{4}$$

and the depth of its centre of gravity O from A is

$$\bar{z} = AE = AO\cos\theta = (h\tan\alpha)\cdot\frac{4\sin\alpha}{\sqrt{1+15\sin^2\alpha}}. \tag{5}$$

Hence, the thrust on the base is

$$F = g\rho S\bar{z} = g\rho\pi h^3\tan^3\alpha\frac{4\sin\alpha}{\sqrt{1+15\sin^2\alpha}}. \tag{6}$$

The weight of the liquid in the cone is

$$W = \frac{1}{3}\pi\left(h^2\tan^2\alpha\right)\cdot hg\rho = \frac{1}{3}\pi g\rho h^3\tan^2\alpha. \tag{7}$$

From (6) and (7), we get

$$\frac{F}{W} = \frac{12\tan\alpha\sin\alpha}{\sqrt{1+15\sin^2\alpha}} = \frac{12\sin^2\alpha}{\cos\alpha\sqrt{1+15\sin^2\alpha}} \textbf{ (Proved)}.$$

EXAMPLE 4.10 *The lighter of two liquids of density ρ rests on the heavier of density δ to a depth of $h/2$. A trapezium is immersed vertically with one parallel side of length 'a' in the surface and the opposite of length b at a depth h. Prove that the thrust on the trapezium is*

$$\frac{gh}{48}[(7a+11b)\rho+(a+5b)\delta].$$

Solution. Let $ABCD$ be the trapezium immersed in the two liquids vertically with its side $AD = a$ on the surface (FIG. 4.15). EF is the surface of separation of two liquids so that depth of each liquid is $\frac{h}{2}$. We join BD. Then, area of the triangle ABD is

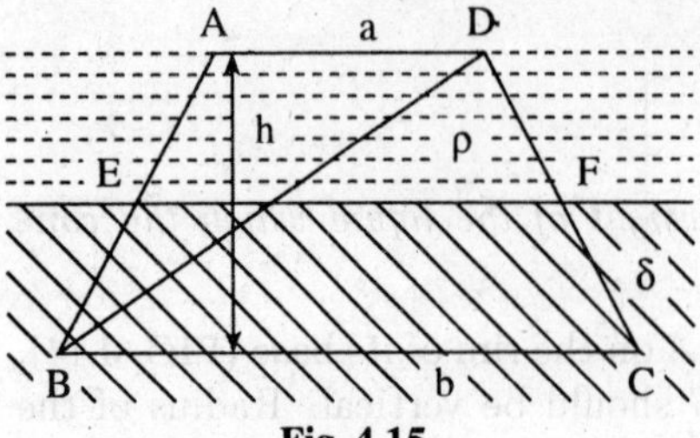

Fig. 4.15

$$S_1 = \frac{1}{2}ah \tag{1}$$

and that of triangle BDC is

$$S_2 = \frac{1}{2}bh. \tag{2}$$

The depth of c.g. of the triangle ABD from the surface is

$$\bar{z}_1 = \frac{1}{3}(0 + 0 + h) = \frac{1}{3}h \tag{3}$$

and that of triangle BDC is

$$\bar{z}_2 = \frac{1}{3}(0 + h + h) = \frac{2}{3}h. \tag{4}$$

Hence, assuming that the whole liquid is of density ρ only, the total thrust on the trapezium would be

$$F_1 = g\rho(S_1\bar{z}_1 + S_2\bar{z}_2)$$

$$= g\rho\left(\frac{1}{2}ah \cdot \frac{1}{3}h + \frac{1}{2}bh \cdot \frac{2}{3}h\right) = \frac{1}{6}g\rho h^2(a + 2b). \tag{5}$$

Now we remove the top liquid of density ρ and assume that the remaining liquid is of density $(\delta - \rho)$. Thrust on area $EBCF$ due to this liquid of density $(\delta - \rho)$ is obtained from (5) by replacing

$$\rho \to (\delta - \rho), h \to \frac{h}{2} \text{ and } a \to \frac{a + b}{2}.$$

Thus the thrust on area $EBCF$ is

$$F_2 = \frac{1}{6}g(\delta - \rho)\frac{h^2}{4}\left(\frac{a + b}{2} + 2b\right), \text{ i.e., } F_2 = \frac{1}{48}g(\delta - \rho)h^2(a + 5b). \tag{6}$$

Hence the total thrust on the trapezium $ABCD$ is

$$F = F_1 + F_2$$

$$\text{i.e., } F = \frac{1}{6}g\rho h^2(a + 2b) + \frac{1}{48}g(\delta - \rho)h^2(a + 5b)$$

$$= \frac{gh^2}{48}[8\rho(a + 2b) + (\delta - \rho)(a + 5b)]$$

$$= \frac{gh^2}{48}[\rho(7a + 11b) + \delta(a + 5b)] \text{ (Proved).}$$

EXAMPLE 4.11 *A triangle of altitude h and base a is immersed in a liquid whose density at depth z is* $(\delta + \lambda z)$, *with the vertex in the surface and base horizontal. Prove that the thrust on it is*

$$a\left[\frac{\pi h}{2} + g\left(\frac{\delta h^2}{3} + \frac{\lambda h^3}{4}\right)\right],$$

where π *is the atmospheric pressure.*

Solution. Let ABC be the triangle immersed in the liquid with its vertex A in the surface and its base $BC = a$ horizontal (FIG. 4.16).

We consider a strip of thickness dz at a depth z from the surface. Then pressure at any point on this strip is

$$p = \pi + g(\delta + \lambda z)z. \quad (1)$$

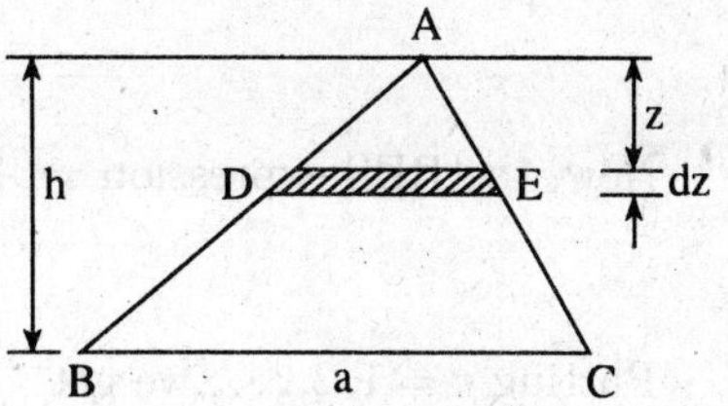

Fig. 4.16

From triangles ADE and ABC, we have

$$\frac{DE}{BC} = \frac{z}{h}, \text{ i.e., } DE = \frac{az}{h}. \quad (2)$$

Hence the thrust on the strip is

$$dF = p \cdot DE \cdot dz = [\pi + g(\delta + \lambda z)z]\frac{az}{h}dz.$$

The total thrust on the triangle is

$$F = \frac{a}{h}\int_0^h \left[\pi z + g(\delta + \lambda z)z^2\right] dz = \frac{a}{h}\left[\pi\frac{h^2}{2} + g\left(\frac{\delta h^3}{3} + \frac{\lambda h^4}{4}\right)\right]$$

$$= a\left[\frac{\pi h}{2} + g\left(\frac{\delta h^2}{3} + \frac{\lambda h^3}{4}\right)\right] \quad \textbf{(Proved)}.$$

EXAMPLE 4.12 *A semi-circle is immersed vertically in a liquid with the diameter in the surface. Show how to divide it into sectors such that thrusts on each is the same.*

Solution. Let ACB be the semi-circle immersed in a liquid with its diameter AB in the surface. We are to divide it into n sectors such that thrusts on each of them is same. Let the sector AOP_r contain r-number of such sectors, where $\angle AOP_r = 2\alpha$ (FIG. 4.17).

Now, for the semi-circle, we have $S = \frac{1}{2}\pi a^2, \bar{z} = \frac{4a}{3\pi}$, so that thrust upon it containing n sectors is

$$F_1 = g\rho\left(\frac{1}{2}\pi a^2\right)\frac{4a}{3\pi} = \frac{2}{3}g\rho a^3. \quad (1)$$

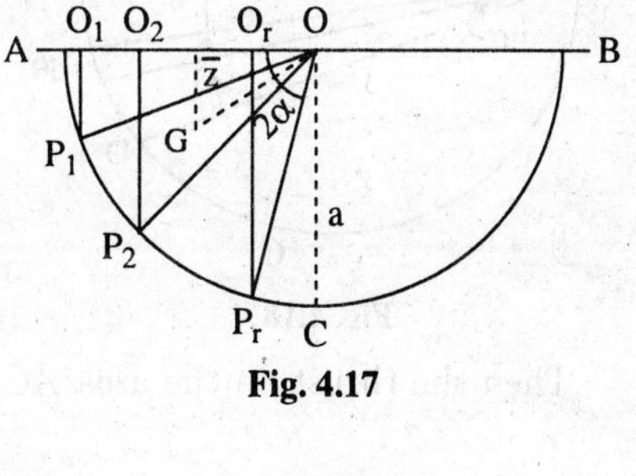

Fig. 4.17

Here ρ is the density of liquid and a = radius of the semi-circle.

Again, for the portion AOP_r containing r sectors,

$$\text{area } S = \alpha a^2 \text{ and } \bar{z} = OG\sin\alpha = \frac{2}{3}a\frac{\sin\alpha}{\alpha}\cdot\sin\alpha = \frac{2a\sin^2\alpha}{3\alpha}$$

so that the thrust upon it is

$$F_2 = g\rho\left(\alpha a^2\right)\frac{2a\sin^2\alpha}{3\alpha} = \frac{2}{3}g\rho a^3\sin^2\alpha. \quad (2)$$

Since the thrust on each sector is to be equal, we must have

$$\frac{F_2}{F_1} = \frac{r}{n}, \text{ or, } \frac{\frac{2}{3}g\rho a^3\sin^2\alpha}{\frac{2}{3}g\rho a^3} = \frac{r}{n},$$

which yields

$$\sin^2\alpha = \frac{r}{n}. \quad (3)$$

This can be written as $\frac{2r}{n} = 2\sin^2\alpha = 1 - \cos 2\alpha$

$$\therefore \frac{2ar}{n} = a - a\cos 2\alpha. \tag{4}$$

Now, the RHS expression $= AO - OO_r = AO_r$. Hence, we get

$$AO_r = \frac{2r}{n} \cdot a. \tag{5}$$

Putting $r = 1, 2, \ldots$, we get

$$AO_1 = \frac{2a}{n}, AO_2 = \frac{4a}{n}, AO_3 = \frac{6a}{n}, \ldots. \tag{6}$$

Eqn (6) shows that distances of the points $O_1, O_2, \ldots$, from A are in arithmetical progression with common difference $2a/n$.

Thus, if we divide the diameter AB into n equal parts each of length $\frac{2a}{n}$, draw perpendiculars on AB from them to meet the circular arc at points $P_1, P_2, \ldots$ and then join the lines $OP_1, OP_2, \ldots$, the sectors obtained will be such that thrust upon each of them is same.

EXAMPLE 4.13 *A semi-circular area is immersed in a liquid vertically with the diameter in the surface. If the line through an extremity of the diameter dividing the semi-circle into two parts, the thrusts on which are equal, makes an angle α with the horizontal, then prove that*

$$\cos^4\alpha = \frac{1}{2}.$$

Solution. Let ACB be a semi-circular area immersed vertically in a liquid with its diameter AB in the surface. Let AD be a line drawn from A, an extremity of the diameter, dividing the semi-circular area into two parts ACD and ABD (FIG. 4.18) so that thrusts on them are equal.

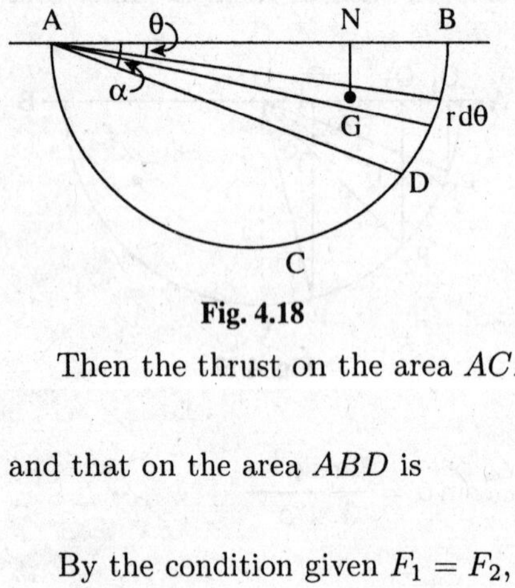

Fig. 4.18

Let R = radius of the semi-circle, ρ = density of the liquid, S_1 = area of the portion ACD, z_1 = depth of c.g. from the surface of the area ACD, S_2 = area of the portion ABD, z_2 = depth of c.g. from the surface of the area ABD.

Then the thrust on the area ACD is

$$F_1 = g\rho S_1 z_1 \tag{1}$$

and that on the area ABD is

$$F_2 = g\rho S_2 z_2. \tag{2}$$

By the condition given $F_1 = F_2$, i.e., $g\rho S_1 z_1 = g\rho S_2 z_2$, which yields

$$S_1 z_1 = S_2 z_2. \tag{3}$$

Now, the area of the semi-circle is $S = \frac{1}{2}\pi R^2$ and depth of its c.g. from AB is

$$z = \frac{4R}{3\pi}.$$

Then, from statics, we have

$$\frac{4R}{3\pi} = \frac{S_1 z_1 + S_2 z_2}{\frac{1}{2}\pi R^2}, \text{ or, } \frac{2}{3}R^3 = S_1 z_1 + S_2 z_2 = 2S_2 z_2 \text{ [using (3)]} \tag{4}$$

Let us now determine the quantity $S_2 z_2$ which is the moment of the area about the diameter AB. With origin at A, the polar equation of the circle is

$$r = 2R\cos\theta. \tag{5}$$

We consider an elementary sector of small vertical angle $d\theta$. Area of this elementary sector is

$$dS = \frac{1}{2}(r)(rd\theta) = \frac{1}{2}r^2 d\theta. \tag{6}$$

Its c.g. G is at a distance of $\frac{2}{3}r$ from A so that

$$GN = AG\sin\theta = \frac{2}{3}r\sin\theta. \tag{7}$$

Hence, the moment of area of the elementary sector about AB is

$$= dS \cdot GN = \left(\frac{1}{2}r^2 d\theta\right)\frac{2}{3}r\sin\theta = \frac{1}{3}r^3\sin\theta d\theta. \tag{8}$$

Hence, the moment of area of the portion ABD about AB is

$$S_2 z_2 = \int \frac{1}{3}r^3 \sin\theta d\theta,$$

which upon using Eqn (5) assumes the form

$$S_2 z_2 = \frac{8R^3}{3}\int_0^{\alpha} \cos^3\theta\sin\theta d\theta. \tag{9}$$

We put $\cos\theta = x$, i.e., $-\sin\theta d\theta = dx$. Then the integral (9) is

$$\begin{aligned} S_2 z_2 &= -\frac{8R^3}{3}\int x^3 dx \\ &= -\frac{8R^3}{3}\left[\frac{x^4}{4}\right] \\ &= -\frac{2R^3}{3}\left(\cos^4\theta\right)\Big|_{\theta=0}^{\theta=\alpha} \\ &= \frac{2R^3}{3}\left(1-\cos^4\alpha\right). \end{aligned} \tag{10}$$

From Eqns (4) and (10), we have

$$\frac{2}{3}R^3 = \frac{4}{3}R^3\left(1-\cos^4\alpha\right),$$

which yields

$$2\cos^4\alpha = 1, \text{ i.e., } \cos^4\alpha = \frac{1}{2} \textbf{ (Proved)}.$$

EXAMPLE 4.14 *A vessel contains n different liquids resting in horizontal layers and of densities* $\rho_1, \rho_2, \ldots, \rho_n$ *starting from the highest liquid. A triangle is held with its base in the upper surface of the highest liquid and with its vertex in the n-th liquid. If* Δ *be the area of the triangle and* $h_1, h_2, \ldots, h_n$ *be the depths of the vertex below the upper surfaces of the 1st, 2nd, ..., n-th liquids respectively, then prove that the thrust on the triangle is*

$$\frac{1}{3}\frac{g\Delta}{h_1^2}\left[\rho_1\left(h_1^3-h_2^3\right)+\rho_2\left(h_2^3-h_3^3\right)+\cdots+\rho_n h_n^3\right].$$

Solution. Let AB_1C_1 be a triangle immersed in n different liquids with its base B_1C_1 in the surface of the uppermost liquid of density ρ_1. Let the length of base of the triangle AB_1C_1 be a_1 and those of other triangles formed with lines of separation be $a_2, a_3, \ldots, a_r, \ldots, a_n$ respectively (FIG. 4.19).

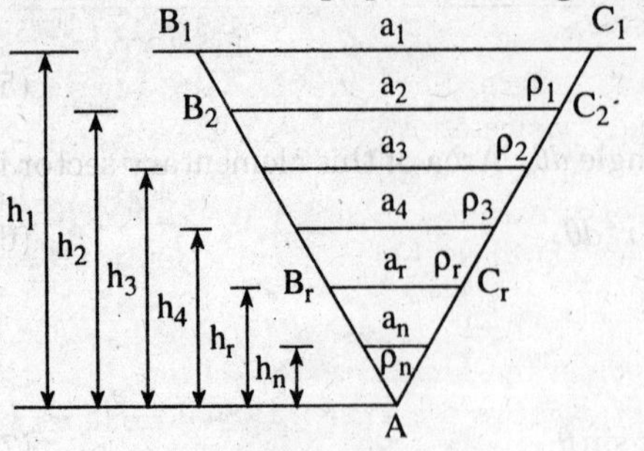

Fig. 4.19

Then, the area of the triangle AB_1C_1 is

$$\Delta = \frac{1}{2}a_1 h_1 \tag{1}$$

and the area of the r-th triangle with base a_r and height h_r is

$$\Delta_r = \frac{1}{2}a_r h_r. \tag{2}$$

Now, from similar triangles AB_1C_1 and AB_rC_r, we have

$$\frac{a_r}{a_1} = \frac{h_r}{h_1}, \text{ i.e., } a_r = \frac{a_1}{h_1}h_r. \tag{3}$$

Putting this value in (2), we get

$$\Delta_r = \frac{1}{2}\left(\frac{a_1}{h_1}h_r\right)h_r = \frac{1}{2}\frac{a_1}{h_1}h_r^2. \tag{4}$$

From (1) and (4),

$$\frac{\Delta_r}{\Delta} = \left(\frac{1}{2}\frac{a_1}{h_1}h_r^2\right)\frac{2}{a_1 h_1} = \frac{h_r^2}{h_1^2},$$

which yields

$$\Delta_r = \frac{h_r^2}{h_1^2}\Delta. \tag{5}$$

Thus, putting $r = 2, 3, \ldots, n$, we get the areas of the triangles $AB_2C_2, \ldots, AB_nC_n$ as

$$\frac{h_2^2}{h_1^2}\Delta, \frac{h_3^2}{h_1^2}\Delta, \ldots, \frac{h_n^2}{h_1^2}\Delta. \tag{6}$$

respectively.

The depths of c.g. of the triangles from their respective bases are

$$\bar{z}_r = \frac{1}{3}h_r, r = 1, 2, \ldots, n. \tag{7}$$

The total thrust on the triangle AB_1C_1 is obtained by using Eqn (4.17). Thus

$$\begin{aligned}
F &= g\left[\rho_1\bar{z}_1\Delta + \sum_{r=2}^{n}(\rho_r - \rho_{r-1})\Delta_r\bar{z}_r\right] \\
&= g\left[\rho_1\Delta\cdot\frac{1}{3}h_1 + \sum_{r=2}^{n}(\rho_r - \rho_{r-1})\frac{h_r^2}{h_1^2}\Delta\cdot\frac{1}{3}h_r\right] \\
&= \frac{g\Delta}{3h_1^2}\left[\rho_1 h_1^3 + \sum_{r=2}^{n}(\rho_r - \rho_{r-1})h_r^3\right] \\
&= \frac{g\Delta}{3h_1^2}\left[\rho_1 h_1^3 + (\rho_2 - \rho_1)h_2^3 + (\rho_3 - \rho_2)h_3^3 + \cdots + (\rho_n - \rho_{n-1})h_n^3\right] \\
&= \frac{g\Delta}{3h_1^2}\left[\rho_1(h_1^3 - h_2^3) + \rho_2(h_2^3 - h_3^3) + \cdots + (\rho_n h_n^3)\right] \text{ (Proved).}
\end{aligned}$$

EXERCISE

1. A square is immersed in a liquid with one side in the surface. Show how to draw a horizontal line in the square dividing it into two portions, the thrusts on which are the same.

[Ans: divides vertical side in ratio $(\sqrt{2}+1) : 1$.]

2. A plane area in the form of a trapezium is immersed vertically in a heavy liquid of density ρ with one of the parallel sides of length a in the surface, and the other of length b at a depth h below the surface. Show that the whole pressure on the area is $\frac{g\rho h^2}{6}(a+2b)$.

3. A square immersed vertically in a liquid with one side in the surface, is divided into two parts by a straight line parallel to a diagonal, so that thrusts on the two portions are equal. If x be the depth of the point in which the dividing line cuts the vertical side a of the square, then prove that

$$2x^3 - 6a^2x + a^3 = 0.$$

4. A parallelogram is immersed in water with one side in the surface. It is divided into n parts by horizontal lines so that the thrusts on each part are the same.

Show that, the depths of the dividing lines are proportional to the square root of natural numbers.

5. A triangle ABC is immersed in a liquid vertically with its side AB in the surface. If O be the centre of the circumscribed circle of the $\triangle ABC$, prove that the thrusts on $\triangle$s OCA and OCB are in the ratio of $\sin 2B : \sin 2A$.

6. A cone full of water is placed on its side on a horizontal table. If W be the weight of the contained water and 2α the vertical angle of the cone, show that the thrust on the base is $3W \sin\alpha$.

7. A trapezoidal plane having its parallel sides of length x and $y (x > y)$ at a distance z apart, is immersed vertically in a liquid of density ρ. The uppermost side is horizontal at a depth d below the free surface.

Show that the total thrust on the surface is $\frac{g\rho}{6}[(2x+y)z + 3d(x+y)]z$.

8. A cylinder is filled with equal volumes of n different liquids which do not mix. The density of the highest liquid is ρ, that of the next is 2ρ, and so on, that of the lowest being $n\rho$. Show that the whole pressure on different portions of the curved surfaces of the cylinder are in the ratios

$$1^2 : 2^2 : 3^2 : \cdots : n^2.$$

9. A triangle completely immersed in a liquid with its corner C in the surface is divided into two parts by a straight line through its vertex A, so that the thrusts on two portions are equal. If x be the depth of the point at which the dividing line cuts BC and α, β be the depths of A and B, then prove that $2x^2 + 2\alpha x - \beta(\alpha+\beta) = 0$.

10. A triangle is immersed vertically in a liquid with the vertex in the surface and the base horizontal. Find how to divide it by horizontal lines into n parts, the thrust on which are equal.

[*Ans.* Depths of parts are proportional tocube roots of natural numbers.]

11. If a parallelogram be immersed in any manner in a homogeneous heavy liquid, prove that the sum of the pressure at the extremities of the diagonal is the same.

12. A triangle ABC is immersed vertically in two liquids with base BC in the surface. If the densities of the upper and lower liquids be δ_1 and δ_2 respectively and the depth of the surface of separation be half the depth of the vertex, show that the ratio of the thrusts on the portions in the two liquids is $\frac{4\delta_1}{3\delta_1+\delta_2}$.

13. Show that the thrusts on each of the plane ends of a circular cylinder of height h and radius of the base R is $g\rho\pi R^2\left(c + \frac{1}{2}h\cos\theta\right)$, where c is the depth of the middle point of the cylinder below the surface of the fluid and θ is the angle made by the axis to the vertical.

14. A triangle ABC is immersed vertically in a liquid with the vertex A in the surface and the median vertical. Prove that the ratio of the thrusts on the two parts into which the triangle is divided by the median is $\frac{3b^2+5c^2-2a^2}{5b^2+3c^2-2a^2}$.

15. A cylindrical tumbler, half-filled with a liquid of density ρ, is filled up with a liquid of density δ which does not mix with the former one. Show that the thrust on the base of the tumbler to the whole pressure on the curved surface as $2R(\rho+\delta) : h(\rho+3\delta)$, where $h =$ height and $R =$ radius of the base of the tumbler.

16. A parallelogram $ABCD$ is immersed in a homogeneous fluid of density ρ, open to the atmosphere (pressure being equal to π) with the side AB in the surface. E is a point in AB such that $AE = \frac{2}{3}AB$. A straight line joins E to a point F in CD. Show that, if the thrusts on $AEFD$ and $EBCF$ are equal, DF is given by $6DF(3\pi + 2g\rho b) = AB(6\pi + 5g\rho b)$, where ρ is the density of fluid and b is the depth of CD.

17. The inclinations of the axis of a submerged solid cylinder to the vertical in two different positions are complimentary to each other. If P and P' be the difference between the pressures on the two ends in the two cases, prove that the weight of the displaced fluid is equal to $\sqrt{P^2 + P'^2}$.

18. A solid triangular prism, the faces of which include angle α, β, γ is placed in any position entirely within a liquid. If P, Q, R be the thrusts on the three faces respectively opposite to angles α, β, γ, then prove that $P \operatorname{cosec} \alpha + Q \operatorname{cosec} \beta + R \operatorname{cosec} \gamma$ is invariable so long as the depth of the centre of gravity of the prism is unchanged.

19. A hollow right circular cone without weight is filled with liquid and suspended from a point in the rim of its base. If the axis of the cone is inclined to the vertical at an angle θ, prove that $\cos 2\theta = \cot 2\alpha - \frac{3}{4} \operatorname{cosec} 2\alpha$, where 2α is the vertical angle of the cone.

20. A parallelogram is immersed in a fluid with a diagonal vertical, an extremity of which is in the surface of the liquid. Through this point lines are drawn, dividing the parallelogram into three equal parts. Compare the pressures on three parts and if P_2 be the pressure on the middle part and P_1, P_3 those on the other two, prove that $16P_2 = 11(P_1 + P_2)$.

CHAPTER 5

Centre of Pressure

5.1 Introduction

We have seen in the earlier chapter that the pressure at any point on a plane surface in a fluid at rest under gravity acts along the normal of the plane surface and its magnitude is proportional to the depth of the point. Thus, the resultant thrust on the plane area is the sum of a system of parallel forces and acts at a particular point on the plane surface. This point is called the ***centre of pressure*** and is denoted by c.p. Thus, ***the centre of pressure of a plane surface in contact with a fluid is the point at which the direction of the resultant fluid thrust on the plane surface meets the surface.***

The concept of centre of pressure is associated with plane surfaces only remaining within heavy fluids. In the case of curved surfaces, all the fluid pressures may not be reduced to a single resultant so that the definition of centre of pressure does not hold good.

In this chapter, we intend to determine the centre of pressure in different cases.

5.2 Position of Centre of Pressure

We consider a plane surface AB immersed in a heavy homogeneous fluid of density ρ inclined at an angle α with the horizontal (FIG. 5.1). The ends of the surface are at depths of z_1, z_2 from the free surface. The pressures at different points vary linearly with depth, so that pressure diagram is a trapezoid which is known as ***pressure prism***. The resultant thrust F is equal to the volume of this pressure prism and will act at its centre of gravity. Thus the centre of pressure remains below the centre of gravity of the plane surface.

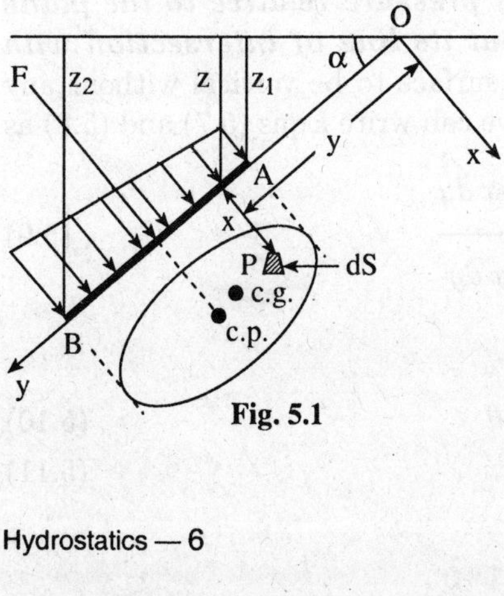

Fig. 5.1

Let us now determine the position of the centre of pressure analytically. Let Ox and Oy be a rectangular Cartesian coordinate system on which lies the plane surface. At any point $P(x, y)$ at a depth z from free surface, we consider an elementary rectangular area $dS = dx\,dy$.

The pressure at the point P is

$$p = g\rho z = g\rho y \sin\alpha. \tag{5.1}$$

Then the thrust on the elementary area is

$$dF = p\,dx\,dy = g\rho y \sin\alpha dx\,dy \tag{5.2}$$

so that the resultant thrust is

$$F = \iint g\rho y \sin \alpha dx\, dy, \tag{5.3}$$

where the integration is taken over the whole area of the plane surface.

The moment of elementary thrust about the y-axis is

$$x dF = g\rho xy \sin \alpha dx\, dy. \tag{5.4}$$

Hence the sum of the moments of all the thrusts about y-axis is

$$= \iint g\rho xy \sin \alpha dx\, dy. \tag{5.5}$$

Let (ξ, η) be the coordinates of the centre of pressure of the plane surface considered here. Then the moment of the resultant thrust about y-axis is

$$\xi \cdot F = \xi \iint g\rho y \sin \alpha dx\, dy. \tag{5.6}$$

For equilibrium of the forces, the sum of the moments of the elementary forces about any line must be equal to the moment of the resultant force about that line. Hence, from Eqns (5.5) and (5.6), we have

$$\xi \iint g\rho y \sin \alpha dx\, dy = \iint g\rho xy \sin \alpha dx\, dy,$$

which yields

$$\xi = \frac{\iint g\rho xy \sin \alpha dx\, dy}{\iint g\rho y \sin \alpha dx\, dy}, \text{ or, } \xi = \frac{\iint \rho xy\, dx\, dy}{\iint \rho y\, dx\, dy}. \tag{5.7}$$

Similarly, by taking moments about x-axis, we get

$$\eta = \frac{\iint \rho y^2 dx\, dy}{\iint \rho y\, dx\, dy}. \tag{5.8}$$

Thus the coordinates of the centre of pressure is independent of the angle α of inclination of the plane surface. Hence, ***the position of the centre of pressure relative to the plane area remains unaltered if the plane area rotates about its line of intersection with the effective surface***. We can, therefore, take the plane surface to be vertical without any loss of generality. Remembering that in this case, $p = g\rho y$, we can write Eqns (5.7) and (5.8) as

$$\xi = \frac{\iint px\, dx\, dy}{\iint p\, dx\, dy}, \eta = \frac{\iint py\, dx\, dy}{\iint p\, dx\, dy}. \tag{5.9}$$

In polar coordinates,

$$x = r\cos\theta, y = r\sin\theta \tag{5.10}$$

$$\text{and } dS = dr(rd\theta) = rd\theta dr. \tag{5.11}$$

Then Eqns (5.7) and (5.8) assume the forms

$$\xi = \frac{\iint \rho r^3 \sin\theta \cos\theta d\theta dr}{\iint \rho r^2 \sin\theta d\theta dr}, \eta = \frac{\iint \rho r^3 \sin^2\theta d\theta dr}{\iint \rho r^2 \sin\theta d\theta dr}. \tag{5.12}$$

Eqns (5.9) in polar coordinates become

$$\xi = \frac{\iint pr^2 \cos\theta d\theta dr}{\iint prd\theta dr}, \eta = \frac{\iint pr^2 \sin\theta d\theta dr}{\iint prd\theta dr}. \tag{5.13}$$

For homogeneous fluids, the density ρ is constant. In this case, Eqns (5.7) and (5.8) are simplified to

$$\xi = \frac{\iint xy\, dx\, dy}{\iint y\, dx\, dy}, \eta = \frac{\iint y^2 dx\, dy}{\iint ydxdy}. \tag{5.14}$$

In polar coordinates, these equations assume the forms

$$\xi = \frac{\iint r^3 \sin\theta \cos\theta d\theta dr}{\iint r^2 \sin\theta d\theta dr}, \eta = \frac{\iint r^3 \sin^2\theta d\theta dr}{\iint r^2 \sin\theta d\theta dr}. \tag{5.15}$$

5.3 Depth of Centre of Pressure

We consider a plane surface of area S immersed vertically in a heavy fluid of density ρ remaining at rest under gravity. Let the plane of the surface meet the effective surface in a line AB (FIG. 5.2).

We take an elementary area dS at a depth z from the effective surface AB. Let G be the centre of gravity of the plane surface and its depth from effecive surface be h.

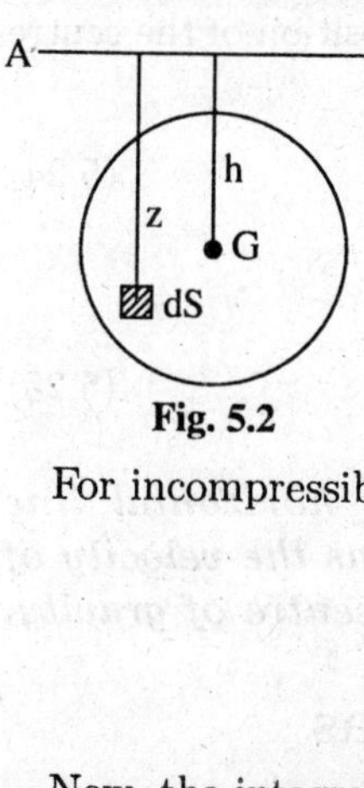

Fig. 5.2

By Eqn (5.14), the depth of centre of pressure below the effective surface is given by

$$\eta = \frac{\int \rho z^2 dS}{\int \rho z\, dS}. \tag{5.16}$$

For incompressible and homogeneous fluids, the density ρ is constant. Hence,

$$\eta = \frac{\int z^2 dS}{\int z\, dS}. \tag{5.17}$$

Now, the integral in the denominator, by statics, represents the first moment of the area S about AB. Hence,

$$\int z\, dS = Sh. \tag{5.18}$$

Again, from dynamics, the integral in the numerator represents the second moment of the area S about AB. Hence,

$$\int z^2 dS = SK^2, \tag{5.19}$$

where K is the radius of gyration of the area S about AB.

By virtue of Eqns (5.18) and (5.19), we get from Eqn (5.17),

$$\eta = \frac{SK^2}{Sh} = \frac{K^2}{h}. \tag{5.20}$$

Let K' be the radius of gyration of the area S about a line passing through the centre of gravity G and parallel to AB. Then, by the theorem of parallel axes, we have

$$SK^2 = SK'^2 + Sh^2, \text{ or, } K^2 = K'^2 + h^2. \tag{5.21}$$

From Eqns (5.20) and (5.21), we get

$$\eta = \frac{K'^2}{h} + h. \tag{5.22}$$

Eqn (5.22) shows that ***the depth of the centre of pressure*** (η) ***is greater than the depth of the centre of gravity*** (h) ***of the plane surface.***

Let ζ denote the difference of the depth of c.p. and c.g. Then

$$\zeta = \eta - h = \frac{K'^2}{h}. \tag{5.23}$$

This shows that ***the vertical distance between the centre of pressure and centre of gravity of a plane surface varies inversely as the depth of the centre of gravity, provided that the plane surface is moved without rotation.***

Eqn (5.23) indicates that ζ decreases with the increase of depth and $\zeta \to 0$ as $h \to \infty$. That is, ***for a plane surface remaining at an infinite depth from the effective surface of the fluid, the centre of pressure and the centre of gravity coincide.***

Let the plane surface be moved down vertically with a velocity v. The position of the centre of gravity of the plane surface being fixed, we write

$$v = \frac{dh}{dt}. \tag{5.24}$$

Differentiating Eqn (5.23) w.r.t. time,

$$\frac{d\zeta}{dt} = -\frac{K'^2}{h^2}\frac{dh}{dt} = -\frac{K'^2}{h^2}v. \tag{5.25}$$

Eqn (5.25) implies that ***the centre of pressure approaches the horizontal line through the centre of gravity with a velocity which directly varies as the velocity of the plane surface and inversely as the square of the depth of the centre of gravity.***

5.4 Depth of C.P. of Some Simple Plane Areas

We now intend to find the position of the centre of pressure of some simple plane areas immersed in a heavy homogeneous fluid remaining at rest under gravity.

(a) **A parallelogram with one side in the surface.** Let $ABCD$ be a parallelogram immersed vertically in a fluid of density ρ with its side AB in the surface (FIG. 5.3).

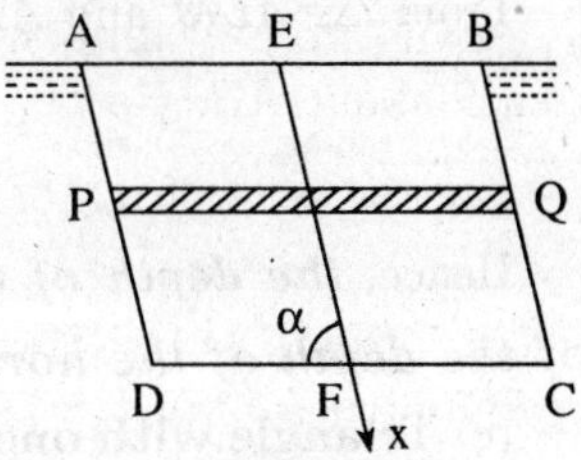

Fig. 5.3

Let E and F be middle points of AB and CD. Join E and F. Let α be the angle which EF makes with the horizontal.

Let $AB = DC = a$ and $AD = BC = l$. Taking the line EF as x-axis, we consider a thin strip PQ of thickness dx at a distance x from the surface along EF.

Then the pressure at any point on this strip is $p = g\rho(x \sin\alpha)$ and the area of the elementary strip is

$$dS = PQ \cdot dx = a\,dx.$$

Hence, the position of the centre of pressure is

$$\xi = \frac{\int xp\,dS}{\int p\,dS} = \frac{\int_0^l x \cdot g\rho x \sin\alpha \cdot a\,dx}{\int_0^l g\rho x \sin\alpha \cdot a\,dx}, \text{ i.e., } \xi = \frac{\int_0^l x^2 dx}{\int_0^l x\,dx} = \frac{2}{3}l = \frac{2}{3}EF. \tag{5.26}$$

Thus, the centre of pressure lies on EF at a distance $\frac{2}{3}$ of depth of DC from free surface.

In the case of a rectangle, $\alpha = \frac{\pi}{2}$. But we get the same result.

(b) **A triangle with vertex in the surface and base horizontal.** Let ABC be a triangle immersed vertically in a liquid with its vertex A in the surface and base BC horizontal. Let its median AD make an angle α with the horizontal (FIG. 5.4).

AE is drawn perpendicular to BC. Let $AE = h$, i.e., the depth of the base of the triangle below the free surface is h. We take x-axis along the median and consider a thin strip PQ at a distance x measured along AD and of a small thickness dx.

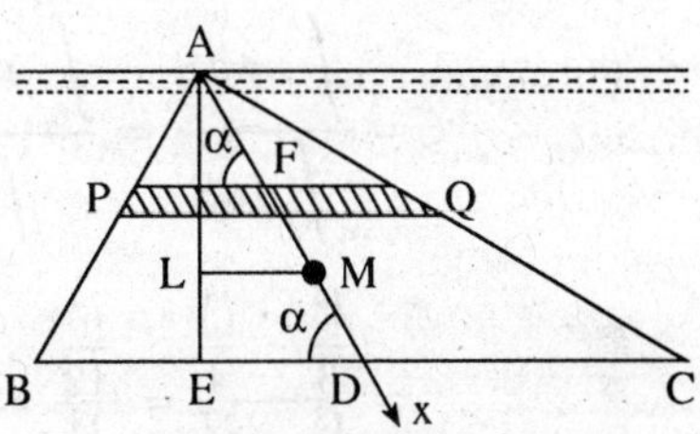

Fig. 5.4

Let $BC = a$, $AD = l$. From the figure, $AF = x$.

Pressure at any point on the strip PQ is

$$p = g\rho(AF \sin\alpha) = g\rho x \sin\alpha$$

and the area of the elementary strip PQ is

$$dS = PQ(dx \cdot \sin\alpha).$$

From Δs APQ and ABC, we have

$$\frac{PQ}{BC} = \frac{AF}{AD}, \text{ i.e., } PQ = \frac{AF}{AD} \cdot BC = \frac{ax}{l}.$$

Then

$$dS = \frac{ax}{l}(dx \cdot \sin\alpha) = \frac{a}{l}x \sin\alpha dx.$$

Hence, the position of the centre of pressure is

$$\xi = \frac{\int xp\,dS}{\int p\,dS} = \frac{\int_0^l x(g\rho x \sin\alpha)\frac{a}{l}x \sin\alpha dx}{\int_0^l (g\rho x \sin\alpha)\frac{a}{l}x \sin\alpha dx} = \frac{\int_0^l x^3 dx}{\int_0^l x^2 dx} = \frac{3}{4}l.$$

Thus, the distance of the centre of pressure M from A is $\frac{3}{4}l$. That is

$$\frac{AM}{AD} = \frac{3}{4}.$$

From $\triangle$s ALM and AED, we have

$$\frac{AL}{AE} = \frac{AM}{AD} = \frac{3}{4}, \therefore AL = \frac{3}{4}h. \tag{5.27}$$

Hence, ***the depth of the centre of pressure from the free surface is equal to*** $\frac{3}{4}$***th of the depth of the horizontal base of the triangle ABC.***

(c) **Triangle with one side in the surface.** Let ABC be a triangle immersed vertically in a fluid with its base BC in the surface. Let the median $AD = l$, and let it make an angle α with the horizontal (FIG. 5.5). AE is drawn perpendicular to BC.

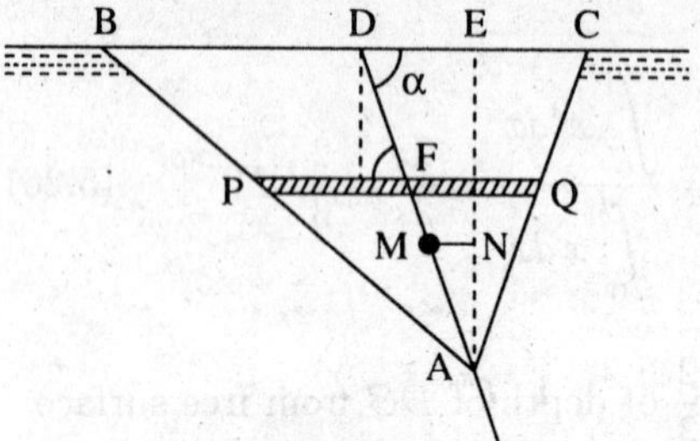

Fig. 5.5

As before, we take x-axis along DA and consider a thin strip PQ of small thickness dx at a distance x.

Let $BC = a$ and $AE = h$. Pressure at any point on the strip PQ is

$$p = g\rho(DF \sin\alpha) = g\rho(x \sin\alpha)$$

and the area of the strip PQ is $dS = PQ(dx \sin\alpha)$.

Now, from the figure we have $\dfrac{PQ}{BC} = \dfrac{AF}{AD} = \dfrac{AD - DF}{AD} = \dfrac{l - x}{l}$.

Hence, $PQ = \dfrac{l-x}{l} \cdot BC = \dfrac{l-x}{l}a$. Then $dS = \left(\dfrac{l-x}{l}a\right)(dx \sin\alpha) = \dfrac{a(l-x)}{l} \sin\alpha dx$.

Hence, the position of the centre of pressure is

$$\xi = \frac{\int xp\, dS}{\int p\, dS} = \frac{\int_0^l x \cdot (g\rho x \sin\alpha) \cdot \frac{a}{l}(l-x) \sin\alpha dx}{\int_0^l (g\rho x \sin\alpha) \frac{a}{l}(l-x) \sin\alpha dx} = \frac{\int_0^l x^2(l-x)dx}{\int_0^l x(l-x)dx}$$

$$= \frac{\frac{l^4}{3} - \frac{l^4}{4}}{\frac{l^3}{2} - \frac{l^3}{3}} = \frac{\frac{l^4}{12}}{\frac{l^3}{6}} = \frac{l}{2}. \tag{5.28}$$

Thus the centre of pressure M lies on the middle of the median.

Since $\dfrac{DM}{AD} = \dfrac{EN}{AE} = \dfrac{1}{2}$, therefore, $EN = \dfrac{1}{2}AE = \dfrac{1}{2}h$. That is, ***the depth of the centre of pressure from the surface is equal to half of the depth of the vertex of the triangle.***

(d) **A totally immersed circular area.** Let a circular area of radius R be immersed vertically in a fluid such that the depth of its centre from the surface is h (FIG. 5.6). With origin at the centre of the circle, we take x-axis vertically downwards and y-axis horizontal. Then equation to the circle is

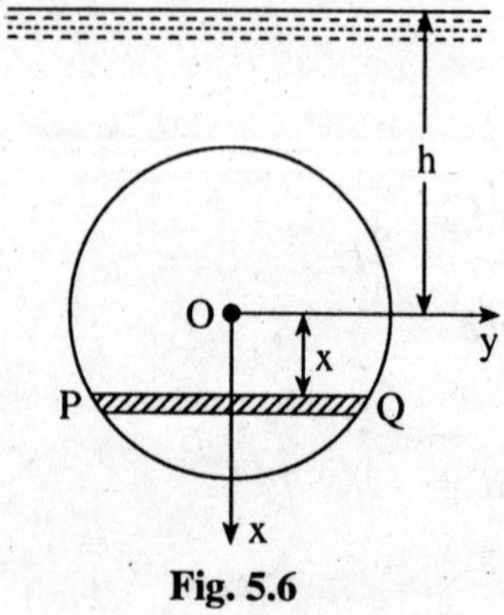

Fig. 5.6

$$x^2 + y^2 = R^2. \tag{5.29}$$

We consider a horizontal strip PQ of small thickness dx at a distance x from O. Then, the pressure at any point on PQ is

$$p = g\rho(h + x)$$

and the elementary area of the strip is $dS = 2y\, dx$.

Hence, the depth of the centre of pressure of the circular area from its centre is

$$\xi = \frac{\int xp\,dS}{\int p\,dS} = \frac{\int_{-R}^{R} x\cdot g\rho(h+x)2y\,dx}{\int_{-R}^{R} g\rho(h+x)2y\,dx} = \frac{\int_{-R}^{R} xy(h+x)dx}{\int_{-R}^{R} y(h+x)dx}$$

To integrate this, we substitute $x = R\cos\theta$, $y = R\sin\theta$, so that $dx = -R\sin\theta d\theta$. Then

$$\xi = \frac{R\int_0^{\pi} \cos\theta\sin^2\theta(h+R\cos\theta)d\theta}{\int_0^{\pi} \sin^2\theta(h+R\cos\theta)d\theta} = \frac{RhI_1+R^2I_2}{hI_3+RI_1}, \tag{5.30}$$

where we have denoted by I_1, I_2, I_3 the following integrals:

$$I_1 = \int_0^{\pi} \cos\theta\sin^2\theta d\theta = \int_0^{\pi} \sin^2\theta d(\sin\theta) = \frac{1}{3}\sin^3\theta\Big|_0^{\pi} = 0$$

$$I_2 = \int_0^{\pi} \cos^2\theta\sin^2\theta d\theta = \frac{1}{4}\int_0^{\pi} 4\cos^2\theta\sin^2\theta d\theta = \frac{1}{4}\int_0^{\pi} \sin^2 2\theta d\theta$$

$$= \frac{1}{8}\int_0^{\pi}(1-\cos 4\theta)d\theta = \frac{1}{8}\left(\theta - \frac{\sin 4\theta}{4}\right)_0^{\pi} = \frac{\pi}{8}$$

$$I_3 = \int_0^{\pi} \sin^2\theta d\theta = \frac{1}{2}\int_0^{\pi}(1-\cos 2\theta)d\theta = \frac{1}{2}\left(\theta - \frac{\sin 2\theta}{2}\right)_0^{\pi} = \frac{\pi}{2}.$$

Using these values in Eqn (5.30), we get

$$\xi = \frac{R^2\dfrac{\pi}{8}}{h\cdot\dfrac{\pi}{2}} = \frac{R^2}{4h}. \tag{5.31}$$

Thus, ***the depth of the centre of pressure of the circular area below the free surface is***

$$\bar{z} = h + \frac{R^2}{4h}. \tag{5.32}$$

If, in particular, the circular area is just immersed in the fluid, then $h = R$. Hence in this case

$$\bar{z} = R + \frac{R^2}{4R} = \frac{5}{4}R. \tag{5.33}$$

Note 1. The plane of the circle may not be vertical. In this case we assume that the plane of the circle makes an angle α to the free surface. Let h be the vertical depth of the centre below the free surface (FIG. 5.7).

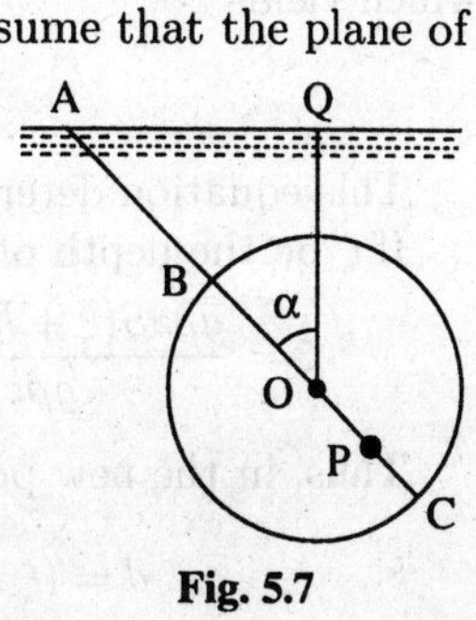

Fig. 5.7

We draw a line ABC in the plane of the circle through its centre intersecting the free surface at A.

Let $OA = h'$. If P be the centre of pressure, then by Eqn (5.31),

$$OP = \frac{R^2}{4h'}. \tag{5.34}$$

In $\triangle AOQ$, $OQ = h$ and $\cos\alpha = \dfrac{OQ}{OA} = \dfrac{h}{h'}$.

Hence,

$$h' = \frac{h}{\cos\alpha}. \tag{5.35}$$

Putting this value in Eqn (5.34), we find the distance of the centre of pressure from the centre of the circle as

$$\xi = OP = \frac{R^2\cos\alpha}{4h}. \tag{5.36}$$

5.5 Effect of Further Immersion

We consider a plane surface of area S immersed vertically in a homogeneous liquid of density ρ. Let G be the centre of gravity at a depth z and P be the centre of pressure at a depth ξ from the free surface (FIG. 5.8).

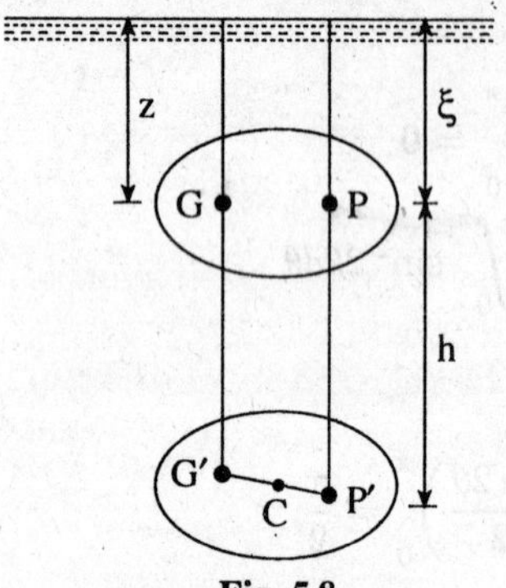

Fig. 5.8

Then the resultant thrust F on the plane surface in this initial position is

$$F = g\rho zS \tag{5.37}$$

acting at the point P.

Now, let the depth be increased either by adding more liquid or by lowering the plane area without rotation by an amount of h. In the new position, G' and P' are the corresponding positions of G and P.

Then the depth of G' is $(z+h)$ and that of P' is $(\xi+h)$. Let the position of the new centre of pressure be at C.

Then, the resultant thrust on the area S in the new position is

$$F' = g\rho S(z+h) \tag{5.38}$$

acting at C.

Due to further immersion, the depth of every point of the area has increased by an amount h. This has caused an extra pressure of $g\rho h$ which is uniform over the area and hence the resultant extra thrust

$$F^* = F' - F = g\rho hS \tag{5.39}$$

will act at G'.

Thus, there are two like parallel forces—the force F acting at P' and the force F^* acting at G'. Their resultant F' acts at C which lies on $P'G'$. Hence,

$$F\cdot P'C = F^*\cdot G'C, \text{ i.e., } g\rho zS\cdot P'C = g\rho hS\cdot G'C,$$

which yields

$$\frac{P'C}{G'C} = \frac{h}{z}. \tag{5.40}$$

This equation determines the location of the new position of the centre of pressure. If ζ be the depth of the new centre of pressure from the surface, then

$$\zeta = \frac{g\rho zS(\xi+h) + g\rho hS(z+h)}{g\rho zS + g\rho hS} = \frac{z(\xi+h)+h(z+h)}{z+h} = \frac{h^2+2hz+\xi z}{z+h}. \tag{5.41}$$

Thus, in the new position, the difference between the depths of P' and C is

$$d = (\xi+h)-\zeta = (\xi+h) - \frac{h^2+2hz+\xi z}{z+h}, \text{ i.e., } d = \frac{(\xi-z)h}{z+h}, \tag{5.42}$$

which is positive since the depth of centre of pressure ξ is greater than that of the centre of gravity z. Thus, depth of P' > depth of C.

This implies that ***the centre of pressure is raised through a vertical distance d given by* Eqn (5.42) *due to a further immersion without rotation of an amount h of the plane surface in the liquid.***

Again, in the new position, the difference in depth between C and G' is

$$\alpha = \frac{h^2 + 2hz + \xi z}{z+h} - (z+h) = \frac{z(\xi - z)}{z+h}. \tag{5.43}$$

The quantities ξ and z being given, Eqn (5.43) shows that α varies inversely as the depth of G', i.e., the depth of the centre of gravity in the new position.

Eqn (5.43) also shows that $\alpha \to 0$ as $h \to \infty$. That is, the depth of the centre of pressure approaches that of the centre of gravity as $h \to \infty$.

From Eqn 5.43), we get by differentiating w.r.t. time

$$\frac{d\alpha}{dt} = -\frac{z(\xi - z)}{(z+h)^2}\frac{dh}{dt}. \tag{5.44}$$

This shows that ***the centre of pressure approaches the horizontal through the centre of gravity with a velocity which is inversely proportional to the square of the depth of its centre of gravity and directly proportional to the velocity of the centre of gravity.***

5.6 Centre of Pressure of Composite Areas

We consider a plane area S with constituent parts of areas $S_1, S_2, \ldots, S_n$ immersed in a fluid, so that

$$S = S_1 + S_2 + \cdots + S_n. \tag{5.45}$$

Let $h_1, h_2, \ldots, h_n$ be the depths of centres of gravity and $\xi_1, \xi_2, \ldots, \xi_n$ the depths of the centres of pressure of the constituent parts.

Then the thrusts on these areas are

$$F_1 = g\rho h_1 S_1, F_2 = g\rho h_2 S_2, \ldots, F_n = g\rho h_n S_n. \tag{5.46}$$

If ξ be the depth of the centre of pressure of the composite area from the free surface, then

$$\xi = \frac{\xi_1 F_1 + \xi_2 F_2 + \cdots + \xi_n F_n}{F_1 + F_2 + \cdots + F_n}. \tag{5.47}$$

For homogeneous fluid, Eqn (5.47) with the use of Eqn (5.46) assumes the form

$$\xi = \frac{\xi_1 h_1 S_1 + \xi_2 h_2 S_2 + \cdots + \xi_n h_n S_n}{h_1 S_1 + h_2 S_2 + \cdots + h_n S_n}. \tag{5.48}$$

If, in particular, we consider a plane area S of which a part S_1 is removed, then the depth of centre of pressure ξ_2 of the remaining part is given by

$$\xi_2 = \frac{\xi h S - \xi_1 h_1 S_1}{hS - h_1 S_1}, \tag{5.49}$$

where, h = depth of c.g. of the whole area S,
ξ = depth of c.p. of the whole area S,
h_1 = depth of c.g. of the removed area S_1,
ξ_1 = depth of c.p. of the removed area S_1.

We can recast Eqn (5.48) with the concept of radius of gyration. Thus, if $K_1, K_2, \cdots, K_n$ be the radii of gyration of the constituent areas $S_1, S_2, \cdots, S_n$ respectively, then

$$\xi_1 = \frac{K_1^2}{h_1}, \xi_2 = \frac{K_2^2}{h_2}, \cdots, \xi_n = \frac{K_n^2}{h_n}. \tag{5.50}$$

With the use of these relations into Eqn (5.48), we get

$$\xi = \frac{K_1^2 S_1 + K_2^2 S_2 + \cdots + K_n^2 S_n}{h_1 S_1 + h_2 S_2 + \cdots + h_n S_n}. \tag{5.51}$$

Eqn (5.49) in this case assumes the form

$$\xi_2 = \frac{K^2 S - K_1^2 S_1}{hS - h_1 S_1} \tag{5.52}$$

where K is the radius of gyration of the whole area S.

5.7 Centre of Pressure of a Triangle in Terms of the Depths of its Vertices

Case 1. *One of the vertices in the free surface.*

Let the vertex A of the triangle ABC be in the free surface and the vertices B, C are at depths b, c below the free surface respectively. We produce BC to meet the free surface at D (FIG. 5.9). Then, $\Delta ABC = \Delta ABD - \Delta ACD$.

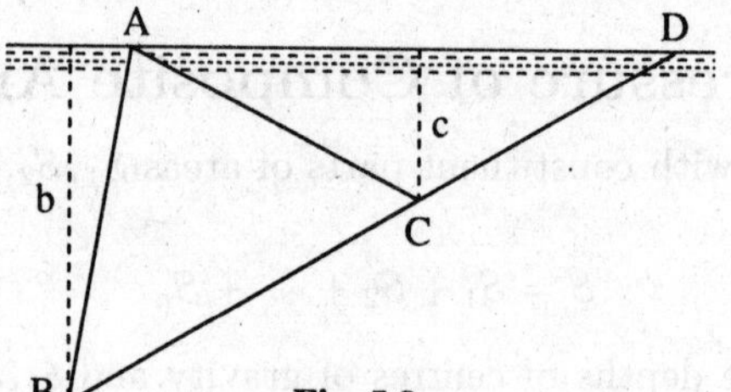

Fig. 5.9

Now, area of the ΔABD is $S = \frac{1}{2} AD \cdot b$, depth of its c.g., $h = \frac{b}{3}$ and depth of its centre of pressure, $\xi = \frac{1}{2} b$.

For the triangle ACD, area $S_1 = \frac{1}{2} AD \cdot c$, depth of c.g., $h_1 = \frac{c}{3}$, depth of c.p., $\xi_1 = \frac{c}{2}$.

Hence, by Eqn (5.49), the depth of the centre of pressure of ΔABC is

$$\xi_2 = \frac{\xi h S - \xi_1 h_1 S_1}{hS - h_1 S_1}$$

$$\text{or, } \xi_2 = \frac{\left(\frac{1}{2} b\right) \frac{b}{3} \cdot \frac{1}{2} AD \cdot b - \left(\frac{1}{2} c\right) \frac{c}{3} \cdot \frac{1}{2} AD \cdot c}{\frac{b}{3} \cdot \frac{1}{2} AD \cdot b - \frac{c}{3} \cdot \frac{1}{2} AD \cdot c}$$

$$\text{or, } \xi_2 = \frac{1}{2} \frac{b^3 - c^3}{b^2 - c^2} = \frac{1}{2} \frac{b^2 + bc + c^2}{b + c} \tag{5.53}$$

Case 2. *All the vertices are within the liquid.*

Let the triangle ABC in case-1 be lowered vertically through a depth α, so that the depths of vertices A, B, C below the free surface are now $\alpha, \alpha + b, \alpha + c$ respectively (FIG. 5.10).

Let S be the area of the $\triangle ABC$. For lowering the triangle by additional depth α, an additional thrust of $F_1 = g\rho\alpha S$ will act at its centre of gravity G, whose depth below the horizontal through A is

$$x_1 = \frac{1}{3}(0+b+c) = \frac{1}{3}(b+c).$$

The other force acting at the centre of pressure P, whose depth from A according to Eqn (5.53) being $x_2 = \frac{1}{2}\frac{b^2+bc+c^2}{b+c}$ is

$$F_2 = \frac{1}{3}(0+b+c)\rho g S = \frac{1}{3}(b+c)\rho g S.$$

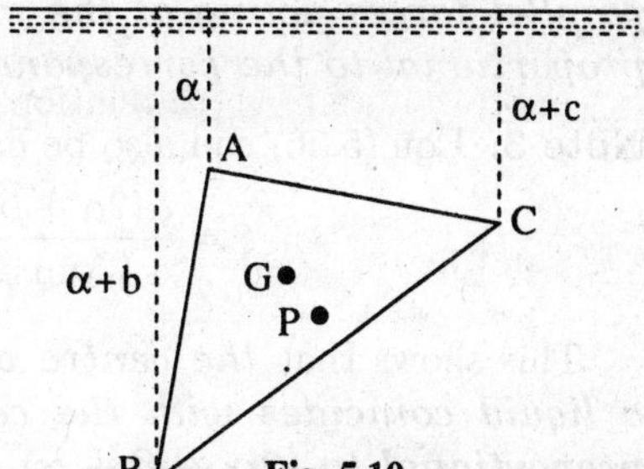

Fig. 5.10

Hence, if $\bar{x}$ be the depth of the new centre of pressure below the horizontal line through A, then

$$\bar{x} = \frac{F_1 x_1 + F_2 x_2}{F_1 + F_2}$$

$$\text{or, } \bar{x} = \frac{g\rho S\left[\frac{\alpha}{3}(b+c) + \frac{1}{3}(b+c)\frac{1}{2}\cdot\frac{b^2+bc+c^2}{b+c}\right]}{g\rho S\left[\alpha + \frac{1}{3}(b+c)\right]}$$

$$\text{or, } \bar{x} = \frac{1}{2}\frac{b^2+bc+c^2+2\alpha(b+c)}{3\alpha+b+c} \tag{5.54}$$

Hence, the depth ξ of the centre of pressure from the free surface of the liquid is

$$\xi = \alpha + \bar{x} = \alpha + \frac{1}{2}\frac{b^2+bc+c^2+2\alpha(b+c)}{3\alpha+b+c}. \tag{5.55}$$

We now take that the depths of vertices B and C below the free surface are β and γ respectively. Then $\beta = b+\alpha, \gamma = c+\alpha$, i.e., $b = \beta-\alpha, c = \gamma-\alpha$.

Then, by Eqn (5.55),

$$\xi = \alpha + \frac{1}{2}\frac{(\beta-\alpha)^2+(\beta-\alpha)(\gamma-\alpha)+(\gamma-\alpha)^2+2\alpha(\beta+\gamma-2\alpha)}{3\alpha+(\beta-\alpha)+(\gamma-\alpha)}$$

$$\text{or, } \xi = \frac{1}{2}\frac{\alpha^2+\beta^2+\gamma^2+\alpha\beta+\beta\gamma+\gamma\alpha}{\alpha+\beta+\gamma} \tag{5.56}$$

Note 2. Eqn (5.56) can be written in the form

$$\xi = \frac{\left(\frac{\alpha+\beta}{2}\right)\left(\frac{\alpha+\beta}{2}\right)+\left(\frac{\beta+\gamma}{2}\right)\left(\frac{\beta+\gamma}{2}\right)+\left(\frac{\gamma+\alpha}{2}\right)\left(\frac{\gamma+\alpha}{2}\right)}{\frac{\alpha+\beta}{2}+\frac{\beta+\gamma}{2}+\frac{\gamma+\alpha}{2}}. \tag{5.57}$$

Now, the depths of vertices of the triangle ABC below the free surface being α, β, γ, the depths of the middle points of the sides of the triangle are

$$\frac{\alpha+\beta}{2}, \frac{\beta+\gamma}{2}, \frac{\gamma+\alpha}{2}.$$

Then, Eqn (5.57) expresses the depth of centre of three parallel forces proportional to

$$\frac{\alpha+\beta}{2}, \frac{\beta+\gamma}{2}, \frac{\gamma+\alpha}{2}$$

acting at middle points of the sides of the triangle. Thus, ***the centre of pressure of a triangular area wholly immersed in a liquid coincides with the centre of three parallel forces acting at the middle points of the sides and whose magnitudes are proportional to the corresponding depths.***

Note 3. Eqn (5.56) can also be expressed as

$$\xi = \frac{\alpha(2\alpha+\beta+\gamma)+\beta(2\beta+\gamma+\alpha)+\gamma(2\gamma+\alpha+\beta)}{(2\alpha+\beta+\gamma)+(2\beta+\gamma+\alpha)+(2\gamma+\alpha+\beta)} \tag{5.58}$$

This shows that ***the centre of pressure of a triangular area wholly immersed in a liquid coincides with the centre of three parallel forces whose magnitudes are proportional to*** $(2\alpha+\beta+\gamma), (2\beta+\gamma+\alpha), (2\gamma+\alpha+\beta)$ ***acting at the vertices of the triangle whose depths below the free surface are*** α, β, γ ***respectively.***

5.8 Plane Surface Immersed in More than One Liquid

We consider a plane surface of area S immersed in n different immiscible liquids of densities $\rho_1, \rho_2, \dots, \rho_n$ in ascending order of magnitude. We intend to find the centre of pressure of the area S.

Let S_1 be the area of the surface in contact with the liquid of density ρ_1, S_2 the area of the surface in contact with the liquid of density ρ_2, and so on. Then,

$$S = S_1 + S_2 + \cdots + S_n. \tag{5.59}$$

The thrust on the whole area remains the same, if we consider that the liquid of density ρ_1 is in contact with the whole area S, a liquid of density $(\rho_2-\rho_1)$ is in contact with the surface area $(S-S_1)$, a liquid of density $(\rho_3-\rho_2)$ is in contact with the surface area $(S-S_1-S_2)$, and so on. Thus, finally a liquid of density $(\rho_n-\rho_{n-1})$ is in contact with the surface area S_n.

Let $F_1, F_2, \dots, F_n$ be the thrusts on areas $S, (S-S_1), \dots, S_n$ due to the above liquids respectively and let $\xi_1, \xi_2, \dots, \xi_n$ be the depths of the centres of pressure of the above areas below the free surface. Then the depth of the centre of pressure ζ of the whole area is given by

$$\zeta = \frac{\xi_1 F_1 + \xi_2 F_2 + \cdots + \xi_n F_n}{F_1 + F_2 + \cdots + F_n}. \tag{5.60}$$

5.9 Locus of the Centre of Pressure

We consider a plane surface of area S immersed vertically in a liquid of density ρ. Let O be a fixed point on the area at a depth h from the free surface (FIG. 5.11). With origin at O, we take a set of ***principal*** Cartesian coordinate axes x and y. Let θ be the angle which the x-axis makes with the vertical. We take a point $P(x, y)$ and consider an elementary area $dx\,dy$ about it. Then the equation of the line of intersection of the plane of the area with the free surface is

$$x\cos\theta + y\sin\theta = h. \tag{5.61}$$

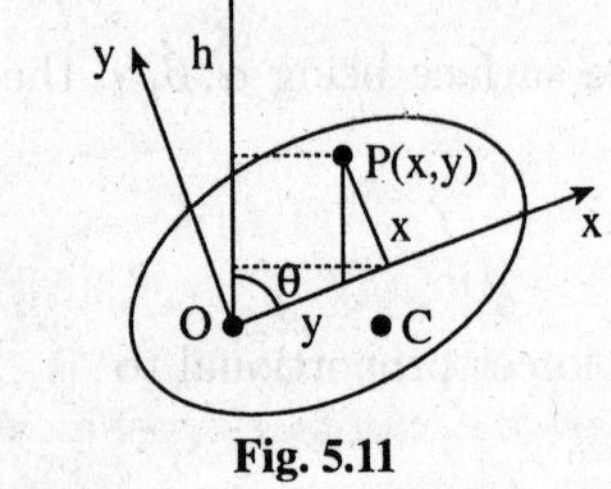

Fig. 5.11

Let in this position, C be the centre of pressure of the area with coordinates (ξ, η) w.r.t. the above coordinate system. If now the area be rotated about a horizontal line through O perpendicular to the plane of the paper, the position of C will change. It will trace a curve on the plane of the surface along with its rotation. We are to determine this curve.

Now, the depth of the elementary area $dx\,dy$ from the free surface is

$$z = (h - x\cos\theta - y\sin\theta) \tag{5.62}$$

so that the pressure on the elementary area is

$$p = g\rho z = g\rho(h - x\cos\theta - y\sin\theta). \tag{5.63}$$

Then,

$$\xi = \frac{\int xp\,dS}{\int p\,dS} = \frac{\iint x\cdot g\rho(h - x\cos\theta - y\sin\theta)dx\,dy}{\iint g\rho(h - x\cos\theta - y\sin\theta)dx\,dy} = \frac{hI_1 - \cos\theta I_2 - \sin\theta I_3}{hI_4 - \cos\theta I_1 - \sin\theta I_5}, \tag{5.64}$$

where we have denoted

$$\left.\begin{aligned} I_1 &= \iint x\,dx\,dy, \quad I_2 = \iint x^2 dx\,dy, \; I_3 = \iint xy\,dx\,dy \\ I_4 &= \iint dx\,dy = S, \; I_5 = \iint y\,dx\,dy, \quad I_6 = \iint y^2 dx\,dy \end{aligned}\right\} \tag{5.65}$$

Similarly, we get

$$\eta = \frac{\iint y(h - x\cos\theta - y\sin\theta)dx\,dy}{\iint (h - x\cos\theta - y\sin\theta)dx\,dy} = \frac{hI_5 - \cos\theta I_3 - \sin\theta I_6}{hI_4 - \cos\theta I_1 - \sin\theta I_5}. \tag{5.66}$$

Now, if $(\bar{x}, \bar{y})$ be the coordinates of the centre of gravity of the area w.r.t. the above coordinate system, then

$$\bar{x} = \frac{\iint x\,dx\,dy}{\iint dx\,dy}, \text{ i.e., } I_1 = S\bar{x}, \; \bar{y} = \frac{\iint y\,dx\,dy}{\iint dx\,dy}, \text{ i.e., } I_5 = S\bar{y} \tag{5.67}$$

Also,

$$I_2 = \iint x^2 dx\,dy = \text{moment of inertia of the area about } y\text{-axis} = Sa^2, \text{ say,}$$

$$I_3 = \iint xy\,dx\,dy = \text{product of inertia of the area about the principal axes} = 0,$$

$$\text{and } I_6 = \iint y^2 dx\,dy = \text{moment of inertia of the area about } x\text{-axis} = Sb^2, \text{ say.}$$

Using these relations, we get from Eqns (5.64) and (5.66),

$$\xi = \frac{hS\bar{x} - \cos\theta\cdot Sa^2 - 0}{hS - \cos\theta\cdot S\bar{x} - \sin\theta\cdot S\bar{y}}, \text{ i.e., } \xi = \frac{h\bar{x} - a^2\cos\theta}{h - \bar{x}\cos\theta - \bar{y}\sin\theta} \tag{5.68}$$

Similarly, we get

$$\eta = \frac{h\bar{y} - b^2\sin\theta}{h - \bar{x}\cos\theta - \bar{y}\sin\theta}. \tag{5.69}$$

Eliminating θ from Eqns (5.68) and (5.69), we get the locus of the centre of pressure which is a conic.

In practice, we take the centre of gravity of the area (which is a fixed point on the area) as the origin of coordinates. In this case, $\bar{x} = \bar{y} = 0$, so that the coordinates of the centre of pressure are

$$\xi = -\frac{a^2 \cos\theta}{h}, \eta = -\frac{b^2 \sin\theta}{h}. \tag{5.70}$$

Eliminating θ from these equations, we get

$$\frac{\xi^2}{\left(\frac{a^2}{h}\right)^2} + \frac{\eta^2}{\left(\frac{b^2}{h}\right)^2} = 1. \tag{5.71}$$

Thus, the locus of the centre of pressure is

$$\frac{x^2}{\left(\frac{a^2}{h}\right)^2} + \frac{y^2}{\left(\frac{b^2}{h}\right)^2} = 1 \tag{5.72}$$

which is an ellipse with the length of semi-major axis equal to a^2/h and that of semi-minor axis equal to b^2/h.

Even if we take the origin of coordinates at the centre of gravity of the plane area, the coordinate axes may not be the principal axes. In this case, the product of inertia does not vanish. Let

$$I_3 = \iint xy\, dx\, dy = Sc^2, \tag{5.73}$$

where S is the area of the plane surface. Then the coordinates of the centre of pressure according to Eqns (5.64) and (5.66) are

$$\left.\begin{aligned} \xi &= \frac{-\cos\theta \cdot Sa^2 - \sin\theta \cdot Sc^2}{hS} = -\frac{1}{h}(a^2 \cos\theta + c^2 \sin\theta) \\ \eta &= \frac{-\cos\theta \cdot Sc^2 - \sin\theta \cdot Sb^2}{hS} = -\frac{1}{h}(c^2 \cos\theta + b^2 \sin\theta) \end{aligned}\right\} \tag{5.74}$$

Solving these equations, we find

$$\cos\theta = \frac{h(b^2\xi - c^2\eta)}{(c^4 - a^2b^2)}, \sin\theta = \frac{h(a^2\eta - c^2\xi)}{(c^4 - a^2b^2)}. \tag{5.75}$$

Hence,

$$h^2\left(b^2\xi - c^2\eta\right)^2 + h^2\left(a^2\eta - c^2\xi\right)^2 = \left(c^4 - a^2b^2\right)^2. \tag{5.76}$$

Thus, the locus of the centre of pressure is

$$h^2\left(b^2x - c^2y\right)^2 + h^2\left(a^2y - c^2x\right)^2 = \left(c^4 - a^2b^2\right)^2, \tag{5.77}$$

which is a second degree equation in x and y. Thus, ***the locus of the centre of pressure is a conic.***

Denoting the quantities

$$A = h^2(b^4 + c^4), B = h^2(c^4 + a^4), H = h^2c^2(a^2 + b^2), \tag{5.78}$$

we write Eqn (5.77) in the form

$$Ax^2 - 2Hxy + By^2 - \left(c^4 - a^2b^2\right)^2 = 0. \tag{5.79}$$

$$\text{Now, } H^2 - AB = h^4c^4\left(a^2+b^2\right)^2 - h^4(b^4+c^4)(c^4+a^4)$$
$$= h^4\left[c^4(a^4+2a^2b^2+b^4) - (b^4c^4+b^4a^4+c^8+c^4a^4)\right]$$
$$= h^4(2a^2b^2c^4 - a^4b^4 - c^8) = -h^4\left(a^2b^2-c^4\right)^2.$$

Hence, $H^2 - AB < 0$.

This implies that the conic represented by Eqn (5.77) is an ellipse.

Thus, ***if a plane surface immersed vertically in a heavy homogeneous liquid is rotated in its own plane about a horizontal axis perpendicular to its plane and passing through its centre of gravity, then the centre of pressume traces an ellipse whose centre is at the centre of gravity.***

We now consider some illustrative examples.

EXAMPLE 5.1 *A quadrilateral is immersed in water with two angular points in the surface and the other two at depths a and b. If x and y are the depths below the surface of the centre of gravity and centre of pressure respectively, then show that* $6xy + ab = 3x(a+b)$.

Solution. Let $ABCD$ be the quadrilateral immersed in water with angular points C, D on the surface (FIG. 5.12). We produce the side BA to meet the surface at E.

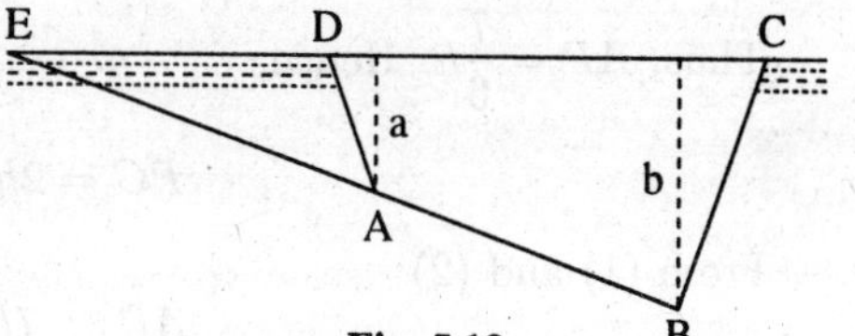

Fig. 5.12

For the triangle EBC,

$$\text{area} = S \text{ (say), depth of c.g., } h = \frac{b}{3}$$
$$\text{and depth of c.p., } \xi = \frac{b}{2}.$$

Similarly, for the triangle EAD,

$$\text{area} = S_1 \text{ (say), depth of c.g., } h_1 = \frac{a}{3} \text{ and depth of c.p., } \xi_1 = \frac{a}{2}.$$

Now, area of the quadrilateral $ABCD = S - S_1$. Depth of its c.g. is

$$x = \frac{S\cdot\frac{b}{3} - S_1\cdot\frac{a}{3}}{S - S_1},$$

which yields

$$3(S-S_1)x = Sb - S_1a, \text{ or, } S(3x-b) = S_1(3x-a). \tag{1}$$

Depth of c.p. of the quadrilateral is

$$y = \frac{S\cdot\frac{b}{3}\cdot\frac{b}{2} - S_1\cdot\frac{a}{3}\cdot\frac{a}{2}}{S\cdot\frac{b}{3} - S_1\cdot\frac{a}{3}},$$

which yields

$$2y(Sb - S_1a) = (Sb^2 - S_1a^2), \text{ or, } S(2by-b^2) = S_1(2ay-a^2). \tag{2}$$

From (1) and (2), we get

$$\frac{3x-b}{2by-b^2} = \frac{3x-a}{2ay-a^2}$$
$$\text{or, } (6axy - 3xa^2 - 2aby + a^2b) = (6bxy - 3xb^2 - 2aby + ab^2)$$
$$\text{or, } 6xy(a-b) - 3x(a^2-b^2) + ab(a-b) = 0$$
$$\text{or, } 6xy - 3x(a+b) + ab = 0, \text{ since } a \neq b$$
$$\text{or, } 6xy + ab = 3x(a+b) \quad \textbf{(Proved)}.$$

EXAMPLE 5.2 *A rhombus is immersed in a liquid with a vertex in the surface and the diagonal through the vertex vertical. Show that the centre of pressure divides the diagonal in the ratio 7 : 5.*

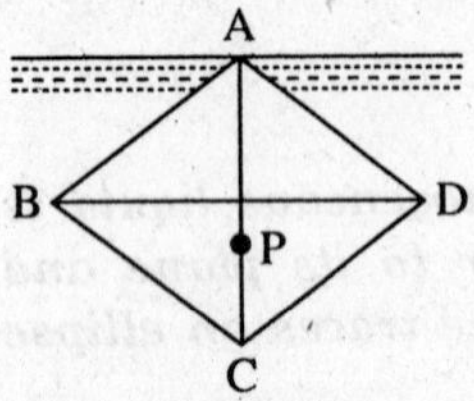

Fig. 5.13

Solution. Let $ABCD$ be a rhombus immersed in a liquid with its vertex in the surface and its diagonal AC vertical (FIG. 5.13).

The rhombus $ABCD$ is symmetrical about its diagonal AC. Hence, the centre of pressure P of the area will lie on AC whose depth will be the same as that of the centre of pressure of $\triangle ABC$ or ADC.

Hence, we are to find the depth of c.p. of the $\triangle ABC$.

Let $AC = 2h$, so that depth of $B = h$. If ξ be the depth of c.p. of the $\triangle ABC$, whose depths of vertices are $\alpha = 0, \beta = h, \gamma = 2h$, then by Eqn (5.56),

$$\xi = \frac{1}{2}\frac{0^2 + h^2 + 4h^2 + 0\cdot h + h\cdot 2h + 0\cdot 2h}{0 + h + 2h} = \frac{1}{2}\frac{7h^2}{3h} = \frac{7}{6}h. \tag{1}$$

Thus, $AP = \frac{7}{6}h$. Hence,

$$PC = 2h - \frac{7}{6}h = \frac{5}{6}h. \tag{2}$$

From (1) and (2)

$$\frac{AP}{PC} = \frac{7h}{6}\cdot\frac{6}{5h} = \frac{7}{5} \quad \textbf{(Proved)}.$$

EXAMPLE 5.3 *ABC is a triangular lamina immersed vertically in water with C in the surface and AB horizontal. Show how to divide the area by a horizontal line PQ into two portions on which the pressures are equal, P and Q being points on AC and BC respectively.*

If h be the length of the perpendicular from C on AB, prove that the height above AB of the centre of pressure of the area APQB in the above case is $\frac{1}{8}h\left[3(4)^{1/3} - 4\right]$.

Solution. The triangular lamina ABC as shown in the figure (FIG. 5.14) is immersed in water with its vertex C in the surface and AB horizontal.

PQ is a horizontal line dividing the $\triangle ABC$ into two parts on which the pressures are equal.

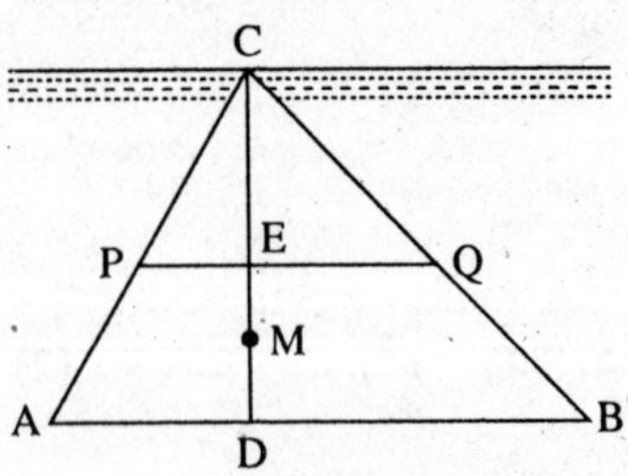

Fig. 5.14

We draw CD perpendicular upon AB which cuts PQ at E. Then $CD = h$, let $CE = H$.

Then, pressure on $\triangle PQC$ is

$$p_1 = g\rho\left(\frac{1}{2}PQ\cdot H\right)\cdot\frac{2}{3}H = \frac{1}{3}g\rho\cdot PQ\cdot H^2. \tag{1}$$

Pressure on the whole triangle ABC is

$$p_2 = g\rho\left(\frac{1}{2}AB\cdot h\right)\cdot\frac{2}{3}h = \frac{1}{3}g\rho\cdot AB\cdot h^2. \tag{2}$$

Since, by condition, pressure on $\triangle PQC$ is equal to the pressure on area $APQB$, we have

$$p_1 = \frac{1}{2}p_2, \text{ i.e., } \frac{1}{3}g\rho\cdot PQ\cdot H^2 = \frac{1}{2}\frac{1}{3}g\rho\cdot AB\cdot h^2$$

which yields

$$\frac{PQ}{AB} = \frac{1}{2}\frac{h^2}{H^2}. \tag{3}$$

Now, from similar triangles ABC and PQC, we have

$$\frac{PQ}{AB} = \frac{CE}{CD} = \frac{H}{h}. \tag{4}$$

From (3) and (4), we get

$$\frac{1}{2}\frac{h^2}{H^2} = \frac{H}{h}, \text{ or, } H^3 = \frac{1}{2}h^3, \text{ i.e., } H = \left(\frac{1}{2}\right)^{1/3} h. \tag{5}$$

Now, for the $\triangle ABC$,

$$\text{area, } S = \frac{1}{2}AB \cdot h$$

$$\text{depth of c.g. from surface, } \bar{x} = \frac{1}{3}(0 + h + h) = \frac{2}{3}h$$

$$\text{depth of c.p. from the surface [by Eqn (5.27)] is } \xi = \frac{3}{4}h$$

Similarly, for the triangle PQC,

$$\text{area, } S_1 = \frac{1}{2}PQ \cdot H = \frac{1}{2}\left(\frac{1}{2} \cdot \frac{h^2}{H^2}AB\right) \cdot H \text{ [using (3)]}$$

$$\text{depth of c.g., } \bar{x}_1 = \frac{2}{3}H$$

$$\text{depth of c.p., } \xi_1 = \frac{3}{4}H.$$

Then, the depth of c.p. of the area $APQB$ from the surface is

$$\xi_2 = \frac{\xi\bar{x}S - \xi_1\bar{x}_1S_1}{\bar{x}S - \bar{x}_1S_1}. \tag{6}$$

Putting here the above values, we get

$$\xi_2 = \frac{\left(\frac{3}{4}h\right)\left(\frac{2}{3}h\right)\left(\frac{1}{2}AB \cdot h\right) - \left(\frac{3}{4}H\right)\left(\frac{2}{3}H\right)\left(\frac{1}{2} \cdot \frac{1}{2}\frac{h^2}{H^2}AB \cdot H\right)}{\frac{2}{3}h\left(\frac{1}{2}AB \cdot h\right) - \frac{2}{3}H\left(\frac{1}{2} \cdot \frac{1}{2}\frac{h^2}{H^2} \cdot AB \cdot H\right)}$$

$$= \frac{\frac{1}{4}h^3 - \frac{1}{8}h^2H}{\frac{1}{3}h^2 - \frac{1}{6}h^2} = \frac{\frac{1}{4}h^2\left(h - \frac{1}{2}H\right)}{\frac{1}{6}h^2},$$

$$\text{i.e., } \xi_2 = \frac{3}{2}h - \frac{3}{4}H. \tag{7}$$

Putting the value of H from (5),

$$\xi_2 = \frac{3}{2}h - \frac{3}{4} \cdot \left(\frac{1}{2}\right)^{1/3} h. \tag{8}$$

Hence, the height of the centre of pressure M from AB is

$$DM = h - \xi = h - \left[\frac{3h}{2} - \frac{3}{4}\left(\frac{1}{2}\right)^{1/3} h\right]$$

$$= h\left[\frac{3}{8} \cdot 2 \cdot \left(\frac{1}{2}\right)^{1/3} - \frac{1}{2}\right] = \frac{h}{8}\left[3(4)^{1/3} - 4\right] \quad \textbf{(Proved)}.$$

EXAMPLE 5.4 *A square lamina of side 'a' has a portion of it in the form of the inscribed circle removed from it and the remaining figure is immersed vertically in the water with one side of the square in the surface. Prove that the depth of the centre of pressure is* $\dfrac{a}{24}\dfrac{64-15\pi}{4-\pi}$.

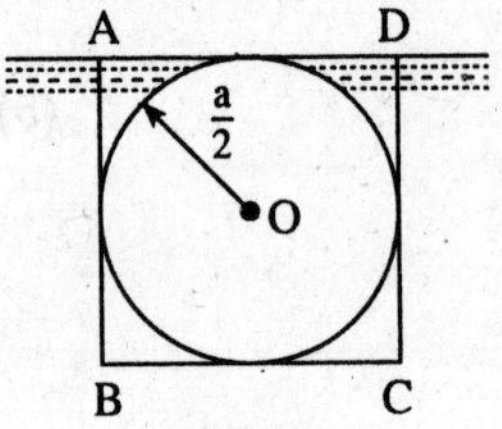

Fig. 5.15

Solution. Let $ABCD$ be a square lamina from which the portion containing the inscribed circle with centre at O has been removed (FIG. 5.15).

For the square $ABCD$,

$$\left.\begin{aligned} \text{area, } S = a^2, \text{ depth of c.g., } h &= \frac{a}{2} \\ \text{depth of c.p., } \xi &= \frac{2}{3}a \end{aligned}\right\} \tag{1}$$

For the circle of radius $\dfrac{a}{2}$,

$$\left.\begin{aligned} \text{area, } S_1 = \pi\left(\frac{a}{2}\right)^2 = \frac{\pi a^2}{4}, \text{ depth of c.g., } h_1 &= \frac{a}{2} \\ \text{depth of c.p., } \xi_1 = \frac{a}{2} + \frac{\frac{a^2}{4}}{4\cdot\frac{a}{2}} &= \frac{5a}{8}. \end{aligned}\right\} \tag{2}$$

If ξ_2 be the depth of centre of pressure of the remaining figure, then

$$\xi_2 = \frac{\xi h S - \xi_1 h_1 S_1}{hS - h_1 S_1} = \frac{\left(\frac{2}{3}a\right)\cdot\left(\frac{a}{2}\right)\cdot a^2 - \left(\frac{5a}{8}\right)\cdot\left(\frac{a}{2}\right)\cdot\frac{\pi a^2}{4}}{\frac{a}{2}\cdot a^2 - \frac{a}{2}\cdot\frac{\pi a^2}{4}}$$

$$= \frac{\frac{1}{3}a^4 - \frac{5}{64}\pi a^4}{\frac{a^3}{2} - \frac{\pi a^3}{8}} = \frac{\left(\frac{1}{3} - \frac{5}{64}\pi\right)a}{\frac{1}{8}(4-\pi)} = \frac{(64-15\pi)}{24(4-\pi)}a \quad \textbf{(Proved)}.$$

EXAMPLE 5.5 *A triangle has its base in the surface of the liquid and its vertex downwards. If the atmospheric pressure be equivalent to a height H of water, prove that the centre of pressure will be higher by a distance* $\dfrac{1}{2}\dfrac{hH}{h+H}$ *than it is when the atmospheric pressure is neglected, where h is the depth of the centre of gravity of the triangle from the surface of water.*

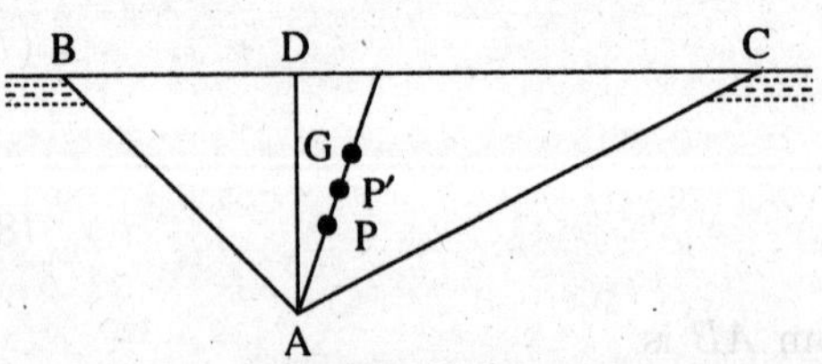

Fig. 5.16

Solution. Let ABC be a triangular area immersed in water with its base BC in the surface and vertex A downwards (FIG. 5.16).

Let P be the position of c.p. when atmospheric pressure is neglected and P', when atmospheric pressure is considered.

Let S = area of $\triangle ABC$,

z = depth of vertex A from the surface.

Then depth of c.g. of the $\triangle ABC$ from surface is $= \dfrac{1}{3}z$.

By condition given,

$$\frac{1}{3}z = h, \text{ i.e., } z = 3h. \tag{1}$$

Then the depth of c.p., when atmospheric pressure is neglected, below the surface is

$$\xi = \frac{1}{2}z = \frac{3}{2}h. \tag{2}$$

In this condition, the total pressure acting at P is

$$F_1 = g\rho Sh. \tag{3}$$

Let us now consider the atmospheric pressure which is equivalent to a height H of water. This causes a total increase of pressure

$$F_2 = g\rho SH \tag{4}$$

acting at c.g. of the area.

Thus, we have two parallel forces—one F_1 acting at P whose depth below BC is $\frac{3}{2}h$, and the other F_2 acting at G whose depth below BC is h.

If ξ^* be the depth of c.p. P' in the new position, then

$$\xi^* = \frac{F_1\frac{3}{2}h + F_2 \cdot h}{F_1 + F_2} = \frac{(g\rho Sh)\frac{3}{2}h + (g\rho SH)h}{g\rho Sh + g\rho SH}$$

$$= \frac{\frac{3}{2}h^2 + hH}{h+H} = \frac{(3h+2H)h}{2(h+H)}$$

Hence, the difference in depths of the original and final positions of the centres of pressure is

$$\xi - \xi^* = \frac{3}{2}h - \frac{(3h+2H)h}{2(h+H)} = \frac{1}{2}\frac{hH}{h+H} \quad \textbf{(Proved)}.$$

Note 4. We could directly use Eqn (5.42). However, it is expected to proceed as above.

EXAMPLE 5.6 *A circular area of radius R is immersed in a liquid with its plane vertical and its centre at a depth c below the free surface. If the area is lowered slowly so that the centre is at a depth $f(t)$ at time t, show that the velocity of the centre of pressure at that instant is*

$$f'(t)\left[1 - \frac{R^2}{4\{f(t)\}^2}\right].$$

Solution. In the initial position when the centre of the circle is at a depth c below the free surface (FIG. 5.17), the depth of the centre of pressure is

$$\xi = c + \frac{R^2}{4c}. \tag{1}$$

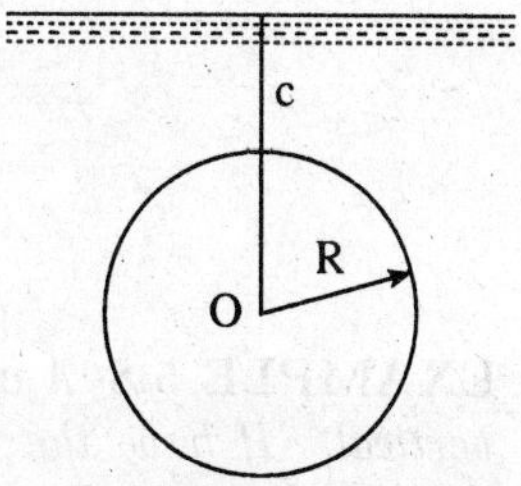

Fig. 5.17

Since the depth of the centre of the circle at any time t after lowering the area is $f(t)$, the depth of the centre of pressure at any time below the free surface is

$$\bar{\xi}(t) = f(t) + \frac{R^2}{4f(t)}. \tag{2}$$

Hence, the velocity of the centre of pressure is

$$\frac{d\bar{\xi}}{dt} = f'(t) + \frac{R^2}{4}\left[\frac{-f'(t)}{\{f(t)\}^2}\right] = f'(t)\left[1 - \frac{R^2}{4\{f(t)\}^2}\right] \quad \textbf{(Proved)}. \tag{3}$$

EXAMPLE 5.7 *When the depth of the liquid is increased by an amount 'a', the depth of the centre of pressure is found to increase by y, and when instead the depth of the liquid is increased by b, that of the centre of pressure is found to increase by z. Show that the depth of the centre of gravity of the area in the original state of the liquid is*

$$\frac{ab(b-a+y-z)}{az-by}.$$

Solution. Let a plane area immersed in a liquid has its c.g. at a depth h below the free surface. If K' be the radius of gyration of the area about an axis through its c.g. parallel to the free surface, then the depth of centre of pressure by Eqn (5.22) in original state is

$$\eta = h + \frac{K'^2}{h}. \tag{1}$$

Since the depth of centre of pressure is increased by y when the depth of liquid becomes $(h+a)$, then

$$\eta + y = (h+a) + \frac{K'^2}{h+a}. \tag{2}$$

Similarly, when the depth of liquid is increased to $(h+b)$, the depth of the centre of pressure below the surface is increased by z.

Hence,

$$\eta + z = (h+b) + \frac{K'^2}{h+b}. \tag{3}$$

From (1) and (2), by subtraction,

$$y = a + K'^2\left(\frac{1}{h+a} - \frac{1}{h}\right), \text{ i.e., } y - a = -\frac{K'^2 a}{h(h+a)},$$

which yields

$$K'^2 = \frac{h}{a}(h+a)(a-y). \tag{4}$$

Similarly, from (1) and (3), we get

$$K'^2 = \frac{h}{b}(h+b)(b-z). \tag{5}$$

Hence, from (4) and (5), we get

$$\begin{aligned}
&\frac{h}{a}(h+a)(a-y) = \frac{h}{b}(h+b)(b-z)\\
\text{or, } &b(h+a)(a-y) = a(h+b)(b-z)\\
\text{or, } &bh(a-y) + ab(a-y) = ah(b-z) + ab(b-z)\\
\text{or, } &h(az-by) = ab(b-z-a+y)\\
\therefore\ &h = \frac{ab(b-a+y-z)}{az-by} \quad \textbf{(Proved)}.
\end{aligned} \tag{6}$$

EXAMPLE 5.8 *A circular disc of radius 'a' is completely immersed in a liquid with its plane vertical. If h be the depth of its centre below the free surface of the liquid, prove that the distance between the centres of pressure of two semi-circles into which the disc is divided by the horizontal diameter is* $\dfrac{6\pi a(4h^2-a^2)}{9\pi^2h^2-16a^2}$.

Solution. The circular disc of radius a is shown in the figure (FIG. 5.18). The horizontal diameter AB divides the disc into two semi-circular parts.

We take polar coordinate system with the centre of the disc O as the origin and the diameter AB as the initial line.

We consider an elementary area $r\,d\theta\,dr$ at a point $P(r,\theta)$ in the upper semi-circle. Then the depth of the area from free surface is

$$z = (h - r\sin\theta) \tag{1}$$

so that the pressure at P is

$$p_1 = g\rho(h - r\sin\theta), \tag{2}$$

where ρ is the density of the liquid.

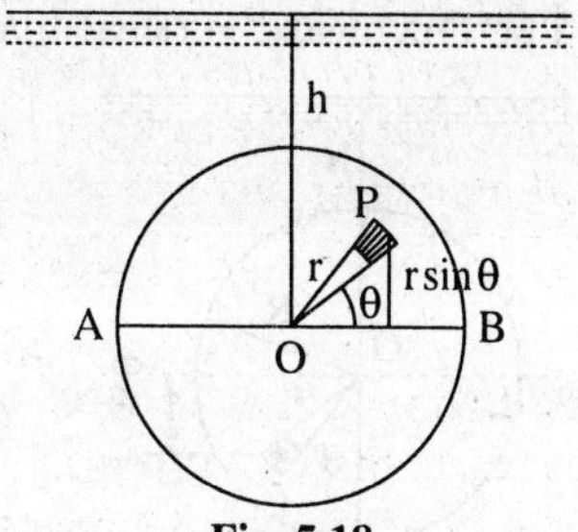

Fig. 5.18

If ξ_1 be the depth of the centre of pressure of the upper circle from the horizontal line AB, then

$$\xi_1 = \frac{\int_0^{\pi}\int_0^{a} r\sin\theta\cdot g\rho(h-r\sin\theta)\cdot r d\theta dr}{\int_0^{\pi}\int_0^{a} g\rho(h-r\sin\theta)\cdot r d\theta dr} = \frac{2\int_0^{\pi/2}\left(h\frac{a^3}{3}\sin\theta - \frac{a^4}{4}\sin^2\theta\right)d\theta}{2\int_0^{\pi/2}\left(h\frac{a^2}{2} - \frac{a^3}{3}\sin\theta\right)d\theta}$$

$$= \frac{h\frac{a^3}{3} - \frac{a^4}{4}\cdot\frac{1}{2}\cdot\frac{\pi}{2}}{h\frac{a^2}{2}\cdot\frac{\pi}{2} - \frac{a^3}{3}}, \text{ i.e., } \xi_1 = \frac{a}{4}\cdot\frac{16h - 3\pi a}{3\pi h - 4a}. \tag{3}$$

If the point $P(r,\theta)$ is taken in the lower circle, then the depth of the point from free surface is

$$z = h + r\sin\theta \tag{4}$$

so that pressure at the point $P(r,\theta)$ is

$$p = g\rho(h + r\sin\theta) \tag{5}$$

Hence in this case, the depth of centre of pressure below the line AB is

$$\xi_2 = \frac{\int_0^{\pi}\int_0^{a} r\sin\theta\cdot g\rho(h+r\sin\theta)\cdot r d\theta dr}{\int_0^{\pi}\int_0^{a} g\rho(h+r\sin\theta)\cdot r d\theta dr} = \frac{a}{4}\cdot\frac{16h+3\pi a}{3\pi h+4a}. \tag{6}$$

Thus, the distance between the two centres of pressure is

$$\xi_2 + \xi_1 = \frac{a}{4}\cdot\frac{16h+3\pi a}{3\pi h+4a} + \frac{a}{4}\cdot\frac{16h-3\pi a}{3\pi h-4a}$$

$$= \frac{a}{4}\left[\frac{(16h+3\pi a)(3\pi h-4a)+(16h-3\pi a)(3\pi h+4a)}{9\pi^2h^2-16a^2}\right]$$

$$= \frac{a}{4}\cdot\frac{48\pi h^2 - 64ah + 9\pi^2ah - 12\pi a^2 + 48\pi h^2 + 64ah - 9\pi^2ah - 12\pi a^2}{9\pi^2h^2-16a^2}$$

$$= \frac{a}{4}\cdot\frac{96\pi h^2 - 24\pi a^2}{9\pi^2h^2-16a^2} = \frac{6\pi a(4h^2-a^2)}{9\pi^2h^2-16a^2} \quad \textbf{(Proved)}.$$

EXAMPLE 5.9 *Show that the centre of pressure of a circular area of radius R immersed in a liquid whose centre is at a depth h below the surface, when the density varies as the depth, is at a depth $\dfrac{2R^2h}{R^2+4h^2}$ below the centre of the circle.*

Solution. Let O be the centre of the circle which is taken as the origin of polar coordinates. The downward vertical line through O is taken as the initial line (FIG. 5.19).

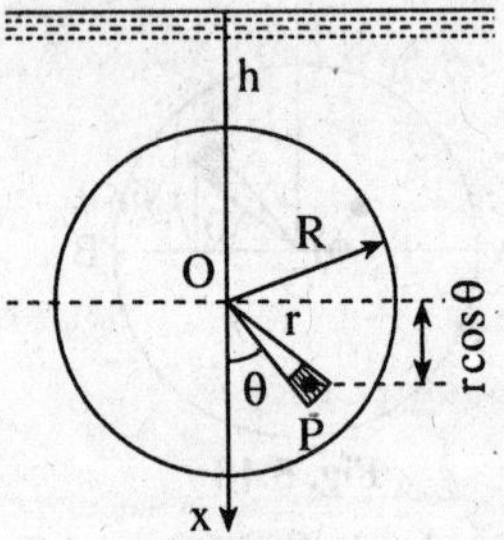

Fig. 5.19

We take a point $P(r, \theta)$ in the circular area and an elementary area $r\,d\theta\,dr$ around it. Then the depth of the point P below the free surface is

$$z = h + r\cos\theta. \tag{1}$$

Since the density ρ of the liquid varies as the depth, its value at P is

$$\rho = \lambda z = \lambda(h + r\cos\theta), \tag{2}$$

where λ is a constant.

Then the pressure at the point P is

$$p = g\rho z = g\lambda(h + r\cos\theta)\cdot(h + r\cos\theta) = g\lambda(h + r\cos\theta)^2. \tag{3}$$

If ξ be the depth of the centre of pressure below the centre of the circle, then

$$\xi = \frac{\displaystyle\int_0^R\int_0^\pi r\cos\theta\cdot g\lambda(h + r\cos\theta)^2\cdot r\,d\theta\,dr}{\displaystyle\int_0^R\int_0^\pi g\lambda(h + r\cos\theta)^2\cdot r\,d\theta\,dr}$$

$$\text{or, } \xi = \frac{\displaystyle\int_0^R\int_0^\pi r^2\cos\theta(h^2 + 2hr\cos\theta + r^2\cos^2\theta)d\theta\,dr}{\displaystyle\int_0^R\int_0^\pi r(h^2 + 2hr\cos\theta + r^2\cos^2\theta)d\theta\,dr} \tag{4}$$

Since $\displaystyle\int_0^\pi \cos^n\theta\,d\theta = 0$, for n is odd, we get from (4)

$$\xi = \frac{2\displaystyle\int_0^R\int_0^{\pi/2} 2hr^3\cos^2\theta\,d\theta\,dr}{2\displaystyle\int_0^R\int_0^{\pi/2} r(h^2 + r^2\cos^2\theta)d\theta\,dr}$$

$$= \frac{2h\dfrac{R^4}{4}\cdot\dfrac{1}{2}\cdot\dfrac{\pi}{2}}{h^2\dfrac{R^2}{2}\cdot\dfrac{\pi}{2} + \dfrac{R^4}{4}\cdot\dfrac{1}{2}\cdot\dfrac{\pi}{2}} = \frac{2hR^2}{R^2 + 4h^2} \quad \textbf{(Proved)}.$$

EXAMPLE 5.10 *A semi-circular lamina is completely immersed in water with its plane vertical, so that the extremity A of its bounding diameter is in the surface and the diameter makes an angle α with the surface. Prove that, if E be the centre of pressure and θ the angle between AE and the diameter, then* $\tan\theta = \dfrac{3\pi + 16\tan\alpha}{16 + 3\pi\tan\alpha}$.

Solution. Let AB be the diameter of the semi-circular lamina which makes an angle α with the liquid surface. We take a polar coordinate system with origin at A and AB as the initial line. Let $E(\xi, \eta)$ be the centre of pressure of the semi-circular lamina so that, by condition given, $\angle BAE = \theta$ (FIG. 5.20). If a be the radius of the semi-circle, then its polar equation is

$$r = 2a\cos\phi. \tag{1}$$

We take a point $P(r, \phi)$ and an elementary area $r\,d\phi\,dr$ around it. Then the depth of the point P from the surface is

$$z = r\sin(\alpha+\phi). \qquad (2)$$

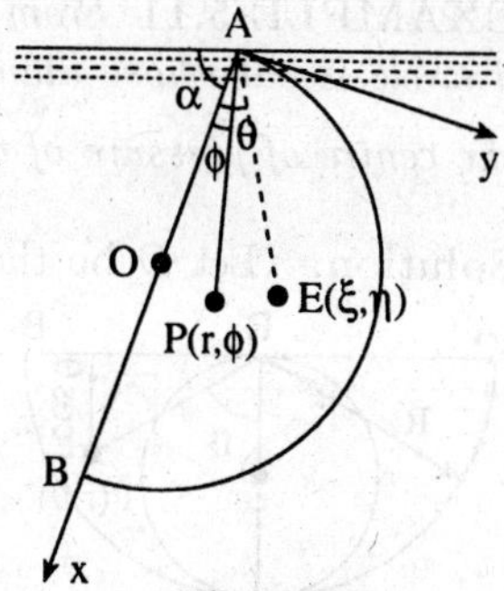

Fig. 5.20

Hence, the pressure at P is

$$p = g\rho r\sin(\alpha+\phi), \qquad (3)$$

where ρ is the density of the liquid. Then the coordinates of the centre of pressure are

$$\xi = \frac{I_1}{I_3},\ \eta = \frac{I_2}{I_3}, \qquad (4)$$

where we have denoted

$$I_1 = \int_0^{\pi/2}\int_0^{2a\cos\phi} r\cos\phi\cdot g\rho r\sin(\alpha+\phi) r\,d\phi\,dr$$

$$I_2 = \int_0^{\pi/2}\int_0^{2a\cos\phi} r\sin\phi\cdot g\rho r\sin(\alpha+\phi) r\,d\phi\,dr$$

$$I_3 = \int_0^{\pi/2}\int_0^{2a\cos\phi} g\rho r\sin(\alpha+\phi) r\,d\phi dr.$$

Then,

$$\tan\theta = \frac{\eta}{\xi} = \frac{I_2}{I_1}. \qquad (5)$$

Now,

$$I_1 = g\rho\int_0^{\pi/2}\frac{(2a\cos\phi)^4}{4}\cos\phi\sin(\alpha+\phi)d\phi$$

$$= 4g\rho\int_0^{\pi/2}\left[\sin\alpha\cdot\cos^6\phi + \cos\alpha\cdot\cos^5\phi\sin\phi\right]d\phi$$

$$\text{or, } I_1 = 4g\rho\left[\sin\alpha\cdot\frac{5}{6}\cdot\frac{3}{4}\cdot\frac{1}{2}\cdot\frac{\pi}{2} + \cos\alpha\cdot\frac{1}{6}\right]$$

$$= \frac{1}{24}g\rho(15\pi\sin\alpha + 16\cos\alpha) \qquad (6)$$

Again,

$$I_2 = g\rho\int_0^{\pi/2}\frac{(2a\cos\phi)^4}{4}\sin\phi\cdot\sin(\alpha+\phi)d\phi$$

$$= 4g\rho\int_0^{\pi/2}\left[\sin\alpha\cdot\sin\phi\cos^5\phi + \cos\alpha\sin^2\phi\cos^4\phi\right]d\phi$$

$$= 4g\rho\left[\sin\alpha\cdot\frac{1}{6} + \cos\alpha\left(\frac{3}{4}\cdot\frac{1}{2}\cdot\frac{\pi}{2} - \frac{5}{6}\cdot\frac{3}{4}\cdot\frac{1}{2}\cdot\frac{\pi}{2}\right)\right]$$

$$= \frac{1}{24}g\rho(16\sin\alpha + 3\pi\cos\alpha) \qquad (7)$$

From (6) and (7),

$$\tan\theta = \frac{I_2}{I_1} = \frac{3\pi\cos\alpha + 16\sin\alpha}{16\cos\alpha + 15\pi\sin\alpha} = \frac{3\pi + 16\tan\alpha}{16 + 15\pi\tan\alpha} \quad \textbf{(Proved).}$$

EXAMPLE 5.11 *From a semi-circle of radius R, whose diameter is in the surface of a liquid, a circle is cut whose diameter is the vertical radius of the semi-circle. Show that the depth of the centre of pressure of the remainder is* $\dfrac{9\pi R}{8(16-3\pi)}$.

Solution. Let O be the centre of the semi-circle of radius R whose diameter AB is in the surface. We take a polar coordinate system with O as the pole and the vertical diameter through O as the initial line (FIG. 5.21). Then equation of the semi-circle is

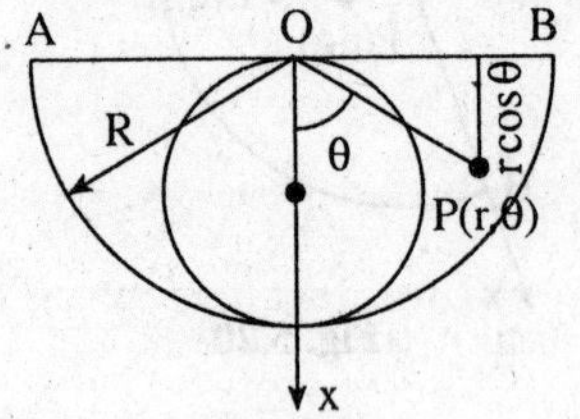

Fig. 5.21

$$r = R, \text{ for } -\frac{\pi}{2} \leq \theta \leq \frac{\pi}{2} \tag{1}$$

and that of the circle is

$$r = R\cos\theta. \tag{2}$$

We take a point $P(r,\theta)$ in the region bounded by the semi-circle of radius R and the circle of radius $R/2$. Its depth from the surface is $r\cos\theta$, so that pressure at P is

$$p = g\rho r\cos\theta, \tag{3}$$

where ρ is the density of the liquid.

We consider an elementary area $r\,d\theta\,dr$ around the point P. Then the thrust on this area is

$$dF = (g\rho r\cos\theta)r\,d\theta\,dr. \tag{4}$$

If ξ be the depth of centre of pressure below the free surface, then, from symmetry

$$\xi = \frac{\displaystyle\int_0^{\pi/2}\int_{R\cos\theta}^{R}(r\cos\theta)(g\rho r\cos\theta)r\,d\theta\,dr}{\displaystyle\int_0^{\pi/2}\int_{R\cos\theta}^{R}(g\rho r\cos\theta)r\,d\theta\,dr}. \tag{5}$$

Numerator is

$$= \left(\int_0^{\pi/2}\cos^2\theta\,d\theta\right)\int_{R\cos\theta}^{R}r^3dr = \int_0^{\pi/2}\cos^2\theta\,d\theta\left[\frac{r^4}{4}\right]_{r=R\cos\theta}^{R}$$

$$= \int_0^{\pi/2}\frac{R^4}{4}\cos^2\theta\,(1-\cos^4\theta)\,d\theta = \frac{R^4}{4}\left(\frac{1}{2}\cdot\frac{\pi}{2}-\frac{5}{6}\cdot\frac{3}{4}\cdot\frac{1}{2}\cdot\frac{\pi}{2}\right) = \frac{9\pi R^4}{384} \tag{6}$$

Denominator is

$$= \int_0^{\pi/2}\cos\theta\,d\theta\int_{R\cos\theta}^{R}r^2dr = \int_0^{\pi/2}\cos\theta d\theta\left[\frac{r^3}{3}\right]_{r=R\cos\theta}^{R}$$

$$= \frac{R^3}{3}\int_0^{\pi/2}\cos\theta(1-\cos^3\theta) = \frac{R^3}{3}\left(1-\frac{3}{4}\cdot\frac{1}{2}\cdot\frac{\pi}{2}\right) = \frac{R^3}{48}(16-3\pi) \tag{7}$$

Hence,

$$\xi = \frac{9\pi R^4}{384}\cdot\frac{48}{R^3(16-3\pi)} = \frac{9\pi R}{8(16-3\pi)} \quad \textbf{(Proved)}.$$

EXAMPLE 5.12 *A quadrant of a circle is just immersed vertically in a liquid with one edge in the surface. Find the position of its centre of pressure if the density of the liquid varies as the depth.*

Solution. Let AOB be a quadrant of a circle of radius R immersed in the liquid with its bounding radius OA in the surface of the liquid. With O as the pole and OA as the initial line, we take a polar coordinate system (FIG. 5.22). We take a point $P(r,\theta)$ whose depth from the free surface is $r\sin\theta$. Since the density of the liquid varies as depth, density at P is

$$\rho = \lambda r\sin\theta,\ \lambda = \text{constant.} \tag{1}$$

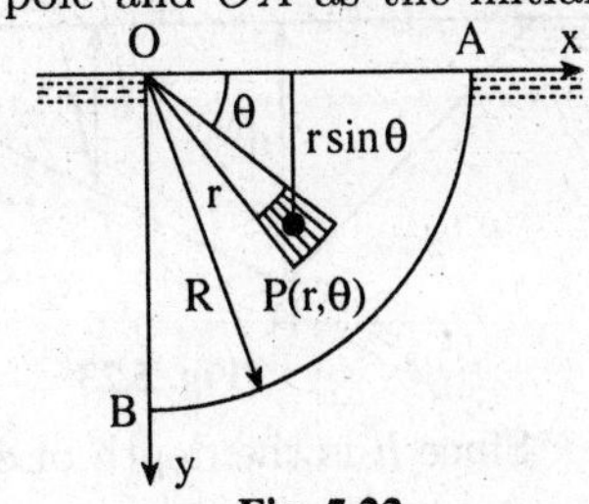

Fig. 5.22

Then the pressure at P is

$$\begin{aligned} p = g\rho z &= g(\lambda r\sin\theta)r\sin\theta \\ &= g\lambda r^2\sin^2\theta. \end{aligned} \tag{2}$$

We consider an elementary area $r\,d\theta\,dr$ about the point P. Then liquid thrust upon it is

$$dF = pr\,d\theta\,dr = g\lambda r^3\sin^2\theta\,d\theta\,dr. \tag{3}$$

If (ξ,η) be the position of the centre of pressure, then

$$\xi = \frac{\displaystyle\int_0^R\int_0^{\pi/2} r\cos\theta\cdot g\lambda r^3\sin^2\theta\,d\theta\,dr}{\displaystyle\int_0^R\int_0^{\pi/2} g\lambda r^3\sin^2\theta\,d\theta\,dr} = \frac{\displaystyle\int_0^{\pi/2}\frac{R^5}{5}\cos\theta\sin^2\theta\,d\theta}{\displaystyle\int_0^{\pi/2}\frac{R^4}{4}\sin^2\theta\,d\theta}$$

$$= \frac{4R}{5}\frac{\left[\dfrac{\sin^3\theta}{3}\right]_0^{\pi/2}}{\dfrac{1}{2}\left[\theta-\dfrac{\sin 2\theta}{2}\right]_0^{\pi/2}} = \frac{4R}{5}\cdot\frac{\frac{1}{3}}{\frac{1}{2}\cdot\frac{\pi}{2}} = \frac{16R}{15\pi} \tag{4}$$

Similarly,

$$\eta = \frac{\displaystyle\int_0^R\int_0^{\pi/2} r\sin\theta\cdot g\lambda r^3\sin^2\theta\,d\theta\,dr}{\displaystyle\int_0^R\int_0^{\pi/2} g\lambda r^3\sin^2\theta\,d\theta\,dr} = \frac{4R}{5}\cdot\frac{\displaystyle\int_0^{\pi/2}\sin^3\theta\,d\theta}{\displaystyle\int_0^{\pi/2}\sin^2\theta\,d\theta}$$

$$= \frac{4R}{5}\frac{\frac{2}{3}\cdot 1}{\frac{1}{2}\cdot\frac{\pi}{2}} = \frac{32R}{15\pi}. \tag{5}$$

Here we have used the property: $\displaystyle\int_0^{\pi/2}\sin^m\theta\,d\theta = \int_0^{\pi/2}\cos^m\theta\,d\theta.$

Thus the position of the centre of pressure is $\left(\dfrac{16R}{15\pi},\dfrac{32R}{15\pi}\right)$.

EXAMPLE 5.13 *A plane quadrilateral ABCD is entirely immersed in water with its side AB in the surface. If the depths of C and D below the surface are c and d respectively and that of its centre of gravity is h, prove that the depth of its centre of pressure is*

$$\frac{c+d}{2}-\frac{cd}{6h}.$$

Solution. The quadrilateral $ABCD$ immersed entirely in water with its side AB in the surface is shown in the figure (FIG. 5.23). The side DC is produced to meet the surface at E. Then, area of the quadrilateral $ABCD$ = area of $\triangle ADE$ − area of $\triangle BCE$.

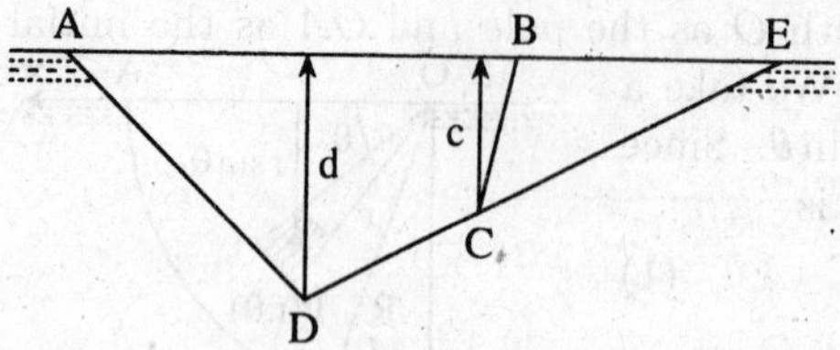

Fig. 5.23

Let $AE = x$ and $BE = y$.

$$\left.\begin{array}{l}\text{Depth of c.g. of } \triangle ADE = \dfrac{d}{3} \\ \text{Depth of c.g. of } \triangle BCE = \dfrac{c}{3}\end{array}\right\} \qquad (1)$$

$$\left.\begin{array}{l}\text{Area of } \triangle ADE = \dfrac{1}{2}xd \\ \text{Area of } \triangle BCE = \dfrac{1}{2}yc\end{array}\right\} \qquad (2)$$

Since h is the depth of c.g. of the area $ABCD$ from the surface,

$$h = \frac{\frac{1}{2}xd \cdot \frac{d}{3} - \frac{1}{2}yc \cdot \frac{c}{3}}{\frac{1}{2}xd - \frac{1}{2}yc} = \frac{xd^2 - yc^2}{3(xd - yc)}. \qquad (3)$$

$$\left.\begin{array}{l}\text{Now, the depth of c.p. of } \triangle ADE = \dfrac{d}{2} \\ \text{and thrust on } \triangle ADE = g\rho\left(\dfrac{1}{2}xd\right)\dfrac{d}{3} = \dfrac{1}{6}g\rho xd^2\end{array}\right\} \qquad (4)$$

$$\left.\begin{array}{l}\text{Similarly, the depth of c.p. of } \triangle BCE = \dfrac{c}{2} \\ \text{and thrust on } \triangle BCE = g\rho\left(\dfrac{1}{2}yc\right)\dfrac{c}{3} = \dfrac{1}{6}g\rho yc^2\end{array}\right\} \qquad (5)$$

Hence, if ξ be the depth of c.p. of the quadrilateral $ABCD$ below the free surface, then

$$\xi = \frac{\frac{1}{6}g\rho xd^2 \cdot \frac{d}{2} - \frac{1}{6}g\rho yc^2 \cdot \frac{c}{2}}{\frac{1}{6}g\rho xd^2 - \frac{1}{6}g\rho yc^2}$$

$$\text{or, } \xi = \frac{xd^3 - yc^3}{2(xd^2 - yc^2)} = \frac{(c+d)(xd^2 - yc^2) + yc^2d - xcd^2}{2(xd^2 - yc^2)}$$

$$= \frac{c+d}{2} - \frac{cd(xd - yc)}{2(xd^2 - yc^2)} = \frac{c+d}{2} - \frac{cd}{6h}, \text{ using (3)} \quad \textbf{(Proved)}.$$

EXAMPLE 5.14 *An area is bounded by two concentric semi-circles with common bounding diameter in the surface of a liquid. Show that the depth of its centre of pressure is*

$$\frac{3\pi}{16}\frac{(a+b)(a^2+b^2)}{a^2+ab+b^2},$$

where a, b $(b > a)$ are the radii of the semi-circles.

Solution. Let O be the centre of the semi-circles with their common bounding diameter AB in the surface of a liquid of density ρ (FIG. 5.24). With origin at O and AB as the initial line, we take a polar coordinate system.

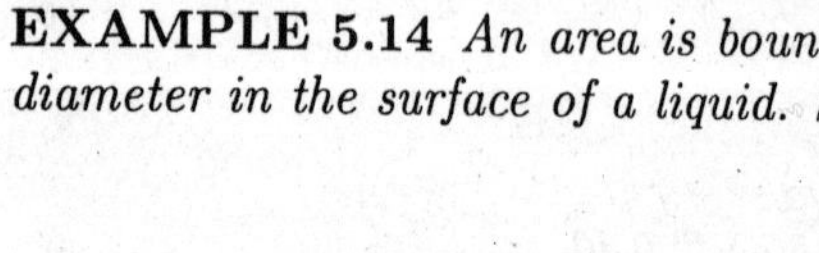

Fig. 5.24

Let $P(r, \theta)$ be any point within the region bounded by the two semi-circles. Then the depth of P from surface is

$$z = r\sin\theta. \qquad (1)$$

We consider an elementary area $r\,d\theta\,dr$ around the point P. Then the liquid thrust on this elementary area is

$$dF = (g\rho r\sin\theta)r\,d\theta\,dr = g\rho r^2\sin\theta\,d\theta\,dr. \qquad (2)$$

The centre of pressure will lie on y-axis due to symmetry. If ξ be the depth of c.p. below the surface, then

$$\xi = \frac{2\int_0^{\pi/2}\int_a^b r\sin\theta\cdot g\rho r^2\sin\theta\,d\theta\,dr}{2\int_0^{\pi/2}\int_a^b g\rho r^2\sin\theta\,d\theta\,dr} = \frac{\int_0^{\pi/2}\sin^2\theta\,d\theta\int_a^b r^3dr}{\int_0^{\pi/2}\sin\theta\,d\theta\int_a^b r^2dr}$$

$$= \frac{\frac{1}{2}\cdot\frac{\pi}{2}\cdot\frac{1}{4}(b^4-a^4)}{1\cdot\frac{1}{3}(b^3-a^3)} = \frac{3\pi}{16}\frac{(b-a)(b+a)(b^2+a^2)}{(b-a)(b^2+ab+a^2)}$$

$$= \frac{3\pi}{16}\frac{(a+b)(a^2+b^2)}{a^2+ab+b^2} \quad \textbf{(Proved).}$$

EXAMPLE 5.15 *Show that, if a lamina totally immersed in a fluid is a quadrant of a circle of radius R, of which the centre is in the surface, the locus of the centre of pressure in the lamina lies on a straight line of length*

$$\frac{3\sqrt{2}(\pi-2)R}{16}.$$

Solution. Let OAB be the lamina in the shape of a quadrant of a circle of radius R immersed in a liquid of density ρ with its centre O in the surface (FIG. 5.25). We take polar coordinate system with O as pole and the bounding radius OB as the initial line.

Let OB make an angle α with the horizontal. We take a point $P(r,\theta)$ within the lamina. Its depth from the surface is

$$z = r\sin(\theta+\alpha) \tag{1}$$

so that fluid pressure at P is

$$p = g\rho z = g\rho r\sin(\theta+\alpha), \tag{2}$$

where ρ is the density of the fluid.

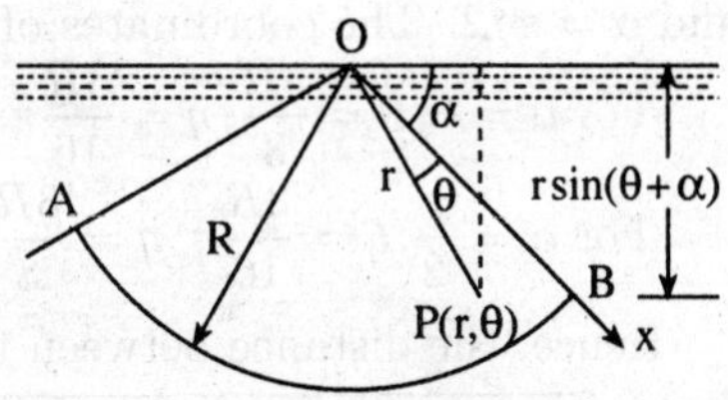

Fig. 5.25

We consider an elementary area $r\,dr\,d\theta$ surrounding the point P. Then the thrust upon the area is

$$dF = pdS = g\rho r\sin(\theta+\alpha)r\,d\theta\,dr. \tag{3}$$

If (ξ,η) be the coordinates of the centre of pressure, then

$$\xi = \frac{\int_0^R\int_0^{\pi/2} r\cos\theta\cdot g\rho r\sin(\theta+\alpha)\cdot r\,d\theta\,dr}{\int_0^R\int_0^{\pi/2} g\rho r\sin(\theta+\alpha)\cdot r\,d\theta\,dr} = \frac{\frac{R^4}{4}\int_0^{\pi/2}(\sin\theta\cos\alpha+\cos\theta\sin\alpha)\cos\theta\,d\theta}{\frac{R^3}{3}\int_0^{\pi/2}(\sin\theta\cos\alpha+\cos\theta\sin\alpha)d\theta}$$

$$= \frac{3R}{4}\cdot\frac{\cos\alpha\int_0^{\pi/2}\sin\theta\cos\theta\,d\theta+\sin\alpha\int_0^{\pi/2}\cos^2\theta\,d\theta}{\cos\alpha\int_0^{\pi/2}\sin\theta\,d\theta+\sin\alpha\int_0^{\pi/2}\cos\theta\,d\theta} = \frac{3R}{4}\cdot\frac{\cos\alpha\cdot\frac{1}{2}+\sin\alpha\cdot\frac{1}{2}\cdot\frac{\pi}{2}}{\cos\alpha+\sin\alpha}$$

$$\text{i.e., } \xi = \frac{3R}{16}\cdot\frac{2\cos\alpha+\pi\sin\alpha}{\cos\alpha+\sin\alpha}. \tag{4}$$

Similarly,

$$\eta = \frac{\int_0^R \int_0^{\pi/2} r\sin\theta \cdot g\rho r \sin(\theta+\alpha)\cdot r\,d\theta\,dr}{\int_0^R \int_0^{\pi/2} g\rho r \sin(\theta+\alpha)\cdot r\,d\theta\,dr}$$

$$= \frac{3R}{4}\frac{\int_0^{\pi/2} \sin\theta(\sin\theta\cos\alpha+\cos\theta\sin\alpha)d\theta}{\int_0^{\pi/2}(\sin\theta\cos\alpha+\cos\theta\sin\alpha)d\theta} = \frac{3R}{4}\frac{\cos\alpha\cdot\frac{1}{2}\cdot\frac{\pi}{2}+\sin\alpha\cdot\frac{1}{2}}{\cos\alpha+\sin\alpha}$$

$$\text{i.e., } \eta = \frac{3R}{16}\cdot\frac{2\sin\alpha+\pi\cos\alpha}{\cos\alpha+\sin\alpha}. \tag{5}$$

From (4) and (5), we have

$$\xi+\eta = \frac{3R}{16}\left[\frac{2\cos\alpha+\pi\sin\alpha}{\cos\alpha+\sin\alpha}+\frac{2\sin\alpha+\pi\cos\alpha}{\cos\alpha+\sin\alpha}\right]$$

$$= \frac{3R}{16}\frac{(2+\pi)(\sin\alpha+\cos\alpha)}{\cos\alpha+\sin\alpha} = \frac{3R}{16}(2+\pi) = \text{constant}. \tag{6}$$

Hence, the locus of the centre of pressure is the straight line

$$x+y = \frac{3R}{16}(2+\pi). \tag{7}$$

The end points of the straight line traversed by the centre of pressure correspond to $\alpha = 0$ and $\alpha = \pi/2$. The coordinates of these end points are:

$$\left.\begin{aligned} &\text{For } \alpha = 0, \xi = \frac{3R}{8}, \eta = \frac{3R}{16}\pi \\ &\text{For } \alpha = \frac{\pi}{2}, \xi = \frac{3R}{16}\pi, \eta = \frac{3R}{8} \end{aligned}\right\} \tag{8}$$

Hence, the distance between these two end points is

$$= \sqrt{\left(\frac{3\pi R}{16}-\frac{3R}{8}\right)^2+\left(\frac{3R}{8}-\frac{3\pi R}{16}\right)^2} = \sqrt{2}\left(\frac{3\pi R}{16}-\frac{3R}{8}\right) = \frac{3\sqrt{2}}{16}(\pi-2)R \quad \textbf{(Proved)}.$$

EXAMPLE 5.16 *A segment of a parabola cut off by a double ordinate at a height h from the vertex, is immersed with this ordinate in the surface of a liquid. Find the position of the centre of pressure on the lamina and also the resultant thrust, neglecting the atmospheric pressure.*

Solution. We take a Cartesian coordinate system with origin at the vertex of the parabola and x-axis vertical (FIG. 5.26). Then equation of the parabola is

$$y^2 = 4ax, \tag{1}$$

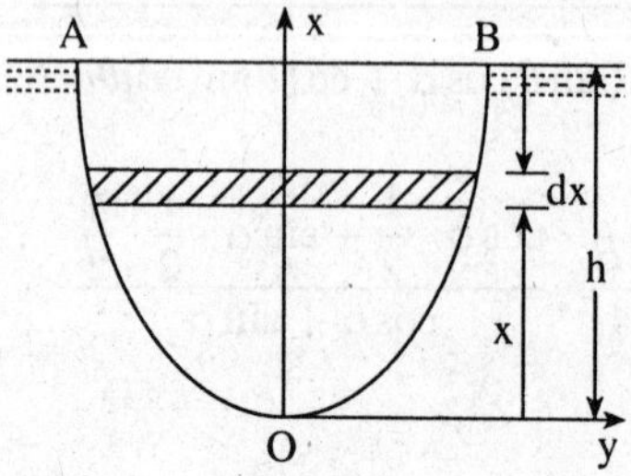

Fig. 5.26

where AB is the double ordinate at a height h from the vertex cutting off the segment AOB and lying in the surface.

We consider a strip of thickness dx at a height x from O. The depth of this strip from the free surface is $(h-x)$. Hence, the liquid pressure at any point on the strip is

$$p = g\rho(h-x), \tag{2}$$

where ρ is the density of the liquid.

The area of the strip is

$$dS = 2ydx = 2\sqrt{4ax}dx = 4\sqrt{ax}dx. \tag{3}$$

Then the thrust on the strip is

$$dF = pdS = g\rho(h-x)4\sqrt{ax}dx. \tag{4}$$

Due to symmetry, the centre of pressure lies on x-axis. If ξ be the height of c.p. from O, then

$$\xi = \frac{\int_0^h x(dF)dx}{\int_0^h dF\,dx} = \frac{\int_0^h 4g\rho\sqrt{a}\left(hx^{3/2} - x^{5/2}\right)dx}{\int_0^h 4g\rho\sqrt{a}\left(hx^{1/2} - x^{3/2}\right)dx}$$

$$= \frac{\left[\frac{2}{5}hx^{5/2} - \frac{2}{7}x^{7/2}\right]_{x=0}^{x=h}}{\left[\frac{2}{3}hx^{3/2} - \frac{2}{5}x^{5/2}\right]_{x=0}^{x=h}} = \frac{\frac{2}{5}h^{7/2} - \frac{2}{7}h^{7/2}}{\frac{2}{3}h^{5/2} - \frac{2}{5}h^{5/2}} = \frac{\frac{4}{35}h^{7/2}}{\frac{4}{15}h^{5/2}} = \frac{3}{7}h. \tag{5}$$

Hence, the depth of c.p. from the surface is

$$= h - \xi = h - \frac{3}{7}h = \frac{4}{7}h. \tag{6}$$

Also, the resultant thrust from (4), is

$$F = \int dF = \int_0^h g\rho(h-x)\cdot 4\sqrt{ax}dx = 4g\rho\sqrt{a}\int_0^h \left(hx^{1/2} - x^{3/2}\right)dx$$

$$= 4g\rho\sqrt{a}\left[\frac{2}{3}hx^{3/2} - \frac{2}{5}x^{5/2}\right]_{x=0}^{x=h} = \frac{16}{15}\sqrt{a}g\rho h^{5/2}.$$

EXAMPLE 5.17 *An ellipse with lengths of semi-axes a and b, is immersed in a liquid vertically so that the axes make equal angles with the vertical. Show that its centre of pressure lies on a line through the centre making with the vertical an angle*

$$\tan^{-1}\left(\frac{a^2-b^2}{a^2+b^2}\right) \text{ and at a depth } h + \frac{a^2+b^2}{8h},$$

h being the depth of the centre of the ellipse.

Solution. Let O be the centre of ellipse at a depth h below the free surface. Since the axes make equal angles with the vertical,

$$\angle COM = \angle AOM = \frac{\pi}{4}. \tag{1}$$

We take OB and OD as x- and y-axes respectively (FIG. 5.27).

Let us take any point $P(x, y)$ in the ellipse. Its depth from free surface is

$$z = h + x\sin\frac{\pi}{4} + y\sin\frac{\pi}{4}$$

$$= h + \frac{1}{\sqrt{2}}(x+y). \tag{2}$$

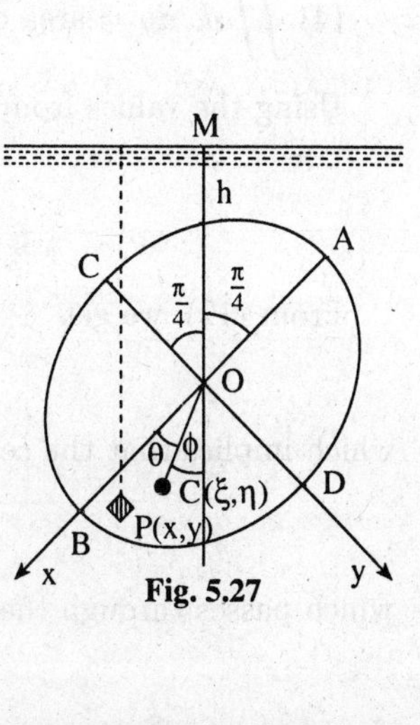

Fig. 5.27

Then pressure due to the liquid of density ρ at the point P is

$$p = g\rho z = g\rho\left[h + \frac{1}{\sqrt{2}}(x+y)\right]. \tag{3}$$

We consider an elementary area $dx\,dy$ surrounding the point P. Then the fluid thrust on the elementary area is

$$dF = p\,dx\,dy = g\rho\left[h + \frac{1}{\sqrt{2}}(x+y)\right]dx\,dy. \tag{4}$$

Let (ξ, η) be the coordinates of the centre of pressure. Then

$$\xi = \frac{\int x\,dF}{\int dF} = \frac{\iint x\cdot g\rho\left[h + \frac{1}{\sqrt{2}}(x+y)\right]dx\,dy}{\iint g\rho\left[h + \frac{1}{\sqrt{2}}(x+y)\right]dx\,dy} = \frac{\iint\left[hx + \frac{1}{\sqrt{2}}(x^2+xy)\right]dx\,dy}{\iint\left[h + \frac{1}{\sqrt{2}}(x+y)\right]dx,dy}. \tag{5}$$

Similarly,

$$\eta = \frac{\iint y\cdot g\rho\left[h + \frac{1}{\sqrt{2}}(x+y)\right]dx\,dy}{\iint g\rho\left[h + \frac{1}{\sqrt{2}}(x+y)\right]dx\,dy} = \frac{\iint\left[hy + \frac{1}{\sqrt{2}}(xy+y^2)\right]dx\,dy}{\iint\left[h + \frac{1}{\sqrt{2}}(x+y)\right]dx\,dy}. \tag{6}$$

Now, since the origin of coordinates has been taken at the centre of ellipse which is also its centre of gravity, we have

$$\iint x\,dx\,dy = 0, \quad \iint y\,dx\,dy = 0. \tag{7}$$

Also,

(1) $\iint x^2 dx\,dy =$ the moment of inertia of the ellipse about y-axis $= S\cdot\frac{a^2}{4} = \frac{\pi a^3 b}{4}$. (8)

(2) $\iint y^2 dx\,dy =$ the moment of inertia of the ellipse about x-axis $= S\cdot\frac{b^2}{4} = \frac{\pi ab^3}{4}$. (9)

(3) $\iint xy\,dx\,dy =$ the product of inertia of the ellipse about x, y-axes $= 0$. (10)

(4) $\iint dx\,dy =$ area of the ellipse $= \pi ab$. (11)

Using the values from (7) to (11), we get from (5),

$$\xi = \frac{\frac{1}{\sqrt{2}}\frac{\pi a^3 b}{4}}{h\pi ab} = \frac{a^2}{4\sqrt{2}h}, \quad \eta = \frac{\frac{1}{\sqrt{2}}\pi ab^3}{\pi ab} = \frac{b^2}{4\sqrt{2}h}. \tag{12}$$

From (12), we get

$$\eta = \frac{b^2}{a^2}\xi, \tag{13}$$

which implies that the centre of pressure of the ellipse lies on the line

$$y = \frac{b^2}{a^2}x, \tag{14}$$

which passes through the origin, i.e., the centre of the ellipse.

Let $C(\xi, \eta)$ be the position of the centre of pressure and let the line OC make angles θ and ϕ with x-axis and the vertical respectively.

Then

$$\theta + \phi = \frac{\pi}{4} \tag{15}$$

$$\text{and } \tan\theta = \frac{\eta}{\xi} = \frac{b^2}{a^2}. \tag{16}$$

Hence, $\tan\phi = \tan\left(\frac{\pi}{4} - \theta\right) = \frac{1 - \tan\theta}{1 + \tan\theta}$, i.e., $\tan\phi = \frac{1 - \frac{b^2}{a^2}}{1 + \frac{b^2}{a^2}} = \frac{a^2 - b^2}{a^2 + b^2}$.

$$\therefore \phi = \tan^{-1}\left(\frac{a^2 - b^2}{a^2 + b^2}\right) \quad \textbf{(Proved)}.$$

Also the depth of c.p. from the surface is

$$= h + \frac{1}{\sqrt{2}}(\xi + \eta) = h + \frac{1}{\sqrt{2}}\left(\frac{a^2}{4\sqrt{2}h} + \frac{b^2}{4\sqrt{2}h}\right) = h + \frac{a^2 + b^2}{8h} \quad \textbf{(Proved)}.$$

EXAMPLE 5.18 *A semi-ellipse bounded by its minor axis is just immersed in a liquid, the density of which varies as the depth. If the minor axis be in the surface, find the eccentricity in order that the focus may be the centre of pressure.*

Solution. Let AB be the minor axis and O be the centre of the ellipse. AB lies on the surface of the liquid (FIG. 5.28). We take the major and minor axes of the ellipse as x- and y-axes respectively. Then the equation of the ellipse is

$$\frac{x^2}{a^2} + \frac{y^2}{b^2} = 1, \tag{1}$$

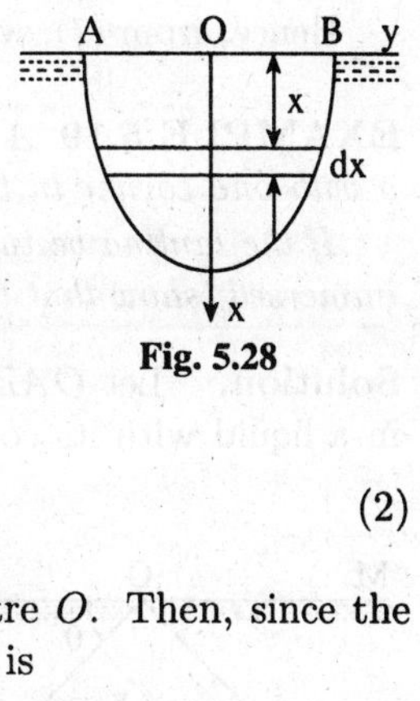

Fig. 5.28

where a, b are the lengths of semi-major and semi-minor axes respectively.

From (1), we have

$$y = \frac{b}{a}\sqrt{a^2 - x^2}. \tag{2}$$

Consider a thin strip of thickness dx at a distance x from the centre O. Then, since the density of the liquid varies as depth, its value at any point on the strip is

$$\rho = \lambda x, \ \lambda = \text{constant}. \tag{3}$$

Then the pressure at any point on the strip is

$$p = g\rho x = g(\lambda x)x = g\lambda x^2. \tag{4}$$

The elementary area of the strip is

$$dS = 2y\,dx = 2\frac{b}{a}\sqrt{a^2 - x^2}dx. \tag{5}$$

Hence the fluid thrust on the strip is

$$dF = p\,dS = 2g\lambda\frac{b}{a}x^2\sqrt{a^2 - x^2}dx. \tag{6}$$

Due to symmetry, the centre of pressure lies on x-axis. If ξ be its depth below the surface,

$$\xi = \frac{\int_0^a x\,dF}{\int_0^a dF} = \frac{\int_0^a 2g\lambda\frac{b}{a}x^3\sqrt{a^2-x^2}dx}{\int_0^a 2g\lambda\frac{b}{a}x^2\sqrt{a^2-x^2}dx} = \frac{\int_0^a x^3\sqrt{a^2-x^2}dx}{\int_0^a x^2\sqrt{a^2-x^2}dx}.$$

Put $x = a\sin\theta$, so that, $dx = a\cos\theta\,d\theta$. The limits of θ are from 0 to $\pi/2$. Hence

$$\xi = \frac{\int_0^{\pi/2} a^3\sin^3\theta\cdot a\cos\theta\cdot a\cos\theta\,d\theta}{\int_0^{\pi/2} a^2\sin^2\theta\cdot a\cos\theta\cdot a\cos\theta}$$

$$= \frac{a\int_0^{\pi/2}\sin^3\theta\cos^2\theta\,d\theta}{\int_0^{\pi/2}\sin^2\theta\cdot\cos^2\theta\,d\theta} = \frac{a\cdot\frac{2}{5}\cdot\frac{1}{3}}{\frac{1}{4}\cdot\frac{1}{2}\cdot\frac{\pi}{2}} = \frac{32}{15\pi}a. \quad (7)$$

Since the centre of pressure coincides with the focus, the eccentricity e of the ellipse is given by

$$\xi = ae, \text{ i.e., } e = \frac{\xi}{a}.$$

Hence, from (7), we get $e = \frac{32}{15\pi}$. **(Ans.)**

EXAMPLE 5.19 *A square lamina is vertically immersed in a homogeneous liquid of density ρ with one corner in the surface. Determine the position of its centre of pressure.*

If the lamina be turned in its own plane about that corner so that it always remains totally immersed, show that the locus of the centre of pressure is a straight line.

Solution. Let $OABC$ be the square lamina with each side equal to $2a$, immersed totally in a liquid with its corner O in the surface. We take the sides OA and OC as x- and y-axes respectively (FIG. 5.29). Let at any time, the side OA make an angle θ with the free surface MN. Then the equation of free surface is

$$y = -x\tan\theta, \text{ or, } x\sin\theta + y\cos\theta = 0. \quad (1)$$

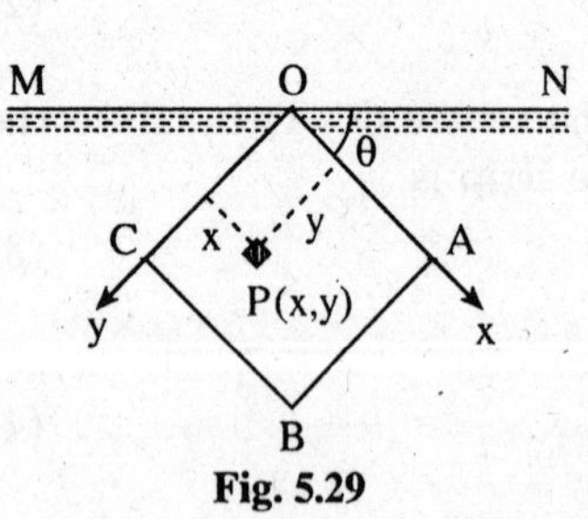

Fig. 5.29

We take a point $P(x, y)$ in the lamina and an elementary area $dx\,dy$ surrounding it. The depth of the point P below free surface is

$$z = x\sin\theta + y\cos\theta \quad (2)$$

so that pressure at the point P is

$$p = g\rho z = g\rho(x\sin\theta + y\cos\theta). \quad (3)$$

Hence, the thrust on the elementary area is

$$dF = p\,dS = g\rho(x\sin\theta + y\cos\theta)dx\,dy. \quad (4)$$

If (ξ, η) be the coordinates of the centre of pressure, then

$$\xi = \frac{\displaystyle\int x \cdot dF}{\displaystyle\int dF} = \frac{\displaystyle\int_0^{2a}\int_0^{2a} x \cdot g\rho(x\sin\theta + y\cos\theta)dx\,dy}{\displaystyle\int_0^{2a}\int_0^{2a} g\rho(x\sin\theta + y\cos\theta)dx\,dy}$$

$$= \frac{\sin\theta \displaystyle\int_0^{2a}\int_0^{2a} x^2 dx\,dy + \cos\theta \int_0^{2a}\int_0^{2a} xy\,dx\,dy}{\sin\theta \displaystyle\int_0^{2a}\int_0^{2a} x\,dx\,dy + \cos\theta \int_0^{2a}\int_0^{2a} y\,dx\,dy}$$

$$= \frac{\sin\theta \cdot \frac{8a^3}{3} \cdot 2a + \cos\theta \cdot \frac{4a^2}{2} \cdot \frac{4a^2}{2}}{\sin\theta \cdot \frac{4a^2}{2} \cdot 2a + \cos\theta \cdot \frac{4a^2}{2} \cdot 2a}$$

$$= \frac{a}{3} \cdot \frac{4\sin\theta + 3\cos\theta}{\sin\theta + \cos\theta}. \tag{5}$$

Similarly,

$$\eta = \frac{\displaystyle\int y \cdot dF}{\displaystyle\int dF} = \frac{\displaystyle\int_0^{2a}\int_0^{2a} y \cdot g\rho(x\sin\theta + y\cos\theta)dx\,dy}{\displaystyle\int_0^{2a}\int_0^{2a} g\rho(x\sin\theta + y\cos\theta)dx\,dy}$$

$$= \frac{\sin\theta \displaystyle\int_0^{2a}\int_0^{2a} xy\,dx\,dy + \cos\theta \int_0^{2a}\int_0^{2a} y^2 dx\,dy}{\sin\theta \displaystyle\int_0^{2a}\int_0^{2a} x\,dx\,dy + \int_0^{2a}\int_0^{2a} y\,dx\,dy}$$

$$= \frac{\sin\theta \cdot \frac{4a^2}{2} \cdot \frac{4a^2}{2} + \cos\theta \cdot \frac{8a^3}{3} \cdot 2a}{\sin\theta \cdot \frac{4a^2}{2} \cdot 2a + \cos\theta \cdot \frac{4a^2}{2} \cdot 2a}$$

$$= \frac{a}{3} \cdot \frac{3\sin\theta + 4\cos\theta}{\sin\theta + \cos\theta}. \tag{6}$$

From (5) and (6), we get

$$\xi + \eta = \frac{a}{3} \cdot \frac{7(\sin\theta + \cos\theta)}{\sin\theta + \cos\theta} = \frac{7a}{3}. \tag{7}$$

Hence, the locus of the centre of pressure is

$$x + y = \frac{7a}{3} \tag{8}$$

which is a straight line.

EXAMPLE 5.20 *A rectangle is immersed in n fluids of densities $\rho, 2\rho, 3\rho, \ldots, n\rho$, the top of the rectangle being in the surface of the first liquid and the area immersed in each being the same. Show that the depth of the centre of pressure of the rectangle is*

$$\frac{3n+1}{2n+1}h,$$

where h is the depth of the lower side.

Solution. Let a rectangle $ABCD$ be immersed in n liquids, its side AD being in the surface (FIG. 5.30). It is given that $AB = DC = h$. Since the area immersed in each liquid is same, the height of each rectangular portion in different liquids is $= \dfrac{h}{n}$.

Let area of each of these rectangular portions be S. Now, the thrust on the whole rectangle $ABCD$ due to all different liquids

= thrust on rectangle $ABCD$ of area nS due to liquid of density ρ only,

+ thrust on rectangle A_1BCD_1 of area $(n-1)S$ due to liquid of density $(2\rho - \rho) = \rho$,

+ thrust on rectangle A_2BCD_2 of area $(n-2)S$ due to liquid of density $(3\rho - 2\rho) = \rho$,

+ ....,

+ thrust on rectangle $A_{n-1}BCD_{n-1}$ of area S due to liquid of density $\{n\rho - (n-1)\rho\} = \rho$.

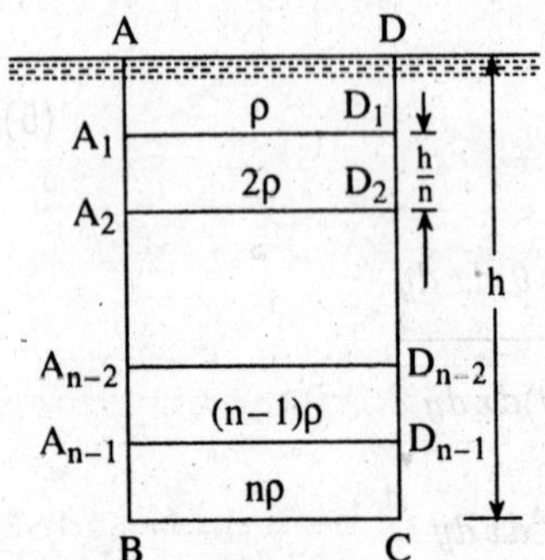

Fig. 5.30

If $F_1, F_2, \ldots, F_n$ be thrusts on these rectangles, then

$$\left.\begin{aligned} F_1 &= g\rho(nS)\cdot\frac{h}{2} = \frac{1}{2}g\rho nSh \\ F_2 &= g\rho(n-1)S\cdot\frac{1}{2}h\left(1-\frac{1}{n}\right) \\ F_3 &= g\rho(n-2)S\cdot\frac{1}{2}h\left(1-\frac{2}{n}\right) \\ &\cdots\cdots\cdots\cdots\cdots\cdots \\ F_n &= g\rho S\cdot\frac{1}{2}\frac{h}{n}. \end{aligned}\right\} \tag{1}$$

If $x_1, x_2, \ldots, x_n$ be the depths of c.p. of the above rectangles below the surface, then

$$x_1 = \frac{2}{3}h,\ x_2 = \frac{2}{3}h\left(1-\frac{1}{n}\right),\ \ldots,\ x_n = \frac{2}{3}\frac{h}{n}. \tag{2}$$

Then the heights of the c.p. of the rectangular portions above the line BC are:

$$\xi_1 = \frac{1}{3}h,\ \xi_2 = \frac{1}{3}h\left(1-\frac{1}{n}\right),\ \xi_3 = \frac{1}{3}h\left(1-\frac{2}{n}\right),\ \ldots,\ \xi_n = \frac{1}{3}\frac{h}{n}. \tag{3}$$

Hence

$$\begin{aligned} F_1\xi_1 &= \frac{1}{2}g\rho nSh\cdot\frac{1}{3}h = \frac{1}{6}(g\rho Sh^2)\cdot n \\ F_2\xi_2 &= g\rho Sh\cdot(n-1)\cdot\frac{1}{2}\frac{n-1}{n}\cdot\frac{1}{3}h\left(1-\frac{1}{n}\right) \\ &= \frac{1}{6}(g\rho Sh^2)\cdot\frac{(n-1)^3}{n^2} \\ F_3\xi_3 &= g\rho Sh\cdot(n-2)\cdot\frac{1}{2}\frac{n-2}{n}\cdot\frac{1}{3}h\left(1-\frac{2}{n}\right) \\ &= \frac{1}{6}(g\rho Sh^2)\cdot\frac{(n-2)^3}{n^2} \\ &\cdots\cdots\cdots\cdots\cdots\cdots\cdots\cdots \\ F_n\xi_n &= g\rho Sh\cdot\frac{1}{2n}\cdot\frac{h}{3n} = \frac{1}{6}(g\rho Sh^2)\frac{1}{n^2} \end{aligned}$$

Thus, if ξ be the height of the centre of pressure of the rectangle $ABCD$ above the line BC, then

$$\xi = \frac{\Sigma F_i \cdot \xi_i}{\Sigma F_i},\ i = 1, 2, \ldots, n,$$

$$\text{i.e., } \xi = \frac{\frac{1}{6} g\rho S h^2 \left[n + \frac{(n-1)^3}{n^2} + \frac{(n-2)^3}{n^2} + \cdots + \frac{1}{n^2} \right]}{\frac{1}{2} g\rho S h \left[n + \frac{(n-1)^2}{n} + \frac{(n-2)^2}{n} + \cdots + \frac{1}{n} \right]}$$

$$= \frac{h}{3n} \cdot \frac{n^3 + (n-1)^3 + (n-2)^3 + \cdots + 1}{n^2 + (n-1)^2 + (n-2)^2 + \cdots + 1}$$

$$= \frac{h}{3n} \cdot \frac{\left\{ \frac{n(n+1)}{2} \right\}^2}{\frac{n(n+1)(2n+1)}{6}} = \frac{h}{3n} \cdot \frac{6}{4} \cdot \frac{n(n+1)}{2n+1}, \text{ i.e., } \xi = \frac{h}{2} \cdot \frac{n+1}{2n+1}. \quad (4)$$

Hence, the depth of the centre of pressure of the rectangle $ABCD$ below the free surface is

$$= h - \xi = h - \frac{h}{2} \frac{n+1}{2n+1} = \frac{3n+1}{2n+1} \cdot \frac{h}{2} \quad \textbf{(Ans.)}$$

EXAMPLE 5.21 *Find the coordinates of the centre of pressure of a square lamina, held vertically in a fluid whose density varies as the depth, with a vertex in the surface, and the diagonal vertical.*

Solution. Let $ABCD$ be a square lamina immersed in a liquid with its vertex A in the surface and the diagonal AC vertical (FIG. 5.31).

Let $AC = BD = 2h$, so that $AO = OC = h$, where O is the centre of the square.

We take a thin strip of thickness dx at a depth x from the surface within the $\triangle ABD$.

Since the density of the fluid varies as depth, its magnitude at any point on the strip is

$$\rho = \lambda x, \ (\lambda = \text{constant}) \quad (1)$$

so that pressure at any point on the strip is

$$p = g\rho x = g(\lambda x)x = g\lambda x^2. \quad (2)$$

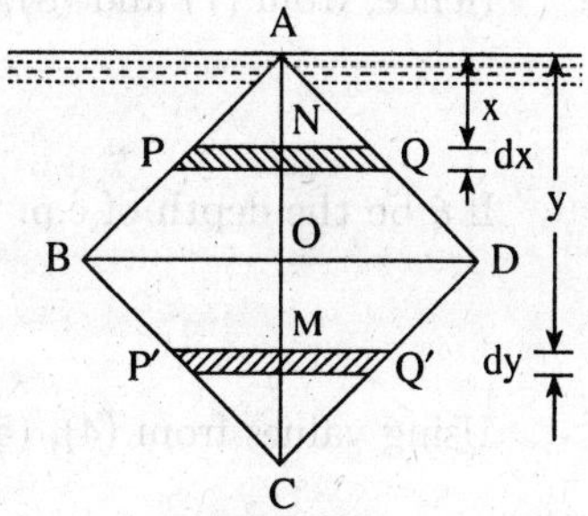

Fig. 5.31

Now, from $\triangle$s APQ and ABD,

$$\frac{PQ}{BD} = \frac{x}{AO}, \text{ i.e., } PQ = \frac{x \cdot BD}{AO} = \frac{x \cdot 2h}{h} = 2x. \quad (3)$$

Hence, the fluid thrust on the $\triangle ABD$ is

$$F_1 = \int_0^h (g\lambda x^2) \cdot 2x\, dx = 2g\lambda \left(\frac{x^4}{4} \right)_0^h = 2g\lambda \cdot \frac{h^4}{4}. \quad (4)$$

Depth of c.p. of $\triangle ABD$ below the surface is

$$\xi_1 = \frac{\int_0^h x \cdot (g\lambda x^2) 2x\, dx}{\int_0^h (g\lambda x^2) 2x\, dx} = \frac{\int_0^h x^4 dx}{\int_0^h x^3 dx} = \frac{h^5}{5} \cdot \frac{4}{h^4} = \frac{4}{5} h. \quad (5)$$

Similarly, we consider a thin strip $P'Q'$ of thickness dy at a depth y from the surface within the $\triangle BCD$. Then, density on the strip is $g\lambda y^2$ and the length of the strip is given by

$$\frac{P'Q'}{BD} = \frac{CM}{CO}, \text{ i.e., } \frac{P'Q'}{2h} = \frac{2h-y}{h}, \text{ i.e., } P'Q' = 2(2h-y). \tag{6}$$

Hence the fluid thrust on the $\triangle BCD$ is

$$F_2 = \int_h^{2h} (g\lambda y^2)2(2h-y)dy = 2g\lambda\int_h^{2h}(2hy^2 - y^3)dy = 2g\lambda\left[2h\frac{y^3}{3} - \frac{y^4}{4}\right]_h^{2h}$$

$$\text{i.e., } F_2 = 2g\lambda\left[2h\left(\frac{8h^3}{3} - \frac{h^3}{3}\right) - \frac{1}{4}(16h^4 - h^4)\right] = 2g\lambda h^4\left(\frac{14}{3} - \frac{15}{4}\right) = 2g\lambda h^4 \cdot \frac{11}{12}. \tag{7}$$

Depth of c.p. of the $\triangle BCD$ below the surface is

$$\xi_2 = \frac{\int_h^{2h} y\cdot(g\lambda y^2)\cdot 2(2h-y)dy}{\int_h^{2h}(g\lambda y^2)\cdot 2(2h-y)dy} = \frac{\int_h^{2h}(2hy^3 - y^4)dy}{\int_h^{2h}(2hy^2 - y^3)dy}.$$

Its numerator is

$$= \left[2h\cdot\frac{y^4}{4} - \frac{y^5}{5}\right]_h^{2h} = 2h\left(\frac{16h^4}{4} - \frac{h^4}{4}\right) - \left(\frac{32h^5}{5} - \frac{h^5}{5}\right)$$

$$= \left(\frac{15}{2} - \frac{31}{5}\right)h^5 = \frac{13}{10}h^5. \tag{8}$$

Hence, from (7) and (8),

$$\xi_2 = \frac{13}{10}h^5 \cdot \frac{12}{11h^4} = \frac{78}{55}h. \tag{9}$$

If ξ be the depth of c.p. of the square lamina $ABCD$ below the free surface, then

$$\xi = \frac{F_1\xi_1 + F_2\xi_2}{F_1 + F_2}. \tag{10}$$

Using values from (4), (5), (7) and (9) in (10), we get

$$\xi = \frac{2g\lambda\cdot\frac{h^4}{4}\cdot\frac{4}{5}h + 2g\lambda h^4\cdot\frac{11}{12}\cdot\frac{78}{55}h}{2g\lambda\cdot\frac{h^4}{4} + 2g\lambda h^4\cdot\frac{11}{12}} = h\cdot\frac{\frac{1}{5}+\frac{78}{60}}{\frac{1}{4}+\frac{11}{12}}$$

$$= h\frac{\frac{12+78}{60}}{\frac{3+11}{12}} = h\frac{\frac{90}{60}}{\frac{14}{12}} = \frac{90}{60}\cdot\frac{12}{14}h = \frac{9}{7}h \quad \textbf{(Ans.)}$$

EXAMPLE 5.22 *Find the coordinates of the centre of pressure of the area enclosed between the curve $\sqrt{x} + \sqrt{y} = \sqrt{a}$ and the axes, taking the axes to be rectangular and one of them in the surface.*

Solution. Let us take the x-axis in the surface and y-axis vertically downwards. With these rectangular coordinates, the shape of the curve is as shown in the figure (FIG. 5.32).

Obviously, the coordinates of the points A and B are $(a, 0)$ and $(0, a)$ respectively.

We consider an elementary strip QR of length x and thickness dy at a depth y from the surface. Then pressure at any point on the strip is

$$p = g\rho y, \tag{1}$$

where ρ is the density of the liquid.

The elementary area, $dS = x\,dy$. Hence the thrust on this elementary area is

$$dF = (g\rho y)x\,dy = g\rho xy\,dy. \tag{2}$$

The centre of pressure of the elementary strip is at the point $\left(\dfrac{x}{2}, y\right)$.

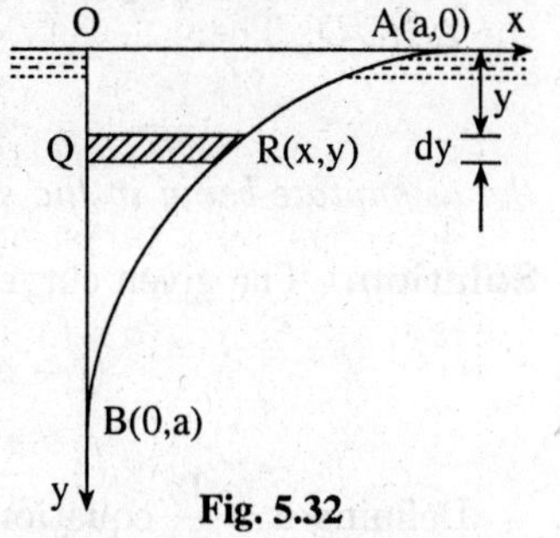

Fig. 5.32

Hence, if (ξ, η) be the coordinates of the centre of pressure of the enclosed area, then

$$\xi = \frac{I_1}{I_3},\ \eta = \frac{I_2}{I_3}, \tag{3}$$

where we have denoted

$$I_1 = \int_0^a \frac{x}{2}(g\rho xy)dy,\ I_2 = \int_0^a y(g\rho xy)dy,\ I_3 = \int_0^a g\rho xy\,dy. \tag{4}$$

In order to find these integrals, we write the parametric equation of the given curve in the form

$$x = a\sin^4\phi,\ y = a\cos^4\phi \text{ so that, } dy = -4a\cos^3\phi\sin\phi\,d\phi. \tag{5}$$

When $y = 0$, $\phi = \dfrac{\pi}{2}$ and when $y = a, \phi = 0$, so that the limits of integration for ϕ are $\dfrac{\pi}{2}$ to zero. Thus,

$$I_1 = \frac{g\rho}{2}\int_{\frac{\pi}{2}}^0 (a^2\sin^8\phi)(a\cos^4\phi)(-4a\cos^3\phi\sin\phi)d\phi = 2g\rho a^4\int_0^{\frac{\pi}{2}} \sin^9\phi\cos^7\phi\,d\phi$$

$$= 2g\rho a^4\cdot\frac{8\cdot6\cdot4\cdot2\cdot6\cdot4\cdot2}{16\cdot14\cdot12\cdot10\cdot8\cdot6\cdot4\cdot2} = \frac{1}{280}g\rho a^4 \tag{6}$$

$$I_2 = g\rho\int_{\frac{\pi}{2}}^0 (a\sin^4\phi)(a^2\cos^8\phi)(-4a\cos^3\phi\sin\phi)d\phi = 4g\rho a^4\int_0^{\frac{\pi}{2}} \sin^5\phi\cos^{11}\phi\,d\phi,$$

$$\text{i.e., } I_2 = 4g\rho a^4\cdot\frac{4\cdot2\cdot1\cdot10\cdot8\cdot6\cdot4\cdot2}{16\cdot14\cdot12\cdot10\cdot8\cdot6\cdot4\cdot2} = \frac{1}{84}g\rho a^4 \tag{7}$$

$$\text{and } I_3 = g\rho\int_{\frac{\pi}{2}}^0 (a\sin^4\phi)(a\cos^4\phi)(-4a\cos^3\phi\sin\phi)d\phi$$

$$= 4g\rho a^3\int_0^{\frac{\pi}{2}} \sin^5\phi\cos^7\phi\,d\phi = 4g\rho a^3\cdot\frac{4\cdot2\cdot6\cdot4\cdot2}{12\cdot10\cdot8\cdot6\cdot4\cdot2} = \frac{1}{30}g\rho a^3 \tag{8}$$

Putting these values in (3),

$$\left.\begin{aligned}\xi &= \frac{(g\rho a^4)}{280}\cdot\frac{30}{g\rho a^3} = \frac{3}{28}a\\ \eta &= \frac{(g\rho a^4)}{84}\cdot\frac{30}{g\rho a^3} = \frac{5}{14}a.\end{aligned}\right\} \tag{9}$$

Thus, the coordinates of the centre of pressure are $\left(\dfrac{3}{28}a, \dfrac{5}{14}a\right)$.

EXAMPLE 5.23 *Show that the depth of the centre of pressure of the area included between the arc and the asymptote of the curve $(r-a)\cos\theta = b$ is*

$$\frac{a}{4}\cdot\frac{3\pi a+16b}{3\pi b+4a},$$

the asymptote being in the surface and the plane of the curve vertical.

Solution. The given curve is

$$(r-a)\cos\theta = b, \text{ i.e., } r = a + \frac{b}{\cos\theta} = \frac{a\cos\theta + b}{\cos\theta}. \tag{1}$$

Defining $u = \dfrac{1}{r}$, equation of the curve can be written as

$$u = \frac{\cos\theta}{a\cos\theta + b} = F(\theta), \text{ say.} \tag{2}$$

The directions in which $r \to \infty$ are given by

$$u = 0, \text{ i.e., } \cos\theta = 0, \text{ i.e., } \theta = (2n+1)\frac{\pi}{2}.$$

We consider the branch corresponding to $\theta = \dfrac{\pi}{2}$. Now,

$$F'(\theta) = \frac{-\sin\theta(a\cos\theta + b) - \cos\theta(-a\sin\theta)}{(a\cos\theta+b)^2} = \frac{-b}{(a\cos\theta+b)^2} \tag{3}$$

For $\theta = \dfrac{\pi}{2}$

$$F'(\theta) = \frac{-b}{b^2} = -\frac{1}{b}. \tag{4}$$

The asymptote of the curve is given by

$$r\sin\left(\theta - \frac{\pi}{2}\right) = \frac{1}{F'(\theta)} = -b, \text{ or, } r\cos\theta = b. \tag{5}$$

The given curve and its asymptote are shown in the figure (5.33).

We consider a point $P(r,\theta)$ and an elementary area $dS = r\,d\theta\,dr$ about it within the region bounded by the curve and the asymptote. The asymptote being in the surface, the depth of the point below the surface is

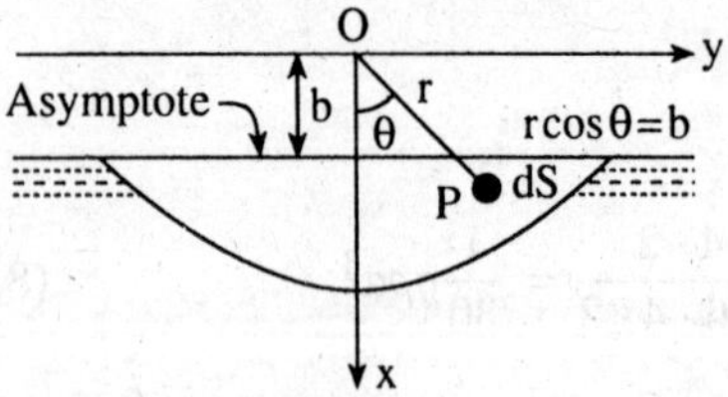

Fig. 5.33

$$z = r\cos\theta - b \tag{6}$$

so that the pressure due to fluid at the point is

$$p = g\rho z = g\rho(r\cos\theta - b), \tag{7}$$

where ρ is the density of the fluid.

From symmetry, the centre of pressure lies on x-axis. If ξ be the depth of c.p. from y-axis, then

$$\xi = \frac{\displaystyle\iint r\cos\theta\cdot g\rho(r\cos\theta - b)r\,d\theta\,dr}{\displaystyle\iint g\rho(r\cos\theta - b)r\,d\theta\,dr} = \frac{\displaystyle\iint (r^3\cos^2\theta - r^2 b\cos\theta)d\theta\,dr}{\displaystyle\iint (r^2\cos\theta - rb)d\theta\,dr}. \tag{8}$$

The limits of integration are

$$\left.\begin{aligned} r &= \text{from } r_1 = \frac{b}{\cos\theta} \text{ to } r_2 = \left(a + \frac{b}{\cos\theta}\right) \\ \text{and } \theta &= \text{from zero to } \frac{\pi}{2} \end{aligned}\right\} \tag{9}$$

Hence, the numerator of (8) is

$$= \int_0^{\frac{\pi}{2}} \left[\cos^2\theta \left(\frac{r^4}{4}\right)_{r_1}^{r_2} - b\cos\theta\left(\frac{r^3}{3}\right)_{r_1}^{r_2}\right] d\theta$$

$$= \int_0^{\frac{\pi}{2}} \left[\frac{\cos^2\theta}{4}\left\{\left(a + \frac{b}{\cos\theta}\right)^4 - \left(\frac{b}{\cos\theta}\right)^4\right\} - \frac{b\cos\theta}{3}\left\{\left(a+\frac{b}{\cos\theta}\right)^3 - \left(\frac{b}{\cos\theta}\right)^3\right\}\right] d\theta$$

$$= \int_0^{\frac{\pi}{2}} \left[\frac{1}{4}(a^4\cos^2\theta + 4ab^3\sec\theta + 6a^2b^2 + 4a^3b\cos\theta) - \frac{b}{3}(a^3\cos\theta + 3a^2b + 3ab^2\sec\theta)\right] d\theta$$

$$= \int_0^{\frac{\pi}{2}} \left[\frac{a^2}{4}\cos^2\theta + \frac{a^2b^2}{2} + \frac{2a^3b}{3}\cos\theta\right] d\theta$$

$$= \frac{a^4}{8}\cdot\frac{\pi}{2} + \frac{a^2b^2}{2}\cdot\frac{\pi}{2} + \frac{2a^3b}{3} = \frac{a^2\pi}{16}(a^2+4b^2) + \frac{2a^3b}{3} \tag{10}$$

Similarly, the denominator of (8) is

$$= \int_0^{\frac{\pi}{2}} \left[\cos\theta\left(\frac{r^3}{3}\right)_{r_1}^{r_2} - b\left(\frac{r^2}{2}\right)_{r_1}^{r_2}\right] d\theta$$

$$= \int_0^{\frac{\pi}{2}} \left[\frac{1}{3}\cos\theta\left\{\left(a+\frac{b}{\cos\theta}\right)^3 - \left(\frac{b}{\cos\theta}\right)^3\right\} - \frac{b}{2}\left\{\left(a+\frac{b}{\cos\theta}\right)^2 - \left(\frac{b}{\cos\theta}\right)^2\right\}\right]$$

$$= \int_0^{\frac{\pi}{2}} \left[\frac{1}{3}\cos\theta(a^3 + 3a^2b\sec\theta + 3ab^2\sec^2\theta) - \frac{b}{2}(a^2 + 2ab\sec\theta)\right]$$

$$= \int_0^{\frac{\pi}{2}} \left(\frac{a^3}{3}\cos\theta + \frac{1}{2}a^2b\right) d\theta = \frac{a^3}{3}\cdot 1 + \frac{a^2b}{2}\cdot\frac{\pi}{2} = \frac{a^2}{12}(4a+3b\pi) \tag{11}$$

From (8) to (11),

$$\xi = \frac{a^2}{48}\left[3(a^2+4b^2)\pi + 32ab\right]\cdot\frac{12}{a^2(4a+3b\pi)} = \frac{1}{4}\frac{3(a^2+4b^2)\pi + 32ab}{4a+3b\pi}. \tag{12}$$

Hence the depth the c.p. from fluid surface is

$$= \xi - b = \frac{1}{4}\frac{3(a^2+4b^2)\pi + 32ab}{4a+3b\pi} - b$$

$$= \frac{3(a^2+4b^2)\pi + 32ab - 16ab - 12b^2\pi}{4(4a+3b\pi)} = \frac{a}{4}\,\frac{3a\pi+16b}{3b\pi+4a} \quad \textbf{(Proved).}$$

EXERCISE

1. The lengths of two parallel sides of a trapezium are a and b, and the distance between them is h. If the trapezium be immersed in water with its plane vertical and the side 'a' in the surface, prove that the centre of pressure will be at a depth of $\dfrac{a+3b}{a+2b}\cdot\dfrac{h}{2}$ below the surface.

2. A lamina in the shape of a quadrilateral $ABCD$ has its side CD in the surface of a liquid and the sides AD, BC vertical and of lengths equal to α, β respectively. Show that the depth of the centre of pressure is $\frac{1}{2}\frac{(\alpha^2+\beta^2)(\alpha+\beta)}{\alpha^2+\alpha\beta+\beta^2}$.

3. A square lamina is just immersed vertically in water and is lowered through a depth h. If 'a' be the length of the edge of the square, prove that the distance of the centre of pressure from the centre of the square is $\frac{a^2}{6a+12h}$.

4. A regular hexagon of side 'a' is immersed in a liquid with one side in the surface. Show that the depth of its c.g. bear a ratio of depth of its c.p. as 18 : 23.

5. A rectangle is immersed vertically in water with two sides horizontal and at depths h and $a+h$ respectively below the effective surface. Prove that the distance of c.p. from the upper side is

$$\frac{a}{h} \cdot \frac{3h+2a}{2h+a}.$$

Show also that the c.p. is below the c.g. of the area and as depth increases, the c.p. approaches to c.g., but never coincides with it.

6. A rectangular area of side a and b has the sides of length a vertical and the centre of the rectangle at a depth h below the surface of water in which it is immersed. Show that the centre of pressure is at a depth $\frac{a^2}{12h}$ below the centre.

7. Proved that, if the density of the liquid varies as the depth below the surface, the depth of the centre of pressure of a parallelogram two of whose sides are horizontal and at depths a and b, is

$$\frac{3}{4} \cdot \frac{a^3+a^2b+ab^2+b^3}{a^2+ab+b^2}.$$

8. If a quadrilateral $ABCD$ be entirely immersed in water, and $\alpha, \beta, \gamma, \delta$, be the depths of its four corners, and h that of its centre of gravity, show that the depth of its centre of pressure is

$$\frac{1}{2}(\alpha+\beta+\gamma+\delta) - \frac{1}{6h}(\alpha\beta+\beta\gamma+\gamma\delta+\gamma\alpha+\alpha\delta+\beta\delta).$$

9. A parallelogram, whose plane is vertical and centre at a depth h below the surface, is totally immersed in a homogeneous fluid. Show that, if a, b be the lengths of the projections of sides on a vertical line, the depth of its centre of pressure will be $h + \frac{a^2+b^2}{12h}$.

10. Show that the depth of the centre of pressure of a semi-circular lamina immersed in a homogeneous liquid with the diameter in the surface is $\frac{3\pi R}{16}$, where R is its radius.

11. A quadrant of a circle of radius R is immersed in a liquid with a bounding radius in the surface. Show that the coordinates of its centre of pressure are $\frac{3}{8}R$ and $\frac{3\pi R}{16}$.

12. Show that the centre of pressure of a circular area of radius R immersed in a liquid with its centre at a depth h below the surface, when the density of the liquid varies as the depth, is at a depth $\frac{2R^2h}{R^2+4h^2}$ below the centre of the circle.

13. A semi-circular area of radius R is immersed vertically with its diameter horizontal at a depth h. If the circumference be below the centre, prove that the depth of its centre of pressure is

$$\frac{3\pi(R^2+4h^2)+32Rh}{4(4R+3h\pi)}.$$

14. A plane lamina consists of a circular disc of radius R and centre O from which a circular portion of radius $\frac{R}{2}$ and centre P has been removed. The lamina is completely immersed in a homogeneous fluid with its plane vertical and P vertically below O. If the centre of pressure of the lamina is at O, prove that O is at a depth $\frac{11}{8}R$ below the fluid.

15. Show that the depth of the centre of pressure of a triangular lamina whose angular points are at depths of α, β, γ from the surface is at a depth

$$\frac{1}{6}\frac{\alpha^2+\beta^2+\gamma^2-\alpha\beta-\beta\gamma-\gamma\alpha}{\alpha+\beta+\gamma}$$

below the centre of gravity of the lamina.

16. Show that the depth of centre of pressure of a rhombus totally immersed with one diagonal vertical and its centre at a depth h is $h+\dfrac{a^2}{24h}$, where a is the length of the vertical diagonal.

17. A parallelogram has its corners at depths h_1, h_2, h_3, h_4 below the surface of a liquid and its centre at a depth of h. Show that the depth of its centre of pressure is $\dfrac{h_1^2+h_2^2+h_3^2+h_4^2+8h^2}{12h}$.

18. An ellipse is just immersed in water with its major axis vertical. Show that, if the centre of pressure coincides with the focus, the eccentricity of the ellipse must be $\dfrac{1}{4}$.

19. A circular disc of radius R is just completely immersed with its plane vertical in a homogeneous liquid. Prove that the distance between the centres of pressure of the two semi-circles into which the disc is divided by its horizontal diameter is $\dfrac{18\pi R}{9\pi^2-16}$.

20. An elliptic disc with semi-axes a, b is completely immersed in a homogeneous liquid with its plane vertical. If h be the depth of the centre of the disc below the free surface and the major axis makes an angle θ with the horizon, show that the depth of the centre of pressure below the centre of the ellipse is $\dfrac{1}{4h}(a^2\sin^2\theta+b^2\cos^2\theta)$.

21. A uniform elliptic lamina whose axes are $2a$ and $2b$, is half immersed in water, the axis $2b$ being in the surface. If the centre of pressure coincides with the focus, show that the eccentricity of the ellipse is $e=\dfrac{3\pi}{16}$.

22. Show that the centre of pressure of a rhombus immersed in two liquids, which do not mix, with a vertex in the upper surface and a diagonal in the common surface, divides the other diagonal in the ratio 5 : 3, if the density of the lower liquid be three times that of the other.

23. The axis of a cylindrical vessel containing liquid is inclined to the vertical at an angle α. Prove that the distance of the centre of pressure of the base from the centre of the base is $\dfrac{R^2\tan\alpha}{4b}$, where R is the radius of the base and $\pi R^2 b$ is the volume of the liquid.

24. Prove that the depth of centre of pressure of a parallelogram, two of whose sides are horizontal and at depths of h, k below the surface of a liquid whose density varies as the depth below the surface is

$$\frac{3}{4}\frac{(h+k)(h^2+k^2)}{h^2+hk+k^2}.$$

25. A vessel contains three fluids of densities $\rho, 2\rho, 3\rho$. A triangle of area S is supported with one side in the surface of the fluid of density ρ and the opposite vertex in the fluid of density 3ρ. If $3h, 2h, h$ are the depths of the vertex below the upper surface of the three fluids, prove that (neglecting atmospheric pressure) the thrust on each face of the triangle is $\frac{4}{3}g\rho Sh$. Also prove that the depth of c.p. of the triangle from the free surface is $\dfrac{59}{36}h$.

26. An isosceles triangle of vertical angle 2α is placed vertically in a liquid with its vertex in the surface and its base inclined at an angle θ to the horizon. Show that the depth of its c.p. is

$$a\frac{3\cos^2\theta\cos^2\alpha+\sin^2\theta\sin^2\alpha}{4\cos\theta\cos\alpha},$$

where a is one of the equal sides of the triangle.

27. A square lamina $ABCD$ is wholly immersed in a homogeneous liquid with its plane vertical, the corner A in the surface and the side AB making an angle θ with the vertical. Show that the depth of c.p. is $\dfrac{a}{3}\cdot\dfrac{2+3\sin\theta\cos\theta}{\sin\theta+\cos\theta}$.

28. The angular points of a triangle immersed in a liquid whose density varies as depth are α, β, γ respectively below the surface. Show that the centre of pressure is at a depth

$$\frac{3}{5}\frac{(\alpha+\beta+\gamma)(\alpha^2+\beta^2+\gamma^2)+\alpha\beta\gamma^2}{\alpha^2+\beta^2+\gamma^2+\alpha\beta+\beta\gamma+\gamma\alpha}.$$

29. An area bounded by the curve $ay^2 = x^3$, the x-axis and the ordinate $x = a$, is immersed in water with the axis in the surface. Show that the coordinates of the centre of pressure are

$$\left(\frac{4}{5}a, \frac{16}{33}a\right)$$

30. The centre of a cube of edge $2l$ is at such a fixed depth h below the surface of the liquid that the cube is completely immersed in any position. For any position of the cube in which two faces remain vertical, P is the centre of pressure of one of the other faces. Show that the greatest distance of P from the centre of the face is $\dfrac{l^2}{3\sqrt{h^2-l^2}}$, the atmospheric pressure being neglected.

31. If the centre of pressure of a triangle ABC, completely immersed in a homogeneous liquid, coincides with the circum-centre, show that the depths of the angular points are as

$$(1-2\cot B\cot C) : (1-2\cot C\cot A) : (1-2\cot A\cot B).$$

32. A sluice-gate is to open automatically by turning about a horizontal axis, as water rises to a fixed level, the upper part turning out and the lower part turning in. If h be the height of the gate and x that of the level of water above the top of the gate when it is to open, show that the axis must be fixed at a height $\dfrac{h}{3}\cdot\dfrac{h+3x}{h+2x}$ from the bottom of the gate.

33. A portion of the side of a vessel in the form of a vertical plane triangle of altitude a, with its vertex uppermost and base horizontal, is movable about a horizontal line in its own plane, whose height above the base is na ($4n$ being less than 1). Prove that this portion will be in equilibrium, if the vessel contains water to a depth $\left[1+n-(1-4n+n^2)^{\frac{1}{2}}\right]a$.

34. One wall of a tank slopes inwards from the bottom at an angle θ to the vertical and contains a triangular trap door of weight W, which is hinged about a horizontal side BB', has the vertex A lower than BB', and can open outwards. The vertical heights of the vertices above the bottom of the tank are a, b, b. Prove that, if water be poured into the tank to a height h so that the trap door is entirely below the surface, it will remain closed, provided that $h < \dfrac{W}{\Delta w}\sin\theta + \dfrac{1}{2}(a+b)$, Δ being the area of the triangle and w, the weight of unit volume of water.

CHAPTER 6

Thrusts on Curved Surfaces

6.1 Introduction

We have seen earlier that the fluid pressure at any point of the surface within the fluid acts along the normal to the surface at that point. Hence, the directions of the pressure at different points on a curved surface change from point to point. Therefore, the fluid thrust on a curved surface cannot be compounded, in general, into a single resultant force.

In discussing the fluid thrust on curved surfaces that are in contact with fluid at rest under gravity, we consider the horizontal and vertical components of the pressure forces and their line of action.

6.2 Resultant Horizontal Thrust

We consider a curved surface immersed in a fluid at rest (FIG. 6.1). The fluid pressure p at any point A acts along the normal to the surface at that point.

We resolve the pressure p along horizontal x-direction and get

$$p_x = p\cos\theta, \tag{6.1}$$

where θ is the angle which the normal to the surface at A makes with the horizontal.

We take an elementary area dS surrounding the point A. Then the fluid thrust along the horizontal x-direction on this small area is

$$p_x dS = p dS \cos\theta. \tag{6.2}$$

Hence, the total horizontal thrust on the whole surface S is

$$F_x = \int p dS \cos\theta, \tag{6.3}$$

the integration being taken over the whole curved surface area.

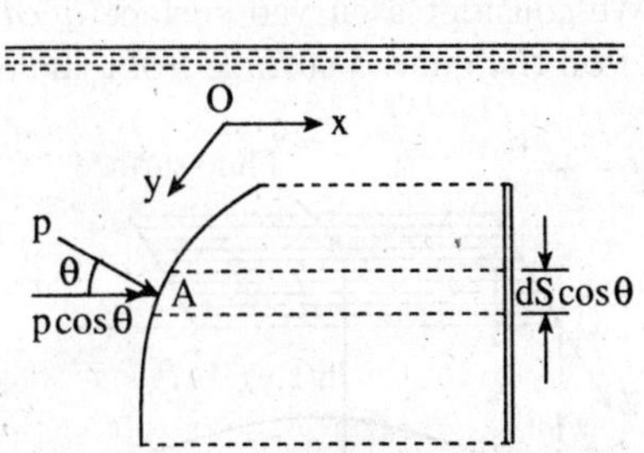

Fig. 6.1

Now, if dS_x be the projection of the area dS on a vertical plane perpendicular to x-direction, then

$$dS_x = dS\cos\theta. \tag{6.4}$$

In view of this, Eqn (6.3) assumes the form

$$F_x = \int p dS_x, \tag{6.5}$$

the integration being taken over the whole of the projected area S_x. This force acts at the centre of pressure of the projected area S_x.

In the same way, if dS_y is the projection of the area dS on a vertical plane perpendicular to y-direction, then the resultant fluid thrust on the surface S in y-direction is

$$F_y = \int p dS_y, \tag{6.6}$$

the integration being taken over the whole of projected area. The resultant force F_y acts at the centre of pressure of the projected area S_y.

Thus, ***the horizontal component of fluid thrust on a given curved surface in a given direction is equal to the total thrust on the projected surface upon a vertical plane perpendicular to the given direction and acts at the centre of pressure of the projected area.***

Eqns (6.5) and (6.6) are precisely identical to the thrusts on plane surfaces and can be expressed as

$$\left.\begin{aligned} F_x &= g\rho S_x \bar{z}_1 \\ F_y &= g\rho S_y \bar{z}_2 \end{aligned}\right\}, \tag{6.7}$$

where $\bar{z}_1, \bar{z}_2$ are the depths of the centres of gravity of the projected areas S_x and S_y below the effective surface respectively.

The two horizontal components F_x and F_y may not, in general, meet at a point, because the centres of pressure of two projected areas may not be at the same depth from the effective surface. However, in case they meet at a point, the resultant of these two horizontal components is given by

$$F = \sqrt{F_x^2 + F_y^2}. \tag{6.8}$$

This force F is known as the ***resultant horizontal thrust.***

6.3 Resultant Vertical Thrust

We consider a curved surface *abcd* immersed in a heavy fluid at rest. The depth of any point A on the curved surface from the effective fluid surface depends on the x- and y-coordinates of the point, so that we write (FIG. 6.2)

$$h = h(x, y). \tag{6.9}$$

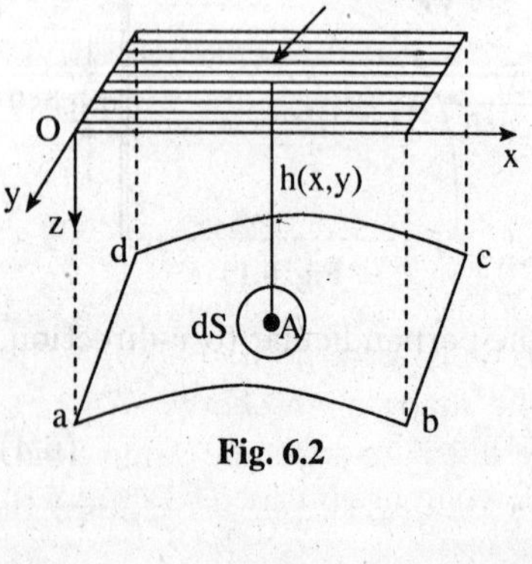

Fig. 6.2

Then the vertical pressure at the point A is

$$p = g\rho h(x, y), \tag{6.10}$$

where ρ is the density of the fluid.

We consider an elementary area dS surrounding the point A. Then the vertical thrust on the top face of the elementary area dS is

$$dF_z = g\rho dS h(x, y). \tag{6.11}$$

Hence, the total thrust on the top face of the curved area S is

$$F_z = \int g\rho dSh(x,y) = g\rho \int dSh(x,y), \tag{6.12}$$

the integration being taken over the whole surface area S.

But the integral $\int dSh(x,y)$ represents the volume of the space between the free surface, the curved surface and the verticals through points a, b, c and d. Denoting this volume by V, we get from Eqn (6.12),

$$F_z = g\rho V. \tag{6.13}$$

The right-hand side of Eqn (6.13) is the weight of the liquid occupying the space of volume V. The fluid within this space is called the ***superincumbent fluid***.

Thus, ***the vertical thrust on any curved surface immersed in a heavy fluid at rest is equal to the weight of the superincumbent liquid and acts vertically downwards through its centre of gravity.***

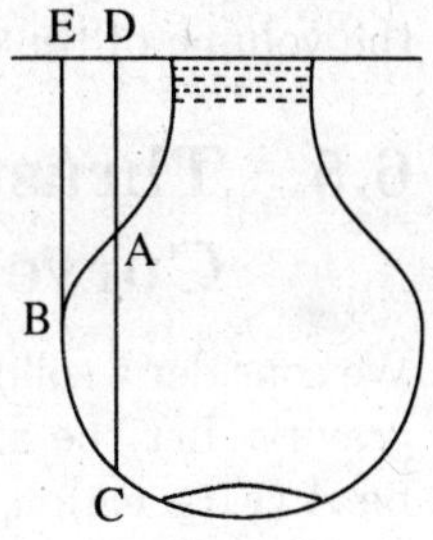

Fig. 6.3

It may happen that some amount of fluid is bounded by a curved surface as shown in the figure (FIG. 6.3).

In this case, the vertical fluid thrust on the curved surface AB is equal to the weight of the superincumbent fluid cylinder $ABED$, but acts in the vertically upward direction.

The vertical fluid thrust on the curved surface BC is equal to the weight of the superincumbent fluid cylinder $CBED$ and acts in the vertically downward direction.

Thus, the vertical resultant thrust on the curved surface ABC is

= weight of fluid cylinder $CBED$—weight of fluid cylinder $ABED$

= weight of fluid cylinder ABC.

Therefore, ***when the fluid is partly above and partly below the curved surface, the resultant vertical thrust is equal to the weight of the fluid contained in that portion and acts vertically downwards through the centre of gravity of the fluid.***

If the horizontal components of thrusts F_x, F_y and the resultant vertical thrust F_z are concurrent, then they can be compounded into a single resultant force R given by

$$R = \sqrt{F_x^2 + F_y^2 + F_z^2} = \sqrt{F^2 + F_z^2}. \tag{6.14}$$

This force R is the **resultant thrust** on the curved surface. If the force R is inclined at an angle ϕ to the horizontal, then

$$\phi = \tan^{-1}\left(\frac{F_z}{F}\right). \tag{6.15}$$

6.4 Principle of Archimedes

We consider a solid body wholly or partially immersed in a heavy fluid at rest under gravity. The solid body remains in equilibrium under the action of two forces—(1) its own weight and (2) the resultant fluid thrust. The resultant fluid thrust on the body does not depend on the material of the composition of the body—it depends only on the shape of the body. We can, therefore, imagine the solid body to be removed and the space filled up with the same fluid. We call this imaginary volume of fluid as ***displaced fluid***. This volume of displaced fluid will remain in equilibrium under the action of its own weight w acting vertically downwards through its centre of gravity and the resultant thrust R of the surrounding fluid upon it. Hence, the resultant thrust R on the volume of the displaced fluid, which is same as the resultant thrust on the solid body, must be equal to the weight of the displaced fluid acting vertically upwards through the centre of gravity of the displaced fluid. This is the ***principle of Archimedes*** which states that:

If a solid body remains wholly or partially immersed in a heavy fluid at rest under gravity, the resultant thrust of the fluid on the solid is equal to the weight of the fluid displaced and acts vertically upwards through the centre of gravity of the fluid displaced.

The resultant thrust on a solid body remaining wholly or partially immersed in a heavy fluid at rest under gravity is called the ***force of buoyancy*** and the centre of gravity of the displaced fluid is known as the ***centre of buoyancy***.

The principle of Archimedes is equally valid in the case of heterogenous fluids. In this case, the volume of the displaced fluid must be of the same varying density as the surrounding one.

6.5 Thrust on Curved Surfaces Bounded by a Plane Curve

We consider a solid body bounded by a plane curve and wholly immersed in a fluid at rest under gravity. Let the area of the curved surface of the body be S and that of the plane boundary be A (FIG. 6.4).

Then, by the principle of Archimedes, the resultant thrust R on the curved surface plus the resultant thrust R_1 on the plane surface must be equal to the weight W of the displaced fluid acting vertically downwards through the centre of gravity of the solid body.

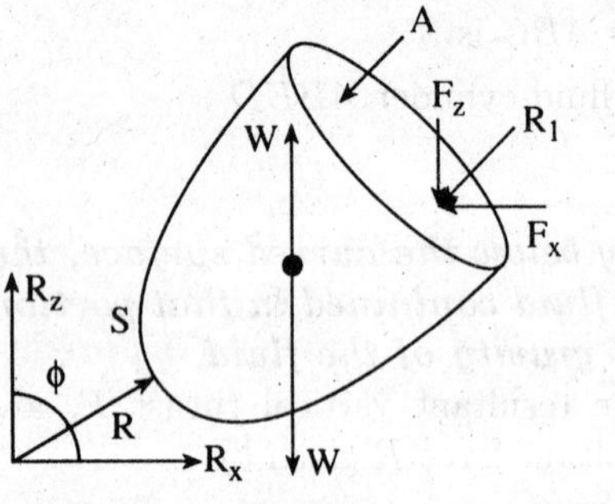

Fig. 6.4

Let the resultant thrust R_1 on the plane area have horizontal and vertical components F_x and F_z respectively and the resultant thrust R on the curved surface S have horizontal and vertical components R_x and R_z respectively. Then, for equilibrium,

$$\left.\begin{aligned} R_x &= F_x \\ R_z - F_z &= W, \text{ i.e., } R_z = W + F_z \end{aligned}\right\} \tag{6.16}$$

Hence, the resultant thrust on the curved surface is

$$R = \sqrt{R_x^2 + R_z^2} = \sqrt{F_x^2 + (W + F_z)^2}. \tag{6.17}$$

If the force R is inclined at an angle ϕ to the vertical, then $\tan\phi = \dfrac{R_x}{R_z} = \dfrac{F_x}{W + F_z}$.

That is

$$\phi = \tan^{-1}\left(\frac{F_x}{W + F_z}\right). \tag{6.18}$$

6.6 Resultant Thrust on any Surface under Arbitrary Forces

We consider a curved surface of area S immersed in a heavy fluid at rest under the action of any given forces (FIG. 6.5).

We take an elementary area dS around a point $P(x, y, z)$. Let p be the pressure at this point in a direction normal to the surface at P. Then the thrust on this elementary area is pdS. Let l, m, n be the direction cosines of the normal to the surface at the point P. Then the components of the thrust pdS along the Cartesian coordinate axes are

$$dX = lpdS, \; dY = mpdS, \; dZ = npdS \tag{6.19}$$

Hence, the components of the resultant thrust along the coordinate axes are

$$\left.\begin{aligned} X &= \int lpdS \\ Y &= \int mpdS \\ Z &= \int npdS \end{aligned}\right\} \tag{6.20}$$

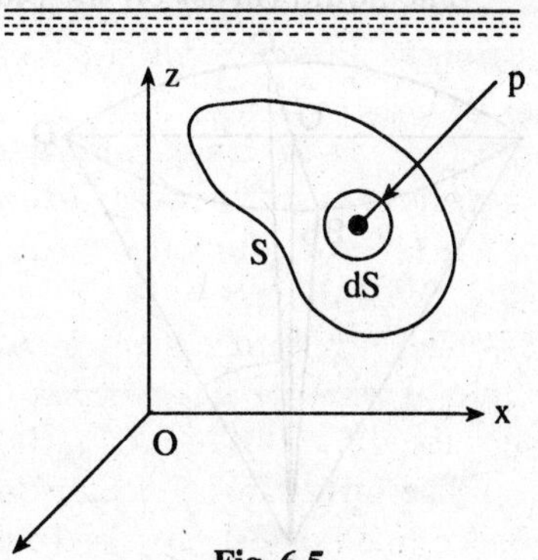

Fig. 6.5

the integrations being taken over the whole surface S.

Similarly, if L, M, N be the couples about x-, y- and z-axes respectively, then we have

$$\left.\begin{aligned} L &= \int (y \cdot npdS - z \cdot mpdS) = \int p(ny - mz)dS \\ M &= \int (z \cdot lpdS - x \cdot npdS) = \int p(lz - nx)dS \\ N &= \int (x \cdot mpdS - y \cdot lpdS) = \int p(mx - ly)dS \end{aligned}\right\} \tag{6.21}$$

The resultant thrust reduces to a single force, if

$$LX + MY + NZ = 0 \tag{6.22}$$

but

$$X^2 + Y^2 + Z^2 \neq 0. \tag{6.23}$$

The system of forces can always be reduced to a wrench, i.e., a single force together with a couple whose axis is along the direction of the force. This axis is called ***Poinsot's Central axis***.

The pitch of the wrench in this case is given by

$$\frac{LX + MY + NZ}{X^2 + Y^2 + Z^2}, \tag{6.24}$$

which shows that the wrench reduces to a single force when pitch is zero and to a couple when the pitch is infinite.

The equation of the central axis is given by

$$\frac{L - yZ + zY}{X} = \frac{M - zX + xZ}{Y} = \frac{N - xY + yX}{Z}. \tag{6.25}$$

We now consider some examples.

EXAMPLE 6.1 *A solid right circular cone is divided into two parts by a plane through its axis and one of these portions is just immersed in a liquid with vertex downwards. Find the resultant thrust on its curved surface and show that it is inclined at an angle of*

$$\tan^{-1}\left(\frac{\pi}{2}\tan\alpha\right)$$

to the horizontal, where α is the semi-vertical angle of the cone.

Solution. Let OA be the axis of the cone and ABC be the vertical plane through the axis dividing the cone into two parts (FIG. 6.6). Let the half-portion of the cone $ABCD$ be immersed in water.

The liquid thrust on the plane surface ABC will be in the direction OD. The solid half-cone being symmetrical about the plane AOD, the horizontal thrust on the curved surfaces ABD and ACD along BC will be equal in magnitude but opposite in directions and hence they nullify each other. Thus, the only horizontal thrust is on the plane surface ABC.

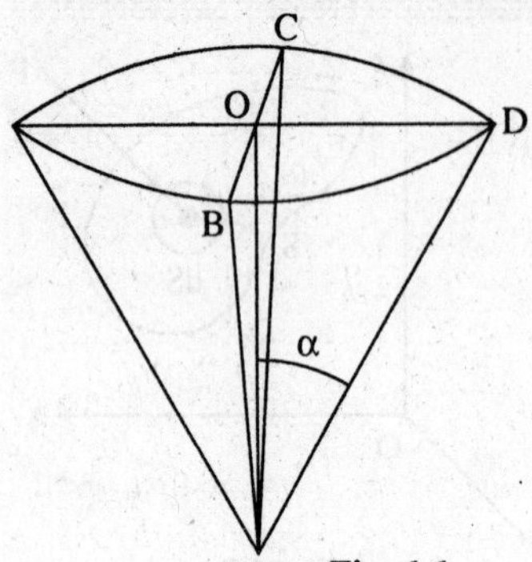

Fig. 6.6

Let $\rho =$ density of the liquid,

$h =$ height of the cone and

$r =$ radius of the circular base.

Then,

$$h = r\cot\alpha, \tag{1}$$

where α is the semi-vertical angle of the cone.

If X be the horizontal thrust on the plane surface ABC, then

$$X = g\rho\cdot\frac{1}{2}(2rh)\cdot\frac{h}{3} = \frac{1}{3}g\rho rh^2, \text{ i.e., } X = \frac{1}{3}g\rho r^3\cot^2\alpha \text{ [by using (1)]}. \tag{2}$$

Let Z be the vertical thrust which is equal to the weight of the liquid displaced. Then

$$Z = g\rho\cdot\frac{1}{2}\cdot\frac{1}{3}\pi r^2 h = \frac{1}{6}\pi g\rho r^3\cot\alpha. \tag{3}$$

Now, the horizontal thrust X acts along the normal to the plane ABC through its centre of pressure which lies on AO. Thus, the line of action of X remains on the plane AOD.

Again, the resultant vertical thrust Z acts vertically upwards through the centre of gravity of the half-cone which remains on the symmetrical plane AOD.

Thus, the two forces X and Z are coplanar and are not parallel. Hence, they must be concurrent and can be compounded into a single resultant force R. Thus,

$$R = \sqrt{X^2+Z^2} = \frac{1}{6}g\rho r^3\cot\alpha\sqrt{\pi^2+4\cot^2\alpha}. \tag{4}$$

If ϕ be the inclination of this resultant thrust R to the horizontal, then

$$\tan\phi = \frac{Z}{X} = \frac{\frac{1}{6}\pi g\rho r^3\cot\alpha}{\frac{1}{3}g\rho r^3\cot^2\alpha} = \frac{\pi}{2}\tan\alpha.$$

Hence, $\phi = \tan^{-1}\left(\frac{\pi}{2}\tan\alpha\right)$ **(Proved)**.

EXAMPLE 6.2 *A hemispherical bowl is filled with water. Find the horizontal fluid thrust on one half of the surface divided by a vetical diametric plane.*

Also show that it is $\frac{1}{\pi}$ *of the magnitude of the resultant fluid thrust on the whole surface.*

Solution. Let the hemispherical bowl is divided into two equal parts by the vertical plane through its diameter AB (FIG. 6.7).

Let $R =$ radius of the hemisphere, $\rho =$ density of water, and $X =$ horizontal thrust along OC. Then,

$$X = \text{thrust on the semi-circle } ABD = g\rho\cdot\frac{1}{2}\cdot\pi R^2\cdot\frac{4R}{3\pi} = \frac{2}{3}g\rho R^3 \tag{1}$$

There will be no horizontal thrust along AB, since the horizontal thrusts on quarter-circle COD from two opposite sides will be equal in magnitude but opposite in directions. Hence, the resultant horizontal thrust is X as given by (1).

Now, if F be the resultant thrust on the whole surface of the hemisphere, then it is equal to the weight of the water contained.

Hence,

$$F = \frac{2}{3} g\rho \cdot \pi R^2 \cdot R = \frac{2}{3} g\rho\pi R^3 \quad (2)$$

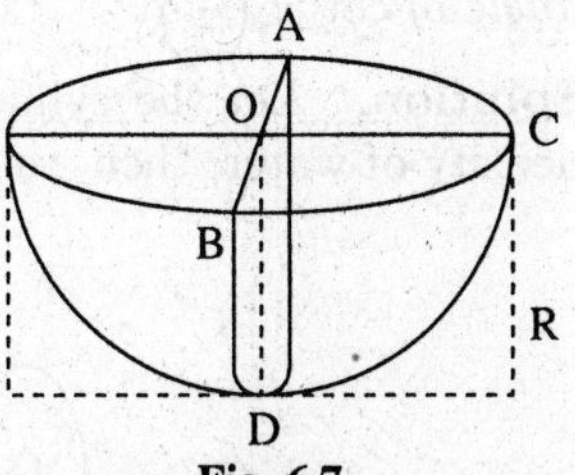

Fig. 6.7

Hence,

$$\frac{X}{F} = \frac{\frac{2}{3} g\rho R^3}{\frac{2}{3} \cdot g\rho\pi R^3} = \frac{1}{\pi} \quad \textbf{(Proved)}.$$

EXAMPLE 6.3 *A solid hemisphere is placed with its base inclined to the surface of a liquid, in which it is just totally immersed, at a given angle* α. *Show that the resultant thrust on the curved portion of the surface will be equal to twice the weight of the liquid displaced, if* $\tan\alpha = 2$.

Solution. Let the hemisphere ABC be just immersed in a fluid of density ρ with its base plane inclined at an angle α (FIG. 6.8). Let R be the radius of the circular base. Then the normal thrust on the plane base is

$$F = g\rho \cdot \pi R^2 \cdot R \sin\alpha, \text{ i.e., } F = g\rho\pi R^3 \sin\alpha \quad (1)$$

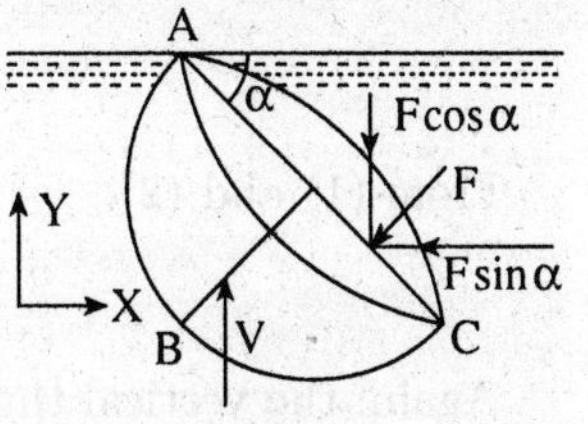

Fig. 6.8

The resultant vertical thrust V on the hemisphere is equal to the weight of the displaced liquid acting vertically upwards and is given by

$$V = \frac{2}{3}\pi R^3 g\rho. \quad (2)$$

Let X = horizontal component and Y = vertical component of the resultant thrust on the curved surface of the hemisphere. Then, for equilibrium,

$$\left.\begin{array}{r} X = F\sin\alpha \\ Y - F\cos\alpha = V, \text{ i.e., } Y = V + F\cos\alpha \end{array}\right\} \quad (3)$$

Hence, the resultant thrust R on the curved surface is

$$R = \sqrt{X^2 + Y^2} = \sqrt{V^2 + F^2 + 2VF\cos\alpha} \quad (4)$$

By condition given,

$$R = 2V, \text{ i.e., } R^2 = 4V^2 \quad (5)$$

From (4) and (5), we get

$$V^2 + F^2 + 2VF\cos\alpha = 4V^2 \text{ or, } F^2 + 2VF\cos\alpha = 3V^2 \quad (6)$$

Putting here the values of F and V from (1) and (2), we get

$$g^2\rho^2\pi^2 R^6 \sin^2\alpha + 2\left(\frac{2}{3}\pi R^3 g\rho\right)(g\rho\pi R^3 \sin\alpha)\cos\alpha = 3\left(\frac{4}{9}\pi^2 R^6 g^2 \rho^2\right)$$

$$\text{or, } 3\sin^2\alpha + 4\sin\alpha\cos\alpha - 4 = 0, \text{ or, } 3\tan^2\alpha + 4\tan\alpha - 4\sec^2\alpha = 0$$

$$\text{or, } 3\tan^2\alpha + 4\tan\alpha - 4(1 + \tan^2\alpha) = 0$$

$$\text{or, } \tan^2\alpha - 4\tan\alpha + 4 = 0, \text{ or, } (\tan\alpha - 2)^2 = 0.$$

Hence, $\tan\alpha = 2$. **(Proved)**

EXAMPLE 6.4 *A cylindrical pipe of circular cross section is half-full of water. If the pipe be imagined to be divided into two halves by a vertical plane along the middle, show that water will tend to push them horizontally with a force W/π, where W is the weight of the water contained. Show that the resultant thrust of water on either half of the pipe makes with the vertical an angle of* $\cot^{-1}\left(\frac{\pi}{2}\right)$.

Solution. Let the cylindrical pipe be of length l and radius of cross section R. If ρ be the density of water, then weight of water contained in the half-full of pipe is (FIG. 6.9)

$$W = g\rho \cdot \frac{1}{2}\pi R^2 \cdot l = \frac{1}{2} g\rho\pi R^2 l \tag{1}$$

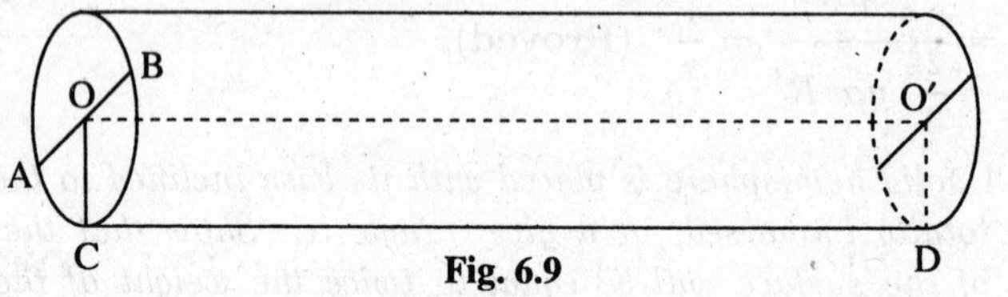

Fig. 6.9

The pipe is imagined to be divided into two halves by the vertical plane $OCDO'$. The horizontal thrust on curved surfaces on either side of this vertical plane $OCDO'$ is equal to the whole pressure on the rectangular area $OCDO'$ of length l and depth R. Hence,

$$X = (l \cdot R) \cdot \left(\frac{1}{2}R\right) \cdot g\rho = \frac{1}{2} g\rho R^2 l. \tag{2}$$

From (1) and (2),

$$X = \frac{W}{\pi} \textbf{ (Proved)}.$$

Again, the vertical thrust Y on the curved surface on either side of the plane $OCDO'$ is

$$Y = \frac{1}{4}\pi R^2 l g\rho = \frac{1}{2} W \text{ [by (1)]} \tag{3}$$

Hence, the resultant thrust on the curved surface is

$$R = \sqrt{X^2 + Y^2} = W\sqrt{\frac{1}{4} + \frac{1}{\pi^2}}, \text{ i.e., } R = \frac{W}{2\pi}\sqrt{4 + \pi^2} \tag{4}$$

If the resultant thrust R makes an angle θ to the vertical, then

$$\tan\theta = \frac{X}{Y} = \frac{W/\pi}{W/2} = \frac{2}{\pi}.$$

Hence,

$$\cot\theta = \frac{\pi}{2}, \text{ or, } \theta = \cot^{-1}\left(\frac{\pi}{2}\right). \quad \textbf{(Proved)}$$

EXAMPLE 6.5 *A solid right circular cone of vertical angle 2α is just immersed in water so that one generator is in the surface of the liquid.*

Prove that the resultant thrust on the curved surface of the cone is to the weight of the fluid displaced by the cone as $\sqrt{1 + 3\sin^2\alpha} : 1$, and that it is inclined to the axis of the cone at an angle of $\cot^{-1}(2\tan\alpha)$.

Solution. Let the right circular cone of height h be just immersed in water of density ρ with a generator on the surface (FIG. 6.10). The thrust on the plane circular base is

$$F = g\rho(\text{area}) \times \text{depth of c.g.}$$
$$= g\rho(\pi h^2 \tan^2\alpha) \cdot h\sin\alpha = g\rho\pi h^3 \tan^2\alpha \sin\alpha. \tag{1}$$

Weight of the displaced liquid is

$$W = g\rho\left(\frac{1}{3}\pi h^2 \tan^2\alpha\right)\cdot h = \frac{1}{3}g\rho\pi h^3 \tan^2\alpha. \quad (2)$$

Let X and Y be the horizontal and vertical components of the resultant thrust on the curved surface. Then by Archimedes principle, we have

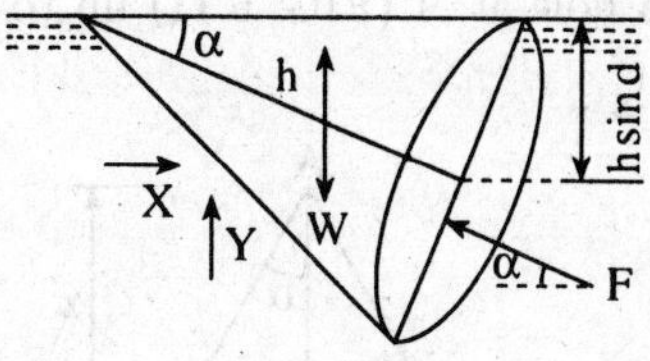

Fig. 6.10

$X = F\cos\alpha = g\rho\pi h^3 \tan^2\alpha \sin\alpha\cos\alpha$

$Y + F\sin\alpha = W$

i.e., $Y = W - F\sin\alpha = \frac{1}{3}g\rho\pi h^3\tan^2\alpha - g\rho\pi h^3 \tan^2\alpha\sin^2\alpha$

i.e., $Y = g\rho\pi h^3 \tan^2\alpha\left(\frac{1}{3} - \sin^2\alpha\right) \quad (3)$

If R be the resultant thrust on the curved surface, then

$$R = \sqrt{X^2+Y^2}$$

$$= g\rho\pi h^3\tan^2\alpha\sqrt{\sin^2\alpha\cos^2\alpha + \left(\frac{1}{3} - \sin^2\alpha\right)^2}$$

$$= g\rho\pi h^3\tan^2\alpha\left[\sin^2\alpha(1-\sin^2\alpha) + \frac{1}{9} - \frac{2}{3}\sin^2\alpha + \sin^4\alpha\right]^{\frac{1}{2}}$$

$$= \frac{1}{3}g\rho\pi h^3\tan^2\alpha\sqrt{1+3\sin^2\alpha} \quad (4)$$

From (2) and (4), we get

$$\frac{R}{W} = \sqrt{1+3\sin^2\alpha}. \quad \textbf{(Proved)}$$

If the resultant thrust R be inclined at an angle ϕ to the horizontal, then

$$\tan\phi = \frac{Y}{X}, \text{ i.e., } \tan\phi = \frac{g\rho\pi h^3\tan^2\alpha\left(\frac{1}{3} - \sin^2\alpha\right)}{g\rho\pi h^3\tan^2\alpha\sin\alpha\cos\alpha} = \frac{1-3\sin^2\alpha}{3\sin\alpha\cos\alpha}. \quad (5)$$

Hence, the inclination of the direction of the thrust R to the axis of the cone is $(\phi+\alpha)$. Then,

$$\tan(\phi+\alpha) = \frac{\tan\phi + \tan\alpha}{1-\tan\phi\tan\alpha}. \quad (6)$$

Putting the value of $\tan\phi$ from (5) into (6), we get

$$\tan(\phi+\alpha) = \frac{\frac{1-3\sin^2\alpha}{3\sin\alpha\cos\alpha} + \frac{\sin\alpha}{\cos\alpha}}{1 - \frac{1-3\sin^2\alpha}{3\sin\alpha\cos\alpha}\cdot\frac{\sin\alpha}{\cos\alpha}} = \frac{\frac{1-3\sin^2\alpha+3\sin^2\alpha}{3\sin\alpha\cos\alpha}}{\frac{3\cos^2\alpha - 1 + 3\sin^2\alpha}{3\cos^2\alpha}}$$

$$= \frac{1}{3\sin\alpha\cos\alpha}\cdot\frac{3\cos^2\alpha}{2} = \frac{1}{2}\frac{\cos\alpha}{\sin\alpha}$$

Hence,

$$\cot(\phi+\alpha) = \frac{2\sin\alpha}{\cos\alpha} = 2\tan\alpha$$

$$\therefore \phi+\alpha = \cot^{-1}(2\tan\alpha). \quad \textbf{(Proved)}$$

EXAMPLE 6.6 *A hollow cone is placed with its vertex upwards on a horizontal table and a liquid is poured in through a small hole in the vertex. If the cone begins to rise when the weight of the liquid poured in it is equal to its own weight, prove that its weight is to the weight of the liquid to fill the cone as* $(9 - 3\sqrt{3}) : 4$.

Solution. Let ABC be the cone placed on a horizontal table and a liquid is poured in through a hole at A (FIG. 6.11) up to the level DE.

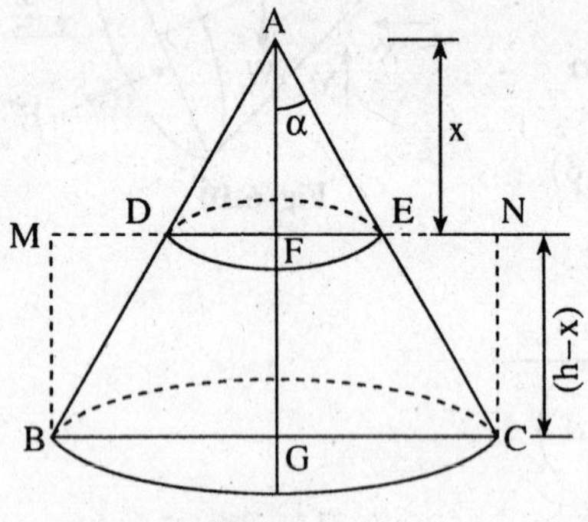

Fig. 6.11

Let h = height of the cone, α = semi-vertical angle of the cone, ρ = density of the liquid.

Then, from the figure, $AG = h$. Let $AF = x, GC = R$ and $EF = r$. Then

$$r = x\tan\alpha,\ R = h\tan\alpha \tag{1}$$

Then, volume of the liquid in the frustrum $DBCE$ is

$$V = \frac{1}{3}\pi\tan^2\alpha(h^3 - x^3) \tag{2}$$

so that its weight is

$$W = V\rho g = \frac{1}{3}\pi\tan^2\alpha(h^3 - x^3)g\rho \tag{3}$$

The vertical thrust P on the curved surface $DBCE$

= weight of the superincumbent liquid

= weight of the liquid in the cylinder $BMNC$ of height $(h - x)$ minus the weight of liquid contained in the frustrum $DBCE$.

Thus,

$$P = \pi h^2\tan^2\alpha(h - x)g\rho - \frac{1}{3}\pi\tan^2\alpha(h^3 - x^3)g\rho. \tag{4}$$

The cone begins to rise when the weight of the liquid poured in is equal to the vertical thrust on the surface of the cone in contact with the liquid. Hence,

$$P = W. \tag{5}$$

Putting values from (3) and (4), we get

$$\pi h^2\tan^2\alpha(h - x)g\rho - \frac{1}{3}\pi\tan^2\alpha(h^3 - x^3)g\rho = \frac{1}{3}\pi\tan^2\alpha(h^3 - x^3)g\rho$$

$$\begin{aligned}
\text{or, }\ & h^2(h - x) - \frac{1}{3}(h^3 - x^3) = \frac{1}{3}(h^3 - x^3)\\
\text{or, }\ & h^2(h - x) = \frac{2}{3}(h^3 - x^3)\\
\text{or, }\ & 3h^3 - 3h^2x + 2x^3 - 2h^3 = 0\\
\text{or, }\ & (h^3 - h^2x) - 2h^2x + 2x^3 = 0\\
\text{or, }\ & h^2(h - x) - 2x(h^2 - x^2) = 0\\
\text{or, }\ & (h - x)[h^2 - 2x(h + x)] = 0\\
\text{or, }\ & 2x^2 + 2xh - h^2 = 0,\ \text{since } h \neq x.
\end{aligned}$$

$$\text{Hence, } x = \frac{-2h \pm \sqrt{4h^2 + 8h^2}}{4} = \frac{-2h \pm 2\sqrt{3}h}{4}.$$

That is,

$$x = \frac{(\sqrt{3}-1)h}{2}, \text{ or, } \frac{x}{h} = \frac{\sqrt{3}-1}{2}, \tag{6}$$

since x cannot be negative.

Now, the weight of the liquid to fill the cone is

$$W' = \frac{1}{3}\pi \tan^2 \alpha (h^3 g\rho) \tag{7}$$

Hence, the ratio

$$\frac{W}{W'} = \frac{\frac{1}{3}\pi \tan^2 \alpha (h^3 - x^3) g\rho}{\frac{1}{3}\pi \tan^2 \alpha (h^3 g\rho)} = \frac{h^3 - x^3}{h^3} = 1 - \left(\frac{x}{h}\right)^3. \tag{8}$$

Putting the value of $\frac{x}{h}$ from (6),

$$\frac{W}{W'} = 1 - \left(\frac{\sqrt{3}-1}{2}\right)^3$$

$$= 1 - \frac{1}{8}(3\sqrt{3} - 3 \cdot 3 \cdot 1 + 3 \cdot \sqrt{3} \cdot 1^2 - 1^3)$$

$$= 1 - \frac{1}{8}(6\sqrt{3} - 10) = \frac{8 - 6\sqrt{3} + 10}{8} = \frac{18 - 6\sqrt{3}}{8} = \frac{9 - 3\sqrt{3}}{4}.$$

Thus, the weight of the liquid poured in is to the weight of liquid to fill the cone as

$$(9 - 3\sqrt{3}) : 4.$$

EXAMPLE 6.7 *A solid sphere rests on a horizontal plane and is just immersed in a liquid. It is then divided into two parts by a plane drawn through its vertical diameter. If ρ is the density of the solid and δ that of the liquid, prove that the two parts will not separate, provided*

$$\delta > \frac{1}{4}\rho.$$

Solution. Let a sphere of radius R resting on a horizontal plane be just immersed in a liquid of density δ. We consider the forces acting on the hemisphere obtained by dividing the sphere by a plane through its vertical diameter (FIG. 6.12).

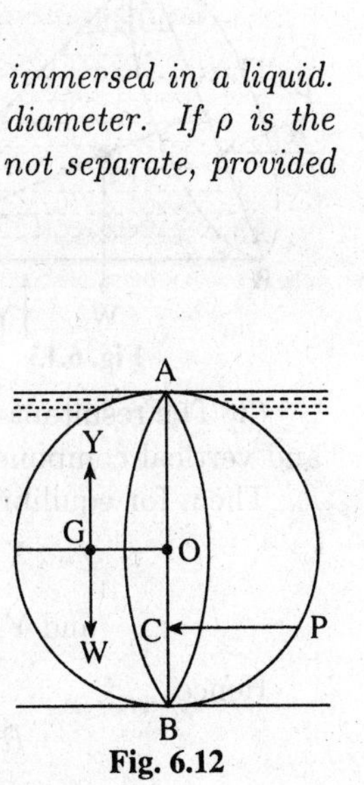

Fig. 6.12

The forces are:

(1) The weight W of the hemisphere is

$$W = \frac{2}{3}\pi R^3 g\rho \tag{1}$$

acting vertically downwards through its centre of gravity G, where

$$OG = \frac{3}{8}R. \tag{2}$$

(2) The horizontal thrust P on the vertical circular area is equal to

$$P = (\pi R^2) \cdot R \cdot g\delta = \pi R^3 g\delta \tag{3}$$

acting at the centre of pressure C of the circle, where

$$OC = \frac{R}{4}, \text{ i.e., } BC = \frac{3}{4}R. \tag{4}$$

(3) The vertical thrust Y on the curved surface which is equal to the weight of the liquid displaced. Thus

$$Y = \frac{2}{3}\pi R^3 g\delta \tag{5}$$

acting vertically upwards through G.

The forces P and Y tend to rotate the hemisphere in clockwise direction which helps to keep the two parts joined; but the force W tends to rotate in anti-clockwise direction which tries to separate the two parts. Hence the two parts will not separate, if the sum of the moments of forces P and Y about B is greater than the moment of W about the same point B. Thus, for no separation,

$$P \cdot BC + Y \cdot OG > W \cdot OG$$

$$\text{i.e.,}\quad (\pi R^3 g\delta)\frac{3}{4}R + \left(\frac{2}{3}\pi R^3 g\delta\right)\frac{3}{8}R \geq \left(\frac{2}{3}\pi R^3 g\rho\right)\frac{3}{8}R$$

$$\text{i.e.,}\quad \frac{3}{4}\delta + \frac{1}{4}\delta \geq \frac{1}{4}\rho,\ \text{i.e.,}\ \delta \geq \frac{1}{4}\rho. \quad \textbf{(Proved)}$$

EXAMPLE 6.8 *A cone whose vertical angle is 2α has its lowest generator horizontal and is filled with a liquid. Prove that the resultant pressure on the curved surface is $\sqrt{1+15\sin^2\alpha}$ times the weight of the liquid.*

Solution. Let ABC be the cone filled with a liquid of density ρ with its lowest generator AB horizontal (FIG. 6.13). Let h be the height of the cone.

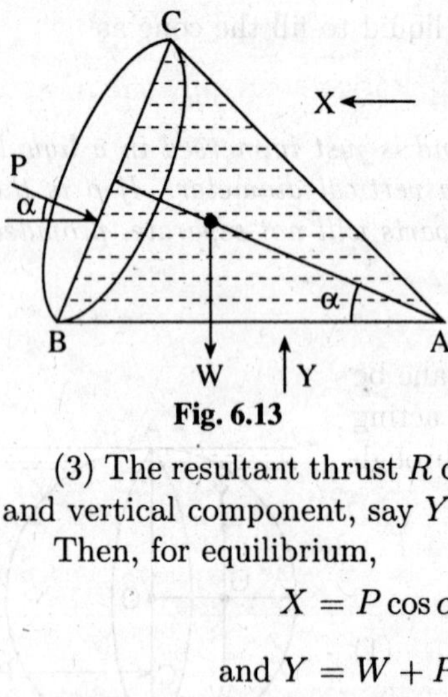

Fig. 6.13

The forces acting on the cone are:

(1) The thrust P on the plane circular face which is equal to

$$P = (\pi h^2 \tan^2\alpha)\cdot(h\sin\alpha)\cdot g\rho$$
$$= \pi g\rho h^3 \tan^2\alpha \sin\alpha \tag{1}$$

acting normally to the surface.

(2) The weight W of the displaced liquid which is equal to

$$W = \frac{1}{3}(\pi h^2 \tan^2\alpha)\cdot h \cdot g\rho = \frac{1}{3}\pi g\rho h^3 \tan^2\alpha. \tag{2}$$

(3) The resultant thrust R on the curved surface which has got horizontal component, say X and vertical component, say Y.

Then, for equilibrium,

$$X = P\cos\alpha = \pi g\rho h^3 \tan^2\alpha \sin\alpha\cos\alpha \tag{3}$$

$$\text{and } Y = W + P\sin\alpha = \frac{1}{3}\pi g\rho h^3\tan^2\alpha + \pi g\rho h^3 \tan^2\alpha\sin^2\alpha. \tag{4}$$

Hence,

$$R = \sqrt{X^2+Y^2}$$
$$= \pi g\rho h^3 \tan^2\alpha\sqrt{\sin^2\alpha\cos^2\alpha + \left(\frac{1}{3}+\sin^2\alpha\right)^2}$$
$$= \pi g\rho h^3 \tan^2\alpha\sqrt{\sin^2\alpha(1-\sin^2\alpha) + \frac{1}{9} + \frac{2}{3}\sin^2\alpha + \sin^4\alpha}$$
$$= \frac{1}{3}\pi g\rho h^3 \tan^2\alpha\sqrt{1+15\sin^2\alpha} \tag{5}$$

From (2) and (5),

$$\frac{R}{W} = \frac{\frac{1}{3}\pi g\rho h^3 \tan^2\alpha\sqrt{1+15\sin^2\alpha}}{\frac{1}{3}\pi g\rho h^3\tan^2\alpha} = \sqrt{1+15\sin^2\alpha}. \quad \textbf{(Proved)}$$

EXAMPLE 6.9 *A hemisphere of radius R is entirely immersed in a liquid of density ρ so that its diametral plane makes an angle φ with the horizontal and its centre is at a depth h from the effective surface. Find the resultant thrust on the curved surface and determine its directions.*

Solution. There are two possible positions of the hemisphere which we consider separately.

Case I : The base of the hemisphere is upwards

This is shown in the figure (FIG. 6.14). The forces acting on the hemisphere are:

(1) Thrust P on the plane circular base which is equal to

$$P = (\pi R^2)\cdot hg\rho = \pi g\rho R^2 h \qquad (1)$$

acting normally at the centre of pressure of the base.

(2) Weight of the displaced liquid

$$W = \left(\frac{2}{3}\pi R^3\right) g\rho = \frac{2}{3}\pi g\rho R^3 \qquad (2)$$

acting vertically downwards through the centre of gravity G.

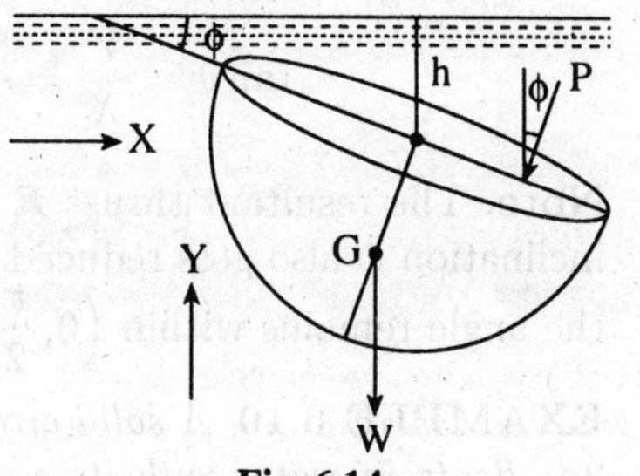

Fig. 6.14

(3) Resultant thrust R on the curved surface which has horizontal and vertical components, say X and Y respectively.

Hence, by Archimedes principle, we have

$$X = P\sin\phi = \pi g\rho R^2 h\sin\phi \qquad (3)$$

$$Y - P\cos\phi = W$$

$$\text{i.e., } Y = W + P\cos\phi = \frac{2}{3}\pi g\rho R^3 + \pi g\rho R^2 h\cos\phi = \frac{1}{3}\pi R^2 g\rho(2R+3h\cos\phi) \qquad (4)$$

Hence, the resultant thrust R on the curved surface is

$$R = \sqrt{X^2+Y^2}$$

$$= \pi g\rho R^2\sqrt{h^2\sin^2\phi + \frac{1}{9}(2R+3h\cos\phi)^2}$$

$$= \pi g\rho R^2\sqrt{h^2\sin^2\phi + \frac{1}{9}(4R^2+12Rh\cos\phi+9h^2\cos^2\phi)}$$

$$= \pi g\rho R^2\sqrt{h^2+\frac{4}{9}R^2+\frac{4}{3}Rh\cos\phi} \qquad (5)$$

If the resultant force R makes an angle ψ with the horizontal, then

$$\tan\psi = \frac{Y}{X} = \frac{\frac{1}{3}\pi g\rho R^2(2R+3h\cos\phi)}{\pi g\rho R^2 h\sin\phi}, \text{ i.e., } \tan\psi = \frac{2R+3h\cos\phi}{3h\sin\phi}$$

$$\therefore \psi = \tan^{-1}\left[\frac{2R+3h\cos\phi}{3h\sin\phi}\right] \qquad (6)$$

Case II: The base of the hemisphere is downwards

This is shown in the figure (FIG. 6.15). We have also shown the forces acting on the hemisphere. In this case, the equations of equilibrium are:

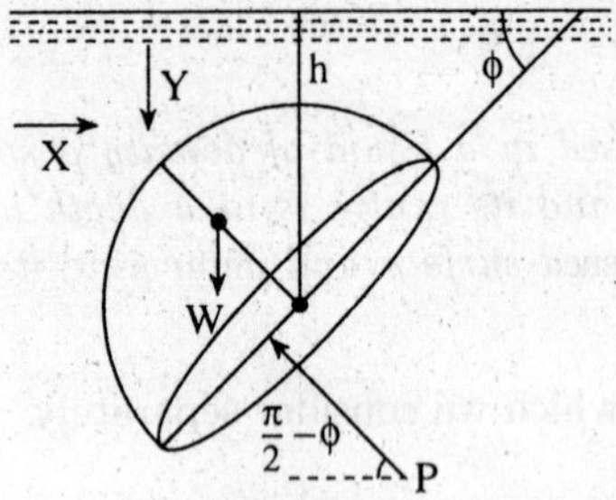

Fig. 6.15

$$X = P\sin\phi = \pi g\rho R^2 h\sin\phi \qquad (7)$$

$$Y = P\cos\phi - W = \pi g\rho R^2 h\cos\phi - \frac{2}{3}\pi g\rho R^3$$

$$= \frac{1}{3}\pi g\rho R^2(3h\cos\phi - 2R) \qquad (8)$$

Hence, the resultant thrust on the curved surface is $R = \sqrt{X^2 + Y^2}$, which, proceeding as before, yields

$$R = \pi g\rho R^2\sqrt{h^2 + \frac{4}{9}R^2 - \frac{4}{3}Rh\cos\phi}. \qquad (9)$$

The angle ψ of inclination of the force R to the horizontal is given by

$$\tan\psi = \frac{Y}{X} = \frac{3h\cos\phi - 2R}{3h\sin\phi}, \text{ i.e., } \psi = \tan^{-1}\left[\frac{3h\cos\phi - 2R}{3h\sin\phi}\right]. \qquad (10)$$

Note. The resultant thrust R on the curved surface becomes less in the Case II. The angle of inclination ψ also gets reduced in this case. This is clear from Eqns (5), (6), (9) and (10), since the angle remains within $\left(0, \frac{\pi}{2}\right)$, for which $\cos\phi$ and $\sin\phi$ assume positive values.

EXAMPLE 6.10 *A solid circular cone of uniform material and height h and of vertical angle 2α, floats in water with its axis vertical and vertex downwards, and a length h′ of the axis is immersed. The cone is bisected by a vertical plane through the axis and the two parts are hinged together at the vertex. Show that the two parts will remain in contact, if* $h' > h\sin^2\alpha$.

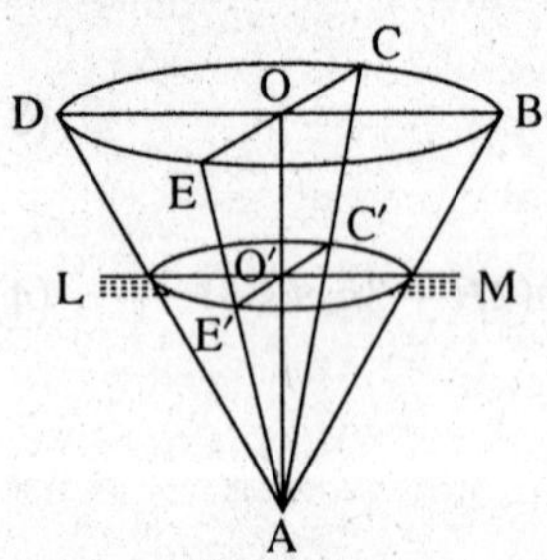

Fig. 6.16

Solution. Let a cone $ABCDE$ of height h float in water with vertex A downwards and axis AO vertical (FIG. 6.16). The cone is divided into two parts by the vertical plane AEC passing through its axis AO.

LM is the free surface of water, then

$$AO = h$$
$$AO' = h'.$$

Let ρ and δ be the densities of the cone and water respectively.

Now, weight of the liquid displaced by the immersed portion of the cone is

$$W_1 = \frac{1}{3}\pi g\delta h'^3\tan^2\alpha. \qquad (1)$$

The weight of the cone is

$$W_2 = \frac{1}{3}\pi g\rho h^3\tan^2\alpha. \qquad (2)$$

Since the cone is floating $W_1 = W_2$,

$$\text{i.e., } \frac{1}{3}\pi g\delta h'^3\tan^2\alpha = \frac{1}{3}\pi g\rho h^3\tan^2\alpha$$

$$\therefore \delta h'^3 = \rho h^3 \qquad (3)$$

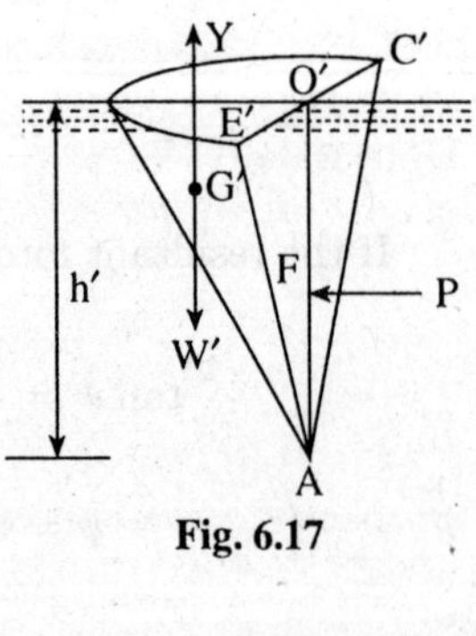

Fig. 6.17

The forces acting on the half-cone are (FIG. 6.17).

(1) The horizontal thrust P on the triangular plane area $AE'C'$ which is equal to

$$P = g\delta \cdot \frac{1}{2}(2h' \tan\alpha \cdot h') \cdot \frac{h'}{3} = \frac{1}{3}h'^3 \delta \tan\alpha \tag{4}$$

acting through the centre of pressure F of the triangle, where

$$O'F = \frac{1}{2}h' = AF. \tag{5}$$

(2) The weight W of the water displaced by the half-cone of height h', which is equal to the vertical thrust Y. Thus

$$Y = W = \frac{1}{6}\pi g\delta h'^3 \tan^2\alpha \tag{6}$$

acting vertically upwards through the centre of gravity G' of the half-cone of height h', so that G' is at a distance of

$$d = \frac{h' \tan\alpha}{\pi} \tag{7}$$

from the axis of the cone.

(3) The weight of the half-cone of height h equal to

$$W' = \frac{1}{6}\pi g\rho h^3 \tan^2\alpha \tag{8}$$

acting vertically through its centre of gravity which is at a distance

$$d' = \frac{h \tan\alpha}{\pi} \tag{9}$$

from the axis of the cone.

The vertical thrust Y and horizontal thrust P tend to keep the two parts of the cone together, whereas the force W' acting vertically downwards tends to separate the two parts.

Thus, in the limit, the two parts will not be separated, if the sum of moments of forces Y and P about the point A is greater than the moment of the force W' about same point A.

Hence,

$$P \cdot \frac{h'}{2} + Y\dot{d} \geq W' \cdot d'$$

or, $\frac{1}{3}h'^3\delta \tan\alpha \cdot \frac{h'}{2} + \frac{1}{6}\pi g\delta h'^3 \tan^2\alpha \cdot \frac{h'\tan\alpha}{\pi} \geq \frac{1}{6}\pi g\rho h^3 \tan^2\alpha \cdot \frac{h\tan\alpha}{\pi}$

or, $h'^3\delta(h' + h'\tan^2\alpha) \geq h^3\rho \cdot h\tan^2\alpha$

or, $h' + h'\tan^2\alpha \geq h\tan^2\alpha$ [using (3)]

or, $h'\sec^2\alpha \geq h\tan^2\alpha$

$\therefore$ $h' \geq h\sin^2\alpha$. **(Proved)**

EXAMPLE 6.11 *A solid octant of a sphere of radius R is immersed in a liquid with one plane side in the surface. Prove that the resultant pressure on the curved surface is*

$$\left(1 + \frac{8}{\pi^2}\right)^{1/2}$$

times the weight of the displaced liquid by the octant. Also show that the depth of the centre of pressure for the curved surface is $\dfrac{\pi R}{\sqrt{\pi^2 + 8}}$.

Solution. Let $OABC$ be the octant of a sphere of radius R immersed in a liquid of density ρ with its plane face AOB on the surface, the other two plane faces AOC and BOC being vertical (FIG. 6.18). We have taken the rectangular coordinate system as shown in the figure.

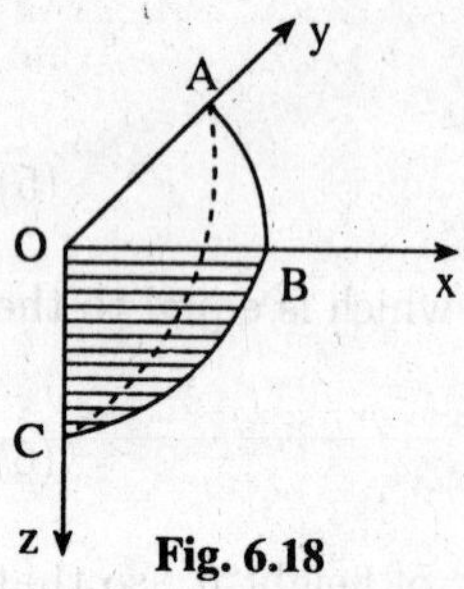

Fig. 6.18

The thrust on the plane face BOC is

$$F_1 = g\rho \times \text{area} \times \text{depth of centre of gravity}$$

$$= g\rho \cdot \frac{1}{4}\pi R^2 \cdot \frac{4R}{3\pi} = \frac{1}{3}g\rho R^3 \tag{1}$$

acting at right angle to the face, i.e., parallel to y-axis.

Similarly, the horizontal thrust on the plane face AOC is

$$F_2 = \frac{1}{3}g\rho R^3 \tag{2}$$

acting parallel to x-axis.

These two horizontal forces F_1 and F_2 meet at a point, their lines of action being at right angles to each other. Hence the resultant horizontal thrust is

$$H = \sqrt{F_1^2 + F_2^2} = \frac{\sqrt{2}}{3}g\rho R^3. \tag{3}$$

Now, the resultant vertical thrust on the whole surface of the solid is equal to the weight W of the displaced liquid acting vertically upwards. Now,

$$W = \frac{1}{8}\left(\frac{4}{3}\pi R^3\right)g\rho = \frac{1}{6}\pi g\rho R^3 \tag{4}$$

If R be the resultant thrust on the curved surface, then

$$R = \sqrt{H^2 + W^2} = g\rho R^3\sqrt{\frac{2}{9} + \frac{\pi^2}{36}} = \frac{1}{6}\pi g\rho R^3\left(1 + \frac{8}{\pi^2}\right)^{1/2}$$

$$= W\left(1 + \frac{8}{\pi^2}\right)^{1/2} \quad \text{[using (4)]}$$

Hence, the resultant thrust R is $\left(1 + \frac{8}{\pi^2}\right)^{1/2}$ times the weight of the displaced liquid.

If the force R makes an angle ϕ with the horizontal, then

$$\tan\phi = \frac{W}{H} = \frac{\frac{1}{6}\pi g\rho R^3}{\frac{\sqrt{2}}{3}g\rho R^3} = \frac{\pi}{2\sqrt{2}} \tag{5}$$

$$\therefore \sin\phi = \frac{\pi}{\sqrt{\pi^2 + 8}}$$

Hence, the depth of the centre of pressure for the curved surface is

$$R\sin\phi = \frac{\pi R}{\sqrt{\pi^2 + 8}}.$$

EXAMPLE 6.12 *A hemispherical bowl is filled with water and two vertical planes are drawn through its central radius, cutting off a semi-lune of the surface. If 2α be the angle between the planes, prove that the angle which the resultant pressure on the surface makes with the vertical is*

$$\tan^{-1}\left(\frac{\sin\alpha}{\alpha}\right).$$

Solution. Let OAC and OBC be the two vertical planes cutting off a semi-lune (FIG. 6.19). Let R be the radius of the hemisphere and ρ be the density of the liquid in the bowl.

The volume of the hemisphere is

$$V = \frac{2}{3}\pi R^3. \quad (1)$$

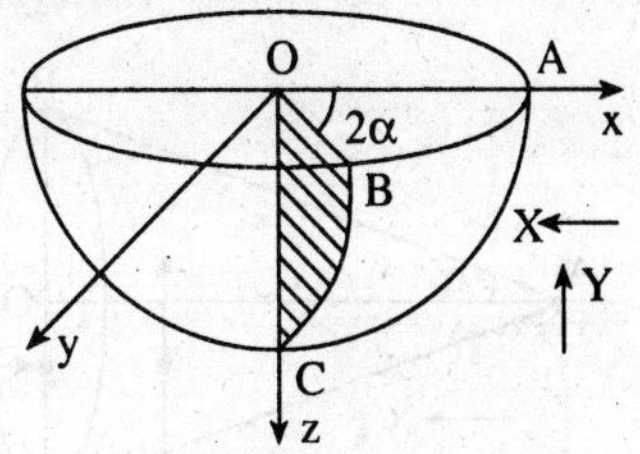

Fig. 6.19

Hence the volume of the solid $OABC$ is

$$V_1 = V \cdot \frac{2\alpha}{2\pi} = \frac{2}{3}\pi R^3 \cdot \frac{2\alpha}{2\pi}$$

$$\text{i.e., } V_1 = \frac{2}{3}\alpha R^3. \quad (2)$$

Now, the horizontal thrust F_1 on the plane AOC is

$$F_1 = g\rho \times \text{area} \times \text{depth of c.g. of the surface}$$
$$= g\rho \times \frac{1}{4}\pi R^2 \times \frac{4R}{3\pi} = \frac{1}{3}g\rho R^3 \quad (3)$$

acting normally on the surface AOC, i.e., parallel to y-axis.

Similarly, the horizontal thrust F_2 on the plane surface OBC is

$$F_2 = \frac{1}{3}g\rho R^3 \quad (4)$$

acting at right angles to the plane OBC. These two forces meet at a point, their lines of action making an angle $(\pi - 2\alpha)$. Hence, the resultant horizontal force is

$$X = \sqrt{F_1^2 + F_2^2 + 2F_1F_2\cos(\pi - 2\alpha)}$$
$$= \frac{1}{3}g\rho R^3\sqrt{1 + 1 - 2\cos 2\alpha} = \frac{1}{3}g\rho R^3 \cdot \sqrt{2} \cdot \sqrt{2}\sin\alpha = \frac{2}{3}g\rho R^3 \sin\alpha \quad (5)$$

The resultant vertical thrust on the curved surface is equal to the weight of the liquid contained, so that its magnitude is

$$Y = V_1 g\rho = \frac{2}{3}g\rho\alpha R^3. \quad (6)$$

Then the resultant thrust on the surface is

$$R = \sqrt{X^2 + Y^2} = \frac{2}{3}g\rho R^3\sqrt{\alpha^2 + \sin^2\alpha}. \quad (7)$$

If ϕ is the angle made by the resultant force R with the vertical, then

$$\tan\phi = \frac{X}{Y} = \frac{\frac{2}{3}g\rho R^3 \sin\alpha}{\frac{2}{3}g\rho R^3\alpha} = \frac{\sin\alpha}{\alpha}$$

$$\text{i.e., } \phi = \tan^{-1}\left(\frac{\sin\alpha}{\alpha}\right). \quad \textbf{(Proved)}$$

EXAMPLE 6.13 *A cone floats with its axis horizontal in a liquid of density double of its own. Find the pressure on its base and prove that, if θ be the inclination to the vertical of the resultant thrust on the curved surface and α be the semi-vertical angle of the cone, then*

$$\tan\theta = \frac{4}{\pi}\tan\alpha.$$

Solution. Since the density of the liquid is double that of the cone, only half of the cone will be immersed in the liquid. Therefore, the axis AO of the cone will be in the surface of the liquid (FIG. 6.20).

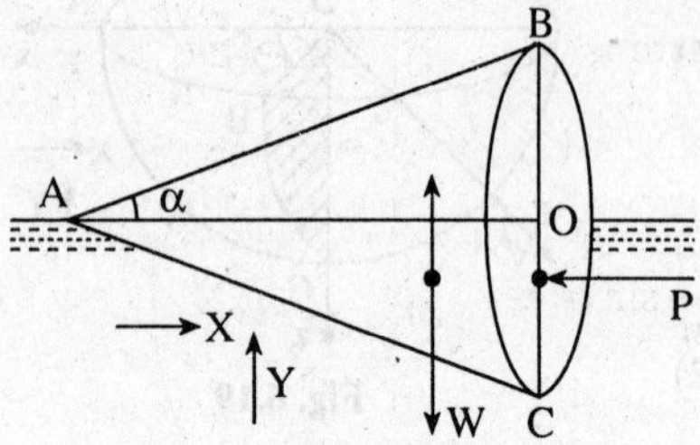

Fig. 6.20

Let R = radius of the circular base of the cone,
h = height of the cone and
ρ = density of the liquid.

Then, the horizontal thrust P on the semi-circular base is

$$P = g\rho \times \text{area} \times \text{depth of c.g.}$$

$$= g\rho \times \frac{1}{2}\pi R^2 \times \frac{4R}{3\pi} = \frac{2}{3}g\rho R^3. \tag{1}$$

The weight W of the liquid displaced is

$$W = \frac{1}{2}\left(\frac{1}{3}\pi R^2 h\right) \cdot g\rho = \frac{1}{6}\pi g\rho R^2 h, \tag{2}$$

which is equal to the resultant vertical thrust on the curved surface acting vertically upwards.

Let X and Y be the horizontal and vertical components of the resultant thrust F on the curved surface. Then, for equilibrium

$$\left.\begin{aligned} X = P = \frac{2}{3}g\rho R^3 \\ Y = W = \frac{1}{6}\pi g\rho R^2 h. \end{aligned}\right\} \tag{3}$$

Hence, the resultant thrust F on the curved surface is

$$F = \sqrt{X^2 + Y^2}$$

$$= g\rho R^2\sqrt{\frac{4}{9}R^2 + \frac{\pi^2}{36}h^2} = \frac{1}{6}g\rho R^2\sqrt{16R^2 + \pi^2 h^2} \tag{4}$$

Since

$$R = h\tan\alpha, h = R\cot\alpha \tag{5}$$

Hence,

$$F = \frac{1}{6}g\rho R^3\sqrt{16 + \pi^2\cot^2\alpha} \tag{6}$$

If the resultant thrust F is inclined to the vertical at an angle θ, then

$$\tan\theta = \frac{X}{Y} = \frac{\frac{2}{3}g\rho R^3}{\frac{1}{6}\pi g\rho R^2 h}, \text{ i.e., } \tan\theta = \frac{4}{\pi}\cdot\frac{R}{h} = \frac{4}{\pi}\tan\alpha. \quad \textbf{(Proved)}$$

EXAMPLE 6.14 *A solid is formed by the revolution of a semi-circle of radius R about its bounding diameter through an angle α, and the solid is immersed in a liquid with one of its plane faces in the surface. Prove that the magnitude of the resultant thrust on the curved surface is*

$$\frac{2}{3}g\rho R^3\left[(\alpha - \sin\alpha\cos\alpha)^2 + \sin^4\alpha\right]^{1/2}.$$

Solution. Let AOB be the solid obtained by the revolution of the semi-circle CAD of radius R about its bounding diameter CD (FIG. 6.21).

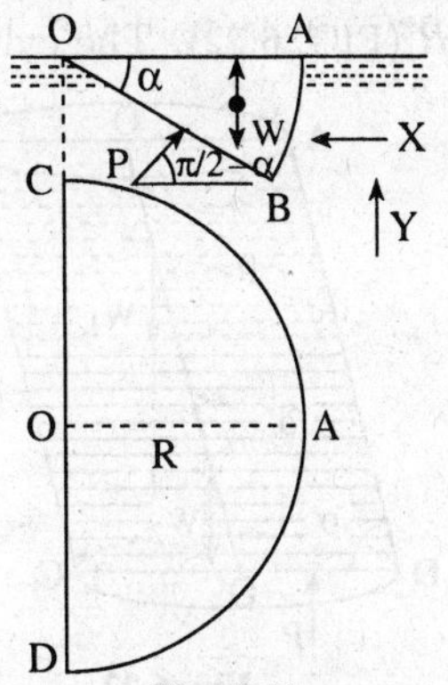

Fig. 6.21

The fluid thrust P on the plane semi-circular face OB is

$$P = g\rho \times \text{area} \times \text{depth of c.g.}$$

The distance of c.g. of the semi-circular area from its centre is

$$d = \frac{4R}{3\pi}. \tag{1}$$

Hence, depth of c.g. from fluid surface is $\bar{z} = d \sin\alpha = \dfrac{4R}{3\pi}\sin\alpha$.

Hence, the fluid thrust P is

$$P = g\rho \cdot \frac{1}{2}\pi R^2 \cdot \frac{4R}{3\pi}\sin\alpha = \frac{2}{3} g\rho R^3 \sin\alpha, \tag{2}$$

where ρ is the density of the liquid.

If the semi-circle is revolved by an angle 2π, we will get a complete sphere of volume

$$\frac{4}{3}\pi R^3.$$

Hence, the volume of the solid obtained by revolving the semi-circle through an angle α is

$$V = \frac{4}{3}\pi R^3 \cdot \frac{\alpha}{2\pi} = \frac{2}{3}R^3\alpha. \tag{3}$$

Then the weight W of the liquid displaced by the solid is

$$W = \frac{2}{3} g\rho R^3 \alpha. \tag{4}$$

Let X and Y be the horizontal and vertical components respectively of the resultant thrust F on the curved surface. Then, for equilibrium

$$\left.\begin{aligned} X &= P\sin\alpha = \frac{2}{3} g\rho R^3 \sin^2\alpha \\ Y + P\cos\alpha &= W \\ \text{i.e., } Y &= W - P\cos\alpha = \frac{2}{3} g\rho R^3 \alpha - \frac{2}{3} g\rho R^3 \sin\alpha\cos\alpha \end{aligned}\right\} \tag{5}$$

Hence the resultant thrust F on the curved surface is

$$F = \sqrt{X^2 + Y^2} = \frac{2}{3} g\rho R^3 \left[\sin^4\alpha + (\alpha - \sin\alpha\cos\alpha)^2\right]^{1/2}. \quad \textbf{(Proved)}$$

Note. If the resultant force F makes an angle ϕ with the horizontal, then

$$\tan\phi = \frac{Y}{X} = \frac{\alpha - \sin\alpha\cos\alpha}{\sin^2\alpha}, \text{ i.e., } \phi = \tan^{-1}\left[\frac{\alpha - \sin\alpha\cos\alpha}{\sin^2\alpha}\right].$$

EXAMPLE 6.15 *A vessel in the form of an oblique cylinder, with its base horizontal, contains two liquids of densities $\rho, \delta (\delta > \rho)$ of equal depth h. Prove that the resultant thrust on the curved surface due to the liquids is equivalent to a couple of moment*

$$\frac{1}{2} W h \cot\alpha \frac{3\rho + \delta}{\rho + \delta},$$

where W is the total weight of the liquid and α is the inclination of the axis of the cylinder to the base.

Solution. Let $ABCD$ be the oblique cylinder of total height $2h$ with its circular base of radius R (FIG. 6.22). The cylinder contains liquid of density ρ in the upper half and liquid of density δ in the lower half-portions. The weights of the liquids are

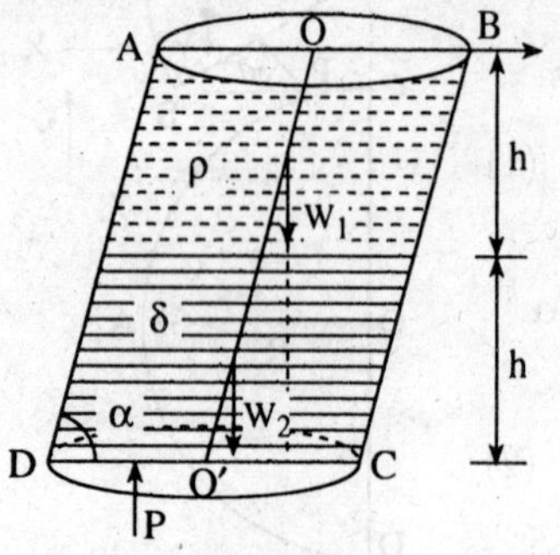

Fig. 6.22

$$W_1 = \pi R^2 hg\rho,\ W_2 = \pi R^2 hg\delta \tag{1}$$

Then the total weight of the liquid is

$$W = W_1 + W_2 = \pi R^2 hg(\rho + \delta), \tag{2}$$

which acts vertically through a point on OO'.

The thrust P on the circular base acts in the vertical direction. Hence, for equilibrium we must have $P = W$. These forces, therefore, form a couple of moment M (say) about the point O'.

The distance of the line of action of the force W_1 from the point O' is

$$d_1 = \left(h + \frac{h}{2}\right) \cot\alpha \tag{3}$$

and that of the force W_2 from the point O' is

$$d_2 = \frac{h}{2} \cot\alpha. \tag{4}$$

Hence, the total moment of the forces about the point O' is

$$M = W_1 d_1 + W_2 d_2 = \pi R^2 hg\rho \left(h + \frac{h}{2}\right) \cot\alpha + \pi R^2 hg\delta \frac{h}{2} \cot\alpha$$

$$= g\pi R^2 h^2 \cot\alpha \left(\frac{3}{2}\rho + \frac{1}{2}\delta\right) = \pi R^2 gh \cdot h\cot\alpha \cdot \frac{1}{2}(3\rho + \delta)$$

$$= \frac{W}{2} \frac{(3\rho + \delta)}{\rho + \delta} h \cot\alpha \text{ [using (2)]} \tag{5}$$

For equilibrium of the whole system, the resultant horizontal thrust on the curved surface must also form a couple of moment $M = \dfrac{W}{2} h \cot\alpha \dfrac{3\rho + \delta}{\rho + \delta}$. **(Proved)**

EXAMPLE 6.16 *A vessel full of homogeneous liquid in the form of an octant of the ellipsoid* $\dfrac{x^2}{a^2} + \dfrac{y^2}{b^2} + \dfrac{z^2}{c^2} = 1$ *is bounded by three principal planes and the z-axis vertically upwards. If the atmospheric pressure be neglected, show that the resultant fluid pressure on the curved surface is*

$$\frac{1}{3} g\rho \left[b^2c^4 + a^2c^4 + \frac{1}{4}\pi^2 a^2 b^2 c^2\right]^{1/2},$$

ρ being the density of the liquid.

Solution. The octant of the ellipsoid as well as the coordinate axes are shown in the figure (FIG. 6.23).

Equation of the ellipsoid is given as

$$\frac{x^2}{a^2} + \frac{y^2}{b^2} + \frac{z^2}{c^2} = 1 \tag{1}$$

Hence, in the figure,

$$OA = a,\ OB = b,\ OC = c. \tag{2}$$

The centre of gravity of the octant of ellipsoid has the coordinates

$$G\left(\frac{3}{8}a,\ \frac{3}{8}b,\ -\frac{3}{8}c\right) \qquad (3)$$

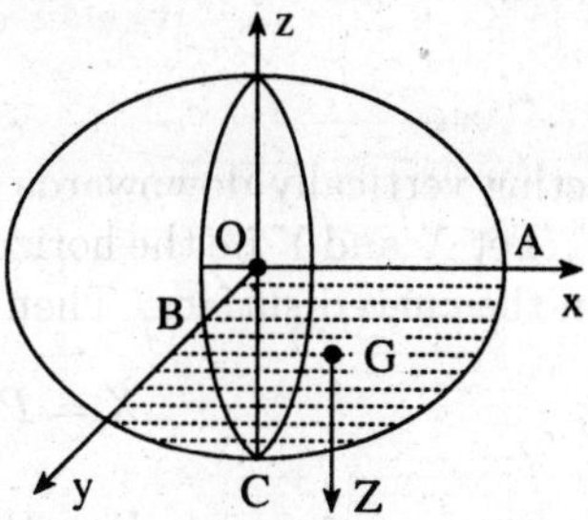

Fig. 6.23

The vertical component Z of the thrust is equal to the weight of the superincumbent liquid and acts vertically downwards through the point G. Thus

$$Z = g\rho \times \frac{1}{8}\text{ the volume of ellipsoid}$$

$$= g\rho \cdot \frac{1}{8} \cdot \frac{4}{3}\pi abc = \frac{1}{6}\pi g\rho abc \qquad (4)$$

The horizontal thrust X on the plane face OBC is

$$X = g\rho \times \text{area} \times \text{depth of c.g.} = g\rho \times \frac{1}{4}\pi bc \times \frac{4c}{3\pi} = \frac{1}{3}g\rho bc^2 \qquad (5)$$

acting at the centre of pressure of the area which has the coordinates $\left(0,\ \frac{3}{8}b,\ -\frac{3}{16}\pi c\right)$.

Similarly, the horizontal thrust Y on the plane face OAC is

$$Y = g\rho \cdot \frac{1}{4}\pi ac \cdot \frac{4c}{3\pi} = \frac{1}{3}g\rho ac^2 \qquad (6)$$

acting parallel to y-axis through the centre of pressure of the area which has the coordinates

$$\left(\frac{3}{8}a,\ 0,\ -\frac{3}{16}\pi c\right).$$

It is seen that all the three forces X, Y, Z meet at the point whose coordinates are

$$\left(\frac{3}{8}a\ ,\ \frac{3}{8}b,\ -\frac{3}{16}\pi c\right).$$

Hence, the forces can be reduced to a single resultant force R whose magnitude is

$$R = \sqrt{X^2+Y^2+Z^2} = \frac{1}{3}g\rho\sqrt{b^2c^4 + a^2c^4 + \frac{1}{4}\pi^2a^2b^2c^2}. \quad \textbf{(Proved)}$$

EXAMPLE 6.17 *A portion of a sphere cut off by two planes through its centre at an angle of 45° is just immersed in a liquid with one face in the surface. Find the resultant thrust on the curved surface and show that it makes an angle ϕ with the horizontal so that* $\tan\phi = \frac{\pi}{2} - 1$.

Solution. Let OAB be the portion of a sphere of radius R cut off by two vertical planes through its centre, the angle between the planes being $\frac{\pi}{4}$. The solid is immersed in a liquid of density ρ (say) with its face OA in the surface of the liquid (FIG. 6.24).

Obviously, the plane faces OA and OB are semi-circles of radius R. Hence, the thrust on the plane surface OB having c.g. at G is

$$P = g\rho \times \text{area} \times \text{depth of c.g.}$$

$$= g\rho \times \frac{1}{2}\pi R^2 \times CG = g\rho \times \frac{1}{2}\pi R^2 \times OG \sin\frac{\pi}{4}$$

$$= g\rho \times \frac{1}{2}\pi R^2 \times \frac{4R}{3\pi} \cdot \frac{1}{\sqrt{2}} = \frac{\sqrt{2}}{3}g\rho R^3 \qquad (1)$$

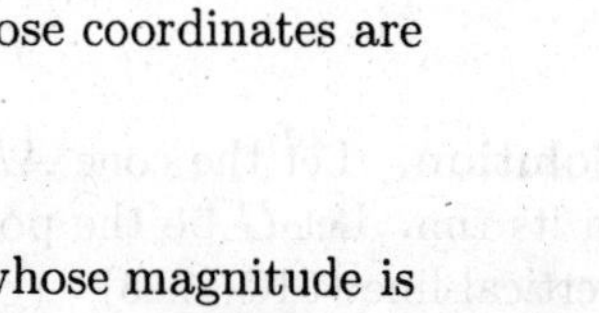

Fig. 6.24

acting normally to the face OB at its centre of pressure.

The weight W of the displaced liquid is

$$W = \frac{2}{3} \cdot g\rho \cdot \frac{\pi}{4} R^3 = \frac{1}{6} g\rho\pi R^3 \tag{2}$$

acting vertically downwards through the c.g. of the displaced liquid.

Let X and Y be the horizontal and vertical components respectively of the resultant thrust F on the curved surface. Then, for equilibrium,

$$X = P \sin\frac{\pi}{4} = \frac{\sqrt{2}}{3} g\rho R^3 \cdot \frac{1}{\sqrt{2}} = \frac{1}{3} g\rho R^3$$

$$Y + P\cos\frac{\pi}{4} = W$$

$$\text{i.e., } Y = W - P\cos\frac{\pi}{4} = \frac{1}{6} g\rho\pi R^3 - \frac{1}{3} g\rho R^3 = \frac{1}{6} g\rho R^3 (\pi - 2) \tag{3}$$

Hence, the resultant thrust F is

$$F = \sqrt{X^2 + Y^2} = \frac{1}{6} g\rho R^3 \sqrt{4 + (\pi - 2)^2} = \frac{1}{6} g\rho R^3 \sqrt{\pi^2 - 4\pi + 8}. \tag{4}$$

If the inclination of the force F be ϕ to the horizontal, then

$$\tan\phi = \frac{Y}{X} = \frac{\frac{1}{6} g\rho R^3(\pi - 2)}{\frac{1}{3} g\rho R^3} = \frac{\pi - 2}{2}, \text{ i.e., } \tan\phi = \frac{\pi}{2} - 1. \quad \textbf{(Proved)}$$

EXAMPLE 6.18 *A hollow cone without weight closed and filled with some liquid is suspended from a point in the rim of its base. If ϕ be the angle which the direction of the resultant pressure makes with the vertical, then show that*

$$\cot\phi = \frac{28\cot\alpha + \cot^3\alpha}{48}.$$

Solution. Let the cone ABC filled with a liquid of density ρ be suspended from the point A in its rim. Let G be the position of c.g. of the cone filled with liquid. Then AG should be a vertical line (FIG. 6.25).

Let h = height of the cone, R = radius of the circular base, α = semi-vertical angle of the cone, θ = angle of inclination of the diameter AB with the horizontal.

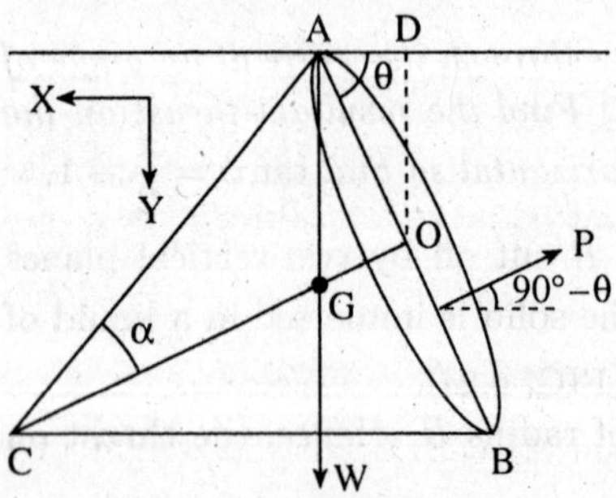

Fig. 6.25

Then,

$$OG = \frac{h}{4}, R = h\tan\alpha, \tan(90° - \theta) = \frac{OG}{OA},$$

$$\text{i.e., } \cot\theta = \frac{h/4}{R} = \frac{h}{4R} = \frac{1}{4}\cot\alpha \tag{1}$$

This yields

$$\left.\begin{aligned} \sin\theta &= \frac{4}{\sqrt{16 + \cot^2\alpha}} \\ \cos\theta &= \frac{\cot\alpha}{\sqrt{16 + \cot^2\alpha}}. \end{aligned}\right\} \tag{2}$$

The thrust on the circular base is

$$P = g\rho \times \text{area} \times \text{depth of c.g.} = g\rho \times \pi R^2 \times R\sin\theta = \pi g\rho R^3 \sin\theta. \tag{3}$$

The weight W of the liquid contained is

$$W = g\rho \cdot \frac{1}{3}\pi R^2 h = \frac{1}{3}\pi g\rho R^3 \cot\alpha \tag{4}$$

Let X and Y be the horizontal and vertical components respectively of the resultant thrust F on the curved surface.

The resultant thrust on all the surfaces is W acting vertically downwards through the c.g. of the cone. Therefore, W is the resultant of forces P, X and Y. Hence, for equilibrium

$$X - P\sin\theta = 0,\ Y - P\cos\theta = W \tag{5}$$

which yields

$$X = P\sin\theta,\ Y = P\cos\theta + W. \tag{6}$$

If the resultant force F is inclined to the vertical at an angle ϕ, then

$$\cot\phi = \frac{Y}{X} = \frac{P\cos\theta + W}{P\sin\theta},\ \text{i.e.,}\ \cot\phi = \cot\theta + \frac{W}{P\sin\theta}. \tag{7}$$

Putting the values of relevant quantities from (1), (2) (3) and (4), we get

$$\cot\phi = \frac{1}{4}\cot\alpha + \frac{\frac{1}{3}\pi g\rho R^3\cot\alpha}{\pi g\rho R^3\sin^2\theta} = \frac{1}{4}\cot\alpha + \frac{1}{3}\cot\alpha\cdot\frac{16+\cot^2\alpha}{16}$$

$$= \frac{1}{48}(28\cot\alpha + \cot^3\alpha).\quad \textbf{(Proved)}$$

EXAMPLE 6.19 *A right circular cone, closed by a plane base is held with its axis horizontal and is full of water. Show that the resultant vertical thrust on the upper half is to that on the lower half of the curved surfaces is*

$$\frac{6-\pi}{6+\pi}.$$

Solution. Let ABC be the cone of height h filled with water of density ρ and held with its axis AO horizontal (FIG. 6.26). We draw the horizontal plane through the axis dividing the cone into two equal parts. The portion $ADEB$ is the upper half and the portion $ADCE$ is the lower half of the curved surfaces.

Then the vertical thrust on upper half of the curved surface $ADEB$ is equal to the weight of the superincumbent liquid contained within the prism $AFGED$ with a triangular base ADE minus the weight of the liquid contained in upper half-portion of the cone. Hence,

$$F_1 = g\rho\left[\frac{1}{2}(2Rh)\cdot R - \frac{1}{2}\cdot\frac{1}{3}\pi R^2\cdot h\right]$$

$$= g\rho\left(R^2h - \frac{1}{6}\pi R^2h\right) = g\rho R^2h\left(1-\frac{\pi}{6}\right), \tag{1}$$

where R is the radius of the circular base of the cone.

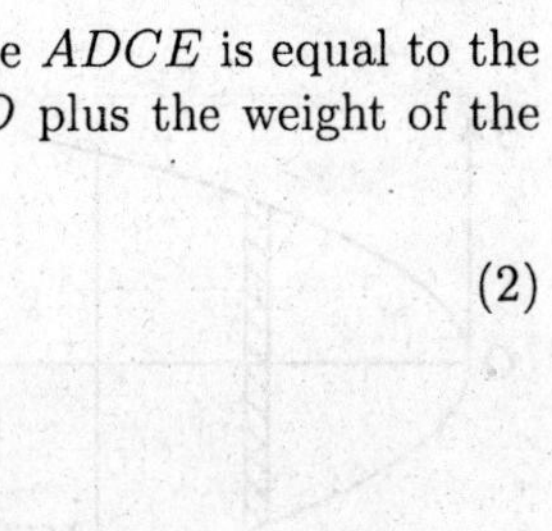

Fig. 6.26

Similarly, the vertical thrust on the lower half of the curved surface $ADCE$ is equal to the weight of the superincumbent liquid contained in the prism $AFGED$ plus the weight of the liquid contained in the lower half of the cone. Thus,

$$F_2 = g\rho\left(R^2h + \frac{1}{6}\pi R^2h\right) = g\rho R^2h\left(1+\frac{\pi}{6}\right). \tag{2}$$

From (1) and (2),

$$\frac{F_1}{F_2} = \frac{1-\frac{\pi}{6}}{1+\frac{\pi}{6}} = \frac{6-\pi}{6+\pi}.\quad \textbf{(Proved)}$$

EXAMPLE 6.20 *A solid is formed by turning a circular area round a tangent line through an angle θ and this solid is just immersed in a liquid of density ρ with one plane face in the surface. Find the resultant thrust on the curved surface and show that it makes an angle ϕ with the horizontal given by*

$$\tan\phi = \frac{\theta - \sin\theta}{\sin^2\theta}.$$

Solution. Let AOB be the solid formed by turning a circular area of radius R about a tangent at O (FIG. 6.27). The solid is just immersed with its circular face on the fluid surface. The volume of the solid generated by the theorem of Pappus is area × arc traversed by c.g. due to turning

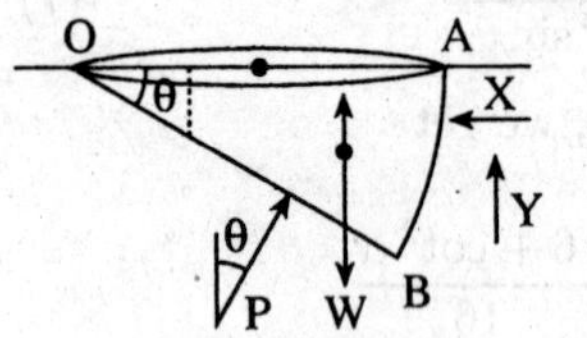

Fig. 6.27

$$\therefore V = \pi R^2 \cdot R\theta = \pi R^3\theta \qquad (1)$$

The thrust P on the circular face OB is

$$P = g\rho \times \text{area} \times \text{depth of c.g.}$$
$$= g\rho \times \pi R^2 \times R\sin\theta = g\rho\pi R^3 \sin\theta \qquad (2)$$

The weight of the liquid displaced is

$$W = g\rho V = g\rho\pi R^3\theta \qquad (3)$$

Let X and Y be the horizontal and vertical components respectively of the resultant thrust F on the curved surface. Then for equilibrium,

$$X = P\sin\theta = g\rho\pi R^3 \sin^2\theta \qquad (4)$$
$$Y + P\cos\theta = W$$
$$\text{i.e., } Y = W - P\cos\theta = g\rho\pi R^3\theta - g\rho\pi R^3 \sin\theta\cos\theta \qquad (5)$$

Hence, the resultant thrust F on the curved surface is

$$F = \sqrt{X^2 + Y^2} = g\rho\pi R^3 \left[\sin^4\theta + (\theta - \sin\theta\cos\theta)^2\right]^{1/2}.$$

If the resultant thrust F is inclined at an angle ϕ with the horizontal, then

$$\tan\phi = \frac{Y}{X} = \frac{\theta - \sin\theta\cos\theta}{\sin^2\theta}. \quad \textbf{(Proved)}$$

EXAMPLE 6.21 *A solid is formed by turning a parabolic area, bounded by the latus rectum, through an angle θ; and this solid is held under water, just immersed, with its lower plane face horizontal. Prove that, if ϕ be the inclination to the horizon of the resultant pressure on the curved surface of the solid,* $3\sin^2\theta\tan\phi = 5\sin\theta - 3\sin\theta\cos\theta - 2\theta$.

Solution. Let us consider a parabola (FIG. 6.28)

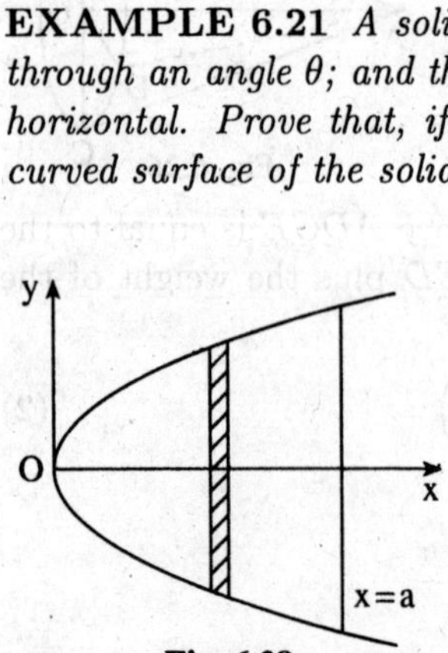

Fig. 6.28

$$y^2 = 4ax \qquad (1)$$

with its latus rectum $x = a$.

The area of the portion bounded by the latus rectum is

$$A = \int_0^a 2y\,dx = \int_0^a 2\sqrt{4ax}\,dx$$
$$= 4\sqrt{a} \cdot \frac{2}{3}a^{3/2} = \frac{8}{3}a^2. \qquad (2)$$

The first moment of the area about y-axis is

$$= \int_0^a x \cdot 2y dx = \int_0^a x \cdot 2\sqrt{4ax} dx = 4\sqrt{a} \cdot \frac{2}{5} a^{5/2} = \frac{8}{5} a^3. \tag{3}$$

Hence, the distance of c.g. of the area from O is

$$\bar{x} = \frac{\frac{8}{5}a^3}{\frac{8}{3}a^2} = \frac{3}{5}a. \tag{4}$$

Then the distance of c.g. from the latus rectum

$$d = a - \bar{x} = a - \frac{3}{5}a = \frac{2}{5}a. \tag{5}$$

Volume of the solid generated by turning the parabola about the latus rectum is

$$V = \text{area} \times \text{arc length traversed by c.g.}$$

$$= \frac{8}{3}a^2 \times \frac{2}{5}a\theta = \frac{16}{15}a^3\theta. \tag{6}$$

Now, let AOB be the above-mentioned solid just immersed in water of density ρ with plane surface OB horizontal (FIG. 6.29). Here $AO = a$.

The thrust P on the parabolic face OB is

$$P = g\rho \times \text{area} \times \text{depth of c.g.}$$

$$= g\rho \times \frac{8}{3}a^2 \times a \sin\theta = \frac{8}{3} g\rho a^3 \sin\theta \tag{7}$$

The thrust Q on the plane face AO is

$$Q = g\rho \times \frac{8}{3}a^2 \times \frac{3}{5}a \sin\theta = \frac{8}{5} g\rho a^3 \sin\theta \tag{8}$$

Weight of the liquid displaced is

$$W = g\rho V = \frac{16}{15} g\rho a^3 \theta \tag{9}$$

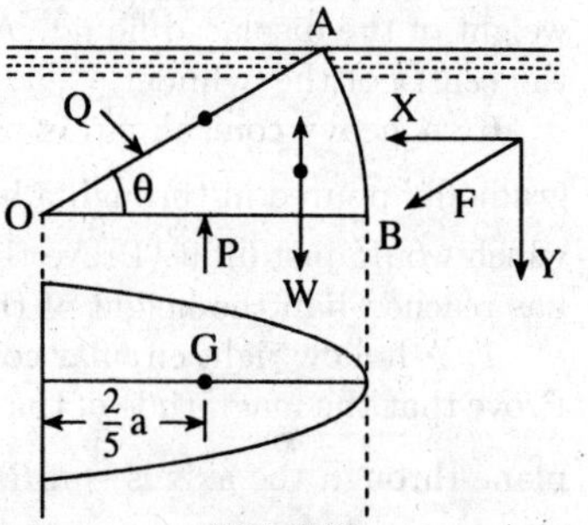

Fig. 6.29

If X and Y be the horizontal and vertical components respectively of the resultant thrust F on the curved surface, then, for equilibrium,

$$X = Q\sin\theta = \frac{8}{5} g\rho a^3 \sin^2\theta \tag{10}$$

$$Y - P + Q\cos\theta = -W,$$

$$\text{i.e., } Y = P - Q\cos\theta - W = \frac{8}{3} g\rho a^3 \sin\theta - \frac{8}{5} g\rho a^3 \sin\theta\cos\theta - \frac{16}{15} g\rho a^3 \theta$$

$$= \frac{8}{15} g\rho a^3 (5\sin\theta - 3\sin\theta\cos\theta - 2\theta) \tag{11}$$

Since ϕ is the inclination of the resultant force F to the horizontal,

$$\tan\phi = \frac{Y}{X} \tag{12}$$

Putting the values of Y and X from (10) and (11),

$$\tan\phi = \frac{5\sin\theta - 3\sin\theta\cos\theta - 2\theta}{3\sin^2\theta},$$

which yields

$$3\sin^2\theta \tan\phi = 5\sin\theta - 3\sin\theta\cos\theta - 2\theta. \quad \textbf{(Proved)}$$

EXERCISE

1. Cylinder is filled with equal volumes of n different liquids which do not mix; the density of the highest is ρ, that of the next is 2ρ, and so on, that of the lowest being $n\rho$. Show that the whole pressures on the different portions of curved surfaces of the cylinder are in the ratio $1^2 : 2^2 : 3^2 : \cdots : n^2$.

2. A vessel in the shape of a hollow hemisphere surmouned by a cone is held with the axis vertical and vertex uppermost. If it be filled with a liquid so as to submerge half the axis of the cone in the liquid and the height of the cone be double the radius of the base, show that the resultant downward thrust of the liquid in the vessel is $\frac{15}{8}$ times the weight of the liquid that the hemisphere can hold.

3. A hemispherical surface of radius R is immersed in a liquid of density ρ with its centre at a depth h and its base inclined at an angle θ to the horizontal. Find the resultant thrust on the curved surface.

Also prove that $(W - V)^2 + H^2$ is constant, V and H being the resultant vertical and horizontal thrusts, and W the weight of the displaced liquid.

4. A double funnel is formed by joining two equal hollow cones at their vertices and stands on a horizontal plane with the common axis vertical. Liquid is poured into the cone until its surface bisects the upper cone. If the liquid be on the point of escaping between the lower cone and the table, prove that the weight of either cone is to that of liquid it can hold as 27 : 16.

5. A right circular cylinder is completely immersed with its parallel plane ends inclined at an angle α to the horizontal. Show that the resultant thrust on the cylindrical surface is $W \sin \alpha$, where $W =$ weight of the displaced liquid. Also show that the line of action of the resultant thrust passes through the centre of the cylinder.

6. A heavy conical cup is placed with vertex upwards on a smooth horizontal table, and water is gradually poured in through a hole at the top. The weight of the cup is $\frac{5}{8}$th of the weight of the water which would just fill it. Prove that the cup will be on the point of rising from the table when the water has reached half the height of the cup.

7. A hollow right circular cone filled with liquid is held with its axis vertical and vertex downwards. Prove that the magnitude of the resultant fluid thrust on half the surface of the cone cut off by a vertical plane through the axis is $\frac{1}{6} g\rho Rh\sqrt{\pi^2 R^2 + 4h^2}$ and that it is inclined at an angle $\tan^{-1}\left(\frac{\pi R}{2h}\right)$ to the vertical. Here $h =$ height of the cone and $R =$ radius of the circular base.

8. A solid hemisphere of radius R is immersed in a liquid of density ρ with its plane face vertical and centre at a depth h below the surface. Show that the resultant thrust is inclined to the horizontal at an angle $\tan^{-1}\left(\frac{2R}{3h}\right)$.

9. A hemisphere of radius R is immersed in a liquid of density ρ. The plane of the base is vertical and its centre at a depth $\sqrt{5}R$ below the surface. Show that the resultant force on the curved surface is $\frac{7}{5}\pi R^3 g\rho$ and that its direction makes with the horizontal an angle ϕ, where $\tan\phi = \frac{2}{\sqrt{45}}$.

10. A closed cylindrical vessel with hemispherical ends is filled with water and placed with its axis horizontal. Find the resultant thrust and show that its line of action passes through the centre of the vessel.

11. A hollow right circular cylinder is filled with liquid and held with its axis horizontal. Show that the resultant thrust on half the curved surface cut off by a vertical plane through the axis is $\frac{1}{2} g\rho R^2 l\sqrt{16 + \pi^2}$ inclined to the horizontal at an angle $\tan^{-1}\left(\frac{\pi}{4}\right)$.

12. A spherical shell formed of two halves in contact along a vertical plane is filled with water. Show that the resultant pressure on either half of the shell is $\sqrt{13}/4$ of the total weight of the liquid.

13. A weightless sphere is divided by a vertical plane into two halves which are hinged together at their lowest point and is just filled with water. Show that the tension of a string which ties together at the highest point of the two halves is three-eighth of the weight of water that the sphere would contain.

14. Two closely fitted hemispheres made of sheet of metal of small uniform thickness are hinged together at a point on their rims and are suspended from the hinge, the rims being greased so that they form a water-tight spherical shell. This shell is now filled with water through a small aperture near the hinge. Prove that the contact will not give way if the weight of the shell exceeds three times that of water it contains.

15. The bottom and top ends of a frustrum of a cone are respectively of radii $a, b (a > b)$. If the frustrum be filled with a liquid of density ρ, prove that the resultant vertical thrust on the curved surface is $\frac{1}{3}\pi g\rho h(2a^2 - ab - b^2)$, where h is the height of the frustrum.

Also prove that, if the frustrum is inverted, the resultant vertical thrust on the curved surface of the frustrum becomes $\frac{1}{3}\pi g\rho h(a^2 + ab - 2b^2)$.

16. A solid hemisphere is immersed in a liquid with highest point of its plane base in the surface and the base inclined at an angle $\tan^{-1} 2$ to the horizon. Show that the resultant thrust on the curved surface is equal to twice the weight of the displaced liquid.

17. A hollow weightless hemisphere with a plane base is filled with water and hung by means of a string, one end of which is attached to a point on the rim of its base. Prove that the whole pressures on the curved surface and the base are in the ratio 19 : 8.

18. A hollow cone of vertical angle 2α is filled with water and placed on its side on a plane rough enough to prevent sliding, which is inclined at an angle β to the horizon. Prove that the resultant horizontal and vertical thrusts on its curved surface are $3W \tan\alpha \cos^2(\alpha + \beta)$ and $W[1 + 3\tan\alpha \sin(\alpha + \beta)\cos(\alpha + \beta)]$ respectively, where W is the weight of the liquid contained.

19. A hollow weightless hemisphere with a plane base is filled with water and hung by means of a string, one end of which is attached to a point on the rim of its base. Show that the inclination to the horizontal of the resultant thrust on its curved surface is $\tan^{-1}(109/96)$.

Also show that the thrust on the plane base is to the weight of the contained liquid as $12 : \sqrt{73}$.

20. A solid cone whose vertical angle 2α is immersed in a liquid with its vertex in the surface and axis vertical. Prove that, if P be the whole pressure on the curved surface and base, and P' the resultant pressure, then $\frac{P}{P'} = \frac{2 + 3\sin\alpha}{\sin\alpha}$.

21. A right cone of vertical angle 2α is just immersed with its slant side of length l in the surface of a liquid. Show that the resultant thrust on the curved surface will cut this slant side at a distance of $\frac{3l}{4(1 - 3\sin^2\alpha)}$ from the vertex. Also show that the magnitude of this thrust is

$$\frac{1}{3}\pi g\rho l^3 \sin^2\alpha \cos\alpha \sqrt{1 + 3\sin^2\alpha}.$$

22. A closed cylinder whose height is equal to the diameter of the base, is filled with water and hangs freely from a string fastened to a point on its upper rim. If the weight of the cylinder be neglected, show that the vertical and horizontal components of thrust on its curved surface are each of them equal to half the weight of the contained water.

23. A spherical shell formed of two halves in contact along a vertical plane is filled with liquid. Show that the resultant pressure on either half of the shell is to the total weight of the liquid as $\sqrt{13} : 4$.

24. A solid cone is just immersed in water with a generating line in the surface. If θ be the inclination to the vertical of the resultant thrust on the curved surface, prove that $(1 - 3\sin^2\alpha)\tan\theta = 3\sin\alpha\cos\alpha$, α being the semi-vertical angle of the cone.

25. A vessel in the shape of the segment of a sphere of radius R cut off by a plane at a distance d from the centre, rests with its base on a table. If the segment is filled with water, show that the resultant vertical thrust on the curved surface will be an upward force, if $d < \frac{R}{2}$.

26. A solid right circular cone is just immersed in water with a generating line in the surface. If the resultant thrust on the curved surface be horizontal, show that the semi-vertical angle of the cone is

$$\sin^{-1}\left(\frac{1}{\sqrt{3}}\right).$$

27. The stem of a funnel has a radius a and its mouth has a radius b, the height of the stem is h and that of its sloped part is l. The mouth is placed on a horizontal plane, being greased so as to be water-tight. If the funnel is filled with water to the top of the stem, prove that the upward pressure on the funnel will be equal to

$$\pi\gamma(b-a)\left[h(b+a)+\frac{1}{3}l(a+2b)\right],$$

where γ is the weight of unit volume of water.

28. A rectangular block whose edges are of length $2a, 2b, 2c$, is divided by a plane through the centre perpendicular to the edges of length $2c$, and the two halves are hinged together along edges parallel to those of length $2a$. The whole is then immersed in a liquid with the line of hinges inclined at an angle θ to the horizon and the dividing plane vertical, the hinges being in the upper face. Prove that the two halves will not separate unless

$$\left[\left(1-\frac{\delta}{\rho}\right)c^2-\frac{2}{3}b^2\right]\cos\theta > 2bd,$$

where d is the depth of c.g. of the block, δ the density of the block and ρ that of liquid.

29. A solid in the form of an oblique circular cylinder is totally immersed in a homogeneous liquid with its ends horizontal. Show that the resultant pressure on its curved surface is a couple of moment $Wd\cot\alpha$, where W is the weight and d the depth of centre of gravity of liquid displaced and α the inclination of the generators to the plane of the ends.

30. A cone is totally immersed in water, the depth of the centre of its base being given. Prove that, P, P', P'' being the resultant pressure on its curved surface, when the sine of the inclination of its axis to the horizon are s, s', s'' respectively,

$$P^2(s'-s'')+P'^2(s''-s)+P''^2(s-s')=0.$$

31. A frustrum of a sphere of radius R, cut off by two parallel planes each distant d from the centre is completely immersed in a liquid with the axis inclined to the vertical at an angle θ and the centre at a depth h.

Show that the resultant thrust on the curved surface is equal to

$$g\rho\left\{\left(\frac{4}{3}\pi d^3\right)^2\cos^2\theta+\left[2\pi\left(R^2d-\frac{d^3}{3}\right)\right]^2\sin^2\theta\right\}^{1/2},$$

where ρ is the density of the liquid.

32. A solid is formed by turning a circular area round a tangent line through an angle θ, and this solid is held under water, just immersed, with its lower plane face horizontal. Prove that, if ϕ be the inclination to the horizon of the resultant pressure on the curved surface of the solid, then $\sin^2\theta\tan\phi = 2\sin\theta-\sin\theta\cos\theta-\theta$.

CHAPTER 7

Equilibrium of Floating Bodies

7.1 Introduction

We know that a big ship made of iron floats in water, whereas a small piece of iron sinks into it. Thus, there are some reasons as to why a body floats or sinks in a fluid. In this chapter, we intend to discuss the conditions of floatation of a body in a fluid at rest under gravity.

7.2 Conditions of Equilibrium of a Floating Body

We consider a solid body of volume V and ***mean density*** ρ floating partially in a fluid of density δ remaining at rest under gravity (FIG. 7.1).

Let V' be the volume of the body immersed, i.e., the volume of the displaced liquid. The only vertical forces acting in the system are:

(1) The weight W of the body acting downwards through its c.g. G, where

$$W = g\rho V. \tag{7.1}$$

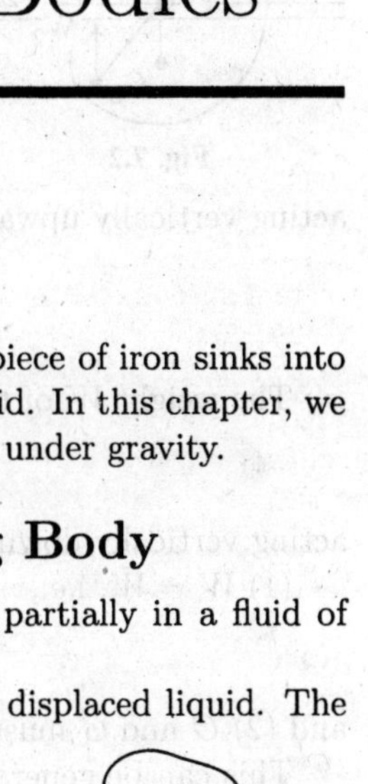

Fig. 7.1

(2) The vertical thrust equal to the weight W' of the displaced liquid acting upwards through the centre of buoyance H, where

$$W' = g\delta V'. \tag{7.2}$$

For equilibrium, these two forces must be equal in magnitude but opposite in directions and must act in the same vertical line.

Hence, we have the following conditions of equilibrium of floating bodies:

(1) ***The weight of the body must be equal to weight of the displaced liquid, and***

(2) ***The centres of gravity of the body and of the fluid displaced must be on the same vertical line.***

By the condition (1), we have from Eqns (7.1) and (7.2),

$$g\rho V = g\delta V', \text{ or, } V' = \frac{\rho}{\delta}V. \tag{7.3}$$

This implies that the volume immersed in the fluid is ρ/δ parts of the volume of the body. Hence

$$\frac{\rho}{\delta} < 1, \text{ i.e., } \rho < \delta. \tag{7.4}$$

That is, the solid can float only if its density is less than that of the fluid.

7.3 Bodies Floating in More than One Liquid

Consider a body of volume V and mean density ρ floating in two fluids of densities δ_1, δ_2 (FIG. 7.2) which do not mix.

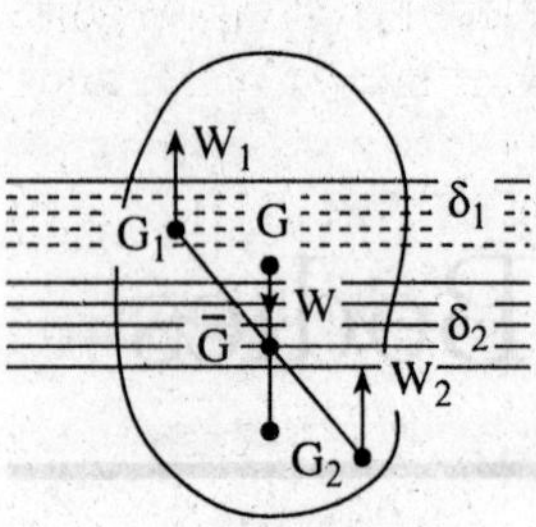

Fig. 7.2

Let V_1 = volume of fluid of density δ_1 displaced, V_2 = volume of fluid of density δ_2 displaced.

Then the weights of the fluids displaced are

$$W_1 = g\delta_1 V_1,\ W_2 = g\delta_2 V_2 \tag{7.5}$$

acting vertically upwards through their centres of gravity G_1 and G_2 respectively.

The resultant of the two forces in Eqn (7.5) is

$$\bar{W} = g(\delta_1 V_1 + \delta_2 V_2) \tag{7.6}$$

acting vertically upwards through the point $\bar{G}$ lying on the line G_1G_2, such that

$$W_1 \cdot \bar{G}G_1 = W_2 \cdot \bar{G}G_2, \text{ i.e., } \frac{\bar{G}G_1}{\bar{G}G_2} = \frac{W_2}{W_1} = \frac{\delta_2 V_2}{\delta_1 V_1}. \tag{7.7}$$

The weight W of the solid body is

$$W = g\rho V \tag{7.8}$$

acting vertically downwards through its c.g. G. Hence, for equilibrium, we must have

(1) $W = \bar{W}$, i.e.,

$$g(\delta_1 V_1 + \delta_2 V_2) = g\rho V \tag{7.9}$$

and (2) G and $\bar{G}$ must be on the same vertical line.

This can be generalised to any number of fluids of different densities which do not mix.

7.4 Equilibrium of Bodies Floating under Constraints

We consider a body floating under some constraints in a fluid at rest under gravity. The constraints imposed will certainly have some effect on the conditions of equilibrium. We consider that the body can turn freely about one fixed point.

Let O be the fixed point about which the body can turn freely (FIG. 7.3). Let W be the weight of the body acting vertically downwards through its c.g. G, and W' that of the fluid displaced acting vertically upwards through its c.g. G'.

From O, we draw a perpendicular on the lines of action of W and W' cutting at points L and L'. Since there is no horizontal force in the system, for equilibrium, (1) the reaction force R at the constraint O should also be vertical. Moreover, the forces R, W and W' should be coplanar. Then

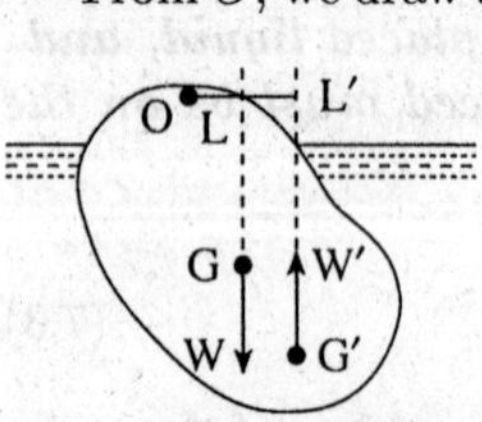

Fig. 7.3

$$R = W - W' \tag{7.10}$$

and (2) the forces W and W' must have equal moments about O, i.e.,

$$W \cdot OL = W' \cdot OL' \tag{7.11}$$

When two points of the body are fixed, the straight line joining these fixed points is also fixed. That is, the body can turn freely about this line which is, therefore, a line of hinges fixed in the body. In this case, for equilibrium, the moments of W and W' about this line must be equal and opposite.

7.5 Floating Body with a String Attached to a Point

Consider a body floating in a fluid with a string attached to a point P on the body (FIG. 7.4). The string will be subjected to a tension T. Thus, the forces acting in the system are:

(1) The weight W of the body acting vertically downwards through its c.g. G, (2) the weight W' of the displaced liquid acting vertically upwards through the centre of buoyancy H, and (3) the tension T in the string acting vertically upwards.

We extend the lines of action of these forces up to the fluid surface meeting at points a, b and c. Obviously, the point b will be in between a and c. Hence, the conditions of equilibrium are:

(1) $W = W' + T$, i.e.,

$$T = W - W'. \tag{7.12}$$

(2) Taking moments about the point P,

$$W \cdot ab = W' \cdot ac. \tag{7.13}$$

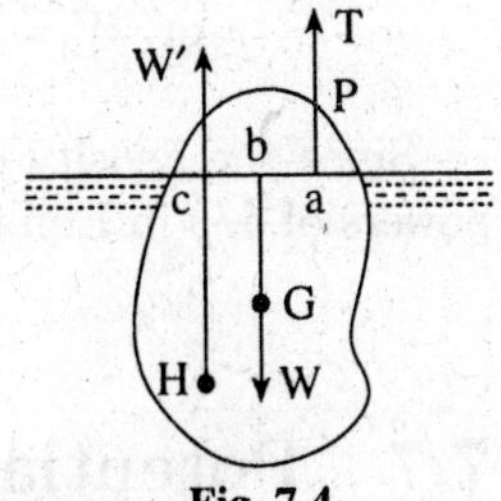

Fig. 7.4

These conditions are also applicable if the body is totally immersed in a fluid whose density is less than that of the body.

If, however, a body be totally immersed in a fluid whose density is greater than that of the body, the weight of the displaced liquid, i.e., the force of buoyancy will be greater than the weight of the body. As such, the body will ascend upwards unless an external vertically downward force is applied to the body.

In the figure (FIG. 7.5), we have shown the body tied down with a string at a point P. Thus the tension T in this case will act vertically downwards. Moreover, the point b will be away from the points a and c.

The conditions of equilibrium are:

(1) $W' = W + T$, i.e.,

$$T = W' - W. \tag{7.14}$$

(2) Taking moments about the point P,

$$W \cdot ab = W' \cdot ac. \tag{7.15}$$

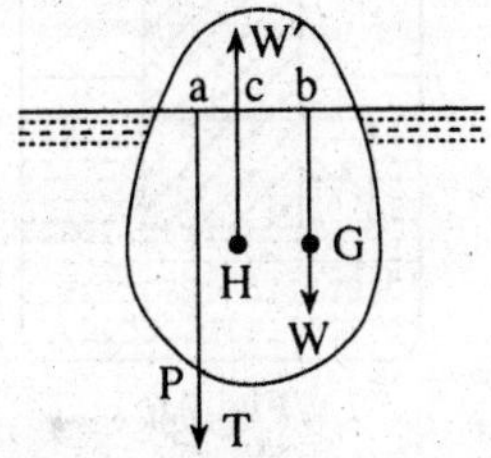

Fig. 7.5

7.6 Weighing a Body Immersed in a Fluid

We will not get the true weight of a body when it is immersed in a fluid and weighed by a balance. This is because the surrounding fluid will exert an upward thrust on the body. The weight thus obtained is known as ***apparent weight*** of the body. Thus, apparent weight of the body is equal to the true weight of the body minus the weight of the displaced fluid.

We, however, can determine the true weight of a body in air in the following manner:

Let W be the true weight of the body of density ρ and w be its corresponding apparent weight which is equal to the sum of the 'weights' of density ρ'. Let the density of air be δ.

If T = tension in the string supporting the body, then

$$T = W - \frac{W}{\rho}\delta, \tag{7.16}$$

where the second term in Eqn (7.16) is the weight of the air displaced by the body.

Similarly, if T' = tension in the string supporting the weights, then

$$T' = w - \frac{w}{\rho'}\delta. \tag{7.17}$$

For a true balance,

$$T = T'. \tag{7.18}$$

This yields

$$W - W\frac{\delta}{\rho} = w - w\frac{\delta}{\rho'}$$

$$\text{or, } W = w\left(1-\frac{\delta}{\rho'}\right)\left(1-\frac{\delta}{\rho}\right)^{-1} = w\left(1-\frac{\delta}{\rho'}\right)\left(1+\frac{\delta}{\rho}+\frac{\delta^2}{\rho^2}+\cdots\right) \tag{7.19}$$

Since the density of air δ is very small compared to ρ and ρ', we may neglect the higher powers of δ/ρ in Eqn (7.18). Hence,

$$W = w\left(1-\frac{\delta}{\rho'}+\frac{\delta}{\rho}\right). \tag{7.20}$$

7.7 Potential Energy of Liquids

As we know, a solid body under gravity possesses potential energy by virtue of its position. Similarly, a mass of heavy liquid at rest under gravity possesses potential energy due to its position. If the liquid is without any friction and is subject to any change in configuration so slowly that its kinetic energy can be neglected, then the loss or gain of potential energy will be equal to the amount of the work done by the gravitational force or by some external agent against the gravitational force respectively.

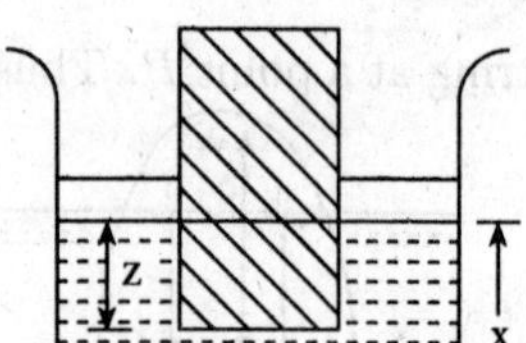

Fig. 7.6

This principle of potential energy and work done can be used to solve problems in hydrostatics when the fluid changes its configuration very slowly.

As an example, we consider a solid body being slowly immersed into a liquid contained in a vessel (FIG. 7.6). In this process, an amount of work is done against the force of buoyancy which is stored up as potential energy due to the gain in elevation of the centre of gravity of the liquid.

Let x = depth of the liquid in the vessel when the depth of the immersed portion of the solid in the liquid is z. Let X and Z be their corresponding areas of cross sections.

Again, let V = volume of the liquid in the vessel and V' = volume of the immersed part of the solid.

Now suppose that the solid is further depressed through a distance dz which causes a corresponding rise of the liquid in the vessel by an amount dx.

Then, extra volume of the solid immersed is

$$= (X - Z)dx \tag{7.21}$$

whereas the volume of the liquid that has risen is

$$= Zdz \tag{7.22}$$

These two quantities must be equal. Hence,

$$(X - Z)dx = Zdz \tag{7.23}$$

Again, the centre of gravity of the solid has moved through a distance $(dz - dx)$. Hence, the increase in potential energy is

$$= V'(dz - dx) \tag{7.24}$$

Eqns (7.23) and (7.24) can be used to determine the total change of potential energy and hence the work done, if the initial and final positions of the solid body be known.

EXAMPLE 7.1 *A ship sailing out of the sea into a river sinks through a distance of x feet. On unloading a cargo of weight P, the ship rises through y feet. Show that the weight of the ship after unloading is*

$$\left[\frac{x\delta}{y(\delta-\rho)}-1\right]P,$$

where δ and ρ are sp. gr. of the sea water and river water respectively. The cross section of the ship near the level of water may be assumed to be uniform.

Solution. Let S = area of cross section of the ship, h = depth in feet of the ship under sea water and W = weight of the ship.

Thus, when in sea, the weight of the ship = weight of water displaced, i.e.,

$$W = Shg\delta. \tag{1}$$

When the ship is in the river water of sp. gr. ρ, it sinks by x feet. Thus, total depth of the ship under river water is $(h+x)$ feet. Then, by above principle,

$$W = S(h+x)g\rho. \tag{2}$$

From (1) and (2), $Shg\delta = S(h+x)g\rho$, which yields

$$h = \frac{x\rho}{\delta-\rho}. \tag{3}$$

Now, when a cargo of weight P is unloaded, the weight of the ship becomes $(W-P)$ and the ship rises by y feet. Thus, the total depth of the ship under river water is $(h+x-y)$.

Hence,

$$W-P = S(h+x-y)g\rho$$

$$\text{or, } W-P = S(h+x)g\rho - Syg\rho = W - Syg\rho \text{ [using (2)]}$$

$$\therefore P = Syg\rho, \text{ i.e., } Sg = \frac{P}{\rho y} \tag{4}$$

Putting the values of h and Sg from (3) and (4) into (1), we get

$$W = \frac{P}{\rho y}\cdot\left(\frac{x\rho}{\delta-\rho}\right)\cdot\delta = \frac{Px\delta}{y(\delta-\rho)} \tag{5}$$

Hence, the weight of the ship after unloading the cargo is

$$W-P = \frac{Px\delta}{y(\delta-\rho)} - P = P\left[\frac{x\delta}{y(\delta-\rho)}-1\right]. \quad \textbf{(Proved)}$$

EXAMPLE 7.2 *A body floating in water has volumes V_1, V_2, V_3 above the surface, when the densities of the surrounding air are respectively ρ_1, ρ_2, ρ_3. Prove that*

$$\frac{\rho_2-\rho_3}{V_1}+\frac{\rho_3-\rho_1}{V_2}+\frac{\rho_1-\rho_2}{V_3}=0.$$

Solution. Let V = volume of the body, W = weight of the body and ρ = density of the liquid.

When a volume of V_1 of the body is above liquid surface and in air of density ρ_1, the volume immersed is $(V-V_1)$.

Since, weight of the body = weight of liquid displaced + weight of air displaced, we write

$$W = (V-V_1)\rho g + V_1 g\rho_1, \text{ i.e., } \frac{W-Vg\rho}{V_1} = g(\rho_1-\rho). \tag{1}$$

We will get similar two such expressions, so that

$$\frac{W - Vg\rho}{V_2} = g(\rho_2 - \rho) \tag{2}$$

$$\frac{W - Vg\rho}{V_3} = g(\rho_3 - \rho) \tag{3}$$

Multiplying (1) by $(\rho_2 - \rho_3)$, (2) by $(\rho_3 - \rho_1)$, (3) by $(\rho_1 - \rho_2)$ and adding, we get

$$(W - Vg\rho)\left[\frac{\rho_2 - \rho_3}{V_1} + \frac{\rho_3 - \rho_1}{V_2} + \frac{\rho_1 - \rho_2}{V_3}\right]$$

$$= g[(\rho_2 - \rho_3)(\rho_1 - \rho) + (\rho_3 - \rho_1)(\rho_2 - \rho) + (\rho_1 - \rho_2)(\rho_3 - \rho)]$$

$$= g[\rho_1(\rho_2 - \rho_3) + \rho_2(\rho_3 - \rho_1) + \rho_3(\rho_1 - \rho_2) - \rho(\rho_2 - \rho_3 + \rho_3 - \rho_1 + \rho_1 - \rho_2)]$$

$$= 0$$

Hence,

$$\frac{\rho_2 - \rho_3}{V_1} + \frac{\rho_3 - \rho_1}{V_2} + \frac{\rho_1 - \rho_2}{V_3} = 0. \quad \textbf{(Proved)}$$

EXAMPLE 7.3 *Prove that, if the volumes U and V of two different substances balance in vacuum and volumes U_1 and V_1 balance when weighted in a liquid, the densities of the substances and liquid are in the ratio of*

$$\frac{U_1 - V_1}{U} : \frac{U_1 - V_1}{V} : \left(\frac{U_1}{U} - \frac{V_1}{V}\right).$$

Solution. Let ρ, δ = densities of two different substances

λ = density of the liquid.

Since the volumes U and V balance in vacuum, their weights are equal. Hence,

$$g\rho U = g\delta V, \text{ i.e., } \frac{\rho}{\delta} = \frac{V}{U}. \tag{1}$$

Again, since the volumes U_1 and V_1 of the same substances balance in a liquid of density λ, their apparent weights are equal.

$$g\rho U_1 - g\lambda U_1 = g\delta V_1 - g\lambda V_1$$

$$\text{or, } \rho U_1 - \delta V_1 = \lambda(U_1 - V_1) \tag{2}$$

$$\text{or, } \frac{\rho}{\delta}U_1 - V_1 = \frac{\lambda}{\delta}(U_1 - V_1) \tag{3}$$

Putting the value of $\frac{\rho}{\delta}$ from (1), we get

$$\frac{V}{U}U_1 - V_1 = \frac{\lambda}{\delta}(U_1 - V_1) \text{ or, } \frac{U_1}{U} - \frac{V_1}{V} = \frac{\lambda}{\delta}\left(\frac{U_1 - V_1}{V}\right) \tag{4}$$

Similarly, we get from (2) dividing by ρ,

$$U_1 - \frac{\delta}{\rho}V_1 = \frac{\lambda}{\rho}(U_1 - V_1)$$

$$\text{or, } U_1 - \frac{U}{V}V_1 = \frac{\lambda}{\rho}(U_1 - V_1) \text{ [using (1)]}$$

$$\text{or, } \frac{U_1}{U} - \frac{V_1}{V} = \frac{\lambda}{\rho}\left(\frac{U_1 - V_1}{U}\right) \tag{5}$$

From (4), we write

$$\delta : \lambda :: \frac{U_1 - V_1}{V} : \left(\frac{U_1}{U} - \frac{V_1}{V}\right) \tag{6}$$

and from (5), we write

$$\rho : \lambda :: \frac{U_1 - V_1}{U} : \left(\frac{U_1}{U} - \frac{V_1}{V}\right) \tag{7}$$

From (6) and (7), we get

$$\rho : \delta : \lambda :: \frac{U_1 - V_1}{U} : \frac{U_1 - V_1}{V} : \left(\frac{U_1}{U} - \frac{V_1}{V}\right). \quad \textbf{(Proved)} \tag{8}$$

EXAMPLE 7.4 *A body of density ρ is weighed by means of weight of density $\rho'(\rho' > \rho)$, the density of air being δ. Now the density of air is increased from δ to δ'. Prove that the body weights less than before by a fraction $\frac{(\rho' - \rho)(\delta' - \delta)}{(\rho - \delta)(\rho' - \delta')}$ of its former weight.*

Solution. Let W = actual weight of the body, V = volume of the body, W_1 = actual weight of the "weights" in the first case, W_2 = actual weight of the "weights" in the second case.

Then,

$$W = g\rho V, \text{ i.e., } V = \frac{W}{g\rho}. \tag{1}$$

Apparent weight of the body in air of density δ

$$= W - Vg\delta = W - W\frac{\delta}{\rho} \tag{2}$$

Apparent weight of the weights in air of density δ

$$= W_1 - \frac{W_1}{g\rho'} \cdot g\delta = W_1 - W_1 \cdot \frac{\delta}{\rho'}. \tag{3}$$

Quantities in (2) and (3) must be equal. Hence,

$$W\left(1 - \frac{\delta}{\rho}\right) = W_1\left(1 - \frac{\delta}{\rho'}\right), \text{ i.e., } W_1 = W\frac{(\rho - \delta)\rho'}{(\rho' - \delta)\rho}. \tag{4}$$

Similarly, when the "weights" are taken in air of density δ', we will have

$$W\left(1 - \frac{\delta'}{\rho}\right) = W_2\left(1 - \frac{\delta'}{\rho'}\right)$$

so that

$$W_2 = W\frac{(\rho - \delta')\rho'}{(\rho' - \delta')\rho}. \tag{5}$$

From (4) and (5),

$$\frac{W_2}{W_1} = \frac{(\rho - \delta')(\rho' - \delta)}{(\rho' - \delta')(\rho - \delta)}. \tag{6}$$

Thus, the body weights less than before by a fraction

$$\xi = 1 - \frac{W_2}{W_1} \tag{7}$$

of its former weight.

From (6) and (7), we get

$$\xi = 1 - \frac{(\rho-\delta')(\rho'-\delta)}{(\rho'-\delta')(\rho-\delta)} = \frac{(\rho'-\delta')(\rho-\delta)-(\rho-\delta')(\rho'-\delta)}{(\rho'-\delta')(\rho-\delta)}$$

$$= \frac{\rho'\rho-\rho'\delta-\rho\delta'+\delta\delta'-(\rho\rho'-\rho\delta-\rho'\delta'+\delta\delta')}{(\rho'-\delta')(\rho-\delta)}$$

$$= \frac{\rho\delta+\rho'\delta'-\rho'\delta-\rho\delta'}{(\rho'-\delta')(\rho-\delta)} = \frac{(\rho'-\rho)(\delta'-\delta)}{(\rho-\delta)(\rho'-\delta')}. \quad \textbf{(Proved)}$$

EXAMPLE 7.5 *A body consists of an alloy of two metals of specific gravities S_1 and S_2 respectively. Its weight in vacuum is W_1 and in water W_2. Show that the proportion of the two metals in the alloy by volume is*

$$\frac{W_2S_2-W_1(S_2-1)}{W_1(S_1-1)-W_2S_1}.$$

Solution. Let V_1 = volume of the metal of sp. gr. S_1, V_2 = volume of the metal of sp. gr. S_2 and ρ = density of water.

Then total weight of the body W_1 is equal to the sum of the weights of the two metals. Hence,

$$W_1 = (V_1S_1+V_2S_2)g\rho \tag{1}$$

Now, the weight of the water displaced by two metals is

$$= V_1g\rho + V_2g\rho = (V_1+V_2)g\rho \tag{2}$$

Hence apparent weight of the body in water

$$W_2 = \text{actual weight} - \text{weight of water displaced}$$
$$= (V_1S_1+V_2S_2)g\rho - (V_1+V_2)g\rho \tag{3}$$

From (1) and (3),

$$\frac{W_1}{W_2} = \frac{V_1S_1+V_2S_2}{(V_1S_1+V_2S_2)-(V_1+V_2)}$$

$$\text{or, } W_2(V_1S_1+V_2S_2) = W_1[V_1S_1+V_2S_2-(V_1+V_2)]$$

$$\text{or, } V_1(W_2S_1-W_1S_1+W_1) = V_2(W_1S_2-W_2S_2-W_1]$$

$$\text{or, } V_1[W_2S_1-W_1(S_1-1)] = V_2[W_1(S_2-1)-W_2S_2]$$

$$\therefore \frac{V_1}{V_2} = \frac{W_2S_2-(S_2-1)W_1}{W_1(S_1-1)-W_2S_1}. \quad \textbf{(Proved)}$$

EXAMPLE 7.6 *A rod of small section and density ρ has a small portion of metal of weight $\frac{1}{n}$th that of the rod attached at one extremity. Prove that the rod will float at any inclination in a liquid of density δ, if $(n+1)^2\rho = n^2\delta$.*

Solution. Let AB be the rod of length $2l$ floating in a liquid of density δ, inclined at an angle θ to the horizontal (FIG. 7.7). The end B is loaded with a weight $\frac{W}{n}$, where W is the weight of the rod. Thus, the forces acting in the rod are:

(1) The weight W of the rod acting vertically downwards through its c.g. G,

(2) The load $\frac{W}{n}$ acting vertically downwards at the end B, and

(3) The weight F of the displaced liquid acting vertically upwards through D, the mid-point of BC.

Hence, for equilibrium,

$$W + \frac{W}{n} = F, \text{ i.e., } W\left(1 + \frac{1}{n}\right) = F \qquad (1)$$

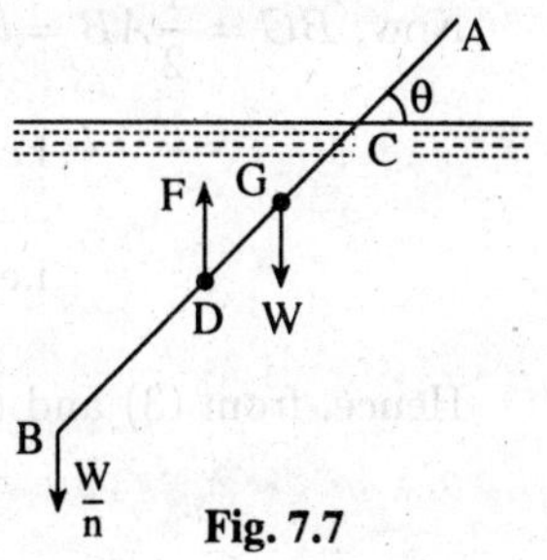

Fig. 7.7

Now, let s = cross sectional area of the rod and $BD = DC = x$.

Then, weight of the rod is

$$W = 2lsg\rho \qquad (2)$$

and weight of the displaced liquid is

$$F = 2xsg\delta. \qquad (3)$$

Hence, from (1), (2) and (3),

$$2lsg\rho\left(1 + \frac{1}{n}\right) = 2xsg\delta, \text{ or, } l\rho\left(1 + \frac{1}{n}\right) = x\delta \qquad (4)$$

Now, taking moments about D, we get

$$\frac{W}{n} \cdot BD = W \cdot GD \text{ or, } \frac{W}{n} \cdot x = W \cdot (l - x) \text{ or, } x\left(\frac{1}{n} + 1\right) = l,$$

which yields

$$x = \frac{ln}{n+1} \qquad (5)$$

From (4) and (5), we get

$$l\rho\left(\frac{n+1}{n}\right) = \frac{ln}{n+1}\delta \qquad (6)$$

Hence, $(n+1)^2\rho = n^2\delta$. **(Proved)**

EXAMPLE 7.7 *A rod of length $2l$ and specific gravity δ whose one end is hinged at a height h above the surface of a liquid of sp. gr. $n\delta (n > 1)$ remains in equilibrium in an inclined position, partly immersed in the liquid. Prove that the inclination of the rod to the vertical is $\cos^{-1}\left(\frac{h}{2l}\sqrt{\frac{n}{n-1}}\right)$ and the length immersed is $2l\left[1 - \sqrt{1 - \frac{1}{n}}\right]$.*

Solution. Let AB be the rod of length $2l$ hinged at its end B at a height h above the liquid surface (FIG. 7.8).

The forces acting on the rod are:

(1) The weight W acting vertically downwards through its c.g. G,

(2) The weight F of the displaced liquid acting vertically upwards through D, the mid-point of AC.

(3) The reaction force R at the hinged end B. This force R must also be acting in the vertical direction, since there is no other force.

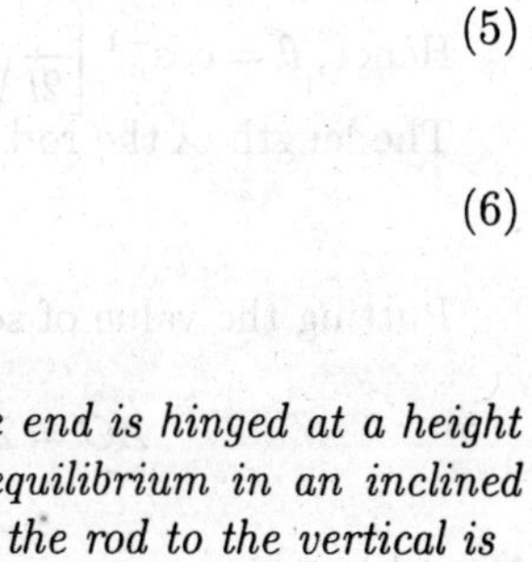

Fig. 7.8

Now, if s = cross sectional area of the rod, then

$$W = 2ls\delta \cdot g\rho \qquad (1)$$

$$\text{and } F = AC \cdot s \cdot n\delta \cdot g\rho = (AB - BC)sn\delta g\rho$$
$$= (2l - h\sec\theta)sn\delta g\rho, \qquad (2)$$

where ρ = density of water.

Taking moments of all forces about B, we get

$$W \cdot BG \sin\theta = F \cdot DB \sin\theta, \text{ i.e., } W \cdot BG = F \cdot DB, \text{ (since } \sin\theta \neq 0) \quad (3)$$

Now, $BG = \frac{1}{2}AB = l, DB = BC + CD,$

$$\text{i.e., } DB = BC + \frac{1}{2}AC = h\sec\theta + \frac{1}{2}(2l - h\sec\theta)$$

$$\text{i.e., } DB = \frac{1}{2}(2l + h\sec\theta) \quad (4)$$

Hence, from (3) and (4),

$$W \cdot l = F \cdot \frac{1}{2}(2l + h\sec\theta) \quad (5)$$

Putting here the values of W and F from (1) and (2),

$$(2ls\delta g\rho) \cdot l = \frac{1}{2}(2l - h\sec\theta)sn\delta g\rho \cdot (2l + h\sec\theta)$$

$$\text{or, } 2l^2 = \frac{1}{2} \cdot n(4l^2 - h^2\sec^2\theta), \text{ or, } 4l^2(n-1) = nh^2\sec^2\theta$$

$$\therefore \sec^2\theta = \frac{4l^2(n-1)}{nh^2}, \text{ i.e., } \cos\theta = \frac{h}{2l}\sqrt{\frac{n}{n-1}} \quad (6)$$

Hence, $\theta = \cos^{-1}\left[\frac{h}{2l}\sqrt{\frac{n}{n-1}}\right]$. **(Proved)**

The length of the rod immersed is

$$= AC = (AB - BC) = 2l - h\sec\theta \quad (7)$$

Putting the value of $\sec\theta$ from (6),

$$AC = 2l - h \cdot \frac{2l}{h}\sqrt{\frac{n-1}{n}} = 2l\left[1 - \sqrt{1 - \frac{1}{n}}\right]. \quad \textbf{(Proved)}$$

EXAMPLE 7.8 *Two uniform straight rods of lengths $2a, 2b$ and densities ρ, δ respectively are joined together to form one straight rod. If it floats freely in water in an inclined position, the rod of length $2a$ and part of the other being immersed, proved that*

$$a^2\rho(\rho - 1) + 2ab\delta(\rho - 1) + b^2\delta(\delta - 1) = 0.$$

Solution. Let the rod AB of length $2a$ and density ρ be joined with the rod BC of length $2b$ and density δ to obtain a longer rod AC. The rod AC floats freely in water with total of AB and a part BD of the rod BC immersed (FIG. 7.9).

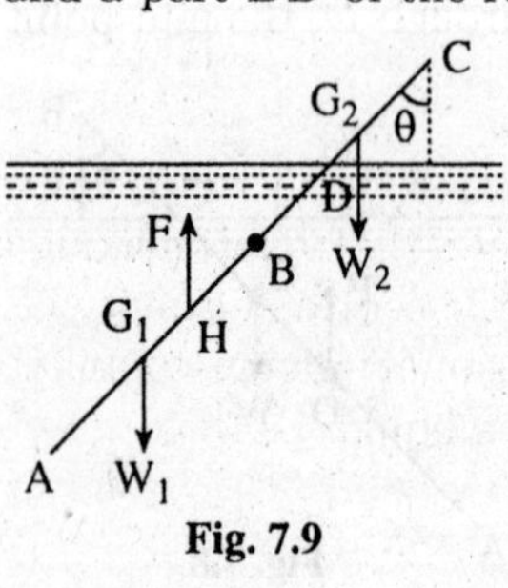

Fig. 7.9

Let $BD = x$. Then

$$\left.\begin{aligned} AB = 2a,\ BC = 2b \\ AD = 2a + x,\ DC = 2b - x \end{aligned}\right\} \quad (1)$$

Let the rod make an angle θ with the vertical. The forces acting on the rod are:

(1) The weight W_1 of the rod AB acting vertically downwards through its c.g. G_1, where

$$W_1 = 2asg\rho \quad (2)$$

(2) The weight W_2 of the rod BC acting vertically downwards through its c.g. G_2, where

$$W_2 = 2bsg\delta \tag{3}$$

(3) The weight F of the water displaced acting vertically upwards at the centre of buoyancy H, where

$$F = AD \cdot sg \text{ (taking density of water to be unity)} = (2a + x)sg \tag{4}$$

We have assumed that both the rods AB, BC have same cross sectional area s.

For equilibrium,

$$F = W_1 + W_2$$

$$(2a + x)sg = (2a\rho + 2b\delta)sg, \text{ or, } 2a + x = 2a\rho + 2b\delta$$

$$\therefore x = 2a(\rho - 1) + 2b\delta. \tag{5}$$

Taking moments of all the forces about the end C,

$$W_1 \cdot G_1C \sin\theta + W_2 \cdot G_2C \sin\theta = F \cdot HC \sin\theta$$

$$\text{i.e., } W_1(G_1B + BC) + W_2\left(\frac{1}{2}BC\right) = F(HD + DC) \; [\because \sin\theta \neq 0]$$

Since, $HD = \frac{1}{2}AD = \frac{1}{2}(2a + x)$, we get

$$W_1(a + 2b) + W_2 b = (W_1 + W_2)\left[\frac{1}{2}(2a + x) + (2b - x)\right]$$

$$\text{or, } W_1(a + 2b) + W_2 b = (W_1 + W_2)\left[(a + 2b) - \frac{x}{2}\right]$$

$$\text{i.e., } W_2(a + b) = \frac{1}{2}x(W_1 + W_2) \tag{6}$$

Putting here the values of W_1, W_2 from (2), (3) and that of x from (5), we get

$$2bsg\delta(a + b) = \frac{1}{2}[2a(\rho - 1) + 2b\delta](2a\rho + 2b\delta)sg$$

$$\text{i.e., } b\delta(a + b) = [a(\rho - 1) + b\delta](a\rho + b\delta)$$

$$\text{or, } ab\delta + b^2\delta = a^2\rho(\rho - 1) + ab\delta(\rho - 1) + ab\rho\delta + b^2\delta^2$$

$$\text{or, } a^2\rho(\rho - 1) + b^2\delta(\delta - 1) + ab\delta(\rho - 1 + \rho - 1) = 0$$

$$\text{or, } a^2\rho(\rho - 1) + 2ab\delta(\rho - 1) + b^2\delta(\delta - 1) = 0. \quad \textbf{(Proved)}$$

EXAMPLE 7.9 *A uniform rod of density δ and length l is freely movable about an end fixed at a depth h below the surface of liquid of density ρ. Show that the rod may rest in a position inclined with vertical with other end outside the liquid, provided*

$$1 < \frac{\rho}{\delta} < \frac{l^2}{h^2}.$$

Solution. Let AB be a rod of length l and density δ movable about the end A which is fixed at a depth h below the liquid surface. The rod rests with end B outside the liquid, making an angle θ with the vertical (FIG. 7.10). Since, in an inclined position, the centre of gravity G of the rod does not coincide with the centre of buoyancy H, the rod cannot remain completely immersed. Hence, the density of the rod must be less than that of the liquid, i.e.,

$$\delta < \rho. \tag{1}$$

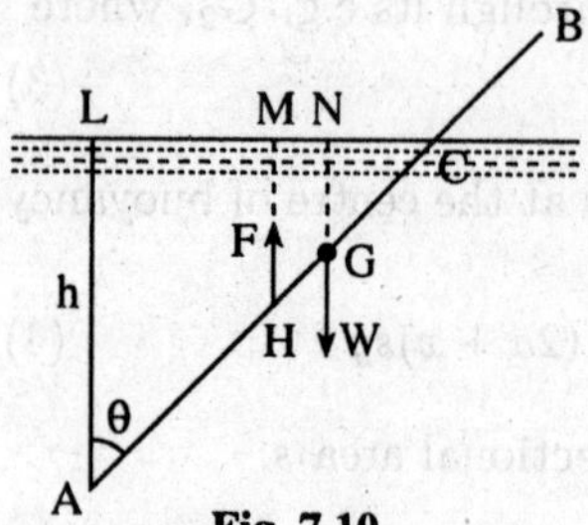

Fig. 7.10

Let $s =$ cross sectional area of the rod. The forces acting on the rod are:

(1) The weight W of the rod acting vertically downwards through its c.g. G, where

$$W = lsg\delta \tag{2}$$

(2) The weight of displaced liquid, i.e., the force of buoyancy F acting vertically upwards through the mid-point H of AC, where

$$F = AC \cdot sg\rho = hsg\rho \sec\theta \tag{3}$$

Taking moments of these forces about A,

$$W \cdot LN = F \cdot ML \tag{4}$$

Now,

$$\left.\begin{aligned} LN &= AG \sin\theta = \frac{1}{2} l \sin\theta \\ ML &= MC = HC \sin\theta = \frac{1}{2} AC \sin\theta = \frac{1}{2} h \sec\theta \sin\theta \end{aligned}\right\} \tag{5}$$

From (2), (3) and (5), we get, by (4)

$$lsg\delta \cdot \frac{1}{2} l \sin\theta = (hsg\rho \sec\theta) \cdot \frac{1}{2} h \sec\theta \sin\theta, \text{ i.e., } \delta l^2 = \rho h^2 \sec^2\theta$$

$$\therefore \cos^2\theta = \frac{\rho h^2}{\delta l^2} \tag{6}$$

For real values of θ, we must have

$$\cos^2\theta < 1 \tag{7}$$

Hence, from (6) and (7)

$$\frac{\rho h^2}{\delta l^2} < 1, \text{ i.e., } \frac{\rho}{\delta} < \frac{l^2}{h^2} \tag{8}$$

Combining (1) and (8), we get $1 < \dfrac{\rho}{\delta} < \dfrac{l^2}{h^2}$. **(Proved)**

EXAMPLE 7.10 *A uniform rod of length $2l$ and density ρ is movable in a plane about one end which is fixed in a liquid of density ρ' at a depth $2h$ below the surface. A liquid of lower density δ is added on the top of the first liquid. If in the oblique position of equilibrium the rod is just covered by the liquid, prove that its inclination to the vertical is*

$$\cos^{-1}\left[\frac{h}{l}\sqrt{\frac{\rho' - \delta}{\rho - \delta}}\right].$$

Solution. Let AB be the rod of length $2l$ with its end A movable in a plane. The rod rests in an inclined position making an angle θ with the vertical, with its portion AC immersed in the liquid of density ρ' and the portion BC in the liquid of density δ (FIG. 7.11). The forces acting on the rod are:

(1) The weight W acting vertically downwards through its c.g. G, where

$$W = 2lsg\rho \tag{1}$$

(2) The force of buoyancy F_1 on the portion AC acting vertically upwards through its mid-point H_1, where

$$F_1 = AC \cdot sg\rho' = 2hsg\rho' \sec\theta \tag{2}$$

(3) The force of buoyancy F_2 on the portion BC acting vertically upwards through its mid-point H_2, where

$$F_2 = BC \cdot sg\delta = (AB - AC) \cdot sg\delta = (2l - 2h\sec\theta)sg\delta \cdot \quad (3)$$

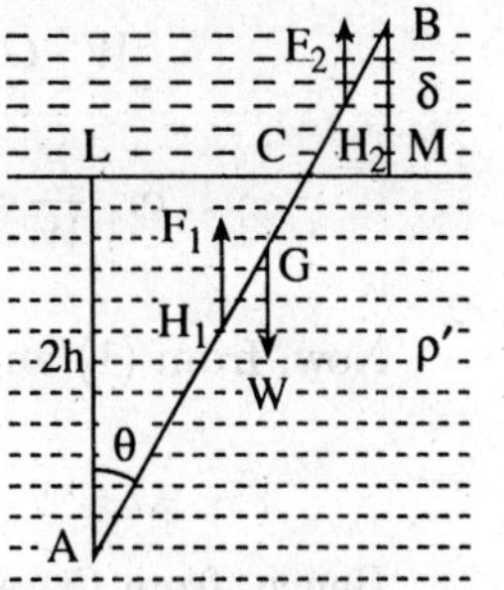

Fig. 7.11

Here we have assumed that, s = cross sectional area of the rod.

(4) The reaction force R at the end A, which also must be vertical. Taking moments of all the forces about A, we get

$$W \cdot AG\sin\theta = F_1 \cdot AH_1 \sin\theta + F_2 \cdot AH_2 \sin\theta$$

$$\text{i.e., } W \cdot AG = F_1 \cdot AH_1 + F_2 \cdot AH_2, \quad (4)$$

since $\sin\theta \neq 0$.

Now, from the figure we have

$$\left.\begin{aligned} AG &= \frac{1}{2}AB = l \\ AH_1 &= \frac{1}{2}AC = \frac{1}{2}\cdot 2h\sec\theta = h\sec\theta \\ AH_2 &= AC + CH_2 = 2h\sec\theta + \frac{1}{2}(2l - 2h\sec\theta), \text{ i.e., } AH_2 = l + h\sec\theta \end{aligned}\right\} \quad (5)$$

Putting the necessary values from (1), (2), (3) and (5) into (4), we get

$$2lsg\rho \cdot l = 2hsg\rho'\sec\theta \cdot h\sec\theta + (2l - 2h\sec\theta)sg\delta \cdot (l + h\sec\theta)$$

$$\text{or, } \rho l^2 = h^2\rho'\sec^2\theta + (l^2 - h^2\sec^2\theta)\delta$$

$$\text{or, } l^2(\rho - \delta) = h^2(\rho' - \delta)\sec^2\theta$$

$$\text{or, } \cos^2\theta = \frac{h^2(\rho' - \delta)}{l^2(\rho - \delta)}.$$

$$\therefore \cos\theta = \frac{h}{l}\sqrt{\frac{\rho' - \delta}{\rho - \delta}}, \text{ i.e., } \theta = \cos^{-1}\left[\frac{h}{l}\sqrt{\frac{\rho' - \delta}{\rho - \delta}}\right]. \quad \textbf{(Proved)}$$

EXAMPLE 7.11 *A thin uniform wooden rod AB is in equilibrium in an inclined position with one end AC immersed in a bowl of water and one point D supported on the edge of the bowl. Show that the sp. gr. δ of the wood is*

$$\frac{AC}{AB} \cdot \frac{2AD - AC}{2AD - AB}$$

and that the greatest fraction of the length of the rod which can remain immersed is $1 - \sqrt{1 - \delta}$.

Solution. Let AB be the rod of cross sectional area s remaining in equilibrium at an angle θ to the vertical (FIG. 7.12).

The forces acting on the rod are:

(1) The weight W acting vertically downwards through its c.g. G, where

$$W = AB \cdot sg\delta. \quad (1)$$

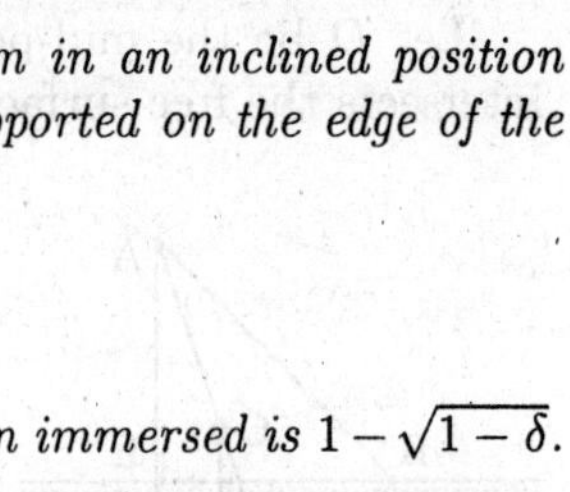

Fig. 7.12

(2) The force of buoyancy F on the portion AC acting vertically upwards through its mid-point H, where

$$F = AC \cdot sg \quad (2)$$

(3) The reaction force R at D. Taking moments of all the forces about D, we get

$$W \cdot GD \sin\theta = F \cdot HD \sin\theta$$

$$\text{or, } \frac{F}{W} = \frac{GD}{HD} = \frac{AD - AG}{AD - AH}, \text{ i.e., } \frac{F}{W} = \frac{AD - \frac{1}{2}AB}{AD - \frac{1}{2}AC} = \frac{2AD - AB}{2AD - AC} \tag{3}$$

Now, from (1) and (2),

$$\frac{F}{W} = \frac{AC \cdot sg}{AB \cdot sg\delta} = \frac{AC}{AB \cdot \delta} \tag{4}$$

Hence, from (3) and (4),

$$\frac{AC}{AB \cdot \delta} = \frac{2AD - AB}{2AD - AC}, \quad \therefore \delta = \frac{AC}{AB} \cdot \frac{2AD - AC}{2AD - AB}. \quad \textbf{(Proved)} \tag{5}$$

The hanging portion BD of the rod has the tendency to uplift the rod out of water. Thus, the greatest part of the rod will remain immersed when BD is minimum. Taking $BD = 0$, the end point B of the rod is just on the rim of the bowl so that $AD = AB$.

Let $\dfrac{AC}{AB} = x$.

Then, from (5), we get $\delta = x \cdot \dfrac{2AB - AC}{2AB - AB} = x(2 - x)$, i.e., $x^2 - 2x + \delta = 0$

$$\therefore x = 1 \pm \sqrt{1 - \delta}$$

Since $x < 1$, we take the value, $x = 1 - \sqrt{1 - \delta}$. **(Proved)**

EXAMPLE 7.12 *A triangular lamina ABC of density ρ floats in a liquid of density δ with its plane vertical, the angle B being in the surface of the liquid and the angle A not immersed. Show that $\rho : \delta = \sin A \sin C : \sin B = (a^2 + b^2 - c^2) : 2b^2$, a, b, c being the lengths of the sides of the triangle.*

Solution. Let ABC be the triangular lamina floating in a liquid of density δ with its plane vertical and vertex B on the surface, but the vertex A above the liquid (FIG. 7.13).

Let D be the mid-point of BC, so that AD is the median of the $\triangle ABC$. The side AC intersects the free surface at E. We join DE. Then DE is the median of the $\triangle BEC$.

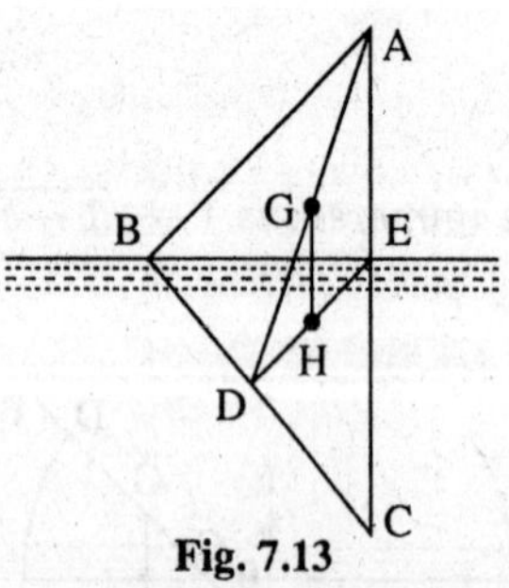

Fig. 7.13

The weight W of the $\triangle ABC$ acts vertically downwards through its c.g. G on AD, such that

$$DG = \frac{1}{3}AD. \tag{1}$$

Similarly, the force of buoyancy F on the $\triangle BEC$ acts vertically upwards through its c.g. H on DE, such that

$$DH = \frac{1}{3}DE \tag{2}$$

Thus, in the $\triangle ADE$, $\dfrac{DG}{AD} = \dfrac{DH}{DE} = \dfrac{1}{3}$.

Hence, GH and AE are parallel. But GH is vertical, so that AE is also vertical. Thus, in equilibrium position of the lamina, the side AC remains vertical.

Now, the weight W of the $\triangle ABC$

$$= \frac{1}{2}BC \cdot AC \sin C \cdot g\rho = \frac{1}{2}abg\rho \sin C \tag{3}$$

The force of buoyancy F on the $\triangle BEC$

$$= \frac{1}{2} BC \cdot EC \sin C \cdot g\delta$$

$$= \frac{1}{2} a \cdot a \cos C \cdot \sin C \cdot g\delta \text{ [since } EC = BC \cos C]$$

$$= \frac{1}{2} a^2 g\delta \sin C \cos C \tag{4}$$

For equilibrium, $W = F$. Hence,

$$\frac{1}{2} abg\rho \sin C = \frac{1}{2} a^2 g\delta \sin C \cos C, \text{ or, } \frac{\rho}{\delta} = \frac{a \cos C}{b}. \tag{5}$$

Now, in the $\triangle ABC$,

$$\cos C = \frac{a^2 + b^2 - c^2}{2ab}. \tag{6}$$

From (5) and (6), we get

$$\frac{\rho}{\delta} = \frac{a}{b} \cdot \frac{a^2 + b^2 - c^2}{2ab} = \frac{a^2 + b^2 - c^2}{2b^2}. \tag{7}$$

Again, in the $\triangle ABC$, we have

$$\frac{a}{b} = \frac{\sin A}{\sin B} \tag{8}$$

From (5) and (8), we get

$$\frac{\rho}{\delta} = \frac{\sin A}{\sin B} \cdot \cos C = \frac{\sin A \cos C}{\sin B} \tag{9}$$

From (7) and (9), we have

$$\frac{\rho}{\delta} = \frac{\sin A \cos C}{\sin B} = \frac{a^2 + b^2 - c^2}{2b^2}. \quad \textbf{(Proved)}$$

EXAMPLE 7.13 *A triangular lamina ABC of specific gravity ρ floats with its plane vertical in water, the vertex A being outside the liquid and BC not horizontal. The angle A is a right angle and $AB = AC$. If θ be the inclination of AB to the vertical, prove that*

$$\sin 2\theta = \frac{2 - 2\rho}{2\rho - 1}, \text{ given } \rho > \frac{3}{4}.$$

Solution. Let ABC be the triangular lamina floating vertically in water with its vertex A outside the surface (FIG. 7.14). The water surface DE divides the $\triangle ABC$ into two parts. Let P, Q be the mid-points of sides DE, BC respectively and G, H be the c.g. of the $\triangle ADE$ and $\triangle ABC$ respectively. We draw AR perpendicular to DE. The line AQ meets DE at S.

We are given,

$$\left.\begin{aligned} AB = AC = x \text{ (say)}, \ \angle BAC = \frac{\pi}{2}, \\ \angle DAR = \theta, \text{ so that, } \angle AER = \theta \end{aligned}\right\} \tag{1}$$

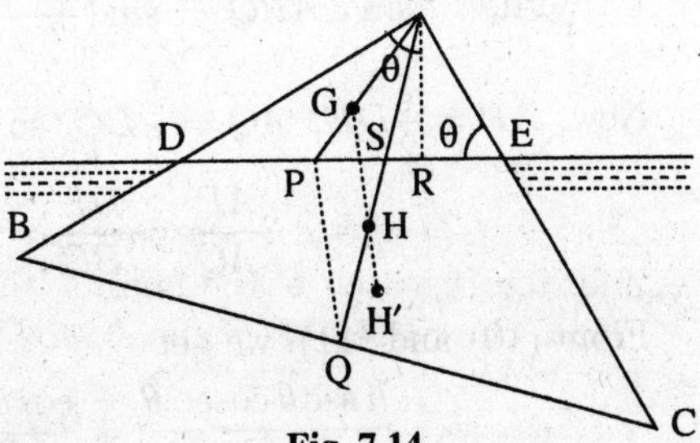

Fig. 7.14

The weight W of the $\triangle ABC$ is

$$W = \frac{1}{2} AB \cdot ACg\rho = \triangle ABC \cdot g\rho \tag{2}$$

acting vertically downwards through H.

The force of buoyancy F on area $BDEC$ is

$$F = \text{area } BDEC \cdot g = (\triangle ABC - \triangle ADE) \cdot g \tag{3}$$

By condition of equilibrium, $W = F$.
Hence,

$$\triangle ABC \cdot g\rho = (\triangle ABC - \triangle ADE) \cdot g, \text{ i.e., } \frac{\triangle ABC}{\triangle ADE} = \frac{1}{1-\rho} \tag{4}$$

Now,

$$\left.\begin{aligned} \triangle ABC &= \frac{1}{2} AB \cdot AC = \frac{1}{2} x \cdot x = \frac{1}{2} x^2 \\ \triangle ADE &= \frac{1}{2} AD \cdot AE = \frac{1}{2} h \sec\theta \cdot h \operatorname{cosec}\theta \end{aligned}\right\} \tag{5}$$

where we have denoted $AR = h$.

Hence, from (5),

$$\frac{\triangle ABC}{\triangle ADE} = \frac{x^2}{h^2 \sec\theta \operatorname{cosec}\theta} = \frac{x^2 \sin 2\theta}{2h^2} \tag{6}$$

From (4) and (6), we get

$$\frac{x^2 \sin 2\theta}{2h^2} = \frac{1}{1-\rho}. \tag{7}$$

Now, the centre of buoyancy for the area $BDEC$ should lie on GH extended. Hence, by condition of equilibrium, GH should be vertical. Thus, GH and AR are parallel. Also, in $\triangle APQ$, $AP : PG = AQ : QH$. Hence, PQ and GH are parallel, that is PQ is also vertical.

Consider the $\triangle APQ$. In this triangle, we find from the figure,

$$\angle AQP = \angle SAR = \angle SAE - \angle RAE = \frac{\pi}{4} - \left(\frac{\pi}{2} - \theta\right) = \theta - \frac{\pi}{4} \tag{8}$$

since, in $\triangle ABC$, $AB = AC$ and AQ, the median, bisects $\angle A$.

Now, since the $\triangle ABC$ is right-angled with hypotenuse BC, and AQ is the median bisecting BC, the median AQ is half of BC. Similarly, in $\triangle ADE$, AP is half of the hypotenuse DE.

Again, since $AP = DP, \angle ADP = \angle DAP$, so that

$$\begin{aligned} \angle APQ &= \angle APS + \angle QPS = \angle ADP + \angle DAP + \frac{\pi}{2} \\ &= 2\left(\frac{\pi}{2} - \theta\right) + \frac{\pi}{2} = \frac{3\pi}{2} - 2\theta \end{aligned} \tag{9}$$

Then,

$$\frac{AP}{AQ} = \frac{\sin \angle AQP}{\sin \angle APQ} = \frac{\sin\left(\theta - \frac{\pi}{4}\right)}{\sin\left(\frac{3\pi}{2} - 2\theta\right)}, \text{ i.e., } \frac{AP}{PQ} = \frac{\sin\theta - \cos\theta}{\sqrt{2}(-\cos 2\theta)} = \frac{\cos\theta - \sin\theta}{\sqrt{2}\cos 2\theta} \tag{10}$$

Now, $AP = \frac{1}{2} DE$, $AQ = \frac{1}{2} BC$, so that

$$\frac{AP}{AQ} = \frac{DE}{BC} = \frac{AD \operatorname{cosec}\theta}{\sqrt{x^2 + x^2}} = \frac{(h \sec\theta) \operatorname{cosec}\theta}{\sqrt{2}x} \tag{11}$$

From (10) and (11), we get

$$\frac{h \sec\theta \operatorname{cosec}\theta}{\sqrt{2}x} = \frac{\cos\theta - \sin\theta}{\sqrt{2}\cos 2\theta}, \text{ i.e., } \frac{2h}{x \sin 2\theta} = \frac{\cos\theta - \sin\theta}{\cos 2\theta}.$$

Squaring both sides,

$$\frac{4h^2}{x^2 \sin^2 2\theta} = \frac{1-\sin 2\theta}{\cos^2 2\theta} = \frac{1-\sin 2\theta}{1-\sin^2 2\theta} = \frac{1}{1+\sin 2\theta}$$

Hence,

$$\frac{x^2 \sin 2\theta}{4h^2} = \frac{1+\sin 2\theta}{\sin 2\theta}, \text{ i.e., } \frac{x^2 \sin 2\theta}{2h^2} = \frac{2(1+\sin 2\theta)}{\sin 2\theta} \qquad (12)$$

From (7) and (12), we get

$$\frac{1}{1-\rho} = \frac{2(1+\sin 2\theta)}{\sin 2\theta}$$

$$\text{or, } \sin 2\theta = 2(1-\rho) + 2(1-\rho)\sin 2\theta, \text{ or, } (2\rho - 1)\sin 2\theta = 2(1-\rho)$$

$$\therefore \sin 2\theta = \frac{2(1-\rho)}{2\rho - 1}. \quad \textbf{(Proved)}$$

The angle θ cannot be equal to $\pi/4$, because, in that case, the side BC should be horizontal, which is contrary to the proposition. In the limiting case, when $\theta = \pi/4$, we have

$$1 = \frac{2(1-\rho)}{2\rho - 1}, \text{ i.e., } \rho = \frac{3}{4}.$$

Thus, ρ should be greater than 3/4, which is given.

EXAMPLE 7.14 *A semi-circular lamina has one of the ends of its diameter smoothly hinged to a fixed point above the surface of a liquid and floats with its plane vertical and its diameter half immersed. If the inclination of the diameter to the horizon is* 45°, *prove that the ratio of the density of the liquid to that of the lamina is* $4(3\pi - 4) : (9\pi - 8)$.

Solution. Let ACB be the semi-circular lamina of radius R with its diametric end A hinged above the liquid surface (FIG. 7.15). We are given that

$$\angle AOD = 45° = \frac{\pi}{4}. \qquad (1)$$

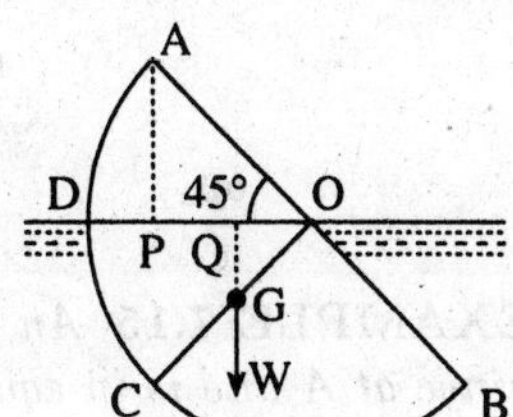

Fig. 7.15

Area of the semi-circle is

$$S = \frac{1}{2}\pi R^2, \qquad (2)$$

so that area of the immersed portion $ODCB$ is

$$= \frac{3}{4}S = \frac{3}{4}\cdot\frac{1}{2}\pi R^2 = \frac{3}{8}\pi R^2 \qquad (3)$$

The distance OG of the centre of gravity of the semi-circle from O is

$$OG = \frac{4R}{3\pi} \qquad (4)$$

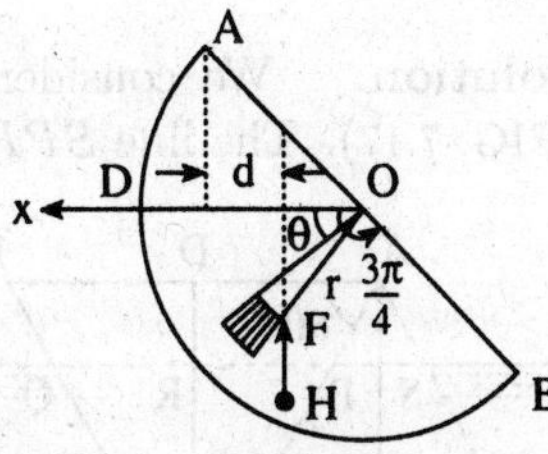

Fig. 7.16

We draw perpendiculars AP and GQ from A and G on OD. Then

$$PQ = OP - OQ = R\cos\frac{\pi}{4} - \frac{4R}{3\pi}\cos\frac{\pi}{4}$$

$$\text{i.e., } PQ = \frac{R}{\sqrt{2}} - \frac{4R}{3\pi}\cdot\frac{1}{\sqrt{2}} = \frac{R(3\pi - 4)}{3\sqrt{2}\pi} \qquad (5)$$

To determine the position of the centre of gravity of the immersed portion $ODCB$, we take x-axis along OD with origin at O and consider a small element $rd\theta dr$ as shown in the figure (FIG. 7.16).

Then

$$\bar{x} = \frac{\int_0^R \int_0^{3\pi/4} r\cos\theta r d\theta dr}{\int_0^R \int_0^{3\pi/4} r d\theta dr} = \frac{\frac{R^3}{3}\cdot\frac{1}{\sqrt{2}}}{\frac{R^2}{2}\cdot\frac{3\pi}{4}} = \frac{8R}{9\sqrt{2}\pi} \tag{6}$$

The horizontal distance of $\bar{x}$ from the fixed point A is

$$d = R\cos\frac{\pi}{4} - \bar{x} = R\cdot\frac{1}{\sqrt{2}} - \frac{8R}{9\sqrt{2}\pi} = \frac{R(9\pi-8)}{9\sqrt{2}\pi} \tag{7}$$

Let ρ and δ be the densities of the lamina and liquid respectively.
The forces acting on the lamina are:
(1) The weight W acting vertically downwards through its c.g. G, where

$$W = \frac{1}{2}\pi R^2 g\rho \tag{8}$$

(2) The force of buoyancy F acting vertically upwards through the centre of buoyancy H, where

$$F = \frac{3}{8}\pi R^2 g\delta \tag{9}$$

(3) The reaction force at the hinged end A. For equilibrium, the sum of moments of all forces about A should vanish. Hence,

$$W\cdot PQ = F\cdot d$$

i.e., $$\frac{1}{2}\pi R^2 g\rho\cdot\frac{R(3\pi-4)}{3\sqrt{2}\pi} = \frac{3}{8}\pi R^2 g\delta\cdot\frac{R(9\pi-8)}{9\sqrt{2}\pi}$$

or, $$\rho(3\pi-4) = \frac{3}{4}\delta\frac{(9\pi-8)}{3}$$

$$\therefore\ \frac{\delta}{\rho} = \frac{4(3\pi-4)}{9\pi-8}. \quad \textbf{(Proved)}$$

EXAMPLE 7.15 *An equilateral triangle ABC of weight W and sp. gr. s is movable about a hinge at A and is in equilibrium when the angle C is immersed in water and AB is horizontal. It is then turned about A in its own plane until the whole of the side BC is in water and horizontal. Prove that the pressure on the hinge in this position is*

$$2\left(\frac{1-\sqrt{s}}{\sqrt{s}}\right)W$$

Solution. We consider the first case, when the side AB is above water and horizontal (FIG. 7.17). The line $SPRQ$ is the water surface.

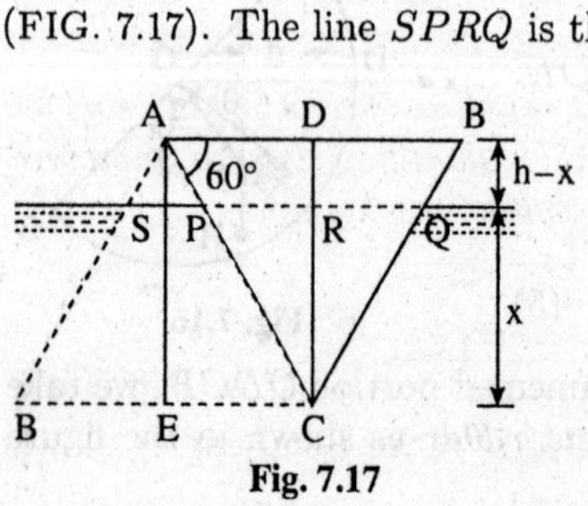

Fig. 7.17

Let $CD = h$ = height of the $\triangle ABC$, and $x = CR$ = depth of the lamina under water.

Then area of the $\triangle ABC$

$$= \frac{1}{2}AB\cdot h = AD\cdot h = h\cot 60^\circ h = \frac{h^2}{\sqrt{3}} \tag{1}$$

Similarly, the area of the $\triangle CPQ$

$$= \frac{1}{2}PQ\cdot x = \frac{x^2}{\sqrt{3}} \tag{2}$$

Let ρ and δ be the densities of the lamina and water.

The forces acting on the lamina are:

(1) The weight W acting vertically downwards through its c.g. which lies on CD, where

$$W = \frac{h^2}{\sqrt{3}} g\rho \tag{3}$$

(2) The force of buoyancy F acting vertically upwards through c.g. of $\triangle PCQ$, which also lies on CD, where

$$F = \frac{x^2}{\sqrt{3}} g\delta \tag{4}$$

(3) There is no reaction force at the hinge A, since it cannot balance the other forces which do not pass through A.

Hence, for equilibrium, we must have

$$W = F, \text{ i.e., } \frac{h^2}{\sqrt{3}} g\rho = \frac{x^2}{\sqrt{3}} g\delta,$$

which yields

$$x = h\sqrt{\frac{\rho}{\delta}} = h\sqrt{s}, \tag{5}$$

since $s = \rho/\delta$.

Now we consider the second case when the $\triangle ABC$ has been turned in its own plane until BC is horizontal. This has been shown in dotted line in the figure.

In this case, the weight W of the lamina remains the same, but acts through its c.g. which lies on AE.

The force of buoyancy F' on the immersed portion is

$$F' = \left[\frac{h^2}{\sqrt{3}} - \frac{(h-x)^2}{\sqrt{3}}\right] g\delta = \frac{1}{\sqrt{3}}(2hx - x^2) g\delta, \tag{6}$$

which acts vertically upwards through a point that lies on AE.

There will be a reaction force R at the hinge A. For equilibrium,

$$W + R = F'$$

$$\text{i.e., } \frac{h^2}{\sqrt{3}} g\rho + R = \frac{1}{\sqrt{3}}(2hx - x^2) g\delta$$

$$\text{i.e., } R = \frac{1}{\sqrt{3}}\left(2h \cdot h\sqrt{s} - h^2 s\right) g\delta - \frac{h^2}{\sqrt{3}} g\rho = \frac{h^2}{\sqrt{3}} g\rho \left[\left(2\sqrt{s} - s\right)\frac{1}{s} - 1\right]$$

$$= W\left(\frac{2}{\sqrt{s}} - 2\right) = 2W\frac{1-\sqrt{s}}{\sqrt{s}}. \quad \textbf{(Proved)}$$

EXAMPLE 7.16 *A rectangle movable about an angular point which is fixed below the surface of a liquid, floats with its sides equally inclined to the vertical and with half of its area immersed in the liquid. Show that the ratio of the density of the body to that of the liquid is*

$$\frac{a-b}{4a}, \ a > b.$$

Solution. Let $ABCD$ be the rectangular lamina of density ρ half-immersed in a liquid of density δ, such that its angular point A is fixed and sides AB, AD make equal angles of 45° with vertical (FIG. 7.18).

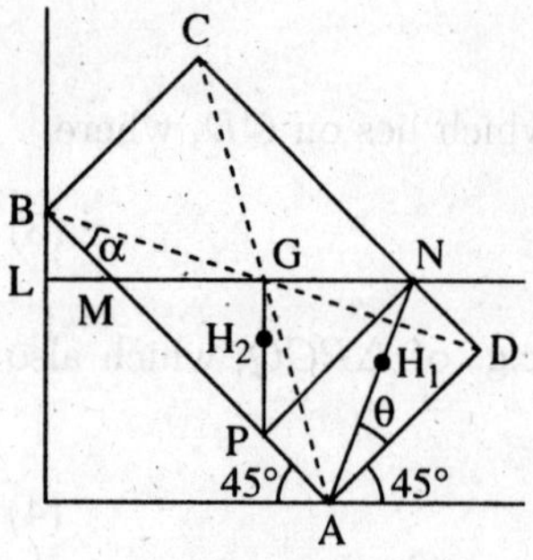

Fig. 7.18

Since the rectangle is half-immersed, the free surface of the liquid LN passes through G, its centre of gravity.

Let

$$\left.\begin{aligned} \angle ABD &= \alpha \\ AB = CD &= a, \\ AD = BC &= b, \\ \text{and } AP = ND &= x \end{aligned}\right\} \tag{1}$$

Since MP and NP are equally inclined to the vertical and hence to the horizontal, we have

$$MP = NP = b \tag{2}$$

Then, area of immersed portion $MNDA$ is

$$= \triangle MNP + \text{rectangle } NPAD = \frac{1}{2}b \cdot b + b\, x = \frac{1}{2}b^2 + bx \tag{3}$$

By condition given,

$$\frac{1}{2}b^2 + bx = \frac{1}{2} \text{ area of } ABCD = \frac{1}{2}ab$$

Hence,

$$x = \frac{1}{2}(a-b) \tag{4}$$

Then,

$$AM = AP + PM = x + b = \frac{1}{2}(a-b) + b = \frac{1}{2}(a+b) \tag{5}$$

The forces acting on the system are:

(1) The weight W of the rectangular lamina $ABCD$ acting vertically downwards through its c.g. G, where

$$W = abg\rho \tag{6}$$

(2) The force of buoyancy F_2 on the immersed $\triangle MPN$ acting vertically upwards through its c.g. H_2, which lies on the vertical median drawn from P, where

$$F_2 = \frac{1}{2}b^2 g\delta. \tag{7}$$

(3) The force of buoyancy F_1 on the rectangle $PADN$ acting vertically upwards through its c.g. H_1, where

$$F_1 = bxg\delta \tag{8}$$

(4) The reaction force R at the fixed corner A. Taking moments of all the forces about A, we get

$$W \cdot AG\cos(\alpha + 45°) = F_2 \cdot PA\cos 45° - F_1 \cdot AH_1\cos(\theta + 45°). \tag{9}$$

Now,

$$\begin{aligned} AG\cos(\alpha + 45°) &= \frac{1}{2}AC \cdot \frac{1}{\sqrt{2}}(\cos\alpha - \sin\alpha) \\ &= \frac{1}{2\sqrt{2}}(AC\cos\alpha - AC\sin\alpha) = \frac{1}{2\sqrt{2}}(a-b) \end{aligned} \tag{10}$$

Similarly,

$$AH_1\cos(\theta+45°)=\frac{1}{2}\cdot AN\cdot\frac{1}{\sqrt{2}}(\cos\theta-\sin\theta)$$
$$=\frac{1}{2\sqrt{2}}(AN\cos\theta-AN\sin\theta)=\frac{1}{2\sqrt{2}}(b-x) \quad (11)$$

Using these values and putting the values of W, F_1 and F_2, we get from (9),

$$abg\rho\cdot\frac{1}{2\sqrt{2}}(a-b)=\frac{1}{2}b^2g\delta\cdot\frac{x}{\sqrt{2}}-bxg\delta\cdot\frac{1}{2\sqrt{2}}(b-x)$$

$$\text{or, } a\rho(a-b)=b\delta x-\delta x(b-x),\ \text{or, } a\rho(a-b)=\delta x^2=\delta\cdot\frac{1}{4}(a-b)^2\ [\text{by (4)}]$$

$$\therefore\ \frac{\rho}{\delta}=\frac{(a-b)^2}{4a(a-b)}=\frac{a-b}{4a}.\quad \textbf{(Proved)}$$

EXAMPLE 7.17 *A uniform isosceles triangular lamina of sp. gr. δ floats in water with its plane vertical, its vertical angle (which is equal to 2α) immersed and the base wholly above the water. Prove that, in the position of equilibrium in which base is not horizontal, the sum of the lengths of the immersed portions of the two sides is $2a\cos^2\alpha$, where a is one of the equal sides. Also prove that $\delta<\cos^4\alpha$, as well as $\cos 2\alpha$.*

Solution. Let ABC be the triangular lamina of sp. gr. δ with $AB=AC=a$, immersed in water with vertex A inside water and side BC wholly above water and not horizontal (FIG. 7.19).

The surface of water cuts the lamina at P and Q.

Let $AP=x, AQ=y$. We are given that $\angle A=2\alpha$.

We take D and E as the mid-points of BC and PQ respectively.

The forces acting on the lamina are:

(1) The weight W acting vertically downwards through its c.g. G, where

$$W=\frac{1}{2}\cdot AB\cdot AC\sin 2\alpha\cdot g\delta=\frac{1}{2}a^2g\delta\sin 2\alpha \quad (1)$$

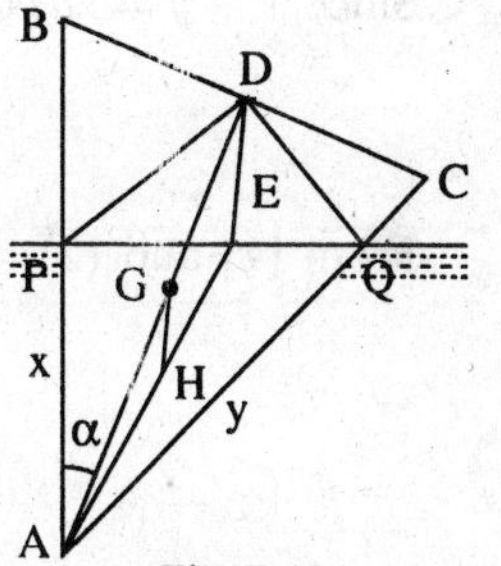

Fig. 7.19

(2) The force of buoyancy F on $\triangle APQ$ acting vertically upwards through its c.g. H, where

$$F=\frac{1}{2}\cdot AP\cdot AQ\sin 2\alpha\cdot g=\frac{1}{2}xyg\sin 2\alpha \quad (2)$$

The conditions of equilibrium are

$$\left.\begin{array}{l}\text{(i) } F=W\\ \text{(ii) the line } GH \text{ should be vertical}\end{array}\right\} \quad (3)$$

Now, in the $\triangle ADE$ and $\triangle AGH$,

$$AG=\frac{2}{3}AD, AH=\frac{2}{3}AE, \text{ so that } \frac{AD}{AG}=\frac{AE}{AH}.$$

Hence, GH and DE are parallel. But GH is a vertical line. Thus, DE also should be vertical. Therefore, in the $\triangle DPQ$, DE is perpendicular to its base PQ and also bisects it. Hence,

$$DP=DQ. \quad (4)$$

Considering the $\triangle APD$ and $\triangle AQD$, we have from cosine formula,

$$\left.\begin{aligned} DP^2 &= x^2 + b^2 - 2bx\cos\alpha \\ \text{and } DQ^2 &= y^2 + b^2 - 2by\cos\alpha \end{aligned}\right\} \tag{5}$$

where we have denoted, $AD = b$.

From (4) and (5), we get

$$x^2 + b^2 - 2bx\cos\alpha = y^2 + b^2 - 2by\cos\alpha$$

$$\text{or, } (x^2 - y^2) - 2b\cos\alpha(x - y) = 0, \text{ i.e., } (x - y)(x + y - 2b\cos\alpha) = 0 \tag{6}$$

Eqn (6) implies that,

$$\text{either } x - y = 0, \text{ i.e., } x = y, \text{ or, } x + y = 2b\cos\alpha \tag{7}$$

But, $x \neq y$, since in that case the side BC is horizontal which is contrary to the proposition. Thus, from (7), we get

$$x + y = 2b\cos\alpha. \tag{8}$$

Now, $b = AD = a\cos\alpha$. Hence,

$$x + y = 2(a\cos\alpha)\cos\alpha = 2a\cos^2\alpha. \quad \textbf{(Proved)}$$

Again, we know from algebra, that the geometric mean of two quantities x and y is less than their arithmetic mean. That is

$$\sqrt{xy} < \frac{x+y}{2}, \text{ i.e., } xy < \frac{(x+y)^2}{4} \tag{9}$$

Since $x + y = 2a\cos^2\alpha$, we get

$$xy < \frac{4a^2\cos^4\alpha}{4}, \text{ i.e., } xy < a^2\cos^4\alpha. \tag{10}$$

From (1) and (2), using the first condition of equilibrium, we have

$$\frac{1}{2}xyg\sin 2\alpha = \frac{1}{2}a^2 g\delta\sin 2\alpha$$

$$\therefore\ xy = a^2\delta, \text{ i.e., } \delta = \frac{xy}{a^2} \tag{11}$$

From (10) and (11), we get

$$\delta < \cos^4\alpha. \quad \textbf{(Proved)} \tag{12}$$

Now, since the side BC should be totally above water, the maximum portion of the $\triangle ABC$ immersed should be such that the vertex C is just on the surface. In this case Q coincides with C, so that

$$AQ = y = a. \tag{13}$$

Then, $x = 2a\cos^2\alpha - y = 2a\cos^2\alpha - a = a\cos 2\alpha$ so that, from (11)

$$\delta = \frac{xy}{a^2} = \frac{(a\cos 2\alpha)a}{a^2} = \cos 2\alpha. \tag{14}$$

Thus, δ must also be less than $\cos 2\alpha$.

EXAMPLE 7.18 *A vertical cylinder of density ρ floats in two liquids, the density of the upper being ρ_1, and that of the lower ρ_2. If the length h of the cylinder be x times the depth of upper liquid, prove that the depth of immersion of the upper face of the cylinder is*

$$\frac{h}{x} - h\frac{\rho_2 - \rho}{\rho_2 - \rho_1}, \text{ provided that } \rho > \rho_2 - \frac{\rho_2 - \rho_1}{x}.$$

Solution. The cylinder of length h and cross sectional area S (say) is shown in the figure (FIG. 7.20). We are given that the depth of lighter liquid of density ρ_1 is h/x.

Let y = length of the cylinder immersed in liquid of density ρ_1. Then the length of the cylinder immersed in liquid of density ρ_2 is $(h - y)$.

Now, the forces acting on the cylinder are: (1) The weight W acting vertically downwards through its c.g., where

$$W = Shg\rho \tag{1}$$

(2) The force of buoyancy due to upper liquid acting vertically upwards, where

$$F_1 = Syg\rho_1, \tag{2}$$

and (3) the force of buoyancy F_2 due to lower liquid acting vertically upwards, where

$$F_2 = S(h - y)g\rho_2. \tag{3}$$

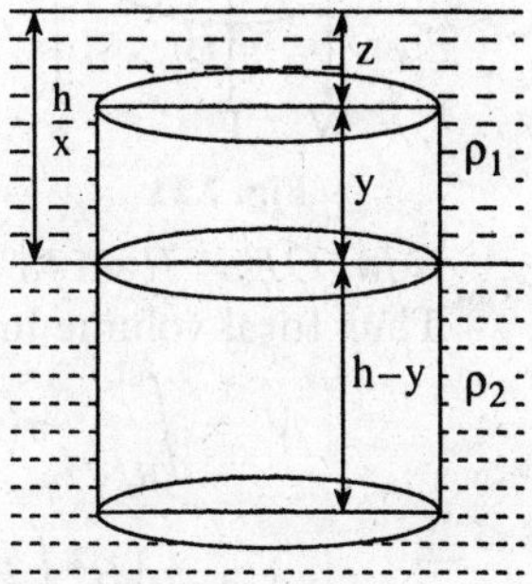

Fig. 7.20

For equilibrium, we must have $W = F_1 + F_2$,

$$\text{i.e., } Shg\rho = Syg\rho_1 + S(h - y)g\rho_2, \text{ or, } h\rho = y\rho_1 + h\rho_2 - y\rho_2$$

$$\therefore y = h\frac{\rho_2 - \rho}{\rho_2 - \rho_1} \tag{4}$$

The depth of immersion of the top surface of the cylinder

$$z = \frac{h}{x} - y = \frac{h}{x} - h\frac{\rho_2 - \rho}{\rho_2 - \rho_1} \tag{5}$$

This depth must be positive. That is,

$$z > 0, \text{ i.e., } \frac{h}{x} - h\frac{\rho_2 - \rho}{\rho_2 - \rho_1} > 0$$

$$\text{or, } \frac{1}{x} > \frac{\rho_2 - \rho}{\rho_2 - \rho_1}, \text{ or, } \rho_2 - \rho_1 > x(\rho_2 - \rho), \text{ or, } \rho x > x\rho_2 - \rho_2 + \rho_1$$

$$\therefore \rho > \rho_2 - \frac{\rho_2 - \rho_1}{x} \tag{6}$$

In order that the cylinder of density ρ floats in the liquid of density ρ_2, we must have

$$\rho < \rho_2 \tag{7}$$

Combining Eqns (6) and (7), we write

$$\rho_2 - \frac{\rho_2 - \rho_1}{x} < \rho < \rho_2. \tag{8}$$

EXAMPLE 7.19 *A hollow hemispherical shell has a heavy particle fixed to its rim and floats in water with the particle just above the surface and with the plane of the rim inclined at an angle of 45° to the surface. Show that the weight of the hemisphere to the weight of the water which it would contain bears the ratio $(4\sqrt{2} - 5) : 6\sqrt{2}$.*

Solution. Let $ABCDA$ be the hemisphere of radius R floating in a liquid of density ρ, such that plane AB makes an angle of 45° to the surface and A is just on the surface (FIG. 7.21).

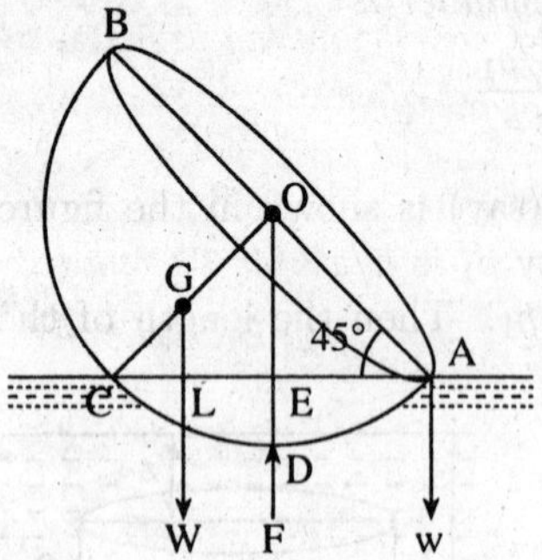

Fig. 7.21

Let W = weight of the hollow hemisphere acting through its c.g. G, which lies on the symmetrical radius OC such that $OG = R/2$, and w = weight of the particle fixed at A.

The portion CDA of the hemisphere is immersed in the liquid. Its volume is determined in the following manner:

Consider a thin circular disc of radius y and thickness dx at a depth x from the centre O (FIG. 7.22).

Then elementary volume is

$$\pi y^2 dx = \pi(R^2 - x^2)dx, \tag{1}$$

since,

$$x^2 + y^2 = R^2 \tag{2}$$

Now, $OE = R\sin 45° = R/\sqrt{2}$.

Thus total volume immersed is

$$V = \int_{R/\sqrt{2}}^{R} \pi(R^2 - x^2)dx$$

$$= \pi\left[R^2\left(R - \frac{R}{\sqrt{2}}\right) - \frac{1}{3}\left(R^3 - \frac{R^3}{2\sqrt{2}}\right)\right]$$

$$= \frac{\pi R^3\,(4\sqrt{2} - 5)}{6\sqrt{2}} \tag{3}$$

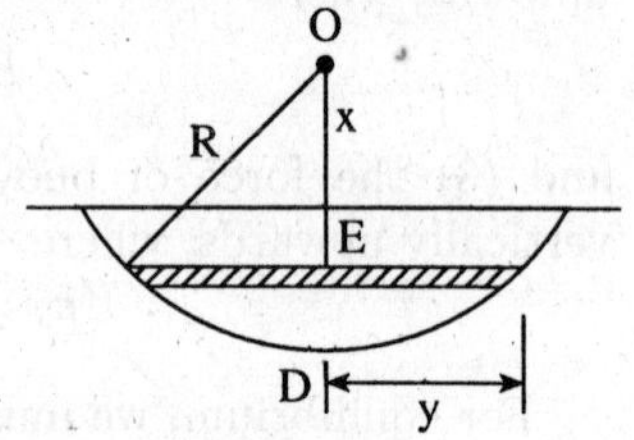

Fig. 7.22

Hence the force of buoyancy F acting vertically along DE is

$$F = Vg\rho = \pi R^3 g\rho \frac{4\sqrt{2} - 5}{6\sqrt{2}}. \tag{4}$$

Taking moments of the forces about A,

$$W \cdot LA = F \cdot EA. \tag{5}$$

Now,

$$\left.\begin{aligned} LA &= LE + EA = OG\cos 45° + OA\cos 45° = \left(\frac{R}{2} + R\right)\frac{1}{\sqrt{2}} = \frac{3R}{2\sqrt{2}} \\ EA &= R \cdot \frac{1}{\sqrt{2}} = \frac{R}{\sqrt{2}} \end{aligned}\right\} \tag{6}$$

From (4), (5) and (6)

$$W \cdot \frac{3R}{2\sqrt{2}} = \pi R^3 g\rho \frac{4\sqrt{2} - 5}{6\sqrt{2}} \cdot \frac{R}{\sqrt{2}}, \text{ or, } W = \frac{2}{3}\pi R^3 g\rho \frac{4\sqrt{2} - 5}{6\sqrt{2}} \tag{7}$$

Now, the weight W' of water which the hemisphere would contain is

$$W' = \frac{2}{3}\pi R^3 g\rho. \tag{8}$$

From (7) and (8), $W = W' \dfrac{4\sqrt{2} - 5}{6\sqrt{2}}$.

Hence, $\dfrac{W}{W'} = \dfrac{4\sqrt{2} - 5}{6\sqrt{2}}$. **(Proved)**

EXAMPLE 7.20 *A solid hemisphere of radius R and weight W is floating in a liquid and at a point on the base at a distance c from the centre rests a weight w. Show that the tangent of the inclination of the axis of the hemisphere to the vertical for the corresponding position of equilibrium assuming the base of the hemisphere to be entirely out of the liquid is* $\frac{8cw}{3RW}$.

Solution. Let $ABPCQ$ be the solid hemisphere of radius R floating in a liquid with its portion PCQ immersed (FIG. 7.23). Let the axis OC of the hemisphere make an angle θ with the vertical. The particle of weight w is at D, such that $OD = c$, where O is the centre of the hemisphere. Ths, the forces acting on the hemisphere are:

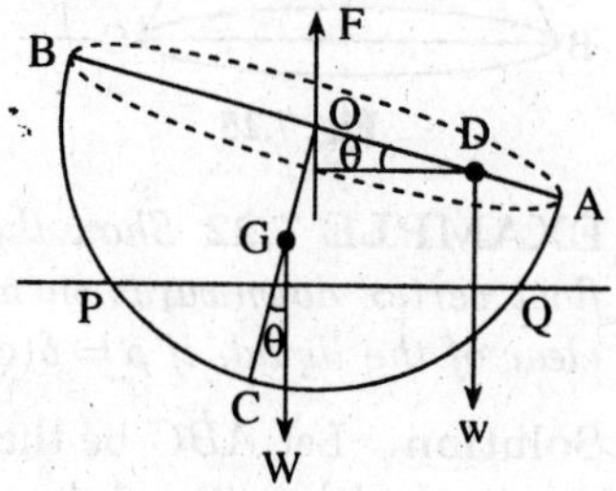

Fig. 7.23

(1) Its weight W acting through its c.g. G vertically downwards, where

$$OG = \frac{3R}{8}. \quad (1)$$

(2) The weight of the particle w acting through D vertically downwards.

(3) The force of buoyancy F on the immersed portion PCQ acting vertically upwards. This force of buoyancy will be along the vertical through the centre O, since the pressure at every point is normal to the surface. Taking moments of all the forces about O, we get

$$W \cdot OG \sin\theta = w \cdot OD \cos\theta, \text{ or, } W \cdot \frac{3R}{8} \sin\theta = wc\cos\theta$$

$$\therefore \tan\theta = \frac{8wc}{3RW}. \quad \textbf{(Proved)}$$

EXAMPLE 7.21 *A solid cone has its axis of length h and is of density ρ. If it floats in a fluid of density δ(> ρ) with its axis vertical, find how much of its axis is out of the fluid.*

Solution. Let the solid cone ABC float in the fluid with vertex downwards (FIG. 7.24).

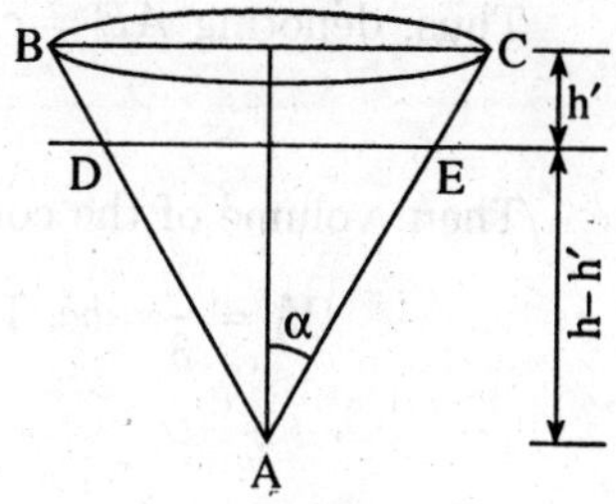

Fig. 7.24

Let α = semi-vertical angle of the cone, h' = length of the axis of the cone outside fluid. Then, weight W of the cone is

$$W = \frac{1}{3}\pi h^3 \tan^2\alpha \cdot g\rho. \quad (1)$$

The force of buoyancy on the portion ADE of the cone is

$$F = \frac{1}{3}\pi(h-h')^3 \tan^2\alpha \cdot g\delta. \quad (2)$$

For equilibrium,

$$F = W$$

i.e., $\frac{1}{3}\pi(h-h')^3\tan^2\alpha \cdot g\delta = \frac{1}{3}\pi h^3\tan^2\alpha \cdot g\rho$, or, $(h-h')^3\delta = h^3\rho$

$$\therefore h' = h\left[1 - \left(\frac{\rho}{\delta}\right)^{1/3}\right] \quad (3)$$

If, however, the vertex A of the cone is upwards and outside the fluid (FIG. 7.25), the force of buoyancy F on the frustrum of the cone $DBCE$ is

$$F = \left(\frac{1}{3}\pi h^3\tan^2\alpha - \frac{1}{3}\pi h'^3\tan^2\alpha\right) g\delta. \quad (4)$$

The weight W remaining same, the condition of equilibrium in this case is

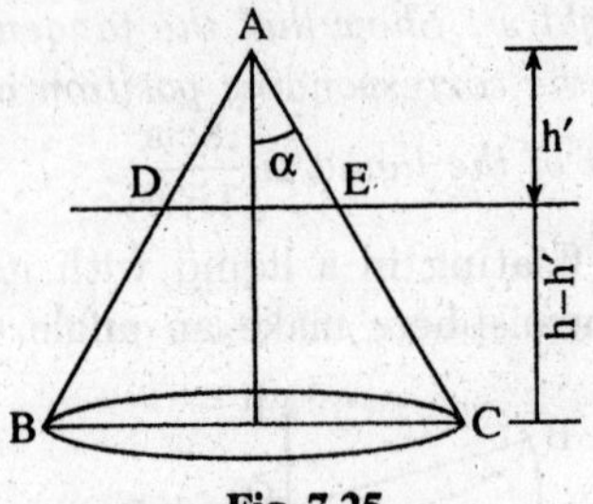

Fig. 7.25

$$\frac{1}{3}\pi h^3 \tan^2 \alpha \cdot g\rho = \left(\frac{1}{3}\pi h^3 \tan^2 \alpha - \frac{1}{3}\pi h'^3 \tan^2 \alpha\right) g\delta$$

$$\text{or, } h^3\rho = h^3\delta - h'^3\delta$$

$$\therefore h'^3\delta = h^3(\delta - \rho)$$

$$\therefore h'^3 = h^3\left(1 - \frac{\rho}{\delta}\right)$$

$$\therefore h' = h\left(1 - \frac{\rho}{\delta}\right)^{1/3}$$

EXAMPLE 7.22 *Show that a right circular cone of density ρ and semi-vertical angle α can float vertex downwards in a liquid of density δ with one generator vertical and the base just clear of the liquid, if $\rho = \delta(\cos 2\alpha)^{3/2}$.*

Solution. Let ABC be the right circular cone of height h and density ρ floating in a liquid of density δ with vertex A downwards, the base plane BC just out of liquid and the generator AB vertical (FIG. 7.26). DC is the liquid surface. The section of the cone by the liquid surface is clearly an ellipse.

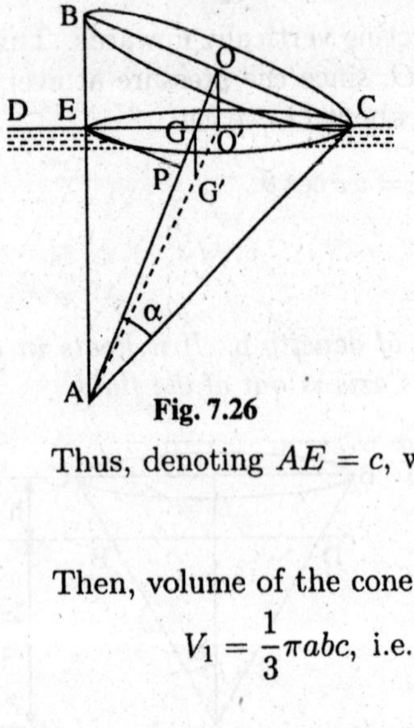

Fig. 7.26

Let EC be the major axis of the ellipse so that $EC = 2a$. We draw EP perpendicular on axis AO; CO is already perpendicular on axis. If b = semi-minor axis of the ellipse, then, by properties of ellipse, we have

$$a = \frac{1}{2}EC = \frac{1}{2}AC \sin 2\alpha$$

$$b = \sqrt{EP \cdot CO} = \sqrt{AC \sin\alpha \cdot AE \sin\alpha}$$

$$\text{i.e., } b = \sin\alpha\sqrt{AC \cdot AC\cos 2\alpha} = AC \cdot \sin\alpha\sqrt{\cos 2\alpha}$$

Thus, denoting $AE = c$, we write,

$$c = AC\cos 2\alpha. \tag{1}$$

Then, volume of the cone AEC immersed in the fluid is

$$V_1 = \frac{1}{3}\pi abc, \text{ i.e., } V_1 = \frac{1}{3}\pi \cdot \frac{1}{2}AC\sin 2\alpha \cdot AC\sin\alpha\sqrt{\cos 2\alpha} \cdot AC\cos 2\alpha$$

$$= \frac{1}{3}\pi(AC)^3 \sin^2\alpha\cos\alpha \cdot (\cos 2\alpha)^{3/2} \tag{2}$$

Thus, the weight of the displaced fluid is

$$W_1 = V_1 g\delta = \frac{1}{3}\pi(AC)^3\sin^2\alpha\cos\alpha(\cos 2\alpha)^{3/2}g\delta \tag{3}$$

Weight of the cone is

$$W = \frac{1}{3}\pi h^3\tan^2\alpha(g\rho) = \frac{1}{3}\pi(AC\cos\alpha)^3\tan^2\alpha \cdot g\rho \tag{4}$$

For equilibrium, $W = W_1$. Hence,

$$\frac{1}{3}\pi(AC\cos\alpha)^3\tan^2\alpha \cdot g\rho = \frac{1}{3}\pi(AC)^3\sin^2\alpha\cos\alpha(\cos 2\alpha)^{3/2}g\delta.$$

Hence, $\rho = \delta(\cos 2\alpha)^{3/2}$. **(Proved)**

We now prove the other condition of equilibrium. If G and G' are the centres of gravity of the cones ABC and AEC, then

$$AG = \frac{3}{4}AO, AG' = \frac{3}{4}AO' \tag{5}$$

Thus $\dfrac{AG}{AO} = \dfrac{AG'}{AO'}$, which implies that GG' and OO' are parallel.

Again, O and O' are middle points of BC and EC respectively. Hence OO' is parallel to BE which is vertical. Then OO' is also vertical, which implies that GG' is a vertical line.

Hence G, G' are on the same vertical line.

EXAMPLE 7.23 *A hollow closed cone of semi-vertical angle* $\sin^{-1}(1/3)$ *and of a metal whose density is* δ*, is made of such uniform thickness that it will float in all the positions wholly submerged in a liquid of density* ρ*. Show that the thickness must be*

$$\frac{h}{4}\left[1-\left(1-\frac{\rho}{\delta}\right)^{1/3}\right]$$

where h is the external height of the cone.

Solution. Let ABC be the cross section of the cone by a vertical plane through its axis (FIG. 7.27). We are given that the semi-vertical angle α of the cone is

$$\alpha = \sin^{-1}\left(\frac{1}{3}\right), \text{ i.e., } \sin\alpha = \frac{1}{3} \tag{1}$$

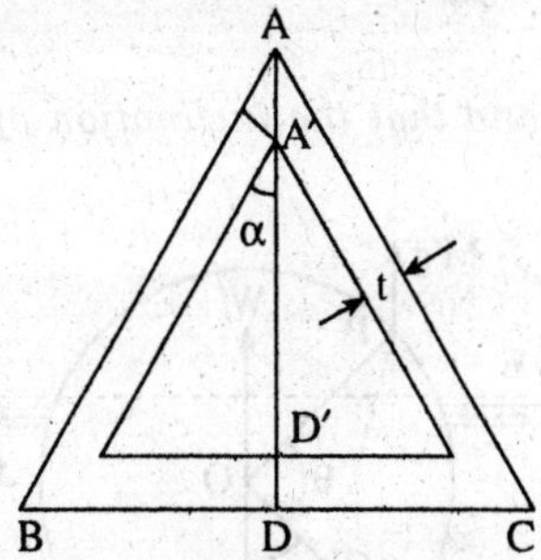

Fig. 7.27

Let h = height (external) of the cone, t = uniform thickness of the metal of the cone.

Then the internal height of the cone is

$$h' = A'D' = AD - AA' - DD'$$
$$= h - t\operatorname{cosec}\alpha - t. \tag{2}$$

The volume of the cone is

$$V = \frac{1}{3}\pi h^2\tan^2\alpha \cdot h = \frac{1}{3}\pi h^3\tan^2\alpha. \tag{3}$$

Hence, the weight of liquid displaced by the cone is

$$W = Vg\rho = \frac{1}{3}\pi h^3 g\rho\tan^2\alpha \tag{4}$$

Volume of the metal in the cone is

$$V' = \frac{1}{3}\pi\tan^2\alpha(h^3 - h'^3) = \frac{1}{3}\pi\tan^2\alpha\left[h^3 - (h - t\operatorname{cosec}\alpha - t)^3\right] \tag{5}$$

Hence, the weight of the metal of the cone is

$$W' = V'g\delta$$
$$\text{i.e., } W' = \frac{1}{3}\pi\tan^2\alpha\left[h^3 - (h - t\operatorname{cosec}\alpha - t)^3\right]g\delta \tag{6}$$

Since the cone floats in all positions submerged in the liquid, we have, for equilibrium

$$W = W'.$$

Thus, from (4) and (6), we get

$$\frac{1}{3}\pi g\rho h^3 \tan^2\alpha = \frac{1}{3}\pi g\delta \tan^2\alpha\,[h^3 - (h - t\,\text{cosec}\,\alpha - t)^3]$$

or, $\rho h^3 = \delta[h^3 - (h - t\,\text{cosec}\,\alpha - t)^3] = \delta[h^3 - (h-4t)^3$, $[\because \text{cosec}\,\alpha = 3]$

or, $\delta(h-4t)^3 = h^3(\delta - \rho)$

or, $(h-4t)^3 = h^3\left(1 - \frac{\rho}{\delta}\right)$

or, $h - 4t = h\left(1 - \frac{\rho}{\delta}\right)^{\frac{1}{3}}$

Hence, $t = \frac{h}{4}\left[1 - \left(1 - \frac{\rho}{\delta}\right)^{1/3}\right]$ **(Proved)**

EXAMPLE 7.24 *A sphere of radius R and mass m is loaded so that its centre of gravity G is at a distance c from its centre O, and is suspended by a string attached to a point P of its surface, GP subtending an angle θ at O. The sphere is partly immersed in a liquid of density ρ, and the tension in the string is M′g. Show that the depth h of O below the surface of the liquid is given by*

$$M - M' = \frac{1}{3}\pi\rho(R+h)^2(2R-h),$$

and that the inclination of GO to the vertical is $\tan^{-1}\left[\frac{M'R\sin\theta}{MC - M'R\cos\theta}\right]$.

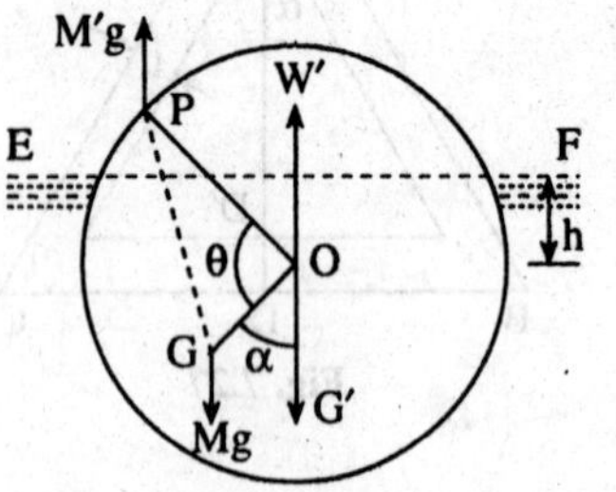

Fig. 7.28

Solution. The sphere of radius R with centre at O immersed in a liquid is shown in the figure (FIG. 7.28), where EF is the surface of the liquid.

Let the inclination of GO to vertical be α. We have the radius of the sphere $OP = R$, $OG = c$, where G is the c.g. of the sphere, and $\angle POG = \theta$.

The forces acting on the sphere are:

(1) Its weight Mg acting vertically downwards through its c.g. G,

(2) The tension $M'g$ acting vertically upwards at P, and

(3) The force of buoyancy W' on the immersed portion of the sphere acting vertically upwards through G', the centre of buoyancy, where

$$W' = Vg\rho, \tag{1}$$

V being the volume of the immersed portion of the sphere. It is clear that the force W' will pass through the centre of the sphere.

Now,

$$V = \frac{2}{3}\pi R^3 + \int_0^h \pi(R^2 - x^2)dx = \frac{2}{3}\pi R^3 + \pi\left(R^2 h - \frac{h^3}{3}\right)$$

$$= \frac{1}{3}\pi\left(2R^3 + 3R^2h - h^3\right) = \frac{1}{3}\pi(R+h)^2(2R-h), \tag{2}$$

where h is the depth of O below the liquid surface.

For equilibrium, we must have

$$Mg - M'g = Vg\rho, \text{ i.e., } M - M' = \frac{1}{3}\pi(R+h)^2(2R-h)\rho \tag{3}$$

Also, for equilibrium, we must have the moments of Mg and $M'g$ about O to be equal. Thus,

$$Mg \cdot GO \sin\alpha = M'g \cdot OP \sin(\alpha+\theta),$$
$$\text{i.e., } Mc\sin\alpha = M'R\sin(\alpha+\theta), \text{ i.e., } Mc\sin\alpha = M'R(\sin\alpha\cos\theta + \cos\alpha\cdot\sin\theta)$$
$$\text{or, } Mc = M'R(\cos\theta + \cot\alpha\cdot\sin\theta), \text{ or, } Mc - M'R\cos\theta = M'R\cot\alpha\cdot\sin\theta.$$

Hence,

$$\tan\alpha = \frac{M'R\sin\theta}{Mc - M'R\cos\theta}, \text{ i.e., } \alpha = \tan^{-1}\left[\frac{M'R\sin\theta}{Mc-M'R\cos\theta}\right]. \quad \textbf{(Proved)}$$

EXAMPLE 7.25 *A cylindrical bucket with water in it balances a mass M over a pulley. A piece of cork of mass m and sp. gr. ρ is then tied to the middle point of the bottom of the bucket so as to be totally immersed.*

Prove that the tension of the string attached to the cork is

$$\frac{2Mmg}{2M+m}\left(\frac{1}{\delta}-1\right).$$

Hence show that the pressure at the lowest point of the curved surface of the bucket will be greater or less than it was originally, according as the volume of the cork $\gtrless \frac{m}{2M}$ *times the volume of water.*

Solution. Initially, the bucket with water balances a mass M over a pulley. When the piece of cork of mass m is attached to the bucket, it will descend with an acceleration f (FIG. 7.29).

Equations of motion are:

$$(M+m)f = (M+m)g - T \tag{1}$$
$$Mf = T - Mg, \tag{2}$$

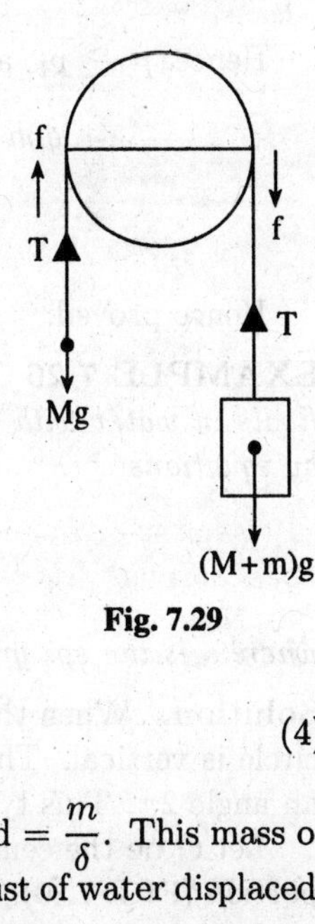

Fig. 7.29

where T is the tension in the string round the pulley.

Eliminating T from (1) and (2), $(2M+m)f = mg$.

Hence,

$$f = \frac{mg}{2M+m}. \tag{3}$$

Now, the forces acting on the cork are:

(1) Its weight mg acting vertically downwards,

(2) The tension T' of the string attached to the cork acting vertically downwards, and

(3) The force of buoyancy R acting vertically upwards.

Thus, the equation of motion of the cork is

$$mf = mg + T' - R. \tag{4}$$

Now, the volume of the cork $= \frac{m}{\delta}$. Hence, the mass of water displaced $= \frac{m}{\delta}$. This mass of water has an acceleration f vertically downwards. Hence the resultant thrust of water displaced, i.e., the force of buoyancy

$$R = \frac{m}{\delta}(g-f). \tag{5}$$

Hence, from (4) and (5),

$$mf = mg + T' - \frac{m}{\delta}(g - f)$$

$$\therefore T' = \frac{m}{\delta}(g - f) - m(g - f) = m\left(\frac{1}{\delta} - 1\right)(g - f) \tag{6}$$

From (3) and (6),

$$T' = m\left(\frac{1}{\delta} - 1\right)\left(g - \frac{mg}{2M + m}\right) = \frac{2Mmg}{2M + 1}\left(\frac{1}{\delta} - 1\right). \quad \textbf{(Proved)} \tag{7}$$

Now, let h = depth of water without cork, h' = depth of water with the cork tied, v = volume of the cork, V = volume of water in the bucket.

Then,

$$\frac{h'}{h} = \frac{V + v}{V}, \text{ i.e., } h' = \frac{V + v}{V}h \tag{8}$$

If ρ be the density of water, then pressure at the lowest point of the curved surface at the initial condition is

$$p_1 = g\rho h. \tag{9}$$

When the bucket is in motion, the pressure at the lowest point is

$$p_2 = (g - f)\rho h'. \tag{10}$$

By (3) and (8), (10) becomes

$$p_2 = \left(g - \frac{mg}{2M + m}\right)\rho \cdot \frac{V + v}{V}h = g\rho h\left(\frac{2M}{2M + m}\right)\frac{V + v}{V} \tag{11}$$

Hence, $p_2 \gtreqless p_1$, according as

$$g\rho h\left(\frac{2M}{2M + m}\right)\frac{V + v}{V} \gtreqless g\rho h, \text{ or, } 2M(V + v) \gtreqless (2M + m)V$$

$$\text{or, } 2Mv \gtreqless mV, \text{ or, } \frac{v}{V} \gtreqless \frac{m}{2M}.$$

Hence proved.

EXAMPLE 7.26 *A life belt in the form of an anchor-ring generated by a circle of radius R floats in water with its equatorial plane horizontal. Show that z, the depth immersed, is given by equations*

$$z = R(1 - \cos\beta)$$

$$2\pi s = (2\beta - \sin 2\beta),$$

where s is the sp. gr. of the material of the belt.

Solution. When the equatorial plane of the life belt is horizontal, the plane of the generating circle is vertical. The solid body is generated by revolving the circle about a vertical axis by an angle 2π. This type of the body is known as ***toroid*** (FIG. 7.30).

Let O be the centre of the vertical generating circle of radius R and ABC be the immersed portion (FIG. 7.30).

Let

$$\angle AOB = \angle BOC = \beta. \tag{1}$$

Then,

$$\left.\begin{array}{r}AO = OC = R,\ AD = DC = R\sin\beta \\ \text{and } OD = R\cos\beta\end{array}\right\} \quad (2)$$

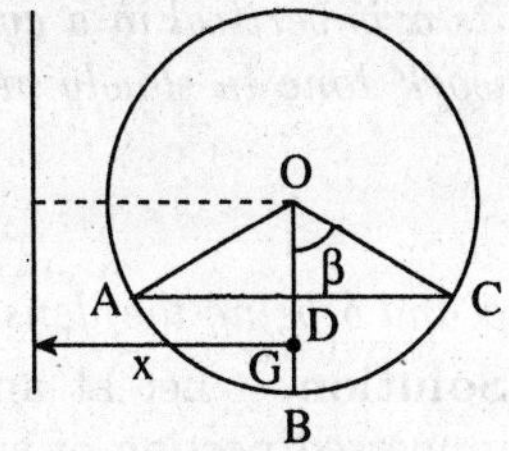

Fig. 7.30

Hence, area of $\triangle OAC$

$$= AD \cdot OD = R\sin\beta \cdot R\cos\beta = R^2 \sin\beta\cos\beta \quad (3)$$

Area of the circle $= \pi R^2$. (4)

Area of the sector $OABC = R^2\beta$. (5)

Hence, area of the immersed portion ABC

$$= \text{area}\, OABC - \text{area}\, \triangle OAC$$
$$= R^2(\beta - \sin\beta\cos\beta) = \frac{1}{2}R^2(2\beta - \sin 2\beta). \quad (6)$$

The depth z of the immersed portion is

$$z = DB = OB - OD = R - R\cos\beta, \text{ i.e., } z = R(1 - \cos\beta) \quad (7)$$

The centre of gravity G of the immersed portion ABC obviously lies on OB.

Let x be the horizontal distance of the line OB from the vertical axis about which the circle revolves to generate the life belt.

Then volume of the life belt by the theorem of Pappus, is

$$V = \text{area of circle} \times \text{arc length moved by its c.g. } O$$
$$= \pi R^2 \cdot 2\pi x = 2\pi^2 R^2 x \quad (8)$$

Similarly, the volume of liquid displaced

$$v = \text{area } ABC \times \text{arc length moved by its c.g. } G$$
$$= \frac{1}{2}R^2(2\beta - \sin 2\beta) \cdot 2\pi x = \pi R^2 x(2\beta - \sin 2\beta) \quad (9)$$

Let ρ and δ be the densities of the material of life belt and the liquid respectively.

Then, the weight of the life belt is

$$W = Vg\rho = 2\pi^2 R^2 x g\rho. \quad (10)$$

Similarly, the weight of the displaced liquid is

$$w = vg\delta = \pi R^2 x(2\beta - \sin 2\beta)g\delta. \quad (11)$$

By condition of floatation, $W = w$.

Hence, from (10) and (11), we get

$$2\pi^2 R^2 x g\rho = \pi R^2 x(2\beta - \sin 2\beta)g\delta, \text{ i.e., } 2\pi\frac{\rho}{\delta} = (2\beta - \sin 2\beta)$$
$$\therefore 2\pi s = (2\beta - \sin 2\beta), \quad (12)$$

where $s = \dfrac{\rho}{\delta}$ is the sp. gr. of the material of the life belt.

From (7) and (12), we conclude that the depth of immersion z is given by the following two equations:

$$\left.\begin{array}{r}z = R(1 - \cos\beta) \\ 2\pi s = (2\beta - \sin 2\beta)\end{array}\right\} \quad \textbf{(Proved)}$$

EXAMPLE 7.27 *A cylindrical piece of wood of length l and sectional area s is floating with its axis vertical in a cylindrical vessel of sectional area S, which contains water. Prove that the work done in slowly pressing down the wood until it is completely immersed is*

$$\frac{1}{2}gl^2 s\left(1-\frac{s}{S}\right)\frac{(\rho-\delta)^2}{\rho},$$

ρ and δ being the densities of water and wood respectively.

Solution. Let at any position, when pressing the wood slowly downwards, x = depth of immersed portion of wood, and h = height of bottom of wood above the bottom of the vessel (FIG. 7.31). Then weight of the wood is

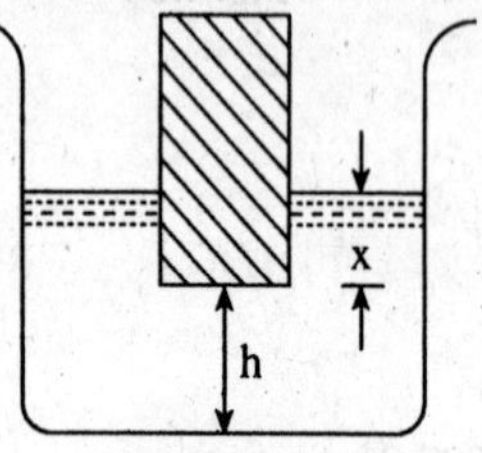

Fig. 7.31

$$W = lsg\delta \tag{1}$$

and that of the displaced liquid is

$$w = xsg\rho \tag{2}$$

Thus, in pressing down the wood slowly, the external force F imposed on the wood is

$$F = w - W = xsg\rho - lsg\delta = gs(x\rho - l\delta). \tag{1}$$

Now, the total volume of water

$$V = Sh + (S-s)x = \text{constant}. \tag{2}$$

Hence, $dV = Sdh + (S-s)dx = 0$, which yields

$$dh = -\left(1-\frac{s}{S}\right)dx \tag{3}$$

Then the work done in pressing down the wood through a depth dh is

$$= -F\cdot dh = gs(x\rho - l\delta)\left(1-\frac{s}{S}\right)dx \tag{4}$$

Now, the portion of wood immersed due to its own weight is

$$l' = \frac{l\delta}{\rho} \tag{5}$$

Thus, the total work done in immersing the wood completely is

$$= \int_{l'}^{l} gs(x\rho - l\delta)\left(1-\frac{s}{S}\right)dx = gs\left(1-\frac{s}{S}\right)\left[\rho\frac{x^2}{2} - l\delta x\right]_{\frac{l\delta}{\rho}}^{l}$$

$$= gs\left(1-\frac{s}{S}\right)\left[\frac{\rho}{2}\left(l^2 - \frac{l^2\delta^2}{\rho^2}\right) - l\delta\left(l - \frac{l\delta}{\rho}\right)\right]$$

$$= \frac{1}{2}gs\left(1-\frac{s}{S}\right)\left[\rho l^2 - \frac{l^2\delta^2}{\rho} - 2l^2\delta\left(1-\frac{\delta}{\rho}\right)\right] = \frac{1}{2}gs\rho l^2\left(1-\frac{s}{S}\right)\left[\left(1-\frac{\delta^2}{\rho^2}\right) - \frac{2\delta}{\rho}\left(1-\frac{\delta}{\rho}\right)\right]$$

$$= \frac{1}{2}gs\rho l^2\left(1-\frac{s}{S}\right)\left(1-\frac{\delta}{\rho}\right)\left[1+\frac{\delta}{\rho}-\frac{2\delta}{\rho}\right] = \frac{1}{2}gs\rho l^2\left(1-\frac{s}{S}\right)\left(1-\frac{\delta}{\rho}\right)^2. \quad \textbf{(Proved)}$$

EXAMPLE 7.28 *A spherical ball of radius r and sp. gr. δ lies at the bottom of a cylindrical vessel of radius R which is filled with water up to a depth of $h(>2r)$. The ball is now slowly lifted just out of water. Show that the work done is*

$$W\left[(\delta-1)h + \frac{2r^3}{3R^2}(1-2\delta) + r\right],$$

where W = weight of water displaced.

Solution. Let the water in the cylinder with the ball at its bottom fill the portion $ABCD$ (FIG. 7.32).

Weight of water displaced is

$$W = \frac{4}{3}\pi r^3 w \qquad (1)$$

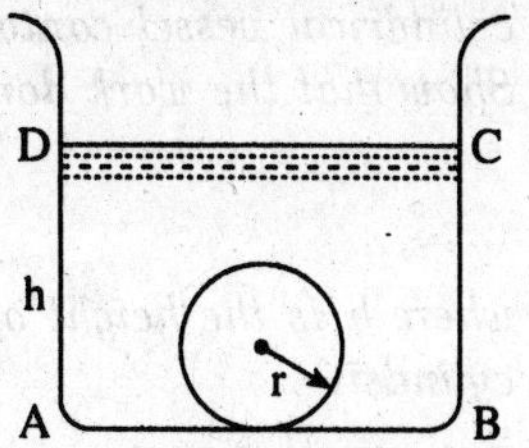

Fig. 7.32

and weight of the ball

$$W' = \frac{4}{3}\pi r^3 \delta w, \qquad (2)$$

where w = weight of unit volume of water.

From (1) and (2),

$$W' = \delta W. \qquad (3)$$

The potential energy at the initial position is P_1 = P.E. of water of portion $ABCD$ − P.E. of water of equal volume of ball + P.E. of the ball.

Thus,

$$P_1 = \pi R^2 hw \cdot \frac{h}{2} - \frac{4}{3}\pi r^3 w \cdot r + \frac{4}{3}\pi r^3 \delta w \cdot r$$

$$\text{i.e., } P_1 = \frac{1}{2}\pi R^2 h^2 w + \frac{4}{3}\pi r^3 w(\delta - 1)r = \frac{3}{8}\frac{R^2 h^2}{r^3}W + W(\delta - 1)r, \text{ [using (1)]} \qquad (4)$$

Let h' = depth of water in the cylinder when the ball has been lifted up just out of water. Since volume of water remains constant,

$$\pi R^2 h' = \pi R^2 h - \frac{4}{3}\pi r^3, \text{ i.e., } h' = h - \frac{4r^3}{3R^2}. \qquad (5)$$

The potential energy in the second case is P_2 = P.E. of water of height h' in the cylinder + P.E. of the ball.

Thus,

$$P_2 = \pi R^2 h' w \cdot \frac{h'}{2} + W'(h' + r) = \frac{3}{8}\frac{R^2 h'^2}{r^3}W + \delta W(h' + r) \qquad (6)$$

Then increase in potential energy is $P = P_2 - P_1$.

From (4) and (6),

$$\begin{aligned} P &= \frac{3}{8}\frac{R^2}{r^3}W(h'^2 - h^2) + \delta W h' + Wr \\ &= \frac{3}{8}\frac{R^2}{r^3}W\left(-\frac{8r^3}{3R^2}h + \frac{16r^6}{9R^4}\right) + \delta W\left(h - \frac{4r^3}{3R^2}\right) + Wr \text{ [by using (5)]} \\ &= -Wh + \frac{2}{3}\frac{r^3}{R^2}W + \delta W h - \frac{4r^3}{3R^2}\delta W + Wr \\ &= W\left[(\delta - 1)h + \frac{2}{3}\frac{r^3}{R^2}(1 - 2\delta) + r\right] \end{aligned} \qquad (7)$$

Work done is equal to increase in potential energy.

Hence the result follows.

Note. If in particular, $\delta = 1$ and $h = 2r$, then from (7),

$$P = W\left[r - \frac{2r^3}{3R^2}\right] = \frac{Wr}{3R^2}(3R^2 - 2r^3).$$

EXAMPLE 7.29 *A right circular cone is slowly lowered, with its axis vertical, into a cylindrical vessel containing liquid of density ρ, until the base of the cone is just immersed. Show that the work done on the liquid is*

$$\frac{1}{36} g\rho h^2 B \frac{3A-2B}{A},$$

where h is the height of the cone, B is its base area and A the area of the cross section of the cylinder.

Solution. Let the cone of base area B and height h be at a depth x below the liquid of density ρ contained in a cylindrical vessel of cross sectional area A (FIG. 7.33). Let y be the distance of the vertex of the cone from the bottom of the cylinder in that position.

If the semi-vertical angle of the cone be α, then

$$B = \pi h^2 \tan^2 \alpha. \tag{1}$$

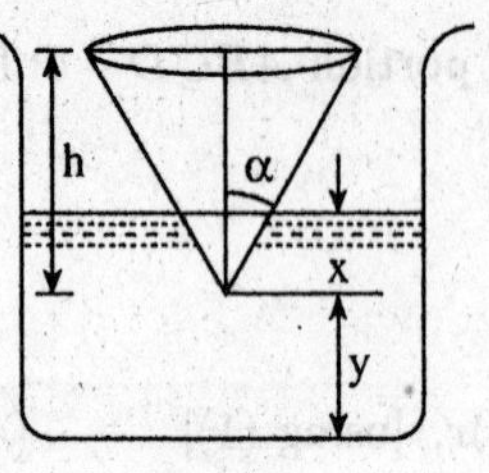

Fig. 7.33

Volume of the cone is

$$V = \frac{1}{3} Bh \tag{2}$$

Volume of the portion under liquid is

$$V_x = \frac{1}{3}\pi x^3 \tan^2 \alpha = \frac{1}{3}\frac{B}{h^2}x^3 \tag{3}$$

Then the force of buoyancy on the immersed portion of the cone is

$$F = V_x g\rho = \frac{1}{3}\frac{B}{h^2}x^3 g\rho \tag{4}$$

Now, volume of water in the vessel is

$$V' = Ay + Ax - V_x$$

$$V' = Ay + Ax - \frac{1}{3}\frac{B}{h^2}x^3 \tag{5}$$

which is constant.

Differentiating (5),

$$dV' \equiv Ady + Adx - \frac{Bx^2}{h^2}dx = 0$$

which yields

$$dy = -\left(1 - \frac{Bx^2}{Ah^2}\right)dx \tag{6}$$

Then work done on the liquid in lowering the cone by a distance of dy is

$$= -Fdy = \frac{1}{3}\frac{B}{h^2}x^3 g\rho\left(1 - \frac{B}{A}\frac{x^2}{h^2}\right)dx \tag{7}$$

Hence the total work done to lower the cone until its base is just immersed, is

$$= \int_0^h \frac{1}{3}\frac{B}{h^2}x^3 g\rho\left(1 - \frac{Bx^2}{Ah^2}\right)dx = \frac{Bg\rho}{3h^2}\int_0^h\left(x^3 - \frac{B}{Ah^2}x^5\right)dx$$

$$= \frac{Bg\rho}{3h^2}\left(\frac{h^4}{4} - \frac{B}{Ah^2}\frac{h^6}{6}\right) = \frac{Bg\rho h^2}{36}\left(3 - \frac{2B}{A}\right)$$

$$= \frac{1}{36}Bg\rho h^2\frac{(3A-2B)}{A}.$$ **(Proved)**

EXAMPLE 7.30 *A cone, whose height is h and vertical angle 2α, has its vertex fixed at a distance c beneath the surface of a liquid; show that it will rest with its base just out of the liquid, if $\delta c^4 \cos^3\alpha \cos\theta = \rho h^4[\cos(\theta+\alpha)\cos(\theta-\alpha)]^{5/2}$, where δ and ρ are the densities of the liquid and cone respectively, and θ is given by the equation $c\cos\alpha = h\cos(\alpha+\theta)$.*

Solution. Let ABC be the cone of height h with its vertex A fixed at a depth c in the liquid and its base BC just out of the liquid (FIG. 7.34). NEC is the liquid surface, O the centre of the base.

Then,

$$AO = h,\ OC = h\tan\alpha \tag{1}$$

ANQ and OE are perpendiculars to the liquid surface.

Then

$$AN = c \tag{2}$$

$$\left.\begin{aligned}\angle CAO = \angle BAO &= \alpha\\ \text{Let } \angle NAO &= \theta\end{aligned}\right\} \tag{3}$$

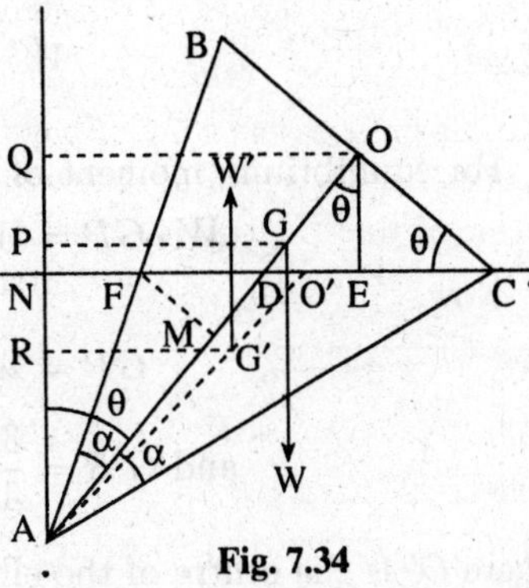

Fig. 7.34

Then, in $\triangle AOQ, \angle AOQ = \dfrac{\pi}{2} - \theta$.

Hence,

$$\angle DOE = \theta, \text{ i.e., } \angle OCE = \theta \tag{4}$$

From the figure,

$$AN = AQ - QN = AQ - OE,$$

$$\text{i.e., } c = h\cos\theta - OC\sin\theta = h\cos\theta - h\tan\alpha\sin\theta$$

$$= \frac{h(\cos\theta\cos\alpha - \sin\theta\sin\alpha)}{\cos\alpha}, \text{ i.e., } c\cos\alpha = h\cos(\theta+\alpha) \tag{5}$$

Now, the section FC of the cone by the liquid surface will be an ellipse of semi-axes a, b (say). We draw FM perpendicular on AO.

Then, $2a = FC = NC - NF$, i.e. $2a = c\{\tan(\theta+\alpha) - \tan(\theta-\alpha)\}$

$$\therefore a = \frac{1}{2}c\frac{\sin 2\alpha}{\cos(\theta+\alpha)\cos(\theta-\alpha)}. \tag{6}$$

By properties of ellipse,

$$b = \sqrt{OC\cdot FM} = \sqrt{h\tan\alpha\cdot AF\sin\alpha},$$

$$\text{i.e., } b = \sqrt{h\tan\alpha\cdot\frac{c\sin\alpha}{\cos(\theta-\alpha)}} = \frac{\sqrt{hc}\sin\alpha}{\sqrt{\cos\alpha\cos(\theta-\alpha)}}. \tag{7}$$

Volume of the cone ABC is

$$V = \frac{1}{3}\pi h^3\tan^2\alpha, \tag{8}$$

so that its weight W acting vertically downwards through its c.g. G is

$$W = Vg\rho = \frac{1}{3}\pi h^3 g\rho\tan^2\alpha \tag{9}$$

Volume of the cone AFC under liquid is

$$V' = \frac{1}{3}\pi abc = \frac{1}{3}\pi\frac{c^2\sqrt{hc}\sin\alpha\sin 2\alpha}{2\cos(\theta+\alpha)\cos(\theta-\alpha)\sqrt{\cos\alpha\cos(\theta-\alpha)}} \tag{10}$$

Hence, force of buoyancy, i.e., weight of displaced liquid acting vertically upwards through its c.g. G' is

$$W' = \frac{1}{3}\pi g\delta \frac{c^2\sqrt{hc}\sin\alpha\sin 2\alpha}{2\cos(\theta+\alpha)\cos(\theta-\alpha)\sqrt{\cos\alpha\cos(\theta-\alpha)}} \tag{11}$$

Using (5), we write (11) in the form

$$W' = \frac{1}{3}\pi g\delta \frac{c^3\sin\alpha\sin 2\alpha}{2[\cos(\theta+\alpha)\cos(\theta-\alpha)]^{3/2}} \tag{12}$$

For equilibrium, moment of these two forces about the fixed point A must be equal. Thus

$$W\cdot GP = W'\cdot G'R \tag{13}$$

Now,

$$GP = AG\cdot\sin\theta = \frac{3}{4}h\sin\theta \tag{14}$$

$$\text{and } G'R = \frac{3}{4}NO' = \frac{3}{4}(NF+FO') = \frac{3}{4}[c\tan(\theta-\alpha)+a] \tag{15}$$

where O' is the centre of the ellipse $FDEC$.

From (6) and (15), we write

$$\begin{aligned} G'R &= \frac{3}{4}c\left[\frac{\sin(\theta-\alpha)}{\cos(\theta-\alpha)} + \frac{\sin 2\alpha}{2\cos(\theta+\alpha)\cos(\theta-\alpha)}\right] \\ &= \frac{3}{4}c\frac{\cos(\theta+\alpha)\sin(\theta-\alpha)+\sin\alpha\cos\alpha}{\cos(\theta+\alpha)\cos(\theta-\alpha)} \\ &= \frac{3}{4}c\frac{\sin\theta\cos\theta}{\cos(\theta+\alpha)\cos(\theta-\alpha)} \end{aligned} \tag{16}$$

Putting the values of W, W' from (9), (12) and $GP, G'R$ from (14), (16), we get from (13),

$$\frac{1}{3}\pi h^3 g\rho\tan^2\alpha\cdot\frac{3}{4}h\sin\theta = \frac{1}{3}\pi g\delta c^3\cdot\frac{\sin^2\alpha\cos\alpha}{[\cos(\theta+\alpha)\cos(\theta-\alpha)]^{3/2}}\cdot\frac{3c}{4}\cdot\frac{\sin\theta\cos\theta}{\cos(\theta+\alpha)\cdot\cos(\theta-\alpha)}$$

$$\text{or, } \frac{h^4\rho}{\cos^2\alpha} = \delta c^4\frac{\cos\alpha\cos\theta}{[\cos(\theta+\alpha)\cos(\theta-\alpha)]^{5/2}}$$

Hence, $\delta c^4\cos^3\alpha\cos\theta = \rho h^4[\cos(\theta+\alpha)\cos(\theta-\alpha)]^{5/2}$. **(Proved)**

EXERCISE

1. A frustrum of a right circular cone, cut off by a plane bisecting the axis perpendicularly, floats with its smaller end in the water and its axis just half-immersed. Prove that the sp. gr. of the cone is 19/56.

2. A thin uniform rod of weight W is loaded at one end with a weight P of insignificant volume. If the rod floats in an inclined position with $\frac{1}{n}$th of its length out of water, prove that

$$(n-1)P = W.$$

3. A hollow conical vessel floats in water with its vertex downwards and a certain depth of its axis immersed. When water is poured into it up to the level originally immersed, it sinks till its mouth is on a level with the surface of the water. What portion of axis was originally immersed? [**Ans.** $\left(\frac{1}{2}\right)^{1/3}$]

4. A uniform rod capable of turning about one of its ends, which is out of water, rests inclined to the vrtical with one-third of its length in water. Prove that its sp. gr. is 5/9.

5. A thin hollow cone with a base floats completely immersed in water wherever it is placed. Show that the vertical angle of the cone is $2\sin^{-1}\left(\frac{1}{3}\right)$.

6. A right circular cylinder of sp. gr. δ floats in water with its axis vertical, two-third being immersed in water. If ρ be the sp. gr. of air, prove that $3\delta = 2 + \rho$.

7. If W_1, W_2 and W_3 be the apparent weights of a given body in fluids whose sp. grs. are s_1, s_2, s_3 respectively, prove that $W_1(s_2 - s_3) + W_2(s_3 - s_1) + W_3(s_1 - s_2) = 0$.

8. A uniform rod of length $2l$ can turn freely about one end which is fixed at a height $h(< 2l)$ above the surface of a liquid. If the densities of the rod and liquid be ρ and δ, show that the rod can rest either in a vertical position or inclined at an angle θ to the vertical such that

$$\cos\theta = \frac{h}{2l}\sqrt{\frac{\delta}{\delta - \rho}}.$$

9. Two solids are each weighed in succession in three homogeneous liquids of different densities. If the weights of the one are w_1, w_2, w_3, and those of the other are W_1, W_2, W_3, prove that

$$w_1(W_2 - W_3) + w_2(W_3 - W_1) + w_3(W_1 - W_2) = 0.$$

10. A thin uniform rod of weight W_1 has a particle of weight W_2 attahced to one end. It is floating in an inclined position in water with this end immsersed. Prove that the length of the rod above water is $\dfrac{W_2}{W_1 + W_2}$ times its whole length, and that the sp. gr. of the rod is $\dfrac{W_1^2}{(W_1 + W_2)^2}$.

11. A heavy uniform rod of small cross section can turn freely in a vertical plane about a hinge placed at a quarter of the length of the rod from one extremity and at a height one-eighth of the length above the surface of a liquid. If d_1 and d_2 $(> d_1)$ be the densities of the rod and the liquid, show that the rod rests at an inclination to the vertical equal to

$$\cos^{-1}\left[\frac{1}{2}\sqrt{\frac{d_2}{9d_2 - 8d_1}}\right]$$

12. A right circular cone of density ρ floats just immersed with its vertex downwards in a vessel containing two liquids of densities δ_1 and δ_2 respectively. Show that the plane of separation of the two liquids cuts off from the axis of the cone a fraction $\sqrt{\dfrac{\rho - \delta_2}{\delta_1 - \delta_2}}$ of its length.

13. AB and BC are two rods of the same cross section, hinged at B, former being heavier and the latter lighter than water. The two rods float in water with the end A freely hinged at a fixed point in the surface of water. Prove that the sp. gr. of the rod AB is equal to $\dfrac{ab + 2by - 2y^2}{ab}$, where $2a, 2b$ are the lengths of the rods AB and BC respectively and $2y$ the length of the immersed portion of the lighter rod.

14. A triangular lamina ABC of which the sides AB, AC are equal, floats in water with BC vertical and three-quarter of length immersed, being kept in equilibrium by means of a string fastened to A and the bottom of the vessel. Find the sp. gr. of the lamina and show that the tension of the string is $\dfrac{1}{27}$th of the weight of the lamina. **[Ans.** $\frac{27}{32}$**]**

15. A prism whose cross section is a triangle ABC is made of unifrom material and floats freely in water with edge C in the surface. Prove that its sp. gr. is either

$$\frac{\sin A\cos B}{\sin C}, \text{ or, } \frac{\sin B\cos A}{\sin C}.$$

16. An equilateral triangular lamina ABC suspended freely from A rests with its side AB vertical and the side AC bisected by the surface of a heavy liquid. Prove that the density of the lamina is to that of the fluid as 15 : 16.

17. A cone of given weight and volume floats in a given liquid with its vertex downwards. Show that the surface of the cone in contact with the liquid is least when the vertical angle of the cone is

$$2\tan^{-1}\left(\frac{1}{\sqrt{2}}\right).$$

18. ABC is an isosceles triangle right angled at A, composed of two heavy rods AB, AC hinged together at A and a light string BC floats with angle A immersed in water. Show that the tension of the string is $\dfrac{W(a-b)}{2a}$, where $2a$ is the length of each rod, $2b$ the length immersed and W the weight of each rod.

19. A square lamina is placed vertically in a liquid of double its density. Prove that it can float only with an edge or diagonal vertical.

20. A uniform prism, whose cross section is an isosceles triangle of vertical angle 2α, floats freely in a liquid with its base just immersed, one edge being in the surface. Show that the ratio of its density to that of the liquid is $2\sin^2\alpha$.

21. A rectangle movable about an angular point floats with half its area immersed in a liquid. If the angular point lies outside the liquid and if the rectangle floats with its sides equally inclined to the vertical, show that the ratio of the density of the rectangle to that of the liquid is $(3b+a) : 4b$, where a and b are the sides of the rectangle $(a < b)$.

22. A sphere of density δ floats just immersed in three liquids. The densities of the liquids in descending order are $\rho, 4\rho, 9\rho$ and the thickness of the two upper liquids are each one-third of the sphere. Prove that $27\delta = 122\rho$.

23. A right circular cone has a plane base in the form of an ellipse. The cone floats in a fluid with its longest generator horizontal. If 2α be the vertical angle of the cone and β the angle between the plane base and the shortest generating line, show that $\cot\beta = \frac{1}{5}\operatorname{cosec} 4\alpha - \cot 4\alpha$.

24. If the height of a right circular cone be equal to the diameter of the base, it will float with its slant side horizontal in any liquid of greater density. Prove this.

25. A hollow cylinder of external and internal radii a, b is open at the top and thickness of its base is c. The cylinder floats in water with its axis vertical and depth h immersed. If the cylinder develops a leak, show that it will never sink if its height is greater than $\dfrac{a^2h-b^2c}{a^2-b^2}$.

26. A solid hemisphere of radius R, floating in a liquid completely immersed with a point of the rim joined to a fixed point by means of a string. Prove that the inclination of the base to the vertical is

$$\tan^{-1}\left(\frac{3}{8}\right).$$

Also prove that the tension of the string is $\frac{2}{3}\pi(\rho-\delta)R^3$, where ρ, δ are the densities of the solid and liquid respectively.

27. A solid cube made of uniform material can turn freely about one edge which is fixed in the surface of water. Prove that, if the cube rests with the face which is not immersed inclined at an angle 30° to the horizon, the density of the cube is to that of water as $(25-7\sqrt{3}) : 18$.

28. A cone of semi-vertical angle α and density ρ has its vertex fixed above the surface of a liquid of density δ. If its axis be inclined to the vertical at an angle θ and it floats with one point of its base in the surface of the liquid, prove that

$$\rho = \delta\left\{1 - \frac{\cos\theta\cos(\theta+\alpha)}{\cos\alpha\cos(\theta-\alpha)}\right\}^{3/2}.$$

29. A sphere of radius a and sp. gr. 0.5 is held completely immersed at the bottom of a circular cylinder of radius b, which is filled with water to a depth h. The sphere is set free and takes up its position of equilibrium. Show that the potential energy lost is

$$W\left[h - \frac{11}{8}a - \frac{a^3}{3b^2}\right],$$

where W is the weight of the sphere.

30. A right circular cone of density δ whose base is an ellipse, floats vertex downwards in a liquid of density ρ with the extremity of the shortest generator on the surface and its base inclined at an angle β to the surface. Prove that the longest generator is vertical and that

$$1 + \tan\alpha\tan\beta = \left(\frac{\rho}{\delta}\right)^{2/3}$$

31. A wooden sphere of radius r is held just immersed in a cylindrical vessel of radius R containing water, and is allowed to rise gently as to be completely out of water. Prove that the loss of potential energy of water is $Wr\dfrac{3R^2 - 2r^2}{3R^2}$, W being the weight of the sphere.

32. To one end of a thin rod floating horizontally in water is attached a string by which the rod is slowly raised from water. Show that until the vertical position is reached, the tension of the string remains constant, and that the work done is then

$$\frac{wl(1-\delta)}{1+\sqrt{1-\delta}},$$

where l, w, δ are the length, weight and sp. gr. of the rod.

33. A solid right cone of density δ, height h and vertical angle 2α can turn freely about its vertex, which is fixed at a height d above the surface of a liquid of density ρ. If it floats with its base wholly immersed and its axis is inclined obliquely at an angle θ with the vertical, show that

$$h^4(\rho - \delta)[\cos(\theta + \alpha)\cos(\theta - \alpha)]^{5/2} = d^4\rho\cos\theta\cos^3\alpha.$$

34. A rectangular lamina floats with its plane vertical and its longer side $2a$ inclined at a small angle θ to the vertical. The lamina is entirely immersed in two fluids of densities δ_1, δ_2 $(\delta_1 < \delta_2)$, its centre is in the common surface and its c.g. is in the line through the centre parallel to the shorter side $2b$ and at a small distance c from the centre. Prove that

$$\theta(\delta_2 - \delta_1)(3a^2 - 2b^2) = 6(\delta_2 + \delta_1)ac.$$

35. A bucket half-full of water is suspended by a string which passes over a pulley small enough to let the other end fall into the bucket. To this end is tied a ball whose sp. gr. δ is greater than 2. Show that, if the ball does not touch the bottom of the bucket and if no water overflows, equilibrium is possible if the weight of the ball lies in between W and $\dfrac{\delta W}{\delta - W}$, where W is the weight of the bucket and water.

CHAPTER 8

Equilibrium of Revolving Fluids

8.1 Introduction

Fluid masses may move in such a way as if it were a rigid body. In these situations, the fluid masses are subjected to accelerations without occurrence of relative motion between fluid particles or between fluid particles and boundaries. The problem of determining pressure in such cases reduces to the statical one, modified to allow for the effects of acceleration.

Let an elementary mass m of the fluid be acted upon by a constant acceleration f whose components are f_x, f_y, f_z, apart from an external force with components X, Y, Z. The resultant effective force per unit mass on the fluid will be

$$\left.\begin{aligned} X' &= X - f_x \\ Y' &= Y - f_y \\ Z' &= Z - f_z, \end{aligned}\right\} \tag{8.1}$$

since the force mf acts in the reversed direction of external force for equilibrium of the system.

8.2 Fluid Masses under Gravity

We consider a fluid mass under gravity subjected to constant acceleration with components f_x, f_z (FIG. 8.1). Since there are no other external force, the components of net forces per unit mass of the fluid are

$$\left.\begin{aligned} X' &= -f_x \\ Z' &= -f_z - g. \end{aligned}\right\} \tag{8.2}$$

We recall the equation of equilibrium from Chapter 3 [ref: Eqn (3.13)]

$$\frac{\partial p}{\partial x} = \rho X,\ \frac{\partial p}{\partial y} = \rho Y,\ \frac{\partial p}{\partial z} = \rho Z, \tag{8.3}$$

which, in this case, assume the forms

$$\frac{\partial p}{\partial x} = -\rho f_x,\ \frac{\partial p}{\partial z} = -\rho(f_z + g). \tag{8.4}$$

These equations characterise the pressure variation through the accelerated mass of the fluid.

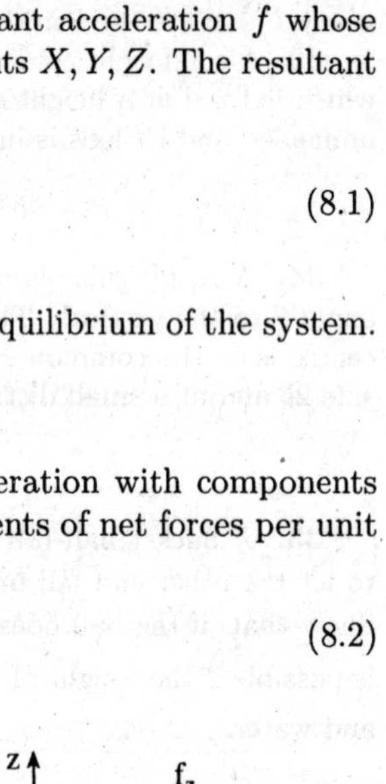

Fig. 8.1

Since, we have

$$dp = \frac{\partial p}{\partial x}dx + \frac{\partial p}{\partial z}dz \tag{8.5}$$

we get, by virtue of Eqn (8.4),

$$dp = -[\rho f_x dx + \rho(f_z + g)dz]. \tag{8.6}$$

Along a line of constant pressure, $dp = 0$. Hence,

$$f_x dx + (f_z + g)dz = 0, \text{ i.e., } \frac{dz}{dx} = -\frac{f_x}{f_z + g}. \tag{8.7}$$

This determines the slope of a line of constant pressure. Its exact position can be obtained with the use of boundary conditions.

Consider, for example, a container of liquid accelerated vertically upwards. Thus, in this case $f_x = 0$, and hence, there is no change of pressure along x-direction. The second of Eqn (8.4) becomes

$$\frac{dp}{dz} = -\rho(f_z + g). \tag{8.8}$$

Since f_z is constant, Eqn (8.8) shows that the pressure variation is linear, but its magnitude will now depend on the value of f_z.

For $f_z = -g$, $\frac{dp}{dz} = 0$, which implies that the pressure will remain constant throughout a freely falling mass of fluid.

If $f_x \neq 0$, the slope of the liquid surface, which is a line of constant pressure, is given by Eqn (8.7). The other lines of constant pressure will be parallel to the surface (FIG. 8.2).

The pressure variation along the direction h, normal to the lines of constant pressure, may be determined thus:

We divide Eqn (8.6) by dh, and get

$$\frac{dp}{dh} = -\rho\left[f_x\frac{dx}{dh} + (f_z + g)\frac{dz}{dh}\right]. \tag{8.9}$$

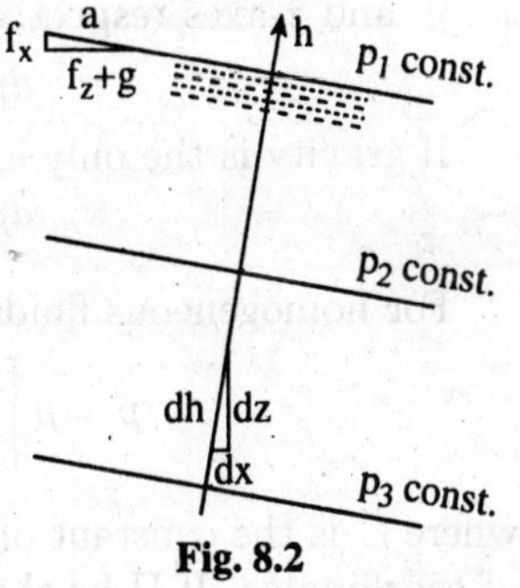

Fig. 8.2

From the similar triangles in the figure, we have

$$\frac{dx}{dh} = \frac{f_x}{a}, \quad \frac{dz}{dh} = \frac{f_z + g}{a}, \tag{8.10}$$

where

$$a = \sqrt{f_x^2 + (f_z + g)^2}. \tag{8.11}$$

Substituting expressions (8.10) into Eqn (8.9), we get

$$\frac{dp}{dh} = -\rho a \tag{8.12}$$

We write Eqn (8.12) in this form

$$\frac{dp}{dh} = -\gamma\frac{a}{g}, \tag{8.13}$$

since the specific weight of the fluid is $\gamma = \rho g$.

This shows that the pressure variation along h is linear and the magnitude of pressure at any depth can be determined by usual method just by taking the specific weight of the fluid equal to

$$\gamma\frac{a}{g} = \frac{\gamma}{g}\sqrt{f_x^2 + (f_z + g)^2}. \tag{8.14}$$

8.3 Fluid Revolving about an Axis

We consider a fluid mass revolving uniformly about an axis without any relative displacement of its particles, i.e., it revolves like a rigid body. We are to determine the pressure at any point in the fluid mass, and the nature of the surfaces of equal pressure.

Let a particle of mass m of the fluid revolve about z-axis with an angular velocity ω at a distance r from the axis (FIG. 8.3).

The particle moves uniformly in a circle of radius r and is subjected to a centripetal force $m\omega^2 r$ towards the centre, i.e., the axis. This force must be equal to the resultant of the external forces acting on the fluid particles and the fluid pressure upon it. Hence the external forces combined with the fluid pressure and the centrifugal force $m\omega^2 r$ away from the axis form a system in statical equilibrium.

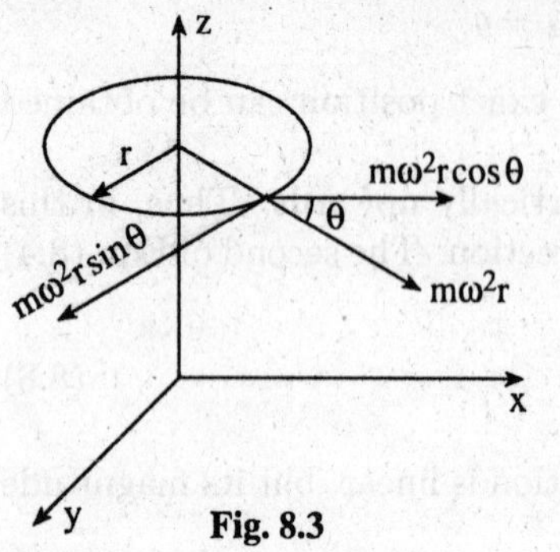

Fig. 8.3

The components of the centrifugal force along x- and y-directions are respectively

$$m\omega^2 r\cos\theta, m\omega^2 r\sin\theta \tag{8.15}$$

If (x, y, z) be the coordinates of the particle in Cartesian system, then

$$x = r\cos\theta, y = r\sin\theta. \tag{8.16}$$

Then the components of the centrifugal force per unit mass of the fluid are

$$\omega^2 x,\ \omega^2 y \tag{8.17}$$

respectively along x- and y-directions.

Let the external force acting per unit mass of the fluid have components X, Y and Z along x-, y- and z-axes respectively. Then the pressure equation for equilibrium is

$$dp = \rho(X dx + Y dy + Z dz + \omega^2 x dz + \omega^2 y dy). \tag{8.18}$$

If gravity is the only external force acting on the fluid mass, then

$$dp = \rho(\omega^2 x dx + \omega^2 y dy - g dz). \tag{8.19}$$

For homogeneous fluids, $\rho =$ constant. In this case, Eqn (8.19) can be integrated to obtain

$$p = \rho\left[\frac{1}{2}\omega^2(x^2+y^2) - gz\right] + C = \rho\left(\frac{1}{2}\omega^2 r^2 - gz\right) + C, \tag{8.20}$$

where C is the constant of integration. Let the vertex of the free surface be taken as the origin of coordinates. If Π be the atmospheric pressure, then we have the boundary condition

$$p = \Pi, \text{ for } r = 0 \text{ and } z = 0. \tag{8.21}$$

This condition when applied to Eqn (8.20), yields $C = \Pi$. Hence,

$$p = \rho\left(\frac{1}{2}\omega^2 r^2 - gz\right) + \Pi \tag{8.22}$$

Surfaces of equi-pressure are given by

$$\rho\left(\frac{1}{2}\omega^2 r^2 - gz\right) + \Pi = \text{constant, i.e., } \frac{1}{2}\omega^2 r^2 - gz = \text{constant,}$$

$$\text{i.e., } r^2 = \frac{2g}{\omega^2} z + \text{constant,} \tag{8.23}$$

which represents paraboloids of revolution having the same latus rectum equal to $2g/\omega^2$.

In the case of heterogeneous fluids, $\rho \neq$ constant. Since dp is an exact differential, ρ must be a function of

$$\left(\frac{1}{2}\omega^2 r^2 - gz\right),$$

so that the same paraboloids are also the surfaces of equal density.

In the case of elastic fluids, $p = k\rho$, where k is a constant. Then the pressure Eqn (8.19) becomes

$$k\frac{dp}{p} = \{\omega^2(x\,dx + y\,dy) - g\,dz\}. \tag{8.24}$$

This, on integration, yields

$$k \log p = \left(\frac{1}{2}\omega^2 r^2 - gz\right) + C. \tag{8.25}$$

Hence, the surfaces of equal pressure and equal density are paraboloids of revolution.

Eqn (8.18) represents the general equation of equilibrium for fluids revolving uniformly about a fixed axis. We can write this equation in the form

$$dp = \rho(X'dx + Y'dy + Z'dz), \tag{8.26}$$

where

$$\left.\begin{aligned} X' &= X + \omega^2 x \\ Y' &= Y + \omega^2 y \\ Z' &= Z. \end{aligned}\right\} \tag{8.27}$$

In order that the fluid is in equilibrium, we must have the necessary and sufficient condition given by

$$X'\left(\frac{\partial Y'}{\partial z} - \frac{\partial Z'}{\partial y}\right) + Y'\left(\frac{\partial Z'}{\partial x} - \frac{\partial X'}{\partial z}\right) + Z'\left(\frac{\partial X'}{\partial y} - \frac{\partial Y'}{\partial x}\right) = 0$$

$$\text{i.e.,}\quad X'\left(\frac{\partial Y}{\partial z} - \frac{\partial Z}{\partial y}\right) + Y'\left(\frac{\partial Z}{\partial x} - \frac{\partial X}{\partial z}\right) + Z'\left(\frac{\partial X}{\partial y} - \frac{\partial Y}{\partial x}\right) = 0$$

$$\text{i.e.,}\quad X\left(\frac{\partial Y}{\partial z} - \frac{\partial Z}{\partial y}\right) + Y\left(\frac{\partial Z}{\partial x} - \frac{\partial X}{\partial z}\right) + Z\left(\frac{\partial X}{\partial y} - \frac{\partial Y}{\partial x}\right)$$

$$= \omega^2\left[x\left(\frac{\partial Z}{\partial y} - \frac{\partial Y}{\partial z}\right) + y\left(\frac{\partial X}{\partial z} - \frac{\partial Z}{\partial x}\right)\right]$$

For fluids under gravity,

$$\left.\begin{aligned} X &= Y = 0 \\ Z &= -g. \end{aligned}\right\} \tag{8.28}$$

In this case, it can be verified that the condition (8.28) is satisfied.

8.4 Revolving Fluids of Mutually Attractive Particles

We consider a fluid mass, the particles of which attract each other according to a definite law, revolving uniformly about a fixed axis. In this case, the fluid particles are in a state of relative equilibrium for a certain form of the free surface. A complete solution of the problem cannot be obtained, since the resultant force of attraction of the fluid mass upon any particle depends on the form of the free surface which is not known.

Let us consider a homogeneous fluid mass, the particles of which attract each other with a force varying directly as the distance, revolving uniformly about an axis through its centre of mass (FIG. 8.4).

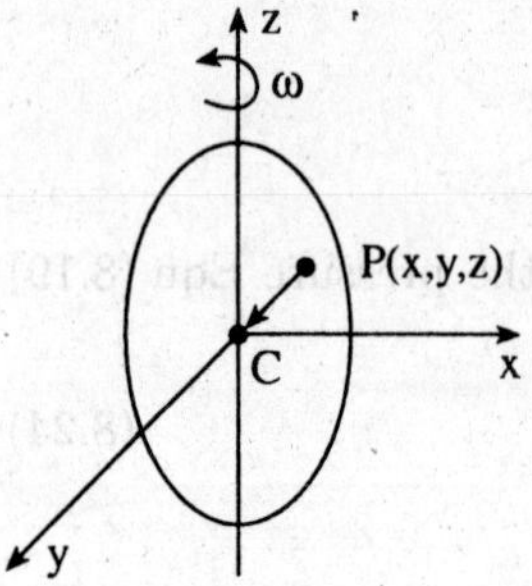

Fig. 8.4

Let $P(x, y, z)$ be the position of any fluid particle w.r.t. a Cartesian coordinate system with its origin at the centre of mass C and axis of rotation as the z-axis. The resultant force of attraction on the particle at P is proportional to its distance from C and directed towards C.

Thus, the components of external force per unit mass at P are

$$X = \mu x, \; Y = \mu y, \; Z = \mu z, \tag{8.29}$$

where μ is a measure of the whole fluid mass.

Then the equation of equilibrium is

$$dp = \rho\{(\omega^2 x - \mu x)dx + (\omega^2 y - \mu y)dy - \mu z dz\}. \tag{8.30}$$

This yields on integration

$$p = \frac{1}{2}\rho\{(\omega^2 - \mu)(x^2 + y^2) - \mu z^2\} + C. \tag{8.31}$$

At the free surface, $dp = 0$. Hence, Eqn (8.30) upon integration yields

$$\frac{1}{2}\rho\{(\omega^2 - \mu)(x^2 + y^2) - \mu z^2\} = \text{constant},$$

$$\text{i.e.,} \quad \left(1 - \frac{\omega^2}{\mu}\right)(x^2 + y^2) + z^2 = D, \tag{8.32}$$

where D is a constant which depends on ω and μ.

Eqn (8.32) represents the form of free surface of the fluid. It is obvious that for small values of ω, the free surface is nearly spherical. As the value of ω increases from zero to μ, the free surface becomes more and more oblate.

When $\omega^2 = \mu$, the free surface consists of two planes defined by

$$z = \pm\sqrt{D}. \tag{8.33}$$

When $\omega^2 > \mu$, the free surface is a hyperboloid of two sheets, which becomes a cone for a certain value of $\omega = \omega^*$.

If $\omega > \omega^*$, the surface is a hyperboloid of one sheet. This surface approximates to the form of a cylinder as ω increases.

8.5 Jacobi's Ellipsoid

Let a fluid mass of density ρ revolve, as if a rigid body, about the axis of z with a constant angular velocity ω. The fluid particles attract each other according to the Newtonian law.

Jacobi discovered that, in this case, an ellipsoid with unequal axes is a possible form of relative equilibrium.

If X, Y, Z be the components of force of attraction on a fluid particle at the point $P(x, y, z)$, then

$$X = Ax, \; Y = By, \; Z = Cz, \tag{8.34}$$

where (Ref. 3)

$$\left.\begin{aligned} A &= 2\pi abc\rho \int_0^\infty \frac{du}{\sqrt{(a^2+u)^3(b^2+u)(c^2+u)}} \\ B &= 2\pi abc\rho \int_0^\infty \frac{du}{\sqrt{(a^2+u)(b^2+u)^3(c^2+u)}} \\ C &= 2\pi abc\rho \int_0^\infty \frac{du}{\sqrt{(a^2+u)(b^2+u)(c^2+u)^3}} \end{aligned}\right\} \tag{8.35}$$

and a, b, c are the semi-axes of the ellipsoid.

Then the equation of equilibrium is

$$dp = \rho\{(A-\omega^2)xdx + (B-\omega^2)ydy + Czdz\}. \tag{8.36}$$

Hence, the equation of free surface is

$$(A-\omega^2)xdx + (B-\omega^2)ydy + Czdz = 0. \tag{8.37}$$

Since a, b, c are the semi-axes of the ellipsoid, its equation is

$$\frac{x^2}{a^2} + \frac{y^2}{b^2} + \frac{z^2}{c^2} = 1. \tag{8.38}$$

Hence,

$$\frac{x}{a^2}dx + \frac{y}{b^2}dy + \frac{z}{c^2}dz = 0. \tag{8.39}$$

Comparing Eqns (8.37) and (8.39), we get

$$A-\omega^2 = \frac{\lambda}{a^2},\ B-\omega^2 = \frac{\lambda}{b^2},\ C = \frac{\lambda}{c^2}, \tag{8.40}$$

where λ is a constant.

From first two relations in (8.40), we get

$$A - \frac{\lambda}{a^2} = B - \frac{\lambda}{b^2}$$

$$\text{i.e., } A - \frac{c^2C}{a^2} = B - \frac{c^2C}{b^2} \text{ [using 3rd of (8.40)]}$$

$$\text{i.e., } a^2b^2(B-A) = (a^2-b^2)c^2C \tag{8.41}$$

Again, from first two relations in (8.40), we get by eliminating λ,

$$\omega^2(a^2-b^2) = Aa^2 - Bb^2. \tag{8.42}$$

Let M be the mass of the fluid ellipsoid. Then

$$M = \frac{4}{3}\pi abc\rho. \tag{8.43}$$

Defining

$$D = [(a^2+u)(b^2+u)(c^2+u)]^{1/2}, \tag{8.44}$$

and using Eqn (8.43), we write Eqns (8.35) in the form

$$A = \frac{3}{2}M\int_0^\infty \frac{du}{D(a^2+u)},\ B = \frac{3}{2}M\int_0^\infty \frac{du}{D(b^2+u)},\ C = \frac{3}{2}M\int_0^\infty \frac{du}{D(c^2+u)}. \tag{8.45}$$

Putting these values in Eqn (8.41), we get

$$\frac{3}{2}Ma^2b^2\int_0^\infty \frac{du}{D}\left[\frac{1}{b^2+u}-\frac{1}{a^2+u}\right]=\frac{3}{2}M(a^2-b^2)\int_0^\infty \frac{c^2du}{D(c^2+u)}$$

$$\text{i.e.,}\quad \int_0^\infty \frac{du}{D}\left[\frac{a^2b^2(a^2-b^2)}{(a^2+u)(b^2+u)}-\frac{c^2(a^2-b^2)}{c^2+u}\right]=0$$

$$\text{i.e.,}\quad (a^2-b^2)\int_0^\infty \frac{du}{D}\left[\frac{a^2b^2}{(a^2+u)(b^2+u)}-\frac{c^2}{c^2+u}\right]=0. \tag{8.46}$$

Since $a \neq b$, we get from this equation

$$\int_0^\infty \frac{du}{D}\left[\frac{a^2b^2}{(a^2+u)(b^2+u)}-\frac{c^2}{c^2+u}\right]=0. \tag{8.47}$$

Eqn (8.47) can be simplified to

$$\int_0^\infty \frac{udu}{D^3}\left(\frac{1}{a^2}+\frac{1}{b^2}-\frac{1}{c^2}+\frac{u}{a^2b^2}\right)=0. \tag{8.48}$$

This equation determines the semi-axis c of the ellipsoid when a, b are given.

The LHS of Eqn (8.48) is negative for $c = 0$ and positive for $c = \infty$. This implies that there is at least one real value of c satisfying the Eqn (8.48).

Again, the expression

$$\frac{1}{a^2}+\frac{1}{b^2}-\frac{1}{c^2}+\frac{u}{a^2b^2} \tag{8.49}$$

is positive for large values of u, and is negative for small u.

Moreover, $\dfrac{u}{D^3}$ is positive. Hence,

$$\frac{1}{c^2}>\frac{1}{a^2}+\frac{1}{b^2}. \tag{8.50}$$

Thus, c is less than the least of the two quantities a, b.

Let us now find the angular velocity ω. Substituting the values of A and B from Eqn (8.45) into Eqn (8.42), we get

$$\omega^2(a^2-b^2)=\frac{3}{2}M\int_0^\infty \frac{du}{D}\left[\frac{a^2}{a^2+u}-\frac{b^2}{b^2+u}\right]$$

$$=\frac{3}{2}M(a^2-b^2)\int_0^\infty \frac{udu}{D(a^2+u)(b^2+u)} \tag{8.51}$$

Since $a \neq b$,

$$\omega^2=\frac{3}{2}M\int_0^\infty \frac{udu}{D(a^2+u)(b^2+u)}. \tag{8.52}$$

The integrand in the RHS is a positive quantity and hence a possible value of ω can be determined.

Thus, an ellipsoid with three unequal semi-axes is a possible form of relative equilibrium of a fluid mass revolving about the smallest axis.

Investigations by Meyer shows that the maximum value of $\omega^2/2\pi\rho$ that will make a Jacobian ellipsoid a possible form of relative equilibrium, is 0.18709. For smaller values of $\omega^2/2\pi\rho$, there is one and only one ellipsoid.

8.6 Elliptic Cylinder

We consider an infinite mass of homogeneous fluid in the form of an elliptic cylinder, revolving, as if rigid, about the axis of the cylinder. The particles of the fluid mass attract each other according to the Newtonian law.

The components of the force of attraction on unit mass at an internal point (x, y) are given by (Ref. 3)

$$X = \frac{4\pi\rho b}{a+b}x, \; Y = \frac{4\pi\rho a}{a+b}y, \tag{8.53}$$

where a, b are the semi-axes of the elliptic section and ρ is the density of the fluid. The pressure equation in this case is

$$\begin{aligned} dp &= \rho\{X dx + Y dy - \omega^2 x dx - \omega^2 y dy\} \\ &= \rho\left\{\left(\frac{4\pi\rho b}{a+b} - \omega^2\right) x dx + \left(\frac{4\pi\rho a}{a+b} - \omega^2\right) y dy\right\}. \end{aligned} \tag{8.54}$$

Hence, the equation of free surface is

$$\left(\frac{4\pi\rho b}{a+b} - \omega^2\right) x dx + \left(\frac{4\pi\rho a}{a+b} - \omega^2\right) y dy = 0. \tag{8.55}$$

The surface of the elliptic cylinder is defined by

$$\frac{x^2}{a^2} + \frac{y^2}{b^2} = 1, \; z = \pm\infty. \tag{8.56}$$

Hence,

$$\frac{x}{a^2}dx + \frac{y}{b^2}dy = 0. \tag{8.57}$$

Comparing Eqns (8.55) and (8.57), we get

$$\begin{aligned} a^2\left(\frac{4\pi\rho b}{a+b} - \omega^2\right) &= b^2\left(\frac{4\pi\rho a}{a+b} - \omega^2\right) \\ \text{i.e.,}\quad 4\pi\rho ab\left(\frac{a-b}{a+b}\right) &= (a^2 - b^2)\omega^2 \\ \text{i.e.,}\quad 4\pi\rho ab\frac{1}{a+b} &= (a+b)\omega^2 \;[\text{since } a \neq b] \\ \therefore\; \omega^2 &= \frac{4\pi\rho ab}{(a+b)^2}. \end{aligned} \tag{8.58}$$

This equation determines ω when a, b and ρ are given.

Thus, an elliptic cylinder is a possible form of relative equilibrium when an infinite mass of homogeneous fluid revolves, as if rigid, about the axis of the cylinder.

Now, from Eqn (8.58), we find that

$$1 - \frac{\omega^2}{\pi\rho} = 1 - \frac{4ab}{(a+b)^2} = \left(\frac{a-b}{a+b}\right)^2,$$

so that

$$\frac{a-b}{a+b} = \sqrt{1 - \frac{\omega^2}{\pi\rho}} \tag{8.59}$$

This implies that the quantity in the RHS must be less than unity. This is possible only if

$$\omega^2 < \pi\rho.$$

Thus, the elliptic cylinder will not be a possible form of relative equilibrium, unless

$$\omega^2 < \pi\rho, \text{ i.e., } \frac{\omega^2}{2\pi\rho} < 0.5. \tag{8.60}$$

8.7 Maclaurin's Theorem

The theorem states:

An oblate spheroid of small ellipticity is a possible form of relative equilibrium of a mass of homogeneous liquid, the particles of which attract each other according to the Newtonian law, rotating uniformly about an axis through its centre of mass.

Proof. Let the equation of the oblate spheroid be

$$\frac{x^2 + y^2}{c^2(1+\lambda^2)} + \frac{z^2}{c^2} = 1 \tag{8.61}$$

rotating uniformly about z-axis with angular velocity ω.

The components of force of attractions towards the origin on a particle of unit mass at the point (x, y, z) are (Ref. 2)

$$\left.\begin{aligned} X &= \frac{2\pi\rho x}{\lambda^3}\left[(1+\lambda^2)\tan^{-1}\lambda - \lambda\right] \\ Y &= \frac{2\pi\rho y}{\lambda^3}\left[(1+\lambda^2)\tan^{-1}\lambda - \lambda\right] \\ Z &= \frac{4\pi\rho z}{\lambda^3}(1+\lambda^2)(\lambda - \tan^{-1}\lambda) \end{aligned}\right\} \tag{8.62}$$

along x-, y- and z-axes respectively. We have assumed that the boundary of the rotating fluid is an oblate spheroid.

The pressure equation is

$$dp = \rho\left[(\omega^2 x - X)\,dx + (\omega^2 y - Y)\,dy - Z dz\right]. \tag{8.63}$$

Hence, the surfaces of equi-pressure are given by

$$(\omega^2 - A)x dx + (\omega^2 - B)y dy - C z dz = 0, \tag{8.64}$$

where

$$\left.\begin{aligned} A = B &= \frac{2\pi\rho}{\lambda^3}\left[(1+\lambda^2)\tan^{-1}\lambda - \lambda\right] \\ C &= \frac{4\pi\rho}{\lambda^3}(1+\lambda^2)\left(\lambda - \tan^{-1}\lambda\right) \end{aligned}\right\} \tag{8.65}$$

Differentiating Eqn (8.61), we have

$$x dx + y dy + (1+\lambda^2) z dz = 0. \tag{8.66}$$

Comparing Eqns (8.64) and (8.66), we get

$$\omega^2 - A = \omega^2 - B = -\frac{C}{1+\lambda^2}. \tag{8.67}$$

Hence,

$$\omega^2 = A - \frac{C}{1+\lambda^2} = \frac{2\pi\rho}{\lambda^3}\left[(1+\lambda^2)\tan^{-1}\lambda - \lambda - 2\left(\lambda - \tan^{-1}\lambda\right)\right]$$

$$\text{or, } \frac{\omega^2}{2\pi\rho} = \frac{1}{\lambda^3}\left[(3+\lambda^2)\tan^{-1}\lambda - 3\lambda\right] \qquad (8.68)$$

This equation determines λ, and hence the ratio of the semi-axes, when ω and ρ are given.

The function in the RHS of Eqn (8.68) is evaluated for different values of λ. This is provided in the Table 8.1.

Table 8.1

λ	0.0	1.0	1.5	2.0	2.5	3.0	3.5	4.0	4.5	5.0	6.0	7.0	8.0	9.0	10.0
$\frac{\omega^2}{2\pi\rho}$	0.0	0.1415	0.1955	0.2187	0.2246	0.2218	0.2148	0.2061	0.1968	0.1876	0.1705	0.1556	0.1424	0.1312	0.1215

With these values, a graph of the function has been drawn which is shown in FIG. 8.5. From the graph, it is found that the function attains a maximum value of 0.2247 for $\lambda = 2.52$.

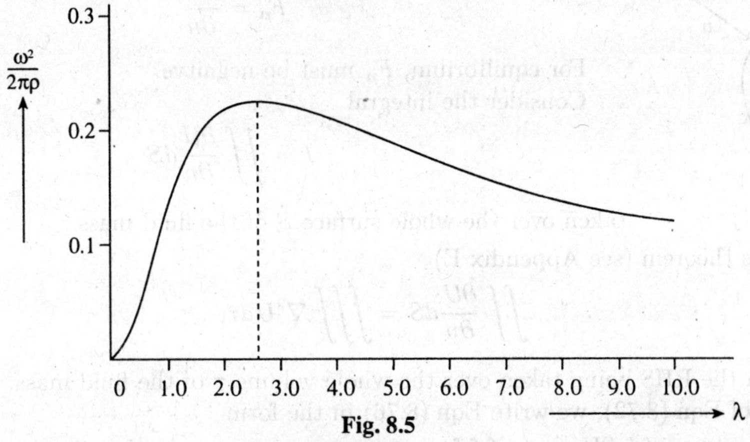

Fig. 8.5

If the value of $\omega^2/2\pi\rho$ is less than 0.2247, there are two real values of λ. Thus, we conclude that:

(1) The oblate spheroid is not a possible form of equilibrium, if

$$\frac{\omega^2}{2\pi\rho} > 0.2247. \qquad (8.69)$$

(2) There are two spheroidal forms possible, if

$$\frac{\omega^2}{2\pi\rho} < 0.2247. \qquad (8.70)$$

8.8 Poincare's Theorem

The theorem states:

If a homogeneous fluid mass of density ρ, the particles of which attract each other according to the Newtonian law, rotates with uniform angular velocity ω about an axis through its centre of mass, such that

$$\frac{\omega^2}{2\pi\rho} > 1, \qquad (8.71)$$

then there is no figure of relative equilibrium possible.

Proof. At every point on the free surface of the fluid mass, there act the force of attraction and the centrifugal force due to rotation. The resultant of these two forces must be directed towards the interior, otherwise the portion would be detached and move away. This is the necessary condition for relative equilibrium of the fluid mass.

Let V be the potential of the attracting forces. Then, by Poisson's equation, at every point at which there is attracting matter of volume density ρ, we have (see Appendix G)

$$\nabla^2 V = -4\pi\rho. \tag{8.72}$$

We define the function

$$U = V + \frac{1}{2}\omega^2 r^2, \tag{8.73}$$

where r is the distance from the axis of rotation.

If n denotes the direction of the outward normal to the surface S of the fluid mass (FIG. 8.6), then the resultant outward normal force acting at any point on the surface of the fluid mass is

$$F_n = \frac{\partial U}{\partial n}. \tag{8.74}$$

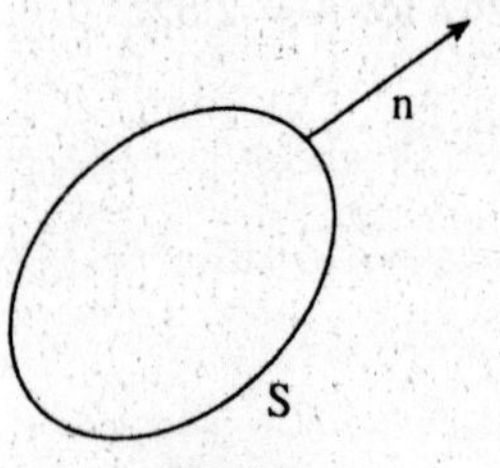

Fig. 8.6

For equilibrium, F_n must be negaitve.

Consider the integral

$$I = \iint \frac{\partial U}{\partial n} dS \tag{8.75}$$

taken over the whole surface S of the fluid mass.

By Green's theorem (see Appendix E)

$$\iint \frac{\partial U}{\partial n} dS = \iiint \nabla^2 U d\tau, \tag{8.76}$$

the integral in the RHS being taken over the whole volume τ of the fluid mass.

By virtue of Eqn (8.72), we write Eqn (8.76) in the form

$$\iint \frac{\partial U}{\partial n} dS = \iiint (-4\pi\rho + 2\omega^2) d\tau = (2\omega^2 - 4\pi\rho)\tau. \tag{8.77}$$

Eqn (8.77) shows that, if

$$\omega^2 > 2\pi\rho, \text{ i.e., } \frac{\omega^2}{2\pi\rho} > 1, \tag{8.78}$$

the resultant force at some points on the surface is directed outwards, and, therefore, equilibrium is impossible. Hence the theorem.

EXAMPLE 8.1 *A conical vessel of height h and vertical angle 2α contains water whose volume is one-half that of the cone. If the vessel and the contained water revolve with uniform angular velocity ω, and no water overflows, show that ω must not be greater than*

$$\sqrt{\frac{2g}{3h}} \cot\alpha.$$

Solution. Let ABC be the cone containing water and revolving about the axis AO (FIG. 8.7). Volume of water in the cone (= volume of empty space) is

$$V_1 = \frac{1}{2} \cdot \frac{\pi}{3} h^3 \tan^2\alpha = \frac{\pi}{6} h^3 \tan^2\alpha. \tag{1}$$

Let D be the vertex of the paraboloid of free surface when the cone revolves such that water reaches the rim of the cone.

Then,

$$OC^2 = \frac{2g}{\omega^2} \cdot OD. \tag{2}$$

But,

$$OC = h \tan \alpha \tag{3}$$

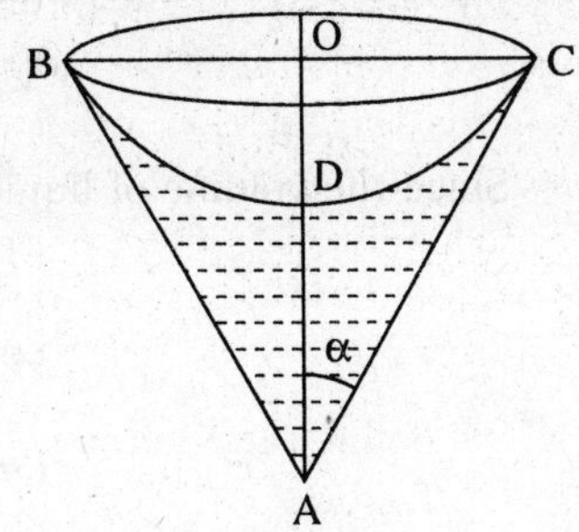

Fig. 8.7

Hence, from (2) and (3), we get

$$\frac{2g}{\omega^2} \cdot OD = h^2 \tan^2 \alpha, \text{ i.e., } OD = \frac{\omega^2 h^2 \tan^2 \alpha}{2g} \tag{4}$$

Volume of the paraboloid BDC is

$$V_2 = \frac{1}{2} \text{ volume of the cylinder of height } OD \text{ and base } BC$$

$$= \frac{1}{2} \cdot \pi h^2 \tan^2 \alpha \cdot \frac{\omega^2 h^2 \tan^2 \alpha}{2g} = \frac{\pi \omega^2 h^4}{4g} \tan^4 \alpha \tag{5}$$

If no water is to overflow, the volume of the empty space of the paraboloid cannot be greater than the volume of empty space before. That is,

$$V_2 \not> V_1, \text{ i.e., } \frac{\pi \omega^2 h^4}{4g} \tan^4 \alpha \not> \frac{\pi}{6} h^3 \tan^2 \alpha, \text{ i.e., } \omega^2 \tan^2 \alpha \not> \frac{2g}{3h}$$

$$\text{i.e., } \omega \not> \sqrt{\frac{2g}{3h}} \cot \alpha. \quad \textbf{(Proved)}$$

EXAMPLE 8.2 *A hollow sphere of radius R half-filled with liquid is made to rotate with angular velocity ω about its vertical diameter. If the lowest point of the sphere is just exposed, show that*

$$\frac{2g}{\omega^2 R} = 2 - \sqrt[3]{4}.$$

Solution. Let OD be the vertical diameter about which the sphere containing water rotates. Since the lowest point is just exposed, AOB is the paraboloid of revolution of the free surface (FIG. 8.8).

With OD as z-axis, equation to this paraboloid is

$$r^2 + (z - R)^2 = R^2, \text{ i.e., } r^2 = (2R - z)z, \tag{1}$$

where $OP = z$, $BP = r$, $BC = R$ and $PC = z - R$.

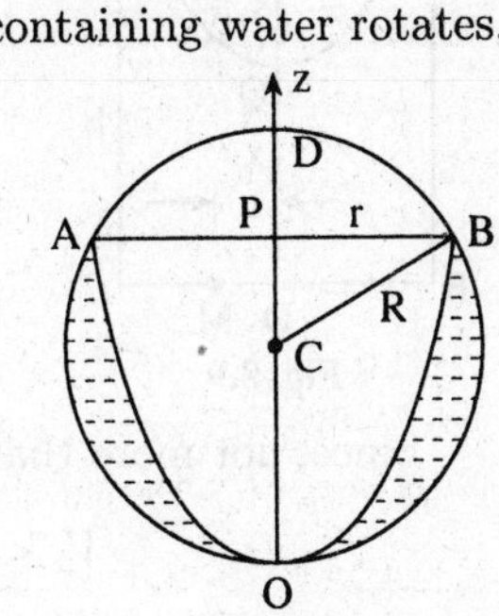

Fig. 8.8

Latus rectum of the paraboloid is given by

$$\frac{2g}{\omega^2} = \frac{BP^2}{OP} = \frac{r^2}{z}, \text{ i.e., } \frac{2g}{\omega^2} = 2R - z \text{ [using (1)]}$$

$$\therefore 2g = \omega^2 (2R - z) \tag{2}$$

Now, the volume of the liquid is

$$V_1 = \frac{2}{3} \pi R^3. \tag{3}$$

Volume of the spherical portion AOB is

$$V_2 = \int_0^z \pi r^2 dz = \int_0^z \pi (2R - z) z dz \text{ [using (1)]}$$

$$= \frac{\pi}{3} z^2 (3R - z) \tag{4}$$

Volume of the empty space $AOBP$ of paraboloid of revolution is

$$V_3 = \frac{1}{2}\pi r^2 \cdot z = \frac{1}{2}\pi(2R - z)z^2 \text{ [using (1)]} \tag{5}$$

Since the volume of liquid remains unaltered, we must have

$$V_1 = V_2 - V_3$$

$$\text{i.e., } \frac{2}{3}\pi R^3 = \frac{\pi}{3}z^2(3R - z) - \frac{\pi}{2}z^2(2R - z)$$

$$\text{i.e., } \frac{2}{3}R^3 = -\frac{1}{3}z^3 + \frac{1}{2}z^3 = \frac{1}{6}z^3$$

$$\therefore\ z = \sqrt[3]{4}R \tag{6}$$

Putting this value of z into (2), we get $2g = \omega^2\,(2 - \sqrt[3]{4})\,R$

$$\therefore\ \frac{2g}{\omega^2 R} = 2 - \sqrt[3]{4}. \quad \textbf{(Proved)}$$

EXAMPLE 8.3 *A right circular cylinder, open at the top, is filled with water and the whole of it revolves with angular velocity ω about its axis. If not more than half the water is spilt, prove that the thrust on the base is*

$$\pi g\rho R^2 h\left(1 - \frac{R^2\omega^2}{4h}\right),$$

where h is the height, R is the radius of the base of the cylinder and ρ is the density of water.

Solution. Let $ABCD$ be the cylinder containing water and revolving about its axis OP. The point Q is the vertex of the paraboloid of revolution of the free surface (FIG. 8.9).

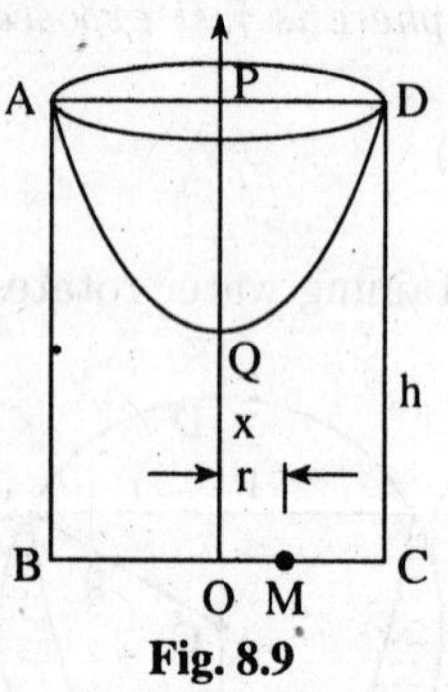

Fig. 8.9

With Q as origin, we take z-axis vertically upwards. In the figure,

$$\left.\begin{aligned} AP = PD &= R \\ AB = CD &= h \\ \text{and, let } OQ &= x \end{aligned}\right\} \tag{1}$$

Now, volume of the paraboloid AQD is

$$V_1 = \frac{1}{2}\pi R^2 \cdot PQ \tag{2}$$

and volume of the cylinder is

$$V_2 = \pi R^2 \cdot OP. \tag{3}$$

Since, not more than half the water is spilt,

$$V_1 < \frac{1}{2}V_2, \text{ i.e., } \frac{1}{2}\pi R^2 \cdot PQ < \frac{1}{2}\pi R^2 \cdot OP, \text{ i.e., } PQ < OP \tag{4}$$

which implies that the vertex Q must remain in between O and P. The figure conforms to this requirement.

The pressure at any point distant r from the axis, by Eqn (8.20), is

$$p = \rho\left(\frac{1}{2}\omega^2 r^2 - gz\right) + C. \tag{5}$$

At the vertex Q, we have the condition $p = 0$. Thus,

$$p = 0, \text{ for } r = 0 \text{ and } z = 0. \tag{6}$$

Using this condition in (5), we find $C = 0$. Hence,

$$p = \rho\left(\frac{1}{2}\omega^2 r^2 - gz\right). \tag{7}$$

Again at A or D, we have $p = 0$. Then

$$p = 0, \text{ for } r = R \text{ and } z = \quad x. \tag{8}$$

Then, from (7), with the use of (8), we get $0 = \rho\left[\frac{1}{2}\omega^2 R^2 - g(h - x)\right]$ which yields

$$gx = \left(gh - \frac{1}{2}\omega^2 R^2\right). \tag{9}$$

Let M be a point on the base of the cylinder at a distance r from the z-axis. Then, the pressure at M by (7) is

$$p = \rho\left[\frac{1}{2}\omega^2 r^2 - g(-x)\right] = \rho\left[\frac{1}{2}\omega^2 r^2 + gh - \frac{1}{2}\omega^2 R^2\right], \text{ [using (9)]}$$

$$= \rho\left[gh - \frac{1}{2}\omega^2(R^2 - r^2)\right] \tag{10}$$

Considering an annular ring at radius $OM = r$ and thickness dr, thrust on the elementary area is

$$dF = (2\pi r dr)p = 2\pi r\rho\left[gh - \frac{1}{2}\omega^2(R^2 - r^2)\right] dr \text{ [using (10)]}$$

Hence, the total thrust on the base of the cylinder is

$$F = \int_0^R 2\pi r\rho\left[gh - \frac{1}{2}\omega^2(R^2 - r^2)\right] dr = 2\pi\rho\left[gh\frac{r^2}{2} - \frac{1}{2}\omega^2\left(R^2\frac{r^2}{2} - \frac{r^4}{4}\right)\right]_0^R$$

$$= 2\pi\rho\left[gh\frac{R^2}{2} - \frac{1}{2}\omega^2\left(\frac{R^4}{2} - \frac{R^4}{4}\right)\right] = \pi\rho\left[R^2 gh - \omega^2\frac{R^4}{4}\right]$$

$$= \pi\rho g h R^2\left(1 - \frac{\omega^2 R^2}{4gh}\right). \quad \textbf{(Proved)}$$

EXAMPLE 8.4 *A closed circular cylinder is just filled with water, and rotates about its axis which is vertical. If the total thrust on the bottom is n times that on the top, show that the angular velocity is* $\frac{2}{R}\sqrt{\frac{gh}{n-1}}$, *where h is the height and R is the radius of the cylinder.*

Solution. Let $ABCD$ be the closed cylinder filled with water and rotates about its axis OP. We take this axis as z-axis vertically upwards with origin at O, the centre of the base of the cylinder (FIG. 8.10). Then, in the figure,

$$\left.\begin{aligned} AB = CD = h \\ AO = OD = R \end{aligned}\right\} \tag{1}$$

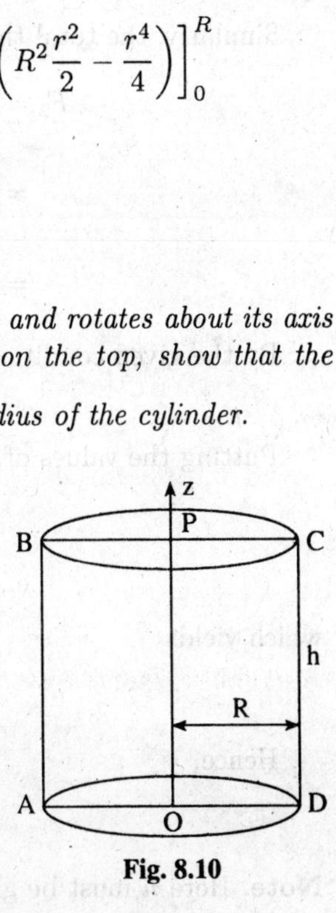

Fig. 8.10

The pressure at any point distant r from the axis is

$$p = \rho\left(\frac{1}{2}\omega^2 r^2 - gz\right) + C, \tag{2}$$

where C is a constant.

At the point P,

$$p = 0, \text{ for } r = 0 \text{ and } z = h. \tag{3}$$

Using this condition in (2), we get

$$0 = \rho(0 - gh) + C, \text{ i.e., } C = g\rho h. \tag{4}$$

Hence, from (2), we get

$$p = \rho\left(\frac{1}{2}\omega^2 r^2 - gz + gh\right). \tag{5}$$

At any point on the top, $z = h$. Hence, the pressure at any point on the top is

$$p_1 = \rho\left(\frac{1}{2}\omega^2 r^2 - gh + gh\right) = \frac{1}{2}\rho\omega^2 r^2 \tag{6}$$

Similarly, at any point on the bottom, $z = 0$, so that the pressure at any point on the bottom is

$$p_2 = \rho\left(\frac{1}{2}\omega^2 r^2 + gh\right). \tag{7}$$

Considering an elementary ring of radius r and thickness dr along which pressure will be same due to symmetry, we find the total thrust F_1 on the top is

$$F_1 = \int_0^R (2\pi r dr)p_1 = \int_0^R (2\pi r dr)\cdot\frac{1}{2}\rho\omega^2 r^2 \text{ [using (6)]}$$
$$= \pi\rho\omega^2\frac{R^4}{4} \tag{8}$$

Similarly, the total thrust on the bottom is

$$F_2 = \int_0^R (2\pi r dr)p_2$$
$$= \int_0^R (2\pi r dr)\cdot\rho\left(\frac{1}{2}\omega^2 r^2 + gh\right), \text{ [using (7)]}$$
$$= 2\pi\rho\int_0^R\left(\frac{1}{2}\omega^2 r^3 + ghr\right)dr = \pi\rho\frac{R^2}{4}(\omega^2 R^2 + 4gh) \tag{9}$$

By the given condition, we have

$$F_2 = nF_1. \tag{10}$$

Putting the values of F_1 and F_2 from (8) and (9) in (10), we get

$$\pi\rho\frac{R^2}{4}(\omega^2 R^2 + 4gh) = n\cdot\pi\rho\omega^2\frac{R^4}{4}$$
$$\text{i.e., } \omega^2 R^2 + 4gh = n\omega^2 R^2, \text{ i.e., } \omega^2 R^2(n-1) = 4gh$$

which yields

$$\omega^2 = \frac{4gh}{R^2(n-1)} \tag{11}$$

Hence,

$$\omega = \frac{2}{R}\sqrt{\frac{gh}{n-1}}. \quad \textbf{(Proved)}$$

Note. Here n must be greater than unity.

EXAMPLE 8.5 *A hollow cone, very nearly filled with water, rotates uniformly about its axis, which is vertical, the vertex being uppermost. If the pressure on the base be equal to six times the weight of the enclosed water, prove that the angular velocity is* $\sqrt{\frac{4g}{R}}\cot\alpha$, *where* R *is the radius of the base and* 2α *the vertical angle of the cone.*

Solution. Let ABC be the hollow cone with O as the centre of the base and the vertex A is uppermost (FIG. 8.11). The cone, very nearly filled with water, rotates uniformly about its axis AO which is vertical.

We take OA as the z-axis along vertically upwards direction.

Let ω = uniform angular velocity, ρ = density of water, h = height of the cone.

Then, in the figure,

$$OA = h, OC = OB = R, \angle OAC = \angle OAB = \alpha. \tag{1}$$

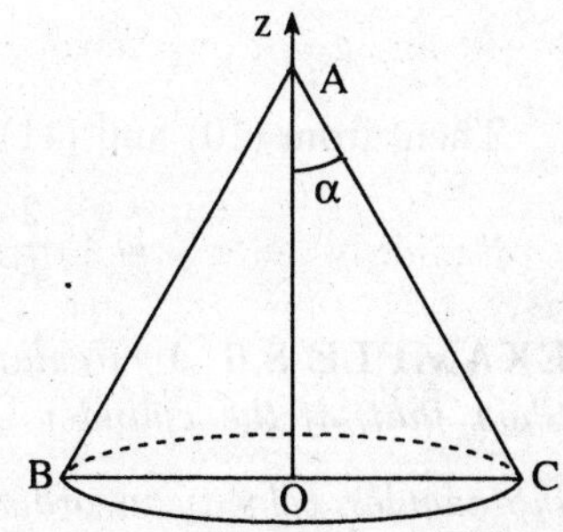

Fig. 8.11

The pressure at any point distant r from the axis is given by

$$p = \rho\left(\frac{1}{2}\omega^2 r^2 - gz\right) + C, \tag{2}$$

where C is the constant of integration.

At the vertex A, the pressure is zero. That is,

$$p = 0, \text{ for } r = 0 \text{ and } z = h. \tag{3}$$

Using this condition in (2), we get

$$0 = \rho(0 - gh) + C, \text{ i.e., } C = g\rho h. \tag{4}$$

From (2) and (4), we get

$$p = \rho\left(\frac{1}{2}\omega^2 r^2 - gz + gh\right). \tag{5}$$

Eqn (5) determines the fluid pressure at any point (r, z).

At the base, $z = 0$. Hence, the pressure at any point $(r, 0)$ on the base is

$$p = \rho\left(\frac{1}{2}\omega^2 r^2 + gh\right). \tag{6}$$

Considering an elementary ring-shaped area of radius r and thickness dr, we find the total thrust on the base equal to

$$F = \int_0^R (2\pi r dr)p = \int_0^R (2\pi r dr)\cdot\rho\left(\frac{1}{2}\omega^2 r^2 + gh\right) \text{ [using (6)]}$$

$$= 2\pi\rho\int_0^R\left(\frac{1}{2}\omega^2 r^3 + ghr\right)dr = 2\pi\rho\left(\frac{\omega^2 R^4}{8} + gh\frac{R^2}{2}\right) = \pi\rho R^2\left(\frac{\omega^2 R^2}{4} + gh\right) \tag{7}$$

The weight of the enclosed liquid is

$$W = \frac{1}{3}\pi R^2 g\rho h. \tag{8}$$

By the given condition, we have

$$F = 6W. \tag{9}$$

Putting the values of F and W from (7) and (8),

$$\pi\rho R^2\left(\frac{\omega^2R^2}{4}+gh\right)=6\cdot\frac{1}{3}\pi R^2 g\rho h,\text{ or, }\frac{\omega^2R^2}{4}+gh=2gh,\text{ or, }\omega^2R^2=4gh$$

which yields

$$\omega=\frac{2}{R}\sqrt{gh}. \tag{10}$$

Now, for the cone,

$$\tan\alpha=\frac{R}{h},\text{ i.e., }h=R\cot\alpha. \tag{11}$$

Then, from (10) and (11), we get

$$\omega=\frac{2}{R}\sqrt{gR\cot\alpha},\text{ i.e., }\omega=\sqrt{\frac{4g}{R}\cot\alpha}.\quad\textbf{(Proved)}$$

EXAMPLE 8.6 *A circular cylinder of radius R and height h contains liquid to a height l. Show that, if the cylinder and the liquid rotate about their common axis which is vertical, the angular velocity in order that the liquid may just not flow out is* $\frac{2}{R}\sqrt{g(h-l)}$ *or,* $\frac{h}{R}\sqrt{\frac{g}{l}}$ *according as l is > or <* $\frac{1}{2}h$.

Solution. Let $ABCD$ be the cylinder with O as its centre of base. We take the z-axis along the vertical through O in the upward direction (FIG. 8.12). Then, $AB=CD=h$.

The cylinder containing water of height l rotates about z-axis with uniform angular velocity. We consider following three cases.

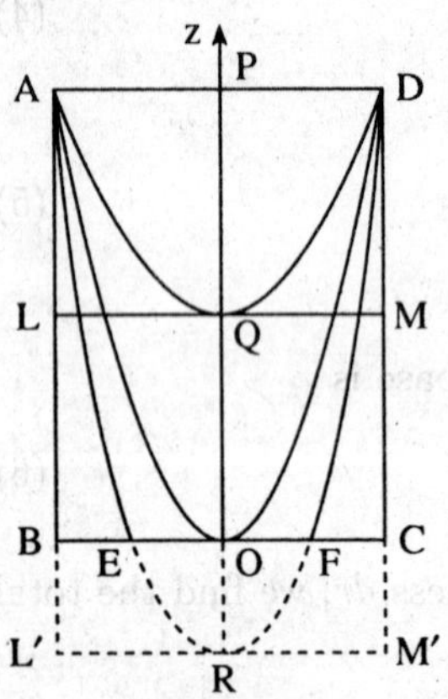

Fig. 8.12

Case 1. When $l=\frac{1}{2}h$, the volume of water is equal to half the volume of the cylinder. In this case, the vertex of the paraboloid of revolution of the free surface just touches the centre O of the base. If angular velocity $=\omega$, then

$$\frac{PD^2}{OP}=\text{latus rectum }\frac{2g}{\omega^2},\text{ i.e., }\frac{R^2}{h}=\frac{2g}{\omega^2},$$

which yields

$$\omega^2=\frac{2gh}{R^2},\text{ i.e., }\omega=\frac{1}{R}\sqrt{2gh}=\frac{1}{R}\sqrt{gl}. \tag{1}$$

Eqn (1) gives the value of ω in order that water may not just flow out, when $l=\frac{1}{2}h$.

Case 2. When $l>\frac{1}{2}h$, volume of water is more than half the volume of the cylinder. In this case, the vertex of the paraboloid of revolution of the free surface will be raised to Q. Let ω_1 be the angular velocity.

Let $PQ=x$, then

$$OQ=h-x. \tag{2}$$

Volume of water originally in the cylinder is

$$V=\pi R^2 l. \tag{3}$$

After rotation, volume of water $=$ volume $LBCM+\frac{1}{2}$ volume $ALMD$.

Hence, $\pi R^2 l = \pi R^2(h-x) + \frac{1}{2}\pi R^2 x$, i.e., $l = h - x + \frac{1}{2}x$.

$$\therefore x = 2(h-l) \tag{4}$$

For the paraboloid AQD, $\frac{PD^2}{PQ}$ = latus rectum $\frac{2g}{\omega_1^2}$, or, $\frac{R^2}{x} = \frac{2g}{\omega_1^2}$, i.e., $\omega_1^2 = \frac{2gx}{R^2}$.

Using (4),

$$\omega_1^2 = \frac{4g(h-l)}{R^2}.$$

Hence,

$$\omega_1 = \frac{2}{R}\sqrt{g(h-l)}. \quad \textbf{(Proved)} \tag{5}$$

Eqn (5) gives the value of the angular velocity so that water may just not flow out.

Case 3. When $l < \frac{1}{2}h$, volume of water is less than half the volume of cylinder. In this case, the vertex R of the paraboloid of revolution of the free surface goes down below the centre O of the base. Let ω_2 be the angular velocity.

Let $PR = y$, then

$$OR = y - h. \tag{6}$$

Volume of water $= \frac{1}{2}$ volume $AL'M'D-$ (volume $BL'M'C-$ volume of paraboloid ERF)

$$\text{i.e., } \pi R^2 l = \frac{1}{2}\pi R^2 y - \left[\pi R^2(y-h) - \frac{1}{2}\pi\cdot OF^2\cdot(y-h)\right] \tag{7}$$

Now, from the figure, we have

$$\frac{OF^2}{OR} = \frac{PD^2}{PR}, \text{ i.e., } OF^2 = \frac{R^2}{y}\cdot OR = \frac{R^2}{y}(y-h). \tag{8}$$

Putting this value in (7), we get

$$\pi R^2 l = \frac{1}{2}\pi R^2 y - \left[\pi R^2(y-h) - \frac{1}{2}\pi R^2\frac{(y-h)^2}{y}\right]$$

$$\text{i.e., } l = \frac{1}{2}y - (y-h) + \frac{1}{2}\frac{(y-h)^2}{y}, \text{ i.e., } l = h - \frac{1}{2}y + \frac{(y-h)^2}{2y}$$

$$\text{i.e., } l = \frac{2yh - y^2 + (y^2 - 2yh + h^2)}{2y}, \text{ i.e., } l = \frac{h^2}{2y}$$

$$\therefore y = \frac{h^2}{2l} \tag{9}$$

Now, for the paraboloid ARD,

$$\frac{PD^2}{PR} = \text{latus rectum } \frac{2g}{\omega_2^2}, \text{ i.e., } \frac{R^2}{y} = \frac{2g}{\omega_2^2}$$

$$\therefore \omega_2^2 = \frac{2gy}{R^2} = \frac{2g}{R^2}\cdot\frac{h^2}{2l} \text{ [using (9)]}$$

$$\text{i.e., } \omega_2^2 = \frac{h^2}{R^2}\frac{g}{l}.$$

Hence, $\omega_2 = \frac{h}{R}\sqrt{\frac{g}{l}}$. **(Proved)**

EXAMPLE 8.7 *A fine circular tube of radius R contains liquid, filling half of its circumference and rotates about the vertical diameter with uniform angular velocity. Prove that the liquid will just separate, if the angular velocity is* $\sqrt{\dfrac{2g}{R}}$.

Solution. Let the circular tube of radius R containing liquid of density ρ rotate about the vertical diameter AOB with uniform angular velocity ω. Since the tube is half-filled with liquid, the free surface will be along the horizontal diameter (FIG. 8.13).

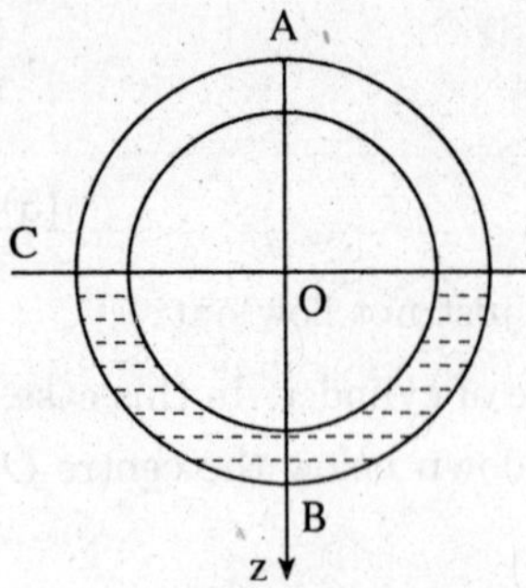

Fig. 8.13

With the centre of the tube as origin, we take z-axis along the vertical diameter in downward direction. Then the pressure equation is

$$dp = \rho(\omega^2 r dr + g dz), \tag{1}$$

which on integration yields

$$p = \rho\left(\frac{1}{2}\omega^2 r^2 + gz\right) + K, \tag{2}$$

where K is the constant of integration.

At points C and D on the free surface, pressure is zero. That is

$$p = 0, \text{ for } r = R \text{ and } z = 0. \tag{3}$$

Condition (3) applied to (2) gives

$$0 = \rho\left(\frac{1}{2}\omega^2 R^2 + 0\right) + K, \text{ i.e., } K = -\frac{1}{2}\omega^2 R^2 \rho \tag{4}$$

Putting this value of K in (2), we get

$$p = \rho\left(\frac{1}{2}\omega^2 r^2 + gz\right) - \frac{1}{2}\rho\omega^2 R^2. \tag{5}$$

This equation determines the pressure at any point (r, z).
At the lowest point B, $r = 0$ and $z = R$.
Hence, pressure at B is

$$p_B = \rho(0 + gR) - \frac{1}{2}\rho\omega^2 R^2 = g\rho R - \frac{1}{2}\rho\omega^2 R^2. \tag{6}$$

The liquid will just separate when the pressure at B vanishes; that is, when

$$0 = g\rho R - \frac{1}{2}\rho\omega^2 R^2. \tag{7}$$

This yields $\omega^2 = \dfrac{2g}{R}$, i.e., $\omega = \sqrt{\dfrac{2g}{R}}$. **(Proved)**

EXAMPLE 8.8 *A cylinder of radius R is filled with water to a depth h. A fine vertical tube is at a distance d (> R) from the axis of the cylinder and communicates with the bottom of the cylinder. If the whole is made to rotate uniformly about the axis of the cylinder, show that the angular velocity ω necessary to make the water rise in the tube to a height x above the bottom of the cylinder is given by*

$$\omega^2\left(d^2 - \frac{1}{2}R^2\right) = 2g(x - h).$$

Solution. The arrangement is shown in the figure (FIG. 8.14). The level of water ABC in the cylinder and the tube initially will be at a height h.

The free surface becomes a paraboloid of revolution COD when the whole system rotates. With origin at O, the vertex of the paraboloid of revolution, we take z-axis vertically upwards.

Then, for the portion of the paraboloid COD,

$$\frac{PD^2}{OP} = \text{latus rectum } \frac{2g}{\omega^2}, \text{ i.e., } \frac{R^2}{OP} = \frac{2g}{\omega^2}.$$

Hence,

$$OP = \frac{R^2\omega^2}{2g}. \quad (1)$$

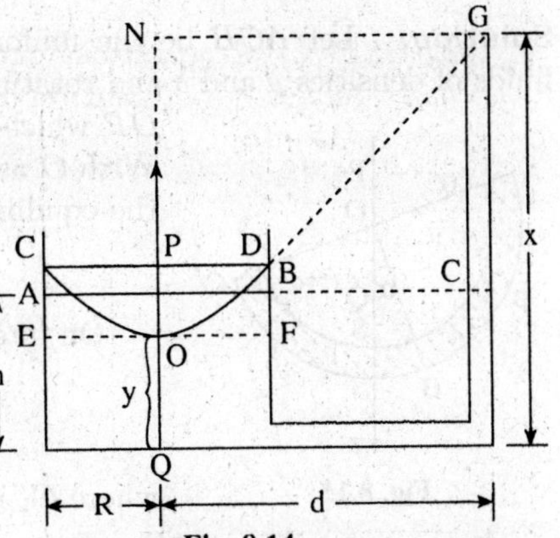

Fig. 8.14

Let $OQ = y$. Neglecting the volume of water in the fine tube, the original volume of water is

$$V_1 = \pi R^2 h. \quad (2)$$

After rotation, volume of water is

$$V_2 = \frac{1}{2}\pi R^2 y + \text{volume of paraboloid} = \frac{1}{2}\pi R^2 y + \frac{1}{2}\pi R^2 \cdot OP. \quad (3)$$

Since the volume of water remains unaltered, $V_1 = V_2$.

From (2) and (3),

$$\pi R^2 h = \pi R^2 y + \frac{1}{2}\pi R^2 \cdot OP, \text{ i.e., } h = y + \frac{1}{2}\cdot OP$$

$$\text{i.e., } h = y + \frac{1}{2}\cdot\frac{R^2\omega^2}{2g} \quad \text{[using (1)]}$$

$$\text{i.e., } y = h - \frac{R^2\omega^2}{4g} \quad (4)$$

The paraboloid of revolution COD continues up to the tube meeting at G. Hence,

$$\frac{GN^2}{ON} = \text{latus rectum } \frac{2g}{\omega^2}$$

$$\text{i.e., } \frac{d^2}{x-y} = \frac{2g}{\omega^2}, \text{ or, } 2g(x-y) = d^2\omega^2, \text{ or, } y = x - \frac{d^2\omega^2}{2g} \quad (5)$$

From (4) and (5), we get

$$h - \frac{R^2\omega^2}{4g} = x - \frac{d^2\omega^2}{2g}, \text{ or, } \frac{\omega^2}{2g}\left(d^2 - \frac{1}{2}R^2\right) = x - h$$

$$\therefore \omega^2\left(d^2 - \frac{1}{2}R^2\right) = 2g(x-h). \quad \textbf{(Proved)}$$

EXAMPLE 8.9 *A uniform semi-circular closed tube of radius R is tightly filled with equal volumes of two liquids of densities ρ and δ respectively, which do not mix and is rotated with angular velocity ω about a vertical radius making an angle α with the line of symmetry. Prove that the pressures at the two ends will be equal, if*

$$\frac{\omega^2 R}{2g}(\delta - \rho) = \frac{\delta}{\cos\alpha + \sin\alpha} - \frac{\rho}{\cos\alpha - \sin\alpha},$$

the fluid of density δ being the lower of the two and the convexity of the tube being downwards.

Solution. Let ACB be the uniform semi-circular closed tube of radius R containing two fluids of densities ρ and δ and rotating with uniform angular velocity ω about the vertical axis OP which makes an angle α with the line of symmetry (FIG. 8.15). With O as origin, we take z-axis vertically downwards. Considering the equilibrium of fluid of density ρ, we write the pressure equation

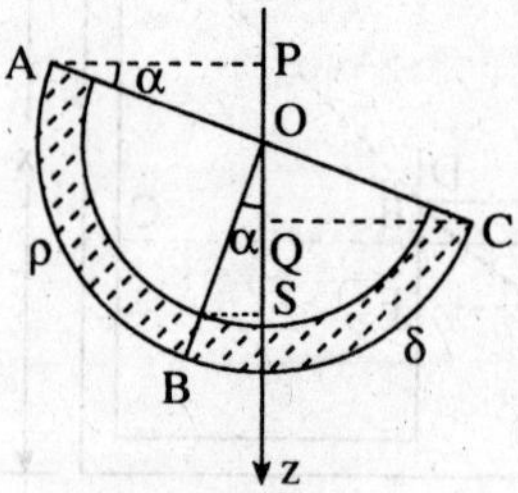

Fig. 8.15

$$dp = \rho(\omega^2 r dr + g dz).$$

On integration, we get

$$p = \rho\left(\frac{1}{2}\omega^2 r^2 + gz\right) + C_1, \tag{1}$$

where C_1 is the constant of integration.

At the point A, $r = AP = R\cos\alpha$, $z = -OP = -R\sin\alpha$. Then pressure at A is given by

$$p_A = \rho\left(\frac{1}{2}\omega^2 R^2\cos^2\alpha - gR\sin\alpha\right) + C_1. \tag{2}$$

Similarly, at the point B,

$$r = BS = R\sin\alpha, z = OS = R\cos\alpha. \tag{3}$$

Hence, pressure at B is given by

$$p_B = \rho\left(\frac{1}{2}\omega^2 R^2\sin^2\alpha + gR\cos\alpha\right) + C_1. \tag{4}$$

From (2) and (4), we get

$$p_A - p_B = \frac{1}{2}\omega^2 R^2\rho(\cos^2\alpha - \sin^2\alpha) - g\rho R(\sin\alpha + \cos\alpha). \tag{5}$$

Now we consider the equilibrium of fluid of density δ.

The pressure at any point in the fluid is given by

$$p = \delta\left(\frac{1}{2}\omega^2 r^2 + gz\right) + C_2, \tag{6}$$

where C_2 is a constant of integration.

Then, using (3), pressure at B is given by

$$p_B = \delta\left(\frac{1}{2}\omega^2 R^2\sin^2\alpha + gR\cos\alpha\right) + C_2. \tag{7}$$

At the point C,

$$r = CQ = R\cos\alpha, z = OQ = R\sin\alpha. \tag{8}$$

Hence, pressure at C is given by

$$p_C = \delta\left(\frac{1}{2}\omega^2 R^2\cos^2\alpha + gR\sin\alpha\right) + C_2. \tag{9}$$

From (7) and (9), we get

$$p_B - p_C = \frac{1}{2}\omega^2 R^2\delta(\sin^2\alpha - \cos^2\alpha) + g\delta R(\cos\alpha - \sin\alpha). \tag{10}$$

By the condition given, $p_A = p_C$. Hence,

$$p_A - p_B = p_C - p_B. \tag{11}$$

From (5) and (10), using (11), we have

$$\frac{1}{2}\omega^2 R^2 \rho(\cos^2\alpha - \sin^2\alpha) - g\rho R(\sin\alpha + \cos\alpha) = \frac{1}{2}\omega^2 R^2 \delta(\cos^2\alpha - \sin^2\alpha) + g\delta R(\sin\alpha - \cos\alpha)$$

or, $$\frac{1}{2}\omega^2 R^2(\rho - \delta)(\cos^2\alpha - \sin^2\alpha) = gR[\delta(\sin\alpha - \cos\alpha) + \rho(\sin\alpha + \cos\alpha)]$$

or, $$\frac{\omega^2 R}{2g}(\delta - \rho) = \frac{\delta}{\cos\alpha + \sin\alpha} - \frac{\rho}{\cos\alpha - \sin\alpha}.$$ **(Proved)**

EXAMPLE 8.10 *A vessel in the form of a paraboloid of revolution of a parabola of latus rectum 4a about its axis, is filled to half its height with a liquid. Show that the greatest angular velocity with which it can revolve about its axis so that no liquid is spilt is* $\frac{1}{4}\sqrt{\frac{6g}{a}}$.

Solution. Let ABC be the paraboloid of revolution of latus rectum $4a$ and height, say h (FIG. 8.16). Then, we have $BD = h$.

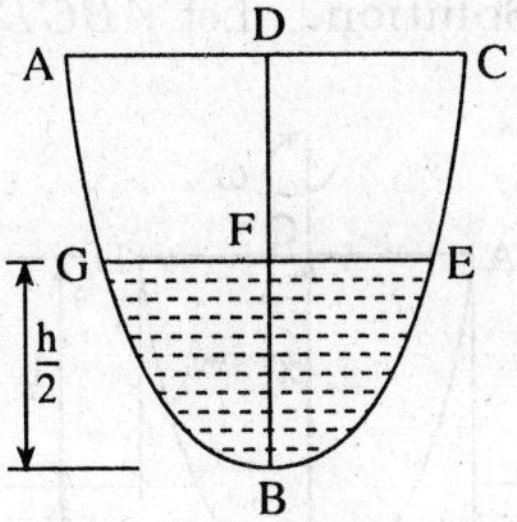

Fig. 8.16

Hence,

$$\frac{CD^2}{BD} = 4a, \text{ i.e., } CD^2 = 4ah. \qquad (1)$$

Initially, liquid is filled to half its height, so that GFE is the free surface. Then,

$$\frac{EF^2}{BF} = 4a, \text{ i.e., } EF^2 = 4a \cdot \frac{h}{2} = 2ah. \qquad (2)$$

Hence the volume of the liquid is

$$V_1 = \frac{1}{2}\pi EF^2 \cdot \frac{h}{2} = \frac{1}{2}\pi \cdot 2ah \cdot \frac{h}{2} \text{ [using (2)]}$$

i.e., $$V_1 = \frac{1}{2}\pi a h^2 \qquad (3)$$

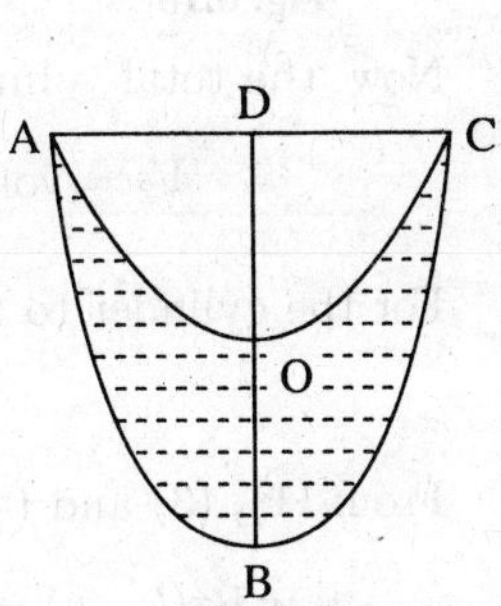

Fig. 8.17

When the liquid rotates with angular velocity ω, the free surface will be a paraboloid of revolution with vertex at O. In the limiting state of the liquid to be spilt out, the free surface will touch points A and C (FIG. 8.17). Then,

$$\frac{CD^2}{OD} = \text{latus rectum } \frac{2g}{\omega^2}, \text{ i.e., } \frac{4ah}{OD} = \frac{2g}{\omega^2} \text{ [using (1)]}$$

i.e., $$OD = \frac{4ah \cdot \omega^2}{2g} = \frac{2ah\omega^2}{g} \qquad (4)$$

The volume of the liquid in rotating condition is

V_2 = volume of the portion $AOCB$

= volume of the paraboloid ABC minus volume of the paraboloid AOC

i.e., $$V_2 = \frac{1}{2}\pi \cdot CD^2 \cdot h - \frac{1}{2}\pi \cdot CD^2 \cdot OD$$

$$= \frac{1}{2}\pi \cdot 4ah \cdot h - \frac{1}{2}\pi \cdot 4ah \cdot \frac{2ah\omega^2}{g} \text{ [using (1) and (4)]}$$

$$= 2\pi a h^2 - 4\pi a^2 h^2 \frac{\omega^2}{g} \qquad (5)$$

Since there is no change in the volume of the liquid, $V_1 = V_2$.

Hence, by (3) and (5), we have

$$\frac{1}{2}\pi ah^2 = 2\pi ah^2 - 4\pi a^2h^2\frac{\omega^2}{g}, \text{ i.e., } \frac{1}{2} = 2 - 4a\frac{\omega^2}{g}, \text{ i.e., } 4a\frac{\omega^2}{g} = 2 - \frac{1}{2} = \frac{3}{2}$$

$$\therefore \omega^2 = \frac{3g}{8a} = \frac{6g}{16a}$$

$$\therefore \omega = \frac{1}{4}\sqrt{\frac{6g}{a}}. \quad \textbf{(Proved)}$$

EXAMPLE 8.11 *A circular cylinder of radius R floats in a liquid at rest with its axis vertical and a length h unimmersed. Show that if the cylinder is sufficiently long, it will float in equilibrium with its upper rim in the surface provided that the liquid is made to rotate with angular velocity* $\frac{2}{R}\sqrt{gh}$.

Solution. Let $ABCD$ be the cylinder of length l with a length h unimmersed in a fluid at rest (FIG. 8.18). Then the volume of the displaced liquid is

$$V_1 = \pi R^2(l-h). \tag{1}$$

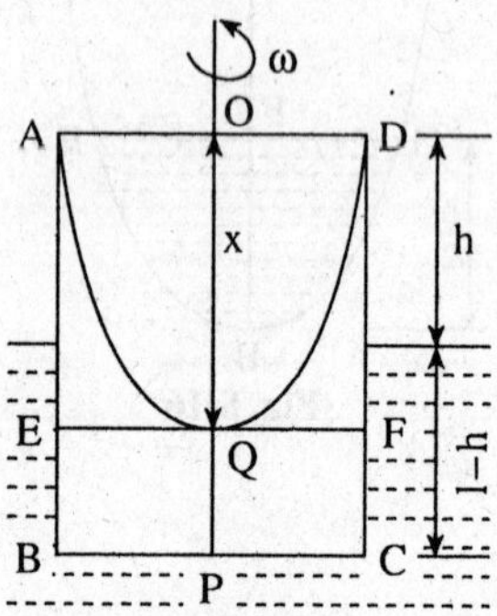

Fig. 8.18

When the liquid rotates with uniform angular velocity ω about the axis of the cylinder OP, the horizontal free surface of the liquid takes the shape of a paraboloid of revolution AQD, assuming that the upper rim of the cylinder is in the surface.

Let $OQ = x$. Then, for the paraboloid AQD,

$$\frac{OD^2}{OQ} = \text{latus rectum } \frac{2g}{\omega^2},$$

$$\text{i.e., } \frac{R^2}{x} = \frac{2g}{\omega^2}, \text{ i.e., } \omega^2 = \frac{2gx}{R^2}. \tag{2}$$

Now, the total volume of displaced liquid is

$$V_2 = \text{volume } EBCF + \frac{1}{2}\text{ volume } AEFD = \pi R^2(l-x) + \frac{1}{2}\pi R^2 x. \tag{3}$$

For the cylinder to float in equilibrium,

$$V_1 = V_2. \tag{4}$$

From (1), (3) and (4), we get

$$\pi R^2(l-h) = \pi R^2(l-x) + \frac{1}{2}\pi R^2 x, \text{ i.e., } l-h = l-x+\frac{1}{2}x, \text{ i.e., } x = 2h. \tag{5}$$

From (2) and (5), we get

$$\omega^2 = \frac{2g}{R^2}\cdot 2h = \frac{4gh}{R^2}$$

$$\therefore \omega = \frac{2}{R}\sqrt{gh}. \quad \textbf{(Proved)}$$

EXAMPLE 8.12 *A hemispherical shell of radius 2R containing water rotates with angular velocity* $\sqrt{2g/3R}$ *about its axis which is vertical. A sphere of radius R rests on the water with its lowest point in contact with the shell without pressing on it. The free surface of the liquid passes through the rim of the shell. Prove that the density of the sphere is to that of water as* $3:8$.

Solution. Let ABC be the hemispherical shell of radius $2R$ containing water and rotating with uniform angular velocity ω about its vertical axis BD (FIG. 8.19).

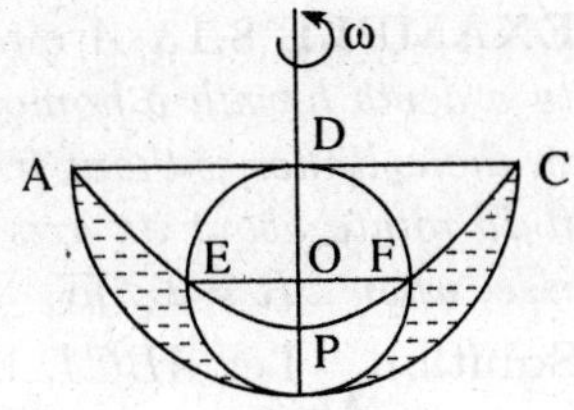

Fig. 8.19

Due to rotation, the free surface takes the shape of a paraboloid of revolution with vertex at P and passing through the rim of the shell.

A solid sphere of radius R with centre at O rests on the shell without exerting any pressure.

For the paraboloid of revolution APC, we have

$$\frac{CD^2}{PD} = \text{latus rectum } \frac{2g}{\omega^2}, \text{ i.e., } \frac{CD^2}{PD} = \frac{2g}{2g} \cdot 3R = 3R, \tag{1}$$

since we are given that $\omega = \sqrt{2g/3R}$.

Now, CD = radius of the hemisphere = $2R$.

Hence, from (1), we get

$$\frac{4R^2}{PD} = 3R, \text{ i.e., } PD = \frac{4}{3}R. \tag{2}$$

Then,

$$OP = PD - OD = \frac{4}{3}R - R = \frac{1}{3}R. \tag{3}$$

For the portion of the paraboloid EPF, we have

$$\frac{OF^2}{OP} = \frac{2g}{\omega^2}, \text{ i.e., } OF^2 = \frac{2g}{\omega^2} \cdot OP = \frac{2g}{2g} \cdot 3R \cdot \frac{1}{3}R = R^2. \tag{4}$$

Hence, $OF = R$.

Let ρ and δ be the densities of the sphere and water respectively.

Then the weight of the sphere is

$$W_1 = \frac{4}{3}\pi R^3 \rho g. \tag{5}$$

Volume of liquid displaced by the portion of the sphere $EBFP$ is

$$\begin{aligned} V &= \frac{2}{3}\pi R^3 - \frac{1}{2}\pi \cdot OF^2 \cdot OP \\ &= \frac{2}{3}\pi R^3 - \frac{1}{2}\pi \cdot R^2 \cdot \frac{1}{3}R \text{ [using (3) and (4)]} \\ &= \frac{2}{3}\pi R^3 - \frac{1}{6}\pi R^3 = \frac{1}{2}\pi R^3 \end{aligned} \tag{6}$$

Hence, weight of the liquid displaced is

$$W_2 = \frac{1}{2}\pi R^3 g\delta. \tag{7}$$

Since the sphere rests on the hemispherical shell without exerting any pressure, the weight of the sphere acting vertically downwards must be balanced by the thrust of the liquid acting vertically upwards. Hence,

$$W_1 = W_2. \tag{8}$$

From (5), (7) and (8), we get

$$\frac{4}{3}\pi R^3 \rho g = \frac{1}{2}\pi R^3 g\delta, \text{ i.e., } \frac{\rho}{\delta} = \frac{3}{8}. \quad \textbf{(Proved)}$$

EXAMPLE 8.13 *A circular cylinder of radius R, whose axis is vertical, is filled with a fluid to a depth h with a homogeneous liquid of density* ρ. *A piston of weight* $\pi\rho g\alpha R^2$, *which works in the cylinder without friction, is placed on the top of the fluid. Show that if the cylinder and fluid rotate about its axis with gradually increasing angular velocity* ω, *the piston will begin to rise, when* $\omega R = 2\sqrt{g\alpha}$.

Solution. Let $ABCD$ be the cylinder containing fluid of density ρ to a depth h. With the centre O of the base as origin, we take z-axis (which is the axis of the cylinder) in the vertically upward direction (FIG. 8.20).

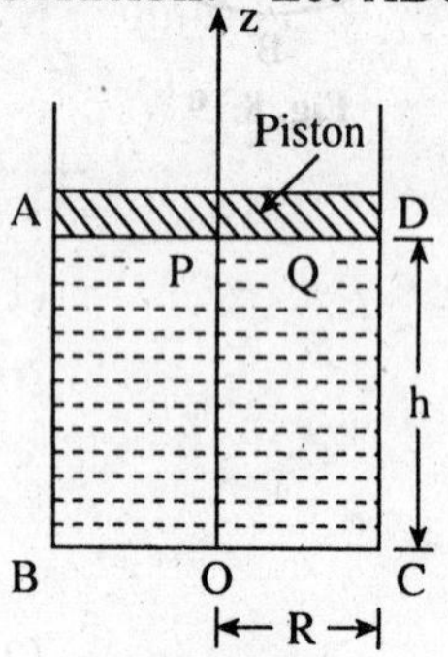

Fig. 8.20

The pressure equation is

$$dp = \rho(\omega^2 r dr - g dz), \tag{1}$$

where r is the horizontal distance from the axis and z is the height from O of the point.

Integrating (1), we get

$$p = \rho\left(\frac{1}{2}\omega^2 r^2 - gz\right) + C, \tag{2}$$

where C is the constant of integration.

At the point P, the centre of the piston, the pressure is zero. That is

$$p = 0, \text{ for } r = 0 \text{ and } z = h. \tag{3}$$

Using this condition in (2), we get

$$0 = -g\rho h + C, \text{ i.e., } C = g\rho h. \tag{4}$$

Hence, from (2) and (4),

$$p = \rho\left(\frac{1}{2}\omega^2 r^2 - gz\right) + g\rho h. \tag{5}$$

This determines the fluid pressure at any point (r, z). Then pressure at any point Q below the piston for which $z = h$ and $r = r$ is, from (5),

$$p = \frac{1}{2}\omega^2 r^2 \rho. \tag{6}$$

Considering an elementary ring of radius r and thickness dr below the piston, we find thrust on it,

$$dF = (2\pi r)dr \cdot p = (2\pi r dr) \cdot \frac{1}{2}\omega^2 r^2 \rho \text{ [using (6)]}$$
$$= \pi\omega^2 \rho r^3 dr. \tag{7}$$

Hence, total thrust on the bottom of the piston acting vertically upwards is

$$F = \int_0^R \pi\omega^2 \rho r^3 dr = \pi\omega^2 \rho \frac{R^4}{4}. \tag{8}$$

The piston will be on the point of rising just when the thrust F on the bottom of the piston will be equal to the weight W of the piston, where, as given,

$$W = \pi g\rho\alpha R^2. \tag{9}$$

Equating (8) and (9), we get

$$\pi\omega^2 \rho \frac{R^4}{4} = \pi g\rho\alpha R^2, \text{ i.e., } \omega^2 R^2 = 4g\alpha$$

$$\therefore \omega R = 2\sqrt{g\alpha}. \quad \textbf{(Proved)}$$

EXAMPLE 8.14 *A tube of small section is in the form of three sides of a square, the middle one of which is horizontal. It is filled with water and revolves about a vertical axis through the middle point of the horizontal side.*

Prove that no liquid escapes unless $\omega > \sqrt{\dfrac{8g}{a}}$, *and that, if this be so, the amount that escapes would fill a length of the tube equal to* $a\sqrt{1 - \dfrac{8g}{\omega^2 a}}$, *where 'a' is the side of the tube and* ω *the uniform angular velocity.*

If the tube revolves about a vertical side, prove that the amount which will flow out would fill a length $\dfrac{\omega^2 a^2}{2g}$, *or,* $a + \sqrt{a^2 - \dfrac{2ga}{\omega^2}}$ *according as* $\omega <$, *or* $> \sqrt{\dfrac{2g}{a}}$.

Solution. Let $ABCD$ be the tube with side BC horizontal and O is the middle point of BC (FIG. 8.21). When the tube revolves about the vertical axis OE, the vertex P of the paraboloid of revolution will remain anywhere on OE.

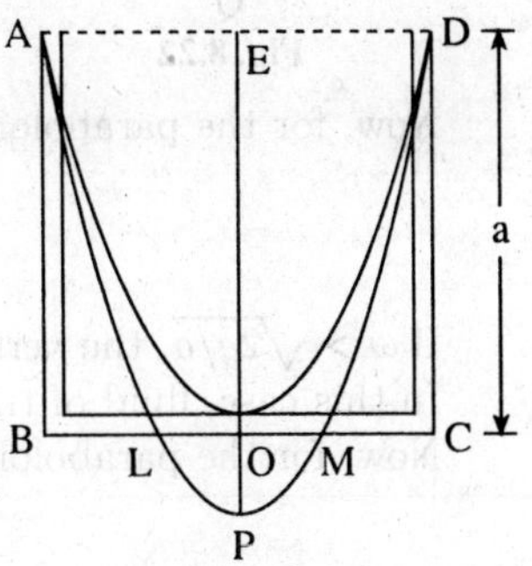
Fig. 8.21

For no escape of the fluid, the lowest position of P will coincide with O, so that AOD is the paraboloid of revolution formed.

In this case,

$$\frac{DE^2}{OE} = \text{latus rectum } \frac{2g}{\omega^2},$$

$$\text{i.e., } \frac{a^2}{4 \cdot a} = \frac{2g}{\omega^2} \quad \left[\text{since } DE = \frac{a}{2}, OE = AB = a\right]$$

$$\text{i.e., } \omega^2 = \frac{8g}{a}, \text{ i.e., } \omega = \sqrt{\frac{8g}{a}}. \tag{1}$$

Thus, no fluid escapes, if

$$\omega < \sqrt{\frac{8g}{a}}. \tag{2}$$

If the angular velocity increases, so that

$$w > \sqrt{\frac{8g}{a}}, \tag{3}$$

the vertex P of the paraboloid of revolution will be below O as shown in the figure.

In this case, fluid of length LM in the tube will escape.

Let $OP = x$.

Then, from the paraboloid APD, we have

$$\frac{DE^2}{EP} = \frac{2g}{\omega^2}. \tag{4}$$

Now, $EP = EO + OP = a + x$. Hence, by (4)

$$\frac{a^2}{4(a+x)} = \frac{2g}{\omega^2}, \text{ i.e., } \omega^2 a^2 = 8g(a+x), \text{ i.e., } x = \frac{\omega^2 a^2}{8g} - a. \tag{5}$$

Again, for the paraboloid LPM,

$$\frac{OM^2}{OP} = \frac{2g}{\omega^2}, \text{ i.e., } \frac{LM^2}{4x} = \frac{2g}{\omega^2} \; [\text{since } LM = 2 \cdot OM, OP = x]$$

$$\text{i.e., } LM^2 = \frac{8gx}{\omega^2} \tag{6}$$

Putting the value of x from (5) into (6)

$$LM^2 = \frac{8g}{\omega^2}\left(\frac{\omega^2 a^2}{8g} - a\right), \text{ i.e., } LM^2 = a^2\left(1 - \frac{8g}{\omega^2 a}\right)$$

$$\therefore LM = a\sqrt{1 - \frac{8g}{\omega^2 a}} \tag{7}$$

Thus, for $\omega^2 > \sqrt{8g/a}$, fluid of tube length LM given by (7) escapes.

Now, if the tube rotates about a vertical side, say AB, the vertex of the paraboloid of revolution will lie on AB (FIG. 8.22). When the vertex falls on B, we have

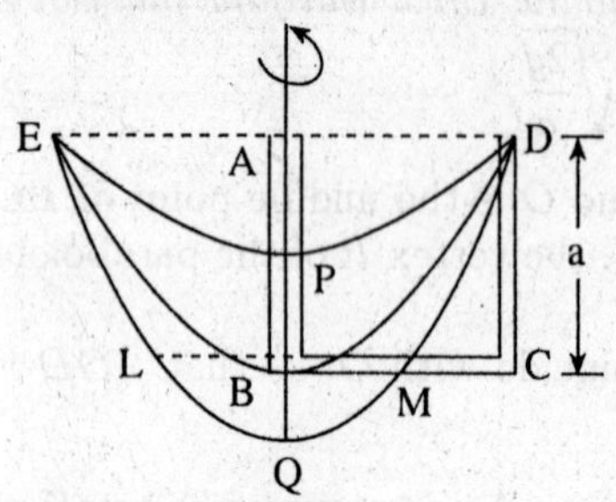

Fig. 8.22

$$\frac{AD^2}{AB} = \frac{2g}{\omega^2}, \text{ i.e., } \frac{a^2}{a} = \frac{2g}{\omega^2},$$

$$\text{i.e., } \omega^2 = \frac{2g}{a}, \text{ i.e., } \omega = \sqrt{\frac{2g}{a}} \tag{8}$$

Thus, when $\omega < \sqrt{2g/a}$, the vertex of the paraboloid of revolution is at P, lying in between A and B. In this case, fluid in the tube of length AP will escape.

Now, for the paraboloid EPD, we have

$$\frac{AD^2}{AP} = \frac{2g}{\omega^2}, \text{ i.e., } AP = \frac{\omega^2 a^2}{2g}. \tag{9}$$

If $\omega > \sqrt{2g/a}$, the vertex of the paraboloid will fall below B as shown in the figure. In this case, fluid of tube length $AB + BM$ will escape.

Now, for the paraboloid EQD, we have

$$\frac{AD^2}{AQ} = \frac{2g}{\omega^2}, \text{ i.e., } \frac{a^2}{a+y} = \frac{2g}{\omega^2}$$

$$\therefore y = \frac{\omega^2 a^2}{2g} - a, \tag{10}$$

where $BQ = y$.

Again, for the paraboloid LQM, we have

$$\frac{BM^2}{BQ} = \frac{2g}{\omega^2}, \text{ i.e., } BM^2 = \frac{2g}{\omega^2}\cdot y = \frac{2g}{\omega^2}\left(\frac{\omega^2 a^2}{2g} - a\right) \text{ [using (10)]}$$

$$\text{i.e., } BM = \sqrt{a^2 - \frac{2ga}{\omega^2}} \tag{11}$$

Hence,

$$AB + BM = a + \sqrt{a^2 - \frac{2ga}{\omega^2}} \tag{12}$$

Thus, when $\omega > \sqrt{2g/a}$, fluid of tube length $a + \sqrt{a^2 - \frac{2ga}{\omega^2}}$ escapes.

EXAMPLE 8.15 *A straight tube making an angle α with the vertical and filled with fluid of density ρ, rotates with uniform angular velocity ω about the vertical axis through its lower end which is closed. Prove that, when the atmospheric pressure is Π, the greatest length of the tube that no fluid may flow out is* $\dfrac{g\rho\cos\alpha + \omega\sin\alpha\sqrt{2\Pi\rho}}{\omega^2\rho\sin^2\alpha}$.

Solution. Let AB be the tube inclined to the vertical at an angle α revolving about the vertical axis AQ. With A as the origin, we take z-axis along AQ in the upward direction (FIG. 8.23). Then the pressure equation is

$$dp = \rho(\omega^2 r dr - g dz), \tag{1}$$

where r = horizontal distance from vertical axis and z = the vertical distance from A of the point P concerned.

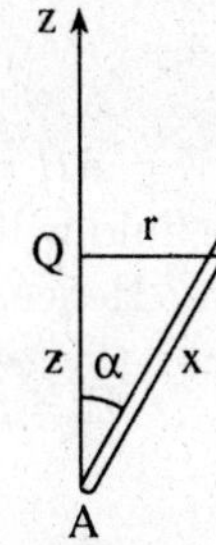

Fig. 8.23

Integrating (1), we get

$$p = \rho\left(\frac{1}{2}\omega^2 r^2 - gz\right) + C, \tag{2}$$

where C is the constant of integration.

If $AP = x$, then, from the figure,

$$r = x \sin\alpha; \quad z = x\cos\alpha. \tag{3}$$

Putting these values in (2), we get

$$p = \rho\left(\frac{1}{2}\omega^2 x^2 \sin^2\alpha - gx\cos\alpha\right) + C. \tag{4}$$

This equation determines the fluid pressure at any point along the length of the tube.

Differentiating (4), we get

$$\frac{dp}{dx} = \rho(\omega^2 x \sin^2\alpha - g\cos\alpha). \tag{5}$$

For maximum or minimum value of p, we must have

$$\frac{dp}{dx} = 0. \tag{6}$$

Thus, from (5) and (6), we get $\rho(\omega^2 x\sin^2\alpha - g\cos\alpha) = 0$, which yields

$$x = \frac{g\cos\alpha}{\omega^2\sin^2\alpha}. \tag{7}$$

Again, differentiating (5), we find

$$\frac{d^2p}{dx^2} = \rho\omega^2\sin^2\alpha,$$

which is positive.

Hence, the pressure at P is minimum and must be equal to zero. Thus,

$$p = 0, \text{ for } x = \frac{g\cos\alpha}{\omega^2\sin^2\alpha} \tag{8}$$

Using this condition in (4), we get

$$0 = \rho\left[\frac{1}{2}\omega^2\cdot\frac{g^2\cos^2\alpha}{\omega^4\sin^4\alpha}\cdot\sin^2\alpha - g\cdot\frac{g\cos\alpha}{\omega^2\sin^2\alpha}\cos\alpha\right] + C$$

$$\text{i.e., } 0 = \rho\left[\frac{1}{2}\frac{g^2\cos^2\alpha}{\omega^2\sin^2\alpha} - \frac{g^2\cos^2\alpha}{\omega^2\sin^2\alpha}\right] + C = -\frac{1}{2}\rho\frac{g^2\cos^2\alpha}{\omega^2\sin^2\alpha} + C$$

Hence,

$$C = \frac{1}{2}\rho\frac{g^2\cos^2\alpha}{\omega^2\sin^2\alpha}. \tag{9}$$

Putting this value of C in (4), we get

$$p = \rho\left[\frac{1}{2}\omega^2 x^2 \sin^2\alpha - gx\cos\alpha + \frac{1}{2}\frac{g^2\cos^2\alpha}{\omega^2\sin^2\alpha}\right] \tag{10}$$

Let $l = AB =$ the length of the tube.

The fluid will escape at the end B when the pressure at B is just more than the atmospheric pressure. Hence, in the limiting condition,

$$\Pi = \rho\left[\frac{1}{2}\omega^2 l^2 \sin^2\alpha - gl\cos\alpha + \frac{1}{2}\frac{g^2\cos^2\alpha}{\omega^2\sin^2\alpha}\right].$$

$$\text{i.e., } \Pi = \frac{1}{2}\rho\omega^2\sin^2\alpha\left[l^2 - \frac{2gl\cos\alpha}{\omega^2\sin^2\alpha} + \frac{g^2\cos^2\alpha}{\omega^4\sin^4\alpha}\right] = \frac{1}{2}\rho\omega^2\sin^2\alpha\left(l - \frac{g\cos\alpha}{\omega^2\sin^2\alpha}\right)^2$$

$$\text{i.e., } \frac{2\Pi}{\rho\omega^2\sin^2\alpha} = \left(l - \frac{g\cos\alpha}{\omega^2\sin^2\alpha}\right)^2, \text{ i.e., } l - \frac{g\cos\alpha}{\omega^2\sin^2\alpha} = \frac{\sqrt{2\Pi/\rho}}{\omega\sin\alpha}$$

$$\therefore\ l = \frac{g\cos\alpha}{\omega^2\sin^2\alpha} + \frac{\sqrt{2\Pi/\rho}}{\omega\sin\alpha} = \frac{g\rho\cos\alpha + \omega\sin\alpha\sqrt{2\Pi\rho}}{\omega^2\rho\sin^2\alpha}.$$ **(Proved)**

EXAMPLE 8.16 *A liquid occupies a portion of a fine tube of radius R subtending an angle $(\pi + \theta)$ at the centre. When the tube rotates uniformly about a vertical tangent, the liquid just reaches the top of the vertical diameter. Prove that the angular velocity is*

$$\sqrt{\frac{g}{R}}\left(\tan\frac{\theta}{2} - \sin^2\frac{\theta}{2}\right)^{-\frac{1}{2}}$$

Solution. Let $ABCE$ be the fine circular tube of radius R containing liquid of density, say ρ. EF is the vertical tangent (FIG. 8.24). P and Q are the fluid levels in the tube such that

$$\text{obtuse } \angle POQ = \pi + \theta.$$

When the tube rotates uniformly with angular velocity ω about the vertical tangent EF, the fluid level P reaches top A of the vertical diameter AC (FIG. 8.25), and the level Q at D, such that $\angle COD = \theta$.

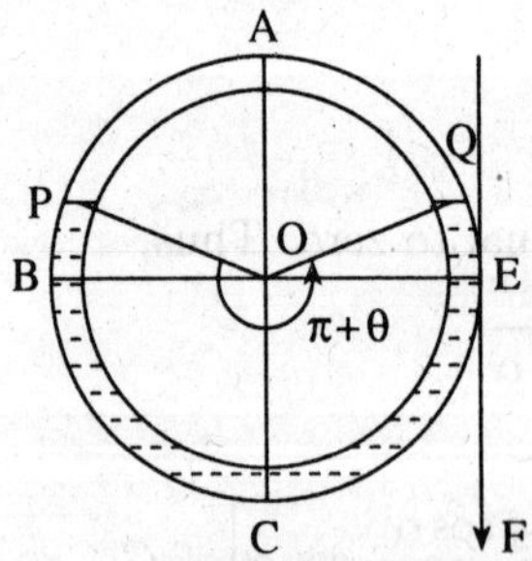

Fig. 8.24

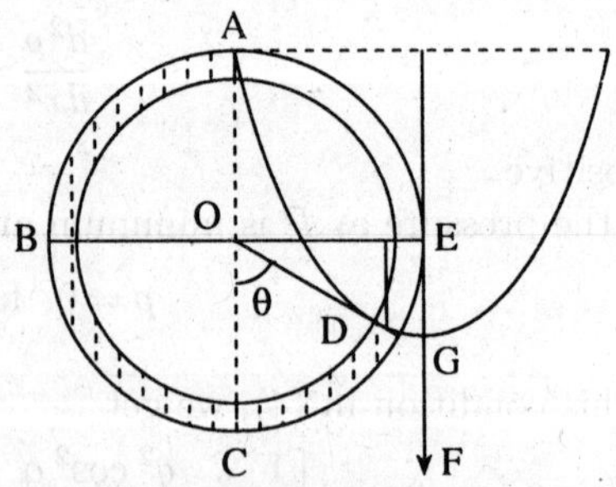

Fig. 8.25

With the point of contact E as origin, we take the z-axis along EF vertically downwards. Then the pressure equation is

$$dp = \rho\left(\frac{1}{2}\omega^2 r\,dr + g\,dz\right), \tag{1}$$

where $r =$ distance from z-axis and z, the distance from E of the point concerned.

Integrating (1), we get

$$p = \rho\left(\frac{1}{2}\omega^2 r^2 + gz\right) + C, \tag{2}$$

where C is the constant of integration.

Now, the pressure at free surface D is zero. The coordinates of D are

$$\left.\begin{aligned} z &= R\sin(90^\circ - \theta) = R\cos\theta \\ r &= R - R\cos(90^\circ - \theta) = R(1 - \sin\theta) \end{aligned}\right\} \tag{3}$$

Thus we have the condition,

$$p = 0, \text{ for } r = R(1 - \sin\theta), z = R\cos\theta \tag{4}$$

Using this condition in (2), we get

$$0 = \rho\left[\frac{1}{2}\omega^2 R^2 (1 - \sin\theta)^2 + gR\cos\theta\right] + C. \tag{5}$$

Hence,

$$C = -\rho\left[\frac{1}{2}\omega^2 R^2 (1 - \sin\theta)^2 + gR\cos\theta\right]. \tag{6}$$

Putting this value in (2), we get

$$p = \rho\left(\frac{1}{2}\omega^2 r^2 + gz\right) - \rho\left[\frac{1}{2}\omega^2 R^2 (1 - \sin\theta)^2 + gR\cos\theta\right] \tag{7}$$

Eqn (7) completely determines the pressure at any point of the tube.

Again, at the point A the pressure is zero. For the point A,

$$r = R, z = -R. \tag{8}$$

Thus, we have another condition,

$$p = 0, \text{ for } r = R \text{ and } z = -R. \tag{9}$$

Using this condition in (7), we get

$$0 = \rho\left(\frac{1}{2}\omega^2 R^2 - gR\right) - \rho\left[\frac{1}{2}\omega^2 R^2 (1 - \sin\theta)^2 + gR\cos\theta\right]$$

i.e., $\frac{1}{2}\omega^2 R^2\left[1 - (1 - \sin\theta)^2\right] = gR(1 + \cos\theta)$

i.e., $\omega^2 R^2 (2\sin\theta - \sin^2\theta) = 2gR(1 + \cos\theta)$

i.e., $\omega^2 R^2\left(4\sin\frac{\theta}{2}\cos\frac{\theta}{2} - 4\sin^2\frac{\theta}{2}\cos^2\frac{\theta}{2}\right) = 4gR\cos^2\frac{\theta}{2}$

i.e., $\omega^2 R\left(\tan\frac{\theta}{2} - \sin^2\frac{\theta}{2}\right) = g$

$$\therefore\ \omega^2 = \frac{g}{R}\left(\tan\frac{\theta}{2} - \sin^2\frac{\theta}{2}\right)^{-1}$$

$$\therefore\ \omega = \sqrt{\frac{g}{R}\left(\tan\frac{\theta}{2} - \sin^2\frac{\theta}{2}\right)^{-\frac{1}{2}}}\ . \quad \textbf{(Proved)}$$

EXAMPLE 8.17 *A hollow circular cylinder of radius R, height h and specific gravity δ, floats in water which is rotating about a vertical axis coincident with the axis of the cylinder which is open at its upper end. Show that the maximum angular velocity ω of the fluid so that the cylinder shall not sink is given by* $\omega^2 R^2 = 4gh(1-\delta)$.

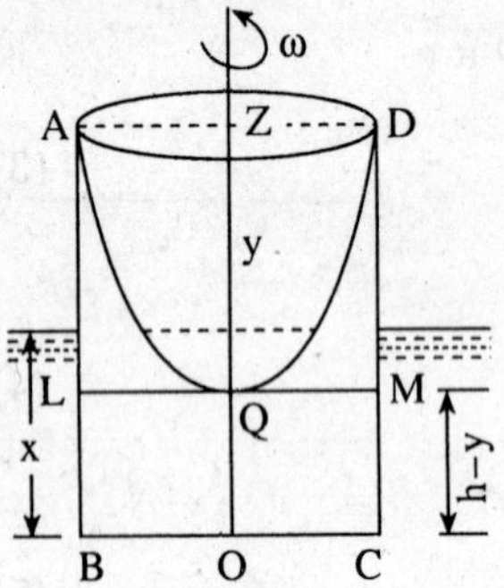

Fig. 8.26

Solution. Let $ABCD$ be the hollow circular cylinder floating in water which is rotating about the axis OZ of the cylinder (FIG. 8.26). Then, we are given,

$$\left.\begin{aligned} AB = CD = h \\ OB = OC = R \end{aligned}\right\} \tag{1}$$

Let x be the depth of the cylinder immersed. Then by the condition of floatation, weight of the cylinder = weight of displaced liquid.

Hence,

$$\pi R^2 hg\delta = \pi R^2 xg, \text{ i.e., } x = \delta h. \tag{2}$$

Due to rotation, the free surface of water will be a paraboloid of revolution. The cylinder will not sink if no water enters into it. For this, it is necessary that the cylinder should float in equilibrium with its upper rim in the surface in its critical situation.

In this situation, the vertex of the paraboloid falls at Q.

Let $QZ = y$. Then the volume of the paraboloid AQD is

$$V_1 = \frac{1}{2} \text{ volume of the cylinder } ALMD = \frac{1}{2}\pi R^2 y. \tag{3}$$

Then volume of the liquid now displaced is

$$V_2 = V_1 + \text{volume } LBCM = \frac{1}{2}\pi R^2 y + \pi R^2 (h-y) \tag{4}$$

Initially, the volume of liquid displaced was

$$V_3 = \pi R^2 x = \pi R^2 \delta h, \text{ [using (2)]} \tag{5}$$

For maintaining equilibrium, we must have

$$V_3 = V_2. \tag{6}$$

From (4), (5) and (6), we get

$$\pi R^2 \delta h = \frac{1}{2}\pi R^2 y + \pi R^2(h-y), \text{ i.e., } \delta h = \frac{1}{2}y + (h-y)$$

$$\therefore y = 2h(1-\delta) \tag{7}$$

Now, for the paraboloid AQD, we have

$$\frac{DZ^2}{QZ} = \text{latus rectum } \frac{2g}{\omega^2}, \text{ i.e., } \frac{R^2}{y} = \frac{2g}{\omega^2}$$

$$\therefore \omega^2 R^2 = 2gy \tag{8}$$

From (7) and (8), we get $\omega^2 R^2 = 4gh(1-\delta)$. **(Proved)**

EXAMPLE 8.18 *A vessel in the form of a hemisphere of radius R is full of liquid and is made to rotate with uniform angular velocity ω about the vertical radius of the bowl. Show that the amount of water spilt is*

$$\frac{\pi}{4}\cdot\frac{\omega^2 R^4}{g} \text{ or } \frac{2\pi}{3}\cdot\frac{R^3\omega^6 - 2g^3}{\omega^6}$$

according as $R\omega^2 < \text{ or } > 2g$.

Solution. Let $ABNCD$ be the hemispherical bowl of radius R completely filled with liquid and rotating about the axis PN (FIG. 8.27).

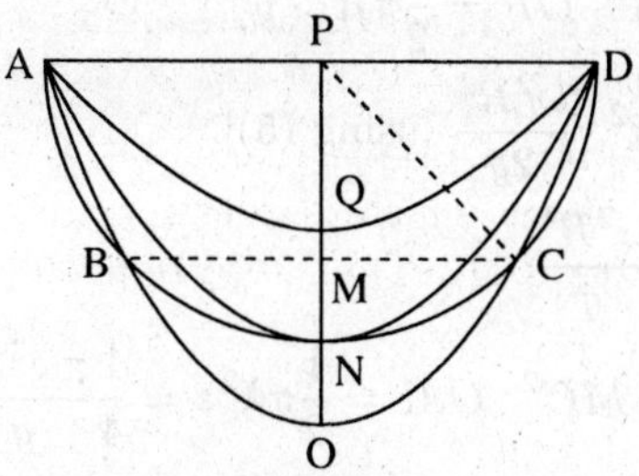

Fig. 8.27

Due to rotation, the free surface takes the shape of a paraboloid of revolution. When the vertex of the paraboloid just touches the lowest point N, then, for the paraboloid AND, we have

$$\frac{PD^2}{PN} = \text{latus rectum } \frac{2g}{\omega^2}, \text{ i.e., } \frac{R^2}{R} = \frac{2g}{\omega^2}, \text{ i.e., } \omega^2 R = 2g. \tag{1}$$

Thus, the vertex of the paraboloid of revolution will be

$$\left.\begin{array}{r}\text{above } N, \text{ if } \omega^2 R < 2g\\ \text{and below } N, \text{ if } \omega^2 R > 2g\end{array}\right\} \tag{2}$$

Let, when $\omega^2 R < 2g$, AQD be the paraboloid of revolution formed.
If $PQ = x$, then

$$\frac{PD^2}{PQ} = \frac{2g}{\omega^2}, \text{ i.e., } \frac{R^2}{x} = \frac{2g}{\omega^2}, \text{ i.e., } x = \frac{R^2\omega^2}{2g}. \tag{3}$$

Then the volume of liquid spilt is equal to the volume of the paraboloid AQD. Hence, volume of liquid spilt is

$$V = \frac{1}{2}\pi R^2 \cdot x = \frac{1}{2}\pi R^2 \cdot \frac{R^2\omega^2}{2g} \text{ [using (3)]}$$

$$\text{i.e., } V = \frac{\pi}{4} \cdot \frac{\omega^2 R^4}{g} \tag{4}$$

Again, when $\omega^2 R > 2g$, let $ABOCD$ be the paraboloid formed due to rotation of the liquid. In this case, the bottom point N of the hemisphere gets exposed.
Let $OP = y$. Then, for the paraboloid $ABOCD$, we have

$$\frac{PD^2}{OP} = \frac{2g}{\omega^2}, \text{ i.e., } \frac{R^2}{y} = \frac{2g}{\omega^2}, \text{ i.e., } y = \frac{\omega^2 R^2}{2g}. \tag{5}$$

Again, let $OM = z, MC = \lambda$, then $PM = OP - OM = y - z$. Then, for the paraboloid BOC, we have

$$\frac{MC^2}{OM} = \frac{2g}{\omega^2}, \text{ i.e., } \frac{\lambda^2}{z} = \frac{2g}{\omega^2}, \text{ i.e., } z = \frac{\lambda^2\omega^2}{2g} \tag{6}$$

Now, volume of the liquid spilt is

$V =$ volume V_1 of paraboloid $ABOCD-$ volume V_2 of the paraboloid BOC
$+$ volume V_3 of the spherical segment BNC.

Here,

$$V_1 = \frac{1}{2}\pi R^2 \cdot OP = \frac{1}{2}\pi R^2 \cdot y$$

$$\text{i.e., } V_1 = \frac{1}{2}\pi R^2 \cdot \frac{\omega^2 R^2}{2g} \text{ [using (5)]}$$

$$\text{i.e., } V_1 = \frac{\pi}{4} \cdot \frac{\omega^2 R^4}{g} \tag{7}$$

$$V_2 = \frac{1}{2}\pi \cdot MC^2 \cdot OM = \frac{1}{2}\pi\lambda^2 z = \frac{1}{4}\frac{\pi\omega^2\lambda^4}{g} \text{ [using (6)]} \tag{8}$$

$$\text{and } V_3 = \int_{y-z}^{R} \pi(R^2-\xi^2)d\xi = \pi\left[R^2\xi - \frac{\xi^3}{3}\right]_{y-z}^{R}$$

$$= \pi\left[R^2\{R-(y-z)\} - \frac{1}{3}\{R^3-(y-z)^3\}\right] \tag{9}$$

Now, in the ΔPMC,

$$PM^2 + MC^2 = PC^2, \text{ i.e., } (y-z)^2 + \lambda^2 = R^2, \text{ i.e., } (y-z)^2 = R^2 - \lambda^2. \tag{10}$$

Putting here the value of y from (5) and that of z from (6), we get

$$\left(\frac{\omega^2 R^2}{2g} - \frac{\omega^2\lambda^2}{2g}\right)^2 = R^2 - \lambda^2, \text{ i.e., } \frac{\omega^4}{4g^2}(R^2-\lambda^2)^2 = R^2 - \lambda^2$$

$$\therefore R^2 - \lambda^2 = \frac{4g^2}{\omega^4}. \tag{11}$$

Using (10), we get from (9)

$$V_3 = \pi\left[R^2\left\{R - \sqrt{R^2-\lambda^2}\right\} - \frac{1}{3}\left\{R^3 - (R^2-\lambda^2)^{3/2}\right\}\right]$$

$$= \pi\left[R^2\left\{R - \frac{2g}{\omega^2}\right\} - \frac{1}{3}\left\{R^3 - \frac{8g^3}{\omega^6}\right\}\right] \text{ [using (11)]}$$

$$= \frac{2}{3}\pi R^3 - \pi\frac{2g}{\omega^2}R^2\left(1 - \frac{4g^2}{3\omega^4 R^2}\right) \tag{12}$$

Using (7), (8) and (12), we get the total volume of liquid spilt as

$$V = V_1 - V_2 + V_3$$

$$= \frac{\pi}{4}\frac{\omega^2 R^4}{g} - \frac{\pi}{4}\frac{\omega^2\lambda^4}{g} + \frac{2}{3}\pi R^3 - \frac{2\pi g R^2}{\omega^2} + \frac{8}{3}\frac{\pi g^3}{\omega^6}$$

$$= \frac{\pi}{4}\frac{\omega^2 R^4}{g} - \frac{\pi}{4}\frac{\omega^2}{g}\left(R^2 - \frac{4g^2}{\omega^4}\right)^2 + \frac{2}{3}\pi R^3 - \frac{2\pi g R^2}{\omega^2} + \frac{8}{3}\frac{\pi g^3}{\omega^6} \text{ [using (11)]}$$

$$= \frac{2}{3}\pi R^3 + \left[\frac{\pi}{4}\frac{\omega^2 R^4}{g} - \frac{\pi}{4}\frac{\omega^2}{g}\left(R^4 - \frac{8R^2 g^2}{\omega^4} + \frac{16g^4}{w^8}\right) - \frac{2\pi g R^2}{\omega^2} + \frac{8}{3}\frac{\pi g^3}{\omega^6}\right]$$

$$= \frac{2}{3}\pi R^3 + \left[\frac{\pi}{4}\frac{\omega^2 R^4}{g} - \frac{\pi}{4}\frac{\omega^2 R^4}{g} + \frac{2\pi g R^2}{\omega^2} - \frac{4\pi g^3}{\omega^6} - \frac{2\pi g R^2}{\omega^2} + \frac{8}{3}\frac{\pi g^3}{\omega^6}\right]$$

$$= \frac{2}{3}\pi R^3 - \frac{4}{3}\frac{\pi g^3}{\omega^6} = \frac{2\pi}{3}\left(\frac{R^3\omega^6 - 2g^3}{\omega^6}\right). \quad \textbf{(Proved)}$$

EXAMPLE 8.19 *A bent tube, of small uniform bore, consists of two straight legs, of which one is horizontal and closed at the end, and the other is vertical and open. If the horizontal leg be filled with mercury and the tube revolves about a vertical axis through the closed end with angular velocity* ω, *prove that mercury will rise to a height d in the vertical leg, when*

$$\omega^2 = \frac{2g(h+d)}{l^2 - d^2}$$

where l is the length of the horizontal leg, and h is the height of the mercury barometer.

Show also that the mercury will not rise at all, unless $\omega^2 l^2 > 2gh$.

Solution. Let ABC be the bent tube with vertical leg AB open to atmosphere and horizontal leg BC of length l closed at C (FIG. 8.28). The leg BC is filled with mercury. When the tube revolves about the vertical axis CZ through closed end C with an angular velocity ω, mercury rises to a height d in the vertical leg.

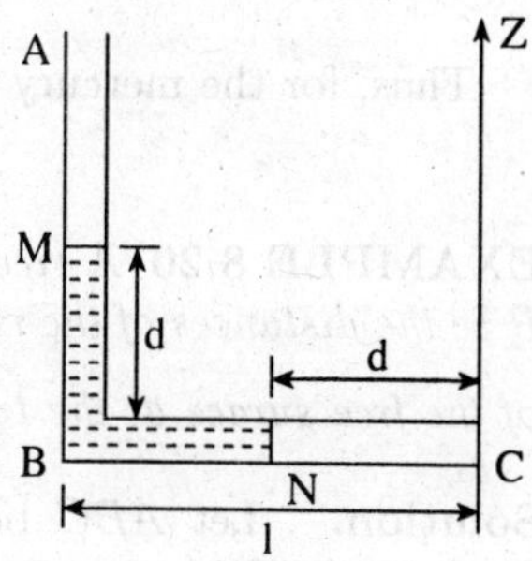

Fig. 8.28

Let ρ = density of mercury.

Since h is the height of mercury barometer, atmospheric pressure is

$$\Pi = g\rho h. \tag{1}$$

With C as origin, we take the z-axis along CZ in the vertical direction.

Then pressure equation is

$$dp = \rho(\omega^2 r dr - g dz), \tag{2}$$

where r is the horizontal distance from z-axis and z the vertical distance from C of the point concerned.

Integrating (2), we get

$$p = \rho\left(\frac{1}{2}\omega^2 r^2 - gz\right) + K, \tag{3}$$

where K is the constant of integration.

Now, at the end N, pressure is zero. Thus we have the condition

$$p = 0, \text{ for } r = d \text{ and } z = 0 \tag{4}$$

Using this condition in (3), we get

$$0 = \rho\left(\frac{1}{2}\omega^2 d^2 - 0\right) + K, \tag{5}$$

which yields

$$K = -\frac{1}{2}\rho\omega^2 d^2. \tag{6}$$

Putting this value of K in (3), we get

$$p = \rho\left(\frac{1}{2}\omega^2 r^2 - gz\right) - \frac{1}{2}\rho\omega^2 d^2. \tag{7}$$

Again, at the end M, the pressure is the atmospheric pressure Π given by (1). Thus we have another condition,

$$p = \Pi = g\rho h, \text{ for } r = l \text{ and } z = d. \tag{8}$$

Using this condition in (7), we get

$$g\rho h = \rho\left(\frac{1}{2}\omega^2 l^2 - gd\right) - \frac{1}{2}\rho\omega^2 d^2, \text{ i.e., } g(h+d) = \frac{1}{2}\omega^2(l^2 - d^2). \tag{9}$$

Hence,

$$\omega^2 = \frac{2g(h+d)}{l^2 - d^2}. \quad \textbf{(Proved)} \tag{10}$$

When mercury does not rise at all, the angular velocity ω is minimum and $d = 0$. Putting $d = 0$ in (10), the minimum value of ω is given by

$$\omega^2 = \frac{2gh}{l^2}, \text{ i.e., } \omega^2 l^2 = 2gh. \tag{11}$$

Thus, for the mercury to rise in the vertical leg,

$$\omega^2 l^2 > 2gh. \quad \textbf{(Proved)} \tag{12}$$

EXAMPLE 8.20 *A river flowing with uniform velocity v makes a circular turning. If r and R be the distances of the river banks from the axis of rotation, prove that the difference in height of the free surace at the two banks is* $\dfrac{v^2}{g}\ln\dfrac{R}{r}$.

Solution. Let ABC be the river bed and AC the free surface of water due to rotation (FIG. 8.29). The nearer bank is at a distance of r and the further bank at a distance of R from the axis of rotation OZ.

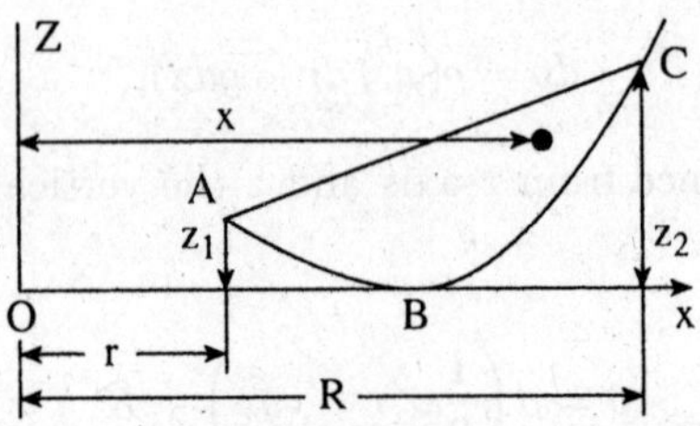

Fig. 8.29

Let ρ = density of river water, z_1 = height of free surface at nearer bank, z_2 = height of free surface at further bank.

Considering a unit mass of water at a distance x from the axis of rotation, we have the components of forces acting on it

$$X = \frac{v^2}{x}, Y = 0, Z = -g. \tag{1}$$

Hence the pressure equation is

$$dp = \rho\left(\frac{v^2}{x}dx - gdz\right). \tag{2}$$

On the free surface, $dp = 0$. Hence, we get

$$\frac{v^2}{x}dx - gdz = 0. \tag{3}$$

Integrating (3), we get

$$v^2 \ln x - gz = K, \; K = \text{constant}. \tag{4}$$

We are given,

$$\left.\begin{aligned} z &= z_1, \text{ when } x_1 = r \\ z &= z_2, \text{ when } x_2 = R \end{aligned}\right\} \tag{5}$$

Using these conditions in (4), we get

$$\left.\begin{aligned} v^2 \ln r - gz_1 &= K \\ v^2 \ln R - gz_2 &= K \end{aligned}\right\} \qquad (6)$$

Eliminating K from these two equations,

$$v^2(\ln R - \ln r) = g(z_2 - z_1), \text{ i.e., } z_2 - z_1 = \frac{v^2}{g} \ln \frac{R}{r}. \quad \textbf{(Proved)}$$

EXAMPLE 8.21 *A spherical shell of radius R is just filled with a liquid of density ρ and the whole rotates with uniform angular velocity ω about the vertical diameter. Show that the maximum value of the pressure on the shell is* $\frac{1}{2}\rho\frac{(\omega^2 R + g)^2}{\omega^2}$.

Solution. Let $ABCD$ be the spherical shell of radius R just filled with a liquid of density ρ rotating with uniform angular velocity ω about the vertical diameter AOC. With origin at the centre O of the shell, we take z-axis along vertically downward direction (FIG. 8.30).

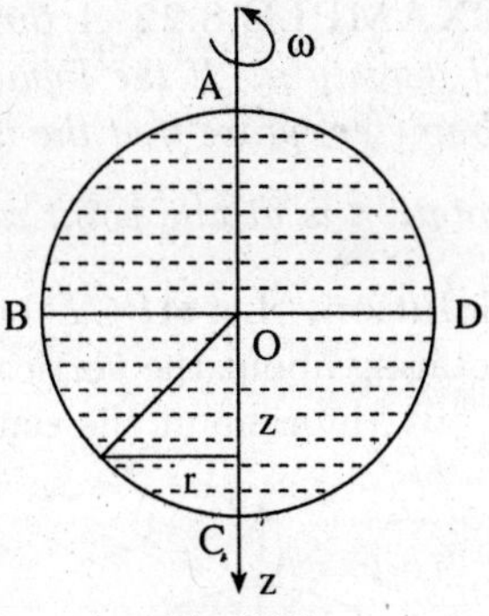

Fig. 8.30

Then the pressure at any point is given by

$$dp = \rho(\omega^2 r dr + g dz), \qquad (1)$$

where r = distance of the point from z-axis, z = depth of the point from O.

Integrating (1),

$$p = \rho\left(\frac{1}{2}\omega^2 r^2 + gz\right) + C, \qquad (2)$$

where C is the constant of integration.

At the point A, the pressure is zero. Thus, we have the condition,

$$p = 0, \text{ for } z = -R \text{ and } r = 0. \qquad (3)$$

Using this condition in (2), we get

$$0 = \rho(0 - gR) + C, \text{ i.e., } C = g\rho R. \qquad (4)$$

From (2) and (4),

$$p = \rho\left[\frac{1}{2}\omega^2 r^2 + g(z + R)\right]. \qquad (5)$$

This determines the pressure at any point on the shell. Now, for any point on the surface of the shell, we have

$$r^2 = R^2 - z^2. \qquad (6)$$

Then, from (5) and (6), pressure on the surface is given by

$$p = \rho\left[\frac{1}{2}\omega^2(R^2 - z^2) + g(z + R)\right]. \qquad (7)$$

For extremum value of this pressure, we must have

$$\frac{dp}{dz} = 0. \qquad (8)$$

Hence, differentiating (7), and using (8), we get $\frac{dp}{dz} \equiv \rho[-\omega^2 z + g] = 0$, which yields

$$z = \frac{g}{\omega^2}. \qquad (9)$$

Again,

$$\frac{d^2p}{dz^2} = \frac{d}{dz}\left[\rho\left(-\omega^2 z + g\right)\right] = -\omega^2\rho, \tag{10}$$

which is negative.

Hence, the pressure on the surface attains its maximum value at the level of z given by (9). Putting the value of z from (9) into (7), we get the maximum value of the pressure. Thus,

$$\begin{aligned} p_{\max} &= \rho\left[\frac{1}{2}\omega^2\left(R^2 - \frac{g^2}{\omega^4}\right) + g\left(\frac{g}{\omega^2} + R\right)\right] = \rho\left[\frac{\omega^2}{2}\cdot\frac{R^2\omega^4 - g^2}{\omega^4} + g\left(\frac{R\omega^2 + g}{\omega^2}\right)\right] \\ &= \rho(R\omega^2 + g)\left[\frac{R\omega^2 - g}{2\omega^2} + \frac{g}{\omega^2}\right] = \rho(R\omega^2 + g)\frac{R\omega^2 + g}{2\omega^2} \\ &= \rho\frac{(R\omega^2 + g)^2}{2\omega^2}. \quad \textbf{(Proved)} \end{aligned}$$

EXAMPLE 8.22 *A thin spherical shell of radius R is just not filled with gravitating liquid of density ρ. If the liquid be rotating in relative equilibrium with angular velocity ω about a diameter, prove that the tension in the shell across the great circle at right angles to the axis of rotation is at any point in that circle equal to* $\dfrac{\omega^2\rho R^3}{8}$.

Solution. Let $ABCD$ be the sphere of radius R containing gravitating liquid of density ρ and rotating about the vertical diameter AC (FIG. 8.31).

With origin at the centre O, we take the z-axis along OA vertically upwards. At any point $P(x, y, z)$ the components of the force of attraction towards the centre on unit mass of liquid are

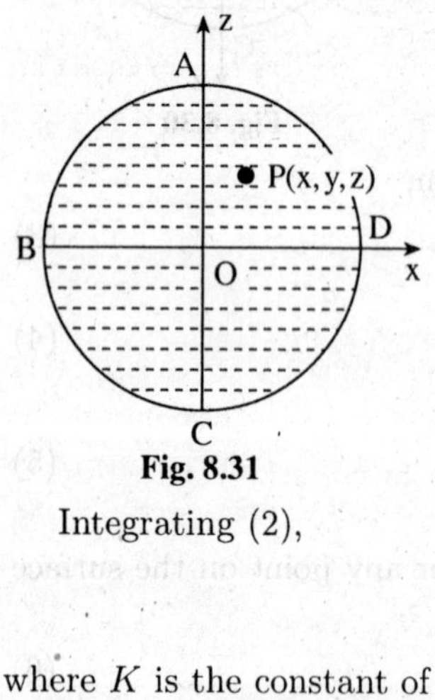

Fig. 8.31

$$X = \frac{4}{3}\pi\gamma\rho x, \; Y = \frac{4}{3}\pi\gamma\rho y, \; Z = \frac{4}{3}\pi\gamma\rho z, \tag{1}$$

where γ is the gravitational constant. Then the pressure at the point P is given by

$$dp = \rho\left[(\omega^2 - \xi)x\,dx + (\omega^2 - \xi)y\,dy - \xi z\,dz\right], \tag{2}$$

where we have denoted

$$\xi = \frac{4}{3}\pi\gamma\rho. \tag{3}$$

Integrating (2),

$$p = \frac{\rho}{2}\left[(\omega^2 - \xi)(x^2 + y^2) - \xi z^2\right] + K, \tag{4}$$

where K is the constant of integration.

Since the sphere is just not filled, the pressure is zero at the top most point A. Hence

$$p = 0, \text{ when } x = y = 0, z = R. \tag{5}$$

Using this condition in (4), we get

$$0 = \frac{\rho}{2}\left[0 - \xi R^2\right] + K, \text{ i.e.. } K = \frac{1}{2}\rho\xi R^2. \tag{6}$$

Putting this value of K in (4), we get

$$p = \frac{\rho}{2}\left[(\omega^2 - \xi)(x^2 + y^2) + \xi(R^2 - z^2)\right]. \tag{7}$$

This equation determines the pressure at any point in the liquid.

Now, for any point on the inner surface of the spherical shell,

$$x^2 + y^2 + z^2 = R^2, \text{ i.e., } x^2 + y^2 = R^2 - z^2. \tag{8}$$

Putting this relation in (7),

$$p = \frac{\rho}{2}[(\omega^2 - \xi)(R^2 - z^2) + \xi(R^2 - z^2)], \text{ i.e., } p = \frac{\rho}{2}\omega^2(R^2 - z^2). \tag{9}$$

This determines the pressure at any point on the inner surface of the shell.

Let us now evaluate the resultant thrust on the lower half of the shell.

At a depth z from origin, we consider a ring-shaped element of radius $R\cos\theta$ and thickness $Rd\theta$ (FIG. 8.32). Due to symmetry, the resultant horizontal thrusts on portions BC and CD will be equal in magnitude and opposite in direction, so that they nullify each other. The resultant vertical thrust on the whole portion BCD will be twice that on either BC or CD.

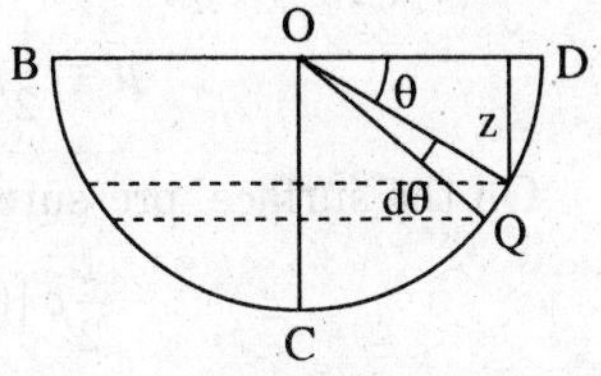

Fig. 8.32

The vertical component of pressure at Q is

$$= p\sin\theta = \frac{\rho}{2}\omega^2(R^2 - z^2)\sin\theta. \tag{10}$$

Thus, the resultant vertical thrust on the hemispherical bowl is

$$F = 2\int_0^{\pi/2} (\pi R\cos\theta)\cdot\frac{\rho}{2}\omega^2(R^2 - z^2)\sin\theta\cdot Rd\theta = \pi R^2\rho\omega^2\int_0^{\pi/2}\cos\theta\cdot R^2\cos^2\theta\cdot\sin\theta d\theta$$

$$= \pi R^4\rho\omega^2\int_0^{\pi/2}\cos^3\theta\sin\theta d\theta = \frac{1}{4}\pi R^4\rho\omega^2 \tag{11}$$

Thus, across the horizontal diameter BOD, equal amounts of vertical thrusts F will act both on the upper and lower hemispheres.

Hence, the tension in the shell across the horizontal diameter is

$$= \frac{F}{2\pi R} = \frac{1}{4}\frac{\pi R^4\rho\omega^2}{2\pi R} = \frac{\omega^2\rho R^3}{8}. \quad \textbf{(Proved)}$$

EXAMPLE 8.23 *A given mass of gravitating fluid of density ρ can rotate in relative equilibrium with angular velocity Ω with its free surface in the form of an ellipsoid with three unequal axes, the greatest semi-axis being a. A rigid vessel in this form is now made and the fluid in it is set rotating with the vessel in relative equilibrium with angular velocity ω about the least axis. Prove that the pressure at any point of the surface is*

$$\frac{1}{2}\rho(\omega^2 - \Omega^2)(x^2 + y^2), \textit{ or, } \frac{1}{2}\rho(\omega^2 - \Omega^2)(x^2 + y^2 - a^2)$$

according as ω is greater or less than Ω.

Solution. In a gravitating fluid in the form of an ellipsoid of semi-axes a, b, c the components of attraction forces on a unit mass at a point $P(x, y, z)$ towards the centre are (FIG. 8.33)

$$X = Ax,\ Y = By,\ Z = Cz \tag{1}$$

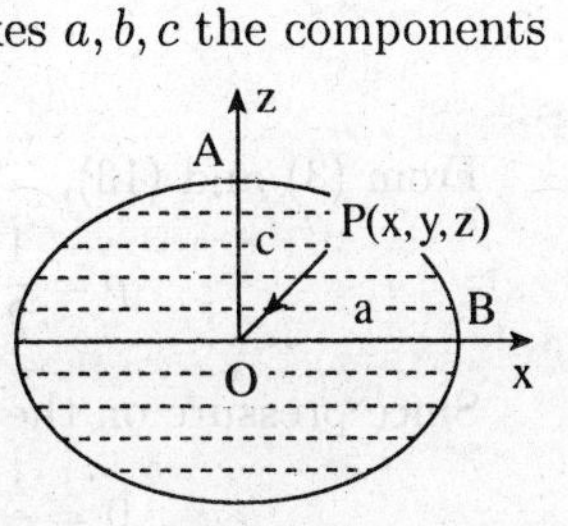

Fig. 8.33

where A, B, C are given by Eqns (8.35).

When the fluid mass rotates with angular velocity Ω, the pressure at any point is given by

$$dp = \rho\left[(\Omega^2 - A)xdx + (\Omega^2 - B)ydy - Czdz\right] \tag{2}$$

Integrating,

$$p = \frac{1}{2}\rho\left[(\Omega^2 - A)x^2 + (\Omega^2 - B)y^2 - Cz^2\right] + D_1. \tag{3}$$

At the topmost point A, pressure is zero. That is

$$p = 0, \text{ when } x = y = 0, z = c. \tag{4}$$

Using this condition in (3)

$$0 = \frac{1}{2}\rho(-Cc^2) + D_1, \text{ i.e., } D_1 = \frac{1}{2}\rho Cc^2. \tag{5}$$

From (3) and (5)

$$p = \frac{1}{2}\rho\left[(\Omega^2 - A)x^2 + (\Omega^2 - B)y^2 - Cz^2\right] + \frac{1}{2}\rho Cc^2. \tag{6}$$

On the surface, pressure is zero. Hence,

$$\frac{1}{2}\rho\left[(\Omega^2 - A)x^2 + (\Omega^2 - B)y^2 - Cz^2 + Cc^2\right] = 0. \tag{7}$$

When the fluid mass in the vessel rotates about Oz with angular velocity $\omega > \Omega$, the pressure equation is

$$dp = \rho\left[(\omega^2 - A)xdx + (\omega^2 - B)ydy - Czdz\right]. \tag{8}$$

Integrating, we get

$$p = \frac{\rho}{2}\left[(\omega^2 - A)x^2 + (\omega^2 - B)y^2 - Cz^2\right] + D_2. \tag{9}$$

At the point A, the pressure is zero. This yields

$$D_2 = \frac{1}{2}\rho Cc^2. \tag{10}$$

Hence pressure at any point is

$$p = \frac{\rho}{2}\left[(\omega^2 - A)x^2 + (\omega^2 - B)y^2 - Cz^2 + Cc^2\right] \tag{11}$$

The pressure at any point on the surface would be the difference of pressures (6) and (11). Thus, for $\omega > \Omega$, the pressure on the surface is

$$p_1 = \frac{1}{2}\rho(\omega^2 - \Omega^2)(x^2 + y^2). \quad \textbf{(Proved)}$$

Again, in Eqn (3), we put the condition that pressure at B is zero. That is

$$p = 0, \text{ when } x = a, y = z = 0 \tag{12}$$

Then we get

$$D_1 = -\frac{1}{2}\rho(\Omega^2 - A)a^2. \tag{13}$$

From (3) and (13),

$$p = \frac{1}{2}\rho\left[(\Omega^2 - A)x^2 + (\Omega^2 - B)y^2 - Cz^2 - (\Omega^2 - A)a^2\right] \tag{14}$$

Since pressure on the surface is zero, we have

$$0 = \frac{1}{2}\rho\left[(\Omega^2 - A)x^2 + (\Omega^2 - B)y^2 - Cz^2 - (\Omega^2 - A)a^2\right] \tag{15}$$

When the fluid mass in the vessel rotates about the axis Oz with angular velocity $\omega < \Omega$, the pressure at the point B vanishes. This condition applied to (9) yields

$$D_2 = -\frac{\rho}{2}(\omega^2 - A)a^2. \tag{16}$$

Hence, in this case, Eqn (9) assumes the form

$$p = \frac{\rho}{2}\left[(\omega^2 - A)x^2 + (\omega^2 - B)y^2 - Cz^2 - (\omega^2 - A)a^2\right]. \tag{17}$$

Then the pressure on the surface from (15) and (17) is

$$p = \frac{1}{2}\rho(\omega^2 - \Omega^2)(x^2 + y^2 - a^2). \quad \textbf{(Proved)}$$

EXAMPLE 8.24 *A mass of gravitating liquid of uniform density ρ is rotating about an axis with a small uniform acceleration ω. The free boundary is an oblate spheroid of small ellipticity ϵ for which the components of force of attraction at a point may be taken as*

$$X = -\frac{4}{3}\pi\rho\left(1 - \frac{2}{5}\epsilon\right)x,\ Y = -\frac{4}{3}\pi\rho\left(1 - \frac{2}{5}\epsilon\right)y,\ Z = -\frac{4}{3}\pi\rho\left(1 + \frac{4}{5}\epsilon\right)z.$$

Prove that $\epsilon = \dfrac{15\omega^2}{16\pi\rho}$.

Solution. If $a, a, a(1-\epsilon)$ are the semi-axes of the oblate spheroid of small ellipticity ϵ, then

$$(1-\epsilon)^2 \simeq (1 - 2\epsilon) \tag{1}$$

so that its equation is

$$x^2 + y^2 + \frac{z^2}{1 - 2\epsilon} = a^2. \tag{2}$$

The pressure equation is

$$dp = \rho\left[(\omega^2 - A)(xdx + ydy) - Czdz\right], \tag{3}$$

where, as given

$$A = \frac{4}{3}\pi\rho\left(1 - \frac{2}{5}\epsilon\right),\ C = \frac{4}{3}\pi\rho\left(1 + \frac{4}{5}\epsilon\right) \tag{4}$$

Integrating (3),

$$p = \frac{\rho}{2}\left[(\omega^2 - A)(x^2 + y^2) - Cz^2\right] + \text{constant}. \tag{5}$$

Since pressure is constant over the boundary, its equation must be of the form

$$(\omega^2 - A)(x^2 + y^2) - Cz^2 = \text{constant}. \tag{6}$$

Comparing (2) and (6), we get

$$\omega^2 - A = -C(1 - 2\epsilon). \tag{7}$$

Putting the values of A and C from (4) into (7), we get

$$\omega^2 = \frac{4}{3}\pi\rho\left[\left(1 - \frac{2}{5}\epsilon\right) - (1 - 2\epsilon)\left(1 + \frac{4}{5}\epsilon\right)\right]$$

$$= \frac{4}{3}\pi\rho\left(1 - \frac{2}{5}\epsilon - 1 - \frac{4}{5}\epsilon + 2\epsilon\right) = \frac{4}{3}\pi\rho\cdot\frac{4}{5}\epsilon.$$

Hence, $\epsilon = \dfrac{15\omega^2}{16\pi\rho}$. **(Proved)**

EXERCISE

1. Fluid is contained within a circular tube of radius R in a vertical plane which can rotate about a vertical axis. If the fluid subtends an angle θ at the centre, the least angular velocity to make the fluid divide into two parts is $\sqrt{\frac{9}{R}\sec\frac{\theta}{4}}$.

2. A circular tube is half-full of liquid and is made to revolve about a vertical tangent line with angular velocity ω. If R be the radius of the tube, prove that the diameter passing through the free surfaces of the liquid is inclined at an angle $\tan^{-1}\left(\frac{\omega^2 R}{g}\right)$ to the horizon.

3. A hollow cone, vertex upwards, is three-quarters full of water, and is set rotating about its axis which is vertical, with an angular velocity of $\sqrt{\frac{8g}{3h}}\cos\alpha$, where α is the semi-vertical angle and h is the height of the cone. Show that the thrust on the base is to the weight of water in the vessel as $10 : 3$.

4. A quantity of water occupies a quadrant of a fine tube of given radius R. When the tube rotates with angular velocity ω about a vertical diameter, the highest of water is at an angular distance of $60°$ from the highest point of the tube. Prove that $\omega^2 = \frac{2g}{R}\left(\sqrt{3}+1\right)$, assuming that water rises on one side only.

5. When a cylinder open at the top and half-full of liquid rotates with angular velocity ω about its axis which is vertical, the liquid just reaches the rim. Prove that the velocity in order that $\frac{1}{n}$th of the liquid may remain in the cylinder is $\omega\sqrt{n}$.

6. Fluid is rotating in a cylinder of radius R, whose bottom is closed by a conical surface of semi-vertical angle α, the vertex being downwards. Prove that the pressure at the surface of the cone is minimum at a point distant $\frac{l}{2}\cot\alpha$ from the axis, where l is the latus rectum of the free surface provided $l < 2R\tan\alpha$.

7. A semicircular tube is filled with water and rotates with uniform angular velocity ω about the diameter joining the two ends, which is vertical. Find where must a hole in the tube be made so that all liquid may escape. [**Ans.** g/ω^2 below centre.]

8. A closed right circular cylinder is very nearly filled with water and is made to rotate about its axis which is vertical. Show that, if the angular velocity is $\frac{\sqrt{2gh}}{R}$, then the whole thrust on the base is half as much again as when the liquid is at rest. Here h is the height and R the radius of the cylinder.

9. A circular tube of small section is half-full of liquid and in the surface at each side floats one of two small equal spheres which just fit the tube. The tube rotates with angular velocity ω about the vertical diameter. Show that the pressure is maximum at a depth g/ω^2 below the centre when ω is greater than a certain quantity.

10. A heavy liquid is contained in a fine circular tube of radius R, the liquid filling one-fourth of the tube. It is set rotating about a vertical diameter with angular velocity ω. If $\omega^2 = \frac{2\sqrt{2}g}{R}$, show that the level of the liquid will just rise to the horizontal diameter whilst the depth of the free surface below the centre is $R/\sqrt{2}$.

11. A rectangular lamina of which the sides are $2a$ and $2b$, is completely immersed in water which is rotating about a vertical axis with uniform angular velocity $(ng/b)^{1/2}$. The plane of the lamina is vertical and one of its sides ($2b$) touches the surface of water at its lowest point, the point of contact being the middle point of the side, and the lamina rotates with the same angular velocity as the water. Show that the depth of the centre of pressure of the lamina is $\frac{a(8a+6h+nb)}{6a+6h+nb}$, where h denotes the height of the water barometer.

12. A solid cylinder floats in a cylindrical vessel, and the system revolves about the common axis with angular velocity ω. Show that the cylinder is depressed by motion through the space $\frac{\omega^2}{4g}(R^2-r^2)$, where R and r denote the radii of the vessel and the cylinder respectively.

13. An elliptic tube half-full of liquid revolves about a fixed vertical axis in its plane with angular velocity ω. Show that the angle which the straight line joining the free surfaces of the liquid makes with the vertical is $\tan^{-1}\left(\frac{g}{c\omega^2}\right)$, where c is the distance of the axis from the centre of the ellipse.

14. A fine straight tube of length l, closed at both ends and inclined to the vertical at an angle α, is just filled with liquid of density ρ. If the tube is rotated with uniform angular velocity ω round a vertical axis through the lower end, prove that the pressure at the highest point is $\frac{1}{2}\rho\omega^2\sin^2\alpha\left(l - \frac{g\cos\alpha}{\omega^2\sin^2\alpha}\right)$, or zero according as $\omega^2 >$ or $< \frac{g\cos\alpha}{l\sin^2\alpha}$.

15. A cylinder of radius R is just full of water, being closed by a heavy lid which can turn about a point on its rim. Prove that the lid will be lifted up when the cylinder and the water are rotating with angular velocity ω about the axis of the cylinder, if its weight be less than $\frac{1}{4}\pi\omega^2 R^4\rho$, where ρ is the density of water.

16. A cubical box, open at the top, whose base is horizontal, is filled with water and made to rotate about a vertical axis through its centre. If the centre of the base be just uncovered, show that the angular velcity is $\sqrt{8g/a}$, where a is the side of the cube.

17. A circular cylinder of radius R contains water and a piston closely fitting the cylinder rests on the surface of the water. The whole is made to rotate with uniform angular velocity ω about the axis of the cylinder, which is vertical. Show that, provided $\omega^2R^2 > 4g(b+h)$, the height x through which the piston rises, is given by the equation $\{4g(x-b-h)+\omega^2R^2\}^2 - 16g\omega^2R^2x = 0$, where h is the height of the water barometer and b the length of that column of water in the cylinder whose weight is equal to that of the piston.

18. A hemispherical bowl, just full of water, is inverted and placed with its mouth downwards on a smooth horizontal table so that no water escapes under the edge of the bowl. The vessel and the contained liquid is then made to rotate with angular velocity ω and the vessel is on the point of rising. Prove that its weight is to the weight of the contained water as $(4g+3\omega^2R) : 8g$, where R is the radius of the bowl.

19. A spherical shell of radius R is just filled with a liquid of density ρ, and the whole rotates with uniform angular velocity ω about the vertical diameter. Show that the ratio of the resultant thrusts on the upper and lower hemispheres is $\frac{4g+3\omega^2R}{20g+3\omega^2R}$.

20. A cylindrical can of radius R with its axis vertical contains water to height h. It is rotated about its axis and given a horizontal acceleration f. Assuming that the spin is just sufficient not to uncover any point of the base and that the can is sufficiently high for none of the water to spill, show that f is less than both $R^2\omega^2$ and $\sqrt{2gh}$ and that the spin ω satisfies the relation $4gh = R^2\omega^2 + \frac{2f^2}{\omega^2}$.

21. A right circular cone of semi-vertical angle α floats, vertex downwards, in water contained in a vertical cylinder, and the surface of the water meets the cone in a circle of radius R. The water is made to rotate with uniform angular velocity ω about the common axis of the cone and the cylinder. Show that the water now meets the cone in a circle of radius r given by $r^3 - \frac{3\omega^2}{4g}r^4\tan\alpha = R^3$.

22. A cup in the form of a paraboloid of revolution cut off by a plane perpendicular to the axis is just filled with liquid and placed with its vertex upwards on a horizontal plane. The whole is made to revolve about the axis of the cup. Prove that the liquid will escape when the angular velocity ω exceeds the value $\sqrt{\frac{g}{2a}\cdot\frac{W-W_1}{W_1}}$, where W, W_1 are the weights of the cup and the liquid respectively and $4a$ is the latus rectum of the parabola.

23. Show that if a liquid of density δ is rotating, as if rigid, with uniform angular velocity about a vertical axis, while the air above it is at rest, the free surface will be a paraboloid of revolution whose latus rectum is to that of a surface of equi-pressure in the liquid as $(\delta-\rho) : \delta$, if the variation in the density ρ of the air be neglected.

24. Water is contained in a cup formed by the revolution of a parabola about its axis, the water and the cup revolve with uniform angular velocity ω about the axis. If the latus rectum of the parabola be less than $2g/\omega^2$, show that all the water would escape through a hole at the lowest point of the cup.

25. A case in the form of a prolate spheroid of small ellipticity ϵ is filled by a fluid nucleus of density $(\rho+\delta)$ surrounded by a fluid of density ρ. Show that, if it rotates round its axis of figure with an angular velocity $\sqrt{\frac{8}{5}\pi\rho\epsilon}$, a possible form of the common surface is a sphere.

26. Two gravitating liquids which do not mix, and whose densities are ρ, δ $(\rho > \delta)$, are enclosed in a rigid spherical envelop, and the whole rotates in relative équilibrium with a small angular velocity ω about a diameter of the sphere. Show that a possible form of common surface of the two liquids is an oblate spheroid of ellipticity $\frac{15}{16} \cdot \frac{\omega^2}{\pi\left(\rho + \frac{3}{2}\delta\right)}$.

27. A uniform gravitating liquid sphere of radius R and density ρ is made to rotate with small uniform angular velocity ω, its surface being free. Show that the pressure at the point $\left(\frac{1}{2}R, \theta\right)$ is

$$\frac{1}{2}\pi G\rho^2 R^2 - \frac{7}{48}\rho\omega^2 R^2 - \frac{5}{16}\rho\omega^2 R^2 \cos^2\theta,$$

where G denotes the constant of gravitation.

28. A mass of liquid of density ρ is surrounded by a mass of liquid of density δ and the whole completely fills a case in the form of an oblate spheroid of small ellipticity ϵ. If the case rotates about its axis with small uniform angular velocity ω, prove that a possible form of common surface is an oblate spheroid of ellipticity ξ given by $\frac{15\omega^2}{16\pi} = \xi\rho + \frac{3}{2}(\xi - \epsilon)\delta$.

CHAPTER 9

Stability of Floating Bodies

9.1 Introduction

We have seen in the earlier chapter that when a body floats partially immersed in a fluid, then the following two conditions are satisfied:

(1) The weight of the body is equal to the weight of the displaced fluid, i.e., the force of buoyancy, and

(2) The centre of gravity of the body and that of the displaced fluid, i.e., the centre of buoyancy are in the same vertical line.

Now, if the floating body be slightly displaced, it may either tend to return to the original position or it may recede further away from the original position. In the former case, the equilibrium is said to be ***stable***, and in the latter case, ***unstable***.

The displacements of the body may be of four different types:

(1) Small vertical displacement
(2) Small horizontal displacement without rotation
(3) Small rotational displacement
(4) Small arbitrary displacement.

Let us first consider the small vertical displacement of a body floating partially immersed in a homogeneous fluid, or immersed wholly or partially in a heterogeneous fluid, the density of which increases with the depth. In this case, if the body is pushed down, the amount of fluid displaced will increase, i.e., the force of buoyancy will increase. This extra force in the vertically upward direction will tend to raise the body to its original position. If the body is elevated, the weight of displaced fluid will diminish and hence the body will tend to sink down to its original position. Thus, ***for small vertical displacements, the equilibrium of a floating body is stable***.

In the case of a small horizontal displacement without rotation of a floating body, there is no resultant horizontal force acting on the body. Hence there is no tendency of the body to move away from its original position. Hence, ***equilibrium is neither stable nor unstable***. We may say that the body is ***neutral*** in respect of stability.

An arbitrary displacement to a floating body will, in general, cause both vertical and angular displacements in the positions of the body. However, if the displacements be small, the effects of the two changes in position can be treated independently.

We, therefore, intend to discuss the effect of a small angular displacement of a floating body in respect to its stability of equilibrium. For this, we make the following assumptions:

(1) The weight of the fluid displaced remains unchanged. Hence, the force of buoyancy in all positions is equal to the weight of the body.

(2) The fluid pressure has no tendency to raise or depress the centre of mass of the body.

9.2 Metacentre

We consider a body floating in a homogeneous fluid. Let G be the centre of gravity of the body and H be the centre of buoyancy (FIG. 9.1) in the original position of equilibrium. In this position, G and H remain on the same vertical line.

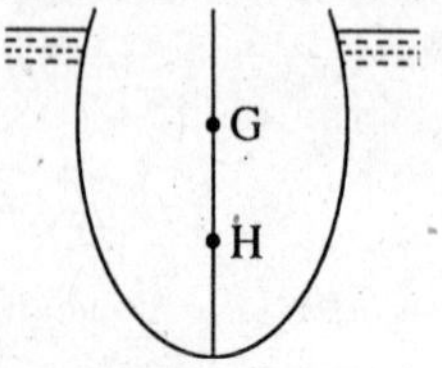

Fig. 9.1

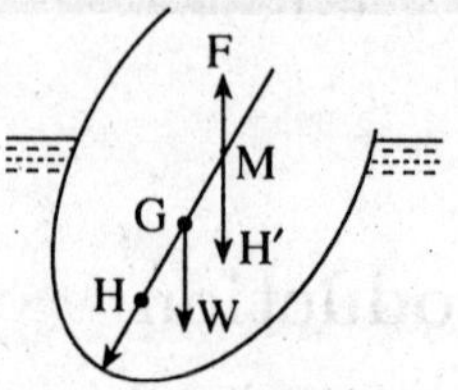

Fig. 9.2

Let now the body be given a small angular displacement. Whereas the position of G remains unaltered, the position of H will be shifted to a new position H' (FIG. 9.2). The force of buoyancy F equal to the weight of the body now acts in the vertically upward direction through H'.

This vertical line will intersect the line GH, if the displacement takes place in a plane of symmetry of the body in which case the two lines are coplanar. Then the point M at which these two lines intersect is called the ***Metacentre***.

Thus, ***if a small angular displacement be given to a body floating freely in a fluid, so that the amount of displaced fluid remains the same, then the point at which the vertical line through the new centre of buoyancy meets the line joining the centre of gravity of the body and the original centre of buoyancy is called the Metacentre.***

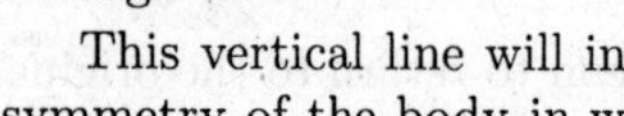

Fig. 9.3

Let W = weight of the floating body acting vertically downwards through G, and F = the force of buoyancy acting vertically upwards through H'. These two forces being equal, they constitute a couple of moment,

$$L = W \cdot GM \sin\theta = W \cdot GM \cdot \theta, \tag{9.1}$$

where θ is the small angular displacement.

Now, two cases may arise. The metacentre M may be above or below the centre of gravity G (FIG. 9.3).

When the metacentre M is above G, the couple tends to restore the body to its original position. In the other case, the couple tends to increase the angular displacement.

Thus, the equilibrium is stable or unstable according as the metacentre M is above or below the centre of gravity G.

The distance of the metacentre M from the centre of gravity G is called the ***metacentric height*** of the floating body.

9.3 Definitions

If a body floats in a homogeneous liquid, the plane in which the body is intersected by the surface of the liquid is called the ***plane of floatation***.

If the floating body moves so that the volume of the liquid displaced remains unchanged, then the envelop of the planes of floatation is called the ***surface of floatation*** and the locus of the centre of buoyancy is called the ***surface of buoyancy***.

Curves of floatation and ***curves of buoyancy*** are the principal normal sections at corresponding points on a surface of floatation and a surface of buoyancy.

9.4 Necessary Condition for Stability

We consider a solid body AQB floating in a homogeneous liquid in equilibrium. With the point O on the plane of floatation AOB as origin, we take a Cartesian coordinate system as shown in the figure (FIG. 9.4). The axes Ox and Oy are in the plane of floatation and Oz vertically downwards.

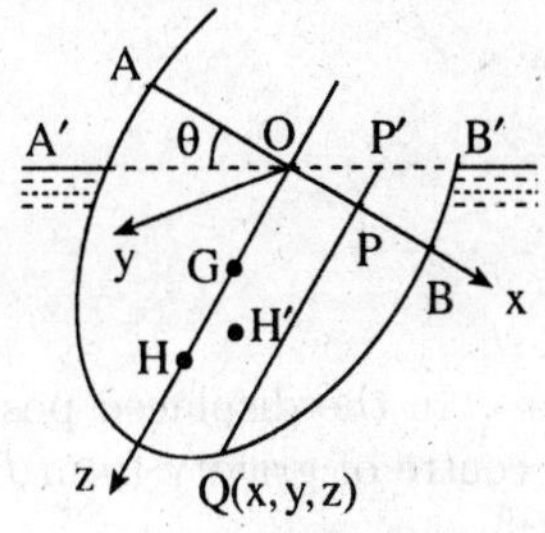

Fig. 9.4

Let the body be given a small angular displacement θ about the y-axis which is perpendicular to the plane of paper, carrying the axes Ox and Oz with it.

Let A = area of the plane of floatation,
V = volume of the immersed portion of the body.

In the original position, we consider an elementary cylinder of cross-sectional area $dxdy$ around the point $P(x, y)$ on the plane of floatation and of length $PQ = z$, the coordinates of the point Q on the body being (x, y, z).

Then the volume immersed in the liquid is

$$V = \iint z dxdy, \tag{9.2}$$

where the integration is carried over the plane of floatation.

In the displaced position, $A'OB'$ is the new plane of floatation and the point P has moved to P' such that

$$PP' = x\theta. \tag{9.3}$$

Hence, the length of the cylinder $P'Q$ is

$$P'Q = z + x\theta \tag{9.4}$$

so that the volume of the elementary cylinder is

$$dV = (z + x\theta)dxdy. \tag{9.5}$$

Then the volume immersed in the liquid in the displaced position is

$$V = \iint (z + x\theta)dxdy, \tag{9.6}$$

where the integration is carried over the area of the new plane of floatation.

Assuming that the volume immersed remains constant, we get from Eqns (9.2) and (9.6),

$$\iint (z + x\theta)dxdy = \iint z dxdy,$$

which yields

$$\iint x dxdy = 0. \tag{9.7}$$

Eqn (9.7) implies that the x-coordinate of the centroid of the plane of floatation is zero, i.e., the centroid of the surface lies on y-axis.

Hence, follows the theorem:

THEOREM 9.1 ***If a plane section of a body cuts off a volume which remains constant for small angular displacements of the plane, then the axis about which the plane turns must pass through the centroid of the section.***

We assume that this condition is satisfied. Let in the original position, $((\bar{x}, \bar{y}, \bar{z})$ be the coordinates of the centre of buoyancy H. The position of the centre of gravity of the elementary cylinder PQ of volume $zdxdy$ being at

$$x,\ y,\ \frac{1}{2}z$$

we have

$$\left.\begin{aligned} V\bar{x} &= \iint xzdxdy \\ V\bar{y} &= \iint yzdxdy \\ V\bar{z} &= \frac{1}{2}\iint z^2 dxdy. \end{aligned}\right\} \tag{9.8}$$

In the displaced position, the length of the element $P'Q$ being $(z + x\theta)$, the distance of its centre of gravity from P is

$$= \frac{1}{2}(z + x\theta) - x\theta = \frac{1}{2}(z - x\theta). \tag{9.9}$$

Hence, for the elementary cylinder $P'Q$ with cross-sectional area $(z + x\theta)dxdy$, the coordinates of the centre of gravity are

$$x,\ y,\ \frac{1}{2}(z - x\theta).$$

If x^*, y^*, z^* be the coordinates of the new position H' of the centre of buoyancy, then we have

$$\left.\begin{aligned} Vx^* &= \iint x(z + x\theta)dxdy \\ Vy^* &= \iint y(z + x\theta)dxdy \\ Vz^* &= \frac{1}{2}\iint (z + x\theta)(z - x\theta)dxdy = \frac{1}{2}\iint (z^2 - x^2\theta^2)dxdy \end{aligned}\right\} \tag{9.10}$$

Since θ is very small, θ^2 can be neglected. Hence

$$Vz^* = \frac{1}{2}\iint z^2 dxdy \tag{9.11}$$

From Eqns (9.8) and (9.11), we find that for small θ,

$$z = z^* \tag{9.12}$$

This implies that the line HH' is parallel to the xy-plane, i.e., the plane of floatation. Since H, H' are two neighbouring points on the surface of buoyancy, we have the following theorem:

THEOREM 9.2 ***The tangent plane at any point on the surface of buoyancy is parallel to the plane of floatation.***

Now, in the displaced position, as shown in the figure (FIG. 9.3), the body is subjected to two equal and opposite parallel forces—its weight W acting vertically downwards through G and the force $F(= W)$ of buoyancy acting vertically upwards through H'. These forces form a couple and the plane of this couple will be at right angles to the axis of rotation, if and only if, the points G and H' lie on a vertical plane perpendicular to the y-axis, i.e., if $\bar{y} = y^*$. Hence, from (9.8) and (9.10), we get

$$\iint y(z + x\theta)dxdy = \iint yzdxdy.$$

This yields

$$\iint xydxdy = 0. \tag{9.13}$$

This implies that the y-axis, i.e., the axis of rotation must be a principal axis of inertia of the section of the body made by the plane of floatation.

When this condition is satisfied, the vertical line through H' intersects the line HG at the point M, the metacentre. Hence, follows the theorem:

THEOREM 9.3 ***If a body floating in equilibrium in a homogeneous liquid be given a small angular displacement, its metacentre will exist, if only the axis of rotation is also a principal axis of the plane of floatation at some point.***

When the axis of rotation is an axis of symmetry of the plane of floation, the above condition is evidently satisfied.

We assume that this condition is also satisfied so that a metacentre of the floating body exists.

Now, from the figure (FIG. 9.3), since θ is small, we have

$$HH' = HM \cdot \theta. \tag{9.14}$$

Also, HG and the vertical through H' remain on a plane parallel to xOz plane, and HH' is parallel to the corresponding plane of floatation, i.e., xOy plane. Hence,

$$HH' = x^* - \bar{x}. \tag{9.15}$$

By virtue of Eqns (9.8) and (9.10), Eqn (9.15) assumes the form

$$HH' = \frac{1}{V}\left[\iint x(z + x\theta)dxdy - \iint xzdxdy\right] = \frac{\theta}{V}\iint x^2dxdy. \tag{9.16}$$

But the integral on the right-hand side of Eqn (9.16) represents the moment of inertia of the area A of the plane of floatation about y-axis. Hence we write,

$$\iint x^2dxdy = Ak^2, \tag{9.17}$$

where k is the radius of gyration of the area A about the axis of rotation y-axis. Hence,

$$HH' = \frac{Ak^2}{V}\theta. \tag{9.18}$$

From Eqns (9.14) and (9.18), we get

$$HM \cdot \theta = \frac{Ak^2}{V} \cdot \theta, \text{ i.e., } HM = \frac{Ak^2}{V}. \tag{9.19}$$

Thus, the distance HM depends only on the area of the surface section and the volume immersed.

The metacentric height GM is given by

$$GM = HM - HG$$
$$= \frac{Ak^2}{V} - HG = \frac{Ak^2 - V \cdot HG}{V}. \tag{9.20}$$

Now, the weight of the floating body being equal to the weight of the displaced liquid,

$$W = g\rho V, \tag{9.21}$$

where ρ is the density of the liquid.

Hence, the couple L tending to restore the body to its original position is

$$L = W \cdot GM \cdot \theta$$
$$= g\rho V \cdot \frac{Ak^2 - HG \cdot V}{V} \cdot \theta = g\rho\theta \left(Ak^2 - HG \cdot V\right). \tag{9.22}$$

Hence, the equilibrium will be stable, if

$$HG < \frac{Ak^2}{V}. \tag{9.23}$$

The plane of floatation has two principal axes passing through its centre of gravity with corresponding radii of gyrations k_1 and k_2. Hence, there are two metacentres corresponding to angular displacements about these two axes. Thus, the equilibrium will be stable for angular displacements about both these axes, if

$$\left.\begin{aligned} & HG < \frac{Ak_1^2}{V} \\ \text{and also} \quad & HG < \frac{Ak_2^2}{V} \end{aligned}\right\} \tag{9.24}$$

These are the necessary conditions for stability of equilibrium.

9.5 Work Done in Producing a Displacement

We consider a body floating in equilibrium in a homogeneous liquid. If the body has been displaced through a small angle θ about either principal axes through the centre of gravity of the surface section, then by Eqn (9.22), the restoring couple L acting on the body is given by

$$L = g\rho(Ak^2 - V \cdot HG)\theta = \xi\theta, \tag{9.25}$$

where we have denoted

$$\xi = g\rho(Ak^2 - V \cdot HG). \tag{9.26}$$

If the angle θ be increased to $\theta + d\theta$ by an external agency, then the work done is

$$dW^* = L \cdot d\theta = \xi\theta d\theta \tag{9.27}$$

Hence, the total work done in producing an angular displacement θ is

$$W^* = \int_0^\theta \xi\theta d\theta = \frac{1}{2}\xi\theta^2. \tag{9.28}$$

9.6 Sufficient Conditions for Stability

Let a body floating in equilibrium in a homogeneous liquid be given a small angular displacement θ about any axis in the plane of floatation through the centre of gravity of the surface section. This small rotation may be compounded of rotations θ_1 and θ_2 about the principal axes of the section. Each of these rotations separately induces a restoring couple, so that

$$\left.\begin{aligned} L_1 &= \xi_1\theta_1 \\ L_2 &= \xi_2\theta_2 \end{aligned}\right\} \tag{9.29}$$

where

$$\left.\begin{aligned} \xi_1 &= g\rho\left(Ak_1^2 - V\cdot HG\right) \\ \xi_2 &= g\rho\left(Ak_1^2 - V\cdot HG\right) \end{aligned}\right\} \tag{9.30}$$

Then work done separately by the external agency in producing angular displacements θ_1, θ_2 are

$$\left.\begin{aligned} W_1^* &= \frac{1}{2}\xi_1\theta_1^2 \\ W_2^* &= \frac{1}{2}\xi_2\theta_2^2. \end{aligned}\right\} \tag{9.31}$$

Hence, the total work done by the external agency is

$$W^* = W_1^* + W_2^* = \frac{1}{2}\left(\xi_1\theta_1^2 + \xi_2\theta_2^2\right). \tag{9.32}$$

It follows from Eqn (9.32), that the conditions

$$\xi_1 > 0 \text{ and } \xi_2 > 0, \text{ i.e., } HG < \frac{Ak_1^2}{V} \text{ and } HG < \frac{Ak_2^2}{V} \tag{9.33}$$

are sufficient to ensure stability for displacements which do not alter the volume of the liquid displaced.

9.7 Metacentric Height of a Ship

The metacentric height of a ship for rolling displacements may be determined by the following method:

Let the weight of the ship together with its cargo be W.

Suppose that a known weight w be shifted from A to B across the deck perpendicular to the ship due to which the ship turns through a small angle θ.

Let the distance $AB = d$. The angle θ can be measured by observing the relative defleciton of a plumb-line (FIG. 9.5).

We introduce two equal and opposite vertical forces w at A. Then the upward vertical force at A and the downward vertical force at B form a couple of moment

$$L_1 = wd\cos\theta. \tag{9.34}$$

The downward vertical force w at A plus the weight of the remaining cargo and the ship is equal to W. This force W acts vertically downward through the centre of gravity G of the ship. The force of buoyancy also equal to W acts vertically upwards through the metacentre M.

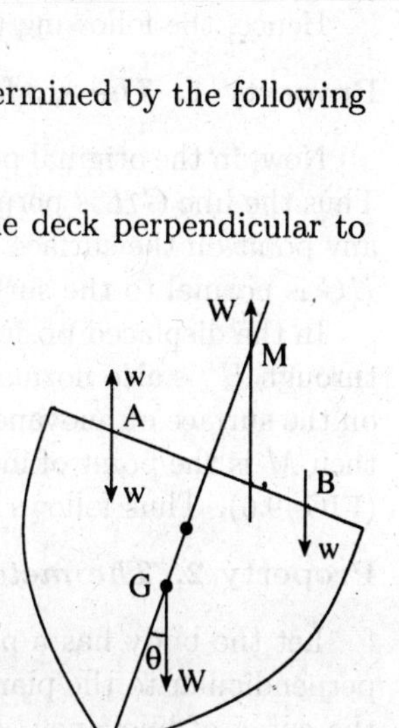

Fig. 9.5

These two equal and opposite parallel forces form a couple of moment

$$L_2 = W \cdot GM \cdot \sin\theta. \tag{9.35}$$

For equilibrium, the above two couples should be equal. Thus

$$L_1 = L_2, \text{ i.e., } wd\cos\theta = W \cdot GM \cdot \sin\theta$$

$$\therefore GM = \frac{wd}{W}\cot\theta = \frac{wd}{W}\theta, \tag{9.36}$$

since θ is very small.

9.8 Some Properties of Surface of Buoyancy

Let $(\bar{x}, \bar{y}, \bar{z})$ be the coordinates of the centre of buoyancy in the original position and (x^*, y^*, z^*) those in the displaced position of a body floating in equilibrium in a homogeneous liquid. Then, from Eqns (9.8) and (9.10), we have

$$\left.\begin{aligned} V\bar{z} &= \frac{1}{2}\iint z^2\,dxdy \\ Vz^* &= \frac{1}{2}\iint (z^2 - x^2\theta^2)dxdy. \end{aligned}\right\} \tag{9.37}$$

These equations yield

$$\bar{z} - z^* = \frac{1}{2}\frac{\theta^2}{V}\iint x^2\,dxdy, \text{ i.e., } \bar{z} - z^* = \frac{1}{2}\frac{Ak^2}{V}\theta^2. \tag{9.38}$$

The right-hand side of Eqn (9.38) being essentially a positive quantity for any small value of θ, we find

$$\bar{z} > z^*. \tag{9.39}$$

This implies that the depth z^* of the new position H' from the plane of floatation is less than that of the original position H.

Hence, the following property.

Property 1. ***The surface of buoyancy is concave upwards.***

Now, in the original position, the plane of floatation is horizontal and the line GH is vertical. Thus the line GH is perpendicular to the plane of floatation. Again, since the tangent plane at any point on the surface of buoyancy is parallel to the corresponding plane of floatation, hence HG is normal to the surface of buoyancy.

In the displaced position when H' is the new position of the centre of buoyancy, the vertical through H' is also normal to the surface of buoyancy. If H and H' are two neighbouring points on the surface of buoyancy and the angular displacement is such that the metacentre M exists, then M is the point of intersection of two normals through H and H', and hence, $HM = H'M$ (FIG. 9.6). Thus follows a property.

Property 2. ***The metacentre is a centre of curvature of the surface of buoyancy.***

Let the body has a plane of symmetry and the angular displacements be given about axes perpendicular to the plane. Then the centres of buoyancy will lie on a curve in the plane called the curve of buoyancy. For such displacements, ***the locus of the metacentre will be the evolute of the curve of buoyancy.***

For small displacement of H, the length GH remains constant since it is a normal. When the position is one of stable equilibrium, G lies between H and M.

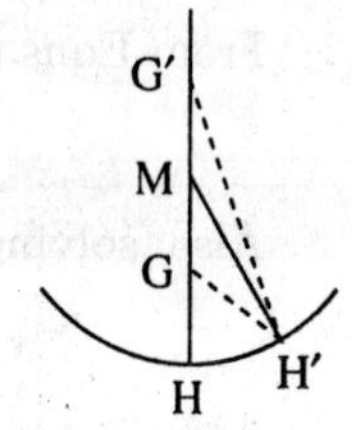

Fig. 9.6

In that case,

$$H'G + GM > H'M$$

i.e., $$H'G + GM > HM \quad (\because H'M = HM)$$

i.e., $$H'G + GM > HG + GM, \text{ i.e., } H'G > HG. \qquad (9.40)$$

Thus, in the case of stable equilibrium, HG is a minimum.

In the case of unstable equilibrium, the position G' of the centre of gravity lies on HM produced. It can be proved in the same manner that

$$HG' > H'G'. \qquad (9.41)$$

Thus, in the case of unstable equilibrium, HG' is a maximum.

9.9 Surface of Buoyancy in General

We consider a body floating in equilibrium in a homogeneous liquid (FIG. 9.7). With the origin of coordinates on the vertical through the centroid Q of the original surface section, we take a rectangular coordinate system as shown in the figure.

Let in the original position, $H(x_0, y_0, z_0)$ be the position of the centre of buoyancy, and $z = c$ be the equation of plane of floatation.

Let the body be given a small angular displacement. In this case, let the equation of the plane of floatation be

$$z = c + lx + my, \qquad (9.42)$$

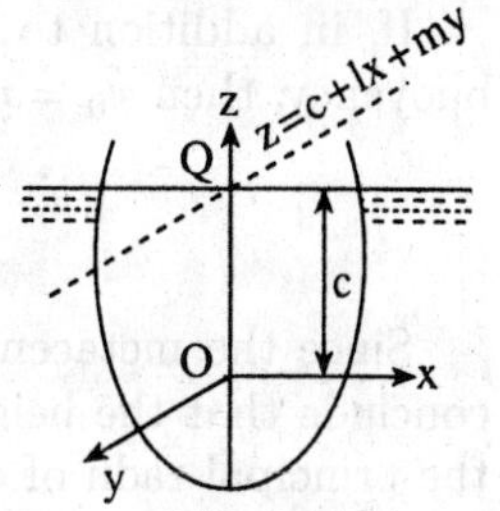

Fig. 9.7

where l, m are constants.

Let $(\bar{x}, \bar{y}, \bar{z})$ be the coordinates of the new position of the centre of buoyancy. Then,

$$V(\bar{x} - x_0) = \iint (z - c)x\,dx\,dy = \iint (lx + my)x\,dx\,dy = al + hm, \qquad (9.43)$$

where the integration is taken over the area of the solid cut by the surface.

Similarly,

$$V(\bar{y} - y_0) = \iint (z - c)y\,dx\,dy = hl + bm \qquad (9.44)$$

and $$V(\bar{z} - z_0) = \iint \frac{1}{2}(z^2 - c^2)\,dx\,dy = \frac{1}{2}(al^2 + 2hlm + bm^2) \qquad (9.45)$$

Here we have denoted

$$\left.\begin{aligned} a &= \iint x^2\,dx\,dy \\ h &= \iint xy\,dx\,dy \\ b &= \iint y^2\,dx\,dy \end{aligned}\right\} \qquad (9.46)$$

where a and b represent the moment of inertia of the plane section of the body at $z = c$ about y- and x-axes respectively, and h represents the product of inertia of that section.

From Eqns (9.43) to (9.45), we find the relation

$$2(\bar{z} - z_0) = l(\bar{x} - x_0) + m(\bar{y} - y_0) \tag{9.47}$$

Also, solving for l and m Eqns (9.42) and (9.43), we get

$$\left.\begin{aligned} l &= \frac{V}{ab - h^2}\left[b(\bar{x} - x_0) - h(\bar{y} - y_0)\right] \\ \text{and } m &= \frac{V}{ab - h^2}\left[a(\bar{y} - y_0) - h(\bar{x} - x_0)\right] \end{aligned}\right\} \tag{9.48}$$

Substituting these values of l and m into Eqn (9.47), we get

$$2(\bar{z} - z_0) = \frac{V}{ab - h^2}\left[b(\bar{x} - x_0)^2 - 2h(\bar{x} - x_0)(\bar{y} - y_0) + a(\bar{y} - y_0)^2\right] \tag{9.49}$$

Eqn (9.49) represents the approximate form of the surface of buoyancy.

If the original axes of x and y are principal axes of the plane section, then the product of inertia $h = 0$. In this case, the approximate form of the surface of buoyancy is given by

$$2(\bar{z} - z_0) = \frac{V}{ab}\left[b(\bar{x} - x_0)^2 + a(\bar{y} - y_0)^2\right] \tag{9.50}$$

If, in addition to the above, the origin of coordinates is taken at the original centre of buoyancy, then $x_0 = y_0 = z_0 = 0$. Hence, the surface of buoyancy is

$$2z = \frac{x^2}{a/V} + \frac{y^2}{b/V}. \tag{9.51}$$

Since the metacentre is a centre of curvature of the surface of buoyancy (property 2), we conclude that the heights h_1, h_2 of the metacentres above the centre of buoyancy are equal to the principal radii of curvature of the surface defined by Eqn (9.51). Thus,

$$h_1 = \frac{a}{V}, \quad h_2 = \frac{b}{V}. \tag{9.52}$$

9.10 Surface of Buoyancy for an Ellipsoid

Let the ellipsoid be defined by the equation

$$\frac{x^2}{a^2} + \frac{y^2}{b^2} + \frac{z^2}{c^2} = 1, \tag{9.53}$$

where a, b, c are the semi-axes of the ellipsoid.

We make the substitutions

$$\left.\begin{aligned} x &= a\xi \\ y &= b\eta \\ z &= c\zeta \end{aligned}\right\} \tag{9.54}$$

which reduces Eqn (9.53) to

$$\xi^2 + \eta^2 + \zeta^2 = 1 \tag{9.55}$$

Thus, our problem is reduced to that of a sphere.

If V denotes the volume of the immersed portion of the ellipsoid, then the corresponding volume of the sphere is

$$V_s = \frac{V}{abc}. \tag{9.56}$$

The plane which cuts off this volume touches a concentric sphere of radius r, such that

$$\int_r^1 \pi(1-x^2)dx = \frac{V}{abc}, \text{ i.e., } \frac{1}{3}\pi(1-r)^2(2+r) = \frac{V}{abc}. \tag{9.57}$$

The centroid of the volume cut off lies on a sphere of radius R, where

$$R\int_r^1 \pi(1-x^2)dx = \int_r^1 \pi x(1-x^2)dx,$$

which yields

$$R = \frac{3}{4}\frac{1+r^2}{2+r} \tag{9.58}$$

We conclude from Eqn (9.57), going back to the original problem, that the surface of floatation is a similar ellipsoid of semi-axes ra, rb and rc, where r is given by

$$(1-r^2)(2+r) = \frac{3V}{\pi abc} \tag{9.59}$$

Similarly, we conclude that the surface of buoyancy is another similar ellipsoid of semi-axes Ra, Rb and Rc, where R is given by Eqn (9.58).

9.11 Surface of Buoyancy for a Cylinder

We consider a cylinder of arbitrary cross section floating in equilibrium in a homogeneous liquid. With origin O at the centre of the base, we take a rectangular coordinate system (FIG. 9.8).

Let A = uniform area of cross section of the cylinder, V = volume immersed of the cylinder, and $z = c$ is the equation of the plane of floatation in the original position.

Then,

$$V = Ac \tag{9.60}$$

In the displaced position, let the equation of the plane of floatation be

$$z = c + lx + my, \tag{9.61}$$

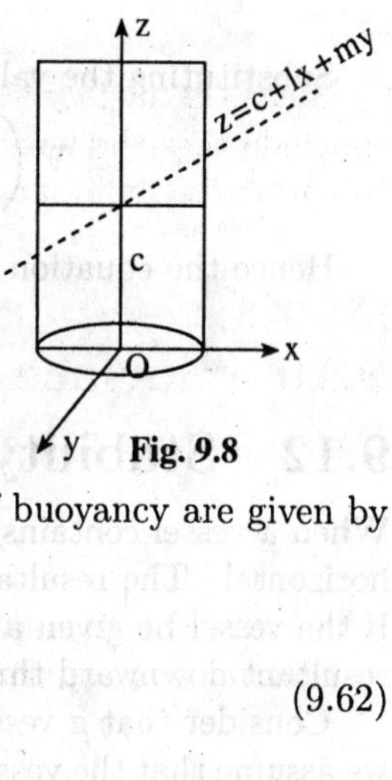

Fig. 9.8

where l, m are constants. The coordinates $(\bar{x}, \bar{y}, \bar{z})$ of the new centre of buoyancy are given by

$$\left.\begin{aligned} V\bar{x} &= \iint xz\,dx\,dy \\ V\bar{y} &= \iint yz\,dx\,dy \\ V\bar{z} &= \frac{1}{2}\iint z^2\,dx\,dy \end{aligned}\right\} \tag{9.62}$$

where all the integrations are carried over the area of the base.

Substituting the value of z from Eqn (9.61), we get

$$V\bar{x} = \iint x(c+lx+my)dx\,dy = al + hm \tag{9.63}$$

$$V\bar{y} = \iint y(c+lx+my)dx\,dy = hl + bm \tag{9.64}$$

$$V\bar{z} = \frac{1}{2}\iint (c+lx+my)^2 dx\,dy = \frac{1}{2}c^2A + \frac{1}{2}(al^2 + 2hlm + bm^2), \tag{9.65}$$

where we have denoted

$$\left.\begin{aligned} a &= \iint x^2 dxdy \\ h &= \iint xydxdy \\ b &= \iint y^2 dxdy \end{aligned}\right\} \tag{9.66}$$

and used the relations

$$\iint xdxdy = 0, \quad \iint ydxdy = 0, \tag{9.67}$$

since the centroid of the base is at origin.

If the axes of x and y are the principal axes of the section, then the product of inertia $h = 0$. In this case,

$$\left.\begin{aligned} V\bar{x} &= al, \text{ i.e., } l = \frac{V\bar{x}}{a} \\ V\bar{y} &= bm, \text{ i.e., } m = \frac{V\bar{y}}{b} \end{aligned}\right\} \tag{9.68}$$

and

$$V\bar{z} = \frac{1}{2}c^2 A + \frac{1}{2}\left(al^2 + bm^2\right)$$

$$\text{i.e., } V\left(\bar{z} - \frac{1}{2}c\right) = \frac{1}{2}\left(al^2 + bm^2\right) \text{ [using Eqn (9.60)]} \tag{9.69}$$

Substituting the values of l and m from Eqn (9.68) into Eqn (9.69), we get

$$V\left(\bar{z} - \frac{1}{2}c\right) = \frac{V^2}{2}\left(\frac{\bar{x}^2}{a} + \frac{\bar{y}^2}{b}\right), \text{ i.e., } \frac{\bar{x}^2}{a} + \frac{\bar{y}^2}{b} = \frac{2\bar{z} - c}{V} \tag{9.70}$$

Hence the equation of the surface of buoyancy is

$$\frac{x^2}{a} + \frac{y^2}{b} = \frac{2z - c}{V} \tag{9.71}$$

9.12 Stability of a Vessel Containing Liquid

When a vessel contains liquid, the volume of the liquid remains constant and its free surface is horizontal. The resultant downward thrust on the vessel is equal to the weight of the liquid. If the vessel be given a small angular displacement, we can determine the line of action of the resultant downward thrust by previous methods.

Consider that a vessel containing liquid is floating in a homogeneous liquid. For simplicity, we assume that the vessel is a body of revolution with its axis vertical, so that there is symmetry about all vertical planes through its axis. Then the centres of mass of the body and of the liquid are in the same vertical line.

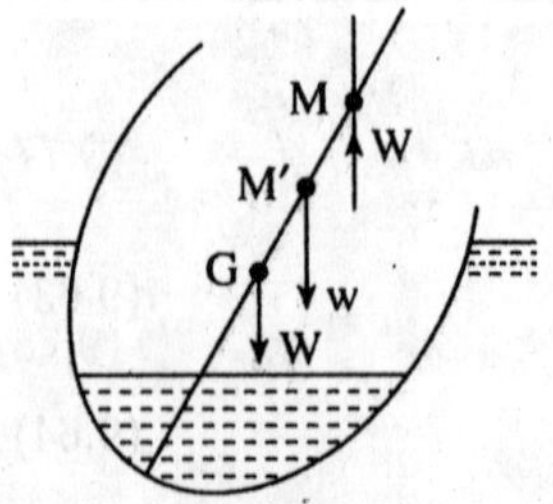

Fig. 9.9

Let the vessel containing liquid and floating in a liquid be given a small angular displacement (FIG. 9.9).

Let W = weight of displaced liquid,

w = weight of the contained liquid,

M = position of metacentre for the displaced liquid, and

M' = position of metacentre for the contained liquid,

G = position of centre of mass of the vessel.

The points G, M and M' are all lying on the axis of the vessel. The forces acting through these points are shown in the figure.

The total moments about G is

$$L = W \cdot GM - w \cdot GM'. \tag{9.72}$$

If $L > 0$, the vessel will tend to come back to its original position, so that equilibrium will be stable. Thus, for stable equilibrium,

$$W \cdot GM - w \cdot GM' > 0, \text{ i.e., } \frac{W}{w} > \frac{GM'}{GM}. \tag{9.73}$$

Similarly, if $L < 0$, the vessel will tend to move further away from its original position, so that equilibrium will be unstable. Thus, when

$$W \cdot GM - w \cdot GM' < 0, \text{ i.e., } \frac{W}{w} < \frac{GM'}{GM}, \tag{9.74}$$

the equilibrium will be unstable.

Let us now consider the case when the vessel is not a body of revolution and that the centres of mass of the displaced liquid and that of the contained liquid are not in the same vertical line.

We assume in this case that the body is symmetrical about the vertical plane through the centres of mass and that the small angular displacement takes place in the direction of that vertical plane.

Let (FIG. 9.10),

G be the centre of mass of the body,
H the centre of mass of the liquid displaced,
H' the centre of mass of the contained liquid,
M the metacentre of the displaced liquid
M' the metacentre of the contained liquid,
W = weight of the displaced liquid,
and w = weight of the contained liquid.

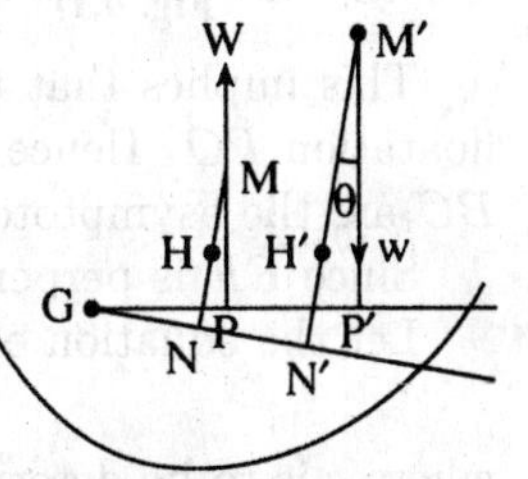

Fig. 9.10

In the original position, GNN' was the horizontal line through G. In the displaced position, let GPP' be the horizontal line through G.

The force W acts through M in the vertically upward direction and the force w acts through M' in the vertically downward direction. Then the total moment about G is

$$L = W \cdot GP - w \cdot GP'. \tag{9.75}$$

Now, from the figure,

$$\left.\begin{aligned} GP &= GN \cos\theta + MN \sin\theta \\ GP' &= GN' \cos\theta + M'N' \sin\theta \end{aligned}\right\} \tag{9.76}$$

Then, from Eqns (9.75) and (9.76), we have

$$L = W(GN \cos\theta + MN \sin\theta) - w(GN' \cos\theta + M'N' \sin\theta). \tag{9.77}$$

But, for equilibrium in the initial position, we must have

$$W \cdot GN = w \cdot GN' \tag{9.78}$$

In view of this, Eqn (9.77) assumes the form

$$L = (W \cdot MN - w \cdot M'N') \sin\theta. \tag{9.79}$$

Hence, the equilibrium is stable or unstable according as

$$L > \text{or} < 0$$

$$\text{i.e., } W \cdot MN - w \cdot M'N' > \text{or} < 0$$

$$\text{i.e., } \frac{W}{w} > \text{or} < \frac{M'N'}{MN} \tag{9.80}$$

9.13 Particular Cases of Curves of Buoyancy

We consider a lamina floating with its plane vertical in a homogeneous liquid having the form of a triangle or a rectangle.

(a) **Triangle.** Let ABC be the triangular lamina of density δ floating in a homogeneous liquid of density ρ (FIG. 9.11). PQ is the line of floatation and H is the centre of buoyancy. E and F are the middle points of PQ and AC respectively. Since HG is vertical, it is perpendicular to PQ. Again, since

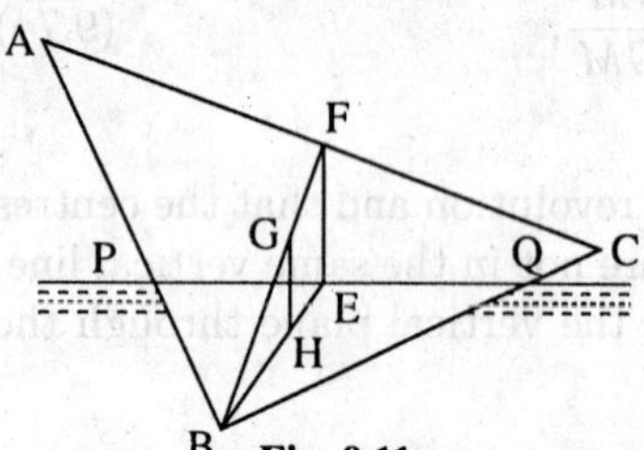

Fig. 9.11

$$BG : GF = BH : HE, \tag{9.81}$$

EF is parallel to GH and, therefore, is perpendicular to PQ. Now, for equilibrium, we must have

$$\Delta ABC \cdot \delta = \Delta BPQ \cdot \rho$$

$$\text{i.e., } \frac{\Delta ABC}{\Delta BPQ} = \frac{\rho}{\delta}. \tag{9.82}$$

This implies that the area of the ΔBPQ remains constant for all positions of the line of floatation PQ. Hence, PQ always touches, at its middle point, a hyperbola of which BA and BC are the asymptotes.

Since EF is perpendicular to PQ, it is normal to the hyperbola at E.

Let the equation of the curve of buoyancy referred to BA and BC as x- and y-axes be

$$xy = c^2, \tag{9.83}$$

where c is to be determined.

Let

$$AB = 2a,\ BC = 2b,\ \angle ABC = \theta. \tag{9.84}$$

Then the coordinates of F, the middle point AC, are (a, b).

Let (ξ, η) be the coordinates of E which lies on the curve (9.83). Then

$$\xi\eta = c^2. \tag{9.85}$$

Then the equation of the normal at E to the curve is

$$y - \eta = \frac{\eta\cos\theta - \xi}{\xi\cos\theta - \eta}(x - \xi) \tag{9.86}$$

Since this normal passes thorugh $F(a, b)$,

$$(b - \eta) = \frac{\eta\cos\theta - \xi}{\xi\cos\theta - \eta}(a - \xi),$$

$$\text{i.e., } (b - \eta)(\xi\cos\theta - \eta) = (a - \xi)(\eta\cos\theta - \xi)$$

$$\text{i.e., } \xi^2 - (a + b\cos\theta)\xi = \eta^2 - (a\cos\theta + b)\eta \tag{9.87}$$

All the points of the hyperbola where the tangents are the lines of floatation can be obtained by solving Eqns (9.85) and (9.87). Writing

$$\eta = \frac{c^2}{\xi} \tag{9.88}$$

and putting it into Eqn (9.87), we get

$$\xi^4 - (a + b\cos\theta)\xi^3 + (a\cos\theta + b)c^2\xi - c^4 = 0 \qquad (9.89)$$

Eqn (9.89) has only one negative root, and one or three positive root by Descarte's rule of signs. Neglecting the negative root, we see that there can be three positions of equilibrium or only one. Now,

$$\triangle BPQ = \frac{1}{2}\cdot BP\cdot BQ\sin\theta = \frac{1}{2}\cdot 2\xi\cdot 2\eta\sin\theta = 2c^2\sin\theta \qquad (9.90)$$

$$\text{and } \triangle ABC = \frac{1}{2}\cdot AB\cdot BC\sin\theta = \frac{1}{2}\cdot 2a\cdot 2b\sin\theta = 2ab\sin\theta \qquad (9.91)$$

Then, for equilibrium we have by Eqn (9.82)

$$\frac{2ab\sin\theta}{2c^2\sin\theta} = \frac{\rho}{\delta}, \text{ i.e., } c^2 = \frac{\delta}{\rho}ab, \qquad (9.92)$$

which determines the value of c.

Note. If the triangle be isosceles, $a = b$. Putting this in Eqn (9.89), we get

$$\xi^4 - a\xi^3(1+\cos\theta) + ac^2\xi(1+\cos\theta) - c^4 = 0$$

$$\text{or, } (\xi^2 - c^2)\left(\xi^2 + c^2 - 2a\xi\cos^2\frac{\theta}{2}\right) = 0 \qquad (9.93)$$

This yields

$$\left.\begin{aligned} \xi &= \pm c \\ \text{and } \xi &= a\cos^2\frac{\theta}{2} \pm \sqrt{a^2\cos^4\frac{\theta}{2} - c^2} \end{aligned}\right\} \qquad (9.94)$$

The negative root $\xi = -c$ is not admissible. If

$$a^2\cos^4\frac{\theta}{2} > c^2 \qquad (9.95)$$

we get three real roots, otherwise only one.

In this case, Eqn (9.92) becomes

$$c^2 = \frac{\delta}{\rho}a^2 \qquad (9.96)$$

In view of this equation, the condition (9.95) becomes

$$\cos^4\frac{\theta}{2} > \frac{\delta}{\rho}, \qquad (9.97)$$

for which there are three positions of equilibrium, otherwise there is only one position of equilibrium.

(b) **Rectangle.** Let AB be the initial plane of floatation and $A'B'$ the position of that plane after a small angular displacement be given to the rectangular lamina (FIG. 9.12). Let H and H' be the positions of the centres of buoyancy corresponding to the two planes of floatation.

Let

$$\left.\begin{aligned} AO = OB &= a \\ OH &= b \\ BB' &= \xi \end{aligned}\right\} \qquad (9.98)$$

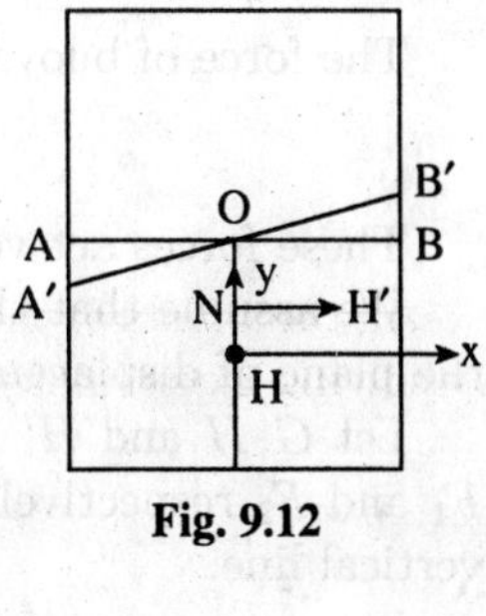

Fig. 9.12

and the area cut off $= S$. With origin at H, we take a rectangular coordinate system. Then, if the coordinates of H' are (x, y),

$$NH' = x,\ NH = y \tag{9.99}$$

Thus, we have

$$\left.\begin{aligned} S \cdot x = S \cdot NH' &= \frac{1}{2}a\xi \cdot \frac{2}{3}a - \frac{1}{2}a\xi\left(-\frac{2}{3}a\right) = \frac{2}{3}a^2\xi \\ \text{and } S \cdot y = S \cdot NH &= \frac{1}{2}a\xi\left(b + \frac{1}{3}\xi\right) - \frac{1}{2}a\xi\left(b - \frac{1}{3}\xi\right) = \frac{1}{3}a\xi^2 \end{aligned}\right\} \tag{9.100}$$

Eliminating ξ between Eqns (9.100), we get the locus of H' as

$$Sx^2 = \frac{4}{3}a^3 y, \tag{9.101}$$

which is a parabola. Thus the curve of buoyancy is a parabola.

The curve of floatation is clearly a point.

Notes. (1) If the lamina be bounded by a parabola, the curves of floatation and buoyancy are equal parabolas.

(2) If the boundary of the lamina be elliptic, the curves of floatation and buoyancy are arcs of similar and similarly situated concentric ellipses.

(3) If the lamina be bounded by a hyperbola, the curves of floatation and buoyancy are arcs of similar hyperbolas.

(4) In the case of a right circular cone floating with its vertex beneath the surface, the surfaces of floatation and byoyancy are hyperboloids of revolution.

9.14 Stability of a Body Floating in Two Liquids

We consider a body floating partially immersed in two liquids which do not mix.

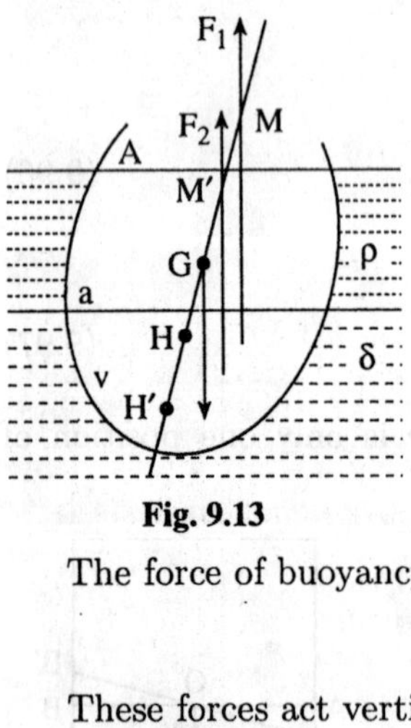

Fig. 9.13

Let (FIG. 9.13)

$V =$ total volume of the body immersed

$v =$ the volume of the body immersed in lower liquid,

$\rho =$ the density of the upper liquid,

$\delta =$ the density of the lower liquid, so that $\delta > \rho$,

$A =$ area of surface section by upper liquid, and

$a =$ area of surface section by lower liquid.

Considering the whole liquid of density ρ, the force of buoyancy is

$$F_1 = g\rho V. \tag{9.102}$$

The force of buoyancy due to the liquid of density $(\delta - \rho)$ on the volume v is

$$F_2 = g(\delta - \rho)v \tag{9.103}$$

These forces act vertically upwards.

We assume that the body is symmetrical with respect to a vertical plane perpendicular to the plane of displacement.

Let G, H and H' be the centre of gravity of the body and the centres of buoyancy forces F_1 and F_2 respectively. Then, due to the symmetry of the body, these points lie on the same vertical line.

Let now the body be given a small angular displacement θ about any horizontal axis in the plane of symmetry. Let M and M' be the positions of metacentres.

Then the total moment L about G of the forces F_1 and F_2 tending to restore equilibrium

$$L = F_1 \cdot GM \cdot \theta + F_2 \cdot GM' \cdot \theta, \tag{9.104}$$

where the positive direction of GM, GM' is upwards.

If M and M' are both above G, by Eqn (9.104) $L > 0$, and hence the equilibrium is stable. If, however, M' is below G, GM' is negative, so that

$$L = F_1 \cdot GM \cdot \theta - F_2 \cdot M'G \cdot \theta. \tag{9.105}$$

Hence, equilibrium is stable, if

$$F_1 \cdot GM \cdot \theta > F_2 \cdot M'G \cdot \theta, \text{ i.e., } \rho V \cdot GM > (\delta - \rho)v \cdot M'G \tag{9.106}$$

Eqn (9.106) takes a different look, if we recall that

$$\left.\begin{aligned} V \cdot GM &= (AK^2 - V \cdot HG) \\ \text{and } v \cdot GM' &= (ak^2 - v \cdot H'G) \end{aligned}\right\}, \tag{9.107}$$

where K is the radius of gyration of the top surface section and k that of bottom surface section about the axis of rotation. Hence, the condition (9.106) in view of relation (9.107) becomes

$$\rho(AK^2 - V \cdot HG) > (\delta - \rho)(v \cdot H'G - ak^2). \tag{9.108}$$

9.15 Metacentric Height in Heterogeneous Liquids

We now consider the case of heterogeneous liquid the density of which depends on the depth from free surface.

Referring to FIG. 9.4, we assume that

$$\rho = f(z), \tag{9.109}$$

where z is measured vertically downwards.

If the body is given a small angular displacement, then the density at the point (x, y, z) is

$$\rho = f(z + x\theta) = \rho + x\theta\frac{d\rho}{dz} \tag{9.110}$$

to the first order of θ.

We assume that the density at the surface is ρ_0, i.e.,

$$\rho_0 = f(z)\big|_{z=0} \tag{9.111}$$

Using the notations of section 9.4, the mass of the liquid displaced in original position is

$$M_0 = \iiint \rho dxdydz \tag{9.112}$$

and the mass of the liquid displaced in final position is

$$M = \iiint \left(\rho + x\theta\frac{d\rho}{dz}\right) dxdydz + \iint \rho_0 x\theta dxdy, \tag{9.113}$$

where the volume integrals are taken over the original volume V displaced, and the surface integral is taken over the surface section due to the wedges on the surface.

Since the mass of the liquid displaced remains constant, $M_0 = M$. Hence,

$$\iiint \left(\rho + x\theta \frac{d\rho}{dz}\right) dxdydz + \iint \rho_0 x\theta dxdy = \iiint \rho dxdydz$$

$$\text{i.e.,} \quad \iiint \left(\frac{d\rho}{dz}\right) xdxdydz + \iint \rho_0 xdxdy = 0 \tag{9.114}$$

Eqn (9.114) will be satisfied, if at all levels

$$\iint xdxdy = 0, \tag{9.115}$$

which implies that the centroids of all the horizontal sections of the volume immersed in the original position remain on the plane yz.

Since the mass of the liquid remains constant,

$$\left.\begin{aligned} M_0\bar{x} &= \iiint \rho xdxdydz \\ M_0\bar{y} &= \iiint \rho ydxdydz \\ M_0\bar{z} &= \iiint \rho zdxdydz \end{aligned}\right\} \tag{9.116}$$

where $\bar{x}, \bar{y}, \bar{z}$ are the coordinates of the centre of buoyancy H in the original position.

Similarly, we have

$$\left.\begin{aligned} M_0x^* &= \iiint \left(\rho + x\theta\frac{d\rho}{dz}\right) xdxdydz + \iint \rho_0 x^2\theta dxdy \\ M_0y^* &= \iiint \left(\rho + x\theta\frac{d\rho}{dz}\right) ydxdydz + \iint \rho_0 yx\theta dxdy \\ M_0z^* &= \iiint \left(\rho + x\theta\frac{d\rho}{dz}\right) zdxdydz \end{aligned}\right\} \tag{9.117}$$

where x^*, y^*, z^* are the coordinates of the centre of buoyancy H' in the displaced condition.

For the existence of the metacentre, the vertical through H' must intersect HG. This requires, as discussed in section 9.4, that $\bar{y} = y^*$. Hence,

$$\iiint \left(\rho + x\theta\frac{d\rho}{dz}\right) ydxdy + \iint \rho_0 yx\theta dxdy = \iiint \rho ydxdydz$$

$$\text{i.e.,} \quad \iiint xy\frac{d\rho}{dz}dxdydz + \iint \rho_0 xydxdydz = 0. \tag{9.118}$$

This equation is satisfied if at all depths, the plane yz intersects the horizontal section in a principal axis of that section.

Let the conditions (9.115) and (9.118) be satisfied. Then, as in section 9.4, we have

$$HM \cdot \theta = HH' = x^* - \bar{x}$$

$$\text{i.e.,} \; HM \cdot \theta = \frac{1}{M_0}\left[\iiint \left(\rho + x\theta\frac{d\rho}{dz}\right) xdxdydz + \iint \rho_0 x^2\theta dxdydz - \iiint \rho xdxdydz\right]$$

$$= \frac{\theta}{M_0}\left[\iiint \frac{d\rho}{dz}(x^2dxdydz) + \iint \rho_0 x^2 dxdydz\right]$$

Hence,

$$HM = \frac{1}{M_0}\left[\int Ak^2\frac{d\rho}{dz}dz + \rho_0 A_0 k_0^2\right], \tag{9.119}$$

where $k =$ the radius of gyration of the section at depth z about its axis in yz-plane,
$A =$ the area of the section at depth z,
$k_0 =$ the radius of gyration of the surface section about its axis in the yz-plane,
$A_0 =$ the area of the surface section.

If the density ρ_0 at the surface vanishes, then

$$HM = \frac{1}{M_0}\int Ak^2\frac{d\rho}{dz}dz \tag{9.120}$$

EXAMPLE 9.1 *A homogeneous circular cylinder of length h, radius R and specific gravity ρ floats in water. Prove that the position with the axis vertical is stable, if*

$$\frac{R^2}{h^2} > 2\rho(1-\rho).$$

Check whether the equilibrium is stable or not, when $R = 1.5$ m, $h = 3$ m and $\rho = 0.8$.

Solution. Let O be the centre of the base of the cylinder floating with axis vertical and G, H, M be the positions of the centre of gravity of the cylinder, centre of buoyancy and metacentre respectively (FIG. 9.14).

Let $x =$ depth of the cylinder immersed. Then the weight W of the cylinder is

$$W = \pi R^2 h\cdot\rho g \tag{1}$$

The force of buoyancy is

$$F = \pi R^2 x g \tag{2}$$

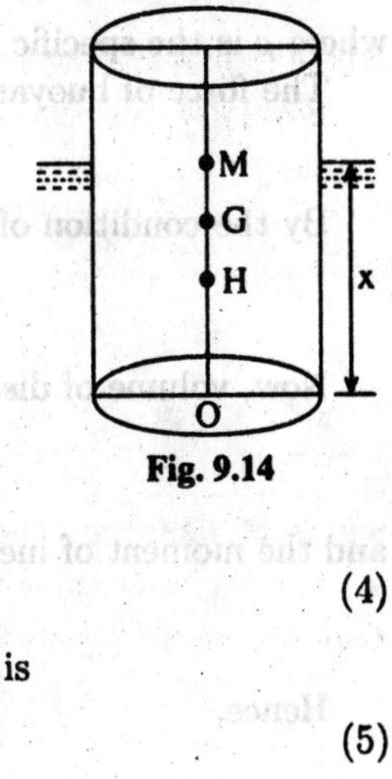

Fig. 9.14

For equilibrium, we must have $W = F$. Hence,

$$\pi R^2 x g = \pi R^2 h\rho g$$

$$\therefore x = h\rho \tag{3}$$

Now, volume V of the immersed portion is

$$V = \pi R^2 x = \pi R^2 h\rho \text{ [using (3)]} \tag{4}$$

and moment of inertia of the surface section about the axis of rotation is

$$Ak^2 = \pi R^2\cdot\frac{R^2}{4} = \frac{\pi}{4}R^4 \tag{5}$$

Hence,

$$HM = \frac{Ak^2}{V} = \frac{\pi}{4}R^4\cdot\frac{1}{\pi R^2 h\rho} = \frac{R^2}{4h\rho} \tag{6}$$

Also

$$HG = OG - OH = \frac{1}{2}(h-x) = \frac{1}{2}h(1-\rho). \tag{7}$$

For equilibrium to be stable, we must have

$$HM > HG, \text{ i.e., } \frac{R^2}{4h\rho} > \frac{1}{2}h(1-\rho) \text{ [from (6) and (7)]}$$

$$\text{i.e., } \frac{R^2}{h^2} > 2\rho(1-\rho) \quad \textbf{(Proved)}$$

In particular we have $R = 1.5\,\text{m}, h = 3.0\,\text{m}, \rho = 0.8$. Then

$$\frac{R^2}{h^2} = \frac{(1.5)^2}{(3.0)^2} = \frac{1}{4} = 0.25 \tag{8}$$

$$\text{and } 2\rho(1-\rho) = 2\cdot(0.8)\cdot(1-0.8) = (1.6)\cdot(0.2) = 0.32. \tag{9}$$

Hence, $\dfrac{R^2}{h^2} < 2\rho(1-\rho)$.

So the equilibrium is unstable.

EXAMPLE 9.2 ***Prove that a circular cylinder of radius R and length $\frac{R}{n}$ cannot float upright in stable equilibrium if its specific gravity lies between***

$$\frac{1}{2}\left[1-\sqrt{1-2n^2}\right] \text{ and } \frac{1}{2}\left[1+\sqrt{1-2n^2}\right]$$

What will happen, if $2n^2 > 1$?

Solution. We refer to the figure (FIG. 9.14) of the previous example.

In this case, $h = \dfrac{R}{n}$.

Proceeding as before, we find

$$W = \pi R^2 \cdot \frac{R}{n} \cdot \rho g = \frac{\pi}{n} R^3 \rho g, \tag{1}$$

where ρ is the specific gravity of the cylinder.

The force of buoyancy is

$$F = \pi R^2 x g. \tag{2}$$

By the condition of floatation, $W = F$. Hence,

$$\frac{\pi}{n} R^2 \rho g = \pi R^2 x g, \text{ i.e., } x = \frac{R}{n}\rho. \tag{3}$$

Now, volume of displaced liquid is

$$V = \pi R^2 x = \pi R^2 \cdot \frac{R}{n}\rho = \frac{\pi}{n} R^3 \rho \tag{4}$$

and the moment of inertia of the surface section about the axis of rotation is

$$Ak^2 = \pi R^2 \cdot \frac{R^2}{4} = \frac{\pi}{4} R^4. \tag{5}$$

Hence,

$$HM = \frac{Ak^2}{V} = \frac{\pi}{4} R^4 \cdot \frac{n}{\pi R^3 \rho}, \text{ i.e., } HM = \frac{nR}{4\rho} \tag{6}$$

Also

$$HG = OG - OH = \frac{1}{2}\frac{R}{n} - \frac{1}{2}x = \frac{1}{2}\frac{R}{n} - \frac{1}{2}\frac{R}{n}\rho = \frac{R}{2n}(1-\rho). \tag{7}$$

The equilibrium is unstable, if

$$HM < HG, \text{ i.e., } \frac{nR}{4\rho} < \frac{R}{2n}(1-\rho) \text{ [using (6) and (7)]}$$

$$\text{i.e., } n^2 < 2\rho(1-\rho), \text{ i.e., } \rho^2 - \rho + \frac{n^2}{2} < 0$$

$$\text{i.e., } \left\{\rho - \frac{1}{2}\left[1-\sqrt{1-2n^2}\right]\right\}\left\{\rho - \frac{1}{2}\left[1+\sqrt{1-2n^2}\right]\right\} < 0$$

Thus, the equilibrium is unstable, if the density ρ of the cylinder lies between

$$\frac{1}{2}\left[1-\sqrt{1-2n^2}\right] \text{ and } \frac{1}{2}\left[1+\sqrt{1-2n^2}\right]$$

If $2n^2 > 1$, the quantities

$$\frac{1}{2}\left[1-\sqrt{1-2n^2}\right] \text{ and } \frac{1}{2}\left[1+\sqrt{1-2n^2}\right]$$

are both complex numbers and so the value of ρ cannot lie between them. Hence, in this case, the upright position is stable for all specific gravities.

EXAMPLE 9.3 *Show that a homogeneous right circular cone of vertical angle 2α cannot float stably with its axis vertical and vertex downwards unless its density as compared with that of the liquid is greater than $\cos^6\alpha$.*

Solution. Let O be the vertex of the cone floating with its axis vertical (FIG. 9.15).

Let h = whole length of the axis of the cone,
x = length of the cone immersed,
ρ = density of the cone,
δ = density of the liquid.

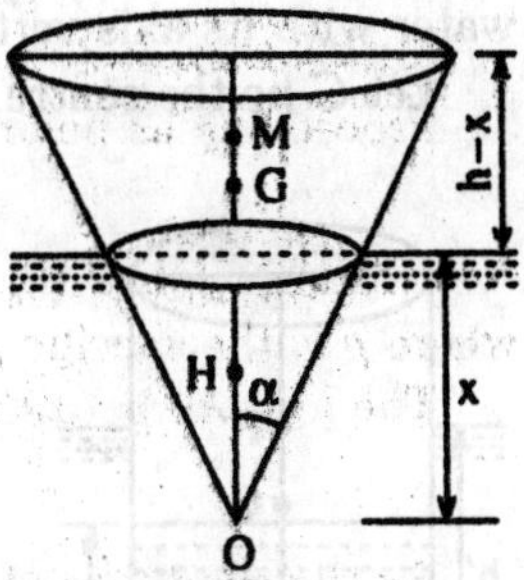

Fig. 9.15

Then the weight W of the cone is

$$W = \frac{1}{3}\pi h^3 \tan^2\alpha g\rho \tag{1}$$

and the force of buoyancy F is

$$F = \frac{1}{3}\pi x^3 \tan^2\alpha g\delta \tag{2}$$

By the condition of floatation, $F = W$.
Hence, from (1) and (2),

$$\frac{1}{3}\pi x^3 \tan^2\alpha g\delta = \frac{1}{3}\pi h^3 \tan^2\alpha g\rho, \text{ i.e., } \left(\frac{x}{h}\right)^3 = \frac{\rho}{\delta}. \tag{3}$$

Let G be the centre of gravity of the cone, H the centre of buoyancy and M the metacentre. Then,

$$OG = \frac{3}{4}h, \; OH = \frac{3}{4}x. \tag{4}$$

Hence,

$$HG = OG - OH = \frac{3}{4}(h - x) \tag{5}$$

The volume V of the displaced liquid is

$$V = \frac{1}{3}\pi x^3 \tan^2\alpha \tag{6}$$

The moment of inertia of the surface section about the axis of rotation is

$$Ak^2 = \pi x^2 \tan^2\alpha \cdot \frac{1}{4}x^2 \tan^2\alpha = \frac{\pi}{4}x^4 \tan^4\alpha \tag{7}$$

Hence,

$$HM = \frac{Ak^2}{V} = \frac{\pi}{4}x^4 \tan^4\alpha \cdot \frac{3}{\pi x^3 \tan^2\alpha} = \frac{3}{4}x \tan^2\alpha \tag{8}$$

The equilibrium is stable, if

$$HM > HG, \text{ i.e., if } \frac{3}{4}x\tan^2\alpha > \frac{3}{4}(h-x) \text{ [using (5) and (8)]}$$

$$\text{i.e., if } x\sec^2\alpha > h$$

$$\text{i.e., if } \left(\frac{x}{h}\right)^3 > \cos^6\alpha \tag{9}$$

Hence, from (3) and (9), we conclude that, for equilibrium to be stable,

$$\frac{\rho}{\delta} > \cos^6\alpha. \quad \textbf{(Proved)} \tag{10}$$

EXAMPLE 9.4 *A thin metal circular cylinder contains water to a depth h and floats in water with its axis vertical, immersed to a depth h′. Show that the vertical position is stable, if the height of the centre of gravity of the cylinder above its base is less than* $\frac{1}{2}(h+h')$.

Solution. Let O be the centre of the base of the cylinder containing liquid and floating in water with its axis vertical (FIG. 9.16).

Let G be the centre of gravity of the cylinder, H the centre of gravity of the contained water and H' the centre of gravity of the displaced liquid.

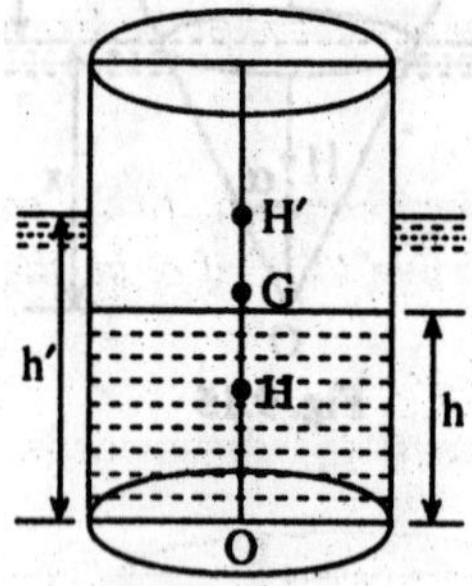

Fig. 9.16

Then

$$OH = \frac{h}{2}, \; OH' = \frac{h'}{2}, \; OG = x \text{ (say)} \tag{1}$$

Let R = radius of the cylinder and ρ = density of liquid. Then the weight of the contained liquid is

$$W = \pi R^2 hg\rho \tag{2}$$

and the weight of the displaced liquid is

$$F = \pi R^2 h' g\rho. \tag{3}$$

Now, let the cylinder be given a small angular displacement (FIG. 9.17). M, M' are the positions of the metacentres of the contained water and the displaced water respectively. Then the force W in (2) acts vertically downwards through M and the force F in (3) acts vertically upwards through M'.

Now, the volume V of the liquid contained is

$$V = \pi R^2 h \tag{4}$$

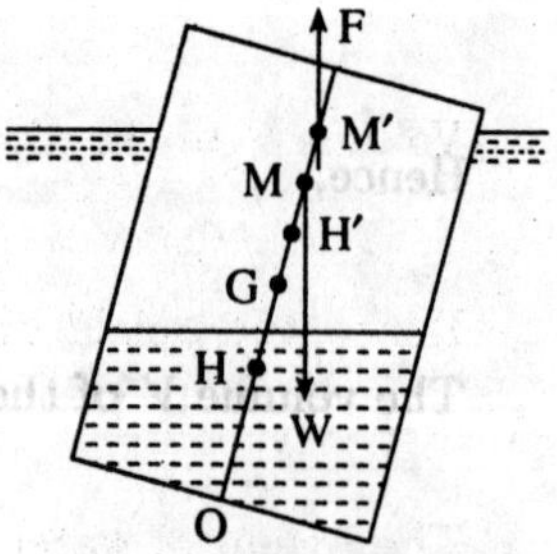

Fig. 9.17

and the moment of inertia of the section about the axis of rotation is

$$Ak^2 = \pi R^2 \cdot \frac{R^2}{4} = \frac{\pi}{4}R^4 \tag{5}$$

Hence,

$$HM = \frac{Ak^2}{V} = \frac{\pi}{4}R^4 \frac{1}{\pi R^2 h} = \frac{R^2}{4h} \tag{6}$$

Similarly,

$$H'M' = \frac{\pi}{4}R^4 \cdot \frac{1}{\pi R^2 h'} = \frac{R^2}{4h'} \tag{7}$$

Now,

$$GM = OM - OG = (OH + HM) - OG = \frac{h}{2} + \frac{R^2}{4h} - x \tag{8}$$

Similarly

$$GM' = OM' - OG = OH' + H'M' - OG = \frac{h'}{2} + \frac{R^2}{4h'} - x \tag{9}$$

The moment L of all the forces about G is

$$L = F \cdot GM' - W \cdot GM \tag{10}$$

The equilibrium will be stable, if

$L > 0$, i.e., if $F \cdot GM' > W \cdot GM$

i.e., if $\pi R^2 h' g\rho \left(\frac{h'}{2} + \frac{R^2}{4h'} - x\right) > \pi R^2 h g\rho \left(\frac{h}{2} + \frac{R^2}{4h} - x\right)$ [using (2), (3), (8) and (9)]

i.e., if $h'\left(\frac{h'}{2} + \frac{R^2}{4h'} - x\right) > h\left(\frac{h}{2} + \frac{R^2}{4h} - x\right)$

i.e., if $\frac{1}{2}\left(h'^2 - h^2\right) > x(h' - h)$

i.e., if $\frac{1}{2}(h' + h) > x$ (11)

Thus, the vertical position of the cylinder is stable, if the height of its centre of gravity from the base is less than $\frac{1}{2}(h + h')$.

EXAMPLE 9.5 *A hollow cone whose weight is half that of the water it can contain, floats in water in stable equilibrium with its axis vertical and vertex downwards. Prove that, if α be the semi-vertical angle of the cone, then $\cos^6 \alpha < \frac{729}{1024}$.*

Solution. Let ABC be the cone floating with its axis AO vertical and vertex A downwards.

Let H = whole length of the axis of the cone, h = length of the immersed portion of its axis, R = radius of the circular base of the cone, r = radius of the circular base at the level of floatation, ρ = density of water.

Then,

$$R = H \tan \alpha \tag{1}$$

$$r = h \tan \alpha \tag{2}$$

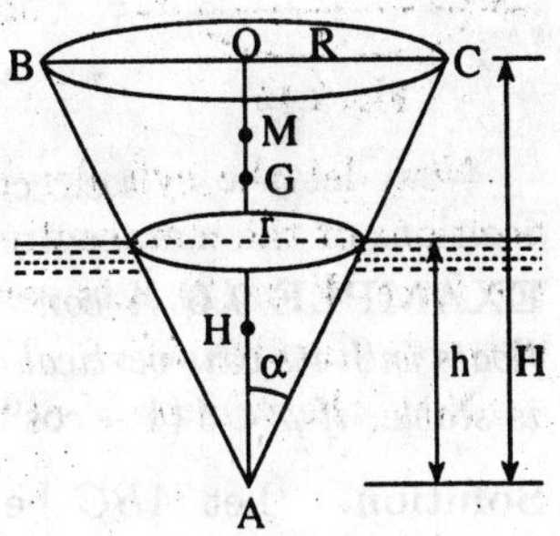

Fig. 9.18

Volume V of the displaced water is

$$V = \frac{1}{3}\pi r^2 h = \frac{1}{3}\pi h^3 \tan^2 \alpha \tag{3}$$

Hence, the weight of the displaced water is

$$W = V g\rho = \frac{1}{3}\pi h^3 \tan^2 \alpha g\rho \tag{4}$$

The weight of water the cone can contain is

$$W' = \frac{1}{3}\pi R^2 H g\rho = \frac{1}{3}\pi H^3 \tan^2 \alpha g\rho \tag{5}$$

Then the weight of the cone is

$$W'' = \frac{1}{2}W' \text{ (by the condition given)}$$

$$= \frac{1}{6}\pi H^3 \tan^2 \alpha g\rho \tag{6}$$

By the condition of floatation,

$$W = W''$$

Hence,

$$\frac{1}{3}\pi h^3 \tan^2 \alpha g\rho = \frac{1}{6}\pi H^3 \tan^2 \alpha g\rho, \text{ i.e., } h^3 = \frac{1}{2}H^3 \tag{7}$$

Let G be the centre of gravity of the cone, H the centre of buoyancy and M the metacentre. Then

$$AH = \frac{3}{4}h, \; AG = \frac{2}{3}H \tag{8}$$

The moment of inertia of the circular section at the surface of floatation about its diameter is

$$Ak^2 = \pi r^2 \cdot \frac{r^2}{4} = \frac{\pi}{4}r^4 = \frac{\pi}{4}h^4 \tan^4 \alpha \tag{9}$$

Then

$$HM = \frac{Ak^2}{V} = \frac{\pi}{4}h^4 \tan^4 \alpha \cdot \frac{3}{\pi h^3 \tan^2 \alpha} = \frac{3}{4}h \tan^2 \alpha \tag{10}$$

From the figure we have

$$HG = AG - AH = \frac{2}{3}H - \frac{3}{4}h \tag{11}$$

For the equilibrium to be stable, we must have

$$HM > HG, \text{ i.e., } \frac{3}{4}h \tan^2 \alpha > \frac{2}{3}H - \frac{3}{4}h \text{ [using (10) and (11)]}$$

$$\text{i.e., } \frac{3}{4}h \sec^2 \alpha > \frac{2}{3}H, \text{ i.e., } \cos^2 \alpha < \frac{9h}{8H}, \text{ i.e., } \cos^6 \alpha < \frac{729}{512}\left(\frac{h}{H}\right)^3$$

$$\therefore \cos^6 \alpha < \frac{729}{512} \cdot \frac{1}{2} \text{ [using (7)]}$$

$$\text{i.e., } \cos^6 \alpha < \frac{729}{1024}. \quad \textbf{(Proved)}$$

EXAMPLE 9.6 *A cone of density ρ, whose height is h and the radius of whose base is R, floats with its axis vertical and vertex upwards in a liquid of density δ. Prove that the equilibrium is stable, if $\rho < \delta\,(1 - \cos^6 \alpha)$, where 2α is the vertical angle of the cone.*

Solution. Let ABC be the cone of height h floating with its axis vertical and vertex A upwards (FIG. 9.19).

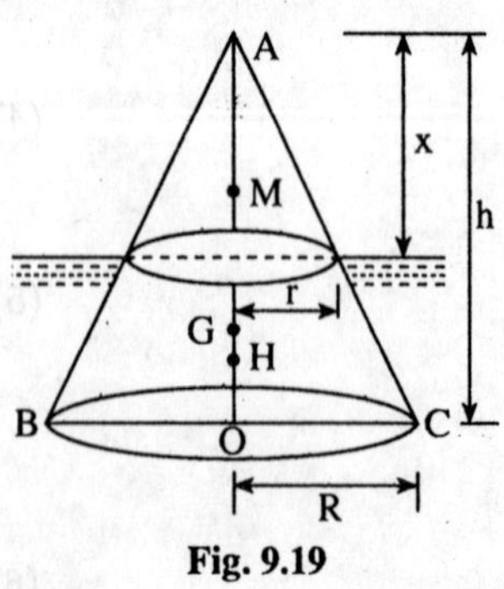

Fig. 9.19

Let r = radius of the circular section at the surface of floatation, x = length of the axis of the cone above the liquid. Then

$$R = h \tan \alpha \tag{1}$$

$$r = x \tan \alpha \tag{2}$$

Volume V_0 of the cone is

$$V_0 = \frac{1}{3}\pi R^2 h = \frac{1}{3}\pi h^3 \tan^2 \alpha \tag{3}$$

Weight W of the cone is

$$W = V_0 \rho g = \frac{1}{3}\pi h^3 \tan^2 \alpha g\rho \tag{4}$$

Volume V of the liquid displaced is

$$V = \frac{1}{3}\pi R^2 h - \frac{1}{3}\pi r^2 x = \frac{1}{3}\pi h^3 \tan^2\alpha - \frac{1}{3}\pi x^3 \tan^2\alpha$$

$$= \frac{1}{3}\pi \tan^2\alpha (h^3 - x^3) \tag{5}$$

Then the weight W' of the liquid displaced is

$$W' = \frac{1}{3}\pi \tan^2\alpha (h^3 - x^3) g\delta \tag{6}$$

By the condition of floatation

$$W = W'$$

i.e., $\frac{1}{3}\pi h^3 \tan^2\alpha g\rho = \frac{1}{3}\pi \tan^2\alpha (h^3 - x^3) g\delta$ [using (4) and (6)]

i.e., $\rho h^3 = \delta(h^3 - x^3)$, i.e., $(\delta - \rho)h^3 = \delta x^3$

$$\therefore \left(\frac{x}{h}\right)^3 = \frac{\delta - \rho}{\delta} = 1 - \frac{\rho}{\delta} \tag{7}$$

Let G be the centre of gravity of the cone, H the centre of buoyancy and M the metacentre. Then from the figure,

$$\left.\begin{aligned} AG &= \frac{3}{4}h \\ \text{and } AH &= \frac{\frac{3}{4}h \cdot h^3 - \frac{3}{4}x \cdot x^3}{h^3 - x^3} = \frac{3}{4}\frac{h^4 - x^4}{h^3 - x^3} \end{aligned}\right\} \tag{8}$$

Hence,

$$HG = AH - AG = \frac{3}{4}\left[\frac{h^4 - x^4}{h^3 - x^3} - h\right] \tag{9}$$

Now, the moment of inertia of the circular section at the surface of floatation is

$$Ak^2 = \pi r^2 \cdot \frac{r^2}{4} = \frac{\pi}{4}r^4 = \frac{\pi}{4}x^4 \tan^4\alpha \tag{10}$$

Hence,

$$HM = \frac{Ak^2}{V} = \frac{\pi}{4}x^4 \tan^4\alpha \cdot \frac{3}{\pi \tan^2\alpha (h^3 - x^3)} \quad \text{[using (5)]}$$

$$= \frac{3}{4}\frac{x^4 \tan^2\alpha}{h^3 - x^3} \tag{11}$$

The equilibrium is stable, if

$$HM > HG$$

i.e., if $\frac{3}{4}\frac{x^4 \tan^2\alpha}{h^3 - x^3} > \frac{3}{4}\left[\frac{h^4 - x^4}{h^3 - x^3} - h\right]$ [using (9) and (11)]

i.e., if $x^4 \tan^2\alpha > hx^3 - x^4$, i.e., if $\tan^2\alpha > \frac{h}{x} - 1$

i.e., if $\sec^2\alpha > \frac{h}{x}$, i.e., if $\cos^6\alpha < \left(\frac{x}{h}\right)^3$

i.e., if $\cos^6\alpha < 1 - \frac{\rho}{\delta}$ [using (7)]

$\therefore \frac{\rho}{\delta} < 1 - \cos^6\alpha$. **(Proved)**

EXAMPLE 9.7 *A solid and homogeneous body consists of a cylinder joining to a hemisphere on the same base and floats with the hemispherical portion partly immersed in water. Find the greatest height of the cylinder consistent with stability.*

Solution. Let $ABOCD$ be the solid body of which the portion $ABCD$ is a cylinder and the portion BOC is a hemisphere on the same base BC (FIG. 9.20).

Let $h =$ the height of the cylinder $ABCD$, $R =$ the radius of the common base, $\rho =$ density of the solid homogeneous body.

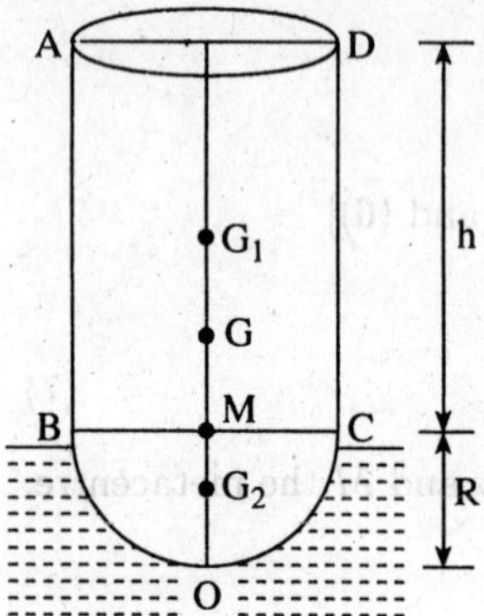

Fig. 9.20

Then weight of the cylindrical part is

$$W_1 = \pi R^2 hg\rho \tag{1}$$

and that of the hemispherical part is

$$W_2 = \frac{2}{3}\pi R^3 g\rho. \tag{2}$$

Let G_1, G_2 and G_3 be the positions of the centres of gravity of the cylinder, hemisphere and the whole body respectively. Then

$$\left.\begin{aligned} OG_2 &= R - \frac{3}{8}R = \frac{5}{8}R \\ OG_1 &= R + \frac{h}{2} \end{aligned}\right\} \tag{3}$$

Hence,

$$OG = \frac{W_1 \cdot OG_1 + W_2 \cdot OG_2}{W_1 + W_2}$$

$$= \frac{\pi R^2 hg\rho\left(R+\frac{h}{2}\right) + \frac{2}{3}\pi R^3 g\rho \cdot \frac{5}{8}R}{\pi R^2 hg\rho + \frac{2}{3}\pi R^3 g\rho} = \frac{3h\left(R+\frac{h}{2}\right)+\frac{5}{4}R^2}{3h+2R} \tag{4}$$

Since the metacentre is the centre of curvature of the surface of buoyancy, the metacentre M of the body then must be at the centre of the base, so that $OM = R$.

For equilibrium to be stable, the centre of gravity of the body cannot be above the metacentre. Hence,

$$OG \ngtr OM, \text{ i.e., } \frac{3h\left(R+\frac{h}{2}\right)+\frac{5}{4}R^2}{3h+2R} \ngtr R$$

$$\text{i.e., } 3h\left(R+\frac{h}{2}\right)+\frac{5}{4}R^2 \ngtr R(3h+2R)$$

$$\text{i.e., } \frac{3}{2}h^2 \ngtr 2R^2 - \frac{5}{4}R^2 = \frac{3}{4}R^2, \text{ i.e., } h^2 \ngtr \frac{R^2}{2}, \text{ i.e., } h \ngtr \frac{R}{\sqrt{2}}$$

Hence, for stability, the greatest height of the cylinder is $\frac{1}{\sqrt{2}}$ times the radius of the hemisphere.

EXAMPLE 9.8 *A thin cylindrical vessel floats when empty with its axis vertical and its base at a depth h below the surface and the equilibrium is unstable. Show that the vessel can float in stable equilibrium with its axis vertical, if a liquid of volume greater than* $\pi R^2\left(c - \frac{h}{2}\right)$ *is poured in, where R is the radius of the vessel and c the distance of its centre of gravity from the base.*

Solution. Let $ABCD$ be the cylindrical vessel with O as the centre of its base (FIG. 9.21).

Let W = the weight of the cylinder and ρ = the density of the liquid.

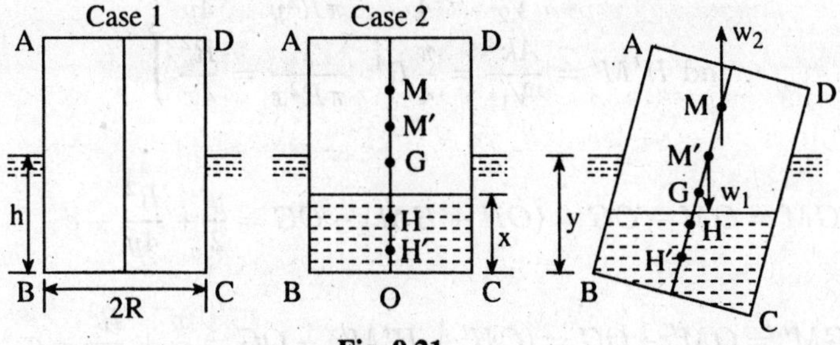

Fig. 9.21

When the vessel is empty, we have for floating

$$W = \pi R^2 hg\rho. \tag{1}$$

Let now x = height of the liquid poured in and y = the depth of the base from the surface of floatation.

Hence, by the condition of floatation,

$$W + \pi R^2 xg\rho = \pi R^2 yg\rho \tag{2}$$

From (1) and (2), we get

$$\pi R^2 hg\rho + \pi R^2 xg\rho = \pi R^2 yg\rho,$$

which yields

$$h + x = y. \tag{3}$$

Let G be the centre of gravity of the vessel with liquid. Then we are given

$$OG = c \tag{4}$$

If H and H' be the centres of gravity of the displaced and contained liquids, then

$$OH = \frac{y}{2},\ OH' = \frac{x}{2} \tag{5}$$

Hence,

$$\left.\begin{aligned} HG = OG - OH &= c - \frac{y}{2} \\ \text{and } H'G = OG - OH' &= c - \frac{x}{2} \end{aligned}\right\} \tag{6}$$

Now, the volume of the contained liquid is

$$V_1 = \pi R^2 x \tag{7}$$

and that of displaced liquid is

$$V_2 = \pi R^2 y \tag{8}$$

The moment of inertia of the circular section at the surfaces of floatation in both the cases is

$$Ak^2 = \pi R^2 \cdot \frac{R^2}{4} = \frac{\pi}{4} R^4. \tag{9}$$

Let M and M' be the metacentres of the displaced and contained liquids respectively.

Then

$$\left.\begin{aligned} HM &= \frac{Ak^2}{V_2} = \frac{\pi}{4}R^4 \cdot \frac{1}{\pi R^2 y} = \frac{R^2}{4y} \\ \text{and } H'M' &= \frac{Ak^2}{V_1} = \frac{\pi}{4}R^4 \cdot \frac{1}{\pi R^2 x} = \frac{R^2}{4x} \end{aligned}\right\} \tag{10}$$

Hence,

$$GM = OM - OG = (OH + HM) - OG = \frac{y}{2} + \frac{R^2}{4y} - c \tag{11}$$

Similarly,

$$GM' = OM' - OG = (OH' + H'M') - OG = \frac{x}{2} + \frac{R^2}{4x} - c \tag{12}$$

The weight of the contained liquid is

$$w_1 = \pi R^2 x g \rho \tag{13}$$

and that of the displaced liquid is

$$w_2 = \pi R^2 y g \rho \tag{14}$$

The force w_1 acts vertically downwards through M' and the force w_2 acts vertically upwards through M.

The total moments of these forces about G is

$$L = w_2 \cdot GM - w_1 \cdot GM' \tag{15}$$

The equilibrium will be stable, if

$$L > 0, \text{ i.e., if } w_2 \cdot GM > w_1 \cdot GM'$$

$$\text{i.e., if } \pi R^2 y g \rho \left(\frac{y}{2} + \frac{R^2}{4y} - c\right) > \pi R^2 x g \rho \left(\frac{x}{2} + \frac{R^2}{4x} - c\right)$$

$$\text{i.e., if } y\left(\frac{y}{2} + \frac{R^2}{4y} - c\right) > x\left(\frac{x}{2} + \frac{R^2}{4x} - c\right)$$

$$\text{i.e., if } \frac{y^2}{2} + \frac{R^2}{4} - cy > \frac{x^2}{2} + \frac{R^2}{4} - cx$$

$$\text{i.e., if } \frac{1}{2}(y^2 - x^2) > c(y - x)$$

$$\text{i.e., if } \frac{1}{2}(y + x) > c \text{ since } y \neq x$$

$$\text{i.e., if } \frac{1}{2}(h + x + x) > c \text{ [using (3)]}$$

$$\text{i.e., if } x > c - \frac{1}{2}h \tag{16}$$

Hence, the volume of the liquid poured in must be greater than

$$\pi R^2 \left(c - \frac{1}{2}h\right)$$

in order to have the equilibrium to be stable.

EXAMPLE 9.9 *A cylinder, floating with its axis horizontal and in the surface, is displaced in the vertical plane through the axis. Determine the metacentre and discuss its stability of equilibrium.*

Solution. Let h = the length of the cylinder and R = radius of the base.

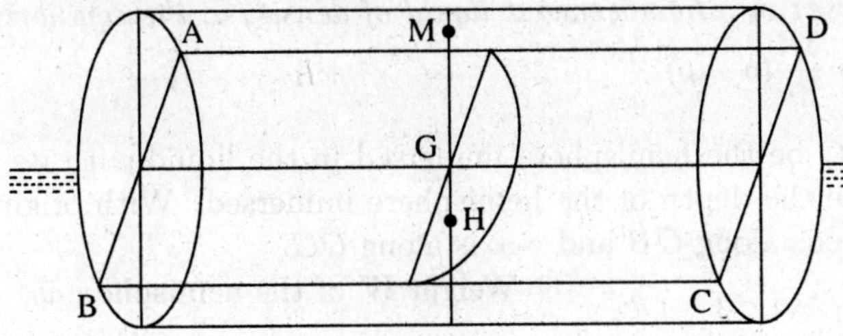

Fig. 9.22

The horizontal axis being in the surface, the plane of floatation is the rectangle $ABCD$ (FIG. 9.22), where

$$AB = 2R, \ AD = h \tag{1}$$

Then the volume of the immersed portion is

$$V = \frac{1}{2}\pi R^2 h \tag{2}$$

and area of surface of floatation is

$$A = 2Rh. \tag{3}$$

The moment of inertia of the surface of floatation is

$$Ak^2 = 2Rh \cdot \frac{1}{3}\left(\frac{h}{2}\right)^2 = \frac{1}{6}Rh^3 \tag{4}$$

Now, let the cylinder be displaced in the vertical plane through its axis. The centre of gravity G being on the axis, the vertical section through G of the portion of the cylinder immersed is a semi-circle.

If H be the centre of buoyancy, i.e., the centre of gravity of the displaced liquid, then

$$GH = \text{the distance of c.g. of a semi-circle from its centre}$$
$$= \frac{4R}{3\pi} \tag{5}$$

Now,

$$HM = \frac{Ak^2}{V} = \frac{1}{6}Rh^3 \cdot \frac{2}{\pi R^2 h} \quad \text{[using (2) and (4)]}$$

$$\text{i.e., } HM = \frac{1}{3}\frac{h^2}{\pi R} \tag{6}$$

Hence

$$GM = HM - GH = \frac{1}{3}\frac{h^2}{\pi R} - \frac{4R}{3\pi} \quad \text{[using (5) and (6)]}$$
$$= \frac{1}{3\pi}\left(\frac{h^2 - 4R^2}{R}\right) \tag{7}$$

This determines the position of the metacentre.

For the equilibrium to be stable, we must have

$$HM > GH$$
$$\text{i.e., } \frac{1}{3}\frac{h^2}{\pi R} > \frac{4R}{3\pi}, \text{ i.e., } h^2 > 4R^2, \text{ i.e., } h > 2R \tag{8}$$

Thus, the equilibrium is stable if the length of the cylinder is greater than its diameter.

EXAMPLE 9.10 *Show that when a uniform hemisphere of radius R and density ρ floats with its plane base immersed in a homogeneous liquid of density δ, the equilibrium is stable and the metacentric height is* $\frac{3R}{8\rho}(\delta - \rho)$.

Solution. Let ABC be the hemisphere immersed in the liquid with its base AB horizontal (FIG. 9.23). Let h be the depth of the hemisphere immersed. With origin at the centre O of the base, we take x-axis along OB and z-axis along OC.

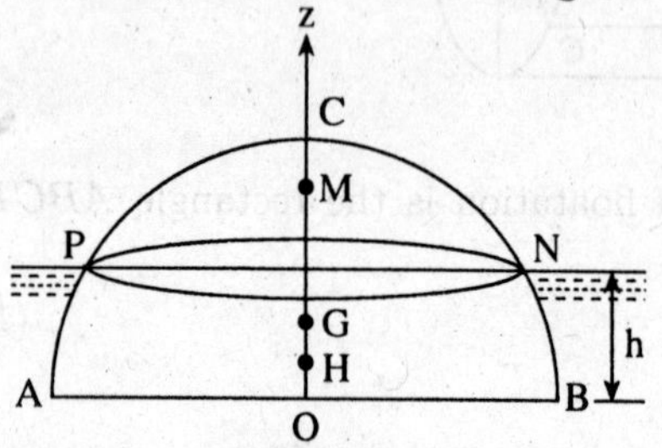

Fig. 9.23

Weight W of the hemisphere is

$$W = \frac{2}{3}\pi R^3 g\rho \tag{1}$$

Volume V of the displaced liquid = volume of the portion $APNB$ of the hemisphere. Thus

$$V = \int_0^h \pi(R^2 - z^2)dz = \frac{\pi h}{3}(3R^2 - h^2) \tag{2}$$

Then the weight W' of the displaced liquid is

$$W' = Vg\delta = \frac{\pi h}{3}(3R^2 - h^2)g\delta \tag{3}$$

By the condition of floatation,

$$W = W' \tag{4}$$

Hence, from (1) and (3)

$$\frac{2}{3}\pi R^3 g\rho = \frac{\pi h}{3}(3R^2 - h^2)g\delta, \text{ i.e., } 2R^3\rho = h(3R^2 - h^2)\delta$$

$$\therefore 3R^2 - h^2 = \frac{2R^3\rho}{h\delta} \tag{5}$$

Let G be the centre of gravity of the hemisphere and H be the centre of gravity of the displaced liquid. Then

$$OG = \frac{3R}{8} \tag{6}$$

$$\text{and } OH = \frac{\int_0^h \pi(R^2 - z^2)\cdot z dz}{V}$$

$$\text{i.e., } OH = \frac{\frac{\pi h^2}{4}(2R^2 - h^2)}{\frac{\pi h}{3}(3R^2 - h^2)} \quad \text{[using (2)]}$$

$$= \frac{3}{4}\cdot\frac{h(2R^2 - h^2)}{3R^2 - h^2} \tag{7}$$

The section is a circle of radius r, where

$$r^2 = R^2 - h^2 \tag{8}$$

The moment of inertia of the surface section about the axis of rotation is

$$Ak^2 = \pi r^2 \cdot \frac{r^2}{4} = \frac{\pi}{4}r^4, \text{ i.e., } Ak^2 = \frac{\pi}{4}(R^2 - h^2)^2 \text{ [using (8)]} \tag{9}$$

If M be the metacentre, then

$$HM = \frac{Ak^2}{V} = \frac{\pi}{4}(R^2 - h^2)^2 \cdot \frac{3}{\pi h(3R^2 - h^2)} \text{ [using (2)]}$$

$$= \frac{3(R^2 - h^2)^2}{4h(3R^2 - h^2)} \quad (10)$$

Hence,

$$OM = OH + HM$$

$$= \frac{3}{4} \cdot \frac{h(2R^2 - h^2)}{3R^2 - h^2} + \frac{3(R^2 - h^2)^2}{4h(3R^2 - h^2)} \text{ [using (7) and (10)]}$$

$$= \frac{3h^2(2R^2 - h^2) + 3(R^4 - 2R^2h^2 + h^4)}{4h(3R^2 - h^2)}$$

$$= \frac{3R^4}{4h(3R^2 - h^2)} \quad (11)$$

From (5), we get

$$4h(3R^2 - h^2) = \frac{8R^3\rho}{\delta} \quad (12)$$

From (11) and (12), we get

$$OM = \frac{3R^4\delta}{8R^3\rho} = \frac{3R\delta}{8\rho} \quad (13)$$

Then

$$GM = OM - OG = \frac{3R\delta}{8\rho} - \frac{3R}{8} \text{ [using (6)]}$$

$$= \frac{3R}{8\rho}(\delta - \rho) \quad (14)$$

Eqn (14) gives the metacentric height. Since the hemisphere of density ρ floats in the liquid of density δ,

$$\delta > \rho, \text{ i.e., } \delta - \rho > 0 \quad (15)$$

This implies that $GM > 0$. That is the metacentre M is above G. Hence, the equilibrium is stable.

EXAMPLE 9.11 *A uniform solid cylinder of length l and radius R, which can turn about its diameter of its upper end as a fixed axis, is held vertically immersed in water to a depth h. Show that this vertical position is not stable unless the ratio of the weight of the cylinder to the water displaced exceeds $2 - \frac{h}{l} - \frac{R^2}{2hl}$.*

Solution. Let $ABCD$ be the cylinder of length l floating vertically in a liquid. The cylinder can turn about a diameter on the top end AD (FIG. 9.24). Let G be the centre of gravity of the cylinder, H the centre of buoyancy and M the metacentre. Then

$$OH = \frac{h}{2}, \; OG = \frac{l}{2}.$$

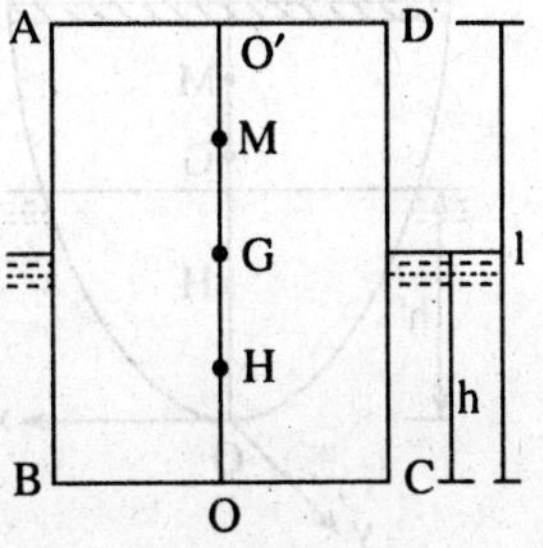

Fig. 9.24

Hence,

$$O'G = \frac{l}{2}, \; O'H = l - \frac{h}{2} \quad (1)$$

Volume of the liquid displaced is

$$V = \pi R^2 h \tag{2}$$

The moment of inertia of the top section about the axis of rotation is

$$Ak^2 = \pi R^2 \cdot \frac{R^2}{4} = \frac{\pi R^4}{4} \tag{3}$$

Hence,

$$HM = \frac{Ak^2}{V} = \frac{\pi R^4}{4} \cdot \frac{1}{\pi R^2 h} = \frac{R^2}{4h} \tag{4}$$

Then

$$O'M = O'O - HM - OH = l - \frac{h}{2} - \frac{R^2}{4h} \tag{5}$$

Let W be the weight of the cylinder acting vertically downwards through G and W' the weight of the displaced liquid acting vertically upwards through M.

The total moment of these two forces about O' is

$$L = W \cdot O'G - W' \cdot O'M$$

$$= W \cdot \frac{l}{2} - W' \cdot \left(l - \frac{h}{2} - \frac{R^2}{4h}\right) \text{ [using (5)]} \tag{6}$$

The equilibrium will be stable, if

$$L > 0, \text{ i.e., if } W \cdot \frac{l}{2} > W'\left(l - \frac{h}{2} - \frac{R^2}{4h}\right)$$

$$\text{i.e., if } \frac{W}{W'} > \frac{2}{l}\left(l - \frac{h}{2} - \frac{R^2}{4h}\right)$$

$$\text{i.e., if } \frac{W}{W'} > 2 - \frac{h}{l} - \frac{R^2}{2lh}. \quad \textbf{(Proved)}$$

EXAMPLE 9.12 *A solid paraboloid of specific gravity δ floats in a liquid of specific gravity ρ with its axis vertical and vertex downwards. Show that the equilibrium will be stable, if*

$$3a > h\left(1 - \sqrt{\frac{\delta}{\rho}}\right),$$

where $4a$ is the latus rectum of the generating parabola and h is the height of the solid.

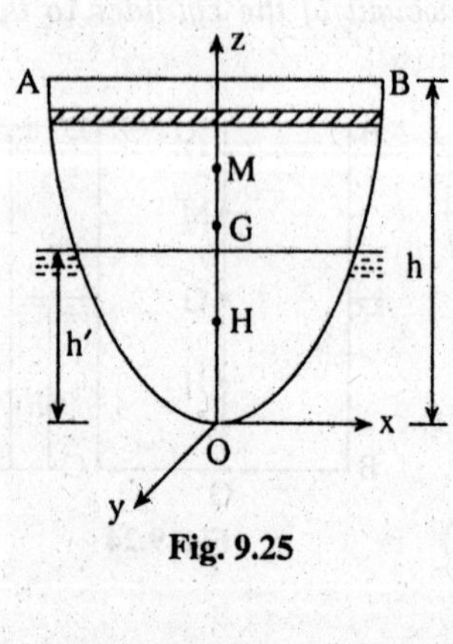

Fig. 9.25

Solution. With origin O at the vertex of the solid, we take the rectangular coordinate system as shown in the figure (FIG. 9.25).

Then the equation of the paraboloid of revolution is

$$x^2 + y^2 = 4az, \tag{1}$$

since $4a$ is the latus rectum of the generating parabola.

Considering a thin circular disc of thickness dz at a height z from origin, we get the volume of the solid

$$V_0 = \int_0^h \pi(x^2 + y^2)dz = \pi \int_0^h 4azdz \text{ [using (1)]}$$

$$= 2\pi a h^2 \tag{2}$$

If h' be the length of the axis immersed, then the volume of liquid displaced is

$$V = 2\pi a h'^2 \tag{3}$$

The weight W of the solid body is

$$W = V_0 g\delta = 2\pi a h^2 g\delta \tag{4}$$

and the weight W' of the displaced liquid is

$$W' = V g\rho = 2\pi a h'^2 g\rho \tag{5}$$

By the condition of floatation,

$$W = W'.$$

Hence, from (4) and (5),

$$2\pi a h^2 g\delta = 2\pi a h'^2 g\rho$$

$$\therefore \left(\frac{h'}{h}\right)^2 = \frac{\delta}{\rho} \tag{6}$$

Let r be the radius at the surface section, where $z = h'$. Then

$$r^2 = x^2 + y^2 = 4ah' \tag{7}$$

The moment of inertia of this surface section about the axis of rotation is

$$Ak^2 = \pi r^2 \cdot \frac{r^2}{4} = \frac{\pi}{4} r^4, \text{ i.e., } Ak^2 = \frac{\pi}{4}(4ah')^2 \text{ [using (7)]}$$

$$= 4\pi a^2 h'^2 \tag{8}$$

Let G be the centre of gravity of the solid, H the centre of buoyancy and M the metacentre. Then

$$OG = \frac{\int_0^h \pi(x^2+y^2) z\,dz}{V_0} = \frac{\int_0^h \pi(4az) z\,dz}{V_0} = \frac{4a\pi \cdot \frac{h^3}{3}}{2\pi a h^2} = \frac{2}{3}h \tag{9}$$

Similarly,

$$OH = \frac{2}{3}h' \tag{10}$$

Then

$$HG = OG - OH = \frac{2}{3}(h - h') \tag{11}$$

Now,

$$HM = \frac{Ak^2}{V} = \frac{4\pi a^2 h'^2}{2\pi a h'^2} \text{ [using (3) and (8)]}$$

$$\text{i.e., } HM = 2a \tag{12}$$

The equilibrium is stable, if

$$HM > HG, \text{ i.e., if } 2a > \frac{2}{3}(h - h'), \text{ i.e., if } 3a > h\left(1 - \frac{h'}{h}\right)$$

$$\text{i.e., if } 3a > h\left(1 - \sqrt{\frac{\delta}{\rho}}\right) \text{ [using (6)]}$$

Hence proved.

EXAMPLE 9.13 *A homogeneous right circular cone of vertical angle 2α floats with its axis vertical and vertex downwards and is in unstable equilibrium.*

Prove that the position becomes stable when a particle of weight w is attached to the vertex, if

$$\frac{w}{W} > \left[\left(\frac{\rho\cos^6\alpha}{\delta}\right)^{1/4} - 1\right],$$

where W is the weight of the cone. It is assumed that the base of the cone is not submerged, that δ is the density of the cone and ρ that of the liquid.

Solution. Let AOB be the cone floating in the liquid with its axis vertical and vertex O downwards, the base AB not submerged (FIG. 9.26). Let h = the whole length of the axis of the cone, x = the length of the axis of the cone immersed, α = the semi-vertical angle of the cone.

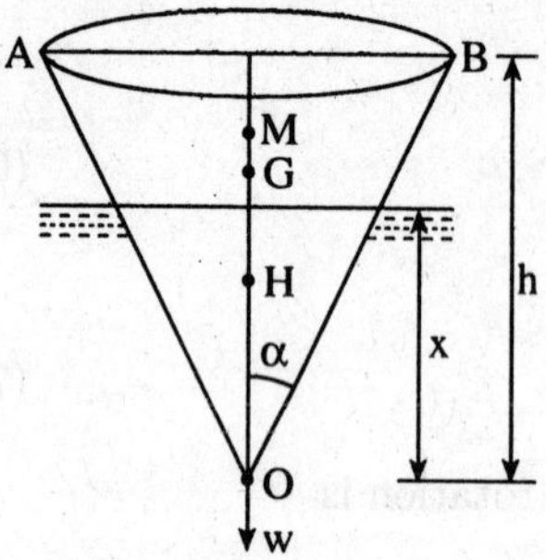

Fig. 9.26

Then, the weight W of the cone is

$$W = \frac{1}{3}\pi h^3 \tan^2\alpha g\delta \tag{1}$$

and the weight W' of the liquid displaced is

$$W' = \frac{1}{3}\pi x^3 \tan^2\alpha g\rho \tag{2}$$

By the condition of floatation, we have

$$W + w = W' \tag{3}$$

From (1) and (2),

$$W + w = \frac{1}{3}\pi x^3 \tan^2\alpha g\rho = \frac{1}{3}\pi h^3 \tan^2\alpha g\delta \cdot \frac{x^3}{h^3}\cdot\frac{\rho}{\delta} = W\cdot\frac{\rho}{\delta}\frac{x^3}{h^3} \quad \text{[using (1)]}$$

Hence,

$$\left(\frac{x}{h}\right)^3 = \frac{W+w}{W}\cdot\frac{\delta}{\rho} \tag{4}$$

Let G be the position of the centre of gravity of the solid and the particle of weight w, H the centre of buoyancy and M the metacentre.

Then

$$(W+w)\cdot OG = W\cdot\frac{3h}{4} \tag{5}$$

since the height of c.g. above O of the cone only is $3h/4$.

From (5), we get

$$OG = \frac{W}{W+w}\cdot\frac{3h}{4} \tag{6}$$

Also

$$OH = \frac{3x}{4} \tag{7}$$

Then

$$HG = OG - OH = \frac{3}{4}\left(\frac{Wh}{W+w} - x\right) \tag{8}$$

Now, the volume of the displaced liquid is

$$V = \frac{1}{3}\pi x^3 \tan^2\alpha \tag{9}$$

and the moment of inertia of the surface section is

$$Ak^2 = \pi r^2 \cdot \frac{r^2}{4} = \frac{\pi}{4} r^4, \tag{10}$$

where r is the radius of the circular section at surface level and

$$r = x \tan \alpha \tag{11}$$

Then

$$Ak^2 = \frac{\pi}{4} x^4 \tan^4 \alpha \tag{12}$$

Hence,

$$HM = \frac{Ak^2}{V} = \frac{\pi}{4} x^4 \tan^4 \alpha \cdot \frac{3}{\pi x^3 \tan^2 \alpha} \text{ [using (9)]}$$
$$= \frac{3}{4} x \tan^2 \alpha \tag{13}$$

For equilibrium to be stable,

$$HM > HG$$

i.e., $\frac{3}{4} x \tan^2 \alpha > \frac{3}{4}\left(\frac{Wh}{W+w} - x\right)$ [using (8) and (13)]

i.e., $x \sec^2 \alpha > \frac{Wh}{W+w}$, i.e., $\left(\frac{x}{h}\right)^3 > \left(\frac{W}{W+w}\right)^3 \cos^6 \alpha$

i.e., $\frac{W+w}{W} \cdot \frac{\delta}{\rho} > \left(\frac{W}{W+w}\right)^3 \cos^6 \alpha$ [using (4)]

i.e., $\left(\frac{W+w}{W}\right)^4 > \frac{\rho}{\delta} \cos^6 \alpha$, i.e., $\frac{W+w}{W} > \left(\frac{\rho}{\delta} \cdot \cos^6 \alpha\right)^{1/4}$

i.e., $1 + \frac{w}{W} > \left(\frac{\rho}{\delta} \cos^6 \alpha\right)^{1/4}$

Hence, $\frac{w}{W} > \left[\left(\frac{\rho}{\delta} \cos^6 \alpha\right)^{1/4} - 1\right]$. **(Proved)**

EXAMPLE 9.14 *Two equal uniform rods of density ρ are joined together at an angle 2α. If they be immersed with the angle downwards in a fluid of density δ, find the position of equilibrium and prove that the curve of buoyancy is a parabola.*

Also prove that the rods cannot float with the line joining their ends inclined to the horizon, unless $\frac{\delta - \rho}{\delta + \rho} > \sin^2 \alpha$. If this condition be fulfilled, determine which of the positions of equilibrium is stable.

Solution. Let AB and AC be the two rods joined at A which is immersed. We are given, $\angle BAC = 2\alpha$ (FIG. 9.27). Let AD be the bisector of the angle BAC.

Let

$$AB = AC = l \tag{1}$$

and the total length of the rods immersed $= 2b$. Then, by the condition of floatation,

$$2lg\rho = 2bg\delta, \text{ i.e., } l\rho = b\delta. \tag{2}$$

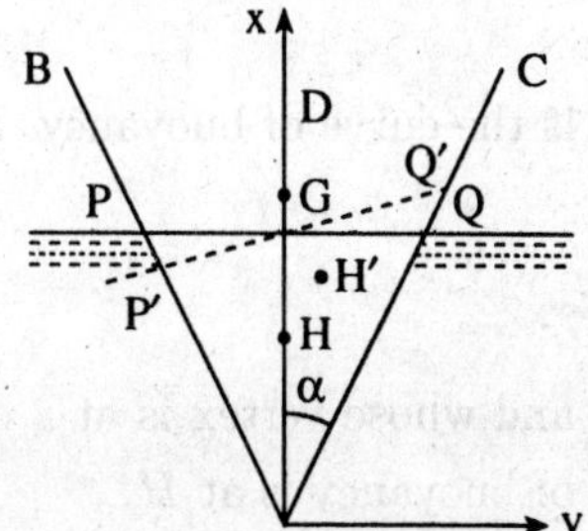

Fig. 9.27

In the symmetric position of floatation, let G be the centre of gravity of the system and H the centre of buoyancy. Then $AP = AQ = b$.

Hence,

$$AG = \frac{1}{2}l\cos\alpha, \ AH = \frac{1}{2}b\cos\alpha \tag{3}$$

Then

$$HG = AG - AH = \frac{1}{2}(l-b)\cos\alpha \tag{4}$$

Also

$$HM = \frac{Ak^2}{V} = \frac{b\sin^2\alpha}{\cos\alpha} \tag{5}$$

In the non-symmetric displaced position, let $P'Q'$ be the surface of floatation and

$$AP' = b - \xi, \ AQ' = b + \xi \tag{6}$$

Let H' be the new centre of buoyancy with coordinates (x, y) with reference to rectangular axes with origin at A.

The coordinates of middle points of AP' and AQ' are

$$\left[\frac{1}{2}(b-\xi)\cos\alpha, -\frac{1}{2}(b-\xi)\sin\alpha\right]$$

$$\text{and} \quad \left[\frac{1}{2}(b+\xi)\cos\alpha, \frac{1}{2}(b+\xi)\sin\alpha\right] \quad \text{respectively.}$$

Then

$$x = \frac{\frac{1}{2}(b-\xi)^2\cos\alpha + \frac{1}{2}(b+\xi)^2\cos\alpha}{2b} = \frac{b^2+\xi^2}{2b}\cos\alpha \tag{7}$$

$$y = \frac{\frac{1}{2}(b+\xi)^2\sin\alpha - \frac{1}{2}(b-\xi)^2\sin\alpha}{2b} = \xi\sin\alpha \tag{8}$$

Eliminating ξ from (7) and (8), we get

$$x = \frac{b}{2}\cos\alpha + \frac{\cos\alpha}{2b}\cdot\frac{y^2}{\sin^2\alpha}, \text{ i.e., } y^2\frac{\cos\alpha}{2b\sin^2\alpha} = x - \frac{b}{2}\cos\alpha$$

Hence,

$$y^2 = \frac{2b\sin^2\alpha}{\cos\alpha}\left(x - \frac{b}{2}\cos\alpha\right) \tag{9}$$

is the curve of buoyancy. It is a parabola whose latus rectum is

$$= \frac{2b\sin^2\alpha}{\cos\alpha} \tag{10}$$

and whose vertex is at a distance of $\frac{b}{2}\cos\alpha$ from A. This means that the vertex of the curve of buoyancy is at H.

There will be one or three normals from the centre of gravity G to the curve of buoyancy according as the distance of G from the vertex H is less or greater than the semi-latus rectum.

Thus, from (4) and (10) we write the condition:

$$\frac{1}{2}(l-b)\cos\alpha < \text{or} > \frac{b\sin^2\alpha}{\cos\alpha}$$

$$\text{i.e.,}\quad (l-b)\cos^2\alpha < \text{or} > 2b\sin^2\alpha$$

$$\text{i.e.,}\quad (l-b)(1-\sin^2\alpha) < \text{or} > 2b\sin^2\alpha$$

$$\text{i.e.,}\quad (l-b) < \text{or} > (l-b)\sin^2\alpha + 2b\sin^2\alpha$$

$$\text{i.e.,}\quad \frac{l-b}{l+b} < \text{or} > \sin^2\alpha,\ \text{i.e.,}\ \frac{\delta-\rho}{\delta+\rho} < \text{or} > \sin^2\alpha\ \text{[using (2)]} \tag{11}$$

Thus, we will get one or three positions of equilibrium, according as

$$\frac{\delta-\rho}{\delta+\rho}\ \text{is less or greater than}\ \sin^2\alpha.$$

In the symmetrical position of equilibrium, the metacentre M is above G, if

$$\frac{\delta-\rho}{\delta+\rho} < \sin^2\alpha, \tag{12}$$

but otherwise not. Hence, if there are three positions of equilibrium, the symmetrical one is unstable. If, however, there is only one position of equilibrium, the symmetrical one is stable.

EXAMPLE 9.15 *A solid of cork bounded by the surface generated by the revolution of a quadrant of an ellipse about the major axis, sinks in mercury up to the focus, its plane face being horizontal and above mercury. If the centre of gravity of the solid cork coincides with the metacentre, prove that the eccentricity e is given by* $2e^4+4e^3+2e^2-e-2=0$.

Solution. Let ABC be the solid generated by the revolution of an octant of an ellipse given by

$$\frac{z^2}{a^2}+\frac{x^2}{b^2}=1 \tag{1}$$

about the major axis, so that $OB=a$ (FIG. 9.28), where O is the centre of the top horizontal base. Then, from (1), we have

$$x^2 = b^2\left(1-\frac{z^2}{a^2}\right) \tag{2}$$

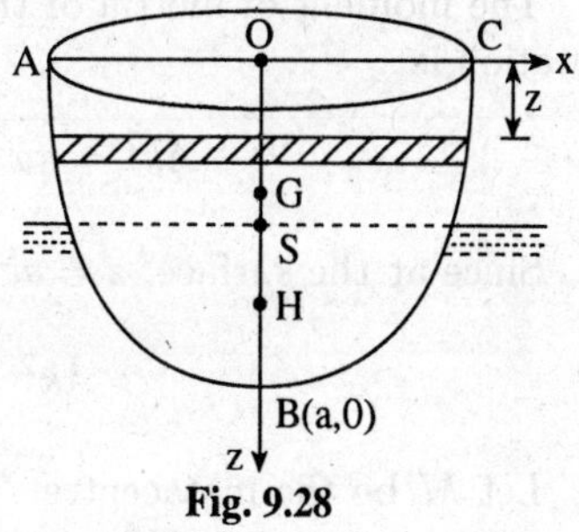

Fig. 9.28

Considering a circular disc of radius x at a depth z and of thickness dz, we find the volume of the solid

$$V_0 = \int_0^a \pi x^2 dz = \int_0^a \pi b^2\left(1-\frac{z^2}{a^2}\right)dz\ \text{[using (2)]}$$

$$= \pi b^2\left(a-\frac{a^3}{3a^2}\right) = \frac{2}{3}\pi ab^2 \tag{3}$$

Since the solid sinks up to the focus S, $SB = ae$.
Hence, the volume of the immersed portion of the solid is

$$V = \int_{ae}^a \pi x^2 dz = \int_{ae}^a \pi b^2\left(1-\frac{z^2}{a^2}\right)dz$$

$$= \pi b^2\left[a(1-e)-\frac{1}{3a^2}\left(a^3-a^3e^3\right)\right] = \pi b^2\cdot\left[a(1-e)-\frac{a}{3}\left(1-e^3\right)\right]$$

$$= \frac{\pi ab^2}{3}(1-e)\left[3-\left(1+e+e^2\right)\right] = \frac{\pi ab^2}{3}(1-e)(2-e-e^2) = \frac{\pi ab^2}{3}(1-e)^2(2+e) \tag{4}$$

Let G be the centre of gravity of the solid and H be the centre of buoyancy. Then

$$OG = \frac{3a}{8} \tag{5}$$

$$\text{and } OH = \frac{\int_{ae}^{a} \pi x^2 dz \cdot z}{V}$$

$$\text{i.e., } V \cdot OH = \int_{ae}^{a} \pi b^2 \left(1 - \frac{z^2}{a^2}\right) z dz = \pi b^2 \int_{ae}^{a} \left(z - \frac{z^3}{a^2}\right) dz$$

$$= \pi b^2 \left[\frac{1}{2}(a^2 - a^2e^2) - \frac{1}{4a^2}(a^4 - a^4e^4)\right]$$

$$= \frac{\pi a^2 b^2}{4}\left[2(1-e^2) - (1-e^4)\right] = \frac{\pi a^2 b^2}{4}\left(e^4 - 2e^2 + 1\right) \tag{6}$$

Hence

$$OH = \frac{\pi a^2 b^2 (e^4 - 2e^2 + 1) \cdot 3}{4 \cdot \pi a b^2 (1-e)^2 (2+e)} = \frac{3a}{4} \cdot \frac{(1-e^2)^2}{(2+e)(1-e)^2} = \frac{3a}{4} \cdot \frac{(1+e)^2}{2+e} \tag{7}$$

Then

$$GH = OH - OG = \frac{3a}{4} \cdot \frac{(1+e)^2}{2+e} - \frac{3a}{8} \text{ [using (5) and (7)]}$$

$$= \frac{3a}{8}\left[\frac{2(1+2e+e^2) - (2+e)}{2+e}\right] = \frac{3a}{8} \cdot \frac{(3+2e)e}{2+e} \tag{8}$$

The moment of inertia of the surface section which is a circle of radius x about the axis of rotation is

$$Ak^2 = \pi x^2 \cdot \frac{x^2}{4} = \frac{\pi}{4} x^4 = \frac{\pi}{4} b^4 \left(1 - \frac{z^2}{a^2}\right)^2, \text{ [using (2)]} \tag{9}$$

Since at the surface, $z = ae$, we get from (9)

$$Ak^2 = \frac{\pi}{4} b^4 \left(1 - \frac{a^2e^2}{a^2}\right)^2 = \frac{\pi}{4} b^4 \left(1 - e^2\right)^2. \tag{10}$$

Let M be the metacentre. Then

$$HM = \frac{Ak^2}{V} = \frac{\pi}{4} b^4 (1-e^2)^2 \cdot \frac{3}{\pi a b^2 (1-e)^2 (2+e)} \tag{11}$$

$$= \frac{3}{4}\frac{b^2}{a}\frac{(1+e)^2}{2+e} = \frac{3}{4}\frac{a^2(1-e^2)}{a} \cdot \frac{(1+e)^2}{2+e} = \frac{3a}{4} \cdot \frac{(1-e)(1+e)^3}{2+e} \tag{12}$$

Since the metacentre coincides with the c.g. of the body, we must have

$$HM = HG.$$

From (8) and (12),

$$\frac{3a}{4} \cdot \frac{(1-e)(1+e)^3}{2+e} = \frac{3a}{8} \cdot \frac{(3+2e)e}{2+e}$$

$$\text{i.e., } 2(1-e)(1+e)^3 = (3+2e)e$$

$$\text{i.e., } 2(1-e)(1+3e+3e^2+e^3) = 3e + 2e^2$$

$$\text{i.e., } 2(1+3e+3e^2+e^3-e-3e^2-3e^3-e^4) = 3e+2e^2$$

$$\therefore\ 2e^4 + 4e^3 + 2e^2 - e - 2 = 0. \quad \textbf{(Proved)}$$

EXAMPLE 9.16 *An oblate spheroid floats half-immersed, with its axis vertical, in a liquid, the density of which varies as the square of depth. Prove that the height of the metacentre above the surface is* $\frac{5}{8}\frac{a^2-b^2}{b}$.

Solution. The oblate spheroid is generated by the revolution of the ellipse,

$$\frac{x^2}{a^2}+\frac{z^2}{b^2}=1, \tag{1}$$

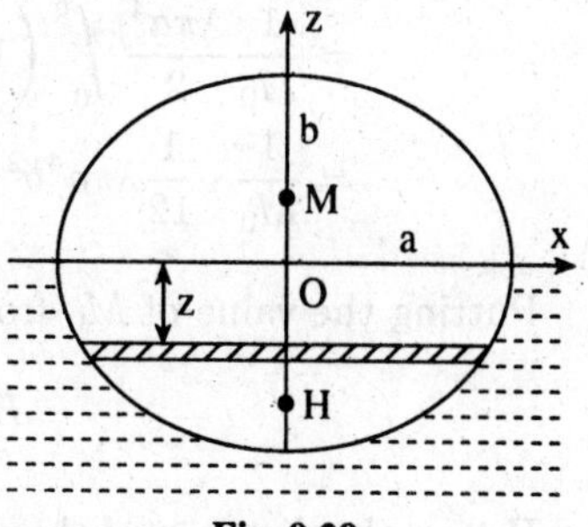

Fig. 9.29

around the z-axis (FIG. 9.29).

Then the equation of the solid body is

$$\frac{x^2+y^2}{a^2}+\frac{z^2}{b^2}=1 \tag{2}$$

which gives

$$x^2+y^2=a^2\left(1-\frac{z^2}{b^2}\right). \tag{3}$$

Thus, for $z=$ constant, we get

$$x^2+y^2=\text{constant}, \tag{4}$$

which implies that horizontal sections of the body are circles.

Since the density of the liquid varies as the square of the depth, we write

$$\rho=\lambda z^2, \tag{5}$$

where λ is a constant. Then, the density ρ_0 on the surface, where $z=0$, is zero.

Hence,

$$A_0k_0^2\rho_0=0 \tag{6}$$

We consider an elementary circular disc at a depth z below the surface and of thickness dz. Then the mass M_0 of the displaced liquid is

$$M_0=\int_0^b \pi(x^2+y^2)\cdot\lambda z^2dz=\lambda\pi\int_0^b a^2\left(1-\frac{z^2}{b^2}\right)z^2dz \text{ [using (3)]}$$

$$=\lambda\pi a^2\left[\frac{b^3}{3}-\frac{1}{b^2}\cdot\frac{b^5}{5}\right]=\frac{2}{15}\lambda\pi a^2b^3 \tag{7}$$

The depth of the centre of gravity H of the displaced liquid from the surface is given by

$$M_0\cdot OH=\int_0^b \pi(x^2+y^2)dz\cdot\lambda z^2\cdot z=\lambda\pi\int_0^b a^2\left(1-\frac{z^2}{b^2}\right)z^3dz$$

$$=\lambda\pi a^2\left[\frac{b^4}{4}-\frac{1}{b^2}\cdot\frac{b^6}{6}\right]=\frac{1}{12}\lambda\pi a^2b^4 \tag{8}$$

Hence,

$$OH=\frac{1}{12}\lambda\pi a^2b^4\cdot\frac{15}{2\lambda\pi a^2b^3} \text{ [using (7)]}$$

$$\text{i.e., } OH=\frac{5}{8}b. \tag{9}$$

Let M be the position of the metacentre.

Then

$$HM = \frac{1}{M_0}\left[\int_0^b Ak^2\frac{d\rho}{dz}dz + A_0k_0^2\rho_0\right] = \frac{1}{M_0}\int_0^b Ak^2\frac{d\rho}{dz}dz \text{ [using (6)]}$$

$$= \frac{1}{M_0}\int_0^b \frac{\pi}{4}(x^2+y^2)^2\cdot 2\lambda z dz = \frac{1}{M_0}\cdot\frac{\lambda\pi}{2}\int_0^b a^4\left(1-\frac{z^2}{b^2}\right)^2 zdz$$

$$= \frac{1}{M_0}\frac{\lambda\pi a^4}{2}\int_0^b\left(1-2\frac{z^2}{b^2}+\frac{z^4}{b^4}\right)zdz = \frac{1}{M_0}\frac{\lambda\pi a^4}{2}\left[\frac{b^2}{2}-\frac{2}{b^2}\cdot\frac{b^4}{4}+\frac{1}{b^4}\cdot\frac{b^6}{6}\right]$$

$$= \frac{1}{M_0}\cdot\frac{1}{12}\lambda\pi a^4b^2 \tag{10}$$

Putting the value of M_0 from (7), we get

$$HM = \frac{15}{2\lambda\pi a^2b^3}\cdot\frac{1}{12}\lambda\pi a^4b^2 = \frac{5}{8}\frac{a^2}{b} \tag{11}$$

Hence, the distance of the metacentre from the surface is

$$OM = HM - OH = \frac{5}{8}\frac{a^2}{b} - \frac{5}{8}b \text{ [using (9) and (11)]}$$

$$= \frac{5}{8}\frac{a^2-b^2}{b}. \quad \textbf{(Proved)}$$

EXAMPLE 9.17 *A right circular solid cone of semi-vertical angle α floats wholly immersed with its vertex upwards and axis vertical, in a liquid the density of which varies as the depth. If h is the height of the cone and b the depth of its vertex below the surface, then show that the distance of the metacentre from the vertex is equal to*

$$\frac{3h}{5}\cdot\frac{5b+4h-h\tan^2\alpha}{4b+3h}.$$

Solution. Let OBC be the cone of height h floating with its axis vertical and the vertex at a depth b from the surface (FIG. 9.30).

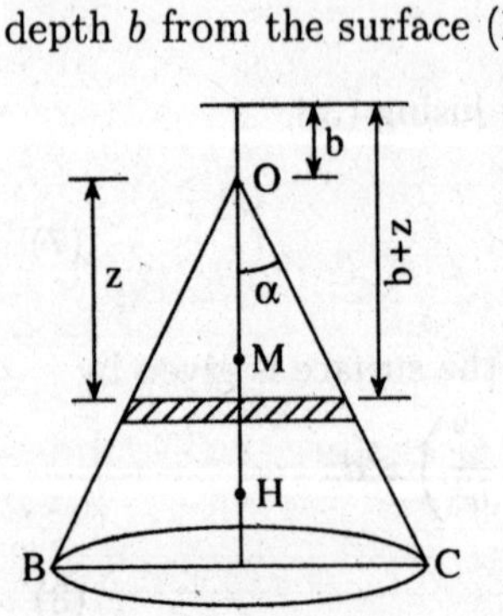

Fig. 9.30

We consider an elementary circular disc of thickness dz at a depth z from the vertex O. Since the density varies as the depth from the surface, we write

$$\rho = \lambda(b+z), \tag{1}$$

where λ is a constant.

Hence, the density at the point O, where $z=0$, is

$$\rho_0 = \lambda b \tag{2}$$

Since area of cross section A_0 at O is zero, we get

$$A_0k_0^2\rho_0 = 0. \tag{3}$$

The mass M_0 of the displaced liquid is

$$M_0 = \int_0^h \pi z^2\tan^2\alpha\cdot\lambda(b+z)dz = \lambda\pi\tan^2\alpha\int_0^h(b+z)z^2dz$$

$$= \lambda\pi\tan^2\alpha\left(b\cdot\frac{h^3}{3}+\frac{h^4}{4}\right) = \frac{1}{12}\lambda\pi\tan^2\alpha h^3(3h+4b) \tag{4}$$

If H be the metacentre, then

$$OH = \frac{1}{M_0}\int_0^h z(\pi z^2 \tan^2\alpha)\cdot\lambda(b+z)\cdot dz = \frac{\lambda\pi}{M_0}\tan^2\alpha\int_0^h z^3(b+z)dz$$

$$= \frac{\lambda\pi}{M_0}\tan^2\alpha\cdot\left(\frac{bh^4}{4}+\frac{h^5}{5}\right) = \frac{\lambda\pi\tan^2\alpha}{20M_0}(4h+5b)h^4 \quad (5)$$

Putting here the value of M_0 from (4), we get

$$OH = \frac{\lambda\pi\tan^2\alpha}{20M_0}(4h+5b)h^4\cdot\frac{12}{\lambda\pi\tan^2\alpha(3h+4b)h^3}$$

$$\text{i.e., } OH = \frac{3h}{5}\frac{4h+5b}{3h+4b} \quad (6)$$

If M be the position of metacentre, then

$$HM = \frac{1}{M_0}\left[\int_0^h Ak^2\frac{d\rho}{dz}dz + A_0k_0^2\rho_0\right] = \frac{1}{M_0}\cdot\int_0^h Ak^2\frac{d\rho}{dz}dz \text{ [using (3)]}$$

$$= \frac{1}{M_0}\int_0^h(\pi z^2\tan^2\alpha)\cdot\frac{1}{4}z^2\tan^2\alpha\cdot\lambda dz = \frac{\lambda\pi}{4M_0}\tan^4\alpha\int_0^h z^4dz$$

$$= \frac{\lambda\pi h^5}{20M_0}\tan^4\alpha = \frac{\lambda\pi h^5\tan^4\alpha}{20}\cdot\frac{12}{\lambda\pi\tan^2\alpha h^3(3h+4b)} \text{ [using (4)]}$$

$$= \frac{3h^2\tan^2\alpha}{5(3h+4b)} \quad (7)$$

Hence, the distance of the metacentre from the vertex is

$$OM = OH - HM$$

$$= \frac{3h}{5}\frac{4h+5b}{3h+4b} - \frac{3h^2\tan^2\alpha}{5(3h+4b)} \text{ [using (6) and (7)]}$$

$$= \frac{3h}{5(3h+4b)}\left[4h+5b-h\tan^2\alpha\right] = \frac{3}{5}h\frac{5b+4h-h\tan^2\alpha}{3h+4b}. \quad \textbf{(Proved)}$$

EXAMPLE 9.18 *A solid paraboloid of revolution floats with its axis vertical and vertex downwards, and focus in the surface of a liquid, the density of which at the depth z is*

$$\mu(a+z), \mu = \text{constant},$$

where 4a is the latus rectum of the generating parabola. Prove that the distance of the metacentre from the vertex is $\frac{21}{8}a$.

Solution. The solid paraboloid is generated by the revolution of the parabola,

$$x^2 = 4az, \quad (1)$$

about the vertical axis Oz, $4a$ being the latus rectum. Then the equation of the paraboloid is (FIG. 9.31)

$$x^2 + y^2 = 4az \quad (2)$$

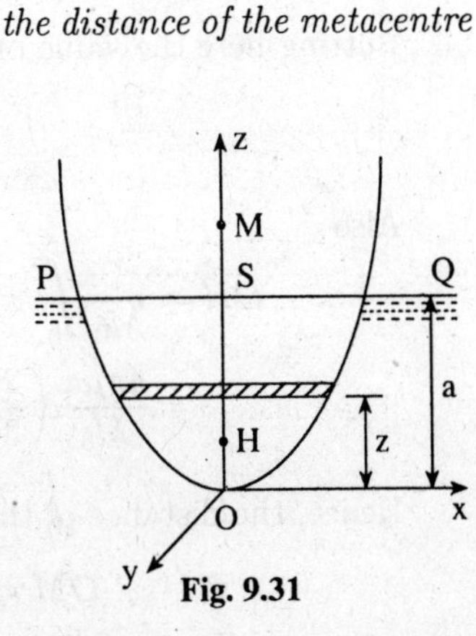

Fig. 9.31

If S be the focus, then $OS = a$. The density of the liquid at a depth z from the surface PQ is given as

$$\rho = \mu(a+z), \ \mu = \text{constant}.$$

Then the density at a height z from O will be given by

$$\rho = \mu(2a - z) \tag{3}$$

so that

$$\rho\big|_{z=0} = 2\mu a \text{ and } \rho\big|_{z=a} = \mu a \tag{4}$$

We consider an elementary circular disc of thickness dz at a height z from O. Then, the mass of the displaced liquid is

$$M_0 = \int_0^a \pi\left(x^2 + y^2\right)\cdot \mu(2a - z)dz$$

$$= \int_0^a \pi \cdot 4az \cdot \mu(2a - z)dz \text{ [using (2)]}$$

$$= 4a\pi\mu \int_0^a \left(2az - z^2\right)dz = 4a\pi\mu\left(2a\cdot\frac{a^2}{2} - \frac{a^3}{3}\right) = \frac{8}{3}\pi\mu a^4 \tag{5}$$

On the surface, when $z = a$, we have from (2)

$$x^2 + y^2 = 4a^2 \tag{6}$$

so that the surface section is a circle of radius $2a$. Hence,

$$A_0k_0^2\rho_0 = \pi(2a)^2 \cdot \frac{1}{4}(2a)^2 \cdot \mu a \text{ [using (4)]}$$

$$= 4\pi\mu a^5 \tag{7}$$

Also

$$\int_0^a Ak^2\frac{d\rho}{dz}dz = \int_0^a \pi(4az)\cdot\frac{1}{4}(4az)\cdot\mu dz = 4\pi\mu a^2\int_0^a z^2dz = \frac{4}{3}\pi\mu a^5 \tag{8}$$

If H be the centre of buoyancy and M the metacentre, then

$$HM = \frac{1}{M_0}\left[\int_0^a Ak^2\frac{d\rho}{dz}dz + A_0k_0^2\rho_0\right]$$

$$= \frac{1}{M_0}\left[\frac{4}{3}\pi\mu a^5 + 4\pi\mu a^5\right] \text{ [using (7) and (8)]}$$

$$= \frac{1}{M_0}\frac{16}{3}\pi\mu a^5 \tag{9}$$

Putting here the value of M_0 from (5), we get

$$HM = \frac{16}{3}\pi\mu a^5 \cdot \frac{3}{8\pi\mu a^4} = 2a \tag{10}$$

Also

$$OH = \frac{1}{M_0}\int_0^a \pi(4az)\cdot\mu(2a - z)\cdot zdz = \frac{1}{M_0}\cdot 4\pi\mu a\int_0^a \left(2az^2 - z^3\right)dz$$

$$= \frac{4\pi\mu a}{M_0}\left(\frac{2}{3}a^4 - \frac{a^4}{4}\right) = \frac{1}{M_0}\cdot\frac{5}{3}\pi\mu a^5 = \frac{3}{8\pi\mu a^4}\cdot\frac{5}{3}\pi\mu a^5 = \frac{5}{8}a \tag{11}$$

Hence, the distance of the metacentre from the vertex is

$$OM = OH + HM = \frac{5}{8}a + 2a = \frac{21}{8}a. \quad \textbf{(Proved)}$$

EXAMPLE 9.19 *A paraboloid of revolution floats with its axis vertical and vertex downwards in a liquid, the density of which varies as the depth. Prove that the equilibrium will be stable or unstable according as*

$$4c < or > 3(m+a),$$

where c is the length of the axis, a the length immersed and m the latus rectum of the generating parabola.

Solution. The paraboloid is generated by the revolution of the parabola,

$$x^2 = mz, \tag{1}$$

about the z-axis, m being the latus rectum. Hence, the equation of the solid body is

$$x^2 + y^2 = mz \tag{2}$$

Let PQ be the surface of the liquid (FIG. 9.32), cutting the axis of the solid at C. Then, as is given,

$$OC = a, OA = c \tag{3}$$

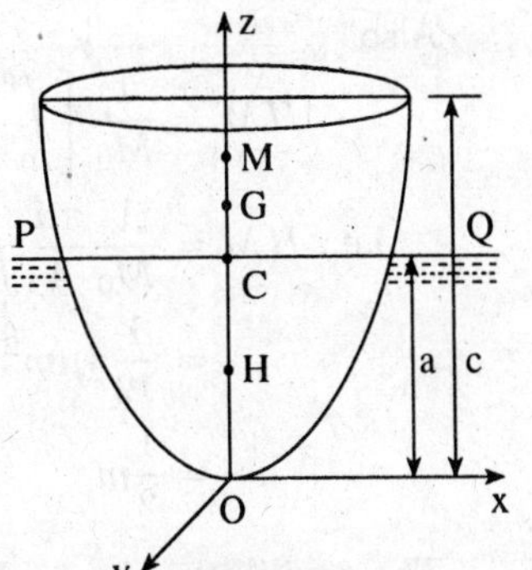

Fig. 9.32

Since the density ρ of the liquid varies as the depth from the surface, at a height z above the vertex O, it will be

$$\rho = \mu(a-z) \tag{4}$$

Hence, the density ρ_0 at the free surface, where $z = a$, vanishes. Thus,

$$A_0 k_0^2 \rho_0 = 0 \tag{5}$$

Let G be the centre of gravity of the paraboloid whose density is, say, δ. Then

$$OG = \frac{\displaystyle\int_0^c \pi(x^2+y^2)z\delta dz}{\displaystyle\int_0^c \pi(x^2+y^2)\delta dz}, \text{ i.e., } OG = \frac{\displaystyle\int_0^c mz \cdot zdz}{\displaystyle\int_0^c mzdz} \text{ [using (2)]}$$

$$= \frac{\frac{c^3}{3}}{\frac{c^2}{2}} = \frac{2}{3}c \tag{6}$$

Let H be the centre of gravity and M_0 the mass of the displaced liquid. Then

$$M_0 = \int_0^a \pi(x^2+y^2)\cdot\mu(a-z)dz = \pi\mu\int_0^a mz(a-z)dz$$

$$= m\pi\mu\left(a\cdot\frac{a^2}{2} - \frac{a^3}{3}\right) = \frac{1}{6}m\pi\mu a^3 \tag{7}$$

The distance of H from the vertex is

$$OH = \frac{1}{M_0}\int_0^a \pi(x^2+y^2)\cdot\mu(a-z)\cdot zdz$$

$$= \frac{1}{M_0}\pi\mu\int_0^a mz(a-z)zdz = \frac{1}{M_0}m\pi\mu\int_0^a (az^2 - z^3)dz$$

$$= \frac{1}{M_0}m\pi\mu\left(a\cdot\frac{a^3}{3} - \frac{a^4}{4}\right) = \frac{1}{12}\frac{m\pi\mu a^4}{M_0} \tag{8}$$

Putting here the value of M_0 from (7), we get

$$OH = \frac{1}{12}m\pi\mu a^4 \cdot \frac{6}{m\pi\mu a^3} = \frac{a}{2} \tag{9}$$

Hence,

$$HG = OG - OH = \frac{2}{3}c - \frac{a}{2} \tag{10}$$

Also

$$HM = \frac{1}{M_0}\left[\int_0^a Ak^2\frac{d\rho}{dz}dz + A_0k_0^2\rho_0\right] = \frac{1}{M_0}\int_0^a \pi(x^2+y^2)\cdot\frac{1}{4}(x^2+y^2)\cdot\mu dz$$

$$\text{i.e., } HM = \frac{1}{M_0}\frac{\pi\mu}{4}\int_0^a mz\cdot mzdz = \frac{1}{M_0}\frac{\pi\mu m^2}{4}\cdot\frac{a^3}{3} = \frac{1}{12}\frac{\pi\mu m^2 a^3}{M_0}.$$

$$= \frac{1}{12}\pi\mu m^2 a^3 \cdot \frac{6}{m\pi\mu a^3} \text{ [using (7)]}$$

$$= \frac{1}{2}m \tag{11}$$

The equilibrium is stable or unstable, according as

$$HM > \text{or} < HG$$

$$\text{i.e., } \frac{m}{2} > \text{or} < \frac{2}{3}c - \frac{a}{2}$$

$$\text{i.e., } \frac{(a+m)}{2} > \text{or} < \frac{2}{3}c$$

$$\text{i.e., } 3(a+m) > \text{or} < 4c \tag{12}$$

Thus, equilibrium is stable or unstable, according as $4c <$ or $> 3(m+a)$. **(Proved)**

EXAMPLE 9.20 *A paraboloid cup, whose latus rectum is $4a$ and whose centre of gravity is at a distance from the vertex equal to $2a$, floats in two liquids of densities δ and ρ ($\delta > \rho$). Prove that the work required to turn the body through a small angle θ about a horizontal axis is*

$$\frac{2}{3}\pi a g\theta^2\left[h^3(\delta-\rho) + (h+h')^3\rho\right],$$

where h, h' are the lengths of the axis immersed.

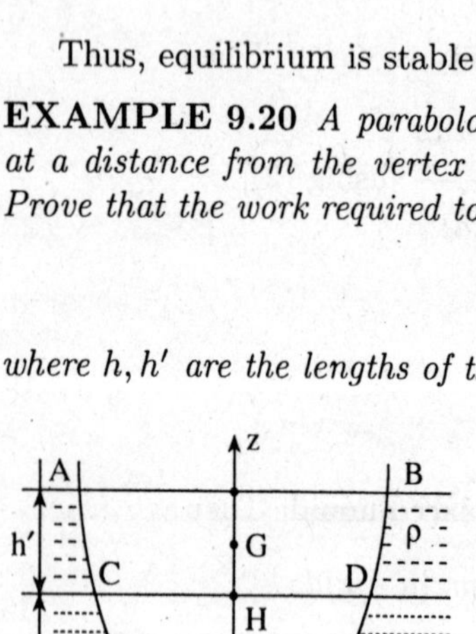

Fig. 9.33

Solution. The paraboloid is generated by the revolution of the parabola,

$$x^2 = 4az, \tag{1}$$

about the z-axis, where $4a$ is the latus rectum. Hence, the equation of the paraboloid is

$$x^2 + y^2 = 4az \tag{2}$$

whose centre of gravity G is at a distance from the vertex equal to $2a$ (FIG. 9.33). That is

$$OG = 2a \tag{3}$$

The volume of the paraboloid up to any height ξ from the vertex is

$$= \int_0^\xi \pi(x^2+y^2)dz = \int_0^\xi \pi(4az)dz = 2a\pi\xi^2 \tag{4}$$

Thus, total volume V of the body immersed in two liquids is

$$V = 2a\pi(h+h')^2 \tag{5}$$

and the volume V' of the body immersed in the lower liquid is

$$V' = 2a\pi h^2 \tag{6}$$

The above system is equivalent to the portion AOB of volume V immersed in a liquid of density ρ plus the portion COD of volume V' immersed in a liquid of density $(\delta - \rho)$.

If H be the centre of buoyancy due to the liquid of density ρ, then

$$OH = \frac{\int_0^{h+h'} \rho\pi(4az)\cdot z\,dz}{\int_0^{h+h'} \rho\pi(4az)dz} = \frac{(h+h')^3}{3}\cdot\frac{2}{(h+h')^2} = \frac{2}{3}(h+h') \tag{7}$$

Similarly, if H' be the centre of buoyancy due to the liquid of density $(\delta - \rho)$, we get

$$OH' = \frac{2}{3}h \tag{8}$$

Hence,

$$\left.\begin{aligned} GH &= OG - OH = 2a - \frac{2}{3}(h+h') \\ GH' &= OG - OH' = 2a - \frac{2}{3}h \end{aligned}\right\} \tag{9}$$

For the upper surface AB, $z = h + h'$, so that from (2),

$$x^2 + y^2 = 4a(h+h') \tag{10}$$

Hence,

$$Ak^2 = \pi(x^2+y^2)\cdot\frac{1}{4}(x^2+y^2) = \frac{\pi}{4}[4a(h+h')]^2 = 4\pi a^2(h+h')^2 \tag{11}$$

Similarly, for the surface CD,

$$A'k'^2 = 4\pi a h^2 \tag{12}$$

Work done W_1 in producing angular displacement θ to the body of volume V immersed in the liquid of density ρ is

$$\begin{aligned} W_1 &= \frac{1}{2}g\rho(Ak^2 - V\cdot GH)\theta^2 \\ &= \frac{1}{2}g\rho\left[4\pi a^2(h+h')^2 - 2a\pi(h+h')^2\left\{2a - \frac{2}{3}(h+h')\right\}\right]\theta^2 \text{ [using (5), (9) and (11)]} \\ &= \frac{1}{2}g\rho\theta^2\left[\frac{4}{3}a\pi(h+h')^3\right] = \frac{2}{3}a\pi g\theta^2\rho(h+h')^3 \end{aligned} \tag{13}$$

Similarly, the work done in producing angular displacement θ to the body of volume V' immersed in the liquid of density $(\delta - \rho)$ is

$$\begin{aligned} W_2 &= \frac{1}{2}g(\delta-\rho)\left[A'k'^2 - V'\cdot GH'\right]\theta^2 \\ &= \frac{1}{2}g(\delta-\rho)\theta^2\left[4\pi a h^2 - 2a\pi h^2\left(2a - \frac{2}{3}h\right)\right] \text{ [using (6), (9) and (12)]} \\ &= \frac{1}{2}g(\delta-\rho)\theta^2\left[\frac{4}{3}a\pi h^3\right] = \frac{2}{3}a\pi g(\delta-\rho)\theta^2 h^3 \end{aligned} \tag{14}$$

Hence, the total work done is

$$W = W_1 + W_2 = \frac{2}{3}a\pi g\theta^2\rho(h+h')^3 + \frac{2}{3}a\pi g(\delta-\rho)\theta^2 h^3$$
$$= \frac{2}{3}\pi a g\theta^2\left[\rho(h+h')^3 + (\delta-\rho)h^3\right]. \quad \textbf{(Proved)}$$

EXAMPLE 9.21 *A solid body of uniform density δ floats partly immersed in a homogeneous liquid of density ρ. Show that a solid of same size and shape and of uniform density $(\rho-\delta)$ can float inverted with the same plane of floatation in the same liquid, and that if equilibrium is stable in one case, it is also stable in the other case for like displacement.*

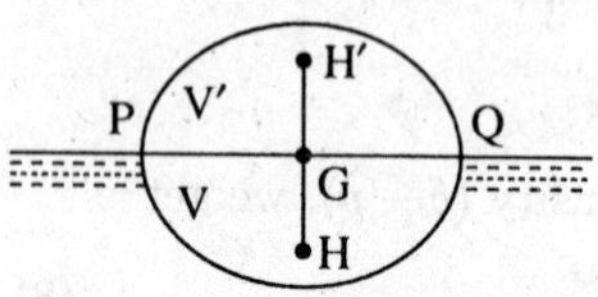

Fig. 9.34

Solution. Let in the first case, V = volume immersed in the liquid, V' = volume of the body above the free surface PQ (FIG. 9.34).

Also, let H be the centre of gravity of the volume V, H' the centre of gravity of the volume V', and G the centre of gravity of the whole body. Then

$$V \cdot HG = V' \cdot H'G \tag{1}$$

and HH' passes through G. Since the centre of gravity of a homogeneous body does not depend on its density, H and H' will also be the centres of buoyancy when the volumes V and V' respectively would be immersed in the liquid of uniform density ρ. Hence, HG is vertical in the first case, and $H'G$ will also be vertical and remains so when the body is inverted.

By the condition of floatation in the first case, we have

$$(V+V')\delta = V\rho, \text{ i.e., } \frac{V}{\delta} = \frac{V+V'}{\rho} = \frac{V'}{\rho-\delta} \tag{2}$$

Eqn (2) yields

$$V'\rho = (V+V')(\rho-\delta) \tag{3}$$

which implies that the weight of liquid of volume V' is equal to the weight of the whole body of density $(\rho-\delta)$. That is, if the body is inverted, it can float in the same liquid with the same plane of floatation, if the density of the body is $(\rho-\delta)$.

Since the plane of floatation remains the same in both the cases, the quantity Ak^2 remains unaltered.

Now, if the equilibrium is stable in the first case, then

$$Ak^2 > V \cdot HG \tag{4}$$

By Eqn (1), we get from (4)

$$Ak^2 > V' \cdot H'G, \tag{5}$$

which is the condition of stability in the second case.

Thus, if the equilibrium is stable in the first case, it is also stable in the second case, when the body is inverted.

EXERCISE

1. A solid cylinder of 3 metres diameter has a height of 3 metres. It is made up of a material whose sp. gr. is 0.8 and is floating in water with its axis vertical. Show that the equilibrium is unstable.

2. A rectangular parallelopiped of sp. gr. 0.5 and edges $2a, 2b, 2c$ floats in water. Show that, if a^2 and b^2 are both greater than $3c^2$, the position of equilibrium in which the edges are vertical is stable.

3. If a solid paraboloid bounded by a plane perpendicular to its axis floats with its axis vertical and vertex immersed, then prove that the height of the metacentre above the centre of gravity of the displaced liquid is equal to half the latus rectum.

4. Show that a thin uniform rod will float in a vertical position in stable equilibrium in a liquid of n times its own density, if a heavy particle be attached to its lower end of weight greater than $(\sqrt{n}-1)$ times its own weight.

5. The radius of a homogeneous circular cylinder is two-third of its height. It floats in water with its axis vertical. Prove that the equilibrium cannot be stable, if the sp. gr. of the cylinder lies between $\dfrac{1}{3}$ and $\dfrac{2}{3}$.

6. A cone whose vertical angle is 60° floats in water with its axis vertical and vertex downwards. Show that its metacentre lies in the plane of floatation and that its equilibrium will be stable, if its sp. gr. is greater than $\dfrac{27}{64}$.

7. A cubical solid floats in stable equilibrium in a fluid with its two faces horizontal. Show that the specific gravity of the cube relative to the liquid cannot lie between 0.211 and 0.788.

8. Prove that a solid right circular cylinder of radius R and height h can float in stable equilibrium with its axis vertical in a liquid of greater density, if $\dfrac{h}{2R} \le 0.707$.

9. A hollow cylindrical vessel of radius R is supported at the ends of a diameter of cross section in which is situated the centre of gravity of the vessel. The vessel contains water to a depth h. Prove that the vertical position is unstable unless $R < \sqrt{2}b$ and h lies between $b \pm \sqrt{b^2 - \frac{1}{2}R^2}$, where b is the distance of the c.g. of the vessel from its base.

10. A solid in the shape of a double cone bounded by two equal circular ends, floats in a liquid of twice its density with its axis horizontal. Prove that the equilibrium is stable or unstable according as the semi-vertical angle is less or greater than 60°.

11. A frustrum of a cone floats with its axis vertical in a liquid of twice its density. Prove that the equilibrium will be stable, if

$$\left\{\frac{a^4+b^4}{(a^3+b^3)^{\frac{4}{3}}}\sqrt[3]{2}-1\right\}h^2 < (a-b)^2,$$

where h is the height of the frustrum and a, b the radii of its ends.

12. A square lamina of density ρ floats in water of density δ with its plane vertical and one angular point below the surface. If $9\delta > 32\rho$, prove that there are three positions of equilibrium in two of which neither diagonal is vertical.

13. A wall-sided ship with a flat bottom draws a depth d of water when its weight, as loaded, is W; also h is the height of the metacentre above the centre of buoyancy. Prove that, if the ship is to be loaded with an extra cargo W', so that the metacentric height is not to be diminished, then the centre of gravity of the extra cargo must not be at a greater height above that of the ship, as originally loaded, than $\dfrac{1}{2}\left[1+\dfrac{W'}{W}\right]d-h$.

14. An elliptic lamina floats half-immersed with its major axis $2a$ vertical in a liquid the density of which varies as the square of the depth. Prove that the depth of the metacentre is $\dfrac{32ae^2}{15\pi}$, where e is the eccentricity.

15. A rectangular parallelopiped with edges $2a, 2b, 2c$ of density d floats with edges $2a$ vertical, partly in a lower liquid of density ρ and partly in an upper liquid of density δ. Prove that for displacements in the planes parallel to $2a, 2b$ the equilibrium is stable, if

$$b^2(\rho-\delta)^2 > 6a^2(\rho-d)(d-\delta).$$

16. A solid body consists of a right cone joined to a hemisphere on the same base and floats with its spherical portion partly immersed. Prove that the greatest height of the cone consistent with stability is $\sqrt{3}$ times the radius of the base.

17. A uniform rod of length l and sp. gr. δ less than that of water has one end fixed at a point nl above a large vessel of water and hangs partially immersed. Prove that the whole potential energy in a position inclined at an angle θ to the vertical differs from that in the vertical position by a quantity proportional to $\{n^2 \sec\theta - (1-\delta)\}(1-\cos\theta)$.

18. A solid in the form of a paraboloid of revolution floats with its axis vertical. If the centre of inertia coincides with the metacentre, prove that the equilibrium is stable.

19. An isosceles triangular lamina ABC floats with its base AB horizontal and above the surface, in a liquid the density of which varies as the depth. If h is the depth of C below the surface, prove that the height of the metacentre above C is $\dfrac{1}{2}h\sec^2\left(\dfrac{C}{2}\right)$.

20. Prove that a circular cylinder of radius R and length λR can float upright in stable equilibrium, if $\lambda^2 < 2$.

21. A uniform solid of revolution of sp. gr. δ and volume V floats in stable equilibrium in water with its axis vertical and a volume V' immersed. Prove that a uniform solid of the same shape and of the same volume but of sp. gr. $(1-\delta)$ can float in water in stable equilibrium with its axis vertical and a volume $(V-V')$ immersed, and with the same plane of floatation.

22. A uniform elliptic disc of eccentricity e and mass M has a particle of mass m attached at one focus and floats with its major axis vertical in a liquid of such density that one-fourth of that axis is immersed. Show that the equilibrium will be stable, if $\left(\dfrac{4\pi}{3\sqrt{3}} - 1 - e\right)m > eM$.

23. If a segment of a sphere of density δ floats in a liquid of density ρ, prove that the position in which the plane face is immersed and horizontal, is a position of stable equilibrium and that the metacentric height is $(\rho-\delta)/\delta$ times the distance of the centre of gravity of the segment from the centre of the sphere.

24. The cross section of a cylindrical ship is two equal arcs of equal parabolas of latus rectum l which touch at the keel, the common vertex of the two parabolas, so that the sides of the ship are concave to the water. The ship is floating upright with its keel at a depth h. Prove that the height of the metacentre above the keel is $h\left(\dfrac{3}{4} + \dfrac{h^2}{l^2}\right)$.

25. A paraboloidal cup, the weight of which is W, standing on a horizontal table, contains a quantity of water, the weight of which is nW. If h be the height of the centre of gravity of the cup and the contained water, the equilibrium will be stable provided the latus rectum of the parabola be $> 2(n+1)h$.

26. A heavy homogeneous cube is completely immersed with two faces horizontal in a fluid whose density $= k$ times the cube of the depth. Prove that the metacentric height is $\dfrac{ka^7}{120M}$, where M is the mass and a the length of an edge of the cube.

27. A cylindrical tube of sheet iron of uniform thickness, of radius a feet and weight w pounds, floats upright in water. Show that its centre of gravity cannot be higher above the lower end than

$$\frac{w}{393a^2} + \frac{49a^4}{w}.$$

Prove also that, whatever be its weight, its metacentre is always more than $0.7a$ feet above the lower end.

28. Prove that the work done in turning a floating body through a small angle θ round its centre of gravity is $\dfrac{1}{2}g\rho\left(Ak^2 + Ab^2 - cV\right)\theta^2$, where c is the distance between the centres of gravity of the body and the liquid displaced, and b is the horizontal distance between the centre of gravity of the body and that of the area of the plane of floatation.

CHAPTER 10

Small Oscillations of Floating Bodies

10.1 Introduction

Suppose that a body is floating in a heavy liquid at rest under gravity in a position of stable equilibrium. If now the body is slightly displaced, it will exhibit small oscillations. In the process, the liquid also will be disturbed and the problem assumes hydro-dynamical in nature.

In this chapter, we intend to discuss small oscillations of floating bodies under the following assumptions:

(1) The motion of the liquid is negligible, although this is not true in practical situations.

(2) The floating body is symmetrical with respect to the vertical plane through its centre.

(3) Oscillations are due to either vertical displacement or rotational displacement; there are no coupled oscillations.

10.2 Oscillations due to Vertical Displacement

We consider a body floating in a liquid in a position of stable equilibrium. Let the body be given a vertical displacement through a depth z (FIG. 10.1). The initial position C of the centre of the plane of floatation will rise to C'. Let G and H be the centre of gravity of the body and the centre of buoyancy respectively. According to our assumptions, C', C, G and H will lie on the same vertical line.

Let A = area of floatation in the initial position, V = volume of the displaced liquid in the initial position, m = mass of the body, and ρ = density of the liquid.

Since the displacement z is very small, we consider the area of floatation A to remain the same in the displaced position also.

Then, by the condition of floatation,

$$mg = Vg\rho, \text{ i.e., } m = V\rho \tag{10.1}$$

The equation for vertical motion of the body is

$$m\frac{d^2z}{dt^2} = mg - Vg\rho - Azg\rho \tag{10.2}$$

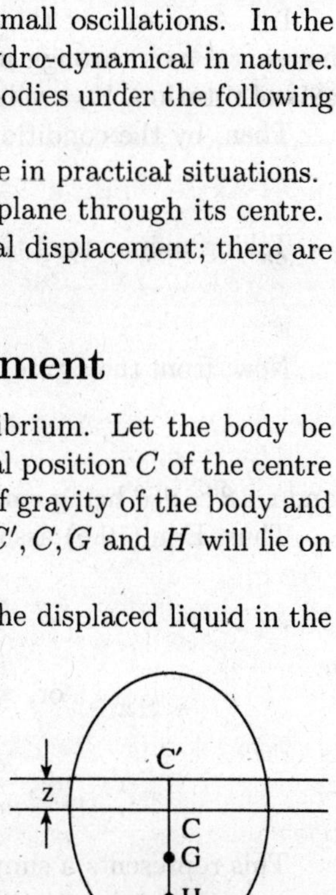

Fig. 10.1

In view of Eqn (10.1), Eqn (10.2) assumes the form

$$\rho V \frac{d^2 z}{dt^2} = -g\rho A z, \text{ i.e., } \frac{d^2 z}{dt^2} = -\frac{gA}{V} z \tag{10.3}$$

This represents a simple harmonic motion with the period of oscillation equal to

$$T = 2\pi\sqrt{\frac{V}{gA}} \tag{10.4}$$

10.3 Oscillations due to Rotational Displacement

We consider, as before, a body of mass m floating in a liquid of density ρ in a position of stable equilibrium. Let the body be rotated through a small angle θ about C, the centre of the plane of floatation (FIG. 10.2). Then the centre of gravity G will be raised by an amount d which depends on θ^2. Since θ is very small, we can neglect the value of d, so that the position of G remains unaltered.

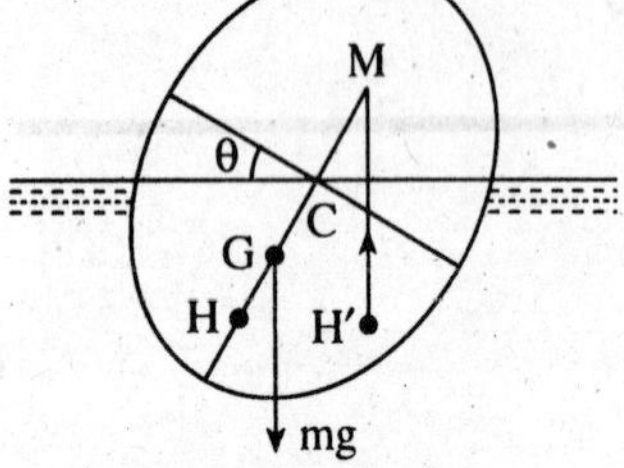

Fig. 10.2

Since the body remains in a position of stable equilibrium, it will exhibit small oscillations about a horizontal axis through C.

G and H are the centre of gravity of the body and the centre of buoyancy respectively initially and H' is the position of the centre of buoyancy in the displaced position.

Let M be the position of the metacentre.

Let $A =$ area of the plane of floatation, $K =$ radius of gyration of the body about the horizontal axis through its c.g. G, $k =$ radius of gyration of the plane of floatation about the axis of rotation, $V =$ volume of the initial liquid displaced, and $GH = h$.

Then, by the condition of floatation,

$$mg = Vg\rho, \text{ i.e., } m = V\rho \tag{10.5}$$

Taking moments of the forces acting in the body while in motion, we have

$$mK^2 \frac{d^2\theta}{dt^2} = -Vg\rho \cdot GM \sin\theta \tag{10.6}$$

Now, from the figure,

$$GM = HM - GH = \frac{Ak^2}{V} - h \tag{10.7}$$

and $\sin\theta \simeq \theta, \theta$ being small.

Thus, Eqn (10.6) assumes the form,

$$mK^2 \frac{d^2\theta}{dt^2} = -Vg\rho\left(\frac{Ak^2}{V} - h\right)\theta$$

$$\text{or, } mK^2 \frac{d^2\theta}{dt^2} = -mg\left(\frac{Ak^2}{V} - h\right)\theta \text{ [using Eqn (10.5)]}$$

$$\text{or, } \frac{d^2\theta}{dt^2} = -\frac{g}{K^2}\left(\frac{Ak^2}{V} - h\right)\theta \tag{10.8}$$

This represents a simple harmonic motion; the period of oscillations is given by

$$T = 2\pi K\sqrt{\frac{V}{g(Ak^2 - Vh)}} \tag{10.9}$$

From Eqn (10.8), it is obvious that, for oscillations,

$$Ak^2 > Vh, \text{ i.e., } h < \frac{Ak^2}{V}, \tag{10.10}$$

which is the criterion for stability.

10.4 Vertical Oscillations in Two Liquids

We consider a body totally immersed in two liquids (which do not mix) of densities ρ_1 and $\rho_2 (> \rho_1)$ (FIG. 10.3).

Let MN be the surface of separation, V_1 = volume of the body immersed in the liquid of density ρ_1, V_2 = volume of the body immersed in the liquid of density ρ_2, A = cross sectional area of the body at the surface of separation, and m = mass of the body.

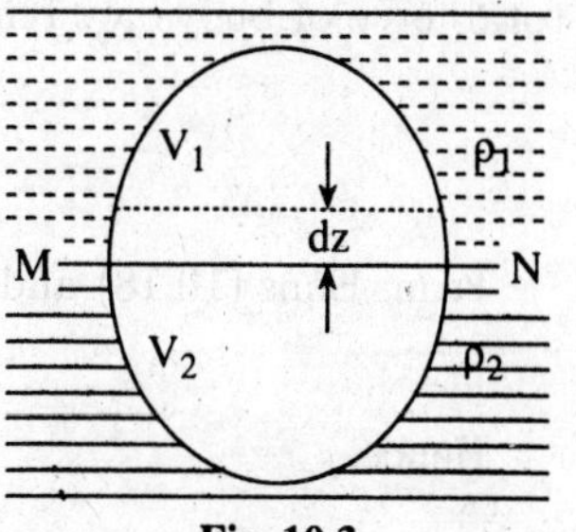

Fig. 10.3

Then by the condition of floatation,

$$mg = V_1 g\rho_1 + V_2 g\rho_2$$

$$\text{i.e., } m = V_1\rho_1 + V_2\rho_2 \tag{10.11}$$

Let the body be pushed down vertically through a small distance z and then released. It will exhibit small oscillations in the vertical directions.

The equation of motion is

$$\begin{aligned} m\frac{d^2z}{dt^2} &= mg - g\rho_1(V_1 - Az) - g\rho_2(V_2 + Az) \\ &= mg - g(V_1\rho_1 + V_2\rho_2) - Ag(\rho_2 - \rho_1)z \\ &= -Ag(\rho_2 - \rho_1)z \text{ [using (10.11)]} \end{aligned} \tag{10.12}$$

Substituting the value of m from Eqn (10.11), we get

$$(V_1\rho_1 + V_2\rho_2)\frac{d^2z}{dt^2} = -Ag(\rho_2 - \rho_1)z, \text{ i.e., } \frac{d^2z}{dt^2} = -\frac{Ag(\rho_2 - \rho_1)}{V_1\rho_1 + V_2\rho_2}z, \tag{10.13}$$

which represents a simple harmonic motion of period

$$T = 2\pi\sqrt{\frac{V_1\rho_1 + V_2\rho_2}{Ag(\rho_2 - \rho_1)}} \tag{10.14}$$

Denoting

$$\xi = \frac{\rho_2}{\rho_1} \tag{10.15}$$

we write Eqn (10.14) in the form

$$T = 2\pi\sqrt{\frac{V_1 + \xi V_2}{Ag(\xi - 1)}} \tag{10.16}$$

If, in particular,

$$V_1 = V_2 = \frac{1}{2}V; \xi = \frac{\rho_2}{\rho_1} = 2,$$

where V is the total volume of the solid, then, from Eqn (10.16), we get

$$T = \sqrt{\frac{3V}{2Ag}} \tag{10.17}$$

10.5 Hydrostatically Equivalent Density

We consider a body of volume V totally immersed in n immiscible liquids of densities $\rho_1, \rho_2, \cdots, \rho_n$.

Let the volume immersed in liquid of density ρ_1 be V_1, that in liquid of density ρ_2 be V_2, and so on. Then the total weight of liquid displaced, i.e., the total force of buoyancy is

$$F = g(V_1\rho_1 + V_2\rho_2 + \cdots + V_n\rho_n) = g\sum_{i=1}^{n} V_i\rho_i \tag{10.18}$$

Let the same body be totally immersed in a homogeneous liquid of density ρ, such that the total force of buoyancy remains unaltered. That is

$$F = g\rho(V_1 + V_2 + \cdots + V_n) = g\rho\sum_{i=1}^{n} V_i \tag{10.19}$$

From Eqns (10.18) and (10.19), we get

$$g\rho\sum V_i = g\sum V_i\rho_i$$

Hence,

$$\rho = \frac{\sum V_i\rho_i}{\sum V_i} \tag{10.20}$$

The density ρ is called the ***hydrostatically equivalent density*** of n different liquids of densities $\rho_1, \rho_2, \cdots, \rho_n$.

Thus, ***the hydrostatically equivalent density of n different immiscible liquids is that uniform density due to which the force of buoyancy upon a solid body totally immesed in it would equal to the sum of the forces of buoyancy upon the same solid when totally immersed in those n liquids.***

10.6 Angular Oscillations in Two Liquids

We consider a solid body of mass m and volume V totally immersed in two immiscible liquids of densities ρ_1 and ρ_2 $(> \rho_1)$.

Let the portion of volume V_1 of the solid be immersed in the liquid of density ρ_1, and the portion of volume V_2 in the liquid of density ρ_2 (FIG. 10.4). Then,

$$V = V_1 + V_2, \tag{10.21}$$

where V_1 = volume of the portion BAD and V_2 = volume of the portion BED.

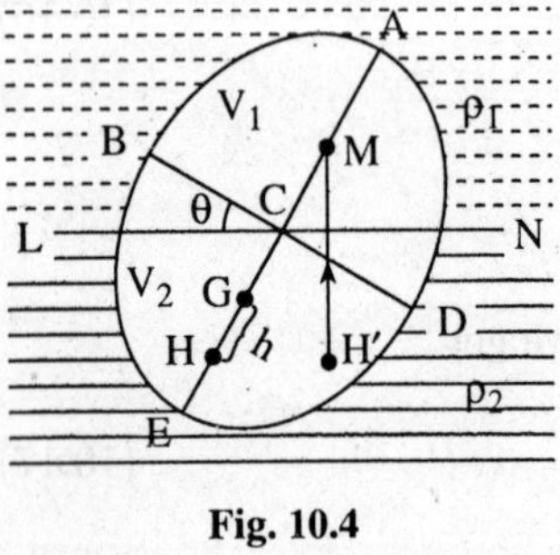

Fig. 10.4

Let LN be the surface of separation of the two liquids. If the body be rotated through a small angle θ about the centre C of the plane of separation and then released, the body will exhibit small oscillations about a horizontal axis through C. During the process, the upper liquid of density ρ_1 will oppose the motion.

The net force of buoyancy is

$$F = g(V_2\rho_2 - V_1\rho_1) \tag{10.22}$$

acting vertically upwards at H', the centre of buoyancy of the lower portion in displaced position.

Let G be the centre of gravity of the body, M the metacentre and $GH = h$. Then taking moments about G of all the forces,

$$mK^2 \frac{d^2\theta}{dt^2} = -F \cdot GM \sin\theta = -F(HM - GH)\theta$$

$$= -g(V_2\rho_2 - V_1\rho_1)\left(\frac{Ak^2}{V_2} - h\right)\theta \tag{10.23}$$

By the condition of floatation,

$$mg = g(V_1\rho_1 + V_2\rho_2), \text{ or, } m = V_1\rho_1 + V_2\rho_2 \tag{10.24}$$

Substituting this value of m into Eqn (10.23), we get

$$\frac{d^2\theta}{dt^2} = -\eta\theta \tag{10.25}$$

where

$$\eta = \frac{g(V_2\rho_2 - V_1\rho_1)}{K^2(V_2\rho_2 + V_1\rho_1)}\left(\frac{Ak^2}{V_2} - h\right) \tag{10.26}$$

This represents a simple harmonic motion with period

$$T = \frac{2\pi}{\sqrt{\eta}} \tag{10.27}$$

That is

$$T = 2\pi K \left[\frac{V_2(V_2\rho_2 + V_1\rho_1)}{g(V_2\rho_2 - V_1\rho_1)(Ak^2 - V_2h)}\right]^{\frac{1}{2}} \tag{10.28}$$

Denoting

$$\xi = \frac{\rho_2}{\rho_1}$$

we write Eqn (10.28) in the form

$$T = 2\pi K \left[\frac{V_2(V_2\xi + V_1)}{g(V_2\xi - V_1)(Ak^2 - V_2h)}\right]^{\frac{1}{2}} \tag{10.29}$$

If $V_1 = V_2 = \frac{1}{2}V$, then

$$T = 2\pi K \left[\frac{V(\xi + 1)}{g(\xi - 1)(2Ak^2 - Vh)}\right]^{\frac{1}{2}} \tag{10.30}$$

If, again, $\xi = 2$, i.e., $\rho_2 = 2\rho_1$, then

$$T = 2\pi K \left[\frac{3V}{g(2Ak^2 - Vh)}\right]^{\frac{1}{2}} \tag{10.31}$$

Note. As a criterion for stability,

$$\frac{Ak^2}{V_2} - h > 0, \text{ i.e., } h < \frac{Ak^2}{V_2}.$$

Also, for oscillatory motion, as we can see in Eqn (10.26),

$$V_2\rho_2 - V_1\rho_1 > 0, \text{ i.e., } \frac{\rho_2}{\rho_1} > \frac{V_1}{V_2}$$

$$\text{or, } \frac{V_1}{V_2} \text{ must be less than } \xi.$$

10.7 Vertical Oscillations in a Heterogeneous Fluid

We consider a solid body of mass m floating completely immersed in a fluid, the density of which varies as the depth from the effective surface.

Let the centre of gravity of the body be at a depth h from the effective surface.

We consider a small volume $d\tau$ of the body at a depth z, where pressure is λz, λ being a constant. Then, for equilibrium,

$$mg = \iiint_V \lambda g z d\tau, \text{ i.e., } m = \iiint_V \lambda z d\tau \tag{10.32}$$

Since h is the depth of centre of gravity, we have, by definition

$$h = \frac{\iiint z d\tau}{\iiint d\tau} = \frac{\iiint z d\tau}{V},$$

which yields

$$\iiint_V z d\tau = Vh \tag{10.33}$$

In view of this equation, we get from Eqn (10.32)

$$m = \lambda V h \tag{10.34}$$

Let now the body be depressed vertically downwards through a small depth η. Then the depth of centre of gravity now becomes $(h+\eta)$, and the depth of the elementary volume becomes $(z+\eta)$, where the fluid pressure is $\lambda(z+\eta)$. The equation of motion is

$$m\frac{d^2\eta}{dt^2} = mg - \iiint_V \lambda g(z+\eta)d\tau \tag{10.35}$$

Similar to Eqn (10.33), we have

$$\iiint_V (z+\eta)d\tau = (h+\eta)V \tag{10.36}$$

By using Eqns (10.34) and (10.36), we get from Eqn (10.35)

$$\lambda V h\frac{d^2\eta}{dt^2} = \lambda g V h - \lambda g V(h+\eta), \text{ or, } \frac{d^2\eta}{dt^2} = -\frac{g}{h}\eta \tag{10.37}$$

This represents a simple harmonic motion of period

$$T = 2\pi\sqrt{\frac{h}{g}}, \tag{10.38}$$

which depends only on the depth of c.g. of the body from the effective surface.

EXAMPLE 10.1 *A cylinder makes vertical oscillations in a liquid contained in another cylinder, the radius of which is n times that of the former. Show that the length of the axis immersed when it is in a position of rest is*

$$\frac{gT^2n^2}{4\pi^2(n^2-1)},$$

where T is the time of a complete oscillation.

Solution. Let R = radius of the inner cylinder, h = length of the axis of the inner cylinder immersed, ρ = density of the liquid, m = mass of the inner cylinder. Then radius of the outer cylinder = nR.

Let z be the depth through which the inner cylinder is depressed and z' be the corresponding rise of water level in the container. Then the total depth gone down is $(z + z')$.

Since the volume of the liquid remains unaltered, we have

$$\pi R^2 z = (\pi n^2 R^2 - \pi R^2)\, z', \text{ i.e., } z = (n^2 - 1)z' \tag{1}$$

By the condition of floatation, we have

$$mg = \pi R^2 h g\rho, \text{ i.e., } m = \pi R^2 h\rho \tag{2}$$

The equation of motion is

$$m\frac{d^2z}{dt^2} = mg - \pi R^2 hg\rho - \pi R^2 g\rho(z + z')$$

$$\text{or, } \pi R^2 h\rho\frac{d^2z}{dt^2} = -\pi R^2 g\rho(z + z') \text{ [using (2)]}$$

$$\text{or, } h\frac{d^2z}{dt^2} = -g\left(z + \frac{z}{n^2-1}\right) \text{ [using (1)]}$$

$$\text{or, } \frac{d^2z}{dt^2} = -\frac{g}{h}\frac{n^2}{n^2-1}z \tag{3}$$

This represents a simple harmonic motion with time period

$$T = 2\pi\sqrt{\frac{h(n^2-1)}{gn^2}} \tag{4}$$

Squaring, $T^2 = 4\pi^2 \dfrac{h}{g}\dfrac{n^2-1}{n^2}$, i.e., $h = \dfrac{gT^2n^2}{4\pi^2(n^2-1)}$. **(Proved)**

EXAMPLE 10.2 *A cube, the length of whose edge is 2a, is floating in a fluid with its centre of gravity at a depth c below the surface. If it receives a small displacement so that two of its faces remain vertical, show that the times of its small vertical and angular oscillations are* $2\pi\sqrt{\dfrac{a+c}{g}}$ *and* $4\pi\sqrt{\dfrac{a^2(a+c)}{g(3c^2-a^2)}}$ *respectively.*

Solution. (i) **Vertical Oscillation.** Let $ABCD$ be the section of the cube by a vertical plane passing through its c.g. G (FIG. 10.5), which is at a depth of c below the free surface. If m be the mass of the cube, then, by the condition of floatation,

$$mg = 4a^2(a + c)g\rho, \text{ i.e., } m = 4a^2(a + c)\rho, \tag{1}$$

where ρ is the density of the liquid.

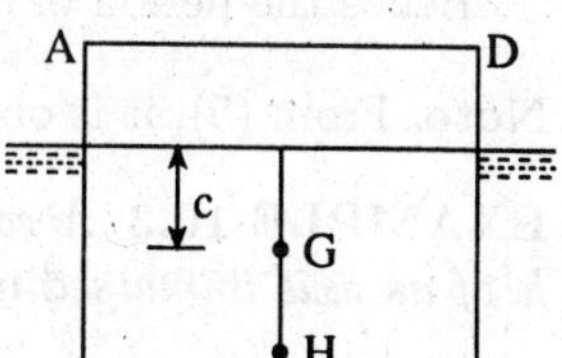

Fig. 10.5

Suppose that the cube is depressed through a depth z vertically and then released.

The equation of motion is

$$m\frac{d^2z}{dt^2} = mg - 4a^2(a + c)g\rho - g\rho \cdot 4a^2 \cdot z$$

$$\text{or, } 4a^2(a + c)\rho\frac{d^2z}{dt^2} = -4a^2 g\rho z \text{ [using (1)]}$$

$$\text{or, } \frac{d^2z}{dt^2} = -\frac{g}{a+c}z \tag{2}$$

This represents a simple harmonic motion with time period

$$T = 2\pi\sqrt{\frac{a+c}{g}}. \quad \textbf{(Proved)}$$

(ii) **Angular Oscillations.** Let the cube be rotated through an angle θ about the centre of the plane of floatation. Taking moments about the centre of gravity G, we have

$$mK^2\frac{d^2\theta}{dt^2} = -g\rho V\left(\frac{Ak^2}{V} - h\right)\theta, \tag{3}$$

where V = volume immersed $= 4a^2(a+c)$

K^2 = radius of gyration of the cube about a horizontal line through its c.g. $G = \frac{2}{3}a^2$

A = area of the plane of floatation $= 4a^2$

k^2 = radius of gyration of the area of floatation about horizontal line through the centre of floatation $= \frac{1}{3}a^2$

and h = distance between the centre of gravity of the cube and the centre of buoyancy $= \frac{1}{2}(a+c) - c = \frac{1}{2}(a-c)$.

Then,

$$\frac{Ak^2}{V} = \frac{4a^2 \cdot \frac{1}{3}a^2}{4a^2(a+c)} = \frac{a^2}{3(a+c)} \tag{4}$$

Putting these values and the value of m from (1) into (3), we get

$$4a^2(a+c)\rho \cdot \frac{2}{3}a^2\frac{d^2\theta}{dt^2} = -4a^2(a+c)g\rho\left[\frac{a^2}{3(a+c)} - \frac{1}{2}(a-c)\right]\theta$$

$$\text{or,} \quad \frac{d^2\theta}{dt^2} = -\frac{3g}{2a^2}\left[\frac{a^2}{3(a+c)} - \frac{1}{2}(a-c)\right]\theta = -\frac{g}{4a^2}\frac{3c^2-a^2}{a+c}\theta \tag{5}$$

Hence the period of oscillation is $T = 4\pi\sqrt{\dfrac{a^2(a+c)}{(3c^2-a^2)}}$. **(Proved)**

Note. From (5), it is obvious that oscillations can occur if $3c^2 > a^2$.

EXAMPLE 10.3 *A cone of vertical angle 2α floats in a cylinder of radius R with a length h of its axis immersed in a liquid. If it be pushed vertically downwards through a small space, show that the time of complete oscillation is $2\pi\sqrt{\dfrac{(R^2 - h^2\tan^2\alpha)h}{3R^2g}}$.*

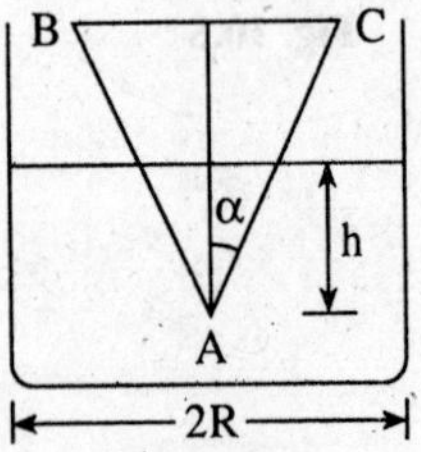

Fig. 10.6

Solution. Let ABC be the cone immersed in a liquid of density ρ contained in a cylinder of radius R (FIG. 10.6).

If m be the mass of the cone, then by the condition of floatation, we have $mg = g\rho V = \frac{1}{3}\pi g\rho h^3\tan^2\alpha$.

Hence,

$$m = \frac{1}{3}\pi h^3\rho\tan^2\alpha \tag{1}$$

Now, let z be the depth through which the cone is depressed vertically downwards, and z' the corresponding rise in water level in the vessel. Since the volume of water in the vessel remains unaltered, we have

$$\frac{1}{3}\pi(h+z)^3\tan^2\alpha - \frac{1}{3}\pi h^3\tan^2\alpha = \pi\left(R^2 - h^2\tan^2\alpha\right)z' \tag{2}$$

Since z is very small, we write

$$(h+z)^3 \simeq h^3 + 3h^2z \tag{3}$$

neglecting the higher powers of z.

Then, from (2) and (3), we get

$$\frac{1}{3}\pi \cdot 3h^2z \cdot \tan^2\alpha = \pi\left(R^2 - h^2\tan^2\alpha\right)z',$$

which yields

$$z' = \frac{h^2\tan^2\alpha}{R^2 - h^2\tan^2\alpha}z \tag{4}$$

The equation of motion is

$$m\frac{d^2z}{dt^2} = mg - g\rho V - g\rho\pi h^2\tan^2\alpha(z+z') = -g\rho\pi h^2\tan^2\alpha(z+z') \tag{5}$$

Substituting the value of m from (1) and that of z' from (4) into (5), we get

$$\frac{1}{3}\pi h^3\rho\tan^2\alpha\frac{d^2z}{dt^2} = -g\rho\pi h^2\tan^2\alpha\left(1 + \frac{h^2\tan^2\alpha}{R^2 - h^2\tan^2\alpha}\right)z$$

$$\text{i.e.,}\quad \frac{d^2z}{dt^2} = -\frac{3gR^2}{h(R^2 - h^2\tan^2\alpha)}z \tag{6}$$

Hence, the period of oscillation is $T = 2\pi\sqrt{\dfrac{h(R^2 - h^2\tan^2\alpha)}{3gR^2}}$. **(Proved)**

EXAMPLE 10.4 *A solid ellipsoid floats in a liquid of twice its own specific gravity with its shortest axis vertical. Find the time of a small vertical oscillation and also the times of small angular oscillations about the two horizontal axes.*

Solution. Since the sp. gr. of the liquid is twice that of the solid ellipsoid, the body will float with half of its volume immersed in the liquid (FIG. 10.7).

Hence, the volume immersed is

$$V = \frac{2}{3}\pi abc, \tag{1}$$

where a, b, c are semi-major axes of the ellipsoid.

The cross section of the ellipsoid by the plane of floatation is an ellipse of semi-major axes a and b. Hence, its area is

$$A = \pi ab. \tag{2}$$

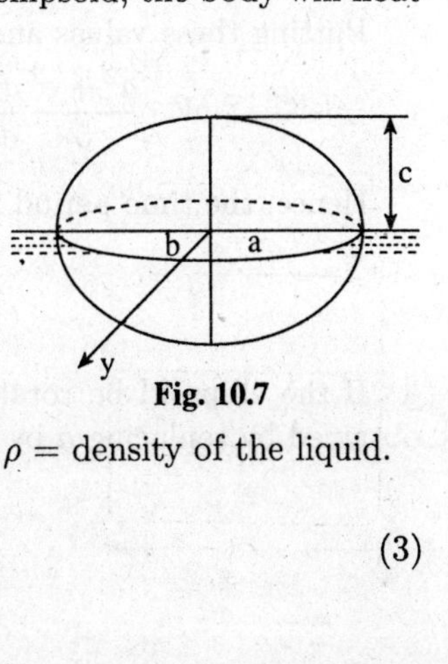

Fig. 10.7

(i) **Vertical Oscillations.** Let m = mass of the ellipsoid, and ρ = density of the liquid. Then by condition of floatation,

$$mg = g\rho V, \text{ i.e., } m = \rho V. \tag{3}$$

If the body be vertically depressed through a depth z, then the equation of motion is

$$m\frac{d^2z}{dt^2} = mg - g\rho V - g\rho Az$$

$$\text{i.e., } \rho V\frac{d^2z}{dt^2} = -g\rho Az \text{ [using (3)]}$$

$$\text{i.e., } \frac{d^2z}{dt^2} = -\frac{gA}{V}z \tag{4}$$

Substituting here the values of A and V from (1) and (2), we get

$$\frac{d^2z}{dt^2} = -g\frac{\pi ab}{\frac{2}{3}\pi abc}z = -\frac{3g}{2c}z \tag{5}$$

Hence, the time period is

$$T = 2\pi\sqrt{\frac{2c}{3g}}. \quad \textbf{(Ans.)} \tag{6}$$

(ii) **Angular Oscillations.** Let the body be rotated about y-axis by a small angle θ, and then released. It will exhibit angular oscillations about y-axis (FIG. 10.8).

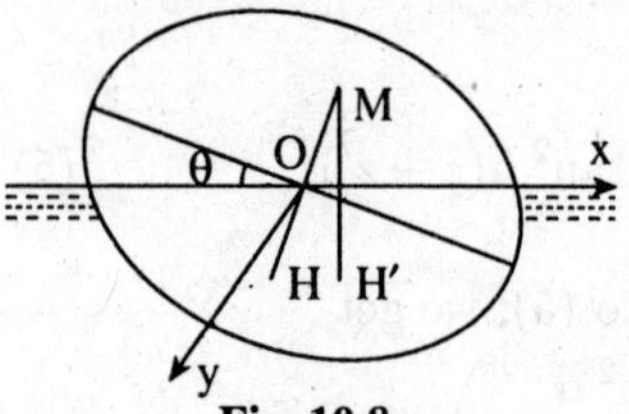

Fig. 10.8

The c.g. of the body is at its centre. Let H and M be the positions of the centre of buoyancy and metacentre respectively.

Then the equation of motion is

$$mK^2\frac{d^2\theta}{dt^2} = -g\rho V\left(\frac{Ak^2}{V} - h\right)\theta, \tag{7}$$

where K = radius of gyration of the body about horizontal axis through its centre, i.e., about y-axis, k = radius of gyration of the plane of floatation (i.e., of the elliptic area) about y-axis, and h = distance of the centre of buoyancy from O.

Here,

$$K^2 = \frac{a^2+c^2}{5}, \; k^2 = \frac{1}{4}a^2, \; h = \frac{3}{8}c \tag{8}$$

Then

$$\frac{Ak^2}{V} - h = \frac{\pi ab\cdot\frac{1}{4}a^2}{\frac{2}{3}\pi abc} - \frac{3}{8}c = \frac{3}{8c}(a^2-c^2) \tag{9}$$

Putting these values and using (3) in (7), we get

$$\rho V\cdot\frac{a^2+c^2}{5}\frac{d^2\theta}{dt^2} = -g\rho V\cdot\frac{3(a^2-c^2)}{8c}\theta, \text{ i.e., } \frac{d^2\theta}{dt^2} = -\frac{15g(a^2-c^2)}{8c(a^2+c^2)}\theta \tag{10}$$

Hence, the time period is

$$T = 2\pi\sqrt{\frac{8c(a^2+c^2)}{15g(a^2-c^2)}}. \quad \textbf{(Ans.)} \tag{11}$$

If the ellipsoid be rotated about x-axis through a small angle θ, the time period can be obtained by replacing a by b in (11). Thus, in this case,

$$T = 2\pi\sqrt{\frac{8c(b^2+c^2)}{15g(b^2-c^2)}}. \quad \textbf{(Ans.)} \tag{12}$$

EXAMPLE 10.5 *An oblate spheroid is completely immersed in two liquids, the specific gravity of the lower being twice that of the upper liquid, and floats with its axis vertical and its centre in the common surface of the liquids. Supposing small displacements to take place, first in a vertical direction and second, about a horizontal line through its centre of gravity, show that the times of small oscillations will be respectively*

$$2\pi\sqrt{\frac{2b}{g}} \text{ and } 2\pi\sqrt{\frac{8b(a^2+b^2)}{5g(a^2-b^2)}},$$

where a and b are semi-axes of the generating ellipse.

Solution. The common surface intersects the solid body in a circle of radius a. Hence,

$$A = \pi a^2 \tag{1}$$

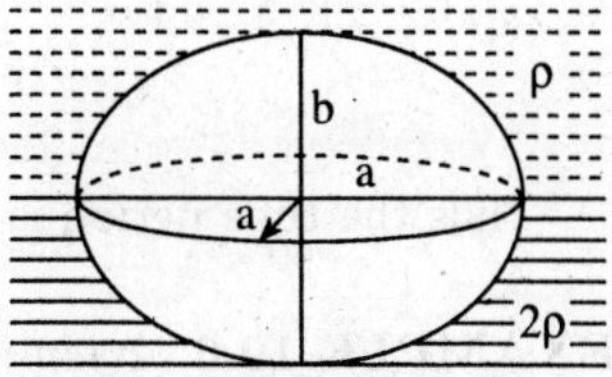

Fig. 10.9

The volumes in two liquids are equal, so that (FIG. 10.9)

$$V_1 = V_2 = \frac{2}{3}\pi a^2 b. \tag{2}$$

If m be the mass of the solid, then, by condition of floatation,

$$mg = \frac{2}{3}\pi a^2 bg(\rho + 2\rho) = 2\pi a^2 bg\rho, \text{ i.e., } m = 2\pi a^2 b\rho \tag{3}$$

For vertical oscillations, the equation of motion (10.12) in this case, is

$$m\frac{d^2z}{dt^2} = -Ag(2\rho - \rho)z$$

$$\text{i.e., } 2\pi a^2 b\rho\frac{d^2z}{dt^2} = -\pi a^2 g\rho z \text{ [using (1) and (3)]}$$

$$\text{i.e., } \frac{d^2z}{dt^2} = -\frac{g}{2b}z \tag{4}$$

Hence, the time period is $T = 2\pi\sqrt{\frac{2b}{g}}$. **(Proved)**

In the case of angular oscillations, the net force of buoyancy is (FIG. 10.10)

$$F = g\,(V_2\rho_2 - V_1\rho_1)$$

$$= g\cdot\frac{2}{3}\pi a^2 b(2\rho - \rho) = \frac{2}{3}\pi a^2 b\rho g \tag{5}$$

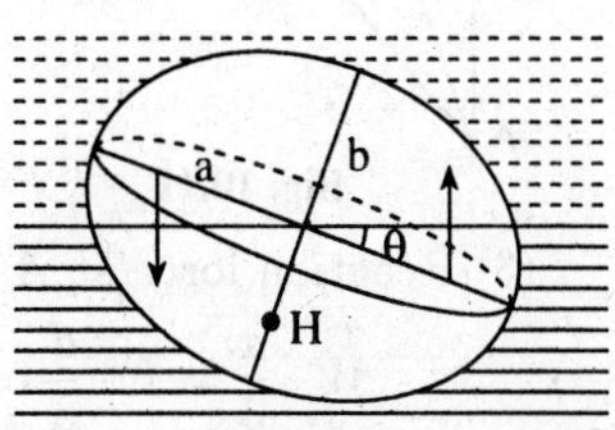

Fig. 10.10

The equation of motion is

$$mK^2\frac{d^2\theta}{dt^2} = -F\left(\frac{Ak^2}{V_2} - h\right)\theta \tag{6}$$

where, K = radius of gyration of the solid about a horizontal axis through its centre, k = radius of gyration of the plane area of floatation about the horizontal axis through its centre and h = distance of the centre of buoyancy from the centre of the solid ellipsoid.

Here,

$$K = \frac{a^2+b^2}{5},\ k = \frac{1}{4}a^2,\ h = \frac{3}{8}b \tag{7}$$

Using (3), (5) and (7) in (6), we get

$$2\pi a^2 b\rho \cdot \frac{a^2+b^2}{5}\frac{d^2\theta}{dt^2} = -\frac{2}{3}\pi a^2 b\rho g\left(\frac{\pi a^2 \cdot \dfrac{a^2}{4}}{\dfrac{2}{3}\pi a^2 b} - \frac{3}{8}b\right)$$

$$\text{or, } \frac{2}{5}(a^2+b^2)\frac{d^2\theta}{dt^2} = -\frac{2}{3}g\left(\frac{3}{8}\frac{a^2}{b} - \frac{3}{8}b\right) \text{ or, } \frac{a^2+b^2}{5}\frac{d^2\theta}{dt^2} = -\frac{1}{8}g\left(\frac{a^2-b^2}{b}\right)$$

Hence,

$$\frac{d^2\theta}{dt^2} = -\frac{5g}{8b}\cdot\frac{a^2-b^2}{a^2+b^2} \tag{8}$$

Thus the time period is $T = 2\pi\sqrt{\dfrac{8b}{5g}\dfrac{a^2+b^2}{a^2-b^2}}$. **(Proved)**

EXAMPLE 10.6 *A lamina of uniform thickness, in the form of an isosceles right-angled triangle, has one of the acute angles fixed below the surface of a liquid, and rests with the side, which is not immersed, horizontal. Prove that the time of a small oscillation in its own plane is* $2\pi\sqrt{\dfrac{a}{g}}$, *where a is the length of each of the equal sides of the triangle.*

Solution. Let ABC be the triangular lamina with its sides $AC = BC = a$ and $\angle C = 90°$. It is fixed at A and rests with sides BC horizontal (FIG. 10.11). Then the side AC is vertical initially.

Let δ = density of the lamina, ρ = density of the liquid and $d = AD$ = depth of the vertex A below the liquid surface.

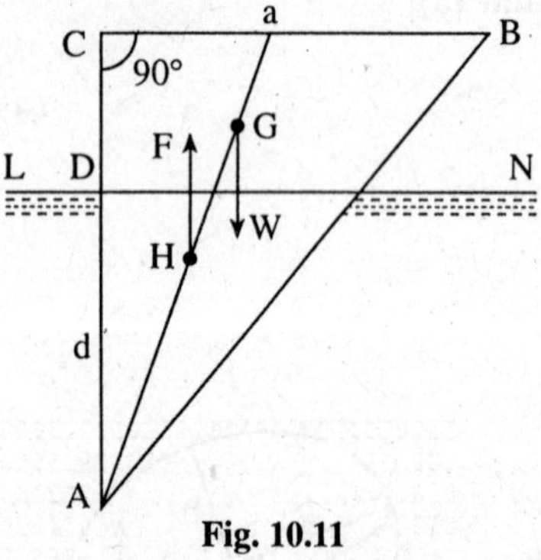

Fig. 10.11

G and H are the positions of centre of gravity of the lamina and the centre of buoyancy respectively.

The forces acting on the lamina are:

(1) Its weight W acting vertically downwards through G, where

$$W = \frac{1}{2}a^2\delta g \tag{1}$$

(2) The force of buoyancy F acting vertically upwards through H, where

$$F = \frac{1}{2}d^2\rho g \tag{2}$$

(3) Reaction force at A. Taking moments about A,

$$W\cdot\frac{a}{3} = F\cdot\frac{d}{3} \text{ or, } \frac{1}{2}a^2\delta g\cdot\frac{a}{3} = \frac{1}{2}d^2\rho g\cdot\frac{d}{3}$$

$$\text{i.e., } a^3\delta = d^3\rho \tag{3}$$

Let now the lamina be slightly displaced so that the side AC now makes an angle θ with the vertical (FIG. 10.12).

Then taking moments about A, we have

$$mK^2\frac{d^2\theta}{dt^2} = mg\cdot GL - F\cdot HN, \tag{4}$$

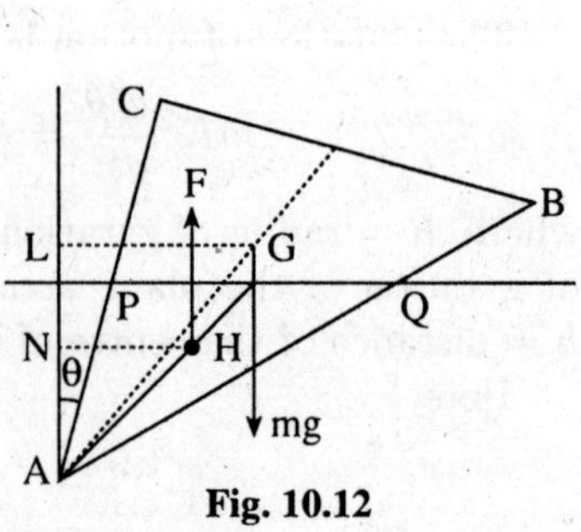

Fig. 10.12

where m = mass of the lamina, H = new position of the centre of buoyancy, K = radius of gyration of the lamina about the horizontal axis through A.

Now, from the figure, we get the distances of A, B, C from the vertical as

$$0,\ \sqrt{2}\,a\sin(45^\circ+\theta),\ a\sin\theta$$

respectively. Hence,

$$GL = \frac{1}{3}\left[\sqrt{2}a\sin(45^\circ+\theta)+a\sin\theta\right]$$
$$= \frac{1}{3}a(2\sin\theta+1) = \frac{1}{3}a(2\theta+1) \tag{5}$$

for small θ.

Also, the distances of points A, P, Q from the vertical line are $0,\ d\tan\theta,\ d\tan(45^\circ+\theta)$ respectively.

Hence,

$$HN = \frac{1}{3}d\left[\tan\theta+\tan(45^\circ+\theta)\right] = \frac{1}{3}d\left(\theta+\frac{1+\theta}{1-\theta}\right), \tag{6}$$

since $\tan\theta \simeq \theta$ for small θ.

The area of the triangle APQ is

$$= \frac{1}{2}PQ\cdot d = \frac{1}{2}d^2[\tan(45^\circ+\theta)-\tan\theta] = \frac{1}{2}d^2\left(\frac{1+\theta}{1-\theta}-\theta\right) \tag{7}$$

Hence, the force of buoyancy F is

$$F = \frac{1}{2}d^2 g\rho\left(\frac{1+\theta}{1-\theta}-\theta\right) \tag{8}$$

From (6) and (8), we get

$$F\cdot HN = \frac{1}{2}d^2 g\rho\left(\frac{1+\theta}{1-\theta}-\theta\right)\left(\frac{1+\theta}{1-\theta}+\theta\right)\cdot\frac{1}{3}d$$
$$= \frac{1}{6}d^3 g\rho\left(\frac{1+\theta^2}{1-\theta}\right)\left(\frac{1+2\theta-\theta^2}{1-\theta}\right)$$
$$= \frac{1}{6}d^3 g\rho\left(\frac{1}{1-\theta}\right)\left(\frac{1+2\theta}{1-\theta}\right),\ [\text{neglecting } \theta^2]$$
$$= \frac{1}{6}d^3 g\rho(1+2\theta)(1-\theta)^{-2}$$
$$= \frac{1}{6}d^3 g\rho(1+2\theta)(1+2\theta+\cdots)$$
$$= \frac{1}{6}d^3 g\rho(1+2\theta)(1+2\theta),\ (\text{neglecting higher powers of } \theta)$$
$$= \frac{1}{6}d^3 g\rho(1+4\theta) \tag{9}$$

Also,

$$K^2 = \frac{1}{3}\left(\frac{a^2}{4}+\frac{5a^2}{4}+\frac{a^2}{2}\right) = \frac{2a^2}{3} \tag{10}$$

Substituting all these values in (4), we get

$$\frac{1}{2}a^2\delta\cdot\frac{2}{3}a^2\frac{d^2\theta}{dt^2}=\frac{1}{2}a^2\delta g\cdot\frac{1}{3}a(2\theta+1)-\frac{1}{6}d^3g\rho(1+4\theta)$$

$$\text{i.e., }\frac{1}{3}a^4\delta\frac{d^2\theta}{dt^2}=\frac{1}{6}a^3\delta g(2\theta+1)-\frac{1}{6}a^3\delta g(1+4\theta)\ [\text{using (3)}]$$

$$\text{i.e., }\frac{1}{3}a^4\delta\frac{d^2\theta}{dt^2}=-\frac{1}{3}a^3\delta g\theta$$

$$\therefore\frac{d^2\theta}{dt^2}=-\frac{g}{a}\theta$$

Hence the time period is $T=2\pi\sqrt{\frac{a}{g}}$. **(Proved)**

EXERCISE

1. A vertical cylinder of height h floats in a liquid of twice its own density contained in a cylindrical vessel. If the radius of the vessel be double that of the cylinder, and the cylinder be slightly displaced in a vertical direction, prove that the time of oscillation is $\pi\sqrt{\frac{3h}{2g}}$.

2. A right cone is floating with its axis vertical and vertex downwards in a fluid, and $\frac{1}{n}$th part of the axis is immersed. A weight equal to the weight of the cone is placed on the base, upon which the cone sinks till its axis is totally immersed, before rising. Show that $n^3+n^2+n=7$.

3. A hollow hemisphere movable about a horizontal diameter is partly filled with liquid. Show that the time of a small oscillation is the same as if there were no fluid in it.

4. A cube, the length of whose edge is $2a$, is completely immersed in two liquids, the sp. gr. of the lower being twice that of the upper liquid and floats with its centre of gravity at a depth $\frac{2}{3}$a below the common surface of the liquids. If it receives small displacement so that two of its faces remain vertical, show that the times of its vertical and angular oscillations are $2\pi\sqrt{\frac{11a}{3g}}$ and $2\pi\sqrt{\frac{20a}{9g}}$ respectively.

5. A solid, the lower portion of whose surface is spherical, floats in a heavy liquid. Show that the time of a small angular oscillation is the same in whatever homogeneous fluid it floats.

6. An oblate spheroid with semi-major axes a, a, c floats in a liquid of twice its own specific gravity with its shortest axis vertical. Show that the time of a small vertical ocillation is $2\pi\sqrt{\frac{2c}{3g}}$.

Also show that the time of a small angular oscillation is $2\pi\sqrt{\frac{8}{15}\frac{a^2+c^2}{a^2-c^2}\cdot\frac{c}{g}}$.

CHAPTER 11

Gases

11.1 Introduction

Gases are compressible fluids. A given mass of gas can be made to expand indefinitely. A gas has neither a definite volume nor a definite shape; it takes the shape of the vessel in which it is contained. Thus, fluids which can be compressed easily and made to expand indefinitely by increasing the volume are called ***gases***.

Air is a gas which has weight and exerts pressure. Atmosphere can be assumed to be a spherical shell of air surrounding the earth's surface. The pressure of the atmosphere is due to the weight of the column of air above. This will not be directly proportional to the height, as in the case of incompressible fluids, on account of the varying density of air. The density of air gradually increases with the increase of height from the surface of the earth.

If we assume the atmosphere to be homogeneous of density ρ and consider a column of air of height H at the earth's surface, such that the pressure $H\rho g$ by this air column is equal to the actual pressure of the atmosphere on the earth's surface, then the height H is called the height of ***homogeneous atmosphere***.

The pressure of air is measured by an apparatus called ***Barometer***.

The Barometer, in its simplest form, is a straight uniform tube AB sealed at the end A and open at the end B; it contains mercury with its lower end immersed in a verssel of mercury (FIG. 11.1).

It is observed that mercury column will remain at rest at a particular height H from the surface of mercury in the vessel. This height is same for all positions of the tube and is known as ***Barometric height***.

The space in the tube is a vacuum and is known as ***Torricellian Vacuum***.

Let δ = density of mercury and Π = atmospheric pressure.

Since pressure at same level are equal, pressure at C and that at Q within the tube are same.

Thus,

$$\Pi = g\delta H \tag{11.1}$$

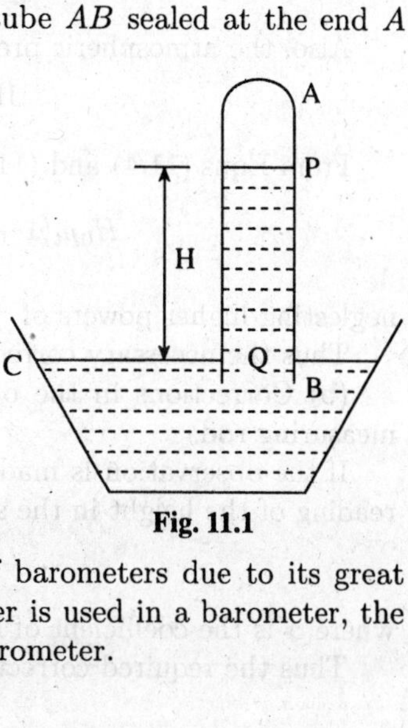

Fig. 11.1

Mercury is the most convenient fluid in the construction of barometers due to its great density which is about 13.568 times that of water. Thus, if water is used in a barometer, the height H would be 13.568 times more than that in a mercury barometer.

There are, however, some corrections to be made to an observation with the mercury barometer as discussed below:

(1) **Correction for capacity.** Whenever there is a change in the position of the mercury level in the tube, the position of the surface of mercury in the vessel will also be changed. Thus, it is necessary to measure the mercury level in the tube correctly in order to get the atmospheric pressure accurately.

Let H = initial height of mercury level in the tube above the surface of mercury in the vessel at any standard position
A = area of cross section of the vessel
a = area of cross section of the tube
x = rise of mercury level in the tube as observed, and
x' = downfall of the mercury level in the vessel.

Since the total volume of mercury remains unaltered, we must have

$$Ax' = ax, \text{ i.e., } x' = \frac{a}{A}x \tag{11.2}$$

Then the correct height of mercury level in the tube above that in the basin is

$$= H + x + \frac{a}{A}x \tag{11.3}$$

Hence the correction is $+\frac{a}{A}x$.

(2) **Correction for temperature.** (a) The barometric height will rise with the increase of temperature due to the thermal expansion of mercury in the tube. Thus the observed atmospheric pressure will be more than the actual, so that necessary corrections are required.

Let H_0 = the required height of mercury level in the tube at 0 °C
H_t = the observed height of mercury level in the tube at t °C
ρ_0 = the density of mercury at 0 °C
ρ_t = the density of mercury at t °C
γ = coefficient of cubical expansion of mercury.

Then we have

$$\rho_0 = \rho_t(1 + \gamma t) \tag{11.4}$$

Also, the atmospheric pressure Π is given by

$$\Pi = g\rho_0 H_0 = g\rho_t H_t, \text{ i.e., } \rho_0 H_0 = \rho_t H_t \tag{11.5}$$

From Eqns (11.4) and (11.5),

$$H_0\rho_t(1 + \gamma t) = \rho_t H_t, \text{ i.e., } H_0 = \frac{H_t}{1 + \gamma t} = H_t(1 - \gamma t), \tag{11.6}$$

neglecting higher powers of γ due to smallness.

Thus the necessary correction in the height is $-H_t\gamma t$.

(b) Corrections in the observed reading are also necessary due to the expansion of the measuring rod.

If the observation is made at t °C above the standard temperature and x be the observed reading of the height in the scale, then the actual reading should be

$$x(1 + \alpha t),$$

where α is the coefficient of linear expansion of the measuring scale.

Thus the required correction is $+x\alpha t$.

(3) **Correction for capillarity.** We have discussed this in the next chapter. On account of surface tension, the mercury in the capillary tube will be below the level of mercury in the vessel by an amount

$$h = \frac{2S\cos\alpha}{g\delta r}, \tag{11.7}$$

where S = surface tension of mercury, α = the angle of contact of mercury with glass, r = radius of the barometer tube, and δ = the density of mercury.

The quantity in Eqn (11.7) is generally obtained experimentally for the barometer to be used and added to the observed height.

(4) **Correction for height above sea level.** Let H be the observed height of the barometer at a place at height z above sea level. The value of H will be less than that observed at sea level. We will see later that the necessary correction is

$$H\left\{\exp\left(\frac{gz}{k}\right) - 1\right\},$$

where k is the constant ratio of pressure and density.

(5) **Correction for gravity.** The value of the acceleration g due to gravity, as we know, is different at different altitudes from the earth's surface. Hence the weight of the mercury column in the barometer will also be different and so will be the barometric height.

The standard position is the sea level at latitude of 45°.

We will discuss this later in this chapter.

11.2 Thermodynamic Parameters

Any thermodynamic system is defined by its temperature, pressure, volume, internal energy, etc. These are called thermodynamic parameters of the system.

If the magnitude of a thermodynamic parameter is independent of path, its change is a perfect differential. For examples, parameters like volume, temperature, potential and kinetic energy are independent of path. Hence their changes are perfect differentials and are written as dV, dT, etc.

Change of parameters, the magnitudes of which depend on the path, are imperfect differentials. For example, work (W) and heat (Q) in a system depend on the path of the process. Hence their changes are imperfect differentials and are written as δW and δQ, but not as dW and dQ.

11.3 Gases under Change of Thermodynamic Parameters

The behaviour of gases under the change of thermodynamic parameters are ruled by the following gas laws:

(a) **Boyle's Law.** It states: ***If the temperature of a given mass of a gas remains constant, the pressure varies inversely as its volume.***

Let V = volume of the gas, ρ its density, p its pressure and m its mass at a certain temperature of t °C. Then, by Boyle's law,

$$p \propto \frac{1}{V}, \text{ i.e., } pV = \text{constant}, \tag{11.8}$$

when temperature remains constant.

Since, $m = V\rho =$ constant, $V = \dfrac{m}{\rho}$ so that, from Eqn (11.8), we have

$$\frac{p}{\rho} = \text{constant}. \tag{11.9}$$

(b) **Charle's Law.** It states: ***If the pressure of a given mass of gas remains constant, the volume of the gas increases by a fraction*** $\left[\frac{1}{273}\right]$ ***of its volume at 0 °C, for every degree centrigrade rise in temperature.***

Thus, if V_0 be the volume of the gas at a pressure p and temperature 0 °C, then its volume V_t at a temperature t °C and at the same pressure p will be

$$V_t = V_0\left[1 + \frac{t}{273}\right] \tag{11.10}$$

Eqn (11.10) shows that, at $t = -273$ °C, the volume of the gas vanishes. In practice, however, most of the gases liquify before this temperature is reached and hence occupies a very little volume. This temperature is called the ***absolute zero*** of the gas thermometer.

The absolute temperature denoted by T corresponding to t °C is $T = t + 273$. Thus Eqn (11.10) may be written as

$$V_t = V_0\frac{t + 273}{273} = \frac{V_0 T}{273}, \tag{11.11}$$

which implies that

$$V \propto T, \tag{11.12}$$

when p is constant.

This is another form of Charle's Law, which states that ***the volume of a given mass of a gas under the same pressure varies directly as its absolute temperature.***

Combining these two laws, we have

$$V \propto \frac{T}{p}, \tag{11.13}$$

when both T and p vary. This result yields

$$\frac{pV}{T} = \text{constant}. \tag{11.14}$$

This constant is generally denoted by R, and is known as ***gas constant***. Thus, we have

$$pV = RT \tag{11.15}$$

This is the ***ideal gas equation*** and the gas which obeys this equation of state is called a ***perfect gas***.

In this equation, p is the absolute pressure, T is the absolute temperature and V is the specific volume. The gas constant R is different for different gases and is determined from $R_u = R \cdot M$, where M is the molar mass, also called molecular weight of the gas and R_u is the universal gas constant. This constant is same for all gases and its value is 8.314 kJ/(kmol. K).

Again, if ρ be the density of the gas at the absolute temperature T, we have

$$\frac{p}{\rho} = RT, \text{ i.e., } p = \rho RT, \tag{11.16}$$

since the mass of the gas is constant.

Also, if ρ_t, ρ_0 are densities of a given mass m of a gas, we have

$$\rho_0 V_0 = \rho_t V_t = m$$

$$\text{i.e.,}\ \ \rho_0 V_0 = V_0 \rho_t (1+\gamma t), \gamma = \frac{1}{273}$$

$$\text{i.e.,}\ \ \rho_0 = \rho_t (1+\gamma t) \tag{11.17}$$

Thus, the thermodynamic state of an ideal gas would be completely known if the pressure and volume of the gas are known. Again, for a given mass of a gas, any one parameter out of the three p, V, T is a function of the other two. These are called ***thermodynamic coordinates***.

11.4 Mixture of Gases

Based on experimental observations, we have the following propositions regarding mixture of gases:

(1) If the pressures of a number of gases at same temperature T and volume V be p_1, p_2, ..., p_n, then the pressure of the mixture of the gases, when combined volume is also V and common temperature is also T, will be $p = p_1 + p_2 + \cdots + p_n$.

Proof. Let the pressure of the second gas be changed from p_2 to p_1. If its volume becomes V_1, then by Boyle's law,

$$p_1 V_1 = p_2 V,\ \text{i.e.,}\ V_1 = \frac{p_2 V}{p_1} \tag{11.18}$$

Similarly, if the pressure of nth gas be changed to p_1, its initial volume V will change to

$$\frac{p_n}{p_1} V.$$

Thus, if the pressure of all gases are p_1, the total volume of the mixture will be

$$V^* = V + \frac{p_2}{p_1} V + \cdots + \frac{p_n}{p_1} V = \frac{V}{p_1}(p_1 + p_2 + \cdots + p_n) \tag{11.19}$$

If now, this volume be changed to V when the pressure of the mixture becomes p, then by Boyle's law,

$$Vp = V^* p_1 = p_1 \cdot \frac{V}{p_1}(p_1 + p_2 + \cdots + p_n)$$

$$\therefore\ p = p_1 + p_2 + \cdots + p_n \tag{11.20}$$

(2) If the pressures of a number of gases at the same temperature T be $p_1, p_2, \ldots, p_n$, while their volumes are $V_1, V_2, \ldots, V_n$, the pressure of the mixture of gases under combined volume V will be $\frac{1}{V}(p_1 V_1 + p_2 V_2 + \cdots + p_n V_n)$, the temperature remaining unaltered.

Proof. Let the pressure of the second gas be changed from p_2 to p_1 while its volume is changed from V_2 to V_2^*. Then by Boyle's law,

$$p_1 V_2^* = p_2 V_2,\ \text{i.e.,}\ V_2^* = \frac{p_2}{p_1} V_2 \tag{11.21}$$

Similarly, for n-th gas, we will have

$$V_n^* = \frac{p_n}{p_1} V_n, \tag{11.22}$$

where V_n^* is the volume of the gas under pressure p_1.

Hence, the total volume V^* of the mixture of gases under pressure p_1 becomes

$$\begin{aligned} V^* &= V_1 + V_2^* + V_3^* + \cdots + V_n^* \\ &= V_1 + \frac{p_2}{p_1}V_2 + \frac{p_3}{p_1}V_3 + \cdots + \frac{p_n}{p_1}V_n \end{aligned}$$

$$\text{i.e., } V^* p_1 = p_1V_1 + p_2V_2 + p_3V_3 + \cdots + p_nV_n \tag{11.23}$$

If the volume V^* be changed to V under same temperature when the pressure becomes p, then by Boyle's law,

$$pV = V^*p_1 = p_1V_1 + p_2V_2 + \cdots + p_nV_n$$

Hence,

$$p = \frac{1}{V}(p_1V_1 + p_2V_2 + \cdots + p_nV_n) \tag{11.24}$$

(3) If $V_1, V_2, \ldots, V_n$ be the volumes of a number of gases at pressures $p_1, p_2, \ldots, p_n$ and absolute temperatures $T_1, T_2, \ldots, T_n$, and if V, p, T denote the volume, pressure and absolute temperature respectively of their mixture, then

$$\frac{pV}{T} = \frac{p_1V_1}{T_1} + \frac{p_2V_2}{T_2} + \cdots + \frac{p_nV_n}{T_n}$$

Proof. Let the volume V_1 and absolute temperature T_1 of the first gas be changed to V and T respectively. Then, its pressure p_1^* is given by

$$\frac{p_1V_1}{T_1} = \frac{p_1^*V}{T}, \text{ i.e., } p_1^* = \frac{T}{V}\cdot\frac{p_1V_1}{T_1} \tag{11.25}$$

We will get similar expressions for other gases, so that for n-th gas we write,

$$p_n^* = \frac{T}{V}\cdot\frac{p_nV_n}{T_n} \tag{11.26}$$

Now, let the gases be mixed in a single vessel of volume V, the temperature remaining constant. Then, by the first proposition, the pressure p of the mixture will be

$$p = p_1^* + p_2^* + \cdots + p_n^* = \frac{T}{V}\left[\frac{p_1V_1}{T_1} + \frac{p_2V_2}{T_2} + \cdots + \frac{p_nV_n}{T_n}\right]$$

Hence,

$$\frac{PV}{T} = \frac{p_1V_1}{T_1} + \frac{p_2V_2}{T_2} + \cdots + \frac{p_nV_n}{T_n} \tag{11.27}$$

11.5 Specific Heats and Heat Capacities

The specific heat of a gas at constant volume is the amount of heat required to raise the temperature of 1 gm of the gas through 1 °C, when its volume is kept constant. It is denoted by c_v.

The specific heat of a gas at constant pressure is the amount of heat required to raise the temperature of 1 gm of the gas through 1 °C, when its pressure is kept constant. It is denoted by c_p.

The ratio of these two quantities c_p, c_v is a pure number which is denoted by γ. Thus

$$\gamma = \frac{c_p}{c_v} \tag{11.28}$$

The value of γ for air is 1.4.

The molar heat capacity of a gas at constant volume is the amount of heat required to raise the temperature of one mole of gas through one degree absolute, when its volume is kept constant. It is denoted by C_v.

Similarly, the molar heat capacity of a gas at constant pressure is the amount of heat required to raise the temperature of one mole of gas through one degree absolute, when its pressure is kept constant. It is denoted by C_p.

If m be the mass of one mole of gas expressed in kg, then

$$c_v = \frac{C_v}{M}, c_p = \frac{C_p}{M}, \text{ i.e., } \frac{C_p}{C_v} = \frac{c_p}{c_v} = \gamma \tag{11.29}$$

When the volume of the gas is kept constant, the heat supplied is used to increase the kinetic energy of the gas molecules resulting in rise of temperature. But no work is done by the gas since the volume does not change.

But in the case when the pressure is kept constant, the volume increases. The heat supplied is used to raise the temperature of the gas as well as in doing work by the gas for the expansion. For increasing the temperature of the gas, the amount of heat required is the same when volume is kept constant. Hence, additional heat energy is required by the gas for doing external work. Thus, C_p is greater than C_v. We will see this later.

11.6 Internal Energy of a Gas

The internal energy stored in a gas is the amount of energy in the form of heat, i.e., the energy due to the vibration of gas molecules. Its magnitude depends on the temperature of the gas and the number of atoms in its molecules.

The internal energy of a unit mass of gas at an absolute temperature is given by

$$U = (T - 273)c_v. \tag{11.30}$$

Thus, if the temperature of a gas changes from T_1 to T_2, then the increase of internal energy per unit mass of gas is

$$U_2 - U_1 = (T_2 - T_1)\, c_v. \tag{11.31}$$

11.7 Enthalpy of Gases

Enthalpy or total heat of a gas at a given temperature is defined as the amount of heat required to raise the temperature of unit mass of the gas from 0 °C to the given temperature, when heated at constant pressure. Hence, for a gas at absolute temperature T, enthalpy is given by

$$H = c_p(T - 273) \tag{11.32}$$

In Eqn (11.32), it is assumed that the value of c_p is constant. However, its value varies with temperature, the increase being very appreciable over a large temperature range.

The total heat can also be found from the sum of the internal energy of the gas and the external work done by the gas due to its expansion. If the total heat is reckoned above absolute zero temperature, and the system is assumed to be open, we write,

$$H = U + pV \tag{11.33}$$

where $U = c_v T$.

11.8 Thermal Conductivity of Gases

The thermal conductivity of a substance is a measure of its capacity to transfer heat by molecular conduction. The steady state heat conduction in one-dimensional heat flow is governed by Fourier's equation

$$q = -kA\frac{dT}{dx}, \tag{11.34}$$

where q = rate of heat conduction in x-direction, A = cross sectional area normal to heat flow, $\frac{dT}{dx}$ = temperature gradient in x-direction, k = thermal conductivity of the medium.

The conduction of heat in gases occurs through the mechanism of diffusion and collision. The molecules of the gas at higher temperature diffuse with those at lower temperature and give up the heat energy after collision.

The thermal conductivity of gases, in general, increases with a rise of temperature. For most gases at atmospheric pressure, the thermal conductivity is independent of pressure.

Since the mechanism of diffusion plays an important role in heat conduction in gases, the thermal conductivity decreases with the increase of molecular weight of the gas. Thus, hydrogen being the lightest gas, it has the highest thermal conductivity.

The values of k for hydrogen at 0 °C and 100 °C are 0.149 kcal/mK and 0.192 kcal/mK respectively.

For air $k = 0.021$ kcal/mK at 0 °C and $k = 0.027$ kcal/mK at 100 °C.

11.9 Thermodynamic Processes

A thermodynamic process in which a change in pressure and volume of a substance occurs without any change in temperature is called an ***isothermal process***.

A thermodynamic process in which a change in pressure and volume of a substance occurs such that no heat is allowed to enter or leave the system is called an ***adiabatic*** process.

A ***reversible*** process is that which can be retraced in the opposite direction so that the substance under the process passes exactly through the same conditions as in the direct process.

Thus, if heat is absorbed in a stage in the direct process, heat will be given out in the reversible process in that stage. Similarly, if work is done ***by*** the substance in the direct process, an equal amount of work will be done ***on*** the substance in the reverse process.

In a reversible process, the work done against internal friction of a gas is zero.

11.10 First Law of Thermodynamics

The law states that, for a closed system, the heat absorbed by a body is equal to the change of the stored energy of the system plus the work done by the body. Thus, if a gas expands due to the heat δQ, which energy appears partly as an increase dU of internal energy and partly as the energy for doing external work, then

$$\delta Q = dU + \delta W \tag{11.35}$$

If the volume of the body expands by an amount dV against a pressure p and external heat, the amount of work done is

$$\delta W = p dV - \delta W_f, \tag{11.36}$$

where δW_f denotes the work done against internal friction of the gas. If the expansion of the gas occurs reversibly, $\delta W_f = 0$.

Thus, the equation of the first law of thermodynamics for a closed system for an ideal gas is written in the form

$$\delta Q = dU + pdV, \tag{11.37}$$

which is valid for an ideal gas.

If we consider an open system when there will be transfer of mass and energy across the boundary, Eqn (11.35) will get modified. We consider a gas of unit mass entering at section 1-1 and leaving at section 2-2 (FIG. 11.2).

Let p_1, p_2 be the pressures and V_1, V_2 be the volumes of the gas at sections 1-1 and 2-2 as shown in the figure. Then the work done on the system to push the mass in $= p_1V_1$ and the work done by the system to push the mass out $= p_2V_2$.

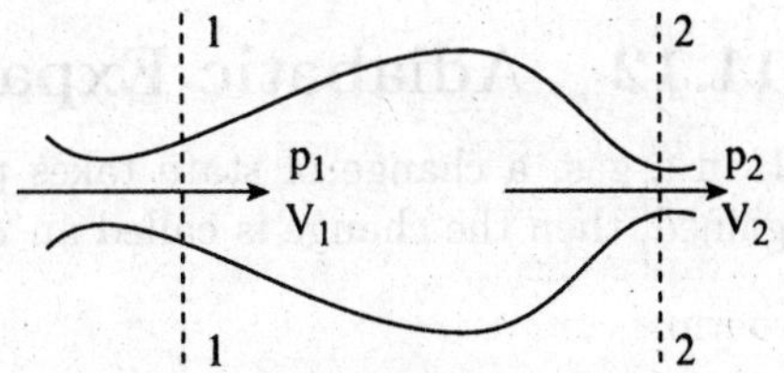

Fig. 11.2

Then the relation (11.35) for steady state of the open system is

$$\begin{aligned} \delta Q &= dU + \delta W + (p_2V_2 - p_1V_1) \\ &= dU + \delta W + d(pV) \\ &= \delta W + dH, \end{aligned} \tag{11.38}$$

where, as defined earlier, H is the enthalpy given by

$$H = U + pV \tag{11.39}$$

so that

$$dH = dU + pdV + Vdp \tag{11.40}$$

11.11 Relation between C_p and C_v

Let a quantity δQ of heat be imparted to a mass of gas, keeping the volume constant. Then we have $dV = 0$.

Thus Eqn (11.35) becomes

$$\delta Q = dU \tag{11.41}$$

But, when volume V is kept constant, we have

$$\delta Q = C_v dT \tag{11.42}$$

Hence, from Eqns (11.41) and (11.42), we have

$$dU = C_v dT \tag{11.43}$$

Thus, we write Eqn (11.37) in view of Eqn (11.43) in the form

$$\delta Q = C_v dT + pdV \tag{11.44}$$

For a perfect gas, we have

$$pV = RT \tag{11.45}$$

so that

$$pdV + Vdp = RdT, \text{ i.e., } pdV = RdT - Vdp \tag{11.46}$$

Substituting this relation into Eqn (11.44), we have

$$\delta Q = C_v dT + RdT - Vdp \tag{11.47}$$

Now, let the amount of gas be imparted a quantity δQ of heat at constant pressure. Then $dp = 0$, and also

$$\delta Q = C_p dT \tag{11.48}$$

With these relations substituted in Eqn (11.47) we have

$$C_p dT = C_v dT + RdT$$

which yields

$$C_p - C_v = R \tag{11.49}$$

Thus, $C_p > C_v$.

11.12 Adiabatic Expansion

If in a gas, a change of state takes place in such a way that neither heat is lost nor heat is gained, then the change is called an ***adiabatic*** change. Thus, in such a process

$$\delta Q = 0 \tag{11.50}$$

In view of this, Eqn (11.44) becomes

$$0 = C_v dT + pdV \tag{11.51}$$

Also, by Eqn (11.46), we have

$$pdV + Vdp = RdT$$

$$\text{i.e., } pdV + Vdp = (C_p - C_v)dT \text{ [by Eqn (11.49)]} \tag{11.52}$$

From Eqn (11.51), we find

$$dT = -\frac{pdV}{C_v} \tag{11.53}$$

Substituting this expression into Eqn (11.52), we get

$$pdV + Vdp = -\frac{C_p - C_v}{C_v} \cdot pdV$$

$$\text{i.e., } pdV + Vdp = \left[1 - \frac{C_p}{C_v}\right] pdV$$

$$\text{i.e., } Vdp + \frac{C_p}{C_v} pdV = 0 \tag{11.54}$$

Denoting

$$\gamma = \frac{C_p}{C_v}, \tag{11.55}$$

we write Eqn (11.54) in the form

$$Vdp + \gamma pdV = 0, \text{ i.e., } \frac{dp}{p} + \gamma \frac{dV}{V} = 0,$$

which on integration yields

$$\log p + \gamma \log V = \text{constant, i.e., } pV^\gamma = \text{constant.} \tag{11.56}$$

This is the equation for a gas in a reversible adiabatic change.

Again, from Eqn (11.45), we have

$$p = \frac{RT}{V} \tag{11.57}$$

Substituting this relation in Eqn (11.56),

$$\frac{RT}{V} \cdot V^{\gamma} = \text{constant, i.e., } TV^{\gamma-1} = \text{constant.} \tag{11.58}$$

This is the equation for a gas in an adiabatic temperature range.

11.13 Work Done in Compressing a Gas

We consider a certain amount of gas in a closed cylinder fitted with a movable piston. Let p_1, V_1, T_1 be the pressure, volume and absolute temperature of the gas in the initial stage. In addition, let there be an external atmospheric pressure Π over the surface of the piston.

Some work would be done on the gas when it is compressed. This process may occur under two conditions — (1) an isothermal condition when the temperature of the gas is maintained constant, and (2) an adiabatic condition when no exchange of heat takes place with the surroundings during the process.

Considering a small change dV in volume of the gas at pressure p, the work done to compress the gas is

$$-(p - \Pi)dV.$$

(1) **Isothermal Process.** Let p_2, V_2 be the pressure and volume of the gas after compression is complete, temperature T_1 remaining unaltered. Then

$$pV = p_1V_1, \text{ i.e., } p = \frac{p_1V_1}{V}. \tag{11.59}$$

Hence, the work done in reducing the volume of the gas from V_1 to V_2 is

$$\begin{aligned} W &= -\int_{V_1}^{V_2} (p - \Pi)dV \\ &= -\int_{V_1}^{V_2} \left[\frac{p_1V_1}{V} - \Pi\right] dV \text{ [using (11.59)]} \\ &= p_1V_1 \log\frac{V_1}{V_2} - \Pi(V_1 - V_2) \end{aligned} \tag{11.60}$$

Again,

$$p_1V_1 = RT_1 \tag{11.61}$$

Hence, Eqn (11.60) can also be written as

$$W = RT_1 \log\frac{V_1}{V_2} - \Pi(V_1 - V_2) \tag{11.62}$$

(2) **Adiabatic Process.** If p_2, V_2, T_2 be the pressure, volume and absolute temperature of the gas after compression, then

$$p_1V_1^{\gamma} = p_2V_2^{\gamma} = pV^{\gamma} = \lambda, \tag{11.63}$$

where λ is a constant.

Then the work done in compressing the gas from volume V_1 to V_2 is

$$W = -\int_{V_1}^{V_2}(p-\Pi)dV = -\int_{V_1}^{V_2}\left[\frac{\lambda}{V^\gamma}-\Pi\right]dV \text{ [using Eqn (11.63)]}$$

$$= -\left[\frac{\lambda}{1-\gamma}\frac{1}{V^{\gamma-1}}-\Pi V\right]_{V_1}^{V_2} = \frac{\lambda}{\gamma-1}\left[\frac{1}{V_2^{\gamma-1}}-\frac{1}{V_1^{\gamma-1}}\right]-\Pi(V_1-V_2)$$

$$= \frac{1}{\gamma-1}\left[\frac{p_2V_2^\gamma}{V_2^{\gamma-1}}-\frac{p_1V_1^\gamma}{V_1^{\gamma-1}}\right]-\Pi(V_1-V_2) \text{ [using Eqn (11.63)]}$$

$$= \frac{1}{\gamma-1}(p_2V_2-p_1V_1)-\Pi(V_1-V_2) \tag{11.64}$$

Again, for the perfect gas

$$p_1V_1 = RT_1,\ p_2V_2 = RT_2 \tag{11.65}$$

Using this relation, we have

$$W = \frac{R}{\gamma-1}(T_2-T_1)-\Pi(V_1-V_2) \tag{11.66}$$

In the adiabatic process, we can have the following interesting relations:

(1) From Eqn (11.63), we have

$$\frac{p_1}{p_2} = \left[\frac{V_2}{V_1}\right]^\gamma \tag{11.67}$$

(2) Since

$$\frac{p_1V_1}{T_1} = \frac{p_2V_2}{T_2},$$

we have

$$\frac{p_1}{p_2} = \frac{V_2}{V_1}\cdot\frac{T_1}{T_2},\ \text{i.e.,}\ \left[\frac{V_2}{V_1}\right]^\gamma = \frac{V_2}{V_1}\cdot\frac{T_1}{T_2},\ \text{[using Eqn (11.67)]}$$

which yields

$$\frac{T_1}{T_2} = \left[\frac{V_2}{V_1}\right]^{\gamma-1} \tag{11.68}$$

(3) Also, we have

$$\frac{T_2}{T_1} = \frac{V_2}{V_1}\cdot\frac{p_2}{p_1} = \left[\frac{T_1}{T_2}\right]^{\frac{1}{\gamma-1}}\cdot\frac{p_2}{p_1} \text{ [using Eqn (11.68)]}$$

$$\text{i.e.,}\ \left[\frac{T_2}{T_1}\right]^{\frac{\gamma}{\gamma-1}} = \frac{p_2}{p_1},\ \text{i.e.,}\ \frac{T_2}{T_1} = \left[\frac{p_2}{p_1}\right]^{\frac{\gamma-1}{\gamma}} \tag{11.69}$$

11.14 Isothermal Atmosphere under Constant Gravity

We consider an isothermal atmosphere under gravity which is assumed to be constant.

Let p = the pressure in the atmosphere at a height z above sea level

ρ = the density of the atmosphere at a height z above sea level

g = acceleration due to gravity assumed to be constant, the height z being measured vertically upwards.

Then the law of pressure is given by

$$dp = -g\rho dz. \tag{11.70}$$

If p_0, ρ_0 denote the pressure and density of the atmosphere, we have, by Boyle's law,

$$p = \lambda\rho, \; p_0 = \lambda\rho_0, \tag{11.71}$$

λ being a constant.

From Eqns (11.70) and (11.71), we have $\dfrac{dp}{p} = -\dfrac{g}{\lambda}dz$, which on integration yields

$$\log p = C - \frac{g}{\lambda}z, \tag{11.72}$$

where C is the constant of integration. Using the condition that $p = p_0$, when $z = 0$, we find

$$\log p_0 = C \tag{11.73}$$

Hence, $\log p = \log p_0 - \dfrac{g}{\lambda}z$ which yields

$$p = p_0 \exp\left(-\frac{g}{\lambda}z\right) \tag{11.74}$$

If, however, we assume that $p = p_0$ at $z = z_0$, then, from Eqn (11.72), we have

$$\log p_0 = C - \frac{g}{\lambda}z_0, \text{ i.e., } C = \log p_0 + \frac{g}{\lambda}z_0 \tag{11.75}$$

Hence,

$$\log p = \log p_0 + \frac{g}{\lambda}z_0 - \frac{g}{\lambda}z, \text{ i.e., } z - z_0 = -\frac{\lambda}{g}\log\frac{p}{p_0} \tag{11.76}$$

This equation may be used to determine the differences of level by observing barometric pressures at two places.

The pressure p is given by

$$\log\frac{p}{p_0} = -\frac{g}{\lambda}(z - z_0), \text{ i.e., } p = p_0 \exp\left[-\frac{g}{\lambda}(z - z_0)\right] \tag{11.77}$$

In view of the relations (11.71), Eqns (11.74) and (11.77) can be expressed in terms of the densities ρ, ρ_0. Thus

$$\rho = \rho_0 \exp\left[-\frac{g}{\lambda}(z - z_0)\right] \tag{11.78}$$

and when $z_0 = 0$,

$$\rho = \rho_0 \exp\left(-\frac{g}{\lambda}z\right) \tag{11.79}$$

Now, suppose H to be the height of a homogeneous atmosphere of density ρ_0 that would produce a pressure p_0. Then

$$p_0 = \lambda\rho_0 = g\rho_0 H, \text{ i.e., } H = \frac{\lambda}{g} = \text{constant}. \tag{11.80}$$

Then Eqns (11.74) and (11.79) become

$$\left.\begin{aligned} p &= p_0 \exp\left(-\frac{z}{H}\right) \\ \rho &= \rho_0 \exp\left(-\frac{z}{H}\right) \end{aligned}\right\} \tag{11.81}$$

Similarly, Eqns (11.77) and (11.78) can be written in the forms:

$$\left.\begin{aligned} p &= p_0 \exp\left[-\frac{z-z_0}{H}\right] \\ \rho &= \rho_0 \exp\left[-\frac{z-z_0}{H}\right] \end{aligned}\right\} \tag{11.82}$$

11.15 Isothermal Atmosphere under Variable Gravity

We consider the atmosphere to be isothermal and the earth to be a sphere of radius R. We also take into account the variation of gravity due to change of distance from the centre of the earth.

We consider a point in the atmosphere at a height z from sea level. If g be the acceleration due to gravity at sea level, its value at the height z will be

$$g_z = g\left[\frac{R}{R+z}\right]^2 \tag{11.83}$$

Let p, ρ be the pressure and density of the atmosphere at a height z from sea level, z being measured vertically upwards.

Then the pressure equation is

$$dp = -g_z \rho dz = -g\left[\frac{R}{R+z}\right]^2 \rho dz \tag{11.84}$$

For isothermal atmosphere, we have

$$p = \lambda\rho, \ \lambda = \text{constant.} \tag{11.85}$$

Hence, from Eqns (11.84) and (11.85), we have

$$\frac{dp}{p} = -\frac{g}{\lambda}\left[\frac{R}{R+z}\right]^2 dz \tag{11.86}$$

Integrating, we have

$$\log p = C + \frac{g}{\lambda}\frac{R^2}{R+z} \tag{11.87}$$

Let, at the sea level where $z = 0$, $p = p_0$. Then, using this condition, we get

$$\log p_0 = C + \frac{gR}{\lambda}, \text{ i.e., } C = \log p_0 - \frac{gR}{\lambda} \tag{11.88}$$

Substituting this value of C in Eqn (11.87),

$$\log p = \log p_0 - \frac{gR}{\lambda} + \frac{g}{\lambda}\frac{R^2}{R+z}$$

$$\text{i.e., } \log p - \log p_0 = -\frac{g}{\lambda}\frac{Rz}{R+z}$$

$$\therefore\ p = p_0 \exp\left[-\frac{g}{\lambda}\frac{Rz}{R+z}\right] \tag{11.89}$$

Similarly, if ρ_0 be the density of the atmosphere at the sea level, we have $p_0 = \lambda\rho_0$. Hence,

$$\rho = \rho_0 \exp\left[-\frac{g}{\lambda}\frac{Rz}{R+z}\right]. \tag{11.90}$$

11.16 Atmosphere under Variable Temperature

We assume that the temperature in the atmosphere is not constant but varies at different heights from sea level. In this case, the relation between the pressure p and density ρ in the atmosphere is given by

$$p = \lambda\rho(1+\gamma t), \tag{11.91}$$

where γ is the coefficient of volume expansion of atmosphere.

Then, the equation of pressure

$$dp = -g\rho dz \tag{11.92}$$

is supplimented by the relation (11.91). Hence, we have

$$\frac{dp}{p} = -\frac{g}{\lambda}\frac{1}{(1+\gamma t)}dz \tag{11.93}$$

Let τ, τ_0 be the temperatures in the atmosphere at heights z, z_0 from the sea level. We assume that, within this region, the temperature is constant having the mean value of

$$t = \frac{1}{2}(\tau + \tau_0) \tag{11.94}$$

so that

$$1 + \gamma t = 1 + \frac{1}{2}\gamma(\tau + \tau_0) = \theta \text{ (say)} \tag{11.95}$$

Then we write Eqn (11.93) in the form:

$$\frac{dp}{p} = -\frac{g}{\lambda\theta}dz \tag{11.96}$$

Integrating,

$$\log p = -\frac{gz}{\lambda\theta} + C \tag{11.97}$$

Assuming $p = p_0$ at $z = z_0$, we find

$$C = \log p_0 + \frac{gz_0}{\lambda\theta} \tag{11.98}$$

so that

$$\log p - \log p_0 = -\frac{g}{\lambda\theta}(z - z_0), \text{ i.e., } z - z_0 = -\frac{\lambda\theta}{g}\log\frac{p}{p_0} \tag{11.99}$$

This determines the differences of levels of two places in the atmosphere in terms of the barometric pressures.

Again, we have from Eqn (11.99),

$$\log\frac{p}{p_0} = -\frac{g}{\lambda\theta}(z - z_0), \text{ i.e., } p = p_0 \exp\left[-\frac{g}{\lambda\theta}(z - z_0)\right], \tag{11.100}$$

which determines the pressure at height z from sea level.

We can, however, look into this problem in a more rigorous way.

For a perfect gas, the equation of state relating density to pressure and temperature is

$$p = \rho RT \tag{11.101}$$

Then, from Eqns (11.92) and (11.101), we have

$$\frac{dp}{p} = -\frac{g}{RT}dz \tag{11.102}$$

In the earth's atmosphere, the layer of air from the surface up to 11 km altitude is called ***troposphere***. The temperature-altitude variation in this zone has been found to be linear and may be expressed as

$$T = T_0 - \xi z, \tag{11.103}$$

where $T_0 = 288\,\text{K}$ is the standard temperature at sea level and ξ is the ***temperature gradient*** or ***lapse rate***. Its value is 6.5 K/km in the international standard of atmosphere.

Substituting for T from Eqn (11.103) into Eqn (11.102),

$$\frac{dp}{p} = -\frac{g}{R}\frac{dz}{T_0 - \xi z} \tag{11.104}$$

Integrating, we have

$$\log p = \frac{g}{\xi R}\log(T_0 - \xi z) + C, \tag{11.105}$$

where C is the constant of integration.

At sea level, $p = p_0, T = T_0$ and $z = 0$. Hence,

$$\log p_0 = \frac{g}{\xi R}\log T_0 + C, \text{ i.e., } C = \log p_0 - \frac{g}{\xi R}\log T_0 \tag{11.106}$$

Substituting this value of C into Eqn (11.105), we have

$$\log p = \frac{g}{\xi R}\log(T_0 - \xi z) + \log p_0 - \frac{g}{\xi R}\log T_0$$

$$\text{i.e., } \log\frac{p}{p_0} = \frac{g}{\xi R}\log\left[1 - \frac{\xi z}{T_0}\right]$$

Hence,

$$\frac{p}{p_0} = \left[1 - \frac{\xi z}{T_0}\right]^{\frac{g}{\xi R}}, \text{ i.e., } p = p_0\left[1 - \frac{\xi z}{T_0}\right]^{\frac{g}{\xi R}} \tag{11.107}$$

This determines the pressure at any height in the troposphere from sea level.

Again, we can write Eqn (11.107) in the form:

$$\left[\frac{p}{p_0}\right]^{\frac{\xi R}{g}} = 1 - \frac{\xi z}{T_0}, \text{ i.e., } z = \frac{T_0}{\xi}\left[1 - \left[\frac{p}{p_0}\right]^{\frac{\xi R}{g}}\right] \tag{11.108}$$

This determines the height from sea level of any place in the troposphere when its barometer reading is known.

The layer of air between 11 km and 20 km above the sea level is known as ***stratosphere***. The temperature in this zone is approximately constant at $-56\,^\circ\text{C}$. The variation of pressure with height, therefore, will be as discussed in section 11.15.

11.17 Convective Equilibrium

The concept of convective equilibrium was introduced as an alternative hypothesis to the constant temperature atmosphere. It implies that if equal masses of air at different levels are interchanged, they will automatically expand or contract according to the local conditions without any loss or gain of heat, i.e., adiabatically. They would merely interchange pressure, density and temperature so that they would always remain in thermal equilibrium in the new surroundings.

In this case, the equation for change of state is

$$pV^{\gamma} = \text{constant.} \tag{11.109}$$

We also have

$$pV = RT, \text{ i.e., } V = \frac{RT}{p} \tag{11.110}$$

With this relation in Eqn (11.109), we have

$$p\left[\frac{RT}{p}\right]^{\gamma} = \text{constant}$$

$$\text{i.e., } p^{1-\gamma}(RT)^{\gamma} = \text{constant}$$

$$\text{i.e., } \frac{p}{T^{\gamma/(\gamma-1)}} = \text{constant} = A \text{ (say)} \tag{11.111}$$

Taking log on both sides,

$$\log p = \frac{\gamma}{\gamma - 1}\log T + \log A \tag{11.112}$$

Differentiating this equation,

$$\frac{dp}{p} = \frac{\gamma}{\gamma - 1}\frac{dT}{T} \tag{11.113}$$

Now, the pressure equation at any height z above see level, where p and ρ are pressure and density, is

$$dp = -g\rho dz \tag{11.114}$$

z being measured upwards and g is assumed to be constant.
Since

$$p = \rho RT, \rho = \frac{p}{RT} \tag{11.115}$$

From Eqns (11.114) and (11.115), we have

$$\frac{dp}{dz} = -g\rho = -\frac{gp}{RT}, \text{ i.e., } \frac{dp}{p} = -\frac{g}{RT}dz \tag{11.116}$$

Equating the two relations (11.113) and (11.116),

$$\frac{\gamma}{\gamma - 1}\frac{dT}{T} = -\frac{g}{RT}dz, \text{ i.e., } \frac{dT}{dz} = -\frac{\gamma - 1}{\gamma}\frac{g}{R}. \tag{11.117}$$

The right hand side of Eqn (11.117) is constant which implies that the temperature in the atmosphere falls uniformly. This rate dT/dz, as we have mentioned earlier, is the lapse rate.
Integrating Eqn (11.117), we have

$$T = C - \frac{\gamma - 1}{\gamma}\frac{gz}{R}, \tag{11.118}$$

where C is the constant of integration. If at sea level, i.e., for $z = 0$, the temperature is T_0, then, from Eqn (11.118),

$$T_0 = C \tag{11.119}$$

so that

$$T = T_0\left[1 - \frac{\gamma - 1}{\gamma}\frac{gz}{RT_0}\right] \tag{11.120}$$

If p_0, ρ_0 be the pressure and density at the sea level and H the height of the homogeneous atmosphere, then

$$p_0 = \rho_0 R T_0 = H\rho_0 g, \text{ i.e., } RT_0 = Hg \tag{11.121}$$

Using this relation in Eqn (11.120), we have

$$T = T_0\left[1 - \frac{\gamma - 1}{\gamma} \cdot \frac{z}{H}\right] \tag{11.122}$$

Thus, ***if T be the absolute temperature in the atmosphere at height z above sea level and T_0 its value at sea level in convective equilibrium of temperature and satisfying the law pV^γ = constant, then***

$$\boldsymbol{T = T_0\left[1 - \frac{\gamma - 1}{\gamma} \cdot \frac{z}{H}\right],}$$

where H is the height of homogeneous atmosphere and the gravity is assumed to be constant.

Let us now assume that gravity is not constant. If its value at sea level be g, then its value at a height z above sea level is

$$g_z = \frac{r^2}{(r+z)^2}g, \tag{11.123}$$

where r is the radius of the earth.

Then the pressure equation becomes

$$dp = -\frac{r^2}{(r+z)^2}g\rho dz, \tag{11.124}$$

where p, ρ denote respectively pressure and density at a height z above sea level.

From Eqn (11.124), we have

$$\frac{dp}{dz} = -\frac{r^2}{(r+z)^2}g\rho = -\frac{r^2}{(r+z)^2}\frac{gp}{RT} \quad \text{[using Eqn (11.115)]}$$

$$\therefore \frac{dp}{p} = -\frac{r^2}{(r+z)^2}\frac{g}{RT}dz \tag{11.125}$$

Equating the relations (11.113) and (11.125),

$$\frac{\gamma}{\gamma - 1}\frac{dT}{T} = \frac{r^2}{(r+z)^2}\frac{g}{RT}dz$$

$$\therefore \frac{dT}{dz} = -\frac{\gamma - 1}{\gamma}\frac{r^2 g}{R(r+z)^2} \tag{11.126}$$

The temperature gradient in this case is not constant, which implies that the temperature in the atmosphere decreases non-uniformly along with the increase in height from sea level.

Integrating Eqn (11.126),

$$T = C + \frac{\gamma - 1}{\gamma}\frac{r^2 g}{R(r+z)}, \tag{11.127}$$

where C is the constant of integration.

If, at sea level where $z = 0$, $T = T_0$, then

$$T_0 = C + \frac{\gamma - 1}{R\gamma} rg, \text{ i.e., } C = T_0 - \frac{\gamma - 1}{R\gamma} rg \tag{11.128}$$

Substituting this value of C in Eqn (11.127),

$$T = T_0 - \frac{\gamma - 1}{R\gamma} rg + \frac{\gamma - 1}{R\gamma} \frac{r^2 g}{(r+z)}$$

$$= T_0 - \frac{\gamma - 1}{R\gamma} \frac{rgz}{r+z} = T_0 \left[1 - \frac{\gamma - 1}{\gamma R T_0} \frac{rgz}{r+z} \right] \tag{11.129}$$

If H be the height of the homogeneous atmosphere, then

$$p_0 = \rho_0 R T_0 = g\rho_0 H, \text{ i.e., } RT_0 = gH \tag{11.130}$$

Substituting this value in Eqn (11.129), we have

$$T = T_0 \left[1 - \frac{\gamma - 1}{H\gamma} \frac{rz}{r+z} \right] \tag{11.131}$$

Thus, ***if T be the absolute temperature at the height z and T_0 its value at the sea level in the atmosphere in convective equilibrium satisfying the law pV^γ = constant, then***

$$\boldsymbol{T = T_0 \left[1 - \frac{\gamma - 1}{H\gamma} \frac{rz}{r+z} \right]},$$

where r is the radius of the earth and H is the height of the homogeneous atmosphere, taking into account the variation of gravity along the height from the surface of the earth.

11.18 Centre of Pressure Exposed to Atmosphere

We consider a lamina exposed to the action of atmosphere (FIG. 11.3). Let z_1, z_2 be the heights above sea level of the bottom and top edges of the plane surface.

The density at any height z from sea level is given by

$$\rho(z) = \rho_0 e^{-\frac{gz}{\lambda}} \tag{11.132}$$

Considering an elementary area dS at a height z from the sea level, we have the height of the centre of pressure of the lamina as

$$\eta = \frac{\int z^2 \rho(z) dS}{\int z\rho(z) dS} = \frac{\int z^2 e^{-\frac{gz}{\lambda}} dS}{\int z e^{-\frac{gz}{\lambda}} dS} \tag{11.133}$$

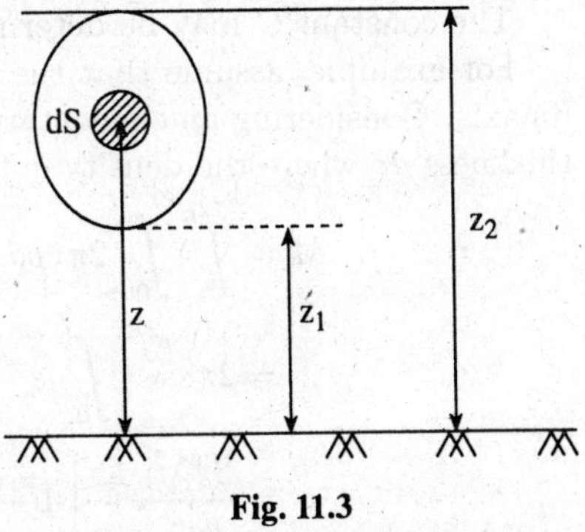

Fig. 11.3

Let in particular, the lamina be a rectangle of height h and breadth b held vertically in the atmosphere at a height H above sea level.

Then, considering an elementary strip of area bdz at a height z above sea level (FIG. 11.4), the height of the centre of pressure

$$\eta = \frac{\int_H^{H+h} b\rho_0 e^{-\frac{gz}{\lambda}} z dz}{\int_H^{H+h} b\rho_0 e^{-\frac{gz}{\lambda}} dz} \tag{11.134}$$

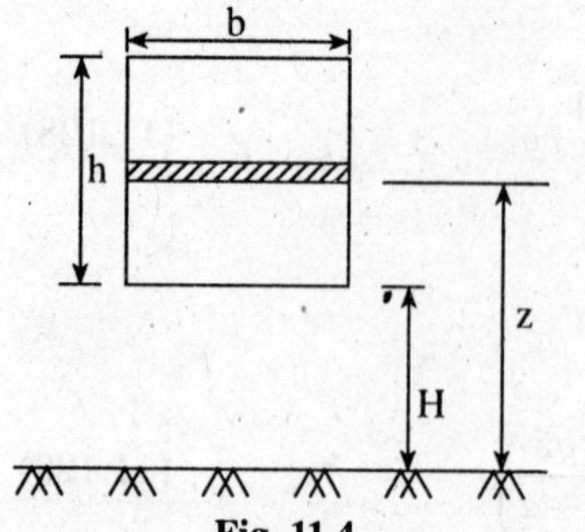

Fig. 11.4

Integrating, we find

$$\eta = \frac{\lambda}{g} + \frac{\left[(H+h)e^{-\frac{gh}{\lambda}} - H\right]}{\left[e^{-\frac{gh}{\lambda}} - 1\right]} \tag{11.135}$$

If the lamina is held at the sea level, then $H = 0$. Hence,

$$\eta = \frac{\lambda}{g} + \frac{he^{-\frac{gh}{\lambda}}}{e^{-\frac{gh}{\lambda}} - 1} = \frac{\lambda}{g} - \frac{h}{e^{\frac{gh}{\lambda}} - 1} \tag{11.136}$$

11.19 Equilibrium of Revolving Gases

We consider a mass of gas under the action of gravity enclosed in a vessel which rotates uniformly about a vertical axis with angular velocity ω. Then the equation of equilibrium is

$$dp = \rho\{\omega^2(xdx + ydy) - gdz\}, \tag{11.137}$$

where

$$p = \lambda\rho, \text{ i.e., } dp = \lambda d\rho \tag{11.138}$$

Hence,

$$\frac{\lambda d\rho}{\rho} = \omega^2(xdx + ydy) - gdz \tag{11.139}$$

Integrating,

$$\lambda \log \rho = \frac{1}{2}\omega^2(x^2 + y^2) - gz + C, \tag{11.140}$$

where C is the constant of integration.

Eqn (11.140) implies that the surfaces of equal pressure and density are paraboloids.

From Eqn (11.140), we write,

$$\rho = e^{\frac{C}{\lambda}} \cdot \exp\left[\frac{\omega^2 r^2 - 2gz}{2\lambda}\right] \tag{11.141}$$

The constant C may be determined if the whole mass M, of the gas be given.

For example, assume that the vessel is a cylinder of radius a and height h rotating about its axis. Considering an elementary ring of gas of radius r, vertical thickness dz and horizontal thickness dr where the density is the same given by Eqn (11.141), we have

$$\begin{aligned} M &= \int_0^h \int_0^a 2\pi r\rho dz dr = \int_0^h \int_0^a 2\pi r e^{\frac{C}{\lambda}} \cdot \exp\left[\frac{\omega^2 r^2 - 2gz}{2\lambda}\right] dz dr \\ &= 2\pi e^{\frac{C}{\lambda}} \left[\int_0^h e^{-\frac{gz}{\lambda}} dz \int_0^a r e^{\frac{\omega^2 r^2}{2\lambda}} dr\right] \\ &= \frac{2\pi\lambda^2}{g\omega^2} e^{\frac{C}{\lambda}} \left[1 - e^{-\frac{gh}{\lambda}}\right]\left[e^{\frac{\omega^2 a^2}{2\lambda}} - 1\right] \end{aligned} \tag{11.142}$$

Hence,

$$e^{\frac{C}{\lambda}} = \frac{g\omega^2 M}{2\pi\lambda^2} \cdot \frac{1}{(1 - e^{-gh/\lambda})(e^{\omega^2 a^2/2\lambda} - 1)}, \tag{11.143}$$

which determines the constant C.

11.20 Rotation of Gravitating Gases

We consider a mass of gas, the particles of which attract each other with a force varying directly as the distance, rotating uniformly about an axis through its centre of mass.

We assume that the gas remains under the action of gravity.

We consider a gas particle about a point $P(x, y, z)$ with reference to a Cartesian coordinate system with the origin at the centre of mass. If μ be a measure of the whole mass of the gas, then the components of attraction forces parallel to the axes on the particle are

$$\left.\begin{aligned} X &= -\mu x \\ Y &= -\mu y \\ Z &= -\mu z \end{aligned}\right\} \tag{11.144}$$

The equation of equilibrium is

$$dp = \rho\left\{\left[\omega^2 x - \mu x\right] dx + \left[\omega^2 y - \mu y\right] dy - \mu z dz\right\}, \tag{11.145}$$

where ω is the uniform angular velocity and z-axis is the axis of rotation.

The pressure-density relation of the gas is

$$p = \lambda\rho, \tag{11.146}$$

λ being a constant. Hence, we have

$$\lambda\frac{dp}{p} = \{(\omega^2 - \mu)xdx + (\omega^2 - \mu)ydy - \mu zdz\} \tag{11.147}$$

Integrating, we have

$$\lambda \log p = C + \frac{1}{2}\left\{(\omega^2 - \mu)(x^2 + y^2) - \mu z^2\right\}, \tag{11.148}$$

where C is the constant of integration.

Surfaces of constant pressure $p = A$ (say) are given by

$$\left[1 - \frac{\omega^2}{\mu}\right](x^2 + y^2) + z^2 = D, \tag{11.149}$$

where we have denoted

$$D = \frac{2}{\mu}(\lambda \log A - C) \tag{11.150}$$

From Eqn (11.149), we observe that:

(1) when ω is very small, the surfaces of constant pressure and constant density are nearly spherical. As ω^2 increases from 0 to $\omega' < \mu$, the spheroidal surface becomes more oblate.

(2) when $\omega^2 = \mu$, the surfaces of constant pressure and constant density are given by

$$z = \pm\sqrt{D} = \pm\left[\frac{2}{\mu}(\lambda \log A - C)\right]^{\frac{1}{2}}, \tag{11.151}$$

which consists of two planes.

(3) when $\omega^2 > \mu$, say, $\omega^2 = \mu + \delta$, then the surfaces of constant pressure and constant density are given by

$$(x^2 + y^2) - \frac{\mu}{\delta}z^2 = -\frac{\mu}{\delta}D, \tag{11.152}$$

which represents hyperboloid of two sheets.

EXAMPLE 11.1 *A cylindrical vessel of depth h and section α is maintained at constant temperature. If ρ_0 and ρ_1 are the densities of air at the top and bottom, show that the total amount of air contained is* $\dfrac{h\alpha(\rho_1 - \rho_0)}{\log \rho_1 - \log \rho_0}$.

Solution. We measure the depth z from the top surface of the cylinder, so that we are given

$$\left.\begin{aligned} \rho &= \rho_0, \text{ when } z = 0 \\ \rho &= \rho_1, \text{ when } z = h \end{aligned}\right\}, \tag{1}$$

where ρ denotes the density of air.

The pressure equation is

$$dp = g\rho dz \tag{2}$$

We also have

$$p = \lambda\rho, \text{ i.e., } dp = \lambda d\rho \tag{3}$$

From (2) and (3), we have

$$\lambda \frac{d\rho}{\rho} = g dz \tag{4}$$

Integrating, we have

$$\lambda \log \rho = gz + C, \tag{5}$$

where C is the constant of integration.

Using the conditions (1) in (5), we have

$$\left.\begin{aligned} \lambda \log \rho_0 &= C \\ \lambda \log \rho_1 &= C + gh \end{aligned}\right\} \tag{6}$$

Eliminating C, we find

$$\lambda \log \rho_1 = \lambda \log \rho_0 + gh, \text{ i.e., } \lambda = \frac{gh}{\log \rho_1 - \log \rho_0} \tag{7}$$

Hence, from (5) and (6), we have

$$\lambda \log \rho = \lambda \log \rho_0 + gz, \text{ i.e., } \rho = \rho_0 e^{gz/\lambda} \tag{8}$$

Mass of air contained in the cylinder is

$$\begin{aligned} &= \int_0^h \rho\alpha dz = \alpha \int_0^h \rho_0 e^{gz/\lambda} dz \\ &= \alpha\rho_0 \frac{\lambda}{g}\left[e^{gz/\lambda}\right]_0^h = \alpha\rho_0 \frac{\lambda}{g}\left(e^{gh/\lambda} - 1\right) \end{aligned} \tag{9}$$

Since $\rho = \rho_1$ at $z = h$, we get from (8),

$$\rho_1 = \rho_0 e^{gh/\lambda} \tag{10}$$

Using this relation in (9), we get the mass of air in the vessel

$$\begin{aligned} &= \alpha\rho_0 \frac{\lambda}{g}\left[\frac{\rho_1}{\rho_0} - 1\right] \\ &= \frac{\alpha}{g}(\rho_1 - \rho_0) \cdot \frac{gh}{(\log \rho_1 - \log \rho_0)}, \text{ using (7)} \\ &= \frac{h\alpha(\rho_1 - \rho_0)}{\log \rho_1 - \log \rho_0}. \quad \textbf{(Proved)} \end{aligned} \tag{11}$$

EXAMPLE 11.2 *If near the earth's surface gravity be assumed to be constant and the absolute temperature in the atmosphere to be given by* $T = T_0\left(1 - \dfrac{z}{nH}\right)$, *where* H *is the height of the homogeneous atmosphere, show that the pressure in the atmosphere will be given by*

$$p = p_0\left(1 - \frac{z}{nH}\right)^n.$$

Solution. We have the pressure equation,

$$dp = -g\rho dz, \tag{1}$$

where p, ρ are the pressure and density at any height z from earth's surface.

We also have

$$p = \rho RT \tag{2}$$

From (1) and (2),

$$\frac{dp}{p} = -\frac{gdz}{RT} = -\frac{g}{RT_0}\frac{dz}{\left[1 - \frac{z}{nH}\right]} \tag{3}$$

Integrating,

$$\log p = \frac{gnH}{RT_0}\log\left(1 - \frac{z}{nH}\right) + C \tag{4}$$

Let p_0, ρ_0 be the pressure and density of the atmosphere at $z = 0$. Then we have

$$p = p_0, \text{ when } z = 0 \tag{5}$$

and also

$$p_0 = \rho_0 gH = \rho_0 RT_0, \tag{6}$$

where H is the height of homogeneous atmosphere.

From (4) and (5), we get

$$C = \log p_0 \tag{7}$$

so that (4) becomes

$$\log\frac{p}{p_0} = \frac{gH}{RT_0}\log\left(1 - \frac{z}{nH}\right)^n \tag{8}$$

From (6), we have

$$RT_0 = gH$$

so that (8) becomes

$$\log\frac{p}{p_0} = \log\left(1 - \frac{z}{nH}\right)^n$$

$$\therefore \frac{p}{p_0} = \left(1 - \frac{z}{nH}\right)^n. \quad \textbf{(Proved)}$$

EXAMPLE 11.3 *In a vertical column of a perfect gas, the pressure and absolute temperature at any height* z *are* p *and* T. *Prove that*

$$z = \frac{p_0}{\rho_0 g T_0}\int_p^{p_0}\frac{T}{p}dp,$$

where p_0, ρ_0, T_0 *are pressure, density and temperature of the gas at the bottom.*

Solution. We have the pressure equation:

$$dp = -g\rho dz \tag{1}$$

We also have, for a perfect gas,

$$p = \rho RT \tag{2}$$

$$\therefore p_0 = \rho_0 RT_0 \tag{3}$$

From (1) and (2), we get

$$\frac{dp}{p} = -\frac{g}{RT}dz, \text{ i.e., } \int_0^z dz = -\frac{R}{g}\int_{p_0}^{p}\frac{T\,dp}{p} \tag{4}$$

From (3), we have

$$R = \frac{p_0}{\rho_0 T_0} \tag{5}$$

Putting this value in (4), we get

$$z = \frac{p_0}{\rho_0 T_0 g}\int_p^{p_0}\frac{T\,dp}{p}. \quad \textbf{(Proved)}$$

EXAMPLE 11.4 *A gas satisfying Boyle's law $p = k\rho$ is acted on by forces*

$$X = -\frac{y}{x^2+y^2},\ Y = \frac{x}{x^2+y^2}$$

Show that the density varies as $e^{\theta/k}$ where $\tan\theta = \dfrac{y}{x}$.

Solution. If p, ρ denote the pressure and density of the gas at any point, the equation of equilibrium is

$$dp = \rho(X\,dx + Y\,dy) \tag{1}$$

Putting here the given forces

$$X = -\frac{y}{x^2+y^2},\ Y = \frac{x}{x^2+y^2}, \tag{2}$$

we write (1) as

$$dp = \frac{\rho}{x^2+y^2}(x\,dy - y\,dx) = \frac{\rho x^2}{x^2+y^2}d\left(\frac{y}{x}\right), \text{ i.e., } dp = \frac{\rho}{1+\left(\frac{y}{x}\right)^2}d\left(\frac{y}{x}\right) \tag{3}$$

We also have

$$p = k\rho, \text{ i.e., } dp = k\,d\rho \tag{4}$$

From (3) and (4), we get

$$k\frac{d\rho}{\rho} = \frac{d\left(\frac{y}{x}\right)}{1+\left(\frac{y}{x}\right)^2} \tag{5}$$

Integrating,

$$k\log\rho = \tan^{-1}\left(\frac{y}{x}\right) + k\log C, \tag{6}$$

where $k\log C$ is the constant of integration.

Since

$$\tan\theta = \frac{y}{x}, \text{ i.e., } \theta = \tan^{-1}\left(\frac{y}{x}\right), \tag{7}$$

we get from (6),

$$\log\frac{\rho}{C} = \frac{\theta}{k}, \text{ i.e., } \frac{\rho}{C} = e^{\frac{\theta}{k}}$$

$$\therefore \rho = Ce^{\frac{\theta}{k}} \tag{8}$$

Equation (8) implies that the density ρ of the gas varies as $e^{\frac{\theta}{k}}$.
Hence proved.

EXAMPLE 11.5 *If the absolute temperature T at a height z is a given function $f(z)$ of the height, show that the ratio of pressures at two heights z_1 and z_2 is given by*

$$\log\frac{p_2}{p_1} = -\frac{g}{R}\int_{z_1}^{z_2}\frac{dz}{f(z)},$$

R being the constant in the equation $p = \rho RT$.

Solution. We are given that

$$T = f(z) \tag{1}$$

Then the perfect gas relation $p = \rho RT$ becomes

$$p = \rho R f(z), \tag{2}$$

where p, ρ, T denote the pressure, density and absolute temperature at a height z above sea level.

The pressure equation is

$$dp = -g\rho dz \tag{3}$$

From (2) and (3), we have

$$\frac{dp}{p} = -\frac{g}{R}\frac{dz}{f(z)} \tag{4}$$

Let p_1, p_2 be the pressures at heights z_1, z_2 respectively above sea level. Then we have

$$\left.\begin{aligned} p = p_1, &\text{ when } z = z_1 \\ p = p_2, &\text{ when } z = z_2 \end{aligned}\right\} \tag{5}$$

Integrating (4) within the limits in (5), we get

$$\left[\log p\right]_{p_1}^{p_2} = -\frac{g}{R}\int_{z_1}^{z_2}\frac{dz}{f(z)}$$

$$\therefore \log\frac{p_2}{p_1} = -\frac{g}{R}\int_{z_1}^{z_2}\frac{dz}{f(z)}. \quad \textbf{(Proved)}$$

EXAMPLE 11.6 *In an atmosphere at rest under gravity, the height H of the equivalent homogeneous atmosphere above any point at height z, measured in terms of density at height z, is given as a function of z. Show that the pressure p at height h is given by*

$$p = p_0 \exp\left(-\int_0^h \frac{dz}{H}\right), \tag{1}$$

where p_0 is the pressure at $z = 0$.

Solution. Let p, ρ denote the pressure and density at a height z above sea level. Then the pressure equation is

$$dp = -g\rho dz \tag{1}$$

Let H be the equivalent height of the homogeneous atmosphere at height z. Then, according to the problem, the pressure of the homogeneous atmosphere at any point at a height z above sea level is given by

$$p = H\rho g \tag{2}$$

From (1) and (2), we have

$$\frac{dp}{p} = -\frac{dz}{H} \tag{3}$$

Integrating,

$$\log p\Big|_{p_0}^{p} = -\int_0^h \frac{dz}{H}, \text{ i.e., } \log\frac{p}{p_0} = -\int_0^h \frac{dz}{H}.$$

Hence,

$$p = p_0 \exp\left(-\int_0^h \frac{dz}{H}\right). \quad \textbf{(Proved)}$$

EXAMPLE 11.7 *A thin uniform circular tube of radius a contains air and rotates with uniform angular velocity ω about an axis in its plane distant c from the centre. Find the pressure at any point neglecting the weight of the air. If $c < a$ and if P and P′ are greatest and least pressures, prove that* $\log\dfrac{P}{P'} = \dfrac{\omega^2}{2k}(a+c)^2$.

Solution. AB is the axis of rotation which is at a distance c from PQ, the axis of the tube (FIG. 11.5).

At any point distant r from the axis of rotation, equation of equilibrium is

$$dp = \rho\omega^2 r dr, \tag{1}$$

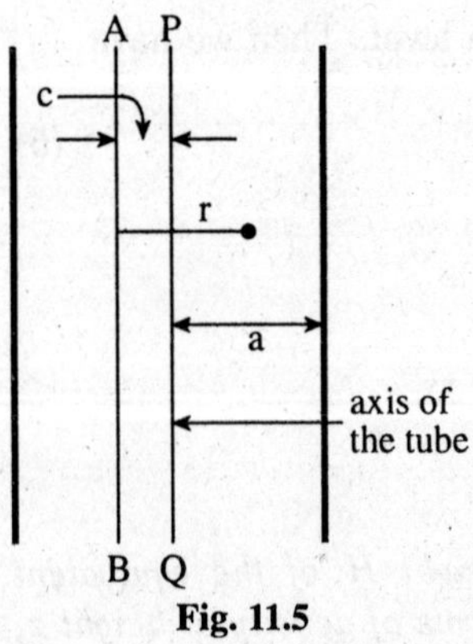

Fig. 11.5

where p, ρ denote the pressure and density at the point.

We also have the perfect gas relation

$$p = k\rho \tag{2}$$

From (1) and (2), we have

$$\frac{dp}{p} = \frac{\omega^2}{k} r dr \tag{3}$$

Integrating, we get

$$\log p = \frac{\omega^2}{2k} r^2 + C, \tag{4}$$

where C is the constant of integration.

From (4), it is obvious that the pressure is greatest at $r = a + c$ and least at $r = 0$. Hence, we are given that

$$\left.\begin{aligned} p &= P, \text{ when } r = a + c \\ p &= P', \text{ when } r = 0 \end{aligned}\right\} \tag{5}$$

Using these conditions in (4), we have

$$\left.\begin{aligned} \log P &= \frac{\omega^2}{2k}(a+c)^2 + C \\ \log P' &= C \end{aligned}\right\} \tag{6}$$

Eleminating C from relations (6), we have

$$\log \frac{P}{P'} = \frac{\omega^2}{2k}(a+c)^2. \quad \textbf{(Proved)}$$

EXAMPLE 11.8 *A hollow gas-tight sphere containing hydrogen requires a force mg to prevent it from rising when the lowest point touches the ground; the total mass of the sphere and hydrogen is M. Show that the sphere can float in equilibrium with its lowest point at a height h above the ground, where* $h = \frac{\lambda}{g}\log\frac{M+m}{M}$, *$\lambda$ being the ratio of the pressure of the atmosphere to its density.*

Solution. Let V = volume of the sphere, ρ_0 = density of hydrogen at the ground, ρ_1 = density of hydrogen at height h.

When the sphere touches the ground, the forces acting are: (1) downward weight Mg of the sphere + hydrogen, (2) downward force mg applied and (3) upward force exerted by the hydrogen which is equal to $V\rho_0 g$.

Hence, for equilibrium, we have

$$V\rho_0 g = Mg + mg, \text{ i.e., } V\rho_0 = M + m \tag{1}$$

When the sphere floats at a height h, we have by the condition of floatation,

$$V\rho_1 g = Mg, \text{ i.e., } V\rho_1 = M \tag{2}$$

From (1) and (2), we have

$$\frac{\rho_0}{\rho_1} = \frac{M+m}{M} \tag{3}$$

The pressure equation is

$$dp = -g\rho dz, \tag{4}$$

where p, ρ denote pressure and density of the gas at a height z above sea level.

We also have the pressure density relation

$$p = \lambda\rho, \text{ i.e., } dp = \lambda d\rho \tag{5}$$

From (4) and (5), we have

$$\lambda\frac{d\rho}{\rho} = -g\,dz \tag{6}$$

Integrating between limits $z = 0$ to $z = h$ when $\rho = \rho_0$ and $\rho = \rho_1$ respectively, we get

$$\lambda \log\rho\Big|_{\rho_0}^{\rho_1} = -gz\Big|_0^h, \text{ i.e., } \lambda\log\frac{\rho_1}{\rho_0} = -gh, \text{ i.e., } \log\frac{\rho_0}{\rho_1} = \frac{g}{\lambda}h$$

$$\therefore h = \frac{\lambda}{g}\log\frac{\rho_0}{\rho_1} \tag{7}$$

From (3) and (7), we get

$$h = \frac{\lambda}{g}\log\frac{M+m}{M}. \quad \textbf{(Proved)}$$

EXAMPLE 11.9 *A box is filled with a heavy gas at a uniform temperature. Prove that, if a is the altitude of the highest point above the lowest, and P, P′ are pressures at these points, the ratio of the pressure to the density at any point is equal to* $\dfrac{ag}{\log\left[\frac{P'}{P}\right]}$.

Solution. Let H = height of the top of the box from sea level (FIG. 11.6).

The pressure equation is

$$dp = -g\rho dz, \tag{1}$$

where p, ρ denote the pressure and density of the gas at a height z above sea level.

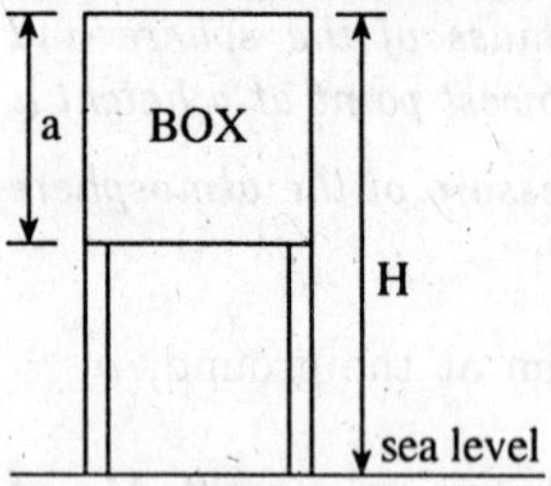

Fig. 11.6

We also have the pressure-density relation

$$p = \lambda\rho \tag{2}$$

From (1) and (2), we have

$$\frac{dp}{p} = -\frac{g}{\lambda}dz, \tag{3}$$

which on integration yields

$$\log p = C - \frac{gz}{\lambda}, \tag{4}$$

where C is the constant of integration. Since P, P' are the pressures at the highest and lowest points of the box, i.e., at height H and $(H - a)$ from the sea level, we get from (4)

$$\log P = C - \frac{gH}{\lambda} \tag{5}$$

$$\text{and } \log P' = C - \frac{g}{\lambda}(H - a) \tag{6}$$

Eliminating C from (5) and (6),

$$\log\frac{P'}{P} = \frac{g}{\lambda}a$$

$$\therefore \lambda = \frac{ga}{\log\left[\frac{P'}{P}\right]}. \quad \textbf{(Proved)}$$

EXAMPLE 11.10 *Taking into account the variation of gravity with height and assuming that the temperature of the air is constant at all heights, prove that at a height x, the pressure p of the air is given by* $\log\dfrac{p}{p_0} = -\dfrac{g_0 ax}{\lambda(a+x)}$, *where a is the radius of the earth,* $\lambda = \dfrac{p_0}{\rho_0}$, *and* p_0, ρ_0, g_0 *are the values of pressure, density and gravity at the earth's surface.*

A balloon carrying a self-registering barometer records pressures equivalent to h and h_1 *inches of mercury when it ascends to heights equal to fractions* α *and* α_1 *of the earth's radius respectively. Prove that* $\dfrac{g_0 a}{\lambda}\cdot\dfrac{\alpha_1 - \alpha}{(1+\alpha_1)(1+\alpha)} = \log\dfrac{h}{h_1} + 2\log\dfrac{1+\alpha_1}{1+\alpha}$.

Solution. The value of gravity g_x at any height x above sea level is

$$g_x = \frac{g_0 a^2}{(a+x)^2} \tag{1}$$

Then the pressure equation is

$$dp = -\frac{g_0 a^2}{(a+x)^2}\rho dx, \tag{2}$$

where p, ρ denote the pressure and density of the air at a height of x above sea level.

We also have the pressure-density relation

$$p = \lambda \rho \tag{3}$$

From (2) and (3), we have

$$\frac{dp}{p} = -\frac{g_0 a^2}{\lambda (a+x)^2} dx \tag{4}$$

Integrating, we get

$$\log p = \frac{g_0 a^2}{\lambda (a+x)} + C, \tag{5}$$

where C is the constant of integration. Since $p = p_0$ at the sea level where $x = 0$, we have

$$\log p_0 = \frac{g_0 a}{\lambda} + C \tag{6}$$

Eliminating C from (5) and (6), we get

$$\log \frac{p}{p_0} = \frac{g_0 a^2}{\lambda (a+x)} - \frac{g_0 a}{\lambda}, \text{ i.e., } \log \frac{p}{p_0} = -\frac{g_0 a x}{\lambda (a+x)}. \quad \textbf{(Proved)} \tag{7}$$

Now, let δ = density of mercury; P, P_1 = pressures when the balloon is at heights of αa and $\alpha_1 a$ respectively, g, g_1 = values of gravity at heights αa and $\alpha_1 a$ respectively.

Since h and h_1 are the barometric heights, we have

$$\left.\begin{aligned} P &= g\delta h \\ P_1 &= g_1 \delta h_1 \end{aligned}\right\} \tag{8}$$

Hence,

$$\frac{P}{P_1} = \frac{gh}{g_1 h_1} \tag{9}$$

By the relation (1), we have at heights corresponding to $x = a\alpha$, and $x = a\alpha_1$, the values of gravity as

$$\left.\begin{aligned} g &= \frac{g_0 a^2}{(a+a\alpha)^2} = \frac{g_0}{(1+\alpha)^2} \\ g_1 &= \frac{g_0 a^2}{(a+a\alpha_1)^2} = \frac{g_0}{(1+\alpha_1)^2} \end{aligned}\right\} \tag{10}$$

Hence,

$$\frac{g}{g_1} = \frac{(1+\alpha_1)^2}{(1+\alpha)^2} \tag{11}$$

From (9) and (11), we get

$$\frac{P}{P_1} = \frac{(1+\alpha_1)^2}{(1+\alpha)^2} \cdot \frac{h}{h_1}$$

$$\therefore \log \frac{P}{P_1} = \log \frac{h}{h_1} + 2 \log \frac{1+\alpha_1}{1+\alpha} \tag{12}$$

Since P, P_1 are pressures at heights $x = a\alpha$ and $x = a\alpha_1$ respectively, we have by relation (7),

$$\log \frac{P}{p_0} = -\frac{g_0 a (a\alpha)}{\lambda (a + a\alpha)} = -\frac{g_0 \alpha a}{\lambda (1+\alpha)} \tag{13}$$

$$\text{and } \log \frac{P_1}{p_0} = -\frac{g_0 a (a\alpha_1)}{\lambda (a + a\alpha_1)} = -\frac{g_0 \alpha_1 a}{\lambda (1+\alpha_1)} \tag{14}$$

Hence,

$$\log\frac{P}{P_1}=\frac{g_0 a}{\lambda}\left[\frac{\alpha_1}{1+\alpha_1}-\frac{\alpha}{1+\alpha}\right]=\frac{g_0 a}{\lambda}\frac{\alpha_1-\alpha}{(1+\alpha_1)(1+\alpha)} \tag{15}$$

From (12) and (15), we get

$$\frac{g_0 a}{\lambda}\cdot\frac{\alpha_1-\alpha}{(1+\alpha_1)(1+\alpha)}=\log\frac{h}{h_1}+2\log\frac{1+\alpha_1}{1+\alpha}. \quad \textbf{(Proved)}$$

EXAMPLE 11.11 *If the law connecting the pressure and density of the air were* $p=\lambda\rho^n$, *prove that, neglecting variations of gravity and temperature, the height of the atmosphere would be* $\dfrac{n}{n-1}$ *times the homogeneous atmosphere.*

Solution. We are given the pressure density relation

$$p=\lambda\rho^n, \tag{1}$$

where p = pressure and ρ = density of air at any height z above the sea level.

The pressure equation is

$$dp=-g\rho dz \tag{2}$$

From (1) we have

$$dp=\lambda n\rho^{n-1}d\rho \tag{3}$$

From (2) and (3), we get

$$-g\rho dz=\lambda n\rho^{n-1}d\rho, \text{ i.e., } -gdz=\lambda n\rho^{n-2}d\rho \tag{4}$$

Integrating, we get

$$C-gz=\frac{\lambda n}{n-1}\rho^{n-1} \tag{5}$$

Let at sea level, i.e., at $z=0$, the density of air $\rho=\rho_0$. Then from (5),

$$C=\frac{\lambda n}{n-1}\rho_0^{n-1} \tag{6}$$

Hence, eliminating C from (5) and (6), we have

$$gz=\frac{\lambda n}{n-1}\left[\rho_0^{n-1}-\rho^{n-1}\right] \tag{7}$$

Let h = height of the total atmosphere. Then

$$\rho=0, \text{ when } z=h \tag{8}$$

Using the condition (8) in (7), we get

$$gh=\frac{\lambda n}{n-1}\rho_0^{n-1} \tag{9}$$

Let p_0 = pressure at the sea level and H = height of the homogeneous atmosphere. Then, we have the relations

$$p_0=g\rho_0 H \text{ and } p_0=\lambda\rho_0^n \tag{10}$$

Hence,

$$g\rho_0 H = \lambda\rho_0^n, \text{ i.e., } gH = \lambda\rho_0^{n-1} \tag{11}$$

Eliminating ρ_0 from (9) and (11), we get

$$\frac{n-1}{\lambda n}gh = \frac{gH}{\lambda}$$

$$\therefore\ h = \frac{n}{n-1}H.$$

Hence proved.

EXAMPLE 11.12 *The absolute temperature T of the atmosphere diminishes according to the law $T = \dfrac{T_0}{1+az+bz^2}$. Show that in an adiabatic change of state, pressure at a height z is given by*

$$p = p_0 \exp\left[-\frac{z}{H}\left[1+\frac{1}{2}az+\frac{1}{3}bz^2\right]\right],$$

where a, b are constants, H is the height of homogeneous atmosphere and p_0, T_0 are the pressure and temperature at $z = 0$.

Solution. Let p, ρ denote the pressure and density of the atmosphere at a height z above sea level. Then the pressure equation is

$$dp = -g\rho dz \tag{1}$$

From the gas relation $pV = RT$, we have

$$p = \rho RT \tag{2}$$

From (1) and (2), we have

$$\frac{dp}{p} = -\frac{g\rho}{\rho RT}dz = -\frac{g}{RT_0}\left[1+az+bz^2\right]dz \tag{3}$$

since we are given

$$T = \frac{T_0}{1+az+bz^2} \tag{4}$$

Integrating (3), we get

$$\log p = -\frac{g}{RT_0}\left[z+\frac{a}{2}z^2+\frac{b}{3}z^3\right]+C, \tag{5}$$

where C is the constant of integration.

Since at the sea level pressure is p_0, we have

$$p = p_0, \text{ when } z = 0 \tag{6}$$

Using this condition in (5), we find

$$C = \log p_0 \tag{7}$$

Hence, from (5) and (7),

$$\log\frac{p}{p_0} = -\frac{gz}{RT_0}\left[1+\frac{a}{2}z+\frac{b}{3}z^2\right] \tag{8}$$

Now, since H is the height of the homogeneous atmosphere,

$$p_0 = \rho_0 R T_0 = H \rho_0 g, \text{ i.e., } RT_0 = Hg \tag{9}$$

Substituting this value in (8), we have

$$\log \frac{p}{p_0} = -\frac{z}{H}\left[1 + \frac{1}{2}az + \frac{1}{3}bz^2\right]$$

Hence,

$$p = p_0 \exp\left[-\frac{z}{H}\left[1 + \frac{1}{2}az + \frac{1}{3}bz^2\right]\right]. \quad \textbf{(Proved)}$$

EXAMPLE 11.13 *Prove that, if the temperature in the atmosphere falls uniformly with the height ascended, the height of a station above sea level is given by*

$$z = a\left\{1 - \left[\frac{h}{h_0}\right]^m\right\},$$

where h, h_0 are readings of the barometer at the station and at sea level respectively and a, m are constants.

Solution. Since the temperature in the atmosphere falls uniformly with the height ascended, we take

$$T = T_0 - \lambda z, \tag{1}$$

where T, T_0 are the absolute temperatures at height z above sea level and at sea level respectively. Here λ is a constant.

Let p, ρ denote the pressure and density of the atmosphere at a height z above sea level. Then the pressure equation is

$$dp = -g\rho dz \tag{2}$$

We also have the gas relation

$$pV = RT, \text{ i.e., } p = \rho RT \tag{3}$$

From (1) and (3), we have

$$p = \rho R(T_0 - \lambda z) \tag{4}$$

From (2) and (4), we have

$$\frac{dp}{p} = -\frac{g}{R}\frac{dz}{(T_0 - \lambda z)} \tag{5}$$

Integrating, we get

$$\log p = \frac{g}{\lambda R}\log(T_0 - \lambda z) + C, \tag{6}$$

where C is the constant of integration.

Let the pressure at sea level be p_0. Then

$$p = p_0, \text{ when } z = 0 \tag{7}$$

Using this condition in (6), we find

$$C = \log p_0 - \frac{g}{\lambda R}\log T_0 \tag{8}$$

Hence, from (6) and (8), we have

$$\log p = \frac{g}{\lambda R}\log(T_0 - \lambda z) + \log p_0 - \frac{g}{\lambda R}\log T_0$$

$$\text{i.e., } \log\frac{p}{p_0} = \frac{g}{\lambda R}\log\left[1 - \frac{\lambda z}{T_0}\right], \text{ i.e., } \log\frac{p}{p_0} = \log\left[1 - \frac{\lambda z}{T_0}\right]^{\frac{g}{\lambda R}}$$

$$\therefore \frac{p}{p_0} = \left[1 - \frac{\lambda z}{T_0}\right]^{\frac{g}{\lambda R}} \qquad (9)$$

Now, let δ be the density of mercury. Since h, h_0 are the barometric heights at the station of height z above sea level and at sea level respectively, we have

$$\left.\begin{aligned} p &= g\delta h \\ p_0 &= g\delta h_0 \end{aligned}\right\} \qquad (10)$$

Hence,

$$\frac{p}{p_0} = \frac{h}{h_0} \qquad (11)$$

Equating the right hand sides of (9) and (11)

$$\frac{h}{h_0} = \left[1 - \frac{\lambda z}{T_0}\right]^{\frac{g}{\lambda R}}, \text{ i.e., } \left[\frac{h}{h_0}\right]^{\frac{\lambda R}{g}} = 1 - \frac{\lambda z}{T_0}, \text{ i.e., } \frac{\lambda z}{T_0} = 1 - \left[\frac{h}{h_0}\right]^{\frac{\lambda R}{g}}$$

$$\therefore z = \frac{T_0}{\lambda}\left[1 - \left[\frac{h}{h_0}\right]^{\frac{\lambda R}{g}}\right] = a\left[1 - \left[\frac{h}{h_0}\right]^{m}\right],$$

where we have denoted

$$a = \frac{T_0}{\lambda},\ b = \frac{\lambda R}{g}, \qquad (12)$$

which are constants. **Hence proved.**

EXAMPLE 11.14 *A gas is compressed to* $\frac{1}{n}$*th of its original volume and in the process, the relation of pressure to volume is expressed by the equation* pV^{γ} = *constant, where* γ *has a constant value. It is then cooled at constant volume to its original temperature. Finally it is expanded at this constant temperature to its original volume.*

Show that the net energy expended on the gas in this cycle is λ *times the work done in compression, where* $\lambda = 1 - \dfrac{(\gamma - 1)\log n}{n^{\gamma-1} - 1}$.

Solution. Let p, V be the original pressure and volume respectively of the gas. When the gas is compressed to $\frac{1}{n}$th of its volume, its volume becomes $\frac{V}{n}$.

By the condition given, the process is adiabatic, so that

$$pV^{\gamma} = \text{constant} = K_1 \text{ (say), i.e., } p = \frac{K_1}{V^{\gamma}} \qquad (1)$$

Hence, the work done on the gas in reducing the volume is

$$W_1 = -\int_V^{V/n} p\,dV = -\int_V^{V/n} \frac{K_1}{V^{\gamma}}dV \text{ [using (1)]}$$

$$= -K_1\left[\frac{V^{1-\gamma}}{1-\gamma}\right]_V^{V/n} = -\frac{K_1}{1-\gamma}\left[\left[\frac{V}{n}\right]^{1-\gamma} - (V)^{1-\gamma}\right] \qquad (2)$$

Now, the gas is cooled at constant volume to its original temperature. In this process, work done is zero since there is no change in volume.

Finally, the gas is expanded from present volume $\dfrac{V}{n}$ to its original volume V at constant temperature. The process is isothermal so that we have

$$pV = \text{constant} = K_2 \text{ (say), i.e., } p = \frac{K_2}{V} \tag{3}$$

Hence, work done in the process is

$$W_2 = \int_{V/n}^{V} p dV = \int_{V/n}^{V} \frac{K_2}{V} dV \text{ [using (3)]}$$

$$\text{i.e., } W_2 = K_2 \big[\log V\big]_{V/n}^{V} = K_2 \log\left[\frac{V}{V/n}\right] = K_2 \log n \tag{4}$$

Thus the net energy expended in the process is

$$\begin{aligned} E = W_1 - W_2 &= -\frac{K_1}{1-\gamma}\left[\left[\frac{V}{n}\right]^{1-\gamma} - V^{1-\gamma}\right] - K_2 \log n \\ &= -\frac{K_1}{1-\gamma} V^{1-\gamma}\left[\frac{1}{n^{1-\gamma}} - 1\right] - K_2 \log n \\ &= \frac{K_1 V^{1-\gamma}}{1-\gamma}\left[1 - n^{\gamma-1}\right] - K_2 \log n \end{aligned} \tag{5}$$

According to the given condition,

$$E = \lambda W_1 \tag{6}$$

Hence,

$$\lambda = \frac{E}{W_1} = \frac{\frac{K_1 V^{1-\gamma}}{1-\gamma}\left[1 - n^{\gamma-1}\right] - K_2 \log n}{\frac{K_1 V^{1-\gamma}}{1-\gamma}\left(1 - n^{\gamma-1}\right)} = 1 - \frac{K_2}{K_1} \frac{1-\gamma}{1 - n^{\gamma-1}} \frac{\log n}{V^{1-\gamma}} \tag{7}$$

From (1) and (3), we have

$$\frac{K_2}{K_1} = \frac{pV}{pV^\gamma} = V^{1-\gamma} \tag{8}$$

Substituting this value in (7), we get

$$\lambda = 1 - V^{1-\gamma} \cdot \frac{1-\gamma}{1 - n^{\gamma-1}} \frac{\log n}{V^{1-\gamma}} = 1 - \frac{(\gamma - 1)\log n}{n^{\gamma-1} - 1}. \quad \textbf{(Proved)}$$

EXAMPLE 11.15 *A piston of weight W is placed at the middle of a horizontal cylinder which it closely fits, the air on each side of it being of atmospheric pressure Π, and the ends of the cylinder closed. The cylinder is then tilted up till its axis is inclined at an angle α to the horizon. Prove that the piston will be in equilibrium again at a distance*

$$l\left[\sqrt{1 + \lambda^2 \operatorname{cosec}^2 \alpha} - \lambda \operatorname{cosec} \alpha\right]$$

from the original position, where $2l$ is the length of the cylinder, A the area of the piston and $\lambda = \Pi A / W$.

Solution. Let AB be the position of the piston when the cylinder of length $2l$ and cross section A was horizontal. In the incined position of the cylinder, let the piston move downwards by a distance z (FIG. 11.7).

Let p_1, p_0 be the pressures of the air at the upper and lower end of the cylinder respectively. Then, by Boyle's law, we have pV = constant, so that

$$p_0(l-z)\cdot A = p_1(l+z)\cdot A = l\Pi A$$

Hence,

$$\left.\begin{aligned} p_0 &= \frac{l\Pi}{l-z} \\ p_1 &= \frac{l\Pi}{l+z} \end{aligned}\right\} \qquad (1)$$

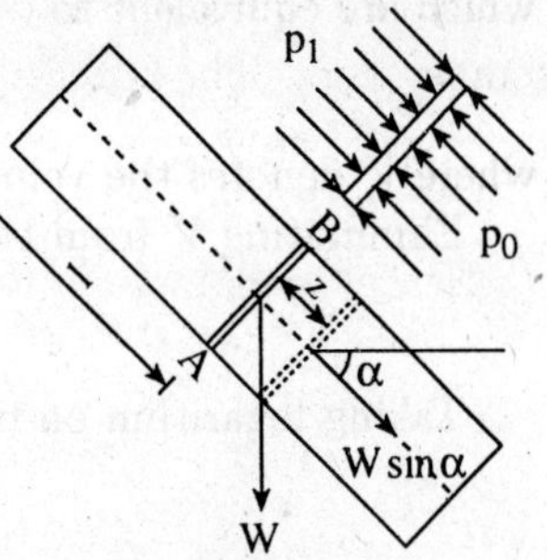

Fig. 11.7

In the inclined position, the component of weight W of the cylinder along its length is

$$F = W \sin\alpha \qquad (2)$$

This downward force is balanced by the thrust due to the net pressure on the piston which is $(p_0 - p_1)$. Hence,

$$F = (p_0 - p_1)A \qquad (3)$$

From (2) and (3), we have

$$W \sin\alpha = (p_0 - p_1)A$$

$$\text{i.e.,}\quad W\sin\alpha = Al\Pi\left[\frac{1}{l-z} - \frac{1}{l+z}\right] \quad \text{[using (1)]}$$

$$\text{i.e.,}\quad W\sin\alpha = Al\Pi\frac{2z}{l^2 - z^2}$$

$$\therefore\ l^2 - z^2 = \frac{2Al\Pi}{W} z \operatorname{cosec}\alpha \qquad (4)$$

We are given that

$$\lambda = \frac{\Pi A}{W} \qquad (5)$$

Hence, from (4) and (5), we have

$$l^2 - z^2 = 2\lambda l z \operatorname{cosec}\alpha$$

$$\text{i.e.,}\quad (z + l\lambda \operatorname{cosec}\alpha)^2 = l^2(1 + \lambda^2 \operatorname{cosec}^2\alpha)$$

$$\text{i.e.,}\quad z + l\lambda\operatorname{cosec}\alpha = l\sqrt{1 + \lambda^2\operatorname{cosec}^2\alpha}$$

$$\therefore\ z = l\left[\sqrt{1+\lambda^2\operatorname{cosec}^2\alpha} - \lambda\operatorname{cosec}\alpha\right]$$

We have ignored negative sign since z cannot be negative.
Hence proved.

EXAMPLE 11.16 *A gaseous atmosphere in equilibrium is such that $p = \lambda\rho^\gamma = R\rho T$, where p, ρ, T are the pressure, density and absolute temperature respectively and λ, γ, R are constants. Prove that the temperature decreases upwards at a constant rate α so that*

$$\frac{dT}{dz} = -\alpha = -\frac{g}{R}\frac{\gamma - 1}{\gamma}.$$

In a certain atmosphere of uniform composition, $T = T_0 - \beta z$, where T_0, β are constants and $\beta < \alpha$. Find the pressure and density and show that they both vanish at height T_0/β.

Solution. We are given the relations

$$p = \lambda\rho^{\gamma},\ p = R\rho T, \tag{1}$$

which are equivalent to

$$pV^{\gamma} = \lambda,\ pV = RT, \tag{2}$$

where V denotes the volume of the gas.

Eliminating V from two relations in (2), we get

$$T^{\gamma} \cdot p^{1-\gamma} = \lambda R^{-\gamma} = \text{constant} = C\,(\text{say}). \tag{3}$$

Taking logarithm on both sides of (3), we get

$$\gamma \log T = (\gamma - 1) \log p + \log C \tag{4}$$

Differentiating (4),

$$\gamma\frac{dT}{T} = (\gamma-1)\frac{dp}{p},\ \text{i.e.,}\ \frac{dp}{p} = \frac{\gamma}{\gamma-1}\frac{dT}{T} \tag{5}$$

Now, the pressure equation is

$$dp = -g\rho dz, \tag{6}$$

where p, ρ are pressure and density of atmosphere at a height z above sea level.

Since $p = R\rho T$, we have from (6),

$$\frac{dp}{p} = -\frac{g\rho}{R\rho T}dz,\ \text{i.e.,}\ \frac{dp}{p} = -\frac{g}{R}\frac{dz}{T} \tag{7}$$

From (5) and (7), we get

$$\frac{\gamma}{\gamma-1}\frac{dT}{T} = -\frac{g}{R}\frac{dz}{T}$$

$$\therefore\ \frac{dT}{dz} = -\frac{g}{R}\frac{\gamma-1}{\gamma}.\quad \textbf{(Proved)} \tag{8}$$

Thus, by the condition of the problem,

$$\alpha = \frac{g}{R}\frac{\gamma-1}{\gamma} \tag{9}$$

Again, we are given

$$T = T_0 - \beta z \tag{10}$$

Since $p = R\rho T$, we have using (10),

$$p = R\rho(T_0 - \beta z) \tag{11}$$

Then from (6) and (11),

$$\frac{dp}{p} = -\frac{g\rho dz}{R\rho(T_0-\beta z)} = -\frac{g}{R}\frac{dz}{T_0-\beta z} \tag{12}$$

Integrating, we get

$$\log p = \frac{g}{R\beta}\log(T_0 - \beta z) + C, \tag{13}$$

where C is the constant of integration. Let the pressure at sea level is p_0. That is

$$p = p_0,\ \text{when}\ z = 0. \tag{14}$$

Using this condition in (13), we get

$$\log p_0 = \frac{g}{R\beta}\log T_0 + C, \text{ i.e., } C = \log p_0 - \frac{g}{R\beta}\log T_0 \tag{15}$$

Hence, from (13) and (15), we have

$$\log p = \log p_0 - \frac{g}{R\beta}\log T_0 + \frac{g}{R\beta}\log(T_0 - \beta z)$$

$$\text{i.e., } \log\frac{p}{p_0} = \frac{g}{R\beta}\log\left[1 - \frac{\beta}{T_0}z\right], \text{ i.e., } \frac{p}{p_0} = \left[1 - \frac{\beta}{T_0}z\right]^{\frac{g}{R\beta}} \tag{16}$$

From (16) it is obvious that $p = 0$ when $z = \dfrac{T_0}{\beta}$. Again, we have

$$p = \rho RT \text{ and } p_0 = \rho_0 RT_0 \tag{17}$$

Then we get

$$\frac{p}{p_0} = \frac{\rho RT}{\rho_0 RT_0} = \frac{\rho}{\rho_0}\cdot\frac{(T_0 - \beta z)}{T_0}$$

$$\therefore \frac{p}{p_0} = \frac{\rho}{\rho_0}\left[1 - \frac{\beta z}{T_0}\right] \tag{18}$$

Hence, from (16) and (18), we get

$$\frac{\rho}{\rho_0}\left[1 - \frac{\beta}{T_0}z\right] = \left[1 - \frac{\beta}{T_0}z\right]^{\frac{g}{R\beta}}$$

$$\therefore \frac{\rho}{\rho_0} = \left[1 - \frac{\beta}{T_0}z\right]^{\frac{g}{R\beta}-1} \tag{19}$$

This shows that $\rho = 0$ when $z = \dfrac{T_0}{\beta}$. **Hence proved.**

EXAMPLE 11.17 *A gaseous atmosphere in equilibrium is such that the temperature gradient is* $\dfrac{dT}{dz} = -\dfrac{g}{R}\cdot\dfrac{\gamma-1}{\gamma}$*, where the symbols have usual meanings.*
Show that the pressure at any height h is given by

$$p = p_0\left[1 + \frac{h}{\xi T_0}\right]^{\frac{\gamma}{\gamma-1}}$$

where p_0, T_0 *are pressure and temperature respectively at sea level, and* $\xi = \dfrac{\gamma}{\gamma-1}\dfrac{R}{g}$.

Solution. We are given that

$$\frac{dT}{dz} = -\frac{g}{R}\frac{\gamma-1}{\gamma} \tag{1}$$

From (1), we have

$$dz = -\frac{\gamma}{\gamma-1}\frac{R}{g}dT, \text{ i.e., } dz = -\xi dT, \tag{2}$$

since we are given

$$\xi = \frac{\gamma}{\gamma-1}\frac{R}{g} \tag{3}$$

Integrating (2), we get

$$\int_0^h dz = -\xi \int_{T_0}^{T} dT, \text{ i.e., } h = \xi(T_0 - T) = \xi T_0 \left[1 - \frac{T}{T_0}\right] \tag{4}$$

From (4), we get

$$\frac{T}{T_0} = 1 - \frac{h}{\xi T_0} \tag{5}$$

Now, following Eqn (11.77), we write

$$\frac{T}{T_0} = \left[\frac{p}{p_0}\right]^{\frac{\gamma-1}{\gamma}} \tag{6}$$

From (5) and (6), we get

$$\left[\frac{p}{p_0}\right]^{\frac{\gamma-1}{\gamma}} = 1 - \frac{h}{\xi T_0}$$

$$\therefore \; \frac{p}{p_0} = \left[1 - \frac{h}{\xi T_0}\right]^{\frac{\gamma}{\gamma-1}}$$

Hence,

$$p = p_0 \left[1 - \frac{h}{\xi T_0}\right]^{\frac{\gamma}{\gamma-1}}. \quad \textbf{(Proved)}$$

Note. The atmospheric pressure at height h becomes zero, when

$$h = \xi T_0 = \frac{\gamma R T_0}{(\gamma - 1)g} \tag{7}$$

This gives the total height of the effective atmosphere from the sea level.

EXAMPLE 11.18 *The height of a balloon is calculated from the barometer pressure reading p on the assumption that the pressure of the air varies as the density. Show that, if the pressure actually varies as the n-th power of the density, there will be an error*

$$h_0 \left\{ \frac{n}{n-1}\left[1 - \left[\frac{p}{p_0}\right]^{\frac{n-1}{n}}\right] - \log \frac{p}{p_0} \right\}$$

in the calculated height, where h_0 is the height of the homogeneous atmosphere and p_0 is the pressure at sea level.

Solution.

Case 1: Pressure varies as density. In this case,

$$p = \lambda \rho, \; \lambda = \text{constant}. \tag{1}$$

The pressure equation is

$$dp = -g\rho dz, \tag{2}$$

where p, ρ are the pressure and density of air at height z above sea level.

From (1) and (2), we have

$$\frac{dp}{p} = -\frac{g\rho dz}{\lambda \rho} = -\frac{g}{\lambda} dz \tag{3}$$

Integrating within proper limits,

$$\int_{p_0}^{p} \frac{dp}{p} = -\frac{g}{\lambda}\int_0^z dz, \text{ i.e., } \log\frac{p}{p_0} = -\frac{g}{\lambda}z \tag{4}$$

Since h_0 is the height of homogeneous atmosphere, we have

$$p_0 = \lambda\rho_0 = h_0\rho_0 g, \text{ i.e., } h_0 = \frac{\lambda}{g} \tag{5}$$

From (4) and (5),

$$\log\frac{p_0}{p} = \frac{z}{h_0}$$

$$\therefore \quad z = h_0 \log\frac{p_0}{p} \tag{6}$$

This gives the height when pressure varies as the density.

Case 2: Pressure varies as n-th power of density. In this case, we have

$$p = \xi\rho^n, \; \xi = \text{constant.} \tag{7}$$

Then from (7),

$$\rho = \left[\frac{p}{\xi}\right]^{\frac{1}{n}} \tag{8}$$

Hence, from (2) and (8), we get

$$dp = -g\left[\frac{p}{\xi}\right]^{\frac{1}{n}} d\bar{z}, \text{ i.e., } \frac{dp}{p^{\frac{1}{n}}} = -\beta d\bar{z}, \tag{9}$$

where we have denoted

$$\beta = \frac{g}{\xi^{\frac{1}{n}}} \tag{10}$$

Integrating (9), we have

$$\frac{p^{1-\frac{1}{n}}}{1-\frac{1}{n}} = -\beta\bar{z} + C, \tag{11}$$

where C is the constant of integration.

Since $p = p_0$, when $\bar{z} = 0$, we find from (11),

$$C = \frac{p_0^{1-\frac{1}{n}}}{1-\frac{1}{n}} \tag{12}$$

Hence from (11) and (12), we have

$$\frac{p^{1-\frac{1}{n}}}{1-\frac{1}{n}} = -\beta\bar{z} + \frac{p_0^{1-\frac{1}{n}}}{1-\frac{1}{n}}, \text{ i.e., } \bar{z} = \frac{1}{\beta}\cdot\frac{1}{1-\frac{1}{n}}\left[p_0^{1-\frac{1}{n}} - p^{1-\frac{1}{n}}\right] \tag{13}$$

Since h_0 is the height of homogeneous atmosphere, we have

$$p_0 = \xi\rho_0^n = h_0\rho_0 g$$

$$\text{i.e., } h_0 = \frac{\xi}{g}\rho_0^{n-1} = \frac{\xi}{g}\left[\frac{p_0}{\xi}\right]^{\frac{n-1}{n}} \quad \text{[using (8)]}$$

$$= \frac{\xi^{\frac{1}{n}}}{g} p_0^{\frac{n-1}{n}} \tag{14}$$

From (13), we write

$$\bar{z} = \frac{\xi^{\frac{1}{n}}}{g} \cdot \frac{n}{n-1} p_0^{\frac{n-1}{n}} \left[1 - \left[\frac{p}{p_0} \right]^{\frac{n-1}{n}} \right] \text{[using (10)]}$$

$$\text{i.e., } \bar{z} = h_0 \frac{n}{n-1} \left[1 - \left[\frac{p}{p_0} \right]^{\frac{n-1}{n}} \right] \text{ [using (14)]}$$

This gives the height when the pressure varies as the n-th power of density. Hence, the error in calculated height is

$$z^* = \bar{z} - z = h_0 \frac{n}{n-1} \left[1 - \left[\frac{p}{p_0} \right]^{\frac{n-1}{n}} \right] - h_0 \log \frac{p_0}{p}$$

$$= h_0 \left\{ \frac{n}{n-1} \left[1 - \left[\frac{p}{p_0} \right]^{\frac{n-1}{n}} \right] - \log \frac{p_0}{p} \right\}. \quad \textbf{(Proved)}$$

EXAMPLE 11.19 *A rectangular area is immersed in a compressible fluid with its plane vertical and one side in the surface where the pressure is zero. Show that, if the density is a linear function of the pressure, the depth of the centre of pressure is*

$$\frac{a}{m} \frac{(m-1)\rho_1 + \left[1 - \frac{1}{2}m^2\right] \rho_0}{\rho_1 - (m+1)\rho_0}$$

where a is the length of the vertical side, ρ_0, ρ_1 are the densities at the top and bottom of the area, and $m = \log \left[\frac{\rho_1}{\rho_0} \right]$.

Solution. Let $ABCD$ be the rectangular area immersed vertically in a compressible fluid (FIG. 11.8).

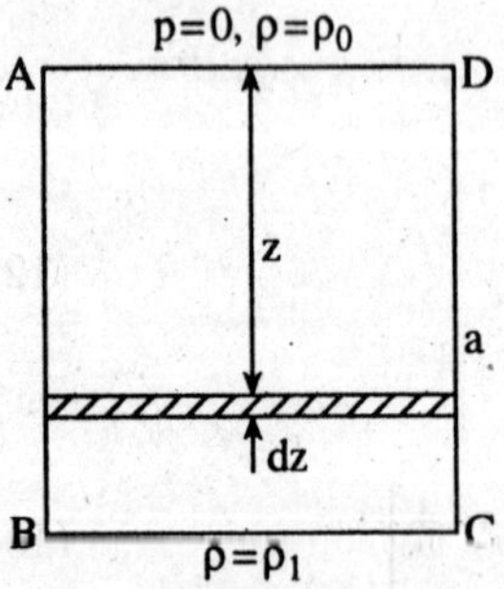

Fig. 11.8

We are given:

$$AB = CD = a \tag{1}$$

$$\left.\begin{aligned} &\text{Pressure on } AD = 0 \\ &\text{Density on } AD = \rho_0 \\ &\text{Density on } BC = \rho_1 \end{aligned}\right\} \tag{2}$$

Since the density is a linear function of pressure, we write:

$$\rho = A + \lambda p \tag{3}$$

where p, ρ denote the pressure and density of the fluid at any point at a depth z below the top surface AD, and A, λ are constants.

From (3) we have

$$d\rho = \lambda dp, \text{ i.e., } dp = \frac{d\rho}{\lambda} \tag{4}$$

The pressure equation is

$$dp = g\rho dz, \text{ i.e., } d\rho = \lambda g\rho dz \text{ [using (4)]}$$

$$\text{i.e., } \frac{d\rho}{\rho} = \lambda g dz \tag{5}$$

Integrating, we have

$$\log \rho = \lambda g z + C, \tag{6}$$

where C is the constant of integration.

We have the conditions:

$$\left.\begin{aligned} \rho &= \rho_0, \text{ when } z = 0 \\ \rho &= \rho_1, \text{ when } z = a \end{aligned}\right\} \tag{7}$$

Using these conditions in (6), we have

$$\left.\begin{aligned} \log \rho_0 &= C \\ \log \rho_1 &= C + \lambda g a \end{aligned}\right\} \tag{8}$$

From (8), we find

$$\left.\begin{aligned} C &= \log \rho_0 \\ \lambda &= \frac{1}{ga} \log \frac{\rho_1}{\rho_0} = \frac{m}{ga} \end{aligned}\right\}, \tag{9}$$

where

$$m = \log \frac{\rho_1}{\rho_0} \tag{10}$$

Substituting the values of C and λ from (9) into (6),

$$\log \rho = \lambda g z + \log \rho_0, \text{ i.e., } \log \frac{\rho}{\rho_0} = \lambda g z = \frac{mz}{a}$$

$$\therefore \rho = \rho_0 e^{\frac{mz}{a}} \tag{11}$$

Since on AD, $p = 0$ and $\rho = \rho_0$, we have from (3),

$$A = \rho_0 \tag{12}$$

so that (3) becomes

$$\rho = \rho_0 + \lambda p, \text{ i.e., } p = \frac{1}{\lambda}(\rho - \rho_0)$$

$$\text{i.e., } p = \frac{ga}{m}\rho_0 \left(e^{\frac{mz}{a}} - 1\right) \text{ [from (11)]} \tag{13}$$

Let $AD = BC = b$.

We consider a strip of thickness dz at a depth z. Then the thrust on the strip is

$$= p \cdot b dz = \frac{ga}{m}\rho_0 b \left(e^{\frac{mz}{a}} - 1\right) dz \tag{14}$$

Hence, the total thrust is

$$\begin{aligned} F &= \frac{ga\rho_0 b}{m} \int_0^a \left(e^{\frac{mz}{a}} - 1\right) dz = \frac{ga\rho_0 b}{m} \left[\frac{a}{m} e^{\frac{mz}{a}} - z\right]_0^a \\ &= \frac{ga\rho_0 b}{m} \cdot \frac{a}{m} (e^m - 1 - m) \end{aligned} \tag{15}$$

If z^* be the depth of centre of pressure from the top surface, then

$$F \cdot z^* = \int p \cdot b dz \cdot z$$

$$\begin{aligned} \text{i.e., } F \cdot z^* &= \int_0^a \frac{ga\rho_0 b}{m} \left(e^{\frac{mz}{a}} - 1\right) z dz = \frac{ga\rho_0 b}{m} \left[z \cdot \frac{a}{m} e^{\frac{mz}{a}} - \frac{a^2}{m^2} e^{\frac{mz}{a}} - \frac{z^2}{2}\right]_0^a \\ &= \frac{ga\rho_0 b}{m} \cdot \frac{a^2}{m^2} \left[m e^m - e^m + 1 - \frac{1}{2} m^2\right] \end{aligned} \tag{16}$$

From (15) and (16), we get

$$z^* = \frac{a}{m} \cdot \frac{me^m - e^m + 1 - \frac{1}{2}m^2}{e^m - 1 - m} \tag{17}$$

Since

$$m = \log \frac{\rho_1}{\rho_0}, \; e^m = \frac{\rho_1}{\rho_0}, \tag{18}$$

hence, (17) becomes

$$z^* = \frac{a}{m} \frac{(m-1)\frac{\rho_1}{\rho_0} + 1 - \frac{1}{2}m^2}{\frac{\rho_1}{\rho_0} - (1+m)} = \frac{a}{m} \frac{(m-1)\rho_1 + \left[1 - \frac{1}{2}m^2\right]\rho_0}{\rho_1 - (1+m)\rho_0}. \quad \textbf{(Proved)}$$

EXAMPLE 11.20 *A hollow closed conical vessel of height h floats partially immersed in water with vertex downwards and axis vertical. A hole is then made very near the vertex and water allowed to enter the vessel so that no air escapes from within. If the vertex was originally at a depth b and H is the height of water barometer, prove that the new depth c of the vertex is given by*

$$c^3 - b^3 = \left\{ c - \frac{H(c^3 - b^3)}{h^3 - (c^3 - b^3)} \right\}^3$$

Solution. Let ABC be the cone partially immersed in water with vertex A downward and axis AE vertical (FIG. 11.9).

Initially, the vertex A was at a depth b below the surface.

Let ρ = density of water, α = semi-vertical angle of the cone. Then the volume of the displaced liquid is

$$V = \frac{1}{3}\pi b^3 \tan^2 \alpha \tag{1}$$

and the weight of displaced liquid is

$$W = V\rho g = \frac{1}{3}\pi g \rho b^3 \tan^2 \alpha \tag{2}$$

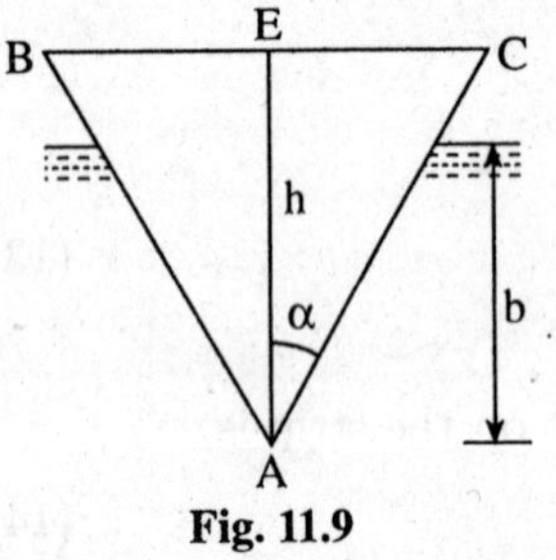

Fig. 11.9

From the condition of floatation, W in (2) will be the weight of the cone.

Now, after the leak, let x be the depth of water inside the vessel, and the depth b changes to c.

Hence, weight of water entered in the vessel is

$$W_1 = \frac{1}{3}\pi g \rho x^3 \tan^2 \alpha \tag{3}$$

and the weight of water displaced is

$$W_2 = \frac{1}{3}\pi g \rho c^3 \tan^2 \alpha \tag{4}$$

By the condition of floatation,

$$W + W_1 = W_2$$

$$\text{i.e., } \frac{1}{3} g\rho\pi \tan^2 \alpha (b^3 + x^3) = \frac{1}{3} g\rho\pi c^3 \tan^2 \alpha$$

$$\text{i.e., } b^3 + x^3 = c^3, \text{ i.e., } x^3 = c^3 - b^3 \tag{5}$$

Originally, the volume of air in the empty cone was

$$V_1 = \frac{1}{3}\pi h^3 \tan^2 \alpha \tag{6}$$

under the atmospheric pressure of

$$P_1 = H\rho g \tag{7}$$

Finally, after leakage of water, volume of air in the cone is

$$V_2 = \frac{1}{3}\pi \left[h^3 - x^3\right] \tan^2 \alpha \tag{8}$$

under the pressure of

$$P_2 = (H + c - x)g\rho \tag{9}$$

According to Boyle's law, we have

$$P_1V_1 = P_2V_2$$

i.e., $H\rho g \cdot \frac{1}{3}\pi h^3 \tan^2 \alpha = (H + c - x)g\rho \cdot \frac{1}{3}\pi(h^3 - x^3) \tan^2 \alpha$

i.e., $h^3H = (H + c - x)(h^3 - x^3)$

i.e., $h^3H = h^3H - Hx^3 + c(h^3 - x^3) - x(h^3 - x^3)$

i.e., $x(h^3 - x^3) = c(h^3 - x^3) - Hx^3$

i.e., $x = c - \dfrac{Hx^3}{h^3 - x^3}$

$$\therefore\ x^3 = \left[c - \frac{Hx^3}{h^3 - x^3}\right]^3 \tag{10}$$

Substituting the value of x^3 from (5) into (10), we get

$$c^3 - b^3 = \left\{c - \frac{H(c^3 - b^3)}{h^3 - (c^3 - b^3)}\right\}^3. \quad \textbf{(Proved)}$$

EXAMPLE 11.21 *A bent tube of uniform bore, the arms of which are at right angles, revolves with constant angular velocity ω about the axis of one of its arms which is vertical and has its extremity immersed in water. Prove that the height to which the water has risen in the vertical tube is*

$$\frac{\Pi}{g\delta}\left[1 - e^{-\frac{\omega^2 a^2}{2k}}\right],$$

where a is the length of the horizontal arm, Π the atmospheric pressure, δ the density of water and k the ratio of the pressure of the atmosphere to its density.

Solution. In the figure (FIG. 11.10), AB is the horizontal arm of length a of the tube and BC the vertical arm with its end C immersed in water.

The pressure equation for air in the horizontal tube due to rotation is

$$dp = \rho\omega^2 r dr, \tag{1}$$

where p = pressure of air, ρ = density of air, r = distance of the point P from the vertical axis.

Also, from Boyle's law, we have

$$p = k\rho \tag{2}$$

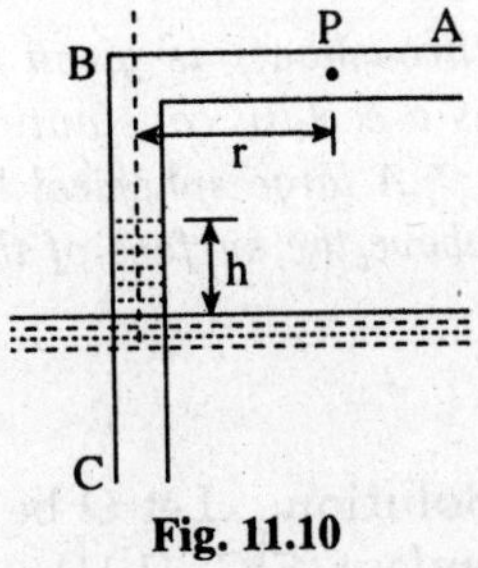

Fig. 11.10

From (1) and (2), we have

$$\frac{dp}{p} = \frac{\omega^2}{k} r dr \tag{3}$$

Integrating,

$$\log p = \frac{\omega^2}{2k} r^2 + C, \tag{4}$$

where C is the constant of integration.

Now, pressure at A is atmospheric, i.e.,

$$p = \Pi, \text{ when } r = a. \tag{5}$$

Using this condition in (4), we get

$$\log \Pi = \frac{\omega^2 a^2}{2k} + C, \text{ i.e., } C = \log \Pi - \frac{\omega^2 a^2}{2k} \tag{6}$$

Hence, from (4) and (6), we have

$$\log p = \frac{\omega^2 r^2}{2k} + \log \Pi - \frac{\omega^2 a^2}{2k}, \text{ i.e., } \log \frac{p}{\Pi} = \frac{\omega^2}{2k}(r^2 - a^2) \tag{7}$$

If pressure at B is p_0, then

$$p = p_0, \text{ when } r = 0. \tag{8}$$

From (7) and (8), we get

$$\log \frac{p_0}{\Pi} = -\frac{\omega^2 a^2}{2k} \tag{9}$$

Hence,

$$p_0 = \Pi e^{-\frac{\omega^2 a^2}{2k}} \tag{10}$$

If h be the height of water of density δ risen in the vertical tube, then for equilibrium,

$$p_0 + g\delta h = \Pi \tag{11}$$

From (10) and (11),

$$\Pi e^{-\frac{\omega^2 a^2}{2k}} + g\delta h = \Pi \tag{12}$$

Hence,

$$h = \frac{\Pi}{g\delta}\left[1 - e^{-\frac{\omega^2 a^2}{2k}}\right]. \quad \textbf{(Proved)}$$

EXAMPLE 11.22 *The pressure at a height z in the atmosphere (temperature constant throughout) is given by $p = p_0 e^{-\frac{gz}{k}}$, where p_0 is the pressure at the earth's surface and k is a certain constant.*

A large spherical balloon of radius r and total weight w floats with its centre at a height h above the surface of the earth. Show that h is given by

$$e^{\frac{gh}{k}} = \frac{4\pi p_0 k}{gw}\left[r \cosh \frac{gr}{k} - \frac{k}{g} \sinh \frac{gr}{k}\right]$$

Solution. Let O be the position of the centre of the balloon at a height h from the earth's surface (FIG. 11.11).

Let W be the weight of displaced air. Then, by the condition of floatation, we must have

$$W = w, \tag{1}$$

w being the weight of the balloon.

We consider a thin circular slice of air at a height z and of thickness dz. Its elementary volume is

$$dV = \pi\{r^2 - (h-z)^2\}dz \tag{2}$$

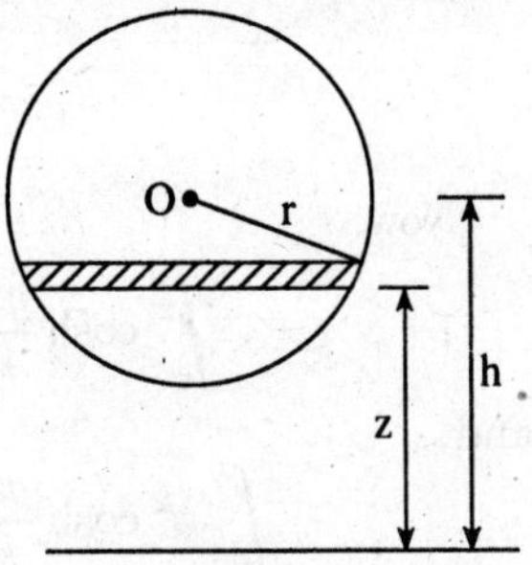

Fig. 11.11

Then the weight of air displaced by this volume is

$$dW = g\rho dV = \pi g\rho\left\{r^2 - (h-z)^2\right\}dz, \tag{3}$$

where ρ is the density of air at height z.

Hence, the total weight of the displaced air is

$$W = \int_{h-r}^{h+r} \pi g\rho\left\{r^2 - (h-z)^2\right\}dz \tag{4}$$

Now, we are given that pressure at a height z in the air is

$$p = p_0 e^{-\frac{gz}{k}} \tag{5}$$

Also we have the relation

$$p = k\rho \tag{6}$$

From (5) and (6), we have

$$\rho = \frac{p_0}{k} e^{-\frac{gz}{k}} \tag{7}$$

Then, from (4) and (7),

$$W = \frac{\pi g p_0}{k}\int_{h-r}^{h+r} e^{-\frac{gz}{k}}\left\{r^2 - (h-z)^2\right\}dz \tag{8}$$

Integral in (8) is split into two like:

$$\begin{aligned} I_1 &= \int_{h-r}^{h} e^{-\frac{gz}{k}}\left\{r^2 - (h-z)^2\right\}dz \\ &= \int_0^r e^{\frac{g}{k}(x-h)}\{r^2 - x^2\}dx \text{ [putting } h - z = x] \\ &= e^{-\frac{g}{k}h}\int_0^r e^{\frac{gx}{k}}(r^2 - x^2)dx \end{aligned} \tag{9}$$

and

$$\begin{aligned} I_2 &= \int_{h}^{h+r} e^{-\frac{gz}{k}}\left\{r^2 - (h-z)^2\right\}dz \\ &= \int_0^r e^{-\frac{g}{k}(x+h)}(r^2 - x^2)dx \text{ [putting } h - z = -x] \\ &= e^{-\frac{g}{k}h}\int_0^r e^{-\frac{g}{k}x}(r^2 - x^2)dx \end{aligned} \tag{10}$$

Hence, using (9) and (10) in (8), we get

$$W = \frac{\pi g p_0}{k} e^{-\frac{gh}{k}} \left\{ \int_0^r \left(e^{\frac{gx}{k}} + e^{-\frac{gx}{k}} \right) (r^2 - x^2) \right\} dx$$

$$= \frac{2\pi g p_0}{k} e^{-\frac{gh}{k}} \int_0^r (r^2 - x^2) \cosh \frac{gx}{k} dx \tag{11}$$

Now,

$$\int_0^r \cosh \frac{gx}{k} dx = \frac{k}{g} \sinh \frac{gx}{k} \Big|_0^r = \frac{k}{g} \sinh \frac{gr}{k} \tag{12}$$

and

$$\int_0^r x^2 \cosh \frac{gx}{k} dx = x^2 \cdot \frac{k}{g} \sinh \frac{gx}{k} \Big|_0^r - \frac{2k}{g} \int_0^r x \sinh \frac{gx}{k} dx$$

$$= \frac{kr^2}{g} \sinh \frac{gr}{k} - \frac{2k}{g} \left[x \cdot \frac{k}{g} \cosh \frac{gx}{k} \Big|_0^r - \int \frac{k}{g} \cosh \frac{gx}{k} \right]$$

$$= \frac{kr^2}{g} \sinh \frac{gr}{k} - \frac{2k^2}{g^2} r \cosh \frac{gr}{k} + \frac{2k^3}{g^3} \sinh \frac{gr}{k} \tag{13}$$

Using (12) and (13) in (11), we get

$$W = \frac{2\pi g p_0}{k} e^{-\frac{gh}{k}} \left[\frac{kr^2}{g} \sinh \frac{gr}{k} - \frac{kr^2}{g} \sinh \frac{gr}{k} + \frac{2k^2 r}{g^2} \cosh \frac{gr}{k} - \frac{2k^3}{g^3} \sinh \frac{gr}{k} \right]$$

$$= 4\pi p_0 e^{-\frac{gh}{k}} \left[\frac{kr}{g} \cosh \frac{gr}{k} - \frac{k^2}{g^2} \sinh \frac{gr}{k} \right] \tag{14}$$

Hence,

$$e^{\frac{gh}{k}} = \frac{4\pi p_0 k}{gw} \left\{ r \cosh \frac{gr}{k} - \frac{k}{g} \sinh \frac{gr}{k} \right\} (\because W = w). \quad \textbf{(Proved)}$$

EXERCISE

1. The readings of a faulty barometer in which there are some air, are a and b, when the true readings are α and β. Show that the true reading when the faulty barometer reads c is

$$c + \frac{(\alpha - a)(\beta - b)(a - b)}{(a - c)(\alpha - a) - (b - c)(\beta - b)}$$

2. A hollow cylinder of height h and open at the top, is inverted and partially immersed so that a length k of it is under water. Prove that the air inside it occupies a length x given by the equation: $x^2 + x(H + k - h) = H \cdot h$, where H is the height of water barometer.

3. A hollow gas-tight balloon containing helium weighs W lbs. When its lowest point touches the ground, it requires a force of w lbs to prevent it from rising. Show that it can float in equilibrium at a height $H \log \left(1 + \frac{w}{W}\right)$, where H is the height of the homogeneous atmosphere.

4. If the pressure of the air varies as $\left(1 + \frac{1}{m}\right)$th power of density, show that, neglecting variation of temperature and gravity, the height of the atmosphere would be equal to $(m + 1)$ times the height of the homogeneous atmosphere.

5. If the absolute temperature diminishes upwards in the atmosphere according to the law $T = \frac{T_0}{1+\beta z}$, where β is a constant, show that the pressure at a height z is given by

$$p = p_0 \exp\left[-\left(\frac{z}{H} + \frac{\beta z^2}{2H}\right)\right]$$

6. A thin conical surface of weight w just sinks to the surface of water. When immersed with its vertex downwards, a weight mw must be placed inside it to make it sink to the same depth as before. If h be the height of a water barometer, prove that the height of the cone is $hm(1+m)^{1/3}$.

7. Masses m and m' of two gases in which the ratios of the pressure to the density are k and k' respectively, are mixed at the same temperature. Prove that the ratio of the pressure and density in the compound is $\frac{mk + m'k'}{m+m'}$.

8. If the law connecting the pressure and density of air were $p = k\rho^2$, prove that, neglecting changes of temperature and gravity, the height of the atmosphere will be twice the height of the homogeneous atmosphere.

9. A closed air-tight cylinder of length $2l$ is half-full of water and half-full of air at atmospheric pressure which is equal to that of a column of height h of water. Water is introduced without letting the air escape so as to fill an additional height k of the cylinder and the pressure of the base is doubled. Prove that $k = l + h - \sqrt{lh + h^2}$.

10. A heavy gas at constant temperature is confined in a vertical cylinder of height h. If ρ be the density at the base, prove that the mean density is $\frac{k\rho}{gh}\left(1 - e^{-\frac{gh}{k}}\right)$.

11. If the temperature of the atmosphere varies inversely as the n-th power of distance from the centre of the earth, show that the pressure of atmosphere at a distance r from the centre of the earth is

$$p = p_0 \exp\left[\frac{g}{RK}\,\frac{r^{n+1} - a^{n+1}}{n+1}\right],$$

where p_0 is the pressure on the surface of the earth whose radius is a.

12. The height of the Torricellian vacuum in a barometer is a inches, and the instrument indicates a pressure of b inches of mercury when the true reading is c inches. If the faulty readings are due to an imperfect vacuum, prove that the true reading corresponding to an apparent reading of d inches is $d + \frac{a(c-b)}{a+b-d}$ inches.

13. As an aeroplane ascends, the temperature and pressure are simultaneously recorded and a curve is drawn plotting the absolute temperature against the logarithm of the pressure. Prove that the height ascended between two readings is $-\frac{k}{g}\int_{x_1}^{x_2} T\,dx$, where $x = \log p$.

14. A closed straight tube of fine uniform bore contains mercury in the middle and air at each end. When the tube is vertical, the portions occupied by air are of lengths α and β respectively, which become α' and β' when the tube is inverted. Prove that the lengths when the tube is horizontal, are

$$\frac{\alpha\alpha'(\beta+\beta')}{\alpha\beta + \alpha'\beta} \text{ and } \frac{\beta\beta'(\alpha+\alpha')}{\alpha\beta+\alpha'\beta'}.$$

15. A mass M of gas at uniform temperature is diffused through all space and at each point (x, y, z), the components of force per unit mass are $-Ax, -By, -Cz$. The pressure and density at the origin are p_0 and ρ_0 respectively. Prove that $ABC\rho_0 M^2 = 8\pi^3 p_0^3$.

16. A piston of weight W rests on a vertical cylinder of transverse section k, being supported by a depth a of air. The piston rod receives a vertical blow P which forces the piston down through a distance h. Prove that

$$(W + \Pi k)\left\{h + a\log\left(1 - \frac{h}{a}\right)\right\} + \frac{gP^2}{2x} = 0$$

17. Two equal vertical tubes are connected at their lower ends by a straight horizontal tube of length a. The upper ends of the tube are open and mercury is poured in. The end of one of the tubes is then closed at a height b above the mercury surface and the system is rotated round the closed tube with angular velocity ω. Prove that, if h be the height of mercury barometer, the mercury surface in the closed tube will sink to a depth z which is the positive root of the equation:

$$4gz^2 + z\{2g(h+2b) - \omega^2 a^2\} - \omega^2 a^2 b = 0.$$

18. One half of a circular tube, radius a, of fine uniform bore, is occupied by mercury and the other half by air at atmospheric pressure. The tube is made to rotate about a vertical diameter with uniform angular velocity ω.

Show that, if the air in the tube remains homogeneous and if $\omega^2 a^2 > 2g(a+h)$, h being the height of the mercury barometer, there will be a vacuum of length $2a\theta$ at the bottom of the tube where

$$\omega^2 a^2 \cos 2\theta = \frac{2\pi gh}{\pi - 2\theta} + 2ga(\sin\theta + \cos\theta).$$

19. Assuming the temperature of the air to diminish uniformly with the height, prove that the difference of level between two stations is $H\dfrac{T_0 - T_1}{273}\dfrac{\log(h_0/h_1)}{\log(T_0/T_1)}$, where H is the height of homogeneous atmosphere at $0\,^\circ\text{C}$, T_0, T_1 are the absolute temperatures at the stations and h_0, h_1 are barometric heights.

Prove also that, if as an approximation the temperature were taken constant and equal to $(T_0+T_1)/2$, the calculated height would be too great by a fraction of its true value equal to

$$\frac{1}{3}\left(\frac{T_0 - T_1}{T_0 + T_1}\right)^2$$

20. A mass M of elastic fluid is rotating about an axis with uniform angular velocity ω, and is acted on by an attraction towards a point in that axis equal to μ times the distance, μ being greater than ω^2. Prove that the equation of a surface of equal density ρ is

$$\mu(x^2+y^2+z^2) - \omega^2(x^2+y^2) = k\log\left\{\frac{\mu(\mu-\omega^2)^2}{8\pi^2}\frac{M^2}{\rho^2 k^2}\right\}$$

21. A given mass M of elastic fluid, for which $p = k\rho$, is forced into a light shell, whose equation is $\dfrac{x^2}{a^2} + \dfrac{y^2}{b^2} + \dfrac{z^2}{c^2} = 1$, and assumes equilibrium under a system of forces, whose force function is

$$\frac{1}{2}k\log\left(\frac{x^2}{a^4} + \frac{y^2}{b^4} + \frac{z^2}{c^4}\right) + \text{constant}.$$

Show that, if p_0 be the pressure at any point of the surface $\dfrac{x^2}{a^4} + \dfrac{y^2}{b^4} + \dfrac{z^2}{c^4} = \dfrac{1}{d^2}$, the mean pressure estimated for equal elements of mass throughout the shell is

$$\frac{4\pi p_0^2 d^2}{15kM}\left\{\frac{bc}{a} + \frac{ca}{b} + \frac{ab}{c}\right\}.$$

CHAPTER 12

Surface Tension and Capillarity

12.1 Introduction

It is our common experience that a small liquid drop tends to assume a spherical shape, and soap films tend to become less extended. If a needle is dropped horizontally with much care in a liquid, it will not sink but float on the surface of the liquid. All these phenomena lead to assume that there is a thin stretched membrane over the liquid surface subjected to a tensile force. We call this force as ***surface tension***.

In an actual stretched membrane, the tension will increase if the membrane is stretched more and more. But in the case of surface tension in a liquid, the tensile force remains constant. Thus the membrane-like behaviour of the liquid surface is not due to the existence of surface tension. It is due to the fact that liquids, not in contact with solids, tend to change the shape of their surfaces such that the surface area is minimum. This is a fundamental property of every liquid.

This property of liquids can be explained on the basis of a molecular hypothesis.

We know that fluids, comprising of small particles of definite shape and size called molecules, are free to move relative to each other. These molecules also attract each other, and the force of attraction between two neighbouring molecules remains effective within a spherical region beyond which it is negligible. We call this region a ***sphere of influence***. Let r_0 be the radius of this sphere of influence.

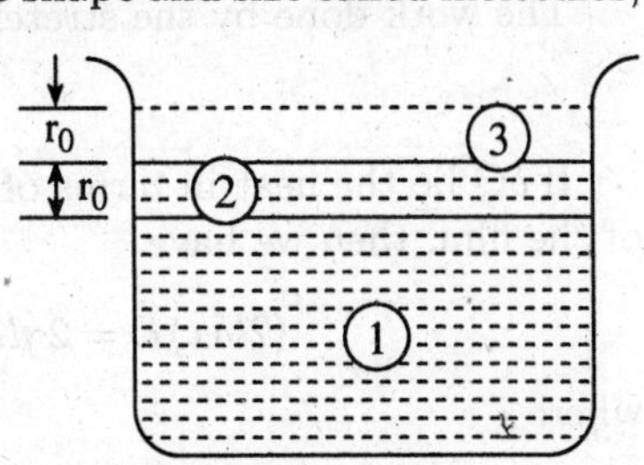

Fig. 12.1

We consider a mass of a fluid and three molecules within it (FIG. 12.1) remaining within their corresponding spheres of influence of radius r_0.

The molecule 1 will be attracted by all other molecules inside the sphere in all directions, so that the resultant force is zero. This molecule is also not influenced by other molecules remaining outside the sphere. The situation however is essentially different in the case of molecules 2 and 3 which are near to the surface. In these cases, a part of the sphere remaining above the liquid will contain gas or vapour. If we neglect the effect of these relatively few molecules, the molecules 2 and 3 can only be attracted by forces which have a resultant component directed towards the interior of the liquid. This downward force will be maximum in the case of molecule 3. Thus the downward force of attraction will be effective to molecules lying within the region of thickness r_0 below the liquid surface. It increases as the molecules approach the surface. The surface layer will be subject to a tension due to the downward force on the surface molecules. Thus the inherent tendency of the liquid surface to contract is explained.

We define the surface tension ***as the force per unit length exerted in the tangential plane to the surface and in a direction perpendicular to an element of a line drawn over the surface.***

12.2 Surface Energy and Surface Tension

The surface of a liquid is caused to contract until it contains a minimum number of molecules. Hence every surface molecule must possess potential energy by virtue of its position. It is manifested very vividly when the soap film is ruptured, the liquid is projected in all directions with a considerable velocity, the potential energy being converted into kinetic energy.

The amount of this energy per unit area of the surface is termed the ***intrinsic surface energy*** or the ***surface energy density*** of the substance.

We consider a liquid film formed in a frame in which the cross-piece AB can move over the two parallel sides (FIG. 12.2) at a distance l apart. The cross-piece AB will be attracted inwards with a force $2\gamma l$, since there are two free faces of the film. In order to hold the film in position, we are to apply a force F in the opposite direction, so that

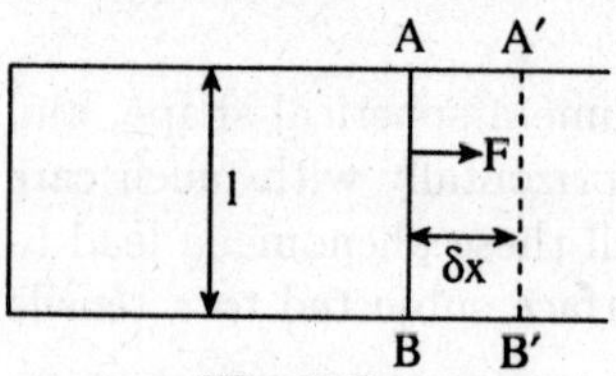

Fig. 12.2

$$F = 2\gamma l \tag{12.1}$$

Here γ is the surface tension of the liquid.

Let AB move a small distance δx to its new position $A'B'$ parallel to its original position. The film will be stretched and as a result, the temperature will fall unless heat is provided from outside. We assume that necessary heat to maintain the temperature of the film is supplied. Then the increase in potential energy of the surface molecules is

$$(2l\delta x)E$$

where E is the intrinsic surface energy of the film.

The work done by the stretching force is

$$F \cdot \delta x$$

If δQ be the heat in terms of thermal energy supplied to maintain the original temperature of the film, then we have

$$(2l\delta x)E = 2\gamma l\delta x + \delta Q, \text{ i.e., } E = \gamma + \frac{\delta Q}{2l\delta x}, \text{ i.e., } E = \gamma + \eta \tag{12.2}$$

where

$$\eta = \frac{\delta Q}{2l\delta x} \tag{12.3}$$

which is the thermal energy supplied per unit increase in the area of the film and is obtained by assuming that the film is subjected to a Carnot cycle.

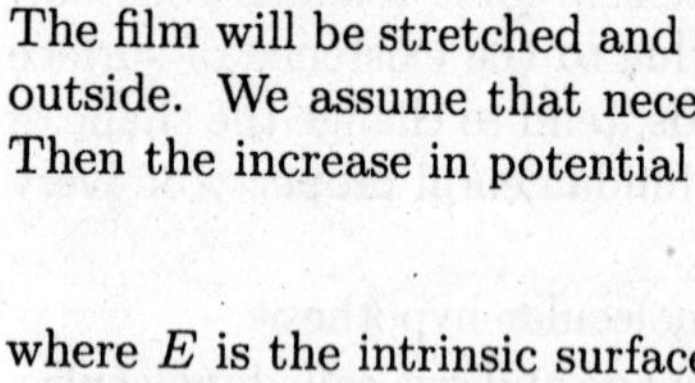

Fig. 12.3

Eqn (12.2) shows that the intrinsic surface energy E of a liquid is actually the sum of two quantities consisting of a mechanical part γ and a thermal part η.

We now consider a thermodynamical system by treating the surface film to a cyclic change of operation in order to determine η (FIG. 12.3).

Let the initial condition of the film be represented by A which is defined by quantities like surface tension γ, absolute temperature T and the area S. We consider that the film is stretched isothermally from condition at A to that represented by B. Since the surface tension is independent of area, work done in this process is

$$\gamma \cdot \delta S$$

and heat absorbed is

$$\eta \cdot \delta S,$$

where δS represents an element of area.

Now we treat the film adiabatically from B to C, where absolute temperature is $(T - \delta T)$. Then from C to D, we allow an isothermal contraction, so that when it expands adiabatically, it finally arrives back at its original state A. Thus the cycle is reversilble.

The work done by the film when it contracts isothermally at absolute temperature $(T-\delta T)$ is

$$\left(\gamma - \frac{\delta\gamma}{\delta T}\delta T\right) \cdot \delta S$$

Hence the net external work done by the film during the whole cycle of operation is

$$\left(\gamma - \frac{\delta\gamma}{\delta T}\delta T\right) \delta S - \gamma \cdot \delta S = -\frac{\delta\gamma}{\delta T}\delta T \delta S \tag{12.4}$$

From the laws of thermodynamics, we write

$$\frac{-\frac{\delta\gamma}{\delta T}\delta T \delta S}{\eta \cdot \delta S} = \frac{\delta T}{T},$$

which yields

$$\eta = -T\frac{d\gamma}{dT} \tag{12.5}$$

Hence, from Eqns (12.2) and (12.5), we write

$$E = \gamma - T\frac{d\gamma}{dT} \tag{12.6}$$

In the case of almost all liquids, the surface tension γ decreases with the rise of temperature so that $\frac{d\gamma}{dT}$ is negative. Hence, E is greater than γ.

The quantity $\frac{d\gamma}{dT}$ is termed the ***temperature coefficient*** of the surface tension.

The surface tension of all liquids at temperature $t\,^\circ$C is given by

$$\gamma = \gamma_0(1 - \alpha t), \tag{12.7}$$

where γ_0 is the surface tension at $0\,^\circ$C and

$$\alpha = \frac{d\gamma}{dt} \tag{12.8}$$

is the temperature coefficient of the surface tension.

According to Van der Waals, the surface tension at absolute temperature T is given by

$$\gamma = \gamma_0 \left(1 - \frac{T}{T_c}\right)^{\frac{3}{2}}, \tag{12.9}$$

where T_c is the critical temperature at which the value of surface tension is zero. It can be observed that the value of α also vanishes at $T = T_c$.

12.3 Pressure Difference Across a Spherical Surface

Let there be a spherical bubble of gas of radius R in a liquid (FIG. 12.4).

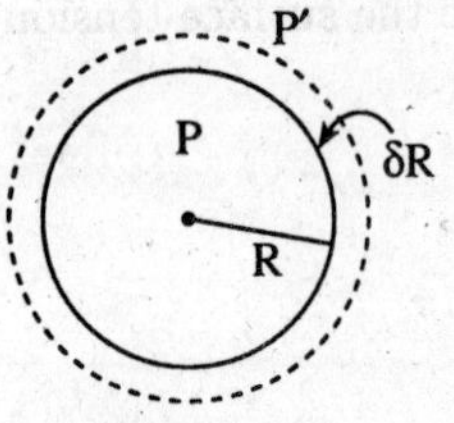

Fig. 12.4

Let P = the pressure inside the bubble, P' = the pressure outside the bubble and γ = the surface tension of the liquid.

It is evident that an excess of pressure acts on the concave side of the surface to counterbalance the effect of surface tension.

We assume that the radius R becomes $R + \delta R$ due to excess of pressure inside, where δR is so small that the pressure inside is not changed. We also assume that the original temperature of the bubble has been maintained by supplying necessary heat from outside.

The surface area of the sphere is

$$S = 4\pi R^2 \tag{12.10}$$

so that

$$\delta S = 8\pi R\delta R. \tag{12.11}$$

Then the increase in total free surface energy is

$$\delta E = 8\pi R\gamma\delta R \tag{12.12}$$

This should be equal to the work done in expanding the bubble.

Let δA be an element of the surface of the bubble. The resultant force acting outwards on it is

$$= P \cdot \delta A - P' \cdot \delta A = p\delta A, \tag{12.13}$$

where $p = P - P'$ = pressure difference.

Then the work done for the expansion is

$$= \left(\int p dA\right) \cdot \delta R = 4\pi R^2 p\delta R \tag{12.14}$$

From Eqns (12.12) and (12.14), we get

$$4\pi R^2 p\delta R = 8\pi\gamma R\delta R, \text{ i.e., } p = \frac{2\gamma}{R} \tag{12.15}$$

For a soap bubble having a double surface,

$$p = \frac{4\gamma}{R} \tag{12.16}$$

12.4 Pressure Difference Across a Cylindrical Surface

It is difficult to produce a cylindrical bubble in a liquid. Hence we assume that it is a soap film having two surfaces.

Let l be the length of the cylinder. Then its total surface area is

$$S = 2(2\pi R)l = 4\pi Rl, \tag{12.17}$$

where R is the radius. Then

$$\delta S = 4\pi l\delta R \tag{12.18}$$

so that the increase in the free surface energy is

$$\delta E = 4\pi l\gamma\delta R \tag{12.19}$$

We assume, as before, that thermal energy has been supplied to restore the original temperature of the film.

Now, the work done in expanding the bubble is

$$= \left(\int p dA \right) \cdot \delta R = 2\pi R l p \cdot \delta R \tag{12.20}$$

Equating the two quantities in Eqns (12.19) and (12.20), we get

$$2\pi R l p \cdot \delta R = 4\pi l \gamma \cdot \delta R, \text{ i.e., } p = \frac{2\gamma}{R} \tag{12.21}$$

12.5 Pressure Difference Across any Curved Surface

We now consider the general case when the curved surface is arbitrary.

Let $ABCD$ be an element of a curved membrane, the surface of which is synclastic (FIG. 12.5). Let the normals on the surface at A and B meet at O, while those at B and C meet at O'. Then, $OA = \rho_1$ and $O'C = \rho_2$ are the principal radii of curvature of the surface.

Let $AB = x$, and $BC = y$.

Let this area be pushed outward normally through a space δz to its new position $A'B'C'D'$, so that

$$A'B' = x + \delta x, \; B'C' = y + \delta y, \; CC' = \delta z.$$

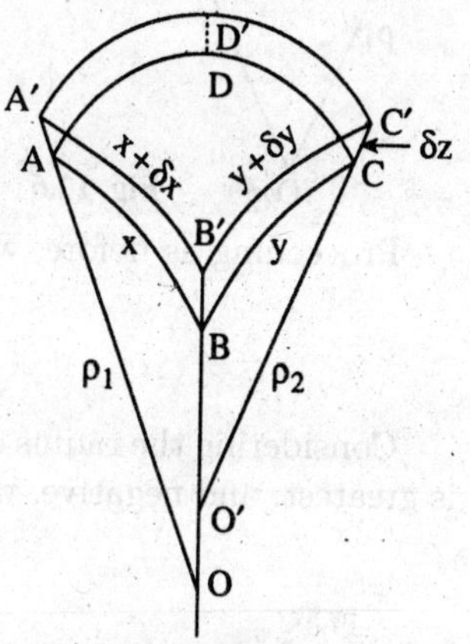

Fig. 12.5

The temperature in this process remains unaltered. Now, from the figure,

$$\frac{x + \delta x}{\rho_1 + \delta z} = \frac{x}{\rho_1}, \text{ i.e., } x + \delta x = x \left(1 + \frac{\delta z}{\rho_1} \right) \tag{12.22}$$

Similarly,

$$y + \delta y = y \left(1 + \frac{\delta z}{\rho_2} \right) \tag{12.23}$$

Let p be the excess of presure on the concave side of the surface $ABCD$ to counterbalance the effect of surface tension. Then the force acting on it is

$$p(\text{area } ABCD) = pxy$$

The work done by this force due to the normal displacement δz is

$$pxy\delta z \tag{12.24}$$

The increase in the free surface energy of the element considered is

$$\begin{aligned}
&\gamma(\text{area } A'B'C'D' - \text{area } ABCD) \\
&= \gamma[(x + \delta x)(y + \delta y) - xy] \\
&= \gamma \left[x \left(1 + \frac{\delta z}{\rho_1} \right) y \left(1 + \frac{\delta z}{\rho_2} \right) - xy \right] \quad \text{[using (12.22) and (12.23)]} \\
&= \gamma \left[xy \left(1 + \frac{\delta z}{\rho_1} + \frac{\delta z}{\rho_2} \right) - xy \right] \\
&= xy\gamma\delta z \left(\frac{1}{\rho_1} + \frac{1}{\rho_2} \right),
\end{aligned} \tag{12.25}$$

where we have neglected the term containing $(\delta z)^2$ due to its smallness.

Now, this increase in free surface energy must be equal to the work done in expanding the surface under isothermal condition. Hence we get, equating the expressions (12.24) and (12.25)

$$p = \gamma\left(\frac{1}{\rho_1} + \frac{1}{\rho_2}\right) \tag{12.26}$$

Let us now consider an anticlastic surface as shown in the figure (FIG. 12.6).

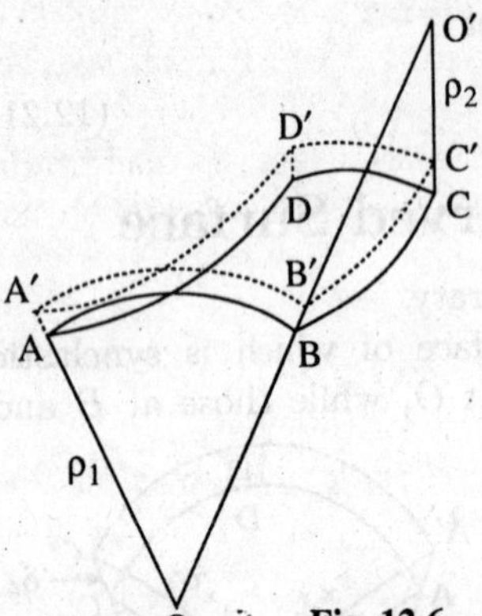

Fig. 12.6

In this case,

$$\frac{B'C'}{BC} = \frac{\rho_2 - \delta z}{\rho_2}, \text{ i.e., } \frac{y+\delta y}{y} = \left(1 - \frac{\delta z}{\rho_2}\right)$$

so that

$$y + \delta y = y\left(1 - \frac{\delta z}{\rho_2}\right) \tag{12.27}$$

Then, area $A'B'C'D'$

$$= (x+\delta x)(y+\delta y) = x\left(1 + \frac{\delta z}{\rho_1}\right)\cdot y\left(1 - \frac{\delta z}{\rho_2}\right)$$

$$= xy\left(1 + \frac{\delta z}{\rho_1} - \frac{\delta z}{\rho_2}\right) \tag{12.28}$$

Proceeding as before, we get

$$p = \gamma\left(\frac{1}{\rho_1} - \frac{1}{\rho_2}\right) \tag{12.29}$$

Considering the radius of curvature to be positive when it lies on the side, where the pressure is greatest, and negative, when it lies on the side of least pressure, we have the general formula

$$p = \gamma\left(\frac{1}{\rho_1} + \frac{1}{\rho_2}\right) \tag{12.30}$$

If the free surface be on both sides of the element as in the case of a film, the pressure difference will be doubled, so that Eqn (12.30) in this case assumes the form

$$p = 2\gamma\left(\frac{1}{\rho_1} + \frac{1}{\rho_2}\right) \tag{12.31}$$

From this equation directly follow the values of pressure difference when the surface is spherical or cylindrical.

12.6 Analytical Method for Surface Tension Equation

We now intend to derive Eqn (12.30) by a rigorous analytical method.

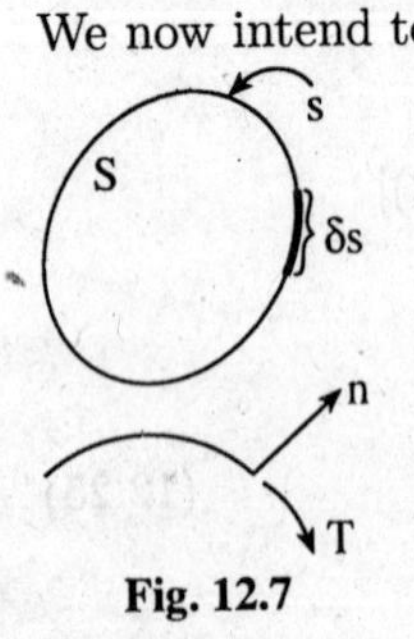

Fig. 12.7

Let the equation of the surface be given by

$$z = f(x, y) \tag{12.32}$$

Then we have the following notations:

$$p = \frac{\partial z}{\partial x},\ q = \frac{\partial z}{\partial y},\ r = \frac{\partial^2 z}{\partial x^2},\ s = \frac{\partial^2 z}{\partial x \partial y},\ t = \frac{\partial^2 z}{\partial y^2} \tag{12.33}$$

We consider the equilibrium of any portion S of the surface bounded by the curve s without any singularity (FIG. 12.7).

The tension γ on the small line element δs will be $\gamma\delta s$ acting at right angles to δs as well as to the normal to the surface.

If l, m, n denote the direction cosines of the tension $\gamma\delta s$, then the resultant of the tensions along z-axis is

$$F_1 = \int \gamma n ds, \tag{12.34}$$

where integration is taken around the boundary of the curve s.

Again, denoting by p^* the difference of pressure on the two sides of the surface, its magnitude on an elementary area δS will be $p^* dS$. If λ, μ, ν be the direction cosines of the normal to the surface, then the resultant of the pressure along z-axis is

$$F_2 = \iint p^* \nu dS \tag{12.35}$$

Since νdS is the projection of the surface area dS on xy-plane which is equal to $dxdy$, we write Eqn (12.35) in the form

$$F_2 = \iint p^* dxdy, \tag{12.36}$$

where the integration is carried over the projection of surface area S on xy-plane.

Now, the direction cosines of the tangent to the line element δs, are

$$x' = \frac{dx}{ds},\ y' = \frac{dy}{ds},\ z' = \frac{dz}{ds} \tag{12.37}$$

Since the normal to the surface S at the point (x, y, z), tangent to the line element δs and normal to the line element δs are three mutually perpendicular lines, we have

$$\frac{l}{\mu z' - \nu y'} = \frac{m}{\nu x' - \gamma z'} = \frac{n}{\lambda y' - \mu x'} = 1 \tag{12.38}$$

Hence, we find

$$n = \lambda y' - \mu x' = \lambda \frac{dy}{ds} - \mu \frac{dx}{ds} \tag{12.39}$$

Using this relation in Eqn (12.34), we get

$$F_1 = \int \gamma(\lambda dy - \mu dx) \tag{12.40}$$

By using Green's theorem, we transform this line integral into a surface integral and write

$$F_1 = \iint \gamma \left(\frac{\partial \lambda}{\partial x} + \frac{\partial \mu}{\partial y}\right) dxdy \tag{12.41}$$

For equilibrium of the portion S of the surface bounded by the curve s, we must have

$$F_1 = F_2$$

$$\text{i.e.,} \quad \iint p^* dxdy - \int \gamma \left(\frac{\partial \lambda}{\partial x} + \frac{\partial \mu}{\partial y}\right) dxdy = 0$$

$$\text{i.e.,} \quad \iint \left[p^* - \gamma \left(\frac{\partial \lambda}{\partial x} + \frac{\partial \mu}{\partial y}\right)\right] dxdy = 0 \tag{12.42}$$

The integral in (12.42) vanishes for an arbitrary region $dxdy$. Hence the integrand must be equal to zero. Thus,

$$p^* = \gamma \left(\frac{\partial \lambda}{\partial x} + \frac{\partial \mu}{\partial y} \right) \tag{12.43}$$

But we have

$$\lambda = \frac{p}{\sqrt{p^2+q^2+1}}, \quad \mu = \frac{q}{\sqrt{p^2+q^2+1}}, \quad \nu = \frac{-1}{\sqrt{p^2+q^2+1}} \tag{12.44}$$

By standard method of partial differentiation, we can have

$$\frac{\partial \lambda}{\partial x} + \frac{\partial \mu}{\partial y} = \frac{r(1+q^2) - 2pqs + t(1+p^2)}{\sqrt{p^2+q^2+1}} \tag{12.45}$$

But the expression on the right-hand side of Eqn (12.45) is equal to

$$\frac{1}{\rho_1} + \frac{1}{\rho_2}$$

so that

$$\frac{\partial \lambda}{\partial x} + \frac{\partial \mu}{\partial y} = \frac{1}{\rho_1} + \frac{1}{\rho_2}, \tag{12.46}$$

where ρ_1, ρ_2 are the principal radii of curvature of the surface at the point (x, y, z). Hence, from Eqns (12.43) and (12.46), we get

$$p^* = \gamma \left(\frac{1}{\rho_1} + \frac{1}{\rho_2} \right) \tag{12.47}$$

Note. In this section, we have denoted the pressure difference by p^* instead of p in order to avoid any confusing with p in Eqn (12.33).

12.7 Interfacial Surface Tensions

The surface tension being due to molecular attraction, it exists in the interfaces of liquids and gases, liquids and solids and solids and gases.

We consider a solid, a liquid and a gas, all in contact and in equilibrium. A vertical section through the system is shown in the figure (FIG. 12.8), where the line of contact is perpendicular to the plane of the paper.

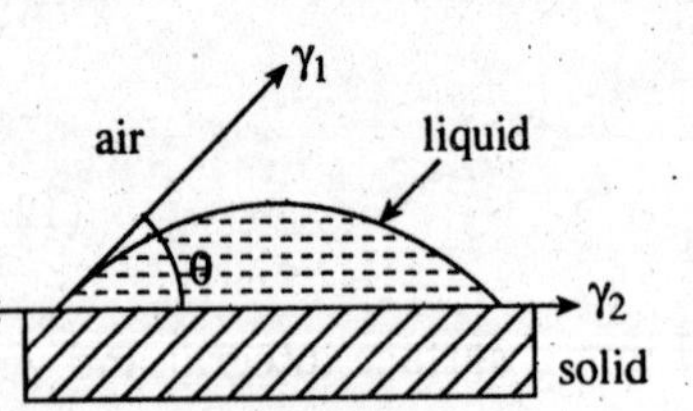

Fig. 12.8

Let γ_1, γ_2 and γ_3 be the interfacial surface tensions between the (1) liquid and air, (2) liquid and solid and (3) solid and air respectively.

Let θ be the angle between the tangent planes to the liquid and to the solid along their line of contact. This angle is called the ***angle of contact*** for the given solid and liquid.

For equilibrium of the liquid drop, we must have

$$\gamma_2 + \gamma_1 \cos\theta - \gamma_3 = 0, \text{ i.e., } \cos\theta = \frac{\gamma_3 - \gamma_2}{\gamma_1} \tag{12.48}$$

Since $\cos\theta$ cannot be greater than unity, we conclude from Eqn (12.48) that, for the equilibrium of the liquid drop, we must have

$$\frac{\gamma_3 - \gamma_2}{\gamma_1} \not> 1. \tag{12.49}$$

From Eqn (12.48), it is also obvious; that when $\gamma_3 > \gamma_2$, $\cos\theta$ is positive, i.e., θ is an acute angle. Similarly, θ is an obtuse angle, when $\gamma_3 < \gamma_2$ (FIG. 12.9).

The solid is only partially wetted by the liquid, when

$$0 < \theta < \frac{\pi}{2} \qquad (12.50)$$

and there will be no wetting, if

$$\frac{\pi}{2} < \theta < \pi \qquad (12.51)$$

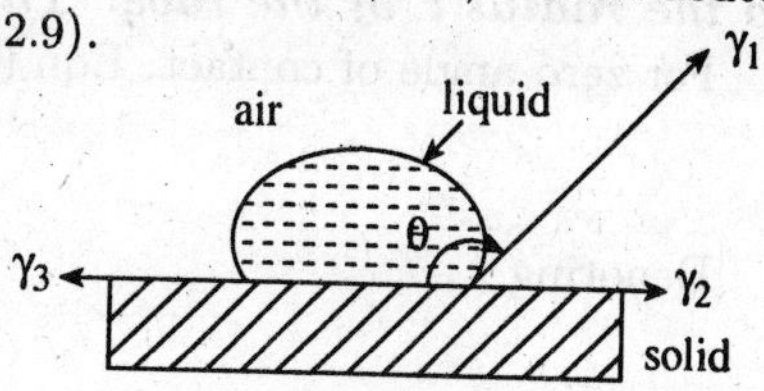

Fig. 12.9

12.8 Capillarity

When a capillary tube is dipped into a liquid, the liquid rises or descends within the tube. This phenomenon of ascent or descent of a liquid in a capillary tube is known as ***capillarity***. The liquid meniscus in the tube is concave in the case of ascent, and convex in the case of descent.

Let us consider the ascent of a liquid in a capillary tube dipped vertically (FIG. 12.10).

Let r = radius of the capillary tube, h = height of the ascent of the liquid in the capillary tube, θ = acute angle of contact, ρ = density of the liquid, γ = surface tension of the liquid.

We assume that the meniscus surface is hemispherical. The line of contact of the liquid in the tube is a circle of radius r. Hence, due to surface tension, a force equal to $2\pi r\gamma$ act along the tangent line to the liquid surface at each point of contact in the inward direction. Its reaction force will be equal in magnitude but opposite in direction. The vertical component F of this reaction force is

$$F = 2\pi r\gamma\cos\theta. \qquad (12.52)$$

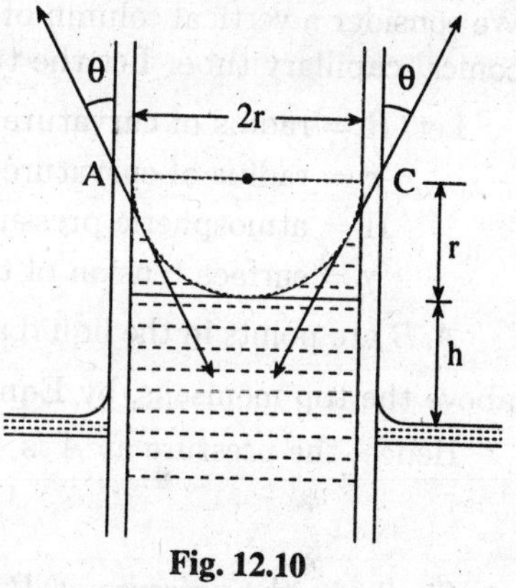

Fig. 12.10

This force is balanced by the weight W of the liquid in the tube which is equal to weight of liquid in the cylinder of radius r and height $(h+r)$ minus the weight of liquid in the hemisphere of radius r.

Hence,

$$W = \pi r^2 h\rho g + \rho g\left(\pi r^3 - \frac{2}{3}\pi r^3\right)$$

$$= \pi r^2 h\rho g + \frac{1}{3}\pi r^3\rho g = \pi r^2\rho g\left(h + \frac{r}{3}\right) \qquad (12.53)$$

For equilibrium,

$$F = W, \text{ i.e., } 2\pi r\gamma\cos\theta = \pi r^2\rho g\left(h + \frac{r}{3}\right), \text{ i.e., } \gamma = \frac{r\rho g\left(h + \frac{r}{3}\right)}{2\cos\theta} \qquad (12.54)$$

In a capillary tube, $h \gg r$, so that the factor $r/3$ in Eqn (12.54) can be neglected. Then

$$\gamma = \frac{rh\rho g}{2\cos\theta} \qquad (12.55)$$

For a given liquid, γ, ρ and θ are constants, so that

$$rh = \text{constant, i.e., } h \propto \frac{1}{r} \qquad (12.56)$$

Thus, ***for a given liquid, the ascent h in a capillary tube is inversely proportional to the radius r of the tube. This is known as Jurin's law.***

For zero angle of contact, Eqn (12.55) yields

$$\frac{2\gamma}{r} = g\rho h \tag{12.57}$$

Denoting

$$\frac{\gamma}{g\rho} = a^2, \tag{12.58}$$

we get from Eqn (12.57)

$$2a^2 = rh \tag{12.59}$$

The quantity a^2 is known as the ***capillary constant*** or ***specific cohesion*** of the liquid. It is also known as ***Laplace's constant***.

12.9 Liquid in a Slightly Conical Capillary Tube

We consider a vertical column of liquid of density ρ and height h in a slightly conical capillary tube. Let the tube taper upwards (FIG. 12.11).

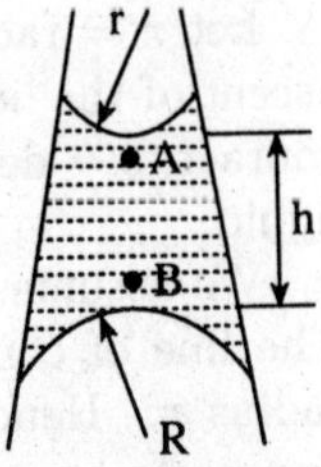

Fig. 12.11

Let R = radius of curvature of lower meniscus
r = radius of curvature of upper meniscus
Π = atmospheric pressure
γ = surface tension of the liquid.

A, B are points in the liquid as shown in the figure. The excess of pressure above the top meniscus, by Eqn (12.15), is $\frac{2\gamma}{r}$.

Hence, the pressure at A is

$$p_A = \Pi - \frac{2\gamma}{r} \tag{12.60}$$

Similarly, the pressure at B is

$$p_B = \Pi - \frac{2\gamma}{R} \tag{12.61}$$

But

$$p_B = p_A + g\rho h, \text{ i.e., } \Pi - \frac{2\gamma}{R} = \Pi - \frac{2\gamma}{r} + g\rho h, \text{ i.e., } g\rho h = 2\gamma\left(\frac{1}{r} - \frac{1}{R}\right) \tag{12.62}$$

Since

$$r < R, \quad \left(\frac{1}{r} - \frac{1}{R}\right) > 0 \tag{12.63}$$

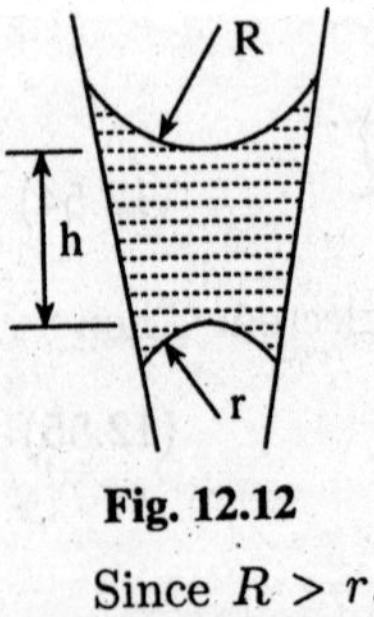

Fig. 12.12

Hence Eqn (12.62) can be satisfied and the liquid takes a position to reach the above height. The equilibrium in this case is stable.

If the narrow end of the tube points downward as shown in the figure (FIG. 12.12), Eqn (12.62) in this case assumes the form

$$g\rho h = 2\gamma\left(\frac{1}{R} - \frac{1}{r}\right) \tag{12.64}$$

Since $R > r$, the right-hand side of Eqn (12.64) is negative, and hence it is impossible to satisfy this equation.

The equilibrium is unstable and the liquid column flows to the lower end of the tube, where a small drop is formed (FIG. 12.13)

In this case, let r_1 = radius of curvature of the top meniscus, R_1 = radius of curvature of the drop, z = the overall length of the drop.

Then pressure at point A is

$$p_A = \Pi - \frac{2\gamma}{r_1} \tag{12.65}$$

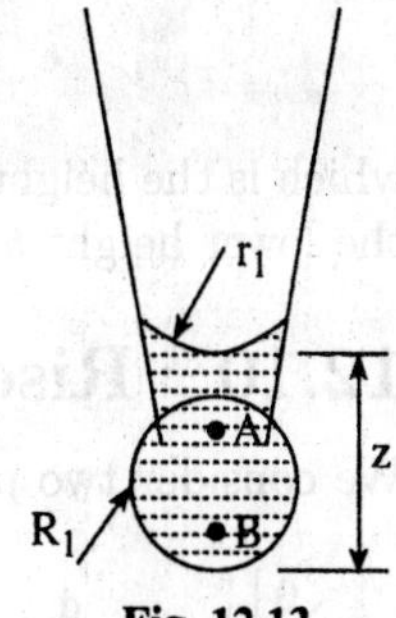

Fig. 12.13

and that at B is

$$p_B = \Pi + \frac{2\gamma}{R_1} \tag{12.66}$$

But

$$p_B = p_A + g\rho z$$

so that

$$\Pi + \frac{2\gamma}{R_1} = \Pi - \frac{2\gamma}{r_1} + g\rho z, \text{ i.e., } g\rho z = 2\gamma\left(\frac{1}{R_1} + \frac{1}{r_1}\right) \tag{12.67}$$

The right-hand side of Eqn (12.67) is always positive whatever may be the value of r_1 and hence the equation can be always satisfied.

Let us now consider a glass conical capillary tube dipped vertically in a liquid, the apex being open to atmosphere (FIG. 12.14).

Let l = the length of the tube, h = height of ascent of the liquid, r = radius of curvature of the meniscus, and R = maximum radius of cross section of the tube.

The angle of contact with glass is zero.

Then, from the figure,

$$\frac{r}{R} = \frac{l-h}{l}, \text{ i.e., } r = \frac{R}{l}(l-h) \tag{12.68}$$

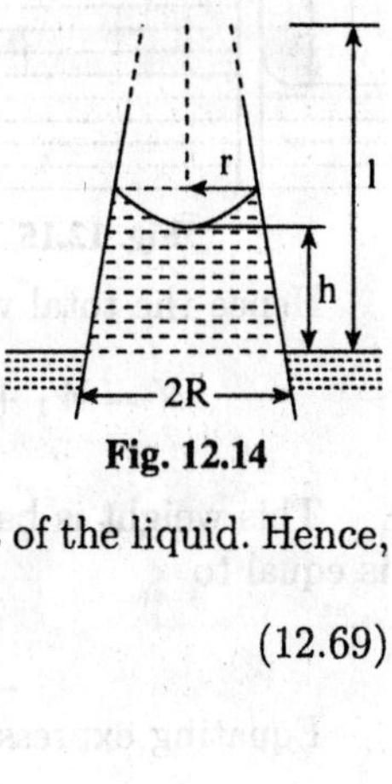

Fig. 12.14

The excess of pressure at the top meniscus $\frac{2\gamma}{r}$ is balanced by the weight of the liquid. Hence,

$$\frac{2\gamma}{r} = g\rho h, \text{ i.e., } r = \frac{2\gamma}{g\rho h} \tag{12.69}$$

From Eqns (12.68) and (12.69),

$$\frac{R}{l}(l-h) = \frac{2\gamma}{g\rho h}, \text{ i.e., } h(l-h) = \frac{2\gamma l}{g\rho R}$$

$$\therefore h^2 - hl + \frac{2\gamma l}{g\rho R} = 0 \tag{12.70}$$

This yields

$$h = \frac{l}{2} \pm \frac{l}{2}\sqrt{1 - \frac{8\gamma}{g\rho R l}} \tag{12.71}$$

For large values of l, we have

$$\sqrt{1 - \frac{8\gamma}{g\rho R l}} \simeq 1 - \frac{4\gamma}{g\rho R l} \tag{12.72}$$

In this case, taking minus sign in Eqn (12.71), we get

$$h_1 = \frac{l}{2} - \frac{l}{2}\left(1 - \frac{4\gamma}{g\rho R l}\right) = \frac{2\gamma}{g\rho R} \tag{12.73}$$

Considering positive sign in Eqn (12.71), we get

$$h_2 = l - \frac{2\gamma}{g\rho R}, \tag{12.74}$$

which is the height of ascent of the liquid in a uniform capillary of radius R. This suggests that the lower height h_1 given by Eqn (12.73) is stable.

12.10 Rise of Liquid between Two Plates

We consider two parallel vertical plates remaining at a small distance d apart, and are dipped in a liquid of density ρ (FIG. 12.15).

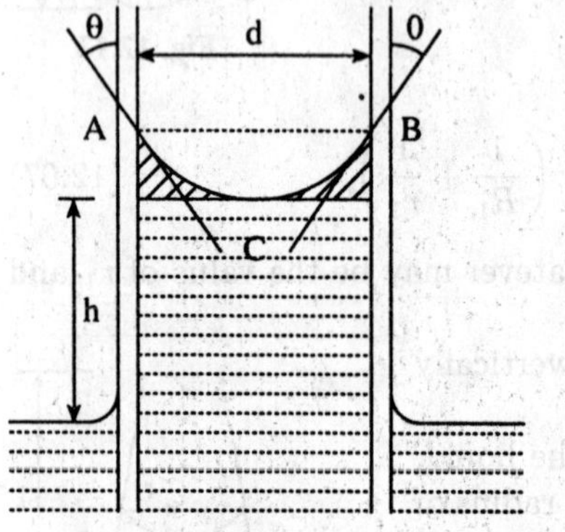

Fig. 12.15

Let θ be the constant angle of contact. Considering unit breadth of the liquid perpendicular to the plane of paper, the weight W_1 of liquid column of height h is

$$W_1 = hd\rho g \tag{12.75}$$

The weight W_2 of the liquid in the shaded portion below the meniscus, which is assumed to be semi-circular, is

$$W_2 = \left(d \cdot \frac{d}{2} - \frac{1}{2}\frac{\pi d^2}{4}\right) g\rho$$

$$= \frac{1}{2}\left(1 - \frac{\pi}{4}\right) d^2 g\rho \tag{12.76}$$

Hence the total weight of the liquid is

$$W = W_1 + W_2 = hdg\rho + \frac{1}{2}\left(1 - \frac{\pi}{4}\right) d^2 g\rho = dg\rho\left[h + \frac{1}{2}\left(1 - \frac{\pi}{4}\right) d\right] \tag{12.77}$$

This weight is balanced by the vertical component of force F due to surface tension which is equal to

$$F = 2\gamma\cos\theta \tag{12.78}$$

Equating expressions (12.77) and (12.78), we get

$$dg\rho\left[h + d\left(\frac{1}{2} - \frac{\pi}{8}\right)\right] = 2\gamma\cos\theta$$

$$\text{i.e., } h = \frac{2\gamma\cos\theta}{dg\rho} - d\left(\frac{1}{2} - \frac{\pi}{8}\right) = \frac{2\gamma\cos\theta}{dg\rho} - 0.107d \tag{12.79}$$

12.11 Force between Two Parallel Plates

We consider two parallel plates separated by a thin layer of liquid (FIG. 12.16), where the area of the plates is large. It is found from experiments that a large force in the direction normal to the plane of the plates is required to separate them. We intend to determine this force in the case of two types of liquids.

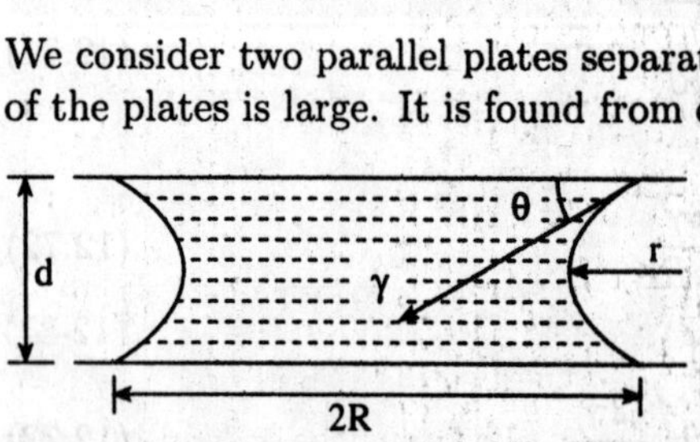

Fig. 12.16

Let d = distance between the plates, R = radius of the liquid drop with the upper plate, γ = surface tension of the liquid, and θ = angle of contact between the liquid and the plates.

In the case when the liquid meniscus is concave of radius of curvature r, the pressure within the liquid would be less than the atmospheric pressure by an amount

$$p = \gamma\left(\frac{1}{r} - \frac{1}{R}\right) \simeq \frac{\gamma}{r} \tag{12.80}$$

Since the plates are very close

$$r = \frac{d}{2\cos\theta} = \frac{1}{2}d\sec\theta \tag{12.81}$$

Thus the pressure difference is

$$p = \frac{\gamma}{\frac{1}{2}d\sec\theta} = \frac{2\gamma}{d\sec\theta} \tag{12.82}$$

Hence the resultant force F on the upper plate acting downwards is

$$F = (\pi R^2)\,p + 2\pi R\gamma\sin\theta$$
$$= \pi R^2\left(\frac{2\gamma}{d\sec\theta}\right) + 2\pi R\gamma\sin\theta = 2\pi\gamma R\left(\frac{R}{d\sec\theta} + \sin\theta\right) \tag{12.83}$$

Hence the force required to separate the plates in the direction normal to their plane is

$$F = 2\pi R\left(\frac{\gamma R}{d\sec\theta} + \gamma\sin\theta\right) \tag{12.84}$$

The term $\gamma\sin\theta$ being very small compared to the other term in Eqn (12.84), can be neglected. Thus

$$F = \frac{2\pi R^2\gamma}{d\sec\theta} \tag{12.85}$$

In the case when the liquid meniscus is convex, as in the case of mercury (FIG. 12.17), the excess of pressure within the mercury is

$$p = \frac{2\gamma}{r} = \frac{2\gamma}{d}\cos(\pi - \theta), \tag{12.86}$$

where r is the radius of curvature.

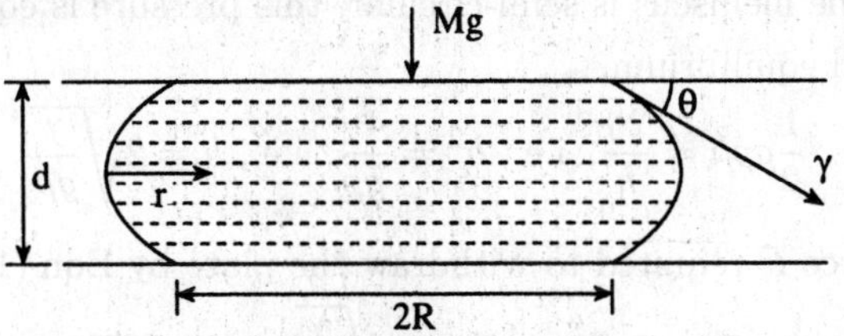

Fig. 12.17

Hence resultant force on the upper plate is

$$F = (\pi R^2)\,p - 2\pi\gamma R\sin(\pi - \theta)$$
$$= \pi R^2\left[\frac{2\gamma}{d}\cos(\pi - \theta)\right] - 2\pi\gamma R\sin(\pi - \theta)$$
$$= 2\pi R\left[\frac{\gamma R}{d}\cos(\pi - \theta) - \gamma\sin(\pi - \theta)\right] \tag{12.87}$$

This force acts in the upward direction and hence the upper plate must be loaded in order to maintain the liquid drop in between the plates.

If M be the required mass on the upper plate, then

$$Mg = F, \text{ i.e. } Mg = 2\pi R\left[\frac{\gamma R}{d}\cos(\pi-\theta) - \gamma\sin\theta\right] \tag{12.88}$$

Neglecting the term $\gamma\sin\theta$, we write

$$Mg = \frac{2\pi\gamma R^2}{d\sec(\pi-\theta)} \tag{12.89}$$

12.12 Force to Pull a Plate from a Liquid Surface

We consider a flat plate placed on the surface of a liquid at rest. If the plate is slightly raised, a layer of liquid will adhere to the plate.

The amount of liquid adhering to the plate will increase along with the increase in force F normal to the plate. The plate will suddenly be separated at a particular value of F (FIG. 12.18).

At this point of situation, let h = the height of the plate above the liquid surface, S = the area of the plate in contact with the liquid, P = the atmospheric pressure, ρ = the density of the liquid.

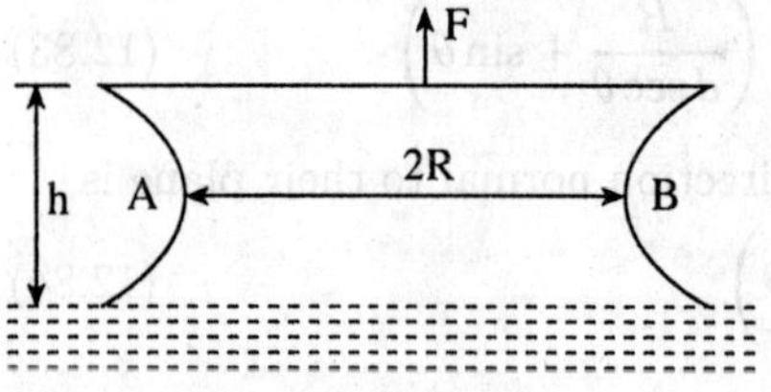

Fig. 12.18

Then the pressure due to the atmosphere on the top of the plate is $P\cdot S$ and that just below the plate is $(P - g\rho h)\cdot S$ so that the force due to the difference of pressure is

$$F = P\cdot S - (P - g\rho h)\cdot S = Sg\rho h \tag{12.90}$$

This will be the force required to withdraw the plate from the liquid surface.

Let the area of plate in contact with the liquid be a circle of radius R. When R is large, the liquid meniscus in a vertical plane is everywhere a semi-circle of radius $h/2$. Then at a point in the liquid along the least horizontal section AB, the pressure will be less than the atmospheric pressure by an amount

$$= \gamma\left(\frac{1}{r} - \frac{1}{R}\right) = \gamma\left(\frac{2}{h} - \frac{1}{R}\right) \simeq \frac{2\gamma}{h}$$

Since the shape of the meniscus is semi-circular, this pressure is equal to $\frac{1}{2}g\rho h$.

Hence, for horizontal equilibrium,

$$\frac{1}{2}g\rho h = \frac{2\gamma}{h}, \text{ i.e., } h^2 = \frac{4\gamma}{g\rho}, \text{ i.e., } h = 2\sqrt{\frac{\gamma}{g\rho}} \tag{12.91}$$

Then the normal force F required to withdraw the plate by Eqn (12.90) is

$$F = Sg\rho h = Sg\rho\cdot 2\sqrt{\frac{\gamma}{g\rho}} = 2S\sqrt{\gamma g\rho} \tag{12.92}$$

12.13 Shape of an Interfacial Surface

We consider an incompressible liquid of density ρ remaining at rest in a vessel under the action of gravity only and in contact with air. We intend to determine the shape of the common surface PQ (FIG. 12.19), with the use of the principle of energy, i.e., the potential energy in the equilibrium state must be stationary.

With O as the origin, we take coordinate axes Ox, Oz and consider an elementary area dS around the point A with coordinates (x, z), S being the area of the surface between the liquid and air.

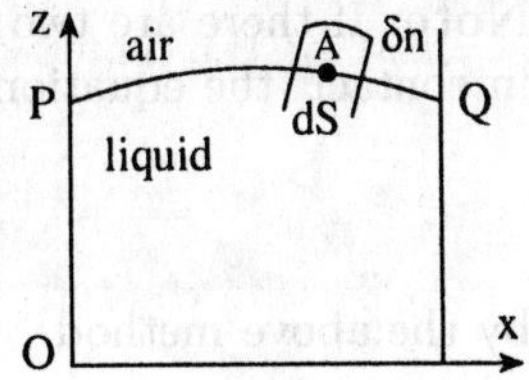

Fig. 12.19

Let the element be slightly displaced normally through a distance δn. Then the variation of the gravitational energy is

$$g\rho z\delta n dS$$

so that the variation of total gravitational energy is

$$= \iint g\rho z\delta n dS \tag{12.93}$$

Let ρ_1, ρ_2 be the principal radii of curvature at the point A of the surface. Then the excess of pressure at the concave side is

$$p = \gamma\left(\frac{1}{\rho_1} + \frac{1}{\rho_2}\right), \tag{12.94}$$

since there is only one surface of separation. The work done by this pressure on elementary area dS due to normal displacement δn is

$$dW = p\delta n dS = \gamma\left(\frac{1}{\rho_1} + \frac{1}{\rho_2}\right)\delta n dS \tag{12.95}$$

Hence the work done due to the small displacement of the entire surface is

$$W = \iint \gamma\left(\frac{1}{\rho_1} + \frac{1}{\rho_2}\right)\delta n dS \tag{12.96}$$

This work done is equal to the negative of the variation of the energy of the surface separating the liquid and air. Hence, by the principle of energy, we require that

$$\iint g\rho z\delta n dS - \iint \gamma\left(\frac{1}{\rho_1} + \frac{1}{\rho_2}\right)\delta n dS = 0$$

$$\text{i.e.,} \iint \left\{g\rho z - \gamma\left(\frac{1}{\rho_1} + \frac{1}{\rho_2}\right)\right\}\delta n dS = 0 \tag{12.97}$$

Since the liquid is incompressible, the variation of its volume is equal to zero. Thus

$$\iint \delta n dS = 0. \tag{12.98}$$

By virtue of this, Eqn (12.97) yields

$$\gamma\left(\frac{1}{\rho_1} + \frac{1}{\rho_2}\right) - g\rho z = \text{constant}, \tag{12.99}$$

which is the equation of the surface of separation.

If we consider the atmospheric pressure Π above the liquid, then the work done by this pressure due to the displacement considered is

$$-\iint \Pi\delta n dS.$$

Eqn (12.97) in this case will assume the form

$$\iint \left\{g\rho z + \Pi - \gamma\left(\frac{1}{\rho_1} + \frac{1}{\rho_2}\right)\right\}\delta n dS = 0, \tag{12.100}$$

which yields

$$\gamma\left(\frac{1}{\rho_1} + \frac{1}{\rho_2}\right) - g\rho z = \Pi + \text{constant} \tag{12.101}$$

as the equation of the surface of separation.

Note. If there are two incompressible and immiscible liquids of densities δ_1 and δ_2 ($\delta_2 > \delta_1$) in contact, the equation of the surface of separation may be determined in the form

$$\gamma\left(\frac{1}{\rho_1}+\frac{1}{\rho_2}\right) - g(\delta_2-\delta_1)z = \Pi + \text{constant} \tag{12.102}$$

by the above method.

12.14 Rise of a Liquid along an Inclined Plate

We consider a perfectly clean glass plate AB dipped into a liquid of density δ making an angle ϕ with the horizon (FIG. 12.20).

The principal section of the free surface is given by the curve APQ, $P(x, z)$ being any point on this curve. The surface being cylindrical

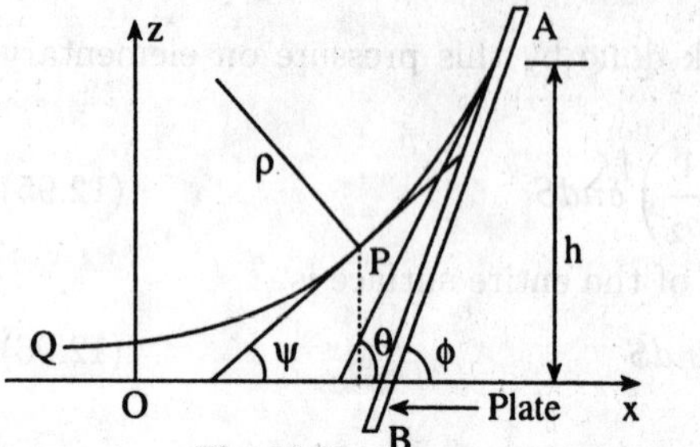

Fig. 12.20

$$\frac{1}{\rho_1} = 0$$

$$\frac{1}{\rho_2} = \frac{1}{\rho} \text{ (say).}$$

Then the general equation of the curve APQ, by Eqn (12.101), is

$$\frac{\gamma}{\rho} - g\delta z = \Pi + \text{constant.} \tag{12.103}$$

We assume that at a great distance from the plate, the curve APQ becomes horizontal so that x-axis becomes an asymptote. That is,

$$\rho \to \infty, \text{ as } z \to 0.$$

This yields from Eqn (12.103) that the constant term is $-\Pi$. Thus we get

$$\frac{\gamma}{\rho} = g\delta z \tag{12.104}$$

We draw a tangent at P, which makes an angle ψ with the x-axis. Then we have the relations

$$\sin\psi = \frac{dz}{ds},\ \frac{1}{\rho} = \frac{d\psi}{ds}, \tag{12.105}$$

where ds is an elementary arc at P.

Using these relations into Eqn (12.104),

$$\gamma\frac{d\psi}{ds} = g\delta z, \text{ i.e., } \gamma\frac{d\psi\sin\psi}{dz} = g\delta z, \text{ i.e., } \gamma\sin\psi d\psi = g\delta z dz \tag{12.106}$$

Integrating,

$$-\gamma\cos\psi = \frac{1}{2}g\delta z^2 + C, \tag{12.107}$$

where C is the constant of integration.

At $z = 0, \psi = 0$, since the curve is horizontal there. Hence, from Eqn (12.107), we get

$$C = -\gamma \tag{12.108}$$

Thus, Eqn (12.107) assumes the final form

$$g\delta z^2 = 2\gamma(1-\cos\psi) \tag{12.109}$$

If the tangent to the liquid surface at the point of contact with the plate makes an angle θ with x-axis, θ is the angle of contact, and

$$\phi = \theta + \psi, \text{ i.e., } \psi = \phi - \theta \tag{12.110}$$

Eqn (12.109) then can be written in the form:

$$g\delta z^2 = 2\gamma\left[1 - \cos(\phi - \theta)\right] = 4\gamma \sin^2\left(\frac{\phi - \theta}{2}\right), \text{ i.e., } z = 2\sqrt{\frac{\gamma}{g\delta}} \sin\frac{\phi - \theta}{2} \tag{12.111}$$

If h be the rise of liquid along the plate, then

$$g\delta h^2 = 2\gamma[1 - \cos(\phi - \theta)] \tag{12.112}$$

If the plate is dipped vertically, $\phi = 90°$. In this case,

$$g\delta h^2 = 2\gamma(1 - \sin\theta), \text{ i.e., } h = \sqrt{\frac{2\gamma}{g\delta}(1 - \sin\theta)} \tag{12.113}$$

For $\theta = 0$, as in water,

$$h = \sqrt{\frac{2\gamma}{g\delta}} \tag{12.114}$$

In terms of the Laplace's constant a^2, we write Eqn (12.114) in the form

$$h = \sqrt{2a^2} = \sqrt{2}a \tag{12.115}$$

If the plate is horizontal and the liquid touches the lower surface of the plate as shown in the figure (FIG. 12.21), then $\phi = 180°$ and Eqn (12.112) yields

$$g\delta h^2 = 2\gamma(1 + \cos\theta)$$

$$\text{i.e., } g\delta h^2 = 4\gamma\cos^2\frac{\theta}{2}$$

$$\text{i.e., } h = 2\sqrt{\frac{\gamma}{g\delta}}\cos\frac{\theta}{2} \tag{12.116}$$

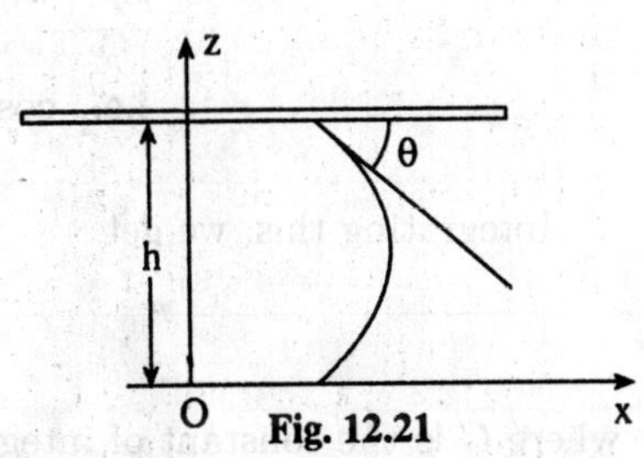

Fig. 12.21

In terms of Laplace's constant, this becomes

$$h = 2a\cos\frac{\theta}{2} \tag{12.117}$$

For $\theta = 0$, i.e., when the angle of contact vanishes,

$$h = 2a \tag{12.118}$$

12.15 Capillary Curve

We consider a liquid of density δ in contact with a vertical wall and assume, as before, that the surface is cylindrical with horizontal generators perpendicular to the plane of paper (FIG. 12.22). The cross section of such a cylindrical surface by a plane perpendicular to the wall will be a curve called ***capillary curve***.

We consider the case when the angle of contact of the liquid with the wall is acute.

Let AB be the wall which the liquid meets at an angle α. We take a horizontal axis Ox at right angles to the wall and at the natural level of the liquid. At this level, the pressure of the liquid will be equal to the atmospheric pressure. We take a point $P(x, y)$ on the capillary curve.

Let ρ = radius of curvature at P on the capillary curve, Π = the atmospheric pressure, γ = the surface tension of the liquid, δ = the density of the liquid.

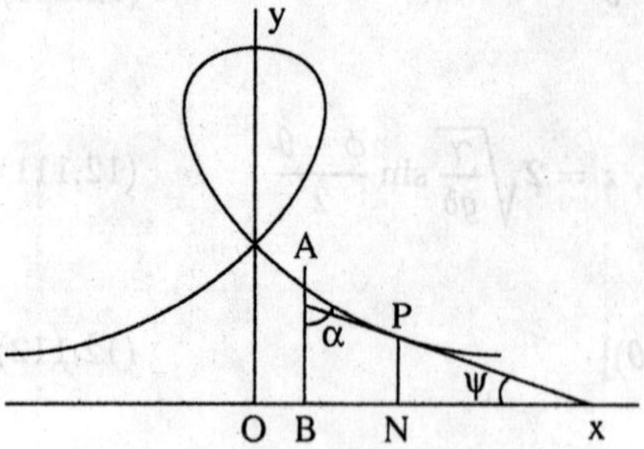

Fig. 12.22

Then, by Eqn (12.30), we have

$$\frac{\gamma}{\rho} = \Pi - p = g\delta y, \tag{12.119}$$

since the other principal radius of curvature of the liquid surface is infinite.

Eqn (12.119) yields

$$\rho y = \frac{\gamma}{g\delta} = a^2, \tag{12.120}$$

where a^2 is the Laplace's constant.

Let the arc length s be measured from the wall, and let the tangent to the curve at P make an acute angle ψ with the x-axis.

Then we have the relations:

$$\rho = -\frac{ds}{d\psi}, \quad \frac{ds}{dy} = -\operatorname{cosec}\psi, \quad \frac{dy}{dx} = -\tan\psi, \tag{12.121}$$

the negative sign being taken, since the angle ψ is made with the negative direction of x-axis.

Then Eqn (12.120) is written as

$$-\frac{ds}{d\psi}y = a^2, \text{ i.e., } -\frac{ds}{dy}\cdot\frac{dy}{d\psi}y = a^2$$

$$\text{i.e., } \operatorname{cosec}\psi\frac{dy}{d\psi}y = a^2, \text{ i.e., } y\,dy = a^2\sin\psi\,d\psi \tag{12.122}$$

Integrating this, we get

$$\frac{y^2}{2} = -a^2\cos\psi + C, \tag{12.123}$$

where C is the constant of integration.

Since $\psi = 0$ when $y = 0$, we have $C = a^2$.

Hence,

$$y^2 = 2a^2(1-\cos\psi) = 2a^2\cdot 2\sin^2\frac{\psi}{2} = 4a^2\sin^2\frac{\psi}{2} \tag{12.124}$$

Thus

$$y = \pm 2a\sin\frac{\psi}{2} \tag{12.125}$$

Since y is positive when ψ is positive, we have to take the upper sign in Eqn (12.125).

Again,

$$\frac{dx}{dy} = -\cot\psi$$

$$\text{i.e., } dx = -\cot\psi\cdot dy = -\cot\psi\cdot\left(a\cos\frac{\psi}{2}d\psi\right) \text{ [from Eqn (12.125)]}$$

$$= -a\cos\frac{\psi}{2}\left[\frac{1-2\sin^2\frac{\psi}{2}}{2\sin\frac{\psi}{2}\cos\frac{\psi}{2}}\right]d\psi = -\frac{a}{2}\left(\operatorname{cosec}\frac{\psi}{2} - 2\sin\frac{\psi}{2}\right)d\psi \tag{12.126}$$

Hence, integrating, we have

$$x = a \log\left(\cot\frac{\psi}{4}\right) - 2a\cos\frac{\psi}{2} \tag{12.127}$$

provided the origin be so chosen that $x = 0$ when $\psi = \pi$.

The capillary curve is represented by the parametric equations:

$$\left.\begin{aligned} x &= a\log\left(\cot\frac{\psi}{4}\right) - 2a\cos\frac{\psi}{2} \\ y &= 2a\sin\frac{\psi}{2} \end{aligned}\right\} \tag{12.128}$$

which is asymptotic to the axis of x and has a loop as shown in the figure.

Since

$$\alpha + \psi = \frac{\pi}{2}, \text{ i.e., } \psi = \frac{\pi}{2} - \alpha, \tag{12.129}$$

the height above the natural level at which the liquid meets the wall, is from Eqn (12.125)

$$= 2a\sin\frac{\psi}{2} = 2a\sin\left(\frac{\pi}{4} - \frac{\alpha}{2}\right) \tag{12.130}$$

Note. It is convenient to measure y downwards in the case of a liquid, such as mercury, for which the angle of contact is obtuse. The figure in this case will be inverted.

12.16 Drop of Liquid

We consider a liquid drop of height H placed on a horizontal plane (FIG. 12.23). The liquid drop has only one surface, so that the equation of equilibrium is

$$p = \gamma\left(\frac{1}{\rho_1} + \frac{1}{\rho_2}\right), \tag{12.131}$$

where p is the difference between the internal pressure and atmospheric pressure.

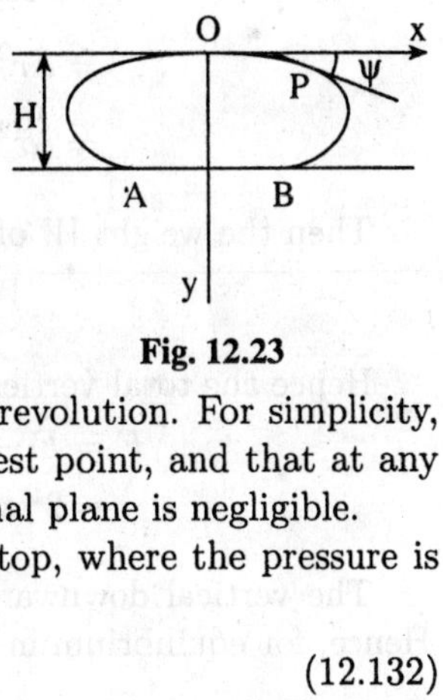

Fig. 12.23

In general, the liquid drop will assume the shape of a surface of revolution. For simplicity, we assume that the liquid drop is so large that it is flat at its highest point, and that at any point on its surface, the curvature in planes other than the meridional plane is negligible.

Taking $\rho_1 = \rho$, $\rho_2 = \infty$, and measuring y downwards from the top, where the pressure is atmospheric, we have

$$\frac{\gamma}{\rho} = p = g\delta y, \text{ i.e., } \rho y = \frac{\gamma}{g\delta} = a^2 \tag{12.132}$$

The vertical section will be a capillary curve. If ψ be the inclination of the tangent at any point P on the capillary curve to the x-axis, then proceeding as before, we will get

$$\left.\begin{aligned} x &= a\log\left(\tan\frac{\psi}{4}\right) + 2a\cos\frac{\psi}{2} + \text{constant} \\ y &= 2a\sin\frac{\psi}{2} \end{aligned}\right\} \tag{12.133}$$

Thus, if α be the angle of contact of the liquid with the plane measured in the liquid, then $\psi = \alpha$, so that the height of the drop is

$$H = 2a\sin\frac{\alpha}{2}. \tag{12.134}$$

12.17 Equilibrium of a Floating Needle

A needle, if carefully dropped, will float on the surface of water. This phenomenon can be explained by the concept of surface tension.

We consider a vertical section of the needle perpendicular to its axis floating on the surface of water (FIG. 12.24).

Let m = mass/unit length of the needle, r = the radius of the circular cross section of the needle, z = height of the axis of the needle from free surface, γ = the surface tension of water, θ = the angle of contact between the water and tangent BD to the needle, δ = the density of water, 2α = the angle made by the arc AB in contact with water at the centre O of the cross section.

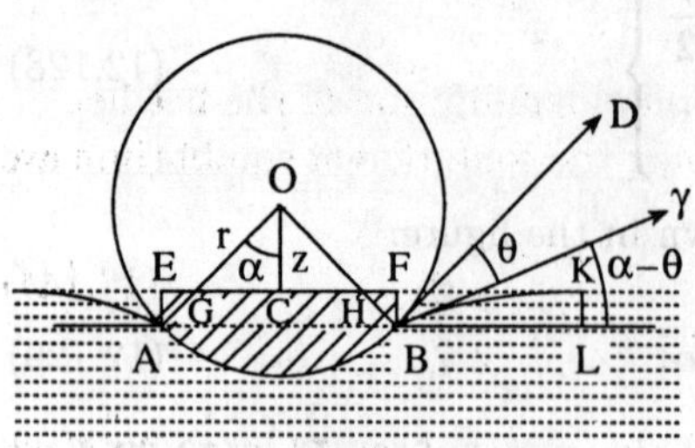

Fig. 12.24

Then the vertical upward force F_1 on the unit length of the needle due to surface tension at A and B is

$$F_1 = 2\gamma \sin(\alpha - \theta) \tag{12.135}$$

Now, from the figure we have

$$\left.\begin{aligned} AB = 2r\sin\alpha,\ AE = BF = r\cos\alpha - z \\ EG = FH = BF\tan\alpha = (r\cos\alpha - z)\tan\alpha \end{aligned}\right\} \tag{12.136}$$

Hence the area of the shaded portion is

$$\begin{aligned} &= \text{area of sector } AOB - \Delta OGH + \Delta AEG + \Delta BFH \\ &= \alpha r^2 - z \cdot z\tan\alpha + 2 \cdot \frac{1}{2}(r\cos\alpha - z)(r\cos\alpha - z)\tan\alpha \\ &= \alpha r^2 - z^2\tan\alpha + (r\cos\alpha - z)^2\tan\alpha \end{aligned} \tag{12.137}$$

Then the weight W of the displaced liquid is

$$W = g\delta\left[\alpha r^2 - z^2\tan\alpha + (r\cos\alpha - z)^2\tan\alpha\right] \tag{12.138}$$

Hence the total vertical upward force F on the needle

$$\begin{aligned} F &= F_1 + W \\ &= 2\gamma\sin(\alpha - \theta) + g\delta\left[\alpha r^2 - z^2\tan\alpha + (r\cos\alpha - z)^2\tan\alpha\right] \end{aligned} \tag{12.139}$$

The vertical downward force is due to the weight of the needle per unit length, i.e., mg. Hence, for equilibrium in the vertical direction, we must have

$$F = mg, \tag{12.140}$$

the force F being given by Eqn (12.139).

Let us now consider the equilibrium in the horizontal direction.

Now, the horizontal component γ' of surface tension along BL is

$$\gamma' = \gamma\cos(\alpha - \theta) \tag{12.141}$$

Horizontal thrust of liquid on the rectangular area $KL = BF$ by unit length is

$$H = \frac{1}{2}g\delta(KL)^2 = \frac{1}{2}g\delta(r\cos\alpha - z)^2 \tag{12.142}$$

Surface tension at K is γ.

For equilibrium in the horizontal direction, we must have

$$\gamma' + H = \gamma, \text{ i.e., } \gamma\cos(\alpha-\theta) + \frac{1}{2}g\delta(r\cos\alpha - z)^2 = \gamma$$

$$\text{i.e., } \frac{1}{2}g\delta(r\cos\alpha - z)^2 = \gamma[1-\cos(\alpha-\theta)]$$

$$\text{i.e., } g\delta(r\cos\alpha - z)^2 = 4\gamma\sin^2\left(\frac{\alpha-\theta}{2}\right) \tag{12.143}$$

Thus, Eqns (12.140) and (12.143) determine the condition of equilibrium of the needle.

If, in particular, $\delta = 1$ and $\theta = 0$, the equations determining the condition of equilibrium are

$$\left.\begin{aligned} 2\gamma\sin\alpha + g[ar^2 + r^2\sin\alpha\cos\alpha - 2rz\sin\alpha] &= mg \\ g(r\cos\alpha - z)^2 &= 4\gamma\sin^2\frac{\alpha}{2} \end{aligned}\right\} \tag{12.144}$$

The condition of equilibrium of the needle in terms of r and α can be obtained by eliminating z from the above two equations.

12.18 Liquid Films and Their Shapes

Liquid films can be produced in various ways. A soap bubble is a very familiar instance of spherical films. Liquid films may be formed by dipping a wire frame into a solution of soap and water. The fact that apparently plane films can be obtained shows that the weight of the film may be neglected in comparison with the tension of the film.

It is observed that a very small tangential action will tear the film, which indicates that the stress across any line is entirely normal to that line. This implies that the tension is the same in all directions.

If the air pressure be the same on both sides of the film, the excess of pressure $p = 0$. Then the equation of equilibrium takes the form

$$\frac{1}{\rho_1} + \frac{1}{\rho_2} = 0, \tag{12.145}$$

that is, the mean curvature is zero. This may occur when $\rho_1 = -\rho_2$, i.e., when the principal curvatures at any point are equal in magnitude but opposite in direction. This can also happen when both ρ_1 are ρ_2 are infinite. In the first case, we get ***catenoidal*** and in the second case, we get ***helicoidal*** forms of liquid films.

In Cartesian coordinates, the condition (12.145) assumes the form

$$r(1+q^2) - 2pqs + t(1+p^2) = 0 \tag{12.146}$$

Writing the expressions for p, q, r, s, t from Eqn (12.33), we get the condition in the form:

$$\frac{\partial^2 z}{\partial x^2}\left\{1+\left(\frac{\partial z}{\partial y}\right)^2\right\} - 2\frac{\partial z}{\partial x}\frac{\partial z}{\partial y}\frac{\partial^2 z}{\partial x\partial y} + \frac{\partial^2 z}{\partial y^2}\left\{1+\left(\frac{\partial z}{\partial x}\right)^2\right\} = 0 \tag{12.147}$$

Let the form of the liquid film be that of a surface of revolution about the x-axis (FIG. 12.25). At any point $P(x, y)$ on the meridian curve, we draw a tangent which makes an angle ψ with x-axis. We also draw a normal to the meridian curve at P which meets the x-axis at N.

If ρ be the radius of curvature of the meridian curve and l be the length PN, then ρ and l are principal radii of curvature of the surface of revolution.

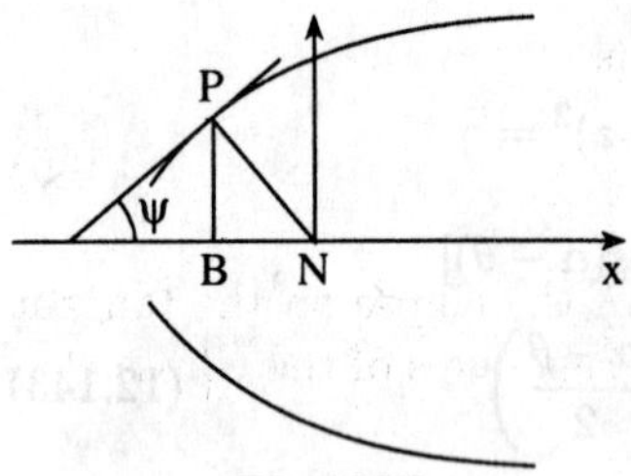

Fig. 12.25

Hence, Eqn (12.145) of equilibrium in this case becomes

$$\frac{1}{\rho}+\frac{1}{l}=0 \tag{12.148}$$

If we consider an element of arc ds at P measured along the positive direction of x-axis, then we have the relations:

$$\rho=-\frac{ds}{d\psi},\ y=l\cos\psi,\ \frac{dy}{ds}=\sin\psi \tag{12.149}$$

Using these relations in Eqn (12.148), we write

$$-\frac{d\psi}{ds}+\frac{\cos\psi}{y}=0,\ \text{i.e.,}\ -\frac{d\psi}{dy}\frac{dy}{ds}+\frac{\cos\psi}{y}=0$$

$$\text{i.e.,}\ -\sin\psi\frac{d\psi}{dy}+\frac{\cos\psi}{y}=0,\ \text{i.e.,}\ \frac{dy}{y}-\frac{\sin\psi d\psi}{\cos\psi}=0 \tag{12.150}$$

Integrating,

$$\log(y\cos\psi)=\text{constant, i.e.,}\ y=c\sec\psi \tag{12.151}$$

Hence,

$$\frac{dy}{ds}=c\sec\psi\tan\psi\frac{d\psi}{ds},\ \text{i.e.,}\ \sin\psi=c\sec\psi\tan\psi\frac{d\psi}{ds},\ \text{i.e.,}\ ds=c\sec^2\psi d\psi$$

$$\therefore\ s=c\tan\psi \tag{12.152}$$

provided we measure s and ψ so that they vanish together.

Eqn (12.152) confirms that the meridian curve is a catenary. Hence, the surface obtained by revolving the curve about its directrix is a catenoid. Therefore, ***a catenoid is the only possible form of the surface of a film when the pressure is the same on both sides***.

When the pressures on two sides of a film are different, the forms of liquid films, are in general, surfaces of revolution. In this case, the condition of equilibrium is

$$p^*=2\gamma\left(\frac{1}{\rho_1}+\frac{1}{\rho_2}\right),\ \text{i.e.,}\ \frac{1}{\rho_1}+\frac{1}{\rho_2}=\frac{p^*}{2\gamma}, \tag{12.153}$$

which implies that the mean curvature of the surface is constant.

Proceeding as above, we write Eqn (12.153) in the form

$$\frac{1}{\rho}+\frac{1}{l}=\frac{p^*}{2\gamma},\ \text{i.e.,}\ -\frac{d\psi}{ds}+\frac{\cos\psi}{y}=\frac{p^*}{2\gamma}$$

$$\text{i.e.,}\ -\frac{d\psi}{dy}\frac{dy}{ds}+\frac{\cos\psi}{y}=\frac{p^*}{2\gamma},\ \text{i.e.,}\ -\sin\psi\frac{d\psi}{dy}+\frac{\cos\psi}{y}=\frac{p^*}{2\gamma}$$

$$\text{i.e.,}\ -y\sin\psi\frac{d\psi}{dy}+\cos\psi=\frac{p^*}{2\gamma}y,\ \text{i.e.,}\ d(y\cos\psi)=\frac{p^*}{2\gamma}y,$$

which upon integration yields

$$y\cos\psi=\frac{p^*}{4\gamma}y^2+\lambda, \tag{12.154}$$

where λ is a constant.

Since $\cos\psi=\dfrac{y}{l}$, Eqn (12.154) assumes the form

$$y^2\left(\frac{1}{l}-\frac{p^*}{4\gamma}\right)=\lambda \tag{12.155}$$

Now, the pedal equation of an ellipse with a focus as the pole is

$$p^2\left(\frac{1}{r}-\frac{1}{2a}\right)=\frac{b^2}{2a}, \tag{12.156}$$

where p is the length of the perpendicular drawn from the focus of the ellipse on the tangent at any point, r is the focal distance of that point, and a, b are the semi-axes of the ellipse.

Comparing Eqns (12.155) and (12.156), we find that P is the focus of the ellipse which touches the x-axis at N (FIG. 12.26). The semi-axes of the ellipse are given by

$$a=\frac{2\gamma}{p^*},\ b=\sqrt{\frac{4\lambda\gamma}{p^*}} \tag{12.157}$$

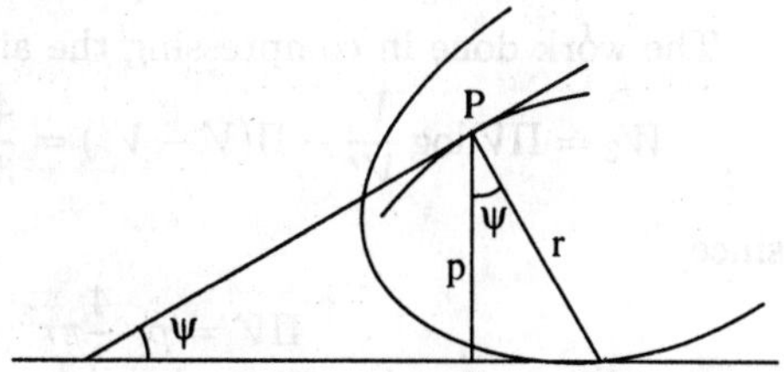

Fig. 12.26

Thus the meridian curve is the path of the focus of an ellipse as it rolls on the straight line along BN. This curve is sinuous and is known as ***unduloid***.

If $p^*=0$, i.e., the pressure is the same on two sides, we have from Eqn (12.155),

$$y^2=\lambda l, \tag{12.158}$$

which shows that the meridian curve is the path of the focus of a parabola as it rolls along the straight line BN. This curve is a catenary and the surface of revolution is a catenoid, which we have seen earlier.

12.19 Energy of Liquid Films

(A) Plane Films. Let AD, BC be the straight wires and CD a movable wire. A rectangular film $ABCD$ is formed by keeping AB in the surface of the liquid and CD drawing out (FIG. 12.27). Then the work done in pulling out the film is

$$W=\tau\cdot AB\cdot AD=\tau ah, \tag{12.159}$$

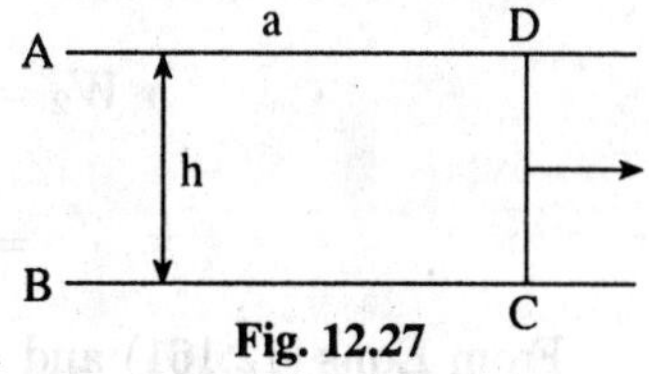

Fig. 12.27

where τ is the tension of the film.

This work expended represents the potential energy of the film.

If S be the superficial energy per unit area, it follows that

$$S=\tau \tag{12.160}$$

It should be observed that here tension of the film is equal to twice the surface tension of either side of the film.

(B) Spherical Films. The energy of a spherical film like a soap bubble is the work done in producing it. This work done consists of two parts: (1) the work done in pulling out the film, and (2) the work done in compressing the air in the bubble.

Let $S=$ the superficial area of the bubble, $\gamma=$ the surface tension.

Then energy of a small plane element $=\gamma dS$. Hence the work done W_1 in pulling out the film is

$$W_1=\gamma S \tag{12.161}$$

Again, let $\Pi=$ atmospheric pressure, $p=$ pressure of air inside the bubble, $r=$ radius of the bubble at that time.

Then

$$p - \Pi = \frac{2\gamma}{r} \tag{12.162}$$

If the bubble contains a mass of air which at atmospheric pressure Π would occupy a volume V, then by Boyle's law,

$$\Pi V = p \cdot \frac{4}{3}\pi r^3 = pV' \text{ (say).} \tag{12.163}$$

The work done in compressing the air isothermally from volume V to V' is

$$W_2 = \Pi V \log \frac{V}{V'} - \Pi(V - V') = \frac{4}{3}\pi r^3 \left\{ \left(\Pi + \frac{2\gamma}{r} \right) \log \left(1 + \frac{2\gamma}{r\Pi} \right) - \frac{2\gamma}{r} \right\} \tag{12.164}$$

since

$$\Pi V = p \cdot \frac{4}{3}\pi r^3 = \left(\Pi + \frac{2\gamma}{r} \right) \cdot \frac{4}{3}\pi r^3 \text{ [using (12.162)]}$$

$$\frac{V}{V'} = \frac{p}{\Pi} = \frac{1}{\Pi}\left(\Pi + \frac{2\gamma}{r} \right) = 1 + \frac{2\gamma}{r\Pi}$$

$$\Pi(V - V') = \Pi V - \Pi V' = pV' - \Pi V' \text{ [using (12.163)]}$$

$$= (p - \Pi)V' = \frac{2\gamma}{r} \cdot \frac{4}{3}\pi r^3$$

If we assume that the difference between the pressures inside and outside the bubble is small compared with the atmospheric pressure Π, we can take $2\gamma/r\Pi$ as small, so that

$$\log \left(1 + \frac{2\gamma}{r\Pi} \right) \simeq \frac{2\gamma}{r\Pi} - \frac{2\gamma^2}{r^2\Pi^2} \tag{12.165}$$

In view of this, we have

$$W_2 = \frac{4}{3}\pi r^3 \left\{ \left(\Pi + \frac{2\gamma}{r} \right)\left(\frac{2\gamma}{r\Pi} - \frac{2\gamma^2}{r^2\Pi^2} \right) - \frac{2\gamma}{r} \right\}$$

$$= \frac{4}{3}\pi r^3 \cdot \frac{2\gamma^2}{r^2\Pi} = \frac{2}{3}\frac{\gamma^2 S}{r\Pi} \text{ (since } S = 4\pi r^2) \tag{12.166}$$

From Eqns (12.161) and (12.166), we find that

$$\frac{W_2}{W_1} = \frac{2\gamma}{3r\Pi} \tag{12.167}$$

Thus, the work done in compressing the air is to that done in pulling out the film as $2\gamma : 3r\Pi$.

EXAMPLE 12.1 *A soap bubble of radius r is blown on a tube with a solution whose surface tension is γ. The atmospheric pressure is P. The tube on which the bubble is blown is then connected to a vessel of volume V filled with air at atmospheric pressure. Show that the bubble will shrink until its radius is $\frac{1}{2}r$, if $2\pi r^2\gamma + \frac{7}{12}\pi r^3 P = \frac{4\gamma V}{r}$.*

Solution. Initially, the volume v of the soap bubble is

$$v = \frac{4}{3}\pi r^3 \tag{1}$$

and pressure inside it

$$= \left(P + \frac{4\gamma}{r} \right) \tag{2}$$

Finally, the volume v' of the soap bubble is

$$v' = \frac{4}{3}\pi\left(\frac{r}{2}\right)^3 = \frac{\pi}{6}r^3 \tag{3}$$

and pressure inside it

$$= P + \frac{4\gamma}{r/2} = P + \frac{8\gamma}{r} \tag{4}$$

When the tube with the bubble is connected to the vessel containing air at atmospheric pressure, the total volume of air is now $(V + v')$ under pressure given by (4).

Hence, according to Boyle's law,

$$\left(P + \frac{4\gamma}{r}\right)v + PV = \left(P + \frac{8\gamma}{r}\right)(v' + V)$$

$$\text{i.e.,}\quad \left(P + \frac{4\gamma}{r}\right)\cdot\frac{4}{3}\pi r^3 + PV = \left(P + \frac{8\gamma}{r}\right)\left(\frac{\pi}{6}r^3 + V\right)$$

$$\text{i.e.,}\quad \frac{4}{3}\pi r^3 P\left(1 - \frac{1}{8}\right) + \frac{16}{3}\pi\gamma r^2 - \frac{4}{3}\pi\gamma r^2 = \frac{8\gamma V}{r}$$

$$\text{i.e.,}\quad \frac{7}{6}\pi r^3 P + 4\pi\gamma r^2 = \frac{8\gamma V}{r}$$

Hence,

$$2\pi\gamma r^2 + \frac{7}{12}\pi r^3 P = \frac{4\gamma V}{r}. \quad \textbf{(Proved)}$$

EXAMPLE 12.2 *If two soap bubbles of radii a and b, are blown from the same liquid, and if they coalesce into a single bubble of radius R, prove that the tension is equal to*

$$\frac{\Pi}{2}\cdot\frac{R^3 - a^3 - b^3}{a^2 + b^2 - R^2},$$

where Π = the atmospheric pressure.

Solution. Let γ = the surface tension of the liquid, p = pressure of air inside the bubble of radius a, p' = pressure of air inside the bubble of radius b, and P = pressure of air inside the bubble of radius R.

Then we have the relations:

$$\frac{2\gamma}{a} = p - \Pi,\ \frac{2\gamma}{b} = p' - \Pi,\ \frac{2\gamma}{R} = P - \Pi \tag{1}$$

By Boyle's law,

$$p\cdot\frac{4}{3}\pi a^3 + p'\cdot\frac{4}{3}\pi b^3 = P\cdot\frac{4}{3}\pi R^3,\ \text{i.e., } pa^3 + p'b^3 = PR^3 \tag{2}$$

Putting here the values of p, p', P from (1), we get

$$\left(\frac{2\gamma}{a} + \Pi\right)a^3 + \left(\frac{2\gamma}{b} + \Pi\right)b^3 = \left(\frac{2\gamma}{R} + \Pi\right)R^3$$

$$\text{or,}\quad 2\gamma(a^2 + b^2 - R^2) = \Pi\,(R^3 - a^3 - b^3)$$

Hence,

$$\gamma = \frac{\Pi}{2}\frac{R^3 - a^3 - b^3}{a^2 + b^2 - R^2}. \quad \textbf{(Proved)}$$

EXAMPLE 12.3 *A solid sphere of radius R is partially immersed in a liquid of density ρ and surface tension γ. Determine the total upward thrust when the centre of the sphere is at a height z from the free surface. Discuss the situation when there will be no effect of surface tension.*

Solution. Let the portion ACB of the sphere with centre at O be immersed in the liquid (FIG. 12.28). Let θ be the angle of contact and $\angle AOC = \phi$. Then the surface tension acts at an angle $(\theta + \phi - \pi)$ to the horizontal.

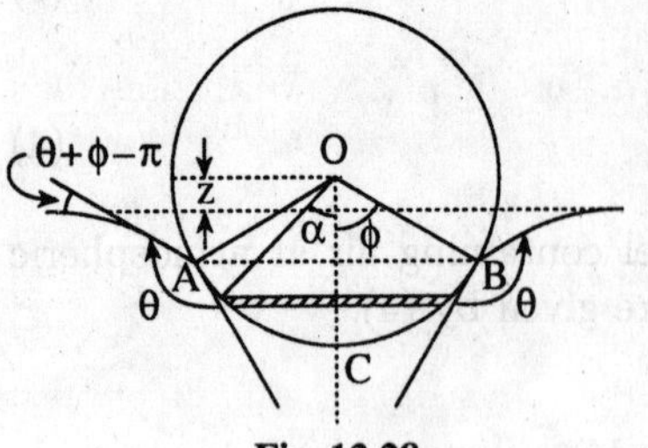

Fig. 12.28

We consider an elementary ring which makes an angle 2α at the centre. Its depth from the free surface is $(R\cos\alpha - z)$. Hence the pressure at all points on the ring is

$$p = g\rho(R\cos\alpha - z) \tag{1}$$

The upthrust due to the liquid on this elementary ring is

$$\begin{aligned} dF &= [(2\pi R\sin\alpha)\cdot R d\alpha \cdot g\rho(R\cos\alpha - z)]\cos\alpha \\ &= 2\pi R^2 g\rho \sin\alpha\cos\alpha(R\cos\alpha - z)d\alpha \end{aligned} \tag{2}$$

Hence the total upthrust on the sphere due to liquid is

$$\begin{aligned} F &= 2\pi R^2 g\rho \int_0^{\phi} \sin\alpha\cos\alpha(R\cos\alpha - z)d\alpha \\ &= 2\pi R^2 g\rho\left[\frac{R}{3}\left(1-\cos^3\phi\right) - \frac{z}{2}\left(1-\cos^2\phi\right)\right] \end{aligned} \tag{3}$$

The vertical force due to the surface tension is

$$F' = 2\pi R\gamma\sin\phi\cdot\cos\left[\frac{\pi}{2} - (\theta+\phi-\pi)\right] = -2\pi R\gamma\sin\phi\sin(\theta+\phi) \tag{4}$$

The total upthrust on the sphere is

$$\begin{aligned} F^* &= F + F' \\ &= 2\pi R^2 g\rho\left[\frac{R}{3}(1-\cos^3\phi) - \frac{z}{2}(1-\cos^2\phi)\right] - 2\pi R\gamma\sin\phi\sin(\theta+\phi) \end{aligned} \tag{5}$$

The effect of surface tension vanishes when $\theta + \phi = \pi$. This occurs when the depth to which the sphere is immersed is such that the liquid surface is horizontal at all points in the neighbourhood of the sphere.

In this case,

$$z = R\cos\phi \tag{6}$$

Then, from (5), we get

$$F^* = 2\pi R^2 g\rho\left[\frac{R}{3}(1-cos^3\phi) - \frac{1}{2}R\cos\phi(1-\cos^2\phi)\right] = \pi R^3 g\rho\left(\frac{2}{3} - \cos\phi + \frac{1}{3}\cos^3\phi\right)$$

EXAMPLE 12.4 *An annular plate of density ρ rests on a liquid of density δ and surface tension γ.*

Determine the vertical upward pull just sufficient to separate the plate from the surface of the liquid when $a, b(a > b)$ are the radii of the plate.

Solution. Let h = rise of the liquid surface due to pull, θ = angle made by the tangent to the profile curve with the horizontal (FIG. 12.29).

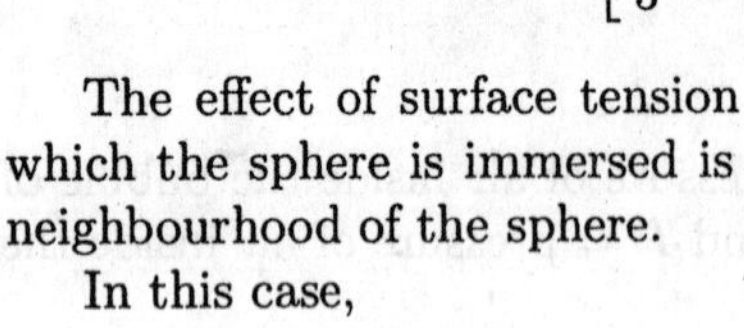

Fig. 12.29

Area of the plate is

$$A = \pi(a^2 - b^2) \tag{1}$$

Then its weight W is

$$W = \pi(a^2 - b^2)g\rho \tag{2}$$

If Π be the atmospheric pressure, then the downward pressure on the top of the plate due to atmosphere is

$$F_1 = \Pi\pi(a^2 - b^2) \tag{3}$$

The pressure on the bottom surface of the plate is $\Pi - g\delta h$, so that the upward force on the plate due to the liquid is

$$F_2 = (\Pi - g\delta h)\pi(a^2 - b^2) \tag{4}$$

The vertical component of the force due to the surface tension is

$$F_3 = 2\pi\gamma(a+b)\sin\theta \tag{5}$$

Resultant vertical downward force

$$\begin{aligned} F &= W + F_1 - F_2 + F_3 \\ &= \pi(a^2 - b^2)g\rho + \pi(a^2 - b^2)[\Pi - (\Pi - g\delta h)] + F_3 \\ &= \pi(a^2 - b^2)g(\rho + \delta h) + F_3 \end{aligned} \tag{6}$$

Hence, the vertical upward pull just sufficient to separate the ring from the surface of the liquid is

$$F = \pi g(a^2 - b^2)(\rho + \delta h) + 2\pi\gamma(a+b)\sin\theta \tag{7}$$

For $\theta = 90°$,

$$\begin{aligned} F &= \pi g(a^2 - b^2)(\rho + \delta h) + 2\pi\gamma(a+b) \\ &= \pi(a+b)[g(a-b)(\rho + \delta h) + 2\gamma] \end{aligned} \tag{8}$$

If a, b are nearly equal, $(a - b) \simeq 0$. In this case,

$$F = 2\pi\gamma(a+b) = 4\pi r\gamma,$$

where $a = b = r$.

EXAMPLE 12.5 *Show that if a light thread with its ends tied together form a part of the internal boundary of a liquid film, the curvature of the thread at every point will be constant.*

If the thread have weight, and if the film be a surface of revolution about a vertical axis, prove that, in the position of equilibrium, the tension of the string is

$$\frac{l}{2\pi}\sqrt{\gamma^2 - w^2},$$

where l is the length and w the weight per unit length of the thread and γ the surface tension of the film.

Solution. First we consider the case when the thread has no weight and is inside the film (FIG. 12.30).

We consider the part of the thread, where its radius of curvature is ρ and tension is T.

Forces acting on a small element ds on this part are: (1) tension dT along the tangent, (2) force γds due to surface tension acting along the normal to the element ds, and (3) $\dfrac{T ds}{\rho}$ due to the tension of the thread acting along the inward normal to the surface.

For equilibrium of the element, we must have

$$dT = 0, \text{ i.e.,, } T = \text{constant} \tag{1}$$

$$\text{and } \frac{T ds}{\rho} = \gamma ds, \text{ i.e., } \rho = \frac{T}{\gamma} = \text{constant.} \tag{2}$$

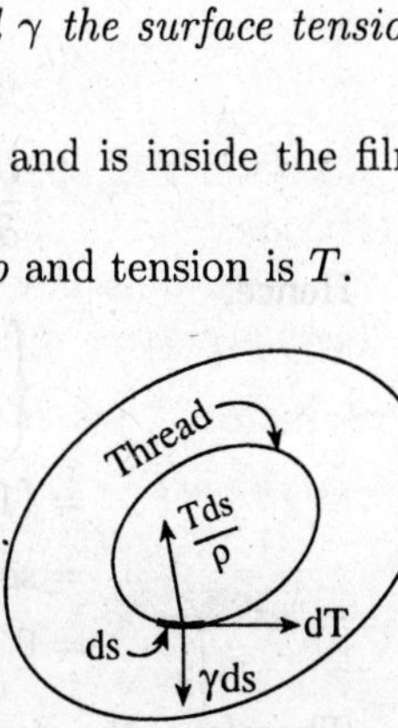

Fig. 12.30

Now we consider the case when the thread has weight and the surface is a surface of revolution. Then the thread assumes the form of a horizontal circle of radius ρ, so that

$$l = 2\pi\rho \tag{3}$$

Considering an element ds of the thread, the forces acting are (FIG. 12.31):

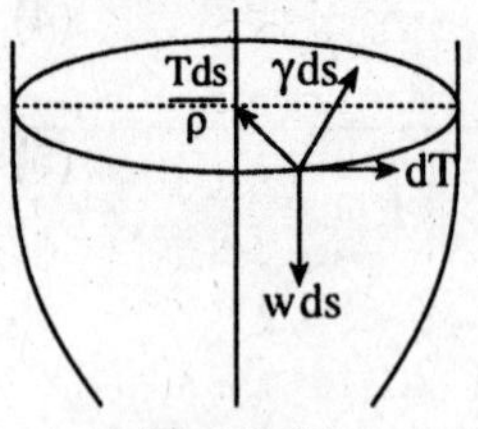

Fig. 12.31

(1) the weight wds acting vertically downwards, (2) the tension dT acting along the tangent, (3) force γds due to surface tension acting at right angles to the element ds, and (4) $T\dfrac{ds}{\rho}$ acting along the inward normal towards the vertical axis of revolution.

Here the force dT only is horizontal and all the other forces are at right angles to each other remaining on the same vertical plane.

For equilibrium in the horizontal direction,

$$dT = 0, \text{ i.e., } T = \text{constant.} \tag{4}$$

The other three forces being at right angles to each other, we have

$$(\gamma ds)^2 = (wds)^2 + \left(T\frac{ds}{\rho}\right)^2$$

$$\text{i.e., } \gamma^2 = w^2 + \left(\frac{T}{\rho}\right)^2, \text{ i.e., } \frac{T^2}{\rho^2} = \gamma^2 - w^2$$

$$\therefore T = \rho\sqrt{\gamma^2 - w^2} = \frac{l}{2\pi}\sqrt{\gamma^2 - w^2} \text{ [using (3)].} \quad \textbf{(Proved)}$$

EXAMPLE 12.6 *Prove that the equation*

$$e^z = \cos y \sec x$$

represents a possible form of a liquid film, the pressure on both sides being the same.

Solution. We are given the equation of surface as

$$e^z = \cos y \sec x$$

$$\text{i.e., } z = \log \cos y + \log \sec x, \text{ i.e., } z = \log \cos y - \log \cos x \tag{1}$$

Then the partial derivatives are

$$\left.\begin{aligned} \frac{\partial z}{\partial x} &= -\frac{1}{\cos x}\cdot(-\sin x) = \tan x \\ \frac{\partial z}{\partial y} &= \frac{1}{\cos y}\cdot(-\sin y) = -\tan y \\ \frac{\partial^2 z}{\partial x^2} &= \sec^2 x, \ \frac{\partial^2 z}{\partial y^2} = -\sec^2 y, \ \frac{\partial^2 z}{\partial x \partial y} = 0 \end{aligned}\right\} \tag{2}$$

Hence,

$$\left\{1 + \left(\frac{\partial z}{\partial y}\right)^2\right\}\frac{\partial^2 z}{\partial x^2} - 2\frac{\partial z}{\partial x}\frac{\partial z}{\partial y}\frac{\partial^2 z}{\partial x \partial y} + \left\{1 + \left(\frac{\partial z}{\partial x}\right)^2\right\}\frac{\partial^2 z}{\partial y^2}$$

$$= (1 + \tan^2 y)\cdot \sec^2 x - 0 + (1 + \tan^2 x)\cdot(-\sec^2 y)$$

$$= \sec^2 y \sec^2 x - \sec^2 x \sec^2 y$$

$$= 0$$

Therefore, the given surface represents a possible form of a liquid film, the pressure on both sides being the same.

EXAMPLE 12.7 *Two soap-bubbles are in contact; if r_1, r_2 be the radii of the outer surfaces, and r be the radius of the circle in which the three surfaces intersect, then*

$$\frac{3}{4r^2} = \frac{1}{r_1^2} + \frac{1}{r_2^2} - \frac{1}{r_1 r_2}$$

Solution. The film surfaces being of the same material, the surface tensions are equal. Hence the angle between them is 120° (FIG. 12.32). The angle between r_1 and r_2 which are perpendicular to the tensions is 60°.

Let the angle between r and r_2 be θ. Then the angle between r and r_1 is $(60° - \theta)$. From the figure, we have

$$\cos\theta = \frac{r}{r_2} \text{ and } \cos(60° - \theta) = \frac{r}{r_1} \qquad (1)$$

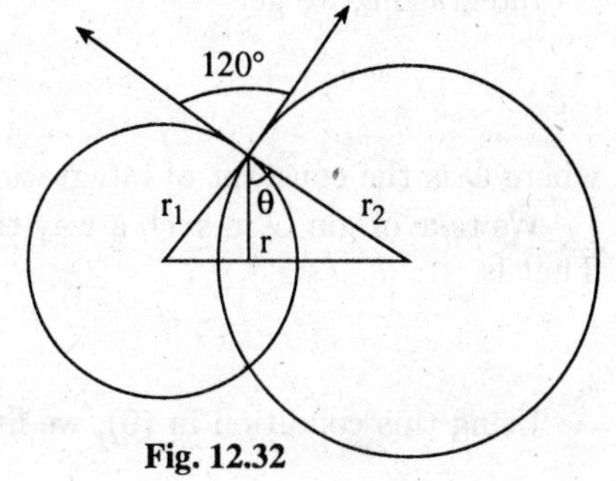

Fig. 12.32

From the second relation, we get

$$\frac{1}{2}\cos\theta + \frac{\sqrt{3}}{2}\sin\theta = \frac{r}{r_1}$$

$$\text{i.e., } \frac{r}{r_2} + \sqrt{3}\sin\theta = \frac{2r}{r_1} \text{ [using (1)]}$$

$$\therefore \sin\theta = \frac{1}{\sqrt{3}}\left(\frac{2r}{r_1} - \frac{r}{r_2}\right). \qquad (2)$$

From (1) and (2),

$$\sin^2\theta + \cos^2\theta = \frac{1}{3}\left(\frac{2r}{r_1} - \frac{r}{r_2}\right)^2 + \frac{r^2}{r_2^2}, \text{ i.e., } 3 = \frac{4r^2}{r_1^2} - \frac{4r^2}{r_1 r_2} + \frac{4r^2}{r_2^2}$$

$$\therefore \frac{3}{4r^2} = \frac{1}{r_1^2} + \frac{1}{r_2^2} - \frac{1}{r_1 r_2}. \quad \textbf{(Proved)}$$

EXAMPLE 12.8 *A liquid of density δ and surface tension γ is contained between two vertical parallel plates of different materials. If h_1, h_2 be the heights of points of contact of the liquid with the plates, where the angles of contact are θ_1, θ_2 respectively, show that*

$$h_1^2 - h_2^2 = \frac{2\gamma}{g\delta}(\sin\theta_2 - \sin\theta_1)$$

Solution. Let AB and CD be the vertical parallel plates and EFG the liquid surface (FIG. 12.33) in between them. The surface will be cylindrical with the horizontal generators perpendicular to the plane of paper.

We take rectangular axes Ox and Oz, with origin on the free surface. For a cylindrical surface, $\rho_1 = \rho$ and $\rho_2 = \infty$. Then by Eqn (12.99), the equation to the curve EFG is

$$\frac{\gamma}{\rho} - g\delta z = \Pi + \text{constant}. \qquad (1)$$

Neglecting the atmospheric pressure, we get

$$\frac{\gamma}{\rho} - g\delta z = \text{constant}. \qquad (2)$$

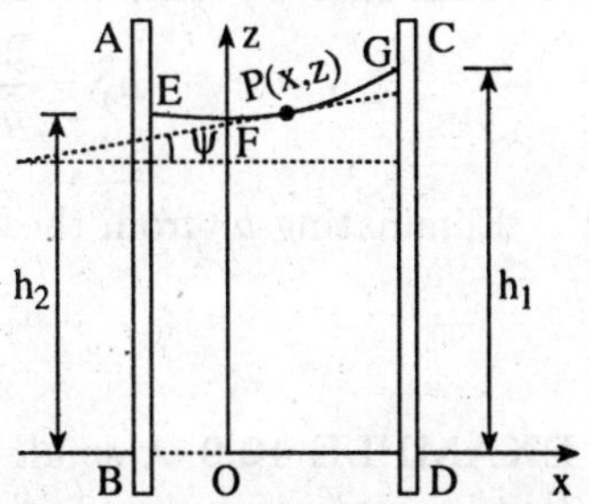

Fig. 12.33

The constant term in (2) may be taken to be zero since the origin lies in the general level such that, for $z = 0$, $\gamma = 0$. Hence, Eqn (2) becomes

$$\frac{\gamma}{\rho} = g\delta z \qquad (3)$$

We consider a point $P(x, z)$ on the profile curve and draw a tangent which makes an angle ψ (say) with x-axis. If we consider a small arc length ds at P, then we have the relations:

$$\rho = \frac{ds}{d\psi}, \text{ i.e., } \frac{1}{\rho} = \frac{d\psi}{ds} \text{ and } \sin\psi = \frac{dz}{ds} \tag{4}$$

Using these relations in (3), we get

$$\gamma\frac{d\psi}{ds} = g\delta z, \text{ i.e., } \gamma\frac{d\psi}{dz}\frac{dz}{ds} = g\delta z, \text{ i.e., } \gamma\sin\psi d\psi = g\delta z dz \tag{5}$$

Integrating, we get

$$-\gamma\cos\psi = \frac{1}{2}g\delta z^2 + C, \tag{6}$$

where C is the constant of integration.

We take origin O in such a way that the tangent to the curve at F at height h is horizontal. That is

$$\psi = 0, \text{ when } z = h \tag{7}$$

Using this condition in (6), we find

$$-\gamma = \frac{1}{2}g\delta h^2 + C, \text{ i.e., } C = -\left(\gamma + \frac{1}{2}g\delta h^2\right) \tag{8}$$

Putting this value of C in (6)

$$-\gamma\cos\psi = \frac{1}{2}g\delta z^2 - \left(\gamma + \frac{1}{2}g\delta h^2\right), \text{ i.e., } \gamma(1-\cos\psi) = \frac{1}{2}g\delta(z^2 - h^2)$$

$$\therefore z^2 = \frac{2\gamma}{g\delta}(1-\cos\psi) + h^2 \tag{9}$$

This is the equation to the profile curve EFG. Since θ_1 is the angle of contact at G and θ_2 at E, the tangents to the curve at these points make angles ψ_1, ψ_2 such that

$$\psi_1 = \frac{\pi}{2} - \theta_1, \; \psi_2 = \frac{\pi}{2} - \theta_2 \tag{10}$$

Thus Eqn (9) takes the form for the points of contact, where $z = h_1$ and $z = h_2$:

$$h_1^2 = \frac{2\gamma}{g\delta}(1-\sin\theta_1) + h^2, \; h_2^2 = \frac{2\gamma}{g\delta}(1-\sin\theta_2) + h^2 \tag{11}$$

Eliminating h^2 from these equations, we get

$$h_1^2 - h_2^2 = \frac{2\gamma}{g\delta}(\sin\theta_2 - \sin\theta_1). \quad \textbf{(Proved)}$$

EXAMPLE 12.9 *A small cube of side a floats with its upper face horizontal in a liquid, such that its angle of contact with the surface of the cube is obtuse and equal to $(\pi - \alpha)$. If δ_1 be the density of the cube and δ_2 that of the liquid and if $g\delta_2 c^2$ is the surface tension, prove that the cube will float if*

$$\frac{\delta_1}{\delta_2} < 1 + 4\frac{c^2}{a^2}\cos\alpha + 2\frac{c}{a}\sin\left(\frac{\pi}{4} - \frac{\alpha}{2}\right)$$

Solution. Let h be the depth of the top face of the cube below the free surface (FIG. 12.34). We are given that the surface tension

$$\gamma = g\delta_2 c^2 \quad (1)$$

The weight W_1 of the cube is

$$W_1 = g\delta_1 a^3 \quad (2)$$

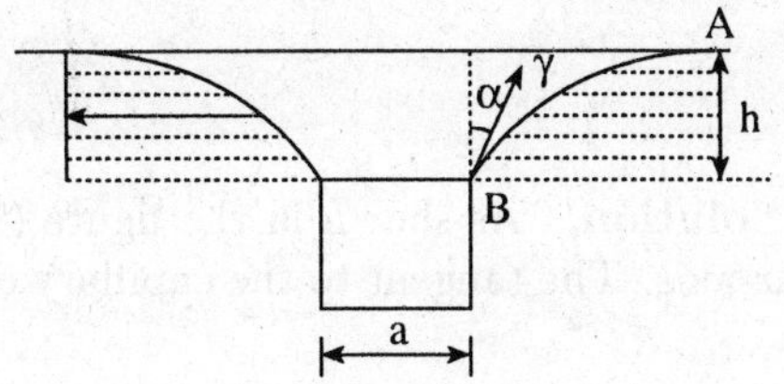

Fig. 12.34

The weight W_2 of the displaced liquid is = weight of the liquid of volume equal to that of the cube + weight of the liquid of height h on the top face of the cube.

Thus

$$W_2 = g\delta_2 a^3 + g\delta_2 a^2 h, \quad (3)$$

which acts vertically upwards.

The surface tension acting along the four edges is

$$= 4a\gamma = 4ag\delta_2 c^2 \quad (4)$$

The resultant surface tension acting vertically upwards

$$F_1 = 4ag\delta_2 c^2 \cos\alpha \quad (5)$$

Thus the total vertically upward force is

$$F = W_2 + F_1 = g\delta_2 a^3 + g\delta_2 a^2 h + 4ag\delta_2 c^2 \cos\alpha \quad (6)$$

The body will float, if

the weight of the cube < the upward forces

i.e., if $W_1 < F$

i.e., if $g\delta_1 a^3 < g\delta_2 a^3 + g\delta_2 a^2 h + 4ag\delta_2 c^2 \cos\alpha \quad (7)$

We now consider the equilibrium in the horizontal direction. For this, we must have, the horizontal component of surface tension at B + horizontal thrust on area of height h and breadth a = surface tension at A.

Thus,

$$a\gamma \sin\alpha + \frac{1}{2}ag\delta_2 h^2 = a\gamma$$

i.e., $g\delta_2 h^2 = 2\gamma(1 - \sin\alpha) = 2g\delta_2 c^2 \left(\sin\frac{\alpha}{2} - \cos\frac{\alpha}{2}\right)^2 = 4g\delta_2 c^2 \sin^2\left(\frac{\pi}{4} - \frac{\alpha}{2}\right)$

i.e., $h = 2c\sin\left(\frac{\pi}{4} - \frac{\alpha}{2}\right) \quad (8)$

Putting this value of h in the condition (7), we get

$$g\delta_1 a^3 < g\delta_2 a^3 + 2g\delta_2 a^2 c\sin\left(\frac{\pi}{4} - \frac{\alpha}{2}\right) + 4ag\delta_2 c^2 \cos\alpha$$

i.e., $\dfrac{\delta_1}{\delta_2} < 1 + 2\dfrac{c}{a}\sin\left(\dfrac{\pi}{4} - \dfrac{\alpha}{2}\right) + 4\dfrac{c^2}{a^2}\cos\alpha.$ **(Proved)**

EXAMPLE 12.10 *A plane plate is partly immersed in a liquid of density ρ and surface tension γ. The angle of capillarity for the liquid and the substance of the plate is β, and the* *plate is inclined at an angle α to the horizontal. Prove that the difference of the heights of the liquid on the two sides of the plate above the undisturbed surface level is*

$$4\left(\frac{\gamma}{g\rho}\right)^{\frac{1}{2}}\cos\frac{\pi-2\beta}{4}\sin\frac{\pi-2\alpha}{4}.$$

Solution. As shown in the figure (FIG. 12.35), the plate AB is inclined at an angle α to the x-axis. The tangent to the capillary curve at the point of contact makes an angle ψ with x-axis. Then

$$\psi = \alpha - \beta \tag{1}$$

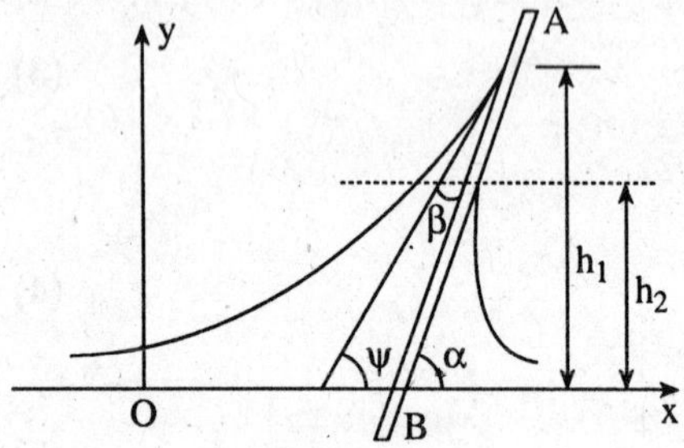

Fig. 12.35

By Eqn (12.109), we have, for any point on the curve

$$g\delta y^2 = 2\gamma(1-\cos\psi) \tag{2}$$

If h_1 be the height of liquid from the undisturbed surface level, then

$$g\delta h_1^2 = 2\gamma[1-\cos(\alpha-\beta)] = 4\gamma\sin^2\frac{\alpha-\beta}{2}$$

$$\therefore h_1 = \sqrt{\frac{4\gamma}{g\delta}}\sin\frac{\alpha-\beta}{2} \tag{3}$$

To determine the height h_2 of the liquid on the other side of the plate, we have to replace α by $(\pi-\alpha)$.

Thus, by (3),

$$h_2 = \sqrt{\frac{4\gamma}{g\delta}}\sin\frac{\pi-\alpha-\beta}{2} \tag{4}$$

Hence, the difference of heights of the liquid on two sides of the plate is

$$\begin{aligned}
&= h_1 - h_2\\
&= \sqrt{\frac{4\gamma}{g\delta}}\sin\frac{\alpha-\beta}{2} - \sqrt{\frac{4\gamma}{g\delta}}\sin\frac{\pi-\alpha-\beta}{2}\\
&= \sqrt{\frac{4\gamma}{g\delta}}\left(\sin\frac{\alpha-\beta}{2} - \sin\frac{\pi-\alpha-\beta}{2}\right) = \sqrt{\frac{4\gamma}{g\delta}}\cdot 2\cos\frac{\pi-2\beta}{4}\sin\frac{\pi-2\alpha}{4}\\
&= 4\sqrt{\frac{\gamma}{g\delta}}\cos\frac{\pi-2\beta}{4}\sin\frac{\pi-2\alpha}{4}. \quad \textbf{(Proved)}
\end{aligned}$$

EXAMPLE 12.11 *A needle floats on water with its axis in the natural level of the surface; if s be the specific gravity of steel referred to water, β the angle of capillarity, and 2α the angle subtended at the axis by the arc of a cross section in contact with the water, prove that*

$$(\pi s - \alpha)\sin\frac{1}{2}(\alpha-\beta) = \cos\alpha\cos\frac{1}{2}(\alpha+\beta)$$

Solution. Let ρ = density of steel, δ = density of liquid, and r = radius of the circular cross section of the needle.

Then mass m per unit length of the needle is

$$m = \pi r^2\rho \tag{1}$$

We refer to the figure (FIG. 12.24) of section (12.17).

Since the needle floats with its axis in the natural level of surface, we have here $z = 0$. Writing β in place of θ and $z = 0$ in Eqn (12.139), we have the total vertical upward force on the needle as

$$F = 2\gamma \sin(\alpha - \beta) + g\delta\,(\alpha r^2 + r^2 \sin\alpha\cos\alpha) \quad (2)$$

Then, for equilibrium, we have

$$F = mg, \text{ i.e., } 2\gamma \sin(\alpha - \beta) + g\delta r^2(\alpha + \sin\alpha\cos\alpha) = \pi r^2 g\rho \quad (3)$$

Similarly, for the equilibrium in the horizontal direction, we have from Eqn (12.143)

$$g\delta r^2 \cos^2\alpha = 4\gamma \sin^2\left(\frac{\alpha - \beta}{2}\right) \quad (4)$$

From (4), we get

$$2\gamma = \frac{g\delta r^2\cos^2\alpha}{2\sin^2\left(\frac{\alpha-\beta}{2}\right)} \quad (5)$$

Putting this value in (3), we get

$$\sin(\alpha - \beta)\cdot\frac{g\delta r^2\cos^2\alpha}{2\sin^2\left(\frac{\alpha-\beta}{2}\right)} + g\delta r^2\,(\alpha + \sin\alpha\cos\alpha) = \pi r^2 g\rho$$

$$\text{i.e., } \cos^2\alpha\frac{\cos\left(\frac{\alpha-\beta}{2}\right)}{\sin\left(\frac{\alpha-\beta}{2}\right)} + (\alpha + \sin\alpha\cos\alpha) = \frac{\pi\rho}{\delta}$$

$$\text{i.e., } \cos^2\alpha\cos\left(\frac{\alpha-\beta}{2}\right) + \sin\alpha\cos\alpha\sin\left(\frac{\alpha-\beta}{2}\right) = (\pi s - \alpha)\cdot\sin\left(\frac{\alpha-\beta}{2}\right)$$

$$\text{i.e., } (\pi s - \alpha)\sin\left(\frac{\alpha-\beta}{2}\right) = \cos\alpha\cdot\left[\cos\alpha\cos\left(\frac{\alpha-\beta}{2}\right) + \sin\alpha\sin\left(\frac{\alpha-\beta}{2}\right)\right]$$

$$\text{i.e., } (\pi s - \alpha)\sin\left(\frac{\alpha-\beta}{2}\right) = \cos\alpha\cos\left(\alpha - \frac{\alpha-\beta}{2}\right)$$

$$\text{i.e., } (\pi s - \alpha)\sin\left(\frac{\alpha-\beta}{2}\right) = \cos\alpha\cos\left(\frac{\alpha+\beta}{2}\right)$$

Here $s = \frac{\rho}{\delta}$ = specific gravity of steel in respect to water.

Hence proved.

EXERCISE

1. If water be introduced between two parallel plates of glass, at a very small distance d from each other, prove that the plates are pulled together with a force equal to $\frac{2A\gamma\cos\alpha}{d} + B\gamma\sin\alpha$.

A being the area of the film, B its periphery, γ the surface tension and α the angle of contact.

2. A soap solution of surface tension γ is used to form a film between a horizontal rod of length l and a length of weightless inextensible thread attached to each end of the rod. A weight W is attached at the midpoint of the thread. Prove that the shape of each half of the thread is circular and that the tension in the thread is equal to $\frac{l\gamma}{\cos\theta_2 - \sin\theta_1}$, where θ_1 is the angle which the tangent to the thread at each end makes with the horizontal, and $2\theta_2$ is the angle between the tangents to the thread at the point, where W is attached.

3. A soap film of surface tension γ has energy E per unit area. If at any subsequent time the mass per unit area of the film is m, show that $\gamma = E - m\dfrac{dE}{dm}$.

4. A drop of water of radius r is formed at the lower end of a vertical capillary tube. The height of water in the tube above the lowest point of the drop is h_1. When the drop and lower end of the tube are immersed in a beaker of water at such a depth that the level in the capillary tube is unaltered, this level is at a height h_2 above the level in the beaker. Show that $\dfrac{2\gamma}{g\rho} = r(h_1 - h_2) - \dfrac{r^2}{3}$, where γ is the surface tension of water and ρ its density.

5. Prove that the equation

$$y = x\tan(az + b)$$

represents a possible form of a liquid film, the pressure on both sides being the same.

6. For a certain large sessile drop of mercury on glass, it was found that the depth of the longest diameter below the highest point in the drop is $\sqrt{3}/2$ times the maximum height of the drop. Show that the angle of contact for mercury on glass is $\pi - \cos^{-1}\left(\dfrac{1}{3}\right)$.

7. If a film under unequal internal and external pressures form a surface of revolution, prove that the inclination ϕ of the tangent plane at a point P to the axis is given by the equation

$$\cos\phi = \frac{x}{a} + \frac{b}{x},$$

x being the perpendicular from P on the axis and a, b constants.

8. A soap bubble is filled with a mass m of a gas whose pressure is $k\times$ its density at the temperature considered. The radius of the bubble is a, when it is first placed in air. The barometer then rises, the temperature remaining unaltered. Show that the radius of the bubble increases or decreases according as the tension of the film is greater or less than $\dfrac{9\,km}{8\pi a^2}$.

9. A film is formed between two horizontal glass rods and two vertical threads connected to the rods at a distance $2a$ apart, the upper rod being clamped and the lower one hanging freely. Prove that the thread takes the form of a circle.

Also show that $\gamma = \dfrac{mg(a-b)}{2(h^2 + a^2 - b^2)}$, where γ is the surface tension, m the mass of the lower rod, $2b$ the distance between the centres of the thread with the film and h is the vertical distance between the rods.

10. A liquid film hangs in the form of a surface of revolution with its axis vertical. The upper boundary of the film is a circular wire held horizontally, the lower boundary is a heavy elastic thread, hanging freely in the form of a horizontal circle of radius r. The natural length of the thread is $2\pi a$, its modulus of elasticity is λ, and its weight is $2\pi aw$. The tension in the film is γ.

Prove that r satisfies the equation: $(\lambda^2 - a^2\gamma^2)r^2 - 2\lambda^2 ar + (\lambda^2 + w^2a^2)a^2 = 0$.

11. A spherical soap bubble of volume V_0 contains a gas at pressure P_0 and absolute temperature T_0. The temperature of the gas is gradually raised. Show that the pressure P and absolute temperature T of the gas at any instant satisfies the relation:

$$P = \sqrt{\frac{C}{T}},$$

where C is a constant.

12. A circular wire of radius a is placed in the surface of soapy water and raised gently, so as to draw after it a film. Prove that, neglecting its weight, the meridian section of the film is a catenary, and investigate the angle at which the film meets the undisturbed surface of the water. Also prove that the parameter of the meridian catenary, when the area of the film is equal to πa^2, is $\frac{a}{z}$, where z is given by $\cosh^{-1} z + z(z^2-1)^{1/2} = z^2$.

CHAPTER 13

Hydrostatic Machines

13.1 Introduction

Hydrostatic machines are mechanical devices which work on the laws of fluid pressure. There are many hydrostatic machines used in practice for various types of activities in our daily life as well as in laboratories. The simplest machine that we are acquainted with is the ***syringe*** used for taking blood from our body for pathological test or for injecting some liquid medicine into our body.

Here we will discuss some of the useful hydrostatic machines.

13.2 Hydraulic or Bramah's Press

This machine works on the principle of multiplication of thrust due to transmission of pressure through fluid.

The hydraulic press, in principle, consists of two vertical cylinders A, B connected by a tube T at their bases (FIG. 13.1). The cylinder B is wider than A and both are fitted with pistons P_2 and P_1 respectively. There is a valve V opening towards the cylinder B. Over the piston P_2, there is a platform D on which the materials like paper to be compressed are placed. The cylinders are full of water.

Fig. 13.1

Let F_1 = force applied on the narrower piston, a = area of the narrower piston P_1 and A = area of the wider piston P_2.

Then the additional pressure developed in water $= \dfrac{F_1}{a}$.

This pressure will get transmitted through water to the piston P_2 which will experience an upward thrust

$$F_2 = \frac{F_1}{a} \cdot A \tag{13.1}$$

Thus, applyinbg a force F_1 on the piston P_1, we can generate a force F_2 which is A/a times bigger on the piston P_2.

In order to get more mechanical advantage, generally a lever system is used.

Let x be the distance of the piston P_1 from the fulcrum O and y be the distance on the handle where a force F is applied. Then the force F_1 induced on the piston P_1 is

$$F_1 = \frac{Fy}{x} \tag{13.2}$$

From Eqns (13.1) and (13.2), we find

$$F = \frac{FAy}{ax} \tag{13.3}$$

Eqn (13.3) shows that even a small force on piston P_1 can produce an enormous force on the piston P_2.

13.3 The Siphon

This is an appliance used for transferring liquid stored in a vessel to another vessel at lower level with the use of a pipeline. The imaginary straight line joining the fluid levels of two vessels remaining in different levels is called the ***hydraulic gradient line***. Thus siphon is actually a pipeline remaining above the hydraulic gradient (FIG. 13.2).

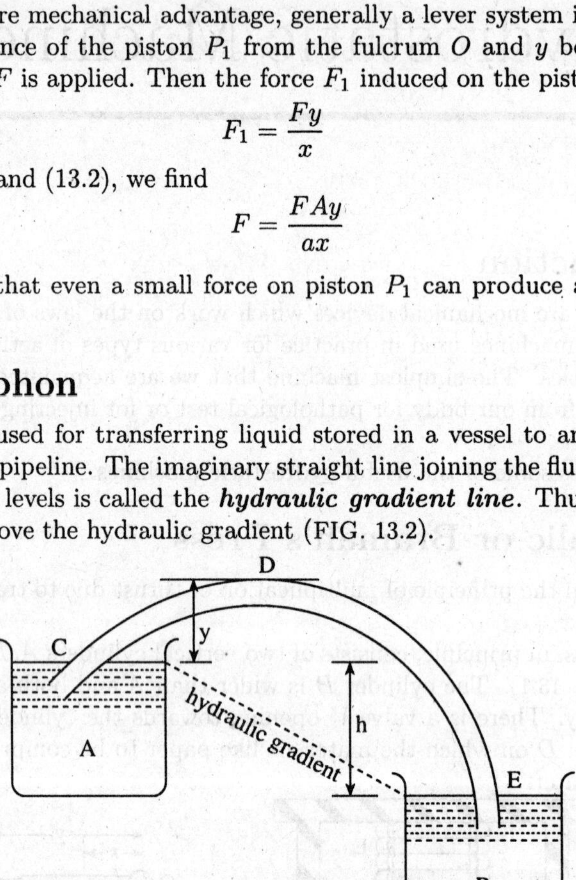

Fig. 13.2

In the laboratory, the pipe is initially filled with the liquid in vessel A by suction or otherwise. Then, keeping both the ends of the pipe temporarily closed, one end is dipped into the vessel A and the other end is placed into the vessel B. If now this end is opened, the liquid from the vessel A will start flowing into the vessel B.

Let $\Pi =$ atmospheric pressure, $y =$ height of the topmost point on the pipe from liquid level in vessel A and $h =$ depth of liquid level B below that of A.

Then fluid pressure at the point $D = \Pi - g\rho y$ and that at the point $E = \Pi + g\rho h$.

Since pressure at $E >$ the atmospheric pressure, the liquid starts flowing from A to B.

It is obvious that for the flow of liquid the pressure at the point D should not be zero. That is, the height of the point D should be less than the barometric height of the liquid.

In practice, siphons are used for transferring water from a reservoir at a high level using long pipeline passing over a hill. In such cases, the velocity of water flowing through the pipe as well as the friction coefficient of the pipe come into play in deciding the maximum height of the point D. Moreover, for continuity of flow, the minimum pressure at the point D should not be less than the separation pressure of the liquid. The ***separation pressure*** head of water is 2.5 m absolute.

Let h = difference of fluid levels,
l_1 = length of the portion CD of the pipe,
l_2 = length of the portion DE of the pipe,
d = uniform diameter of the pipe,
v = average velocity of the fluid flow,
f = friction coefficient of the pipe, and
g = acceleration due to gravity.

Then the average velocity of fluid in the pipe is given by (Ref. 5)

$$v = \sqrt{\frac{2gdh}{fl}}, \tag{13.4}$$

where $l = l_1 + l_2$.

If H be the barometric height of the liquid and H_s be the separation pressure head, then the maximum height of y is given by the pressure equation

$$H = H_s + \frac{v^2}{2g} + y_{\max} + \frac{fl_1v^2}{2gd} \tag{13.5}$$

$$\text{i.e., } y_{\max} = (H - H_s) - \left(1 + \frac{fl_1}{d}\right)\frac{v^2}{2g} \tag{13.6}$$

The last term in Eqn (13.5) represents the head loss due to friction and the second term represents the pressure head due to velocity.

Note. If the pressure at any point in a pipe through which water is flowing reaches 2.5 m absolute, water commences to vaporise and cavities of dissolved gases and vapours are formed. This causes the water in the pipe to separate and flow in sections due to which the flow is no longer continuous. Vibrations and "knocking" will occur in the pipe. This phenomenon is known as ***separation*** or ***cavitation*** and the pressure of 2.5 m absolute is called the ***separation pressure*** of water. For other liquids, the value of the separation pressure will be different.

13.4 The Diving Bell

A diving bell is a heavy rigid chamber used to transport divers in the ocean for underwater works. The most common types of the diving bells are ***wet bell*** and ***closed*** or ***dry bells***.

The ***wet bell*** consists of a heavy hollow vessel closed at the top and open at the bottom. The bell is lowered into the water by cables from a crane attached to a ship or dock. The bell is ballasted so as to remain upright in the water and to be negatively buoyant so that it sinks even when completely full of air. At the bottom, there are structural arrangements where the divers can stand or sit with their heads out of water. The air space inside the bell is kept at ambient pressure at all times.

There are two hoses, fed by pumps on the surface; one is used for providing fresh breathing gas into the bell and the other for leaking out the excess gas. As the bell is lowered, increasing pressure from the water compresses the air in the bell. If the air pressure inside the bell were not raised by adding gas to compensate for the outside water pressure, the bell would gradually fill with water as the air was compressed. Adding pressurised gas ensures that the air space within the bell remains at constant volume as the bell descends.

A ***closed*** or ***dry*** bell is a sealed chamber for human occupation which is lowered into the sea to the work place. The pressure inside is kept equalised to the environmental pressure. The internal pressure requires a strong structure and so a sphere or a cylinder with spherical end is a most efficient structure for this purpose. When it is under water, it must be possible for the divers to get in or out without flooding the entire interior. This requires a pressure hatch at the bottom. The hatch opens inward so that the internal pressure will hold it closed.

We will consider the mechanics of a wet bell having a cylindrical shape (FIG. 13.3).

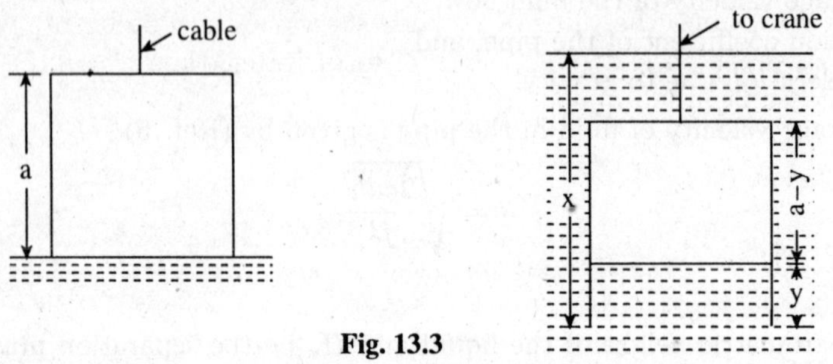

Fig. 13.3

In the wet bell, the greatest structural loads are usually the self weight of the bell and the buoyancy of air space within it. The cable through which the bell is suspended will experience tension.

Initially, the open end of the bell just touches the water surface. Then the bell is made to sink gradually to the desired depth.

Let A = area of cross-section of the bell,
H = height of water barometer,
a = height of the bell,
x = depth of the bottom of the bell from surface,
y = height of water risen inside the bell,
ρ = density of water, and
W = total weight of the bell.

Then the height of the compressed air inside the bell is $(a-y)$ and its pressure is now $(H+x-y)$ compared to its initial pressure H. Hence, by Boyle's law, we have

$$aH = (a-y)(H+x-y)$$

$$\text{or, } aH = aH + ax - ay - yH - yx + y^2, \text{ or, } y^2 - y(H+x+a) + ax = 0$$

$$\text{or, } y = \frac{1}{2}\left[(H+x+a) \pm \sqrt{(H+x+a)^2 - 4ax}\right] \tag{13.7}$$

Eqn (13.7) determines the value of y when x is known. The force of buoyancy due to the volume of compressed air inside the bell $= g\rho A(a-y)$. If F be the force of tension in the cable in this position, then

$$F = W - g\rho A(a-y) \tag{13.8}$$

Suppose that the height of water inside the bell should be kept at a constant value $z = na$, n = constant, by providing air from surface through the pump. Let V be the volume of air pumped in at the atmospheric pressure for this purpose.

Then total volume of air that has entered the bell at atmospheric pressure

$$= V + Aa \tag{13.9}$$

This quantity of air is compressed to a volume

$$= A(a - z) = Aa(1 - n) \tag{13.10}$$

at a pressure of $(H + x - z) = (H + x - na)$.

Then by Boyle's law,

$$(V + aA)H = Aa(1 - n)(H + x - na)$$

$$\text{or, } V = \frac{Aa}{H}\left[x - n(H + a + x) + n^2 a\right]. \tag{13.11}$$

This equation determines the value of V when the values of n and x are provided.

If it is desired to keep the bell completely clear of water, then $z = 0$, i.e., $n = 0$. In this case, Eqn (13.11) assumes the form

$$V = \frac{Aa}{H}x. \tag{13.12}$$

Since Aa/H is constant, Eqn (13.12) implies that the ***volume of air pumped at atmospheric pressure into the bell to keep it clear of water varies as the depth to which the bell has been lowered.***

13.5 The Common Pump

The common pump is used to raise water from below the surface. It consists of a vertical cylinder AB connected to the pipe BC which is dipped below the surface of the liquid to be raised. At the junction of the cylinder and the pipe, there is a ***valve*** V which opens only upwards. A ***valve*** is a mechanical arrangement which allows a fluid to move in one direction only and prevents its motion in the opposite direction.

The cylinder AB is tightly fitted with a piston P having a valve N which also opens only upwards. The piston is connected with a lever for getting more mechanical advantage. The upper part of the cylinder is covered leaving sufficient space for the movement of the piston rod. The piston can move upwards up to the point L just below the sprout through which the raised liquid is collected (FIG. 13.4).

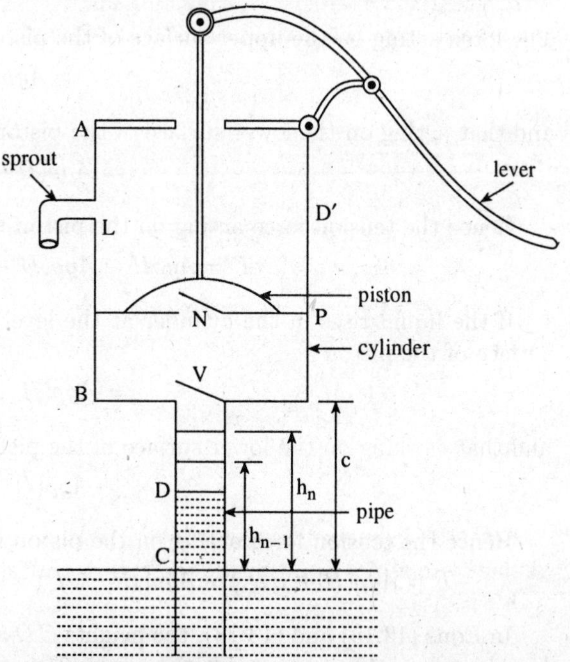

Fig. 13.4

Assuming the piston to be at its lowest point, we consider its upward motion. As the piston moves up, the air remaining below the piston expands causing a fall in pressure. The valve V opens and air in the pipe flows into the cylinder.

Thus the pressure in the pipe BC is reduced. The liquid surface outside the pipe being under atmospheric pressure, liquid is forced to rise in the pipe. The upward journey of the piston stops at the point L so that the cylinder contains air. This air gets compressed during the downward journey of the piston when the valve V gets closed and the valve N opens through which the air escapes. The piston again comes at its lowest position, and we say that one stroke is complete.

After a few strokes, air in the cylinder completely escapes when the liquid enters the cylinder during upward motion. In the next downward motion of the piston, the valve V gets closed and the valve N opens when the liquid passes through and remains on the top of the piston. During the next upward stroke, more liquid enters the cylinder through the valve V when some liquid over the piston flows down the sprout. In each of the following strokes, this process continues.

The atmospheric pressure plays the main role behind the working of the pump. Thus the height BC must be less than the barometric height of the liquid. Since the liquid at level L should stay for next stroke, for efficient working of the pump, the height CL should be less than the barometric height of the liquid.

We assume that, at any instant of time, the liquid level in the pipe is below B and at D.

Let A = cross-sectional area of the cylinder,
a = cross-sectional area of the pipe,
H = barometric height of the liquid,
ρ = density of the liquid,
l = the height BL,
c = the height BC, and
Π = atmospheric pressure.

Tension in the Piston rod

The force acting on the upper surface of the piston due to air pressure

$$= Ag\rho H, \tag{13.13}$$

and that acting on the lower surface of the piston

$$= Ag\rho(H - CD). \tag{13.14}$$

Hence the tension force acting on the piston rod is

$$T = Ag\rho H - Ag\rho(H - CD) = Ag\rho \cdot CD \tag{13.15}$$

If the liquid rises in the cylinder at the level D', then the total force acting on the upper surface of the piston

$$= Ag\rho(H + PD'), \tag{13.16}$$

and that working on the lower surface of the piston

$$= Ag\rho(H - CP). \tag{13.17}$$

Hence the tension force acting on the piston rod is

$$T = Ag\rho(H + PD') - Ag\rho(H - CP) = Ag\rho(PD' + CP) = Ag\rho \cdot CD' \tag{13.18}$$

In Eqns (13.15) and (13.18), the heights CD and CD' represent the differences between the liquid levels within and outside the pump. Hence, ***the tension in the piston rod is always equal to area of the cylinder cross-section multiplied by the weight of the liquid column whose height is the difference between liquid levels within and outside the pump.***

Rise of liquid in the pipe

Let h_{n-1} = height of the liquid surface in the pipe at the end of $(n-1)$th stroke, and h_n = same quantity at the end of nth stroke.

Then, for equilibrium at the height h_{n-1},

$$\Pi = \Pi_1 + g\rho h_{n-1}, \tag{13.19}$$

where Π_1 is the air pressure above the height h_{n-1}.

Similarly, for equilibrium at the height h_n,

$$\Pi = \Pi_2 + g\rho h_n, \tag{13.20}$$

where Π_2 is the air pressure above the height h_n.

Now, volume of the air under pressure Π_1

$$= a(c - h_{n-1}), \tag{13.21}$$

since the piston is at the lowest position now. At the end of nth stroke, the piston has come to the position L. Hence the volume of air above the liquid surface in the pipe

$$= a(c - h_n) + Al. \tag{13.22}$$

By Boyle's law,

$$\Pi_1\{a(c - h_{n-1})\} = \Pi_2\{a(c - h_n) + Al\} \tag{13.23}$$

Inserting here the values of Π_1 and Π_2 from Eqns (13.19) and (13.20), we get

$$(\Pi - g\rho h_{n-1})\{a(c - h_{n-1})\} = (\Pi - g\rho h_n)\{a(c - h_n) + Al\}$$

$$\text{or, } (g\rho H - g\rho h_{n-1})\{a(c - h_{n-1})\} = (g\rho H - g\rho h_n)\{a(c - h_n) + Al\}$$

$$\therefore (H - h_{n-1})\{a(c - h_{n-1})\} = (H - h_n)\{a(c - h_n) + Al\} \tag{13.24}$$

This equation determines the height h_n when the value of h_{n-1} is given.

Rise of liquid in the cylinder

We assume that the liquid enters the cylinder at the end of $(n+1)$th stroke.

Let h_n = height of liquid surface in the pipe at the end of nth stroke, y = height of liquid surface in the cylinder above B at the end of $(n+1)$th stroke, and Π_1, Π_2 = air pressures above the liquid surfaces in the above two cases respectively. Then

$$\left.\begin{aligned} \Pi &= \Pi_1 + g\rho h_n \\ \Pi &= \Pi_2 + g\rho(c + y) \end{aligned}\right\} \tag{13.25}$$

Volume of air above the liquid surface in the pipe $= a(c - h_n)$ and that above the liquid surface in the cylinder $= A(l - y)$.

By Boyle's law,

$$\Pi_1\{a(c - h_n)\} = \Pi_2\{A(l - y)\}$$

$$\text{or, } (\Pi - g\rho h_n)\{a(c - h_n)\} = \{\Pi - g\rho(c + y)\}\{A(l - y)\} \text{ [using Eqn (13.25)]}$$

$$\text{or, } a(H - h_n)(c - h_n) = A(H - c - y)(l - y) \tag{13.26}$$

Eqn (13.26) determines the height y of the liquid in the cylinder when the value of h_n is given. The liquid will escape through the sprout at the next stroke.

13.6 The Force Pumps

Pumps which lift water by means of pressure are known as ***Force pumps***. A diagrammatic view of a force pump is shown in FIG. 13.5. It consists of a cylinder C within which a tight piston P moves due to the rotation of the crank A. The cylinder is connected to a suction pipe S one end of which is dipped into water to be raised and the other end to a delivery pipe D whose end is free through which water escapes. At the junction of the cylinder with the pipes, there are two valves a, b, both opening outwards. When the piston moves to the right, suction occurs causing a vacuum in the cylinder. The atmospheric pressure on the water surface forces the water up the suction pipe S. The valve 'a' opens and the water enters the cylinder. On the return stroke of the piston, the water pressure closes the suction valve 'a' and opens the delivery valve 'b'. Water is then forced up the delivery pipe D and so raised to the required height.

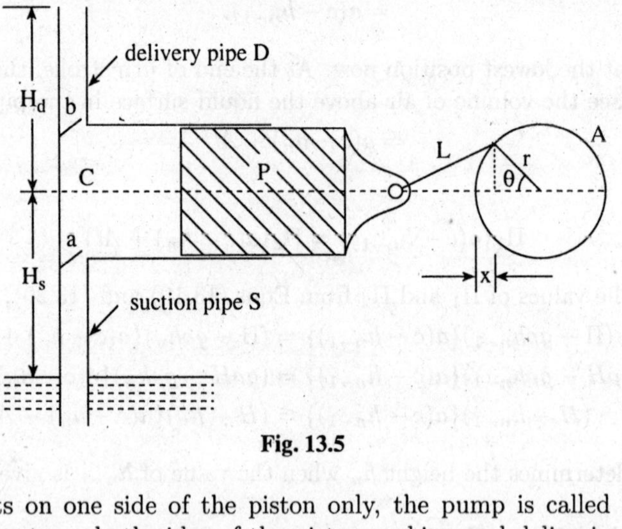

Fig. 13.5

If the water acts on one side of the piston only, the pump is called ***single-acting***. If, however, the water acts on both sides of the piston sucking and delivering during one stroke, the pump is said to be ***double-acting***.

Let $A =$ cross-sectional area of the piston,

$r =$ radius of the crank,

$L =$ length of stroke so that $L = 2r$,

$H_s =$ height of centre of the cylinder above water surface,

$H_d =$ height to which water is raised above the centre of the cylinder, and

$w =$ weight of water actually lifted per second.

We assume that the total lift is small, so that we can neglect the velocity head of water in the delivery pipe. Then the work done is

$$W = w(H_s + H_d). \tag{13.27}$$

Due to the frictional resistance of water and the mechanical parts, and due to leakage, the actual volume of water discharged will be less than the volume swept through by the piston. The difference between these two volumes is called ***slip***.

Hence the actual horse-power of the pump required would be greater than the value obtained by taking the work done from Eqn (13.27).

Acceleration of Piston

The piston will have an acceleration at the beginning and a retardation at the end of each stroke. This will cause an acceleration and a retardation to the water in the suction and the delivery pipes. Hence there will be a variation of pressure in the cylinder.

For simplicity, we assume that the piston moves with simple harmonic motion. Actually, this would be the case when the length L of the connecting rod is very long compared with the length $2\pi r$ of the crank.

Let ω = angular velocity of the rotating crank,
v = velocity of the piston,
f = acceleration of the piston,
θ = angle of rotation of the crank in t seconds, and
ρ = density of water.

Since the motion is simple harmonic,

$$\theta = \omega t \tag{13.28}$$

Displacement of the piston from end of stroke

$$x = r - r\cos\omega t. \tag{13.29}$$

Hence the velocity of the piston

$$v = \frac{dx}{dt} = \omega r \sin\omega t, \tag{13.30}$$

and its acceleration

$$f^* = \frac{d^2x}{dt^2} = \omega^2 r\cos\omega t. \tag{13.31}$$

Let v' and f' are the velocity and acceleration of water in the pipe. Since the volume of water flowing from the pipe per second is equal to the volume of water flowing into the cylinder per second, we have

$$av' = Av, \text{ i.e., } v' = \frac{A}{a}v$$

Hence

$$v' = \frac{A}{a}\omega r \sin\omega t. \tag{13.32}$$

Similarly, acceleration of water in the pipe

$$f' = \frac{A}{a}f^* = \frac{A}{a}\omega^2 r\cos\omega t. \tag{13.33}$$

Let l = length of the pipe through which water is flowing, p_a = intensity of pressure due to acceleration of water in the pipe, and H_a = acceleration pressure in terms of water head.

Then, from Newton's laws of motion,

$$ap_a = \rho al \cdot \frac{A}{a}f^*, \text{ or, } p_a = \frac{A\rho l f^*}{a} \tag{13.34}$$

Hence the acceleration pressure in the pipe in terms of water head is

$$H_a = \frac{p_a}{\rho g} = \frac{Alf^*}{ga} \tag{13.35}$$

Substituting the value of f^* from Eqn (13.31), we get

$$H_a = \frac{Al}{ga}\omega^2 r\cos\omega t = \frac{Al}{ga}\omega^2 r\cos\theta. \tag{13.36}$$

This equation determines the variation of pressure due to acceleration acting on the piston. Thus, at the beginning of the stroke when $\theta = 0$,

$$H_a = \frac{Al}{ga}\omega^2 r. \tag{13.37}$$

At the middle of the stroke, $\theta = 90°$. Hence,

$$H_a = 0 \tag{13.38}$$

At the end of the stroke, $\theta = 180°$, so that

$$H_a = -\frac{Al}{ga}\omega^2 r. \tag{13.39}$$

If the simple harmonic motion is not assumed, it can be proved that, at the beginning of the stroke

$$H_a = \frac{Al}{ga}\omega^2 r\left(1 + \frac{r}{L}\right), \tag{13.40}$$

and at the end of the stroke,

$$H_a = \frac{Al}{ga}\omega^2 r\left(1 - \frac{r}{L}\right). \tag{13.41}$$

Acceleration in suction pipe

During the suction stroke, the piston moves along the cylinder and must produce a vacuum sufficient to lift the water up the height H_s and also to accelerate the water. Hence the vacuum pressure in the cylinder must be equal to $(H_s + H_a)$.

If l_s = length of the suction pipe and a_s = cross-sectional area of the suction pipe, then

$$H_a = \frac{Al_s}{ga_s}\omega^2 r\cos\theta. \tag{13.42}$$

At the centre of the stroke, $H_a = 0$. At the beginning of the suction stroke H_a must be added to the suction head as the piston is accelerating the water. At the end of the stroke, the piston is retarding the water which causes a positive pressure on the piston. The vacuum pressure in the cylinder is thereby reduced, and therefore, H_a is subtracted. ***The inertia of water does not affect the net work done, it only causes a variation of the pressure in the cylinder.***

It is to be noted that $(H_s + H_a)$ must not fall below the cavitation pressure of water. That is, if H be the barometric height, then we must have

$$H_s + H_a \not> H - 2.5, \tag{13.43}$$

where the cavitation pressure of water is taken as 2.5 m absolute. This is taken care of by varying H_s, l_s, the ratio A/a_s or the speed of the pump.

Acceleration in the delivery pipe

At the beginning of the delivery stroke, the water column in the delivery pipe will be accelerated. The water column will be retarded at the end of the stroke. The delivery pipe, in general, is much longer than the suction pipe and hence the acceleration head is very large in this case.

If l_d = length of the delivery pipe and a_d = its cross-sectional area, then

$$H_a = \frac{Al_d}{ga_d}\omega^2 r \cos\theta. \tag{13.44}$$

The minimum pressure head in the cylinder

$$= H_d - H_a, \tag{13.45}$$

above atmospheric. That is, in absolute units, the minimum pressure head in the cylinder

$$= H + H_d - H_a. \tag{13.46}$$

This amount must not be less than the separation pressure of water, otherwise cavitation will take place at the end of the stroke. In the limiting condition, therefore, we must have

$$H + H_d - H_a = 2.5\,\text{m}. \tag{13.47}$$

The maximum pressure head in the cylinder in absolute units

$$= H + H_d + H_a. \tag{13.48}$$

The work done is not affected by the acceleration of the water which causes only the variation of the pressure in the cylinder.

Friction in pipes

The flow in a pipe gets resistance due to friction. ***The loss of pressure head due to friction is directly proportional to the square of velocity of fluid flow***. Thus we write,

$$H_f = \lambda v^2, \tag{13.49}$$

where the constant of proportionality λ depends on the length and diameter of the pipe and the coefficient of friction.

From Eqn (13.32), we have the velocity of fluid in the pipe

$$v' = \frac{A}{a}\omega r \sin\theta \tag{13.50}$$

Thus, the head loss in suction pipe is

$$H_{fs} = \lambda_s\left(\frac{A}{a_s}\omega r \sin\theta\right)^2 \tag{13.51}$$

and that in the delivery pipe is

$$H_{fd} = \lambda_d\left(\frac{A}{a_d}\omega r \sin\theta\right)^2 \tag{13.52}$$

Here the suffix d refers to the delivery pipe and the suffix s to the suction pipe.

At the two ends of the stroke, $\sin\theta = 0$. Hence there will be no loss of heads in the pipes. At the middle of the stroke, $\sin\theta = 1$ and hence

$$\left.\begin{aligned} H_{fs} &= \lambda_s\left(\frac{A}{a_s}\omega r\right)^2 \\ H_{fd} &= \lambda_d\left(\frac{A}{a_d}\omega r\right)^2 \end{aligned}\right\} \tag{13.53}$$

The equations for H_{fs} and H_{fd} are parabolas and hence their mean ordinates are equal to 2/3 rd of their maximum ordinates. Then the work done per second against friction during suction stroke

$$= \frac{2}{3}H_{fs} \cdot w = \frac{2}{3}\lambda_s\left(\frac{A}{a_s}\omega r\right)^2 w \tag{13.54}$$

Similarly, the work done per second against friction during delivery stroke

$$= \frac{2}{3}\lambda_d\left(\frac{A}{a_d}\omega r\right)^2 w. \tag{13.55}$$

Therefore, total work done during suction stroke is

$$W_s = \left(H_s + \frac{2}{3}H_{fs}\right)w, \tag{13.56}$$

and the total work done during delivery stroke is

$$W_d = \left(H_d + \frac{2}{3}H_{fd}\right)w. \tag{13.57}$$

Thus the total work done by the pump per second is

$$W = W_s + W_d = \left[H_s + H_d + \frac{2}{3}(H_{fs} + H_{fd})\right]w. \tag{13.58}$$

13.7 The Common Air-pump

This is also known as Hawksbee's air pump and is used to exhaust the air from a vessel. The pump consists of a cylindrical barrel AD having a tightly fitting piston P. There is a valve 'a' in the piston which opens only outwards. The barrel is connected by a tube or pipe to the receiver R from which the air is to be pumped out. At the junction of the barrel and the pipe, there is a valve 'b' which opens into the barrel (FIG. 13.6).

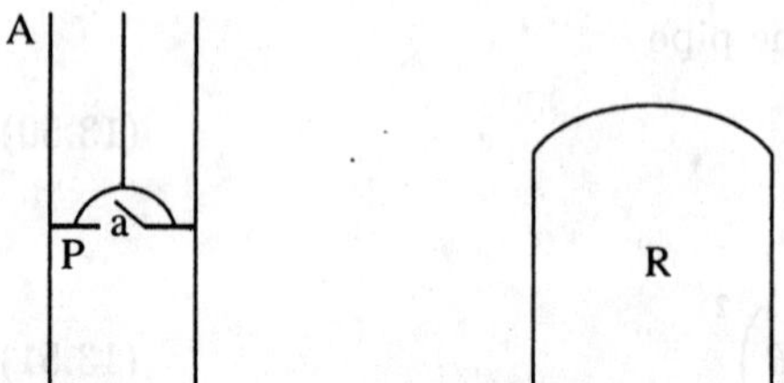

Fig. 13.6

The space in the barrel from the bottom up to the valve 'b' shown in dotted line remains untraversed by the piston and is called the ***clearance***.

During the down-stroke of the piston, the valve 'b' gets closed and so the air in the barrel is compressed. The valve 'a' then opens when the compressed air escapes.

During the up-stroke of the piston, the valve 'a' gets closed due to the atmospheric pressure. A vacuum is created within the barrel where the air pressure falls below that of the air in the receiver. This causes the valve 'b' open when the air from the receiver rushes into the barrel. The process continues and the air in the receiver gets exhausted gradually.

In practice, it is not possible to pump out all the air in the receiver due to following reasons:

(1) Since the piston cannot move down up to the bottom of the barrel, there remains some air below the piston in all its up-stroke. Thus the efficiency of the pump cannot be utilised completely.

(2) When the difference of pressure on the opposite sides of the valve 'b' is small, the valve 'b' will not open and the pump stops working.

Let B = volume of the barrel up to the dotted line,
C = volume of the clearance,
V = volume of the receiver,
ρ = density of the air, and
ρ_n = density of air at the end of nth up-stroke.

Then the density of air in the receiver initially is ρ.

Consider the first up-stroke of the piston. In the beginning, the volume of air below the piston is $(C+V)$ so that its mass is

$$= (C+V)\rho. \tag{13.59}$$

At the end of the first up-stroke, the volume of air below the piston is $(B+C+V)$. This air is rarefied and its density is ρ_1. Then mass of air below the piston

$$= (B+C+V)\rho_1. \tag{13.60}$$

The mass of air remaining the same, we have

$$(B+C+V)\rho_1 = (C+V)\rho. \tag{13.61}$$

We define,

$$\xi = \frac{C}{B+C+V}, \quad \eta = \frac{V}{B+C+V} \tag{13.62}$$

and write Eqn (13.61) in the form:

$$\rho_1 = \xi\rho + \eta\rho. \tag{13.63}$$

During the next downward stroke, the piston valve 'a' opens so that air in the barrel including clearance is of density ρ. Since the valve 'b' gets closed during this stroke, air in the receiver is of density ρ_1.

Now consider the second up-stroke of the piston. In the beginning of the stroke, there remains air in the clearance of density ρ and air in the receiver of density ρ_1. Hence, mass of air below the piston

$$= C\rho + V\rho_1. \tag{13.64}$$

At the end of the second up-stroke, the volume of air below the piston is $(B+C+V)$ and its density is ρ_2. Then the mass of air

$$= (B+C+V)\rho_2. \tag{13.65}$$

The mass of the air remaining the same, we have

$$(B+C+V)\rho_2 = C\rho + V\rho_1. \tag{13.66}$$

Using notations in (13.62), we write

$$\rho_2 = \xi\rho + \eta\rho_1. \tag{13.67}$$

Continuing the process as above, we get the series of equations like

$$\left.\begin{aligned} \rho_1 &= \xi\rho + \eta\rho, \\ \rho_2 &= \xi\rho + \eta\rho_1, \\ \rho_3 &= \xi\rho + \eta\rho_2, \\ &\ldots\ldots\ldots\ldots\ldots, \\ \rho_n &= \xi\rho + \eta\rho_{n-1} \end{aligned}\right\} \tag{13.68}$$

Thus

$$\begin{aligned}\rho_n &= \xi\rho + \eta\rho_{n-1} = \xi\rho + \eta(\xi\rho + \eta\rho_{n-2}) \text{ [using value of } \rho_{n-1} \text{ from above]}\\ &= \xi\rho + \eta\{\xi\rho + \eta(\xi\rho + \eta\rho_{n-3})\} \text{ [using value of } \rho_{n-2} \text{ from above]}\\ &\cdots\cdots\\ &= \xi\rho\,(1+\eta+\eta^2+\cdots+\eta^{n-1}) + \eta^n\rho\\ &= \rho\left\{\frac{\xi(1-\eta^n)}{1-\eta} + \eta^n\right\}\end{aligned} \tag{13.69}$$

If the clearance were negligible, then $C = 0$. The relations in (13.62) assume the forms

$$\xi = 0,\ \eta = \frac{V}{B+V}. \tag{13.70}$$

Hence

$$\rho_n = \rho\left(\frac{V}{B+V}\right)^n. \tag{13.71}$$

13.8 Smeaton's Air-pump

Hawksbee's air-pump was modified by Smeaton in the middle of 18th century. In this case, the barrel is closed at the top having a valve '*c*' which opens outwards (FIG. 13.7). Smeaton's air-pump was found to be easier in working and considered to be the best in construction and invention in those days. In this case, there is an extra clearance at the top of the barrel.

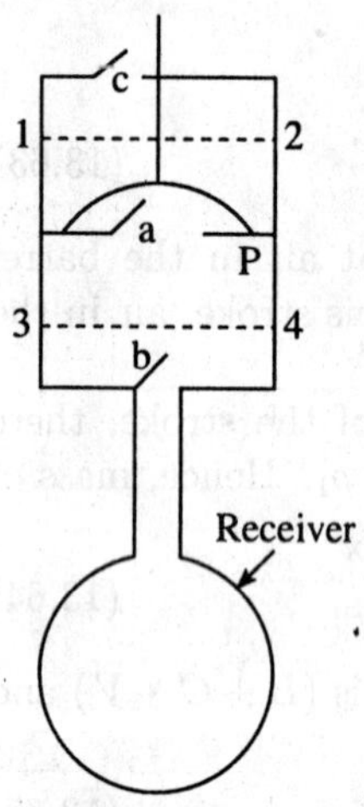

Fig. 13.7

In an up-stroke, the air from inside the barrel escapes through the valve '*c*'. Due to this, there is less resistance in opening the piston valve '*a*' in the downward stroke. Moreover, less effort is required to raise the piston, since the pressure of air above the piston during up-strokes is less than the atmospheric pressure. These are the reasons why Smeaton's pump has got advantage over other pumps.

Let B = volume of the barrel excluding clearances at the top and bottom,
C = volume of the clearance at the bottom,
D = volume of the clearance at the top,
V = volume of the receiver,
ρ = density of the atmospheric air, and
ρ_n = density of air in the receiver after n complete strokes.

At the end of any up-stroke, the piston reaches up to the level 1-2. The air above this is atmospheric with density ρ.

Consider the situation at the end of $(n-1)$th stroke. The density of air in the barrel below the piston as well as in the receiver is ρ_{n-1}. Then mass of air in the whole barrel

$$= (B+C)\rho_{n-1} + D\rho. \tag{13.72}$$

Let now the piston move down-wards up to the level 3-4. The air in whole barrel gets mixed with a common density δ (say). Then mass of air in the barrel is

$$= (B+C+D)\delta. \tag{13.73}$$

Since mass of air in the barrel remains constant, we have from Eqns (13.72) and (13.73)

$$(B+C+D)\delta = (B+C)\rho_{n-1} + D\rho. \tag{13.74}$$

This yields

$$\delta = \frac{(B+C)\rho_{n-1} + D\rho}{B+C+D} \tag{13.75}$$

In this situation mass of air below the piston remaining in the receiver and clearance of volume C is

$$= \delta C + \rho_{n-1} V. \tag{13.76}$$

Let now the piston move upwards to complete the nth stroke. The density of air below the piston in the barrel and the receiver is ρ_n. Hence mass of air below the piston is

$$= (B+C+V)\rho_n. \tag{13.77}$$

Since mass of air below the piston remains constant during up-stroke, we have from Eqns (13.76) and (13.77)

$$(B+C+V)\rho_n = \delta C + \rho_{n-1} V. \tag{13.78}$$

Inserting here the value of δ from Eqn (13.75), we get

$$(B+C+V)\rho_n = V\rho_{n-1} + \frac{C(B+C)\rho_{n-1} + CD\rho}{B+C+D}$$

$$\text{or, } (B+C+D)(B+C+V)\rho_n = [(B+C+D)V + C(B+C)]\rho_{n-1} + CD\rho \tag{13.79}$$

Denoting

$$\left.\begin{aligned} \xi &= \frac{CD}{(B+C+D)(B+C+V)} \\ \text{and } \eta &= \frac{V(B+C+D) + C(B+C)}{(B+C+D)(B+C+V)} \end{aligned}\right\} \tag{13.80}$$

we write Eqn (13.79) in the form

$$\rho_n = \xi\rho + \eta\rho_{n-1} \tag{13.81}$$

Putting successively the value of $n = 1, 2, 3, \ldots, n$, we get a series of relations like

$$\left.\begin{aligned} \rho_1 &= \xi\rho + \eta\rho \; (\because \rho_0 = \rho) \\ \rho_2 &= \xi\rho + \eta\rho_1 \\ \rho_3 &= \xi\rho + \eta\rho_2 \\ &\cdots\cdots\cdots\cdots \\ \rho_n &= \xi\rho + \eta\rho_{n-1} \end{aligned}\right\} \tag{13.82}$$

Proceeding in the same way as in the previous article, we get

$$\rho_n = \rho\left\{\frac{\xi(1-\eta^n)}{1-\eta} + \eta^n\right\} \tag{13.83}$$

If the clearances were neglected, then $C = D = 0$. In this case,

$$\xi = 0 \text{ and } \eta = \frac{V}{B+V} \tag{13.84}$$

and hence Eqn (13.83) assumes the simple form

$$\rho_n = \rho\left(\frac{V}{B+V}\right)^n. \tag{13.85}$$

13.9 The Hydraulic Intensifier

The hydraulic intensifier is used for increasing the intensity of pressure of water by means of the energy of a larger quantity of water at low pressure. This is necessary when water pressure supplied to a machine is not of sufficient intensity.

The hydraulic intensifier consists of a fixed ram through which the water at high pressure flows to the machine. There is a hollow sliding ram, mounted externally on the fixed ram (FIG. 13.8). The sliding ram contains the water at high pressure. There is a fixed cylinder, encasing the sliding ram and containing the low-pressure water. The low pressure water from the supply exerts pressure on the end of the sliding ram, forcing it downwards on to the fixed ram. This causes an increase in the pressure of water in the sliding ram.

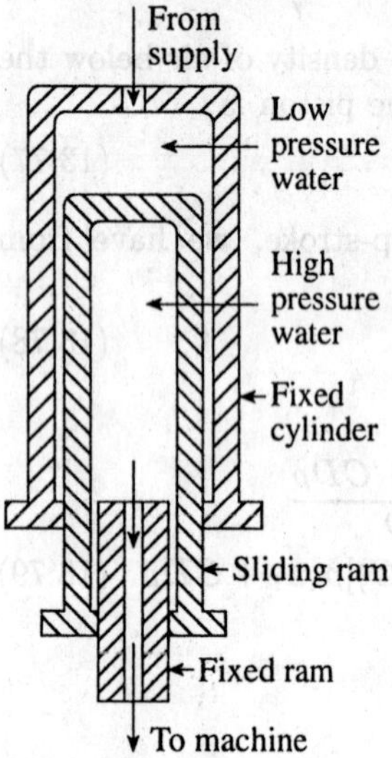

Fig. 13.8

Let A = external area of the end of sliding ram,
a = area of the end of fixed ram,
P = intensity of pressure of low-pressure water in the fixed cylinder, and
p = intensity of pressure of high-pressure water in the sliding ram.

Then the total upward force = pa, and the total downward force = PA. These two forces must be equal. Hence $pa = PA$, which yields

$$p = \frac{PA}{a} \tag{13.86}$$

13.10 The Air-condenser

The air-condenser is used to increase the pressure of air contained in a receiver. The instrument consists of a cylindrical barrel, open at the top, and tightly fitted with a piston. There is a valve 'a' in the piston opening outwards. The piston at its lowest position, leaves a clearance at the bottom of the barrel.

The barrel is connected with the receiver. At the end of the barrel, there is a valve 'b' which opens towards the receiver (FIG. 13.9).

Initially, the receiver is full of air at atmospheric pressure and the piston is at its lowest position. At the up-stroke of the piston, a vacuum is created in the space below the piston, causing a fall of pressure. The air pressure above the piston forces the valve 'a' to open and the air pressure within the receiver forces to close the valve 'b'. Due to this, air rushes into the barrel. At the next down-stroke of the piston, the air in the barrel below piston gets compressed, the piston valve 'a' gets closed and the valve 'b' at the bottom of the barrel opens. The compressed air within the barrel enters into the receiver. More and more air is forced into the receiver by continuing this process.

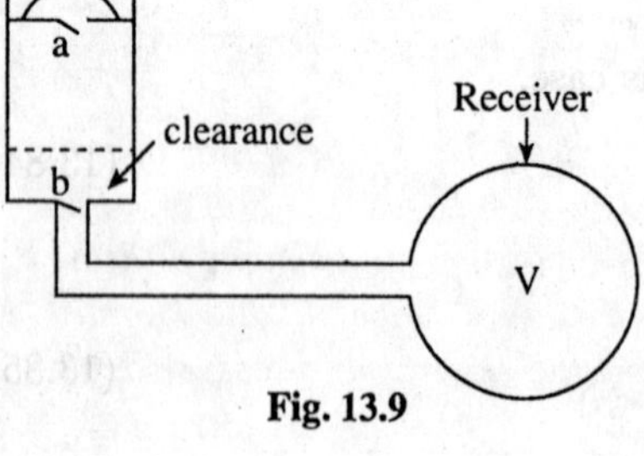

Fig. 13.9

Thus the purpose of the machine is compression, instead of exhaustion and hence the process is just the reverse of an air pump.

Let B = volume of the barrel,
C = volume of the clearance at the bottom,
V = volume of the receiver,
ρ = density of atmospheric air as well as of the initial air in the receiver and
ρ_n = density of the air in the receiver after nth down-stroke.

Consider the situation at the end of $(n-1)$th up-stroke. The air in the receiver is of density ρ_{n-1} and that in the barrel as well as in the clearance is of density ρ. Hence mass of total air

$$= (B+C)\rho + V\rho_{n-1}. \tag{13.87}$$

Now let the piston move downwards up to the dotted line. The volume of air is $(C+V)$ of density ρ_n. Hence the mass of total air

$$= (C+V)\rho_n. \tag{13.88}$$

Since mass of air remains same during a down-stroke, we have

$$(C+V)\rho_n = (B+C)\rho + V\rho_{n-1} \tag{13.89}$$

We define

$$\xi = \frac{B+C}{C+V}, \quad \eta = \frac{V}{C+V} \tag{13.90}$$

and write Eqn (13.89) in the form

$$\rho_n = \xi\rho + \eta\rho_{n-1} \tag{13.91}$$

Putting the value of $n = 1, 2, 3, \ldots, n$, we get the following series of relations

$$\left.\begin{aligned} \rho_1 &= \xi\rho + \eta\rho \;\; (\because \rho_0 = \rho) \\ \rho_2 &= \xi\rho + \eta\rho_1 \\ \rho_3 &= \xi\rho + \eta\rho_2 \\ &\cdots\cdots\cdots\cdots \\ \rho_n &= \xi\rho + \eta\rho_{n-1} \end{aligned}\right\} \tag{13.92}$$

Proceeding in the same way as in article 13.7, we find

$$\rho_n = \rho\left\{\xi\frac{1-\eta^n}{1-\eta} + \eta^n\right\} \tag{13.93}$$

If the clearance is negligible, $C = 0$. Then, from Eqn (13.90)

$$\xi = \frac{B}{V}, \quad \eta = 1. \tag{13.94}$$

In this case, Eqn (13.93) assumes the simple form

$$\begin{aligned} \rho_n &= \rho\left\{\frac{B}{V}\left(1+\eta+\eta^n+\cdots+\eta^{n-1}\right) + \eta^n\rho\right\} \\ &= \rho\frac{B}{V}n + \rho = \rho\left(1+\frac{nB}{V}\right) \end{aligned} \tag{13.95}$$

EXAMPLE 13.1 *Two reservoirs containing water are connected by a pipeline going over a hill. The length of the pipe up to the summit is* 200 m *and on the downstream is* 300 m. *The pipeline is of uniform diameter of* 30 cm. *The reservoirs are open to the atmosphere and the difference of water levels in the reservoirs is* 10 m. *Find the maximum height of the summit point, if the absolute pressure head at that point does not fall below the separation pressure head.*

Take the coefficient of friction of the pipeline as $f = 0.025$.

Solution. Here we are given

$$l_1 = 200\,\text{m},\ l_2 = 300\,\text{m},\ d = 30\,\text{cm} = 0.3\,\text{m}$$
$$h = 10\,\text{m},\ H_s = 2.5\,\text{m},\ f = 0.025$$

Then

$$l = l_1 + l_2 = 200 + 300 = 500\,\text{m}. \tag{1}$$

Taking the atmospheric pressure to be equal to that of 76 cm of mercury, we have the barometric height of water

$$H = \frac{76 \times 13.6}{100} = 10.34\,\text{m (of water)}. \tag{2}$$

The velocity of fluid flow by Eqn (13.4)

$$v = \sqrt{\frac{2gdh}{fl}} = \sqrt{\frac{2 \times 9.81 \times 0.3 \times 10}{0.025 \times 500}} = 2.17\,\text{m/second}. \tag{3}$$

Then the maximum height at summit by Eqn (13.6)

$$\begin{aligned} y_{\max} &= (H - H_s) - \left(1 + \frac{fl_1}{d}\right)\frac{v^2}{2g} \\ &= (10.34 - 2.5) - \left(1 + \frac{0.025 \times 200}{0.3}\right) \times \frac{(2.17)^2}{2 \times 9.81} \\ &= 7.84 - 4.24 = 3.60\,\text{m} \quad \textbf{(Ans.)} \end{aligned}$$

The value of H_s, the separation pressure of water, has been taken as 2.5 m absolute.

EXAMPLE 13.2 *A single-acting pump has a stroke of* 30 cm *and a piston diameter of* 15 cm. *The centre of the pump is* 5 m *above water level in sump and* 33 m *below delivery water level. The lengths of the suction and delivery pipes are* 6 m *and* 40 m *respectively, and their diameters are* 8 cm.

If the pump is working at 30 r.p.m., *find the pressure head on the piston at the beginning, middle, and end of both strokes and find the total work done in driving the pump.*

Ignore the velocity head of the discharge water and take $\lambda = \dfrac{4fl}{2gd}, f = 0.01.$

Solution. Here we are given

L = 30 cm = 0.3 m, piston diameter = 0.15 m

H_s = 5.0 m, H_d = 33.0 m, l_s = 6.0 m, l_d = 40.0 m

d_s = d_d = 0.08 m, rotation per minute = 30

f = 0.01 and $\lambda = \dfrac{4fl}{2gd}$

Then $r = \dfrac{L}{2} = 0.15\,\text{m}, \omega = \dfrac{2\pi(30)}{60} = 3.14.$

(1) **Suction stroke.** At the ends of the stroke

$$\begin{aligned} H_a &= \frac{l_s}{g}\frac{A}{a_s}\omega^2 r = \frac{6 \cdot 0}{9 \cdot 81} \times \left(\frac{0.15}{0.08}\right)^2 (3.14)^2 \times (0.15) \\ &= 3.18\,\text{m of water.} \end{aligned} \tag{1}$$

At middle stroke

$$H_{fs} = \frac{4fl_s}{2gd_s}\left(\frac{A}{a_s}\omega r\right)^2 = \frac{4 \times 0.01 \times 6.0}{2 \times 9.81 \times 0.08}(3.5156 \times 3.14 \times 0.15)^2$$

$$= 0.419 \text{ m of water.} \tag{2}$$

Pressure at the beginning of stroke

$$= H_s + H_a = 5.0 + 3.18 = 8.18 \text{ m of water.} \tag{3}$$

Pressure at the end of the stroke

$$= H_s - H_a = 5.0 - 3.18 = 1.82 \text{ m of water.} \tag{4}$$

Pressure at the middle of stroke

$$= H_s + H_{fs} = 5.0 + 0.419 = 5.419 \text{ m of water.} \tag{5}$$

(2) **Delivery stroke.** At the ends of the strokes

$$H_a = \frac{l_d}{g}\frac{A}{a_d}\omega^2 r = \frac{40}{9.81}\left(\frac{0.15}{0.08}\right)^2 (3.14)^2 \times (0.15)$$

$$= 21.2 \text{ m of water.} \tag{6}$$

At the middle of stroke

$$H_{fd} = \frac{4fl_d}{2gd_d}\left(\frac{A}{a_d}\omega r\right)^2 = \frac{4 \times 0.01 \times 40}{2 \times 9.81 \times 0.08}(3.5156 \times 3.14 \times 0.15)^2$$

$$= 2.795 \text{ m of water.} \tag{7}$$

Pressure at the beginning of stroke

$$= H_d + H_a = 33.0 + 21.2 = 54.2 \text{ m of water.} \tag{8}$$

Pressure at the end of the stroke

$$= H_d - H_a = 33.0 - 21.2 = 11.8 \text{ m of water.} \tag{9}$$

Pressure at the middle of the stroke

$$= H_d + H_{fd} = 33.0 + 2.795 = 35.795 \text{ m of water.} \tag{10}$$

(3) **Work done.** Weight of water per stroke = weight of water/m^3 × area × length of stroke

$$\therefore W = 1000 \times \frac{\pi}{4}(0.15)^2 \times 0.3 = 5.298 \text{ kg.}$$

Work done during suction stroke

$$= W\left(H_s + \frac{2}{3}H_{fs}\right) = 5.298\left(5.0 + \frac{2}{3} \times 0.419\right) = 27.97 \text{ kg.m.} \tag{11}$$

Work done during delivery stroke

$$= W\left(H_d + \frac{2}{3}H_{fd}\right) = 5.298\left(33.0 + \frac{2}{3} \times 2.795\right) = 184.706 \text{ kg.m} \tag{12}$$

Total work done per second = $27.97 + 184.706 = 212.68$ kg.m.

Total work done per revolution = $212.68 \times \frac{30}{60} = 106.34$ kg.m.

Since 1 metric Horse Power (H.P.) = 75 kg.m/second, the power required for the pump to do the total amount of work = $\frac{106.34}{75} = 1.418$ metric H.P.

EXAMPLE 13.3 *A cylindrical diving bell of height 'a' is furnished with a barometer and lowered into a fluid, the heights of the mercury in the barometer before and after immersion being h and h′ respectively. Show that the depth of the bottom of the bell below the surface of the fluid is equal to* $\left(\frac{\theta}{\beta}+\frac{a}{h'}\right)(h'-h)$, *where θ is the sp. gr. of mercury and β that of the fluid.*

Solution. Let x be the depth of the bottom of the bell from surface and y be the height of water risen. Then the length occupied by the air inside the bell is $(a-y)$. Hence, by Boyle's law,

$$ah=(a-y)h', \text{ i.e., } \frac{h}{h'}=1-\frac{y}{a}. \tag{1}$$

Since h' is the height of the mercury barometer at a depth of $(x-y)$, we must have

$$h'\theta=h\theta+(x-y)\beta. \tag{2}$$

From (2),

$$\beta x=h'\theta-h\theta+\beta y, \text{ i.e., } x=(h'-h)\frac{\theta}{\beta}+y. \tag{3}$$

Inserting here the value of

$$y=a\left(1-\frac{h}{h'}\right), \tag{4}$$

we get

$$x=(h'-h)\frac{\theta}{\beta}+a\left(1-\frac{h}{h'}\right)=\left(\frac{\theta}{\beta}+\frac{a}{h'}\right)(h'-h). \quad \textbf{(Proved)}$$

EXAMPLE 13.4 *Prove that, in a force pump, the maximum vacuum pressure in the suction pipe will occur when the angle θ of the crank is zero.*

Solution. Let H be the total vacuum pressure in the suction pipe for any angle θ of the crank. Then

$$H=H_s+H_a+\frac{v_s^2}{2g}+H_{fs} \tag{1}$$

Using Eqns (13.42), (13.50) and (13.51), we get

$$\begin{aligned}H&=H_s+\frac{Al_s}{ga_s}\omega^2r\cos\theta+\frac{1}{2g}\left(\frac{A}{a_s}\omega r\sin\theta\right)^2+\lambda_s\left(\frac{A}{a_s}\omega r\sin\theta\right)^2\\&=H_s+\frac{Al_s}{ga_s}\omega^2r\cos\theta+\left(\frac{A}{a_s}\right)^2\omega^2r^2\sin^2\theta\left(\lambda_s+\frac{1}{2g}\right)\end{aligned} \tag{2}$$

Differentiating Eqn (2) with respect to θ

$$\frac{dH}{d\theta}=-\frac{Al_s}{ga_s}\omega^2r\sin\theta+2\left(\frac{A}{a_s}\right)^2\omega^2r^2\sin\theta\cos\theta\left(\lambda_s+\frac{1}{2g}\right). \tag{3}$$

For maximum value of H, we must have

$$\frac{dH}{d\theta}=0. \tag{4}$$

This yields, by Eqn (3)

$$\omega^2r\frac{A}{a_s}\sin\theta\left[-\frac{l_s}{g}+2r\left(\frac{A}{a_s}\right)\left(\lambda_s+\frac{1}{2g}\right)\cos\theta\right]=0. \tag{5}$$

Hence,

$$\text{either } \sin\theta = 0, \text{ or, } \cos\theta = \frac{l_s a_s}{Ar(1+2g\lambda_s)} \quad (6)$$

The value of $\cos\theta$ obtained from Eqn (6) when applied to actual pumps is always greater than unity. That is, in actual pumps

$$l_s a_s > Ar(1+2g\lambda_s) \quad (7)$$

Hence we discard this value of $\cos\theta$. Therefore, $\sin\theta = 0$, i.e., $\theta = 0$. **(Proved)**

Note. We conclude from above that ***the maximum vacuum pressure occurs just inside the suction pipe at the section where it enters the cylinder.***

EXAMPLE 13.5 *A diving bell is in the form of a cylinder with a hemispherical top; c is the length and a is the radius of the cylinder. Find how far the bell must be sunk in order that the hemisphere may be the only part containing air. Show that in that position, the volume of air at atmospheric pressure which must be forced in, to clear the whole bell of water, must be*

$$\left(\frac{c}{H} + \frac{3}{2}\frac{c}{a}\right)V,$$

where H is the height of the water barometer and V is the volume of the bell.

Solution. Let x be the depth of the bell from the surface when the hemispherical part only is occupied by air (FIG. 13.10).

Volume of the bell

$$V = \pi a^2 c + \frac{1}{2}\cdot\frac{4}{3}\pi a^3 = \pi a^2\left(c + \frac{2}{3}a\right). \quad (1)$$

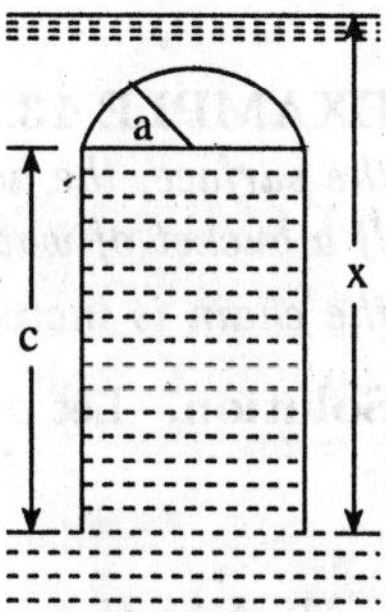

Fig. 13.10

Initially, when the bell was above the surface, it was full of air under atmospheric pressure head H. When under water, volume of air at the top

$$= \frac{2}{3}\pi a^3, \quad (2)$$

under the pressure of $(H + x - c)$.

Then by Boyle's law,

$$V \cdot H = \frac{2}{3}\pi a^3 (H + x - c)$$

$$\text{i.e., } \pi a^2\left(c + \frac{2}{3}a\right)H = \frac{2}{3}\pi a^3(H + x - c) \text{ or, } cH = \frac{2}{3}a(x - c),$$

which yields

$$x = c\left(1 + \frac{3}{2}\frac{H}{a}\right). \quad (3)$$

Let V' be the volume of air at atmospheric pressure which is required to force into the bell to clear it of water. Then

$$(V + V')H = V(H + x), \text{ i.e., } V' = \frac{V}{H}x. \quad (4)$$

Inserting here the value of x from Eqn (3), we get

$$V' = \frac{V}{H}\left(c + \frac{3}{2}\frac{cH}{a}\right) = V\left(\frac{c}{H} + \frac{3}{2}\frac{c}{a}\right). \quad \textbf{(Proved)}$$

EXAMPLE 13.6 *If a diving bell descends from the surface with a uniform velocity v, show that the water will ascend a height b in the bell in time*

$$\frac{b}{v}\left(1+\frac{H}{l-b}\right),$$

where l is the length of the bell and H is the height of water barometer.

Solution. Let A = uniform cross-sectional area of the bell, t = time of descent of the bell, x = depth to which the bell has descended.

Since the bell has descended with uniform velocity v, the distance traversed in time t is

$$x = vt. \tag{1}$$

Initially, when the bell is above water surface, the volume of air within it is Al and its pressure is H. After time t, length of air column within it is $(l-b)$ under pressure of

$$H + x - b = H + vt - b. \tag{2}$$

Then by Boyle's law, we have

$$AlH = A(l-b)(H + vt - b)$$

$$\text{i.e., } b^2 - b(l+H) = (b-l)vt \text{ or, } b - \frac{bH}{b-l} = vt$$

$$\therefore\ t = \frac{b}{v}\left(1+\frac{H}{l-b}\right) \quad \textbf{(Proved)}$$

EXAMPLE 13.7 *A diving bell is immersed in water so that its top is at a depth 'a' below the surface, the height of air within the bell being x and the height of water barometer being H. If a bucket of water of small weight w be now drawn up into the bell, prove that the tension of the chain is increased by* $\dfrac{wx}{H+a+2x}$ *approximately.*

Solution. Let ρ be the density of water. Then the volume of the water drawn up into the bell

$$= \frac{w}{g\rho}. \tag{1}$$

If A be the cross-sectional area of the cylindrical bell, then the volume of the air initially was Ax under a pressure head of $(H + a + x)$ (FIG. 13.11).

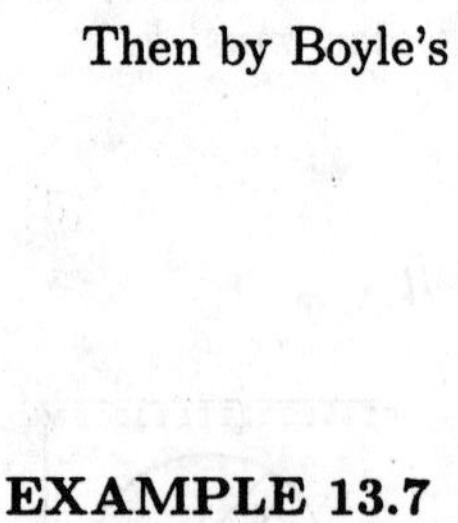

Fig. 13.11

Let after drawing up the bucketful of water, the water level within the bell rises by z. Then the volume of air within the bell

$$= A(x-z) - \frac{w}{g\rho} = A(x - z - \alpha), \tag{2}$$

where we have denoted

$$\alpha = \frac{w}{g\rho A} \tag{3}$$

The pressure of this air is now $(H + a + x - z)$.

Hence by Boyle's law,

$$Ax(H+a+x) = A(x-z-\alpha)(H+a+x-z)$$

$$\text{or, } x(H+a+x) = x(H+a+x) - xz - (z+\alpha)(H+a+x-z)$$

$$\text{or, } xz + z(H+a+x) - z^2 + \alpha(H+a+x) - \alpha z = 0$$

$$\text{or, } z^2 - z(H+a+2x-\alpha) - \alpha(H+a+x) = 0$$

$$\therefore\ z = \frac{1}{2}(H+a+2x-\alpha) \pm \frac{1}{2}\left\{(H+a+2x-\alpha)^2 + 4\alpha(H+a+x)\right\}^{1/2} \tag{4}$$

The expression under radical sign in Eqn (4)

$$= (H+a+2x)^2 + \alpha^2 - 2\alpha(H+a+2x) + 4\alpha(H+a+x)$$
$$= (H+a+2x)^2 + \alpha^2 + 2\alpha(H+a). \tag{5}$$

Neglecting α^2 in Eqn (5), we get from Eqn (4)

$$z = \frac{1}{2}(H+a+2x-\alpha) - \frac{1}{2}\left\{(H+a+2x)^2 + 2\alpha(H+a)\right\}^{1/2}$$
$$= \frac{1}{2}(H+a+2x-\alpha) - \frac{1}{2}(H+a+2x)\left\{1 + \frac{2\alpha(H+a)}{(H+a+2x)^2}\right\}^{1/2}$$

Expanding the last term binomially and taking only two terms,

$$z = \frac{1}{2}(H+a+2x-\alpha) - \frac{1}{2}(H+a+2x)\left\{1 + \frac{\alpha(H+a)}{(H+a+2x)^2}\right\}$$
$$= \frac{1}{2}(H+a+2x-\alpha) - \frac{1}{2}(H+a+2x) - \frac{1}{2}\frac{\alpha(H+a)}{H+a+2x}$$
$$= -\frac{\alpha}{2}\left(1 + \frac{H+a}{H+a+2x}\right) \tag{6}$$

Initially, the tension in the chain is

$$T_0 = W - g\rho Ax, \tag{7}$$

where W is the weight of the bell.

Finally, the tension in the chain becomes

$$T_1 = w + W - g\rho A(x - z). \tag{8}$$

Hence the increase in the tension in the chain

$$T = T_1 - T_0 = w + W - g\rho A(x-z) - W + g\rho Ax$$
$$\text{i.e., } T = w + g\rho Az. \tag{9}$$

Inserting the value of z from (6) and using (3), we find

$$T = w + \frac{w}{\alpha}z = w - \frac{w}{2}\left(1 + \frac{H+a}{H+a+2x}\right)$$
$$= w\left(1 - \frac{1}{2} - \frac{1}{2}\cdot\frac{H+a}{H+a+2x}\right) = \frac{wx}{H+a+2x}. \quad \textbf{(Proved)}$$

EXAMPLE 13.8 *A cylindrical diving bell of height b is immersed in water with its highest point at a depth 'a' below the surface. If the barometer rises so that the increase of the pressure on its top is P, show that the alteration in the tension of the chain is*

$$\frac{P}{2}\left[1 - \frac{a+H}{\sqrt{(a+h)^2 + 4bH}}\right] \text{ approximately.}$$

Solution. Let A = cross-sectional area of the bell,

y = height of water inside the bell when depth of top = a,

z = increase in the height of water inside the bell, when the barometer rises,

ρ = density of water, and

g = acceleration due to gravity.

When the bell is on the surface of water, volume of air inside it $V = Ab$ under barometric pressure $P_1 = H$. So, we have

$$P_1V_1 = AbH. \tag{1}$$

When the bell is below water with its top at a depth $= a$, the volume of air inside it $V_2 = A(b-y)$ under a pressure of $P_2 = (H + a + b - y)$.

Hence,

$$P_2V_2 = A(b-y)(H+a+b-y). \tag{2}$$

Now, the barometer rises, causing an increase of pressure head of

$$h = \frac{P}{g\rho A}, \tag{3}$$

on the top. The volume of air inside the bell $V_3 = A(b-y-z)$ under a pressure head $P_3 = (H+a+h+b-y-z)$. Hence,

$$P_3V_3 = A(b-y-z)(H+a+h+b-y-z). \tag{4}$$

By Boyle's law, we must have

$$P_1V_1 = P_2V_2 = P_3V_3 \tag{5}$$

From Eqns (1) and (2), we get

$$\begin{aligned} AbH &= A(b-y)(H+a+b-y) \\ \text{or,}\quad ab + b^2 - by - yH - ay - by + y^2 &= 0 \\ \text{or,}\quad y^2 - y(H+a+2b) + b(a+b) &= 0 \end{aligned} \tag{6}$$

From Eqns (2) and (4), we have

$$\begin{aligned} A(b-y)(H+a+b-y) &= A(b-y-z)(H+a+h+b-y-z) \\ \text{or,}\quad (b-y)(H+a+b-y) &= (b-y)(H+a+b-y) + (b-y)(h-z) \\ &\quad -z(H+a+h+b-y-z) \\ \text{or,}\quad z^2 - z(H+a+h+2b-2y) + h(b-y) &= 0 \\ \therefore\quad z &= \frac{1}{2}(H+a+h+2b-2y) \pm \frac{1}{2}\sqrt{X}, \end{aligned} \tag{7}$$

where we have denoted

$$X = (H+a+h+2b-2y)^2 - 4h(b-y). \tag{8}$$

We write Eqn (8) in the form

$$\begin{aligned} X &= (H+a+2b-2y)^2 + h^2 + 2h(H+a+2b-2y) - 4h(b-y) \\ &= (H+a+2b-2y)^2 + 2h(a+H), \text{ (neglecting } h^2) \\ &= (H+a+2b-2y)^2 \left\{1 + \frac{2h(a+H)}{(H+a+2b-2y)^2}\right\} \end{aligned} \tag{9}$$

Hence,

$$\sqrt{X} = (H+a+2b-2y)\left\{1 + \frac{h(a+H)}{(H+a+2b-2y)^2}\right\} \tag{10}$$

approximately.

Inserting this value in Eqn (7), we get

$$z = \frac{1}{2}(H+a+h+2b-2y) - \frac{1}{2}(H+a+2b-2y)\cdot\left\{1+\frac{h(a+H)}{(H+a+2b-2y)^2}\right\}$$

$$= \frac{h}{2} - \frac{h}{2}\frac{a+H}{(H+a+2b-2y)} \tag{11}$$

Now,

$$\begin{aligned}(H+a+2b-2y)^2 &= (a+H)^2+4bH+4b^2+4y^2-4yH-4ay+4ab-8yb\\ &= (a+H)^2+4bH+4[y^2-y(H+a+2b)+b(a+b)]\\ &= (a+H)^2+4bH \text{ [by virtue of Eqn (6)]}\end{aligned} \tag{12}$$

Hence, from Eqns (11) and (12), we write

$$z = \frac{h}{2}\left[1-\frac{a+H}{\sqrt{(a+H)^2+4bH}}\right] \tag{13}$$

If W be the weight of the bell, the tensions in the chain in two cases are

$$\left.\begin{aligned}T_0 &= W - g\rho A(b-y)\\ T_1 &= W - g\rho A(b-y-z)\end{aligned}\right\} \tag{14}$$

Increase in tension is

$$T = T_1 - T_0 = \{W - g\rho A(b-y-z)\} - \{W - g\rho A(b-y)\}$$

$$= g\rho Az = \frac{P}{h}z, \text{ using Eqn (3)} \tag{15}$$

From Eqns (13) and (15), we get

$$T = \frac{P}{2}\left[1-\frac{a+H}{\sqrt{(a+H)^2+4bH}}\right] \text{ approximately. } \textbf{(Proved)} \tag{16}$$

EXAMPLE 13.9 *A circular cone, hollow but of great weight is lowered into the sea by means of a rope attached to the vertex. If h be the height of the cone, c the depth of the vertex below the free surface, H the height of the water barometer and T and T′ the absolute temperatures of the air at the surface and of the water, prove that the depth of the water surface inside, below the vertex of the cone is given by* $x^4 + x^3(H+c) - Hh^4\frac{T'}{T} = 0$.

Solution. Let 2α be the vertical angle of the cone and x be the depth of water surface below the vertex of the cone when it is inside the water (FIG. 13.12).

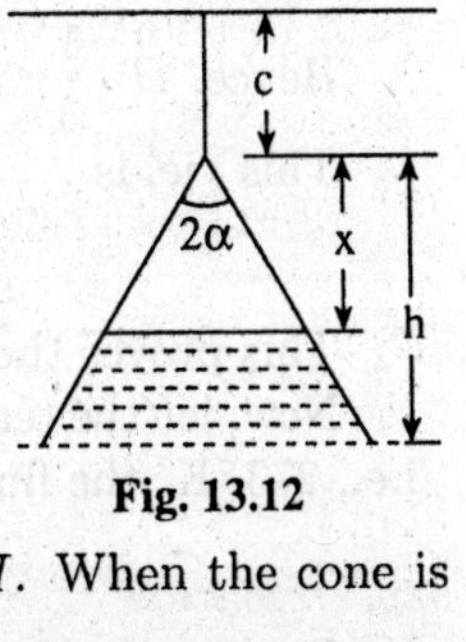

Fig. 13.12

If r be the radius of the cone, then

$$r = h\tan\alpha. \tag{1}$$

Volume of air inside the cone when above water surface

$$V_1 = \frac{1}{3}\pi r^2 h = \frac{1}{3}\pi(h\tan\alpha)^2 h = \frac{1}{3}\pi h^3\tan^2\alpha. \tag{2}$$

Temperature of this air as given $= T$ and the pressure of this air $= H$. When the cone is below the sea, the volume of the air inside it

$$V_2 = \frac{1}{3}\pi x^3\tan^2\alpha. \tag{3}$$

Temperature of this gas $= T'$ and its pressure is $(H+c+x)$.

Then by gas law,

$$\frac{HV_1}{T} = \frac{(H+c+x)V_2}{T'} \tag{4}$$

Inserting here the values of V_1, V_2 from Eqns (2) and (3),

$$\frac{H}{T} \cdot \frac{1}{3}\pi h^3 \tan^2\alpha = \frac{H+c+x}{T'} \cdot \frac{1}{3}\pi x^3 \tan^2\alpha, \text{ or, } \frac{Hh^3T'}{T} = (H+c+x)x^3.$$

Hence, $x^4 + x^3(H+c) - Hh^3\frac{T'}{T} = 0.$ **(Proved)**

EXAMPLE 13.10 *If a diving bell in the shape of a cone of height 'a' be lowered till its vertex is at a depth d, prove that the height x of the part of the bell occupied by the air is given by the equation $x^4 + x^3(H+d) = a^3H$, where H is the height of the water barometer.*

If the temperature of the air inside be now raised through t °C, prove that the tension of the supporting chain is diminished by

$$\frac{3\alpha tHw}{3(H+d)+4x} \textit{ (nearly)},$$

where w is the weight of water the cone could contain and α is the coefficient of expansion, the square of α being neglected.

Solution. Let 2θ = vertical angle of the cone, W = weight of the cone, ρ = density of water, and x = height of air inside the cone when it is inside the water.

We are given that a = height of the bell, and d = depth of the vertex of the cone.

When the bell is above the water surface, the volume of air inside it is

$$V_1 = \frac{1}{3}\pi a^3 \tan^2\theta. \tag{1}$$

Its pressure,

$$p_1 = H. \tag{2}$$

When the bell is inside the water, volume of air inside it

$$V_2 = \frac{1}{3}\pi x^3 \tan^2\theta \tag{3}$$

Its pressure is

$$p_2 = H + d + x \tag{4}$$

By Boyle's law, $p_1V_1 = p_2V_2$.

Hence, $H \cdot \frac{1}{3}\pi a^3 \tan^2\theta = (H+d+x) \cdot \frac{1}{3}\pi x^3 \tan^2\theta.$

This yields

$$x^4 + x^3(H+d) = a^3H. \tag{5}$$

This proves the first part of the problem.

Now, let the temperature of air inside be raised by t °C. If the initial temperature were 0 °C, i.e., 273 °K, the final temperature is then

$$T_3 = 273 + t. \tag{6}$$

Due to rise of temperature, the height of air inside will become $x(1+\alpha t)$. Then the volume of air inside is

$$V_3 = \frac{1}{3}\pi x^3(1+\alpha t)^3, \text{ i.e., } V_3 = \frac{1}{3}\pi x^3(1+3\alpha t) \tag{7}$$

Here we have neglected α^2. The pressure of this gas becomes

$$p_3 = (H + d + x + x\alpha t) \tag{8}$$

By the combined gas law,

$$\frac{p_2 V_2}{T_2} = \frac{p_3 V_3}{T_3}, \ (T_2 = 273\,°\text{K}) \tag{9}$$

Inserting here the necessary values, we get

$$\frac{(H + d + x)x^3}{T_2} = \frac{(H + d + x + x\alpha t)x^3(1 + 3\alpha t)}{T_3}$$

$$\text{or, } \{x^4 + x^3(H + d)\}T_3 = T_2[\{x^4 + x^3(H + d)\} + 3\alpha t x^3(H + d + x) + x^4\alpha t(1 + 3\alpha t)] \tag{10}$$

Using relation (5),

$$a^3 H(T_3 - T_2) = T_2 x^3 \alpha t[3(H + d + x) + x], \text{ neglecting } \alpha^2$$

$$\text{or, } a^3 Ht = T_2 x^3 \alpha t[3(H + d) + 4x]$$

$$\text{or, } \frac{x^3}{a^3} = \frac{H}{T_2\alpha[3(H + d) + 4x]} = \frac{H}{3(H + d) + 4x}, \tag{11}$$

since $T_2\alpha = 273\alpha = 1$.

Since w is the weight of water the cone could contain

$$w = g\rho \cdot \frac{1}{3}\pi a^3 \tan^2\theta. \tag{12}$$

This yields

$$\frac{1}{3}\pi \tan^2\theta = \frac{w}{g\rho a^3}. \tag{13}$$

If F_1, F_2 be the tensions in the chain before and after raising of temperature, then

$$\left.\begin{aligned} F_1 &= W - g\rho \cdot \frac{1}{3}\pi x^3 \tan^2\theta \\ F_2 &= W - g\rho \cdot \frac{1}{3}\pi x^3(1 + 3\alpha t) \tan^2\theta \end{aligned}\right\} \tag{14}$$

By virtue of relation (13), Eqn (14) become

$$\left.\begin{aligned} F_1 &= W - w\frac{x^3}{a^3} \\ F_2 &= W - w\frac{x^3}{a^3}(1 + 3\alpha t) \end{aligned}\right\} \tag{15}$$

Hence the difference in tension of the chain is

$$F = F_2 - F_1 = -3\alpha t w \cdot \frac{x^3}{a^3} \tag{16}$$

Using here Eqn (11),

$$F = -\frac{3\alpha t w H}{\{3(H + d) + 4x\}} \tag{17}$$

This shows that the tension in the chain is diminished. Hence proved.

EXAMPLE 13.11 *Prove that if h, h' be the heights at which the water stands in the lower cylinder of a common pump before and after a stroke, then*

$$(h'-h)(h'+h-H-a)+nb(H-h')=0,$$

where a, b are the lengths of the lower and upper cylinders, n is the ratio of the sectional area of the latter to that of the former, and H is the height of the water barometer.

Solution. Let A = cross-sectional area of the barrel, and A' = cross-sectional area of the pipe. Then by question,

$$\frac{A}{A'}=n, \text{ i.e., } A=nA'. \tag{1}$$

We are given, a = length of the pipe, b = length of the barrel, h = height of water level in the pipe before a stroke, and h' = same after a stroke.

Let Π = atmospheric pressure, Π_1 = air pressure above the height h, and Π_2 = air pressure above the height h'.

For equilibrium at these heights,

$$\left.\begin{aligned}\Pi=\Pi_1+g\rho h, \text{ i.e., } \Pi_1=\Pi-g\rho h\\ \Pi=\Pi_2+g\rho h', \text{ i.e., } \Pi_2=\Pi-g\rho h'\end{aligned}\right\} \tag{2}$$

Volume of air under pressure Π_1

$$=A'(a-h). \tag{3}$$

At the end of the stroke, volume of air above liquid surface

$$=A'(a-h')+Ab. \tag{4}$$

By Boyle's law,

$$\Pi_1 A'(a-h)=\Pi_2\{A'(a-h')+Ab\} \tag{5}$$

Using here Eqns (1) and (2),

$$(\Pi-g\rho h)(a-h)=(\Pi-g\rho h')\{(a-h')+nb\} \tag{6}$$

Since H is the height of water barometer,

$$\Pi=g\rho H. \tag{7}$$

Inserting this into Eqn (6), we get

$$(H-h)(a-h)=(H-h')(a-h'+nb)$$

$$\text{or, } aH-Hh-ah+h^2=aH-Hh'-ah'+h'^2+nb(H-h')$$

$$\text{or, } (h'^2-h^2)-H(h'-h)-a(h'-h)+nb(H-h')=0$$

$$\text{or, } (h'-h)(h'+h-H-a)+nb(H-h')=0. \quad \textbf{(Proved)}$$

EXAMPLE 13.12 *The mass of air in the receiver of an air pump is m at the beginning and it becomes m' after n complete strokes of the piston. If V and B denote the volumes of the receiver and the barrel, prove that* $n=\dfrac{\log m-\log m'}{\log(B+V)-\log V}$.

Solution. Let ρ = density of air initially in the receiver, and ρ_n = density of the air in the receiver after complete n strokes.

We are given that V = volume of the receiver, B = volume of the barrel, m = initial mass of air in the receiver, and m' = mass of air in the receiver after n complete strokes.

Then $m=V\rho$ and $m'=V\rho_n$, so that

$$\frac{m}{m'}=\frac{\rho}{\rho_n}. \tag{1}$$

At the beginning of the first upstroke, volume of air below the piston is V and its density is ρ. At the end of this stroke, the volume of air is $(B+V)$ and its density is ρ_1. Hence, by Boyle's law,

$$(B+V)\rho_1 = V\rho. \tag{2}$$

During the next down-stroke, the air in the barrel is of density ρ, while that in the receiver is of density ρ_1. At the end of the second up-stroke, volume of air below the piston is $(B+V)$ and its density is ρ_2. Hence,

$$(B+V)\rho_2 = V\rho_1. \tag{3}$$

Proceeding in the same way, we have, at the end of n-th stroke

$$(B+V)\rho_n = V\rho_{n-1}. \tag{4}$$

Denoting

$$\xi = \frac{V}{B+V}, \tag{5}$$

we write the above relations in the form

$$\left.\begin{aligned} \rho_1 &= \xi\rho, \\ \rho_2 &= \xi\rho_1, \\ \rho_3 &= \xi\rho_2, \\ &\cdots\cdots\cdots\cdots \\ \rho_n &= \xi\rho_{n-1} \end{aligned}\right\} \tag{6}$$

These equations yield

$$\rho_n = \xi^n\rho = \left(\frac{V}{B+V}\right)^n \rho. \tag{7}$$

From Eqns (1) and (7), we find

$$\frac{m}{m'} = \frac{\rho}{\rho_n} = \left(\frac{B+V}{V}\right)^n \tag{8}$$

Taking log on both sides,

$$\log m - \log m' = n[\log(B+V) - \log V] \tag{9}$$

Hence,

$$n = \frac{\log m - \log m'}{\log(B+V) - \log V}. \quad \textbf{(Proved)} \tag{10}$$

EXAMPLE 13.13 *If the volume of a receiver of a condenser is A and that of the barrel B, and at the end of each stroke there is a volume C between the piston and the valve, prove that if ρ be the density of the atmosphere and also the initial density in the receiver, the density after n strokes is*

$$\rho\left[1 + \frac{B-C}{C}\left\{1 - \left(\frac{A}{A+C}\right)^n\right\}\right]$$

Solution. Here total volume of the barrel including the space between the piston and the valve at the lowest position is B. Therefore, the effective volume of the barrel to be considered is $(B-C)$.

Following the discussion in article 13.10, we find the density of air in the condenser after n strokes

$$\rho_n = \rho\left\{\xi\frac{1-\eta^n}{1-\eta} + \eta^n\right\}, \tag{1}$$

where

$$\xi = \frac{(B-C)+C}{C+A} = \frac{B}{C+A} \tag{2}$$

$$\text{and } \eta = \frac{A}{C+A}. \tag{3}$$

Then

$$1-\eta = 1-\frac{A}{C+A} = \frac{C}{C+A} \tag{4}$$

Eqn (1) can, therefore, be written as

$$\rho_n = \rho\left\{\frac{B}{C+A}\cdot\frac{C+A}{C}(1-\eta^n)+\eta^n\right\}$$

$$\text{i.e., } \rho_n = \rho\left\{\frac{B}{C}(1-\eta^n)+\eta^n\right\}$$

$$\text{i.e., } \rho_n = \rho\left\{\frac{B}{C}+\left(1-\frac{B}{C}\right)\eta^n\right\} = \rho\left[1+\frac{B-C}{C}(1-\eta^n)\right] \tag{5}$$

Inserting here the value of η from Eqn (3),

$$\rho_n = \rho\left[1+\frac{B-C}{C}\left\{1-\left(\frac{A}{A+C}\right)^n\right\}\right]. \quad \textbf{(Proved)}$$

EXAMPLE 13.14 *If a Hawksbee's single-barrelled air pump be perfect except that the piston valve will not open unless the pressure in the barrel exceeds the atmospheric pressure by p, prove that the extra work to be done in completely exhausting the receiver of volume V on account of this defect is equal to that done in compressing a volume V of air at atmospheric pressure until the pressure is increased by p.*

Solution. Let B = volume of the barrel, Π = pressure of the atmospheric air and Π_n = pressure of air in the receiver after n strokes.

Since the pressure of air in the receiver is directly proportional to its density, we write following Eqn (13.71),

$$\Pi_n = \Pi\left(\frac{V}{B+V}\right)^n. \tag{1}$$

Let, during the next down-stroke, the piston valve open so that the pressure in the barrel is $(\Pi+p)$. If the volume of air below the piston in this situation is v, then by Boyle's law,

$$(\Pi+p)v = \Pi_n B. \tag{2}$$

This yields

$$v = \frac{\Pi_n}{\Pi+p}B. \tag{3}$$

The work done to compress the air from volume B to volume v is

$$W_n = -\int_B^v p\,dv \tag{4}$$

Now, from (2), we have

$$\Pi + p = \frac{\Pi_n B}{v}, \text{ i.e., } p = \frac{\Pi_n B}{v} - \Pi \tag{5}$$

Inserting this value of p in Eqn (4),

$$W_n = -\int_B^v \left(\frac{\Pi_n B}{v} - \Pi\right) dv$$

$$\text{or, } W_n = -\Pi_n B \log v + \Pi v \Big|_B^v$$

$$\text{or, } W_n = -\Pi_n B \log \frac{v}{B} + \Pi(v - B) \tag{6}$$

Inserting here the value of v from Eqn (3), we get

$$W_n = -\Pi_n B \log\left(\frac{\Pi_n}{\Pi + p}\right) + \Pi B\left(\frac{\Pi_n}{\Pi + p} - 1\right) \tag{7}$$

If the pump were without any defect, then $p = 0$. In this case, the work done

$$W_n' = -\Pi_n B \log\left(\frac{\Pi_n}{\Pi}\right) + \Pi B\left(\frac{\Pi_n}{\Pi} - 1\right) \tag{8}$$

Hence the extra work done is

$$\begin{aligned} W_n^* &= W_n - W_n' \\ &= -\Pi_n B\left[\log\left(\frac{\Pi_n}{\Pi + p}\right) - \log\left(\frac{\Pi_n}{\Pi}\right)\right] + \Pi B\left(\frac{\Pi_n}{\Pi + p} - \frac{\Pi_n}{\Pi}\right) \\ &= \Pi_n B\left[\log\left(1 + \frac{p}{\Pi}\right) - \frac{p}{\Pi + p}\right] \end{aligned} \tag{9}$$

Using Eqn (1), we write Eqn (9) in the form

$$W_n^* = \Pi B\left(\frac{V}{B + V}\right)^n \left[\log\left(1 + \frac{p}{\Pi}\right) - \frac{p}{\Pi + p}\right]. \tag{10}$$

Putting here the values $n = 1, 2, 3, \ldots, \infty$ and summing the series, we get the work done for complete exhaustion

$$W^* = \Pi B\left[\log\left(1 + \frac{p}{\Pi}\right) - \frac{p}{\Pi + p}\right] \sum_{n=1}^{\infty} \left(\frac{V}{B + V}\right)^n \tag{11}$$

This is a geometric series with 1st term $\dfrac{V}{B + V}$ and common ratio less than unity. Hence,

$$\begin{aligned} W^* &= \Pi B\left[\log\left(1 + \frac{p}{\Pi}\right) - \frac{p}{\Pi + p}\right] \frac{\dfrac{V}{B + V}}{1 - \dfrac{V}{B + V}} \\ &= \Pi V\left[\log\left(1 + \frac{p}{\Pi}\right) - \frac{p}{\Pi + p}\right\}. \end{aligned} \tag{12}$$

Now let us find the work done in compressing air of volume V at atmospheric pressure Π until the pressure is increased to $(\Pi + p)$. Let the volume change to v'. Then by Boyle's law,

$$\Pi V = (\Pi + p)v', \text{ i.e., } v' = \frac{\Pi}{\Pi + p} V \tag{13}$$

The work done in producing this compression

$$W_1 = -\int_V^{v'} pdv'$$

$$= -\int_V^{v'} \left(\frac{\Pi V}{v'} - \Pi\right) dv'$$

$$\text{or, } W_1 = -\Pi V \log v' + \Pi v' \Big|_V^{v'}$$

$$= -\Pi V \log \frac{v'}{V} + \Pi(v' - V) \tag{14}$$

Inserting here the value of v' from Eqn (13), we get

$$W_1 = \Pi V \left[\log\left(1 + \frac{p}{\Pi}\right) - \frac{p}{\Pi + p}\right] \tag{15}$$

From Eqns (12) and (15), we find

$$W^* = W_1. \tag{16}$$

Hence the result.

EXAMPLE 13.15 *In a Smeaton's air pump, if the piston works only between two points, distant a and b respectively, from the top and the bottom of the barrel and at a distance c apart, while the receiver is equal to a length l of the barrel, find the density of the air in the receiver after two strokes.*

Solution. Let A be the cross-sectional area of the barrel. Then following the notations of the article 13.8, we get

$$\left.\begin{aligned} &\text{Volume of the barrel, } B = Ac, \\ &\text{Volume of the clearance at the top, } D = Aa, \\ &\text{Volume of the clearance at the bottom, } C = Ab \\ \text{and } &\text{Volume of receiver, } V = Al. \end{aligned}\right\} \tag{1}$$

Then, by Eqns (13.80),

$$\xi = \frac{CD}{(B + C + D)(B + C + V)} = \frac{ab}{(a + b + c)(b + c + l)} \tag{2}$$

$$\eta = \frac{V(B + C + D) + C(B + C)}{(B + C + D)(B + C + V)} = \frac{l(a + b + c) + b(b + c)}{(a + b + c)(b + c + l)} \tag{3}$$

If ρ be the density of atmospheric air in the receiver at the beginning, the density of air after two strokes, by Eqn (13.83), is

$$\rho_2 = \rho\left\{\frac{\xi(1 - \eta^2)}{1 - \eta} + \eta^2\right\} = \rho\{\xi(1 + \eta) + \eta^2\}. \tag{4}$$

Inserting here the values of ξ and η from Eqns (2) and (3), we get

$$\frac{\rho_2}{\rho} = \frac{ab[(a + b + c)(b + c + 2l) + b(b + c)] + [l(a + b + c) + b(b + c)]^2}{(a + b + c)^2(b + c + l)^2} \tag{5}$$

The numerator of this equation

$$
\begin{aligned}
&= ab(a+b+c)(b+c+2l) + ab^2(b+c) + l^2(a+b+c)^2 \\
&\qquad + 2lb(b+c)(a+b+c) + b^2(b+c)^2 \\
&= l^2(a+b+c)^2 + 2abl(a+b+c) + 2lb(b+c)(a+b+c) \\
&\qquad + ab(a+b+c)(b+c) + ab^2(b+c) + b^2(b+c)^2 \\
&= l^2(a+b+c)^2 + 2lb(a+b+c)(a+b+c) \\
&\qquad + (b+c)[ab(a+b+c) + ab^2 + b^2(b+c)] \\
&= (a+b+c)^2(l^2+2lb) + b(b+c)[a(a+b+c) + ab + b(b+c)] \\
&= (a+b+c)^2(l^2+2lb) + b(b+c)(a+b)(a+b+c) \qquad (6)
\end{aligned}
$$

From Eqns (5) and (6), we get

$$
\begin{aligned}
\frac{\rho_2}{\rho} &= \frac{(a+b+c)^2(l^2+2lb) + b(b+c)(a+b)(a+b+c)}{(a+b+c)^2(b+c+l)^2} \\
&= \frac{l(l+2b)}{(b+c+l)^2} + \frac{b(b+c)(a+b)}{(a+b+c)(b+c+l)^2} \quad \textbf{(Ans)}.
\end{aligned}
$$

EXERCISE

1. Two reservoirs containing water are connected by a pipeline passing over a hill. The length of the pipe up to the summit is l_1 and on the downstream is l_2. The reservoirs are open to atmosphere and the difference of water levels in them is h. If H and H_s denote the pressure heads on the water surface and summit point respectively, show that

$$H - H_s - y = \frac{h}{l_1 + l_2}\left(\frac{d}{f} - l_1\right),$$

where y is the height of summit point from water level in the top reservoir, d is the uniform diameter of the pipeline and f is the coefficient of friction of the pipe material.

2. A single-acting pump has a stroke of 1.0 ft and a piston diameter of 6.0 inches. The centre of the pump is 15 ft above the level of water in the sump and 100 ft below delivery water level. The lengths of the suction and delivery pipes are 30 ft and 120 ft respectively, and their diameters are 3.0 inches. The coefficient of friction for these pipes is 0.015. If the pump is working at 30 r.p.m., find the horse power required to drive the pump, when the velocity head of discharge water is neglected. Take weight of water = 62.4 lbs/cft and 1 H.P. = 550 ft-lbs. [**Ans.** 1.43 H.P.]

3. In a hydraulic press, the radii of the cylinders are 3 inches and 6 ft respectively. The power is applied at the end of a lever whose length is 2 ft, the piston being attached at a distance 2 inches from the fulcrum. If a body weighing 10 tons be placed on the larger piston, find the power that must be applied to the lever.

If the materials of the press will only bear a pressure of 150 lbs wt. per square inch, find what is the greatest weight that can be lifted. [**Ans.** $3\frac{13}{54}$ lbs wt., $1091\frac{1}{49}$ tons wt.]

4. A piston is slowly pushed along a cylinder against water under pressure and to prevent leakage the piston is fitted with a leather collar of breadth b which the water pushes against the walls of the cylinder. If μ is the coefficient of friction between the leather and the cylinder and r is the radius of a section of the cylinder, prove that the fraction of the work done which is spent in overcoming friction is

$$\frac{2\mu b}{r + 2\mu b}.$$

5. A siphon with vertical arms filled with mercury of density ρ and closed at both ends, is inserted in a basin of water of density δ. Prove that when the stoppers are removed, provided the barometer is high enough, (1) if k, the whole length of the outside arm be $> h$, the whole length of the inside arm, the mercury will flow out followed by water; (2) If h be $> k$, the end of the immersed tube must be at a depth below the free surface $> (h-k)\dfrac{\rho}{\delta}$, in order that the mercury may not flow back into the basin.

6. A cylindrical diving bell of height a is lowered till its top is at a depth h below the surface of the water. If the bell be now half-full of water, and air be pumped in till all the water is expelled, prove that the bell must be lowered a further distance $(4H - 2h)$, before the bell is again half-full of water, H being the height of the water barometer.

7. A cylindrical diving bell of height $h/4$ is sunk into water till the lower end is at a depth nh below the surface. If the water fills $\dfrac{1}{5}$th of the bell, prove that the bell contains air, whose volume at atmospheric pressure would be

$$\frac{4}{5}\left(n+\frac{19}{20}\right)V,$$

where V is the volume of the bell and h is the height of the water barometer.

8. A diving bell is in the form of a cylinder of length a surmounted by a cone of height h. If no air is pumped in when it is immersed, find how far it must be lowered for all the air to be forced into the conical part. Show that the volume of air at atmospheric pressure which must now be pumped in, that the bell may be filled, is $\left(\dfrac{a}{H}+\dfrac{3a}{h}\right)V$, where H is the height of the water barometer and V is the volume of the bell.

9. A cylindrical diving bell, whose cross-sectional area is A, is suspended in water with its flat top at a distance a below the surface, the air inside the bell then occupying a length b of the bell. A body of volume $A\alpha$ and sp. gr. ρ falls from the platform inside the bell into the enclosed water and floats. Show that (1) the level of water inside the bell rises, but that (2) the amount of water inside the bell is less than before. Find also the change in tension of the supporting chain. (The weight of the air displaced by the body may be neglected). [**Ans.** Tension in the chain is diminished.]

10. In a diving bell, a soda-water bottle is opened which in the external air would liberate a volume V of gas. Show that the tension of the chain is diminished by

$$\frac{wHV}{\sqrt{(a+H)^2+4bH}},$$

where squares of $\dfrac{V}{HA}$ are neglected, w being the weight of unit volume of water, A the cross-section, a the depth of the top, b the height of the bell and H the height of the water barometer.

11. The weight in water of the material of a cylindrical diving bell is W, and the tension in the chain is P. A length a of the bell is occupied by air. Show that, if a weight w of water is drawn up in a bucket into the air space, and the bell is raised through a distance $\dfrac{wa}{W-P}$, the tension of the chain will be the same as before.

12. If a cylindrical diving bell of height a and whose chamber could contain a weight W of water, be lowered so that the depth of the highest point is d, prove that when the temperature is raised through $t\,°$C, the tension of the supporting chain is diminished by

$$\frac{WHt}{T\sqrt{(H+d)^2+4aH}} \text{ (nearly)},$$

where H is the height of the water barometer and T is the original absolute temperature.

13. A cube of density ρ_0 floats on a liquid of density ρ, and a cylindrical diving bell of length l and cross-section A, is placed over it and lowered until the cube just touches the top. Show that the height to which the liquid rises inside the bell is

$$\frac{\{A(\rho-\delta)-\rho a^2\}\{l(\rho-\delta)-a(\rho-\rho_0)\}}{\rho(\rho-\delta)(A-a^2)},$$

where δ is the density of the atmosphere and a is the length of an edge of the cube.

14. In the common pump, the heights of the barrel and the pipe leading to it are b, a respectively, and n is the ratio of the sectional area of the former to that of the latter, and there is a length c at the bottom of the barrel which the piston does not traverse; show that if the water just rises into the barrel after two strokes, then

$$h^2\left(\frac{a}{nb}+\frac{2c}{b}-1\right)\left(\frac{a}{nb}+\frac{2c}{b}-2\right)+h\left\{a\left(\frac{3a}{nb}+\frac{6c}{b}-4\right)-n(b-c)\right\}+a(2a+nb)=0,$$

where h is the height of water barometer.

15. If of the volume B of the cylinder of a condenser only C is traversed by the piston, prove that the pressure in the receiver cannot be made to exceed $\dfrac{B}{B-C}$ atmospheres.

16. Prove that if the temperature remains constant, the work required (1) to increase q-fold the density of the atmospheric air of pressure P in a receiver of volume V is $PV(q\log q-q+1)$; (2) to diminish to $1/q$th is

$$PV\left(1-\frac{1}{q}-\frac{1}{q}\log q\right).$$

17. The barrel of an air pump is of volume B, but of this a volume b is not traversed by the piston. Show that the density in the receiver never becomes as small as b/B of the atmospheric density.

18. Show that, if the piston valve of a conderser does not open until the pressure on the outside exceeds that on the inside by k times the atmospheric pressure, the number of strokes which can be made before the valve ceases to act is the greatest integer in

$$1+\log\left\{\frac{A}{(A+B)k}\right\}\Big/\log\left\{\left(\frac{B+C}{C}\right)\right\},$$

where C is the volume of the receiver, A that of the portion of the cylinder traversed by the piston and B that of the portion not traversed by it.

19. If the greatest and the least volumes between the valves of a conderser be V and v, and if the valves be open or close according as the difference of pressures on the two sides be greater or less than p, show that the limiting pressure in the receiver will be $(\Pi-p)\dfrac{V}{v}-p$, where Π is the pressure of the atmosphere.

20. If air be taken from a vessel of volume A and condensed in a vessel of volume A' by a piston working in a cylinder of volume B of which the clearances C and C' at the ends are not traversed by the piston, show that neglecting the effect of the weights of the several valves, the limiting ratio of the pressure in A to that in A' is

$$\frac{CC'}{(B-C)(B-C')}.$$

21. Air is pumped from a vessel of volume A and forced into a vessel of volume C by means of a Smeaton's pump of volume B. Show that if the whole of the air have originally a density ρ, and if the piston be originally at the end of the cylinder nearest to the vessel A, then after the piston has made $(2n-1)$ strokes, reaching the end C for the nth time, the density of the air in the two vessels will be

$$\left(\frac{A}{A+B}\right)^n\rho \text{ and } \left[A+B+C-\frac{A^n}{(A+B)^{n-1}}\right]\frac{\rho}{C}.$$

22. If h be the range of the piston in a Smeaton's air pump, a its distance from the top of the barrel in its highest position, b its distance from the bottom of the barrel in its lowest position, and ρ the density of the atmosphere, prove that the limiting density of the air in the receiver will be

$$\frac{ab}{(h+a)(h+b)}\rho.$$

23. Air is pumped into a vessel of volume V by means of a compressing pump in which the volume of the barrel varies between v_1 and v_2. The pressure of the air in the vessel at the outset is the atmospheric pressure Π. Assuming the process to be isothermal, show that if p_n is the pressure in the vessel after the nth stroke, $(V+v_1)p_{n+1} = Vp_n + \Pi v_2$.

Show that the pressure of the air in the vessel cannot be raised beyond the value P, where

$$P = \frac{\Pi v_2}{v_1}, \text{ and that } p_n = P - \left(\frac{V}{V+v_1}\right)^n (P - \Pi).$$

APPENDIX A

Volume and Surface of Solids

A.1 Parallelopipeds

If a, b, c be the length, breadth and height of a solid rectangular parallelopiped, then

$$\text{its volume } V = abc \tag{A.1}$$

$$\text{and its whole surface} = 2(ab + bc + ca) \tag{A.2}$$

When $a = b = c$, we get a ***cube***. In this case,

$$\text{its volume } V = a^3 \tag{A.3}$$

$$\text{and its whole surface } S = 6a^2 \tag{A.4}$$

A.2 Cylinder

If R be the radius of the base and h the height of the cylinder, then

$$\text{its volume } V = \pi R^2 h \tag{A.5}$$

$$\text{and its whole surface } S = 2\pi R(R + h) \tag{A.6}$$

A.3 Cone

Let R = radius of the base of the cone, h = height of the cone, α = semi-vertical angle of the cone, and l = slant height of the cone.

Then

$$\left.\begin{aligned} R &= h \tan\alpha \\ l &= h \sec\alpha \end{aligned}\right\} \tag{A.7}$$

Its volume

$$V = \frac{1}{3}\pi R^2 h = \frac{1}{3}\pi h^3 \tan^2\alpha \tag{A.8}$$

Its curved surface

$$S = \pi R l = \pi h^2 \tan\alpha \sec\alpha \tag{A.9}$$

For a ***frustum of a cone*** having R_1 and R_2 as the radii of circular ends and h the perpendicular distance between them,

$$\text{Volume } V = \frac{1}{3}\pi h(R_1^2 + R_1R_2 + R_2^2) \tag{A.10}$$

$$\text{Its curved surface } S = \pi(R_2 + R_1)\sqrt{h^2 + (R_2^2 - R_1^2)} \tag{A.11}$$

A.4 Sphere

If R be the radius of the sphere, then

$$\text{its volume } V = \frac{4}{3}\pi R^3 \tag{A.12}$$

$$\text{and its surface area } S = 4\pi R^2 \tag{A.13}$$

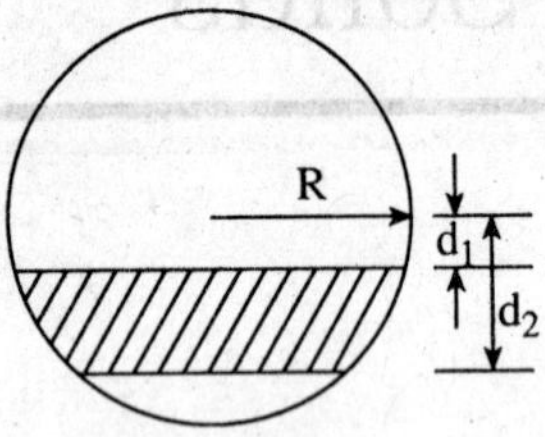

Fig. A.1

For a ***portion of the sphere*** between two parallel planes at distances d_1 and d_2 from the centre,

$$\text{Volume } V = \frac{\pi}{3}(d_2 - d_1)\{3R^2 - (d_1^2 + d_1d_2 + d_2^2)\} \tag{A.14}$$

$$\text{and its curved surface } S = 2\pi R(d_2 - d_1) \tag{A.15}$$

A.5 Paraboloid of Revolution

If R be the radius of its circular base and h its height, then

$$\text{its volume } V = \frac{1}{2}\pi R^2 h, \tag{A.16}$$

which is half the volume of a cylinder with same base and same height.

A.6 Ellipsoid

If a, b, c be the semi-major axes of the ellipsoid, then its volume

$$V = \frac{4}{3}\pi abc \tag{A.17}$$

If $a = b$, we call it a ***spheroid***. The spheroids are of two types—***oblate spheroids*** when $a > c$ and ***Prolate spheroids*** when $a < c$. Spheroids are formed by the revolution of an ellipse about one of its major axes.

APPENDIX **B**

Centre of Gravity of Solids

B.1 Semi-circular Lamina

If R be the radius of the circular lamina, then

$$\bar{z} = \frac{4R}{3\pi} \tag{B.1}$$

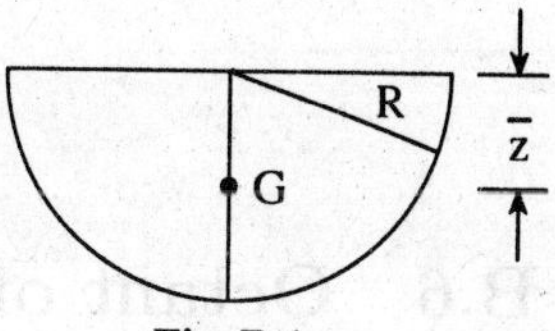

Fig. B.1

B.2 Sector of a Circle

If R be the radius of the sector of a circle and 2θ the angle subtended at the centre, then

$$\bar{z} = \frac{2R}{3}\frac{\sin\theta}{\theta}. \tag{B.2}$$

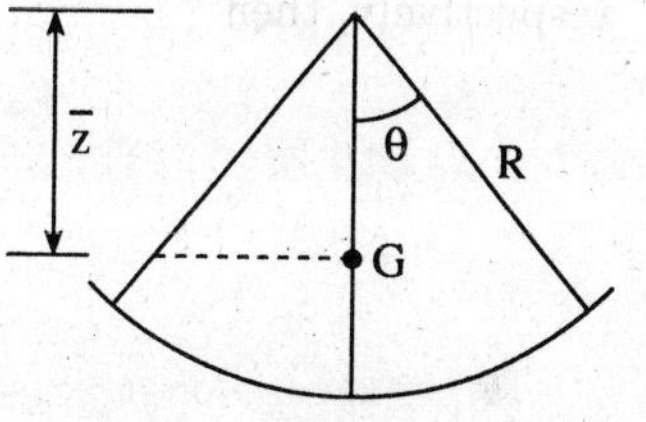

Fig. B.2

B.3 Arc of a Circle

If R be the radius of the arc and 2θ the angle subtended by it at the centre, then

$$\bar{z} = \frac{R\sin\theta}{\theta} \tag{B.3}$$

For a semi-circular arc, $\theta = \frac{\pi}{2}$. Hence

$$\bar{z} = \frac{2R}{\pi} \tag{B.4}$$

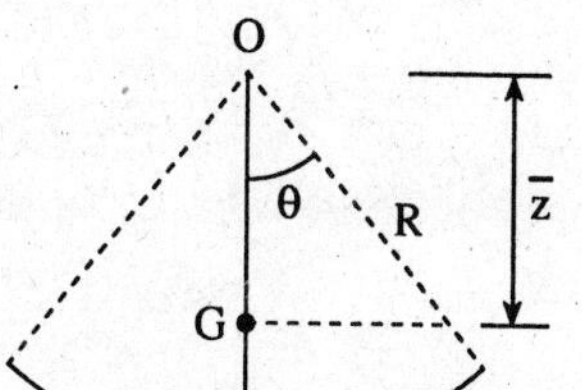

Fig. B.3

B.4 Hemisphere

If R be the radius of the hemisphere, then

$$\bar{z} = \frac{3R}{8} \tag{B.5}$$

For a thin hemispherical shell,

$$\bar{z} = \frac{R}{2} \tag{B.6}$$

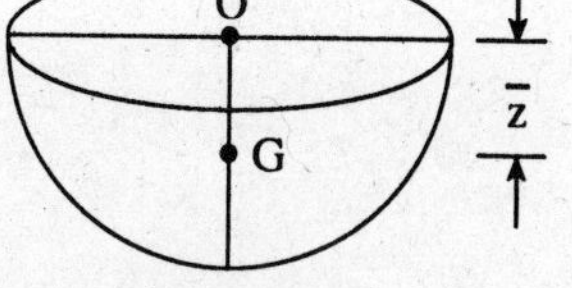

Fig. B.4

B.5 Cone

For a solid cone of height h,

$$\bar{z} = \frac{3}{4}h \qquad \text{(B.7)}$$

For a thin hollow cone of height h,

$$\bar{z} = \frac{3}{4}h \qquad \text{(B.8)}$$

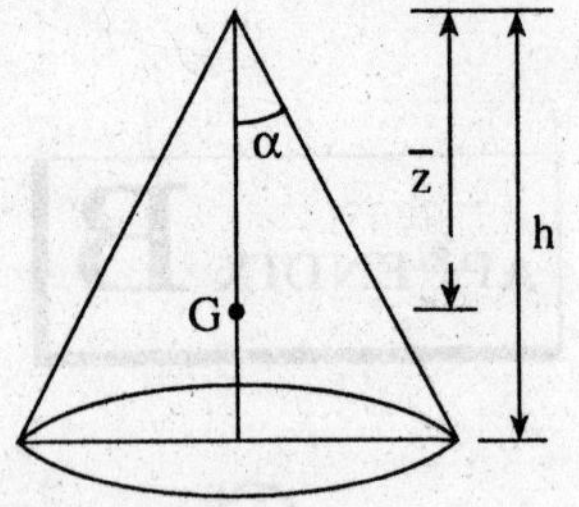

Fig. B.5

For a solid half-cone of circular base of radius R,

$$\bar{z} = \frac{R}{\pi} \qquad \text{(B.9)}$$

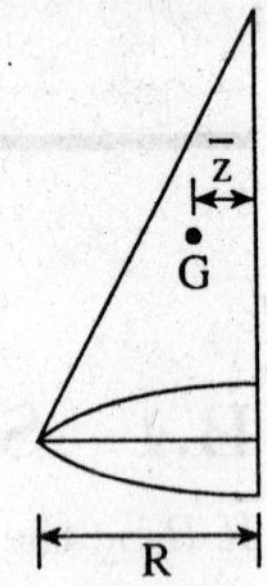

Fig. B.6

B.6 Octant of Ellipsoid

If a, b, c be the lengths of semi-axes of the ellipsoid along x-, y- and z-axes of coordinates respectively, then

$$\left.\begin{aligned} \bar{x} &= \frac{3}{8}a \\ \bar{y} &= \frac{3}{8}b \\ \bar{z} &= \frac{3}{8}c \end{aligned}\right\} \qquad \text{(B.10)}$$

APPENDIX C

Radius of Gyration of Solids

C.1 Thin Rod

If l be the length of the rod, the radius of gyration about an axis passing through the middle point of the rod at right angle to the length is

$$K = \frac{l}{2\sqrt{3}} \tag{C.1}$$

C.2 Rectangular Lamina

If a, b are the lengths of the sides of the rectangular lamina, then the radius of gyration about an axis passing through its centre and parallel to side a is

$$K = \frac{b}{2\sqrt{3}} \tag{C.2}$$

When the axis passes through the centre and is parallel to side b,

$$K = \frac{a}{2\sqrt{3}} \tag{C.3}$$

C.3 Circular Lamina

If R be the radius of the circular lamina, then the radius of gyration about a diameter is

$$K = \frac{R}{2} \tag{C.4}$$

C.4 Elliptic Lamina

If a, b are the lengths of the semi-major and semi-minor axes of the elliptic lamina, then

$$K = \frac{a}{2} \tag{C.5}$$

about the minor axis, and

$$K = \frac{b}{2} \tag{C.6}$$

about the major axis.

C.5 Triangular Lamina

ABC is the triangular lamina and AX is the axis of rotation. If the distances of B and C from the axis of rotation are α, β respectively, then

$$K = \left[\frac{1}{6}\left(\alpha^2 + \alpha\beta + \beta^2\right)\right]^{\frac{1}{2}} \tag{C.7}$$

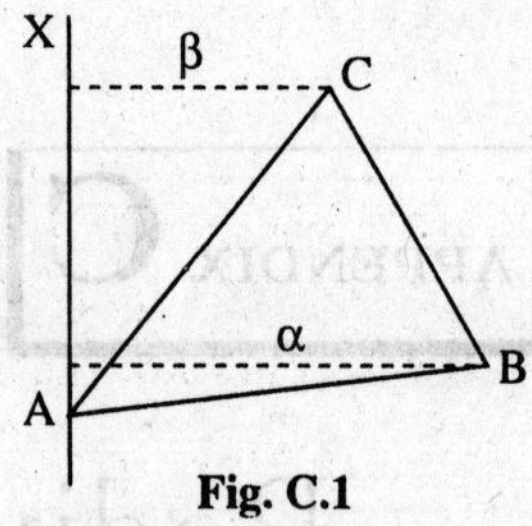

Fig. C.1

APPENDIX **D**

Some Useful Integrals

D.1 Odd and Even Functions

A function $f(x)$ is said to be an ***odd function*** if

$$f(-x) = -f(x) \tag{D.1}$$

in its domain of definition.

In this case,

$$\int_{-a}^{a} f(x)dx = 0 \tag{D.2}$$

A function $f(x)$ is said to be an ***even function*** if

$$f(-x) = f(x) \tag{D.3}$$

In this case,

$$\int_{-a}^{a} f(x)dx = 2\int_{0}^{a} f(x)dx \tag{D.4}$$

Thus, when n is odd, we have

$$\cos^n(\pi - x) = -\cos^n x \tag{D.5}$$

Hence

$$\int_0^{\pi} \cos^n x dx = 0 \tag{D.6}$$

when n is odd.

When n is even,

$$\int_0^{\pi} \cos^n x dx = 2\int_0^{\pi/2} \cos^n x dx \tag{D.7}$$

D.2 Integrals Containing $\sin x$ and $\cos x$

We have

$$\int_0^{\pi} \sin^m x \cos^n x dx = 0, \tag{D.8}$$

if n is odd, regardless of the value of m.

Similarly,

$$\int_0^{\pi} \sin^m x \cos^n x dx = 2\int_0^{\pi/2} \sin^m x \cos^n x dx, \tag{D.9}$$

when n is even.

D.3 Integrals of Powers of $\sin x$ and $\cos x$

We have

$$\int_0^{\pi/2} \sin^n x dx = \int_0^{\pi/2} \cos^n x dx \tag{D.10}$$

for any value of n.

The value of the integral is

$$= \frac{(n-1)(n-3)\cdots}{n(n-2)(n-4)\cdots}\left(\frac{\pi}{2}\right) \tag{D.11}$$

The factor $\frac{\pi}{2}$ will appear ***only when n is even***.

Thus,

$$\int_0^{\pi/2} \sin^5\theta d\theta = \int_0^{\pi/2} \cos^5\theta d\theta = \frac{4\cdot 2}{5\cdot 3\cdot 1} = \frac{8}{15}$$

$$\text{and} \int_0^{\pi/2} \sin^6\theta d\theta = \int_0^{\pi/2} \cos^6\theta d\theta = \frac{5\cdot 3\cdot 1}{6\cdot 4\cdot 2}\cdot\frac{\pi}{2} = \frac{15}{96}\pi$$

Again, we have

$$\int_0^{\pi/2} \sin^m\theta\cos^n\theta d\theta = \frac{\{(m-1)(m-3)\cdots 2 \text{ or } 1\}\{(n-1)(n-3)\cdots 2 \text{ or } 1\}}{(m+n)(m+n-2)\cdots 2 \text{ or } 1}\cdot\frac{\pi}{2} \tag{D.12}$$

The factor $\frac{\pi}{2}$ appears only when ***both m and n are even***.

Thus,

$$\int_0^{\pi/2} \sin^5\theta\cos^7\theta d\theta, (m+n=12) = \frac{(4\cdot 2)(6\cdot 4\cdot 2)}{12\cdot 10\cdot 8\cdot 6\cdot 4\cdot 2} = \frac{1}{120}$$

$$\text{and} \int_0^{\pi/2} \sin^6\theta\cos^4\theta d\theta, (m+n=10) = \frac{(5\cdot 3\cdot 1)(3\cdot 1)}{10\cdot 8\cdot 6\cdot 4\cdot 2}\cdot\frac{\pi}{2} = \frac{3\pi}{512} \left[\begin{array}{l}\text{since both } m, n\\ \text{are even}\end{array}\right]$$

APPENDIX E

Green's Theorem

The theorem states:

If u, v, w are functions of x, y, z having continuous derivatives with respect to x, y, z respectively throughout a singly-connected region bounded by a closed surface S free from singularities and l, m, n are the direction cosines of the normal drawn outwards from a point on an elementary area dS of the surface, then

$$\iiint \left(\frac{\partial u}{\partial x} + \frac{\partial v}{\partial y} + \frac{\partial w}{\partial z}\right) dxdydz = \iint (lu + mv + nw)dS,$$

where the surface integral is taken over the boundary and the volume integral throughout the region enclosed.

Proof. Let S be a closed surface free from singularities. We take a prism of cross section $dxdy$ parallel to z-axis. It intersects the surface S in two elements dS_1, dS_2. Let the outward-drawn normals to these elementary areas have direction cosines (l_1, m_1, n_1) and (l_2, m_2, n_2) respectively.

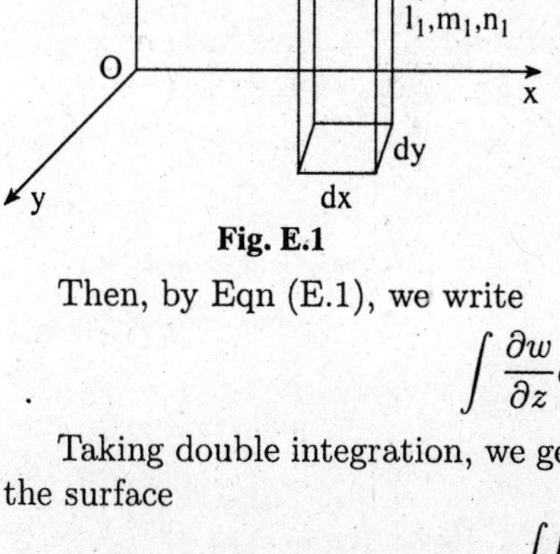

Fig. E.1

Since the projections of elements dS_1, dS_2 on xy-plane are equal to the area $dxdy$, we have with due care of signs

$$dxdy = -n_1 dS_1 = n_2 dS_2. \tag{E.1}$$

Hence, for this particular prism, we get by integration with respect to z,

$$\begin{aligned}\int \frac{\partial w}{\partial z} dxdydz &= \int \left(\frac{\partial w}{\partial z} dz\right) dxdy \\ &= (wdxdy)_2 - (wdxdy)_1,\end{aligned} \tag{E.2}$$

where the suffixes denote the top and bottom intersections of the prism with the surface respectively.

Then, by Eqn (E.1), we write

$$\int \frac{\partial w}{\partial z} dxdydz = n_2 w_2 dS_2 + n_1 w_1 dS_1 \tag{E.3}$$

Taking double integration, we get by summing for all similar parallel prisms which intersect the surface

$$\iiint \frac{\partial w}{\partial z} dxdydz = \iint nwdS \tag{E.4}$$

Similarly, by considering prisms along x- and y-axes, we can prove that

$$\iiint \frac{\partial w}{\partial x} dxdydz = \iint ludS \tag{E.5}$$

$$\text{and} \quad \iiint \frac{\partial w}{\partial y} dxdydz = \iint mvdS \tag{E.6}$$

Hence

$$\iiint \left(\frac{\partial w}{\partial x} + \frac{\partial w}{\partial y} + \frac{\partial w}{\partial z}\right) dxdydz = \iint (lu + mv + nw)dS \tag{E.7}$$

Thus the theorem is proved. We find that Green's theorem connects a volume integral with a surface integral over the boundary of the enclosed volume.

APPENDIX **F**

Gravitational Attraction and Potential

F.1 Newton's Law of Gravitation

The law is as follows:

Every particle in the universe attracts every other particle with a force which is directly proportional to the product of the masses of the particles and inversely proportional to the square of the distance between them.

Thus, if m_1, m_2 are the masses of two particles placed at a distance r apart, the force of attraction between them is

$$F = \gamma \frac{m_1 m_2}{r^2}, \tag{F.1}$$

where γ is known as the ***gravitation constant***. Its value in C.G.S. unit is

$$\gamma = \frac{1}{15,500,000} \text{ (approx.)} \tag{F.2}$$

F.2 Attraction of a System of Particles

Let particles of masses $m_1, m_2, \ldots, m_n$ be situated at points $P_1, P_2, \ldots, P_n$ whose coordinates referred to rectangular axes are $(a_1, b_1, c_1), (a_2, b_2, c_2), \ldots, (a_n, b_n, c_n)$. We are required to determine the components X, Y, Z of force of attraction of the system of particles at any point $P(x, y, z)$.

Consider the particle at $P_k(a_k, b_k, c_k)$. If the distance $PP_k = r_k$, then

$$r_k^2 = (a_k - x)^2 + (b_k - y)^2 + (c_k - z)^2 \tag{F.3}$$

and the force of attraction at P along PP_k is

$$F_k = \frac{\gamma m_k}{r_k^2}. \tag{F.4}$$

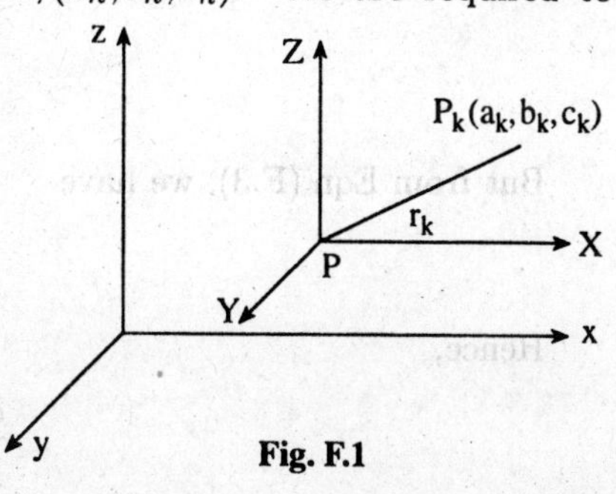

Fig. F.1

The direction cosines of the line PP_k are

$$\left.\begin{aligned}\lambda_k &= \frac{a_k - x}{r_k}\\ \mu_k &= \frac{b_k - y}{r_k}\\ \nu_k &= \frac{c_k - z}{r_k}\end{aligned}\right\} \tag{F.5}$$

Hence, the components of force F_k parallel to coordinate axes are

$$\left.\begin{aligned}X_k &= \frac{\gamma m_k(a_k - x)}{r_k^3}\\ Y_k &= \frac{\gamma m_k(b_k - y)}{r_k^3}\\ Z_k &= \frac{\gamma m_k(c_k - z)}{r_k^3}\end{aligned}\right\} \tag{F.6}$$

Therefore, by the principle of superposition, the components of force of attraction of the system of particles at the point P are

$$\left.\begin{aligned}X &= \sum_{k=1}^{n} X_k = \sum_{k=1}^{n} \frac{\gamma m_k(a_k - x)}{r_k^3}\\ Y &= \sum_{k=1}^{n} Y_k = \sum_{k=1}^{n} \frac{\gamma m_k(b_k - y)}{r_k^3}\\ Z &= \sum_{k=1}^{n} Z_k = \sum_{k=1}^{n} \frac{\gamma m_k(c_k - x)}{r_k^3}\end{aligned}\right\} \tag{F.7}$$

F.3 Potential

We define a function V such that

$$V = \sum_{k=1}^{n} \frac{\gamma m_k}{r_k}, \tag{F.8}$$

where r_k is as defined by Eqn (F.3).

Thus, V is a function of the coordinates (x, y, z) of the point P, and obviously, it is a single valued function in the sense that it cannot have more than one value at a particular point $P(x, y, z)$.

Differentiating both sides of Eqn (F.8),

$$\frac{\partial V}{\partial x} = -\sum_{k=1}^{n} \frac{\gamma m_k}{r_k^2} \frac{\partial r_k}{\partial x} \tag{F.9}$$

But from Eqn (F.3), we have

$$r_k \frac{\partial r_k}{\partial x} = -(a_k - x) \tag{F.10}$$

Hence,

$$\frac{\partial V}{\partial x} = \sum_{k=1}^{n} \frac{\gamma m_k(a_k - x)}{r_k^3} \tag{F.11}$$

Comparing Eqns (F.7) and (F.11), we find

$$\frac{\partial V}{\partial x} = X \tag{F.12}$$

Similarly,

$$\left.\begin{aligned}\frac{\partial V}{\partial y} &= Y\\ \frac{\partial V}{\partial z} &= Z\end{aligned}\right\} \tag{F.13}$$

Thus, the partial derivatives of V with respect to x, y, z give the components of attraction force at P along the direction of coordinate axes.

Since the directions of axes can be arbitrarily chosen, the derivative of V in any direction gives the component of attraction in that direction.

The function V defined by Eqn (F.8) is called the ***potential of the attracting particles.***

F.4 Attraction and Potential of Some Solids

(1) **Uniform straight rod.** Components of attraction at P are

$$\left.\begin{aligned}X &= \frac{2m\gamma}{p}\sin\frac{\alpha+\beta}{2}\sin\frac{\alpha-\beta}{2}\\ Y &= \frac{2m\gamma}{p}\sin\frac{\alpha-\beta}{2}\cos\frac{\alpha+\beta}{2}\end{aligned}\right\}, \tag{F.14}$$

where m = mass per unit length of the rod AB, p = perpendicular distance of P from AB, $\angle MPA = \alpha$ and $\angle MPB = \beta$

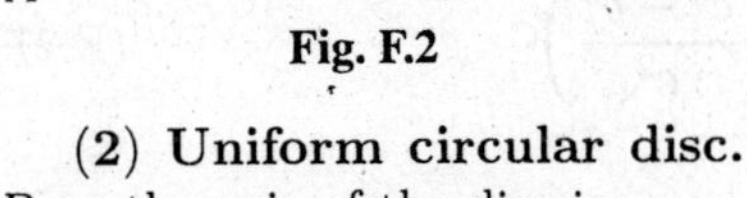

Fig. F.2

Potential at P is

$$V = \gamma m \log\left(\cos\frac{\theta}{2}\cot\frac{\phi}{2}\right), \tag{F.15}$$

where

$$\left.\begin{aligned}\theta &= \angle PAB = 90^\circ - \alpha\\ \phi &= \angle PBA = 90^\circ + \beta\end{aligned}\right\} \tag{F.16}$$

(2) **Uniform circular disc.** Attraction at a point P on the axis of the disc is

$$F = 2\pi\gamma m(1-\cos\alpha), \tag{F.17}$$

where m = mass per unit area of the disc, α = angle that OA makes at P.

Potential of the disc at P is

$$V = 2\pi\gamma m\left\{\sqrt{p^2+a^2}-p\right\}, \tag{F.18}$$

where p = perpendicular distance OP, a = radius of the circular disc.

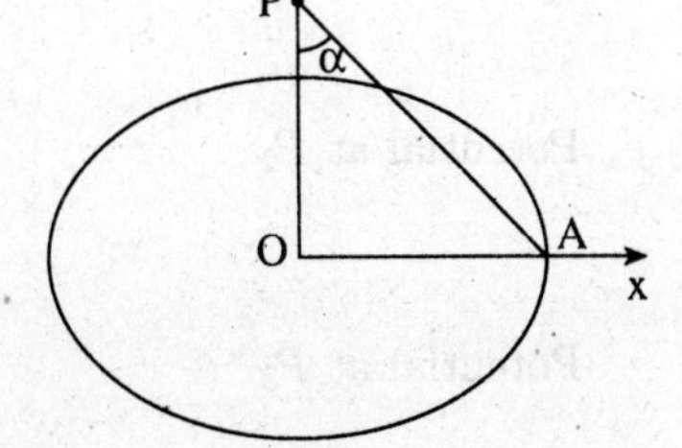

Fig. F.3

(3) Thin spherical shell. When P is inside the shell, the attraction at P is zero. When P is outside the shell, the attraction is

$$F = \gamma \frac{M}{c^2}, \tag{F.19}$$

where M = mass of the shell, c = distance of P from centre O.

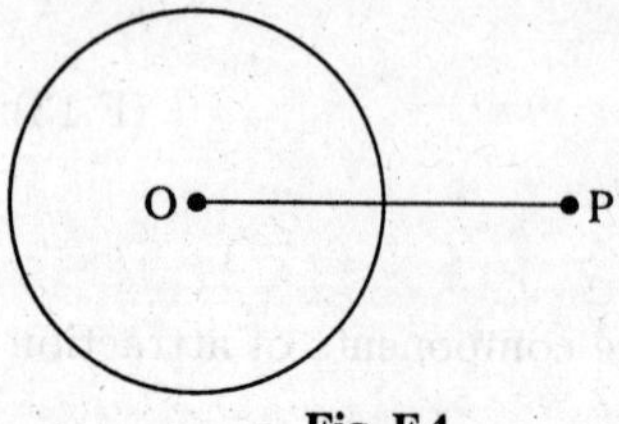

Fig. F.4

The potential at a point P inside the shell is

$$V = \frac{\gamma M}{a}, \tag{F.20}$$

where a= radius of the shell.

The potential at a point P outside the shell is

$$V = \frac{\gamma M}{c} \tag{F.21}$$

(4) Uniform solid sphere. Attraction at an internal point P is

$$F = \frac{4}{3}\pi\gamma\rho c, \tag{F.22}$$

where ρ = density of the sphere, c = distance of the point P from the centre of the sphere.

Attraction at an external point P is

$$F = \frac{\gamma M}{c^2} \tag{F.23}$$

Potential at an internal point is

$$V = \frac{2}{3}\pi\gamma\rho(3a^2 - c^2) \tag{F.24}$$

Potential at an external point is

$$V = \frac{\gamma M}{c} \tag{F.25}$$

(5) Thick spherical shell. Let a, b be the external and internal radii of the shell and c the distance of the point from the centre.

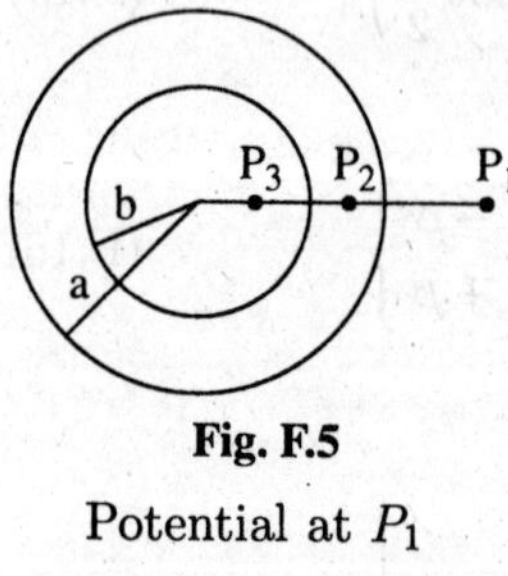

Fig. F.5

Attraction at P_1

$$F = \frac{4}{3}\pi\gamma\rho\left(\frac{a^3 - b^3}{c^2}\right) \tag{F.26}$$

Attraction at P_2

$$F = \frac{4}{3}\pi\gamma\rho\left(\frac{c^3 - b^3}{c^2}\right) \tag{F.27}$$

Attraction at P_3 is zero.

Potential at P_1

$$V = \frac{4}{3}\pi\gamma\rho\left(\frac{a^3 - b^3}{c}\right) \tag{F.28}$$

Potential at P_2

$$V = \frac{2}{3}\pi\gamma\rho\left(3a^2 - c^2 - \frac{2b^3}{c}\right) \tag{F.29}$$

Potential at P_3

$$V = 2\pi\gamma\rho(a^2 - b^2). \tag{F.30}$$

APPENDIX **G**

Laplace's and Poisson's Theorems

G.1 Laplace's Theorem

The theorem states:

If V be the potential of a system of attracting particles at a point (x, y, z) not in contact with the particles, then

$$\frac{\partial^2 V}{\partial x^2} + \frac{\partial^2 V}{\partial y^2} + \frac{\partial^2 V}{\partial z^2} = 0$$

Proof. Let m be the mass of the particle at the point $Q(a, b, c)$ and $P(x, y, z)$ any point not in contact with any of the attracting particles. Then the potential at P is

$$V = \sum \frac{\gamma m}{r}, \tag{G.1}$$

where

$$r^2 = (x-a)^2 + (y-b)^2 + (z-c)^2 \tag{G.2}$$

Hence,

$$\frac{\partial V}{\partial x} = \sum \left(-\frac{\gamma m}{r^2} \frac{\partial r}{\partial x} \right) = -\sum \frac{\gamma m(x-a)}{r^3} \tag{G.3}$$

Then, differentiating once more,

$$\frac{\partial^2 V}{\partial x^2} = -\sum \left\{ \frac{\gamma m}{r^3} - \frac{3\gamma m}{r^4} \frac{\partial r}{\partial x}(x-a) \right\} = -\gamma \sum \left\{ \frac{m}{r^3} - \frac{3m}{r^5}(x-a)^2 \right\} \tag{G.4}$$

Similarly,

$$\frac{\partial^2 V}{\partial y^2} = -\gamma \sum \left\{ \frac{m}{r^3} - \frac{3m}{r^5}(y-b)^2 \right\} \tag{G.5}$$

$$\frac{\partial^2 V}{\partial z^2} = -\gamma \sum \left\{ \frac{m}{r^3} - \frac{3m}{r^5}(z-c)^2 \right\} \tag{G.6}$$

Adding all these,

$$\frac{\partial^2 V}{\partial x^2} + \frac{\partial^2 V}{\partial y^2} + \frac{\partial^2 V}{\partial z^2} = -\gamma \left\{ \frac{3m}{r^3} - \frac{3m}{r^5}[(x-a)^2 + (y-b)^2 + (z-c)^2] \right\}$$

$$= -\gamma \left\{ \frac{3m}{r^3} - \frac{3m}{r^5} \cdot r^2 \right\} = 0$$

Thus

$$\nabla^2 V = 0 \tag{G.7}$$

This is known as ***Laplace's equation*** which is satisfied by the potential of an attracting system of particles at every point at which there is no matter.

G.2 Poisson's Theorem

The theorem states:

If V be the potential of a system of attracting particles at any point $P(x, y, z)$ at which the volume density is ρ, then

$$\frac{\partial^2 V}{\partial x^2} + \frac{\partial^2 V}{\partial y^2} + \frac{\partial^2 V}{\partial z^2} = -4\pi\gamma\rho.$$

Proof. We describe a sphere of small radius ϵ with centre at (a, b, c) and enclosing the point $P(x, y, z)$. The sphere being very small, we can regard the density ρ of the matter inside it as uniformly distributed.

Let V_1 = potential at P due to the matter outside the small sphere, and
V_2 = potential at P due to the matter inside the small sphere.

Since the point P is not in contact with matter which produces potential V_1, we have, by Laplace's theorem

$$\nabla^2 V_1 = 0 \tag{G.8}$$

The potential V_2 at $P(x, y, z)$ inside the small sphere of radius ϵ, is, by Eqn (F.24) given by

$$V_2 = \frac{2}{3}\pi\gamma\rho(3\epsilon^2 - r^2), \tag{G.9}$$

where

$$r^2 = (x-a)^2 + (y-b)^2 + (z-c)^2 \tag{G.10}$$

Hence,

$$\begin{aligned}\frac{\partial V_2}{\partial x} &= -\frac{4}{3}\pi\gamma\rho(x-a)\\ \frac{\partial^2 V_2}{\partial x^2} &= -\frac{4}{3}\pi\gamma\rho\end{aligned} \tag{G.11}$$

Similarly,

$$\left.\begin{aligned}\frac{\partial^2 V_2}{\partial y^2} &= -\frac{4}{3}\pi\gamma\rho\\ \frac{\partial^2 V_2}{\partial z^2} &= -\frac{4}{3}\pi\gamma\rho\end{aligned}\right\} \tag{G.12}$$

Thus

$$\nabla^2 V_2 \equiv \frac{\partial^2 V_2}{\partial x^2} + \frac{\partial^2 V_2}{\partial y^2} + \frac{\partial^2 V_2}{\partial z^2} = -4\pi\gamma\rho \tag{G.13}$$

Since $V = V_1 + V_2$, we get

$$\frac{\partial^2 V}{\partial x^2} + \frac{\partial^2 V}{\partial y^2} + \frac{\partial^2 V}{\partial z^2} = -4\pi\gamma\rho \tag{G.14}$$

This is known as ***Poisson's equation*** which is satisfied by the potential of a system of attracting particles at every point where there exists attracting matter of volume density ρ.

APPENDIX **H**

Theorems of Pappus

H.1 Theorem on Volume

The theorem states:

If any plane area revolves through any angle about an axis in its own plane, then the volume generated by the area is equal to the product of area and the length of the path described by the centroid of the area.

Proof. Let A be the area and $\bar{x}$ the distance of its centroid from y-axis (FIG. H.1).

Let the area A revolve about y-axis by an angle θ.

We consider a small area dA at a point P at a distance x from y-axis. Then the length of the arc described by the point P is $= x\theta$.

Hence the volume generated by the elemental area is

$$dV = x\theta dA \tag{H.1}$$

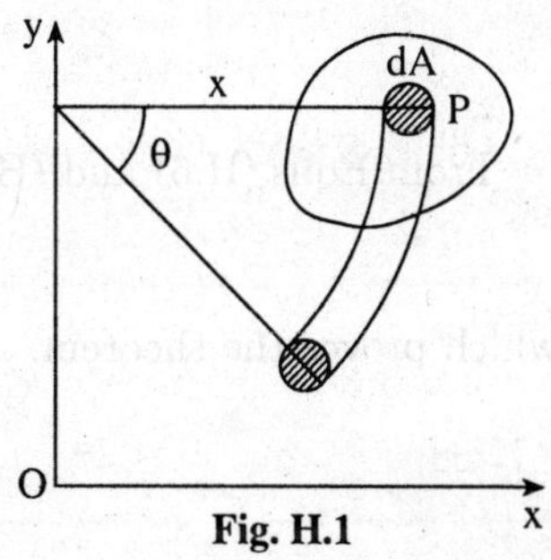

Fig. H.1

The whole volume V generated by the area is

$$V = \int x\theta dA = \theta \cdot \int x dA \tag{H.2}$$

By definition,

$$\bar{x} = \frac{\int x dA}{\int dA} = \frac{\int x dA}{A},$$

so that

$$\int x dA = A\bar{x} \tag{H.3}$$

From Eqns (H.2) and (H.3), we get

$$V = A \cdot \theta\bar{x}, \tag{H.4}$$

which proves the theorem.

H.2 Theorem on Surface Area

The theorem states:

If any plane curve revolves through any angle about an axis on its plane, then the area of the surface generated by the curve is equal to the product of the length of the curve and the length of the path described by the centroid of the curve.

Proof. Let l be the length of the curve which revolves by any angle θ about y-axis. Let $\bar{x}$ be the distance of the centroid of the curve from y-axis (FIG. H.2).

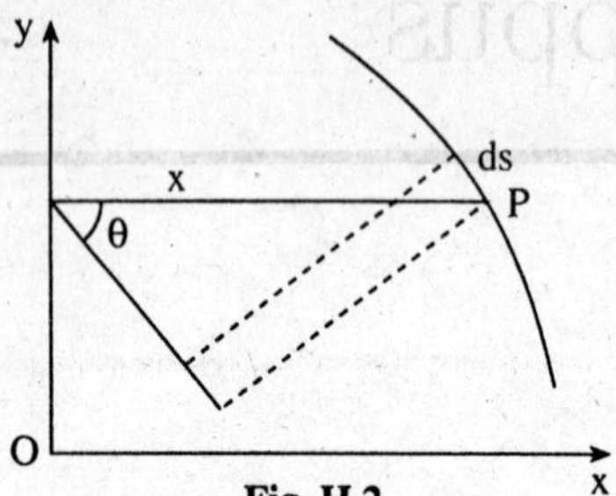

Fig. H.2

We consider a small arc length ds of the curve at a point P. Then the length of the arc described by the point P is $= x\theta$.

Hence the surface generated by the elemental arc is

$$dS = x\theta \cdot ds \tag{H.5}$$

The whole surface area S generated by the curve is

$$S = \int x\theta \cdot ds = \theta \int x ds \tag{H.6}$$

Now, by definition, $\bar{x} = \dfrac{\int x ds}{\int ds} = \dfrac{\int x ds}{l}$ so that

$$\int x ds = l\bar{x} \tag{H.7}$$

From Eqns (H.6) and (H.7), we get

$$S = l \cdot \bar{x}\theta, \tag{H.8}$$

which proves the theorem.

APPENDIX I

Measurement of Static Fluid Pressure

Pressure in a fluid at rest can be measured by simple instruments called ***pressure gauges***. These are described below:

1. Piezometer Tube. It consists of a glass tube with open ends. The pressure in a pipe or vessel, full of a liquid, can be measured by inserting the piezometer tube vertically into the liquid. The liquid will rise in the tube to a height equal to the equivalent static head of the pressure in the vessel.

2. U-Tube This pressure gauge in the form of a U-tube contains a heavier fluid which does not mix with the fluid of which the pressure is to be measured.

Suppose we are to measure the pressure in a pipe full of water with the U-tube. The left limb of the U-tube is kept open to the atmosphere and the right limb, containing water, is connected to the pipe. The pressure in the pipe will force the heavy liquid in the right limb downwards. This, in turn, will cause the heavy liquid in the left limb to rise by the same amount (FIG. I.1)

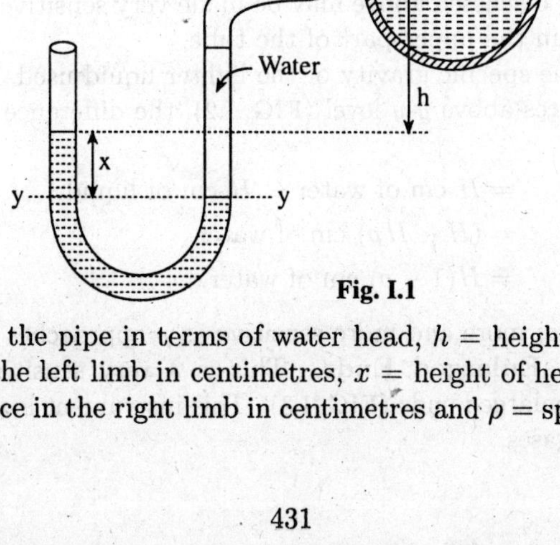

Fig. I.1

Let H = pressure in the pipe in terms of water head, h = height of the centre of the pipe above liquid surface in the left limb in centimetres, x = height of heavy liquid in the left limb above the common surface in the right limb in centimetres and ρ = specific gravity of the liquid in the U-tube.

The liquid below the level of the common surface (shown by dotted line y-y) being homogeneous, the fluid pressures at this level on both the limbs should be equal.

Now the pressure in the left limb at y-y level

$$= x\rho + \Pi \tag{I.1}$$

and that in the right limb at y-y level

$$= x + H + h, \tag{I.2}$$

where Π is the atmospheric pressure in terms of water head in centimetres.

By condition,

$$x + H + h = x\rho, \text{ i.e., } H = x(\rho - 1) - h. \tag{I.3}$$

If, in particular, the heavy liquid in the U-tube is mercury, then $\rho = 13.6$. In this case, Eqn (I.3) assumes the form

$$H = x(13.6 - 1) - h = 12.6x - h \tag{I.4}$$

of water head in centimetres.

Mercury is used as the heavy liquid for measuring high pressure. For small pressures, the liquid used should be a little heavier than water.

3. Inverted U-Tube. This instrument is used for measuring the difference of pressure between two sections of a pipe containing water. The upper part of the tube contains air, while the water from the two sections of the pipe for which the pressure difference is being measured passes into the left and right limbs respectively (FIG. I.2).

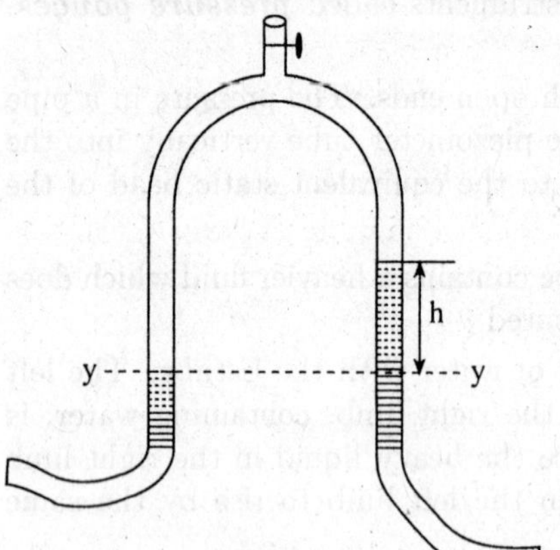

Fig. I.2

There is a valve at the top which is used to adjust the heights of water columns by letting out air through it.

Since the air trapped in the upper part of the tube is under constant pressure, the difference of pressure between the two sections of the pipe should be equal to the difference in heights of the two water columns. If this difference in heights of water columns be H centimetres, the difference of pressures between two sections of the pipe should be H cm of water.

4. Differential Gauge. Gauges which are made more sensitive by using liquids of different specific gravity are known as ***differential gauges***.

The inverted U-tube as discussed above may be made very sensitive by using a liquid lighter than water in place of air in the upper part of the tube.

In this case, let ρ be the specific gravity of the lighter liquid used.

Considering the pressures above y-y level (FIG. I.2), the difference of pressure between the two sections of the pipe

$$\begin{aligned} &= H \text{ cm of water} - H \text{ cm of liquid} \\ &= (H - H\rho) \text{ cm of water} \\ &= H(1 - \rho) \text{ cm of water.} \end{aligned} \tag{I.5}$$

The instrument becomes more and more sensitive as ρ approaches unity.

5. Oil Gauge with Enlarged Ends. This is a very sensitive type of gauge which consists of a U-tube with enlarged ends (FIG. I.3). This instrument is used for measuring small differences of pressure of gases.

Water and oil are taken in the two limbs of the U-tube such that the free surface of each liquid is in the enlarged ends having same cross-sectional area.

Let A = cross-sectional area of enlarged ends, a = cross-sectional area of the tube, ρ = specific gravity of oil used.

We assume that both the ends of the U-tube are exposed to the same pressure. In this case, let h cm be the height of free surface of oil above the common surface m-m. Then height of free surface of water above m-m level is

$$H = \rho h \tag{I.6}$$

Now, let the surface of oil be subjected to an additional pressure equivalent to x cm of water. Due to this, the common surface m-m will fall by an amount, say y cm to the level n-n. The level of oil in the enlarged end will consequently fall by an amount

$$z = \frac{a}{A}y. \tag{I.7}$$

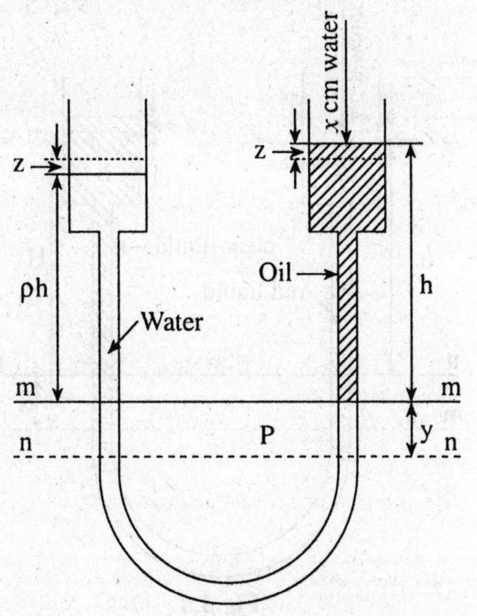

Fig. I.3

Hence, the level of water in the other limb will rise by the same amount as in Eqn (I.7). Let us now consider the equilibrium in both the limbs at the new common surface n-n. Height of oil surface above n-n

$$= h + y - z = h + y - \frac{a}{A}y = h + y\left(1 - \frac{a}{A}\right) \tag{I.8}$$

Height of water surface above n-n

$$= \rho h + y + z = \rho h + y + \frac{a}{A}y = \rho h + y\left(1 + \frac{a}{A}\right) \tag{I.9}$$

Since the pressures in both limbs at n-n level are equal, we must have

$$\rho h + y\left(1 + \frac{a}{A}\right) = \rho\left[h + y\left(1 - \frac{a}{A}\right)\right] + x$$

$$\text{i.e.,}\quad y\left(1 + \frac{a}{A}\right) = \rho y\left(1 - \frac{a}{A}\right) + x$$

$$\text{or,}\quad x = y\left[1 + \frac{a}{A} - \rho\left(1 - \frac{a}{A}\right)\right] \tag{I.10}$$

EXAMPLE I.1 *A pressure gauge consists of two cylindrical bulbs A and B, each of* 2.40 cm^2 *cross-sectional area, which are connected by a U-tube with vertical limbs, each of* 0.06 cm^2 *cross-sectional area. A red liquid of sp. gr.* 0.95 *is filled into A and a clear liquid of sp. gr.* 0.9 *is filled into B, the surface of separation being in the limb attached to B. Find the displacement of the surface of separation when the pressure on the surface in A is greater than that in B by an amount equivalent to* 2.0 cm *head of water.*

Solution. The arrangement is shown in FIG. I.4. We consider the case when pressure in bulb A is same as that in B.

If h be the height of liquid in A above the common surface m-m, then the height of liquid in B above m-m is

$$H = \frac{0.95}{0.9}h \tag{1}$$

Now, let the pressure in A be increased by 2.0 cm of water head. This will cause the common surface to rise by x cm (say). Then the surface of liquid in A will fall by an amount

$$z = \frac{a}{A}x = \frac{0.06}{2.40}x = \frac{x}{40} \text{ cm} \tag{2}$$

and that in B will rise by the same amount. We consider equilibrium of liquid columns above new common surface n-n.

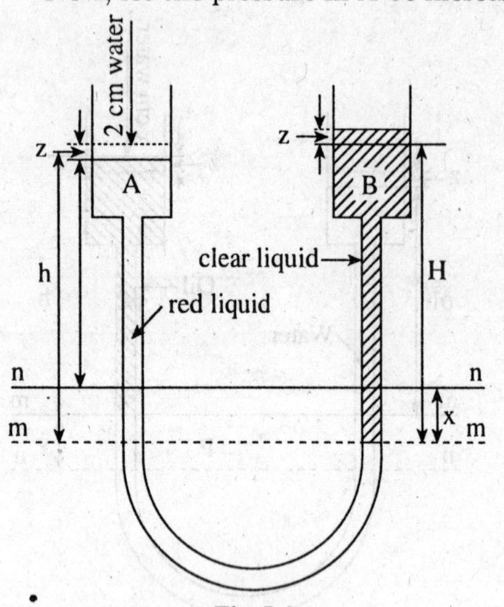

Fig. I.4

Height of liquid in bulb A above n-n level

$$= h - x - z$$

$$= h - x - \frac{x}{40} \tag{3}$$

Height of liquid in bulb B above n-n level

$$= H - x + z$$

$$= \frac{0.95}{0.9}h - x + \frac{x}{40} \tag{4}$$

Then

$$\left.\begin{aligned} \text{total pressure in } A &= 0.95\left(h - x - \frac{x}{40}\right) + 2 \\ \text{and total pressure in } B &= 0.9\left(\frac{0.95}{0.9}h - x + \frac{x}{40}\right) \end{aligned}\right\} \tag{5}$$

Since total pressure in A = total pressure in B, we have from (5)

$$0.95\left(h - x - \frac{x}{40}\right) + 2 = 0.9\left(\frac{0.95}{0.9}h - x + \frac{x}{40}\right)$$

$$\text{or,}\quad -0.95x\left(1 + \frac{1}{40}\right) + 2 = -0.9x\left(1 - \frac{1}{40}\right)$$

$$\text{or,}\quad 2 = x\left\{0.95\frac{41}{40} - 0.9\frac{39}{40}\right\} = \frac{x}{4}(3.895 - 3.51)$$

$$\text{or,}\quad x = \frac{8}{0.385} = 20.8\,\text{cm}$$

$\therefore$ the displacement of the surface of separation is **20.8 cm.**

APPENDIX **J**

Density Measurement of Liquids

Density measurement of liquids at a particular temperature may be made by the following methods:

(1) weighing a known volume of liquid, (2) hydrostatic weighing, (3) Westphal balance, (4) Nicholson's and common hydrometer and (5) U-tube.

For determining accurately the weight of a known volume of liquid, a device called ***pycnometer*** is used. This is usually a glass vessel whose weight, volume and the variation of volume with temperature have been accurately determined. If the weight of the empty pycnometer be W_0 and its weight containing a volume V of the liquid at a particular temperature be W_1, then

$$\rho V = W_1 - W_0, \text{ i.e., } \rho = \frac{W_1 - W_0}{V}, \tag{J.1}$$

where ρ is the density of the liquid at that temperature. Thus the density can be directly determined from Eqn (J.1).

For determining the density by hydrostatic weighing, we essentially use a ***plummet*** whose weight in air as well as in the liquid whose density is to be determined are known (FIG. J.1).

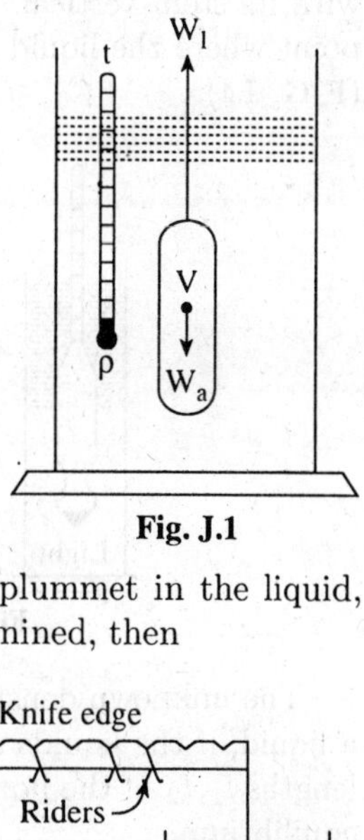

Fig. J.1

Thus, if W_a = weight of the plummet in air, W_l = weight of the plummet in the liquid, V = volume of the plummet, and ρ = density of the liquid to be determined, then

$$W_l + \rho V - W_a = 0,$$

which yields

$$\rho = \frac{W_a - W_l}{V} \tag{J.2}$$

The Westphal balance consists of a scale beam with special riders placed at special points (FIG. J.2). The scale is prepared by utilising the buoyant force on the plummet as a measure of density. The plummet dipped within the liquid is connected to the scale beam and is balanced with the special riders. This gives a direct and precise value of the density from the reading in the scale.

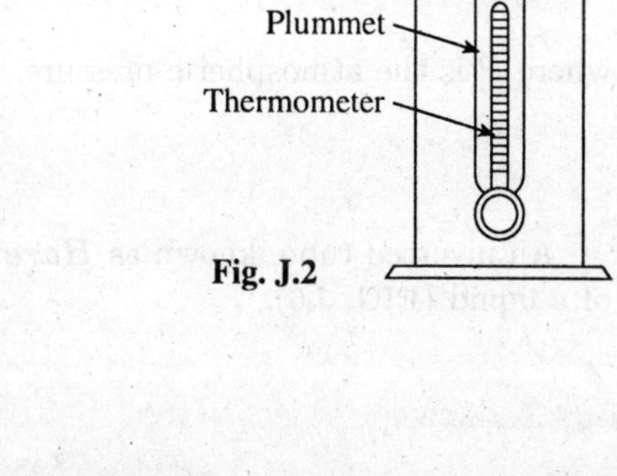

Fig. J.2

The use of hydrometer is probably the most common method for obtaining the density of a liquid. ***Nicholson's hydrometer*** consists of a hollow metal vessel which supports a thin stem with a pan at upper end. There is a well-defined mark on the stem and the instrument is made to sink up to this mark always. There is a small heavy cup at the lower end which helps the hydrometer to float with its stem vertical (FIG. J.3).

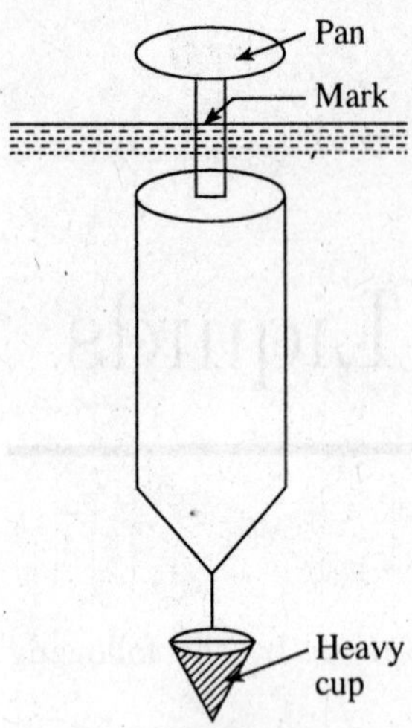

Fig. J.3

Let W_0 = weight of the hydrometer

W_1, W_2 = weights required to be placed on upper pan to sink the hydrometer up to the mark in the liquid and in water respectively.

Then the weight of the displaced liquid $= W_0 + W_1$ and the weight of the same volume of water $= W_0 + W_2$.

Hence, the specific gravity of the liquid is

$$s = \frac{W_0 + W_1}{W_0 + W_2} \tag{J.3}$$

The ***common hydrometer*** consists of a thin glass tube of uniform cross section ending to a bulb. The lower part of the bulb contains mercury such that the instrument can float with its stem vertical. The stem is graduated to read the specific gravity of the liquid at the point where the liquid surface intersects the stem when the instrument is floated in the liquid (FIG. J.4).

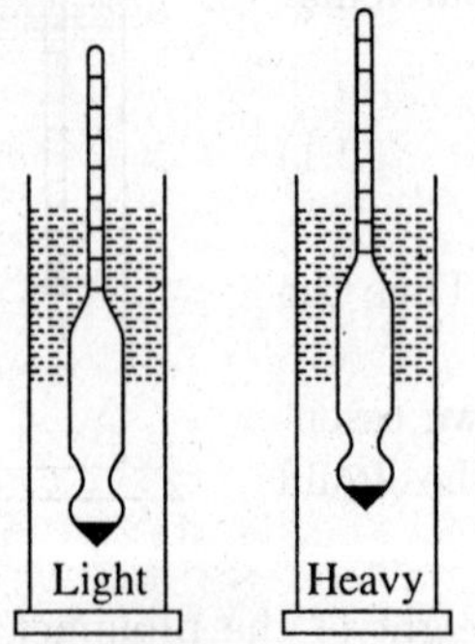

Fig. J.4

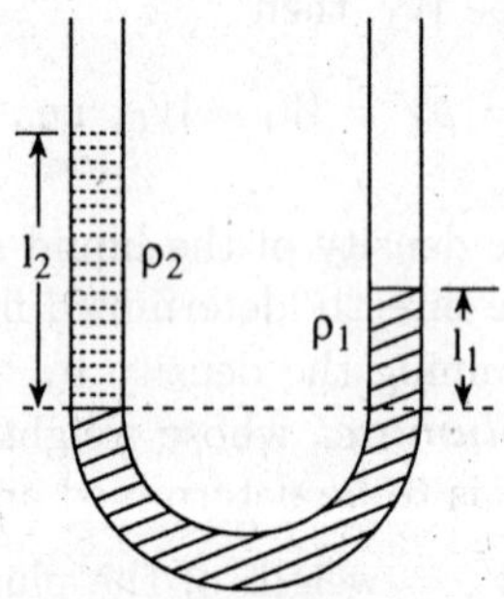

Fig. J.5

The unknown density of a liquid can be obtained approximately from the known density of a liquid, if the liquids are not miscible, by placing them in an open U-tube and measuring the lengths l_1, l_2 of the liquid columns (FIG. J.5). If ρ_1, ρ_2 be the corresponding densities, then for equilibrium,

$$P + g\rho_1 l_1 = P + g\rho_2 l_2, \tag{J.4}$$

where P is the atmospheric pressure. Hence, we get

$$\frac{\rho_1}{\rho_2} = \frac{l_2}{l_1} \tag{J.5}$$

An inverted tube, known as ***Hare's hydrometer***, is also used for determining the density of a liquid (FIG. J.6).

In the curved portion of the U-tube, there is an outlet pipe along with a stopcock through which a quantity of air can be drawn out from the instrument by suction.

Then open ends of the U-tube are immersed in two liquids kept in separate vessels. Thus, here we do not have the condition of immiscibility of the liquids.

Let P = atmospheric pressure, P^* = pressure inside the tube after suction, l_1, l_2 = heights of the liquid columns in the two arms of the U-tube, and ρ_1, ρ_2 = densities of the two liquids.

Then, for equilibrium, we have

$$\left.\begin{aligned} P &= P^* + g\rho_1 l_1 \\ \text{and } P &= P^* + g\rho_2 l_2 \end{aligned}\right\} \tag{J.6}$$

This yields

$$\frac{\rho_1}{\rho_2} = \frac{l_2}{l_1}. \tag{J.7}$$

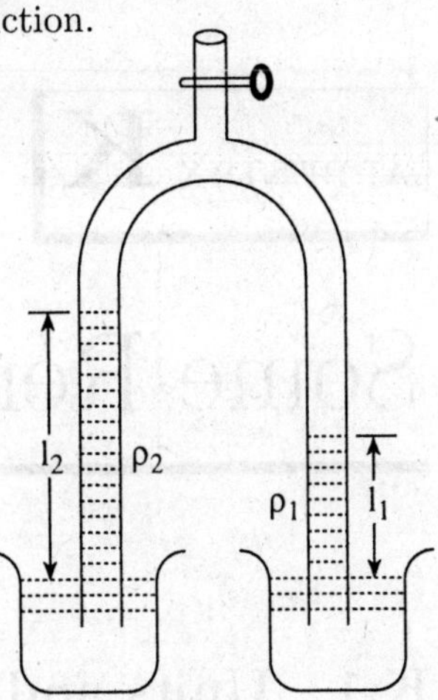

Fig. J.6

APPENDIX **K**

Some Remarks on Surface Tension

K.1 Units and Dimensions

In S.I. system, the unit of surface tension is newton/metre, and in C.G.S., the unit is dyne/centimetre.

The dimension of surface tension is

$$[\gamma] = \frac{[\mathrm{M}][\mathrm{L}]}{[\mathrm{T}^2]} \cdot \frac{1}{[\mathrm{L}]} = \mathrm{MT}^{-2} \tag{K.1}$$

K.2 Surface Tension for Some Liquids

Surface tension for some liquids under liquid-air interfaces are given below:

Liquid	Surface Tension
Water	73 dynes/cm
Salt water	75 "
Soapy water	20-30 "
Ether	17 "
Alcohol	23 "
Carbon Tetrachloride	27 "
Lubricating oil	25-35 "
Mercury	480 "

K.3 Work Done to Increase the Surface Area

When a liquid drop of radius R is broken into n numbers of smaller drops each of radius r, the surface area increases and some ***work is done against the surface tension***. If the increase in surface area is dS, and the surface tension is γ, then the work done is γdS. Here increase in surface area

$$= n \cdot 4\pi r^2 - 4\pi R^2 = 4\pi(nr^2 - R^2) \tag{K.2}$$

Since volume of liquid remains the same, we must have

$$n \cdot \frac{4}{3}\pi r^3 = \frac{4}{3}\pi R^3,$$

which yields

$$r = \frac{R}{n^{1/3}}. \tag{K.3}$$

Inserting this value into Eqn (K.2), we have the increase in surface area

$$= 4\pi\left(n\frac{R^2}{n^{2/3}} - R^2\right) = 4\pi R^2(n^{1/3} - 1) \tag{K.4}$$

Therefore, work done is

$$W = 4\pi R^2\gamma(n^{1/3} - 1) \tag{K.5}$$

K.4 Increase of Temperature due to Decrease in Surface Area

When the surface area decreases, the surface energy also decreases. This causes the release of some energy W which is spent for increasing the temperature of liquid surface.

If n identical liquid drops, each of radius r, coalesce to form a single liquid drop of radius R, then decrease in surface area

$$= n \cdot 4\pi r^2 - 4\pi R^2 = 4\pi(nr^2 - R^2) \tag{K.6}$$

Hence the energy released is

$$W = 4\pi\gamma(nr^2 - R^2) \tag{K.7}$$

Let ρ = density of the liquid, s = specific heat of the liquid, and θ = rise in temperature of the single drop of liquid. Then the quantity of heat absorbed by the drop is

$$Q = \text{mass} \times \text{sp. heat} \times \text{temperature rise, i.e., } Q = \frac{4}{3}\pi R^3\rho s\theta \tag{K.8}$$

Now, by the first law of thermodynamics, we have $W = JQ$, where J = mechanical equivalent of heat. Hence, from (K.7) and (K.8), we have

$$4\pi\gamma(nr^2 - R^2) = J \cdot \frac{4}{3}\pi R^3\rho s\theta, \text{ i.e., } \theta = \frac{3\gamma}{J\rho s}\left(\frac{nr^2}{R^3} - \frac{1}{R}\right) \tag{K.9}$$

Again, since the total volume of liquid remains the same, we must have, by (K.3)

$$nr^3 = R^3, \text{ i.e., } \frac{nr^2}{R^3} = \frac{1}{r}$$

Thus the temperature rise θ is given by

$$\theta = \frac{3\gamma}{J\rho s}\left(\frac{1}{r} - \frac{1}{R}\right) \tag{K.10}$$

K.5 Surface Tension and Evaporation

We consider a liquid drop which is free to evaporate.

Let r = radius of the liquid drop, ρ = density of the liquid, L = latent heat of evaporation, and E = surface energy per unit area.

We know that

$$E = \gamma + \eta, \tag{K.11}$$

where γ = surface tension and η = thermal energy supplied per unit increase in the surface area.

Let the radius of the liquid drop be diminished by a small quantity δr due to evaporation. Then decrease in surface area

$$= 4\pi r - 4\pi(r - \delta r)^2 = 8\pi r \delta r.$$

Hence, loss of surface energy

$$= 8\pi E r \delta r \tag{K.12}$$

Now, volume of liquid evaporated

$$= \frac{4}{3}\pi r^3 - \frac{4}{3}\pi(r - \delta r)^3 = 4\pi r^2 \delta r \tag{K.13}$$

Then the mass of liquid evaporated

$$= 4\pi r^2 \rho \delta r \tag{K.14}$$

Hence, the heat energy necessary for the evaporation of above mass of liquid

$$= 4\pi r^2 \rho L \delta r \tag{K.15}$$

Evaporation will discontinue when loss in surface energy will be less or equal to the heat energy given by (K.15). That is,

$$\frac{8\pi E r \delta r}{J} \leq 4\pi r^2 \rho L \delta r, \text{ i.e., } r \geq \frac{2E}{\rho J L} \tag{K.16}$$

If, however, the liquid drop is prevented from receiving heat from the surroundings, we have $\eta = 0$ in Eqn (K.11), so that $E = \gamma$. In this case, Eqn (K.16) assumes the form

$$r \geq \frac{2\gamma}{\rho J L} \tag{K.17}$$

K.6 Surface Tension and Boiling

In the normal process of boiling, a liquid bubble is formed. This bubble contains vapour as well as some gas.

Let r = radius of the bubble, p_1 = pressure of vapour inside the bubble, p_2 = pressure of gas inside the bubble, γ = surface tension, and p = external pressure. Then, for the equilibrium of the bubble, we must have

$$p_1 + p_2 = p + \frac{2\gamma}{r} \tag{K.18}$$

Assuming that the gas in the bubble obeys Boyle's law, we can write:

$$p_2 = \frac{\lambda}{r^3} \quad (\lambda = \text{constant}) \tag{K.19}$$

From (K.18) and (K.19), we have

$$P \equiv p_1 - p = \frac{2\gamma}{r} - \frac{\lambda}{r^3} \tag{K.20}$$

For minimum value of P,

$$\frac{dP}{dr} \equiv -\frac{2\gamma}{r^2} + \frac{3\lambda}{r^4} = 0,$$

which yields

$$\lambda = \frac{2}{3}\gamma r^2 \tag{K.21}$$

With this value of λ, we get from (K.20),

$$P \equiv p_1 - p = \frac{2\gamma}{r} - \frac{2}{3}\gamma r^2 \cdot \frac{1}{r^3} = \frac{4\gamma}{3r} \tag{K.22}$$

Thus, boiling begins when the excess pressure P inside the bubble of radius r is equal to $\frac{4\gamma}{3r}$.

K.7 Eötvös' Law

We consider a liquid of density ρ and molecular weight M.

Let γ = surface tension at absolute temperature T and T_c = critical temperature of the liquid.

Assuming that the molecules of the liquid are symmetrical in shape, the surface area of one gm-molecule of the liquid is proportional to $\left(\frac{M}{\rho}\right)^{2/3}$.

This surface is called ***molar surface***.

The surface energy in the molar surface is, therefore, proportional to $\gamma\left(\frac{M}{\rho}\right)^{2/3}$. This surface energy is called ***molar energy***.

Eötvös' law states:

$$\gamma\left(\frac{M}{\rho}\right)^{2/3} = K(T_c - T) \tag{K.23}$$

The constant K is known as ***Eötvös' constant*** and has the value of 2.2 approximately. The relation (K.23) was corrected by ***Ramsey and Shields*** giving the following form:

$$\gamma\left(\frac{M}{\rho}\right)^{2/3} = K(T_c - T - \delta) \tag{K.24}$$

Here δ is a constant which assumes a value between 6 and 8 in the case of most liquids.

From Eqn (K.24), it is obvious that the surface tension vanishes when $T = T_c - \delta$. From Eqn (K.23), the critical temperature T_c of the liquid is obtained as

$$T_c = T + \frac{\gamma}{K}\left(\frac{M}{\rho}\right)^{2/3} \tag{K.25}$$

Taking log on both sides, we find the relation:

$$\log(T_c - T) = \log\left(\frac{\gamma}{K}\right) + \frac{2}{3}\log\left(\frac{M}{\rho}\right) \tag{K.26}$$

The constant K in Eqn (K.24) may be obtained by differentiating the expression in the form:

$$K = -\frac{d}{dT}\left[\gamma\left(\frac{M}{\rho}\right)^{2/3}\right] \tag{K.27}$$

APPENDIX L

Vector Analysis

L.1 Vectors and Scalars

Physical quantities are of two types. A quantity which can be specified completely by its magnitude with corresponding unit is called a ***scalar***. For examples, mass, volume, density, energy and temperature are scalar quantities.

A physical quantity which can be completely and meaningfully specified by its magnitude and direction is called a ***vector***. For examples, velocity, acceleration, force, displacement and torque are vector quantities. A vector quantity is graphically represented by a straight line with an arrow at the tip; the direction of the quantity is indicated by the head of the arrow and the magnitude of the quantity is given by the length of the straight line. Unit vector has a length of unity. We will denote vectors by boldface letters in italics.

L.2 Cartesian System

In Cartesian coordinate system, we take unit vectors $\boldsymbol{i}, \boldsymbol{j}, \boldsymbol{k}$ along the coordinate axes. Consider a point $P(x, y, z)$ (FIG. L.1). The ***position*** or ***radius vector*** $\boldsymbol{r}$ of the point is the vector drawn from the origin of coordinate system to the point P. Thus

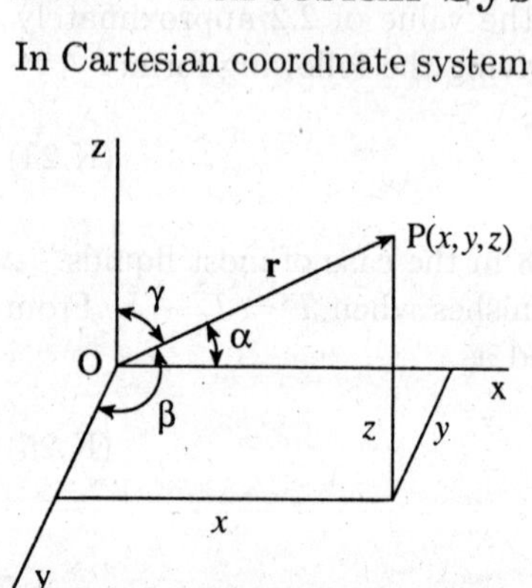

Fig. L.1

$$\boldsymbol{r} = \boldsymbol{i}x + \boldsymbol{j}y + \boldsymbol{k}z. \tag{L.1}$$

The magnitude of the vector $\boldsymbol{r}$ is given by

$$r^2 = |\boldsymbol{r}|^2 = x^2 + y^2 + z^2 \tag{L.2}$$

If α, β, γ are the angles that the vector $\boldsymbol{r}$ makes with coordinate axes, then the direction cosines are defined as

$$l = \cos\alpha,\ m = \cos\beta,\ n = \cos\gamma \tag{L.3}$$

The coordinates of the point P are related as

$$x = rl,\ y = rm,\ z = rn \tag{L.4}$$

Eqns (L.2) and (L.4) yield

$$l^2 + m^2 + n^2 = 1 \tag{L.5}$$

Let us consider an arbitrary vector $\boldsymbol{F}$ having projections F_x, F_y, F_z along the coordinate axes. Then

$$\boldsymbol{F} = \boldsymbol{i}F_x + \boldsymbol{j}F_y + \boldsymbol{k}F_z \tag{L.6}$$

Its magnitude is given by

$$|\boldsymbol{F}| = \sqrt{F_x^2 + F_y^2 + F_z^2} \tag{L.7}$$

The unit vector along $\boldsymbol{F}$ is

$$\frac{\boldsymbol{F}}{|\boldsymbol{F}|} = \boldsymbol{i}\cos\alpha + \boldsymbol{j}\cos\beta + \boldsymbol{k}\cos\gamma \tag{L.8}$$

L.3 Algebraic Operations with Vectors

We consider two vectors $\boldsymbol{E}$ and $\boldsymbol{F}$ given by

$$\left.\begin{aligned} \boldsymbol{E} &= \boldsymbol{i}E_x + \boldsymbol{j}E_y + \boldsymbol{k}E_z \\ \boldsymbol{F} &= \boldsymbol{i}F_x + \boldsymbol{j}F_y + \boldsymbol{k}F_z \end{aligned}\right\} \tag{L.9}$$

(a) **Addition:** Addition (or subtraction) of two vectors $\boldsymbol{E}$ and $\boldsymbol{F}$ is obtained simply by adding (or subtracting) similar components. Thus,

$$\begin{aligned} \boldsymbol{E} \pm \boldsymbol{F} &= (\boldsymbol{i}E_x + \boldsymbol{j}E_y + \boldsymbol{k}E_z) \pm (\boldsymbol{i}F_x + \boldsymbol{j}F_y + \boldsymbol{k}F_z) \\ &= \boldsymbol{i}(E_x \pm F_x) + \boldsymbol{j}(E_y \pm F_y) + \boldsymbol{k}(E_z \pm F_z) \end{aligned} \tag{L.10}$$

(b) **Scalar (or Dot) Product:** The ***scalar product*** of two vectors $\boldsymbol{E}$ and $\boldsymbol{F}$ is a scalar and is defined as

$$\boldsymbol{E} \cdot \boldsymbol{F} = |\boldsymbol{E}|\,|\boldsymbol{F}|\cos\theta, \tag{L.11}$$

where θ is the angle between the directions of the vectors. Then

$$\left.\begin{aligned} \boldsymbol{i}\cdot\boldsymbol{i} = \boldsymbol{j}\cdot\boldsymbol{j} = \boldsymbol{k}\cdot\boldsymbol{k} = 1 \\ \boldsymbol{i}\cdot\boldsymbol{j} = \boldsymbol{j}\cdot\boldsymbol{k} = \boldsymbol{k}\cdot\boldsymbol{i} = 0 \end{aligned}\right\} \tag{L.12}$$

With these relations, Eqn (L.11) can be written as

$$\boldsymbol{E} \cdot \boldsymbol{F} = E_xF_x + E_yF_y + E_zF_z \tag{L.13}$$

This implies that

$$\boldsymbol{E} \cdot \boldsymbol{F} = \boldsymbol{F} \cdot \boldsymbol{E} \tag{L.14}$$

That is, the scalar product is commutative.

(c) **Vector (or Cross) Product:** The ***vector or cross product*** of the two vectors is a vector and is defined by

$$\boldsymbol{E} \times \boldsymbol{F} = |\boldsymbol{E}|\,|\boldsymbol{F}|\sin\theta\boldsymbol{n} = \boldsymbol{n}EF\sin\theta, \tag{L.15}$$

where $\boldsymbol{n}$ is a unit vector along the normal to the plane containing the vectors $\boldsymbol{E}$ and $\boldsymbol{F}$. The positive direction of $\boldsymbol{n}$ is determined by the right-hand screw rule when it is turned from $\boldsymbol{E}$ to $\boldsymbol{F}$ (FIG. L.2).

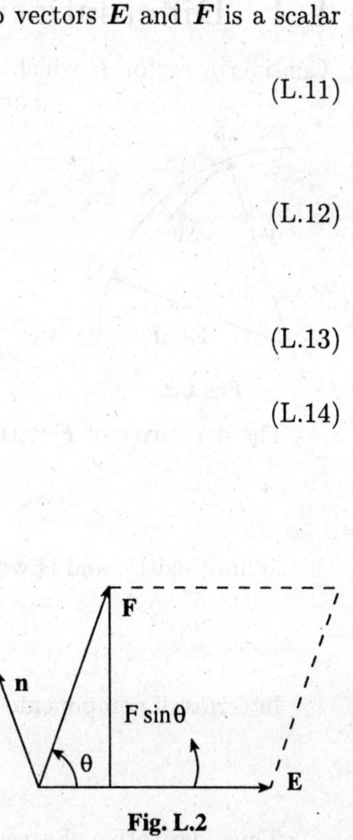

Fig. L.2

Hence,

$$\boldsymbol{E} \times \boldsymbol{F} = -\boldsymbol{F} \times \boldsymbol{E} \tag{L.16}$$

Moreover, by Eqn (L.15), we have

$$\left.\begin{aligned} \boldsymbol{i}\times\boldsymbol{i}=\boldsymbol{j}\times\boldsymbol{j}=\boldsymbol{k}\times\boldsymbol{k}=0\\ \boldsymbol{i}\times\boldsymbol{j}=\boldsymbol{k},\ \boldsymbol{j}\times\boldsymbol{k}=\boldsymbol{i},\ \boldsymbol{k}\times\boldsymbol{i}=\boldsymbol{j}\end{aligned}\right\} \tag{L.17}$$

With the use of Eqn (L.9), Eqn (L.15) can be written as

$$\begin{aligned} \boldsymbol{E}\times\boldsymbol{F} &= (\boldsymbol{i}E_x+\boldsymbol{j}E_y+\boldsymbol{k}E_z)\times(\boldsymbol{i}F_x+\boldsymbol{j}F_y+\boldsymbol{k}F_z)\\ &= \boldsymbol{i}\,(E_yF_z-E_zF_y)+\boldsymbol{j}\,(E_zF_x-E_xF_z)+\boldsymbol{k}\,(E_xF_y-E_yF_x)\\ &= \begin{vmatrix} \boldsymbol{i} & \boldsymbol{j} & \boldsymbol{k}\\ E_x & E_y & E_z\\ F_x & F_y & F_z\end{vmatrix}\end{aligned} \tag{L.18}$$

Note. The area of the parallelogram in (FIG. L.2) is $EF\sin\theta$. Hence, Eqn (L.15) implies that we can represent the area of a parallelogram by a vector. In this situation, the area of the parallelogram is shown by means of a vector drawn normally to the plane of the parallelogram. The direction of the vector is determined by the right-hand screw rule.

In fact, any plane area can be represented by a vector whose length is proportional to the magnitude of the area and whose direction is normal to the plane of the area. The tip of the arrow-head is determined by rotating the right-hand screw in the direction in which the boundary curve of the plane area is traversed.

L.4 Differentiation of Vector Functions

Consider a vector $\boldsymbol{F}$ which depends on a scalar variable t such that for each value of t, there corresponds a value of $\boldsymbol{F}$. Then $\boldsymbol{F}$ is a function of t and we write,

$$\boldsymbol{F}=\boldsymbol{F}(t) \tag{L.19}$$

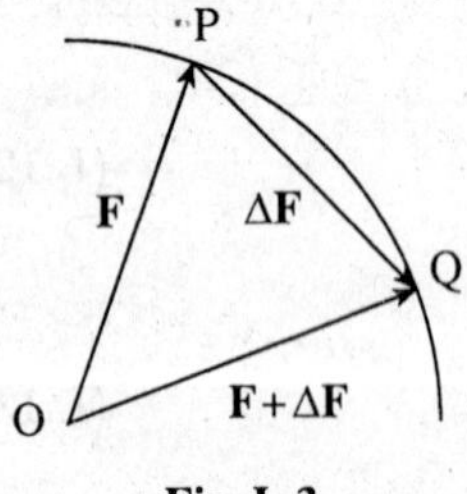

Fig. L.3

Let the scalar variable t change to $t+\Delta t$. Then (FIG. L.3)

$$\boldsymbol{F}+\Delta\boldsymbol{F}=\boldsymbol{F}(t+\Delta t) \tag{L.20}$$

Hence,

$$\frac{\Delta\boldsymbol{F}}{\Delta t}=\frac{\boldsymbol{F}(t+\Delta t)-\boldsymbol{F}(t)}{\Delta t} \tag{L.21}$$

The derivative of $\boldsymbol{F}$ with respect to t is defined as

$$\lim_{\Delta t\to 0}\frac{\Delta\boldsymbol{F}}{\Delta t}, \tag{L.22}$$

if the limit exists, and is written as $d\boldsymbol{F}/dt$. Thus,

$$\frac{d\boldsymbol{F}}{dt}=\lim_{\Delta t\to 0}\frac{\boldsymbol{F}(t+\Delta t)-\boldsymbol{F}(t)}{\Delta t} \tag{L.23}$$

In terms of components of $\boldsymbol{F}$, we get

$$\frac{d\boldsymbol{F}}{dt}=\boldsymbol{i}\frac{dF_x}{dt}+\boldsymbol{j}\frac{dF_y}{dt}+\boldsymbol{k}\frac{dF_z}{dt} \tag{L.24}$$

Thus, derivative of a vector is a vector along the direction of $\Delta\boldsymbol{F}$.

We can easily prove the following relations:

$$\left.\begin{aligned}
&\text{(i)} \quad \frac{d}{dt}(\boldsymbol{E}+\boldsymbol{F}) = \frac{d\boldsymbol{E}}{dt} + \frac{d\boldsymbol{F}}{dt} \\
&\text{(ii)} \quad \frac{d}{dt}(\boldsymbol{E}\cdot\boldsymbol{F}) = \boldsymbol{E}\cdot\frac{d\boldsymbol{F}}{dt} + \frac{d\boldsymbol{E}}{dt}\cdot\boldsymbol{F} \\
&\text{(iii)} \quad \frac{d}{dt}(\boldsymbol{E}\times\boldsymbol{F}) = \boldsymbol{E}\times\frac{d\boldsymbol{F}}{dt} + \frac{d\boldsymbol{E}}{dt}\times\boldsymbol{F} \\
&\text{(iv)} \quad \frac{d}{dt}(\lambda\boldsymbol{F}) = \lambda\frac{d\boldsymbol{F}}{dt}, \ \lambda = \text{a constant scalar}
\end{aligned}\right\} \tag{L.25}$$

In a special case, if $\boldsymbol{F}$ is a vector of constant magnitude, we have

$$\frac{d\boldsymbol{F}}{dt} = 0, \text{ i.e., } \boldsymbol{F}\cdot\frac{d\boldsymbol{F}}{dt} = 0 \tag{L.26}$$

This shows that the vectors $\boldsymbol{F}$ and $\dfrac{d\boldsymbol{F}}{dt}$ are perpendicular to each other.

L.5 Partial Differentiation

Consider a vector $\boldsymbol{F}$ which is a function of the rectangular coordinates x, y, z of a point in space. That is,

$$\boldsymbol{F} = \boldsymbol{F}(x, y, z) \tag{L.27}$$

Keeping the variables y and z constant, we increase x to $x + \Delta x$. Then, by definition,

$$\frac{\partial \boldsymbol{F}}{\partial x} = \lim_{\Delta x \to 0} \frac{\boldsymbol{F}(x+\Delta x, y, z) - \boldsymbol{F}(x, y, z)}{\Delta x}, \tag{L.28}$$

when the limit exists.

Similarly, we define the partial derivatives $\dfrac{\partial \boldsymbol{F}}{\partial y}$ and $\dfrac{\partial \boldsymbol{F}}{\partial z}$. The partial derivatives represent the rate of increase of the vector with respect to one variable when the other two remain constant.

A vector differential operator $\boldsymbol{\nabla}$ is defined by

$$\boldsymbol{\nabla} = \boldsymbol{i}\frac{\partial}{\partial x} + \boldsymbol{j}\frac{\partial}{\partial y} + \boldsymbol{k}\frac{\partial}{\partial z} \tag{L.29}$$

This operator called ***del*** (***nabla***) ***operator*** is of immense importance in physics. The scalar operator ∇^2 is obtained from Eqn (L.29) as

$$\nabla^2 = \boldsymbol{\nabla}\cdot\boldsymbol{\nabla} = \frac{\partial^2}{\partial x^2} + \frac{\partial^2}{\partial y^2} + \frac{\partial^2}{\partial z^2} \tag{L.30}$$

This is called the ***Laplacian operator***.

The ***divergence*** of a vector $\boldsymbol{F}$ is defined as

$$\text{div}\,\boldsymbol{F} = \boldsymbol{\nabla}\cdot\boldsymbol{F} \tag{L.31}$$

In terms of the components of $\boldsymbol{F}$,

$$\text{div}\,\boldsymbol{F} = \left(\boldsymbol{i}\frac{\partial}{\partial x} + \boldsymbol{j}\frac{\partial}{\partial y} + \boldsymbol{k}\frac{\partial}{\partial z}\right)\cdot(\boldsymbol{i}F_x + \boldsymbol{j}F_y + \boldsymbol{k}F_z) = \frac{\partial F_x}{\partial x} + \frac{\partial F_y}{\partial y} + \frac{\partial F_z}{\partial z} \tag{L.32}$$

Thus, ***divergence of a vector is a scalar quantity***.

The ***curl*** of a vector $\boldsymbol{F}$ is defined as

$$\text{curl}\,\boldsymbol{F} = \boldsymbol{\nabla} \times \boldsymbol{F} \tag{L.33}$$

In terms of the components of $\boldsymbol{F}$,

$$\text{curl}\,\boldsymbol{F} = \begin{vmatrix} \boldsymbol{i} & \boldsymbol{j} & \boldsymbol{k} \\ \dfrac{\partial}{\partial x} & \dfrac{\partial}{\partial y} & \dfrac{\partial}{\partial z} \\ F_x & F_y & F_z \end{vmatrix} \tag{L.34}$$

If $\text{curl}\,\boldsymbol{F} = 0$, the vector $\boldsymbol{F}$ is called an ***irrotational vector***. The gravitational and electrostatic fields are irrotational.

If the vector $\boldsymbol{F}$ represents a conservative force field, i.e., when the work done by the force on an object to move it from position A to position B is independent of the path, then $\boldsymbol{F} = \boldsymbol{\nabla}\phi$, where ϕ is a scalar function. In this case also $\text{curl}\,\boldsymbol{F} = 0$.

L.6 Physical Meaning of Divergence

Consider an elementary parallelopiped of sides dx, dy and dz (FIG. L.4). Then the elementary volume is

$$d\tau = dx\,dy\,dz \tag{L.35}$$

Let $\boldsymbol{V}$ be a vector point function at the origin of coordinates O. We take this vector to be associated with fluid flow where

$$\boldsymbol{V} = \rho\boldsymbol{v} \tag{L.36}$$

Here ρ = density of the fluid and $\boldsymbol{v}$ = velocity vector of the fluid. Then Eqn (L.36) represents the rate of mass flow of the fluid through unit area normal to the velocity vector $\boldsymbol{v}$. This is called the ***flux of vector V***.

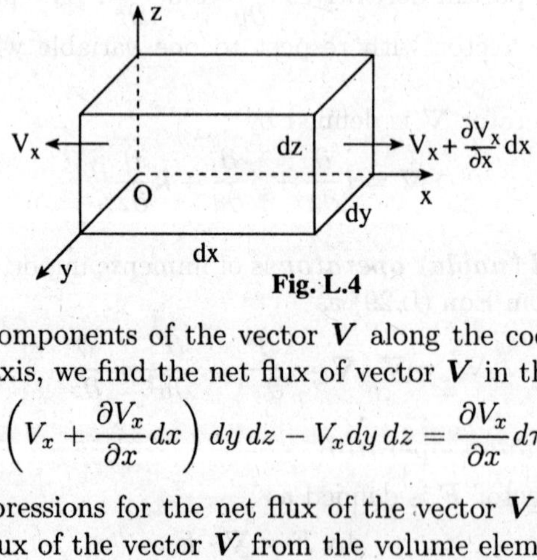

Fig. L.4

Let V_x, V_y, V_z be the components of the vector $\boldsymbol{V}$ along the coordinate axes. Considering the surfaces normal to x-axis, we find the net flux of vector $\boldsymbol{V}$ in the direction of x-axis is

$$\left(V_x + \frac{\partial V_x}{\partial x} dx\right) dy\,dz - V_x dy\,dz = \frac{\partial V_x}{\partial x} d\tau \tag{L.37}$$

We will get similar expressions for the net flux of the vector $\boldsymbol{V}$ in the directions of y- and z-axes. Hence, the total flux of the vector $\boldsymbol{V}$ from the volume element is

$$\left(\frac{\partial V_x}{\partial x} + \frac{\partial V_y}{\partial y} + \frac{\partial V_z}{\partial z}\right) d\tau \tag{L.38}$$

Hence, the amount of outward flux of the vector $\boldsymbol{V}$ per unit volume is

$$\frac{\partial V_x}{\partial x} + \frac{\partial V_y}{\partial y} + \frac{\partial V_z}{\partial z} \tag{L.39}$$

By Eqn (L.32), this is the same as div $\boldsymbol{V}$. Thus, ***divergence of a vector V represents the amount of outward flux of that vector per unit volume***. This is the physical meaning of divergence of a vector.

If div $\boldsymbol{V} > 0$, then occurs outflow of the flux.

If div $\boldsymbol{V} < 0$, then occurs inflow of the flux.

If div $\boldsymbol{V} = 0$, then there is no inflow or outflow of the flux. A vector for which divergence is zero is called a ***solenoidal*** vector.

L.7 Gradient of a Scalar Function

We consider a scalar point function $\phi = \phi(x, y, z)$ and draw two neighbouring surfaces S and S' over which the magnitude of the scalar point function remains constant (FIG. L.5).

Let the function assume the values ϕ and $\phi + \Delta\phi$ on the surfaces S, S' respectively. We take points P, P' on the neighbouring surfaces S, S' respectively and draw PM perpendicular to S'. Then

$$\overline{OP} = \boldsymbol{r};\ \overline{OP'} = \boldsymbol{r} + \Delta\boldsymbol{r} \tag{L.40}$$

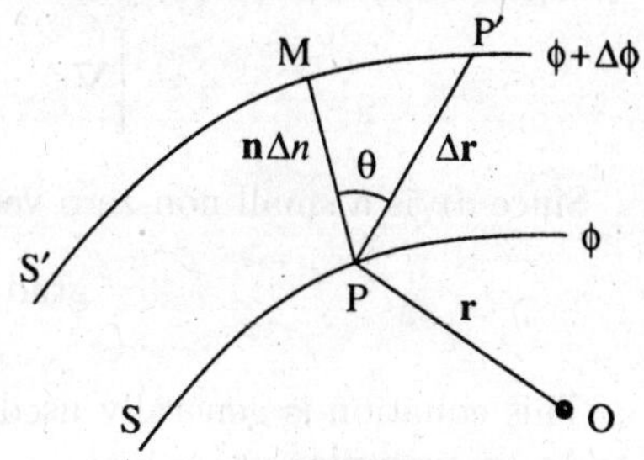

Fig. L.5

Let $PM = \Delta n, PP' = \Delta s$ and $\angle MPP' = \theta$, then

$$\Delta n = \Delta s \cos\theta. \tag{L.41}$$

Hence, the rate of change of ϕ w.r.t. s as we go from the point P on the surface S to the point P' on the surface S' is

$$\frac{\Delta\phi}{\Delta s} = \frac{\Delta\phi}{\Delta n}\frac{\Delta n}{\Delta s} = \cos\theta\frac{\Delta\phi}{\Delta n} \tag{L.42}$$

In the limit as $P' \to P$ along $\overline{P'P}$, we get

$$\frac{\partial\phi}{\partial s} = \cos\theta\frac{\partial\phi}{\partial n} \tag{L.43}$$

Hence,

$$\max\left|\frac{\partial\phi}{\partial s}\right| = \left|\frac{\partial\phi}{\partial n}\right| \tag{L.44}$$

at a chosen point P on S. This maximum rate of change takes place when $\theta = 0$, i.e., along the direction of unit normal vector $\boldsymbol{n}$.

Thus, the maximum rate of change of ϕ occurs in the direction of $\boldsymbol{n}$ and is a vector denoted by

$$\boldsymbol{n}\frac{\partial\phi}{\partial n}. \tag{L.45}$$

This vector is termed the gradient of the scalar point function $\phi(x, y, z)$ at the point P and is denoted by grad ϕ or $\boldsymbol{\nabla}\phi$ so that

$$\boldsymbol{\nabla}\phi = \text{grad}\,\phi = \boldsymbol{n}\frac{\partial\phi}{\partial n} \tag{L.46}$$

Hence, the gradient of a scalar function ϕ at any point is the vector having a magnitude equal to the maximum rate of change of ϕ at that point in the direction of the normal to the surface $\phi(x, y, z)$ = constant at that point.

Now, $\phi = \phi(x, y, z)$, and so

$$d\phi = \frac{\partial \phi}{\partial x}dx + \frac{\partial \phi}{\partial y}dy + \frac{\partial \phi}{\partial z}dz \tag{L.47}$$

Since

$$\boldsymbol{dr} = \boldsymbol{i}dx + \boldsymbol{j}dy + \boldsymbol{k}dz \tag{L.48}$$

Eqn (L.47) can be written as

$$d\phi = \left(\boldsymbol{i}\frac{\partial \phi}{\partial x} + \boldsymbol{j}\frac{\partial \phi}{\partial y} + \boldsymbol{k}\frac{\partial \phi}{\partial z}\right) \cdot \boldsymbol{dr} \tag{L.49}$$

But

$$d\phi = \frac{\partial \phi}{\partial n}dn = \frac{\partial \phi}{\partial n}\boldsymbol{n} \cdot \boldsymbol{dr} = \boldsymbol{\nabla}\phi \cdot \boldsymbol{dr} \tag{L.50}$$

Eqns (L.49) and (L.50) yield

$$\left[\boldsymbol{\nabla}\phi - \left(\boldsymbol{i}\frac{\partial \phi}{\partial x} + \boldsymbol{j}\frac{\partial \phi}{\partial y} + \boldsymbol{k}\frac{\partial \phi}{\partial z}\right)\right] \cdot \boldsymbol{dr} = 0 \tag{L.51}$$

Since $\boldsymbol{dr}$ is a small non-zero vector chosen arbitrarily, we get

$$\text{grad}\,\phi = \boldsymbol{\nabla}\phi = \boldsymbol{i}\frac{\partial \phi}{\partial x} + \boldsymbol{j}\frac{\partial \phi}{\partial y} + \boldsymbol{k}\frac{\partial \phi}{\partial z} \tag{L.52}$$

This equation is generally used to evaluate the gradient of a scalar function. As an example, let

$$\phi(x, y, z) = 3x^2 + 6xy + y^2 - 9zx + z^2$$

Then

$$\frac{\partial \phi}{\partial x} = 6x + 6y - 9z, \frac{\partial \phi}{\partial y} = 6x + 2y, \frac{\partial \phi}{\partial z} = -9x + 2z$$

Hence,

$$\text{grad}\,\phi = \boldsymbol{i}(6x + 6y - 9z) + \boldsymbol{j}(6x + 2y) + \boldsymbol{k}(-9x + 2z)$$

At the point $(1, 0, 0)$,

$$\text{grad}\,\phi = 6\boldsymbol{i} + 6\boldsymbol{j} - 9\boldsymbol{k}.$$

L.8 Integration of Vectors

Integration of vectors are performed by the usual procedures of integral calculus. Let $\phi(x, y, z)$ be a scalar point function and $\boldsymbol{F}$ be a vector in space.

(a) **Line Integral:** Line integrals are performed along some defined contour γ. Thus

(1) $$\int_\gamma \phi \boldsymbol{dr} = \int_\gamma \phi(\boldsymbol{i}dx + \boldsymbol{j}dy + \boldsymbol{k}dz) = \boldsymbol{i}\int_\gamma \phi dx + \boldsymbol{j}\int_\gamma \phi dy + \boldsymbol{k}\int_\gamma \phi dz \tag{L.53}$$

(2) $$\int_\gamma \boldsymbol{F} \cdot \boldsymbol{dr} = \int_\gamma (\boldsymbol{i}F_x + \boldsymbol{j}F_y + \boldsymbol{k}F_z) \cdot (\boldsymbol{i}dx + \boldsymbol{j}dy + \boldsymbol{k}dz)$$

$$= \int_\gamma F_x dx + \int_\gamma F_y dy + \int_\gamma F_z dz \tag{L.54}$$

(3) $$\int_\gamma \boldsymbol{F} \times \boldsymbol{dr} = \int_\gamma (\boldsymbol{i}F_x + \boldsymbol{j}F_y + \boldsymbol{k}F_z) \times (\boldsymbol{i}dx + \boldsymbol{j}dy + \boldsymbol{k}dz)$$

$$= \boldsymbol{i}\int_\gamma (F_y dz - F_z dy) + \boldsymbol{j}\int_\gamma (F_z dx - F_x dz) + \boldsymbol{k}\int_\gamma (F_x dy - F_y dx) \tag{L.55}$$

Here F_x, F_y, F_z are the components of the vector $\boldsymbol{F}$ along the coordinate axes. The above integrals can be evaluated by the usual methods of integation.

(b) Surface Integral: We can represent a plane surface area by a vector called ***area vector*** (FIG. L.6).

Let $d\sigma$ be an elementary surface taken from the surface σ. Then

$$\int_\sigma \boldsymbol{F} \cdot d\sigma = \int \boldsymbol{F} \cdot \boldsymbol{n} d\sigma, \tag{L.56}$$

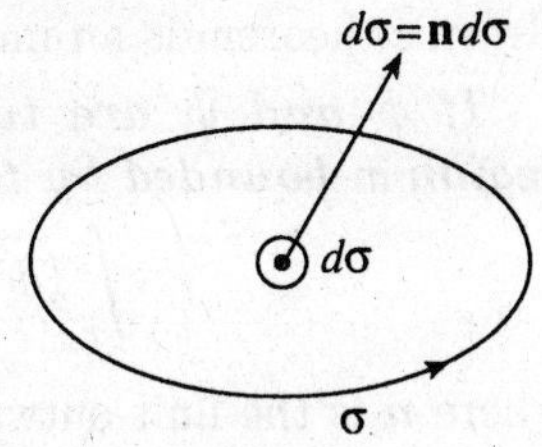

Fig. L.6

where $\boldsymbol{n}$ is the unit normal vector on the surface.

The integral in the L.H.S. of Eqn (L.56) is defined as the flux of the vector $\boldsymbol{F}$ through the surface σ.

(c) Volume Integral: We consider a volume element $d\tau$ which is a scalar quantity. Hence, the volume integral of a vector $\boldsymbol{F}$ over the volume τ is

$$\int_\tau \boldsymbol{F} d\tau = \boldsymbol{i} \int F_x d\tau + \boldsymbol{j} \int F_y d\tau + \boldsymbol{k} \int F_z d\tau \tag{L.57}$$

This is simple to evaluate by usual methods.

L.9 Gauss's Divergence Theorem

Gauss's divergence theorem relates a volume integral with the surface integral. It states that:

The volume integral of the divergence of the vector point function F taken over any volume τ is equal to the surface integral of the vector F taken over the closed surface surrounding the volume τ.

Thus

$$\int_\tau \operatorname{div} \boldsymbol{F} d\tau = \int_\sigma \boldsymbol{F} \cdot \boldsymbol{n} d\sigma, \tag{L.58}$$

where $\boldsymbol{n}$ is the unit outward-drawn normal vector (proof is not given).

As an example of its application, let us evaluate the integral

$$I = \int_\sigma (ax\boldsymbol{i} + by\boldsymbol{j} + cz\boldsymbol{k}) \cdot \boldsymbol{n} d\sigma,$$

where σ is the surface of the sphere $x^2 + y^2 + z^2 = 1$.

Here, $\boldsymbol{F} = ax\boldsymbol{i} + by\boldsymbol{j} + cz\boldsymbol{k}$.

Hence, $\operatorname{div} \boldsymbol{F} = \frac{\partial}{\partial x}(ax) + \frac{\partial}{\partial y}(by) + \frac{\partial}{\partial z}(cz) = a + b + c$.

By Gauss's theorem, $\int_\sigma \boldsymbol{F} \cdot \boldsymbol{n} d\sigma = \int_\tau \operatorname{div} \boldsymbol{F}.d\tau$.

Hence, the given integral is $\int_\tau (a + b + c) d\tau = (a + b + c)\tau$.

Volume of the sphere $\tau = \frac{4}{3}\pi r^3 = \frac{4}{3}\pi \ (\because \ r = 1)$.

Thus, the value of the given integral is $\frac{4}{3}\pi(a + b + c)$.

L.10 Green's Theorem

Green's theorem is an important corollary of divergence theorem. It states that:

If ϕ and ψ are two scalar functions having continuous derivatives within a region τ bounded by the closed surface σ, then

$$\int_\tau (\phi\nabla^2\psi - \psi\nabla^2\phi)\, d\tau = \int_\sigma (\phi\nabla\psi - \psi\nabla\phi)\cdot \boldsymbol{n} d\sigma, \tag{L.59}$$

where $\boldsymbol{n}$ is the unit outward-drawn normal vector to the surface.

L.11 Stokes' Theorem

Stokes' theorem relates a surface integral with a line integral. It states that:

The surface integral of the curl of a vector point function F taken over any surface σ is equal to the line integral of the vector point function F taken around the periphery of the surface.

Thus

$$\int_\sigma (\nabla\times \boldsymbol{F})\cdot \boldsymbol{n} d\sigma = \int \boldsymbol{F}\cdot d\boldsymbol{r}, \tag{L.60}$$

where $\boldsymbol{n}$ is the outward-drawn unit normal vector to the surface (proof is not given).

L.12 Green's Theorem in Plane

It states that:

If M and N are continuous functions of x and y having continuous derivatives in a region σ bounded by a closed curve C, then

$$\int_C (M dx + N dy) = \int_\sigma \left(\frac{\partial N}{\partial x} - \frac{\partial M}{\partial y}\right) dx dy \tag{L.61}$$

As an application of this theorem, let us evaluate the integral

$$I = \int_C (e^{-x}\sin y\, dx + e^{-x}\cos y\, dy),$$

where C is a rectangle with vertices $(0,0), (\pi,0), (\pi,\pi/2), (0,\pi/2)$.

Here, $M = e^{-x}\sin y; N = e^{-x}\cos y$.

Hence, $\dfrac{\partial M}{\partial y} = e^{-x}\cos y; \dfrac{\partial N}{\partial x} = -e^{-x}\cos y$.

$$\text{Then } I = \int_C (M dx + N dy) = \int_{x=0}^{x=\pi}\int_{y=0}^{y=\pi/2}\left(\frac{\partial N}{\partial x} - \frac{\partial M}{\partial y}\right)$$

$$= \int_{x=0}^{x=\pi}\int_{y=0}^{y=\pi/2} -2e^{-x}\cos y\, dx dy$$

$$= \int_{x=0}^{x=\pi} (-2e^{-x}dx)\int_{y=0}^{y=\pi/2}\cos y\, dy = 2\left(e^{-\pi}-1\right).$$

L.13 Directional Derivative

The directional derivative of any scalar point function ϕ along any direction is obtained by multiplying a unit vector in that direction scalarly by grad ϕ.

Thus, if $\boldsymbol{a}$ be a vector defined by

$$\boldsymbol{a} = \boldsymbol{i}a_x + \boldsymbol{j}a_y + \boldsymbol{k}a_z, \tag{L.62}$$

then unit vector along the direction of $\boldsymbol{a}$ is

$$\boldsymbol{n} = \frac{\boldsymbol{i}a_x + \boldsymbol{j}a_y + \boldsymbol{k}a_z}{\sqrt{a_x^2 + a_y^2 + a_z^2}} \tag{L.63}$$

Hence, the directional derivative of $\phi(x, y, z)$ along the direction of $\boldsymbol{a}$ is

$$\boldsymbol{n} \cdot \text{grad}\, \phi = \boldsymbol{n} \cdot \boldsymbol{\nabla}\phi \tag{L.64}$$

L.14 Some Useful Identities

Let ϕ, ψ be two scalar point functions and $\boldsymbol{A}, \boldsymbol{B}$ be two vector point functions. Then we have the following identities:

$$(1)\ \text{grad}\,(\phi\psi) = \phi\,\text{grad}\,\psi + \psi\,\text{grad}\,\phi \tag{L.65}$$

$$(2)\ \text{grad}\,(\boldsymbol{A} \cdot \boldsymbol{B}) = \boldsymbol{A} \times \text{curl}\,\boldsymbol{B} + \boldsymbol{B} \times \text{curl}\,\boldsymbol{A} + (\boldsymbol{A} \cdot \boldsymbol{\nabla})\boldsymbol{B} + (\boldsymbol{B} \cdot \boldsymbol{\nabla})\boldsymbol{A} \tag{L.66}$$

$$(3)\ \text{div}\,(\phi\boldsymbol{A}) = \phi\,\text{div}\,\boldsymbol{A} + \text{grad}\,\phi \cdot \boldsymbol{A} \tag{L.67}$$

$$(4)\ \text{div}\,(\boldsymbol{A} \times \boldsymbol{B}) = \boldsymbol{B} \cdot \text{curl}\,\boldsymbol{A} - \boldsymbol{A} \cdot \text{curl}\,\boldsymbol{B} \tag{L.68}$$

$$(5)\ \text{curl}\,(\phi\boldsymbol{A}) = \text{grad}\,\phi \times \boldsymbol{A} + \phi\,\text{curl}\,\boldsymbol{A} \tag{L.69}$$

$$(6)\ \text{curl}\,(\boldsymbol{A} \times \boldsymbol{B}) = \boldsymbol{A}\,\text{div}\,\boldsymbol{B} - \boldsymbol{B}\,\text{div}\,\boldsymbol{A} + (\boldsymbol{B} \cdot \boldsymbol{\nabla})\boldsymbol{A} - (\boldsymbol{A} \cdot \boldsymbol{\nabla})\boldsymbol{B} \tag{L.70}$$

$$(7)\ \boldsymbol{\nabla} \times \boldsymbol{\nabla} \times \boldsymbol{A} = \boldsymbol{\nabla}(\boldsymbol{\nabla} \cdot \boldsymbol{A}) - \nabla^2\boldsymbol{A} \tag{L.71}$$

EXAMPLE L.1 *If* $\boldsymbol{v} = \boldsymbol{\omega} \times \boldsymbol{r}$, *prove that* curl $\boldsymbol{v} = 2\boldsymbol{\omega}$.

Solution. Let

$$\boldsymbol{\omega} = \boldsymbol{i}\omega_1 + \boldsymbol{j}\omega_2 + \boldsymbol{k}\omega_3 \tag{1}$$

We have

$$\text{curl}\,\boldsymbol{v} = \boldsymbol{\nabla} \times \boldsymbol{v} = \boldsymbol{\nabla} \times (\boldsymbol{\omega} \times \boldsymbol{r}) \tag{2}$$

Thus

$$\begin{aligned}
\text{curl}\,\boldsymbol{v} &= \boldsymbol{\nabla} \times \begin{vmatrix} \boldsymbol{i} & \boldsymbol{j} & \boldsymbol{k} \\ \omega_1 & \omega_2 & \omega_3 \\ x & y & z \end{vmatrix} \\
&= \boldsymbol{\nabla} \times [\boldsymbol{i}(\omega_2 z - \omega_3 y) + \boldsymbol{j}(\omega_3 x - \omega_1 z) + \boldsymbol{k}(\omega_1 y - \omega_2 x)] \\
&= \begin{vmatrix} \boldsymbol{i} & \boldsymbol{j} & \boldsymbol{k} \\ \dfrac{\partial}{\partial x} & \dfrac{\partial}{\partial y} & \dfrac{\partial}{\partial z} \\ \omega_2 z - \omega_3 y & \omega_3 x - \omega_1 z & \omega_1 y - \omega_2 x \end{vmatrix} \\
&= 2(\boldsymbol{i}\omega_1 + \boldsymbol{j}\omega_2 + \boldsymbol{k}\omega_3) \\
&= 2\boldsymbol{\omega}\ \text{[by Eqn (1)]}
\end{aligned}$$

EXAMPLE L.2 *Show that a necessary and sufficient condition that* $F_x dx + F_y dy + F_z dz$ *be an exact differential is that* $\boldsymbol{\nabla} \times \boldsymbol{F} = 0$, *where* $\boldsymbol{F} = \boldsymbol{i}F_x + \boldsymbol{j}F_y + \boldsymbol{k}F_z$.

Solution. Let there exists a function ϕ such that

$$F_x dx + F_y dy + F_z dz = d\phi \tag{1}$$

But

$$\frac{\partial \phi}{\partial x}dx + \frac{\partial \phi}{\partial y}dy + \frac{\partial \phi}{\partial z}dz = d\phi \tag{2}$$

Comparing Eqns (1) and (2), we get

$$\left.\begin{aligned} F_x &= \frac{\partial \phi}{\partial x} \\ F_y &= \frac{\partial \phi}{\partial y} \\ F_z &= \frac{\partial \phi}{\partial z} \end{aligned}\right\} \tag{3}$$

Hence,

$$\boldsymbol{F} = \boldsymbol{i}F_x + \boldsymbol{j}F_y + \boldsymbol{k}F_z = \boldsymbol{i}\frac{\partial \phi}{\partial x} + \boldsymbol{j}\frac{\partial \phi}{\partial y} + \boldsymbol{k}\frac{\partial \phi}{\partial z} = \boldsymbol{\nabla}\phi \tag{4}$$

Thus,

$$\boldsymbol{\nabla} \times \boldsymbol{F} = \boldsymbol{\nabla} \times \boldsymbol{\nabla}\phi = 0 \tag{5}$$

Conversely, if $\boldsymbol{\nabla} \times \boldsymbol{F} = 0$, then we can express $\boldsymbol{F}$ in terms of a scalar function ϕ in the form

$$\boldsymbol{F} = \boldsymbol{\nabla}\phi \tag{6}$$

Hence,

$$\boldsymbol{F} \cdot d\boldsymbol{r} = \boldsymbol{\nabla}\phi \cdot d\boldsymbol{r}$$

$$\text{i.e., } (\boldsymbol{i}F_x + \boldsymbol{j}F_y + \boldsymbol{k}F_z) \cdot (\boldsymbol{i}dx + \boldsymbol{j}dy + \boldsymbol{k}dz) = \left(\boldsymbol{i}\frac{\partial \phi}{\partial x} + \boldsymbol{j}\frac{\partial \phi}{\partial y} + \boldsymbol{k}\frac{\partial \phi}{\partial z}\right) \cdot (\boldsymbol{i}dx + \boldsymbol{j}dy + \boldsymbol{k}dz)$$

$$\begin{aligned} \text{i.e., } F_x dx + F_y dy + F_z dz &= \frac{\partial \phi}{\partial x}dx + \frac{\partial \phi}{\partial y}dy + \frac{\partial \phi}{\partial z}dz \\ &= d\phi \end{aligned}$$

Thus, $F_x dx + F_y dy + F_z dz$ is an exact differential.

EXAMPLE L.3 *If $\boldsymbol{F} = r^n \boldsymbol{r}$, where $r^2 = x^2 + y^2 + z^2$, find the condition when $\boldsymbol{F}$ is solenoidal.*

Solution. We are given

$$\boldsymbol{F} = r^n(\boldsymbol{i}x + \boldsymbol{j}y + \boldsymbol{k}z) = \boldsymbol{i}F_x + \boldsymbol{j}F_y + \boldsymbol{k}F_z, \tag{1}$$

where

$$F_x = xr^n, F_y = yr^n, F_z = zr^n \tag{2}$$

For $\boldsymbol{F}$ to be solenoidal, we must have

$$\text{div } \boldsymbol{F} = 0, \text{ i.e., } \frac{\partial F_x}{\partial x} + \frac{\partial F_y}{\partial y} + \frac{\partial F_z}{\partial z} = 0 \tag{3}$$

Since $r^2 = x^2 + y^2 + z^2$, we have

$$\frac{\partial r}{\partial x} = \frac{x}{r}, \frac{\partial r}{\partial y} = \frac{y}{r}, \frac{\partial r}{\partial z} = \frac{z}{r} \tag{4}$$

From Eqn (2), we have

$$\frac{\partial F_x}{\partial x} = r^n + nxr^{n-1} \cdot \frac{x}{r} = r^n + nr^{n-2}x^2$$

Similarly,

$$\frac{\partial F_y}{\partial y} = r^n + nr^{n-2}y^2$$

$$\frac{\partial F_z}{\partial z} = r^n + nr^{n-2} \cdot z^2$$

Putting these values in Eqn (3), we find

$$3r^n + nr^{n-2}(x^2 + y^2 + z^2) = 0, \text{ i.e., } (3+n)r^n = 0,$$

which yields $n = -3$ for the vector $\boldsymbol{F}$ to be solenoidal.

EXAMPLE L.4 *Determine the constants a, b, c so that the vector*

$$\boldsymbol{F} = \boldsymbol{i}(2x + 3y + az) + \boldsymbol{j}(bx + 2y + 3z) + \boldsymbol{k}(2x + cy + 3z)$$

becomes irrotational.

Solution. The vector $\boldsymbol{F}$ becomes irrotational when curl $\boldsymbol{F} = 0$. (1)

Thus, in the present case

$$\begin{vmatrix} \boldsymbol{i} & \boldsymbol{j} & \boldsymbol{k} \\ \dfrac{\partial}{\partial x} & \dfrac{\partial}{\partial y} & \dfrac{\partial}{\partial z} \\ 2x + 3y + az & bx + 2y + 3z & 2x + cy + 3z \end{vmatrix} = 0$$

$$\text{or, } \boldsymbol{i}(c-3) + \boldsymbol{j}(a-2) + \boldsymbol{k}(b-3) = 0$$

This yields $a = 2, b = 3, c = 3$.

EXAMPLE L.5 *Find the directional derivative of the function $\phi = xy + yz + zx$ at the point $(3, 1, 2)$ in the direction of the vector $2\boldsymbol{i} + 3\boldsymbol{j} + 6\boldsymbol{k}$.*

Solution. Unit vector in the direction of the given vector is

$$\boldsymbol{n} = \frac{2\boldsymbol{i} + 3\boldsymbol{j} + 6\boldsymbol{k}}{\sqrt{4 + 9 + 36}} = \frac{2}{7}\boldsymbol{i} + \frac{3}{7}\boldsymbol{j} + \frac{6}{7}\boldsymbol{k} \tag{1}$$

Since $\phi = xy + yz + zx$, we have

$$\text{grad } \phi = \boldsymbol{i}\frac{\partial \phi}{\partial x} + \boldsymbol{j}\frac{\partial \phi}{\partial y} + \boldsymbol{k}\frac{\partial \phi}{\partial z} = \boldsymbol{i}(y + z) + \boldsymbol{j}(z + x) + \boldsymbol{k}(x + y) \tag{2}$$

At the point $(3, 1, 2)$,

$$\text{grad } \phi = 3\boldsymbol{i} + 5\boldsymbol{j} + 4\boldsymbol{k} \tag{3}$$

Hence the required directional derivative

$$= \boldsymbol{n} \cdot \text{grad } \phi$$

$$= \left(\frac{2}{7}\boldsymbol{i} + \frac{3}{7}\boldsymbol{j} + \frac{6}{7}\boldsymbol{k}\right) \cdot (3\boldsymbol{i} + 5\boldsymbol{j} + 4\boldsymbol{k})$$

$$= \frac{6}{7} + \frac{15}{7} + \frac{24}{7} = \frac{45}{7} = 6\frac{3}{7}$$

EXAMPLE L.6 *Find the unit vector normal to the surface $x^2 + 2y^2 + z^2 = 7$ at the point* $(1, -1, 2)$.

Solution. The unit vector $\boldsymbol{n}$ normal to the surface $\phi(x, y, z) =$ constant is given by

$$\boldsymbol{n} = \frac{\text{grad } \phi}{|\text{ grad } \phi|} \tag{1}$$

Here $\phi = x^2 + 2y^2 + z^2 - 7$ so that

$$\text{grad } \phi = \boldsymbol{i}\frac{\partial \phi}{\partial x} + \boldsymbol{j}\frac{\partial \phi}{\partial y} + \boldsymbol{k}\frac{\partial \phi}{\partial z} = \boldsymbol{i}2x + \boldsymbol{j}4y + \boldsymbol{k}2z \tag{2}$$

At the point $(1, -1, 2)$,

$$\text{grad } \phi = 2\boldsymbol{i} - 4\boldsymbol{j} + 4\boldsymbol{k} \tag{3}$$

Hence,

$$|\text{ grad } \phi| = \sqrt{4 + 16 + 16} = 6 \tag{4}$$

Thus, the unit vector normal to the given surface at the given point is

$$\boldsymbol{n} = \frac{1}{6}(2\boldsymbol{i} - 4\boldsymbol{j} + 4\boldsymbol{k}) = \frac{1}{3}(\boldsymbol{i} - 2\boldsymbol{j} + 2\boldsymbol{k})$$

EXAMPLE L.7 *Find* $\int_C \boldsymbol{F} \cdot d\boldsymbol{r}$, *where*

$$\boldsymbol{F} = \boldsymbol{i}yz + \boldsymbol{j}zx^2 - \boldsymbol{k}xy$$

and the curve C is given by $x = t, y = t^2, z = t^3$ from $P(0, 0, 0)$ to $Q(2, 4, 8)$.

Solution. The radius vector of any point on the curve is

$$\boldsymbol{r} = \boldsymbol{i}x + \boldsymbol{j}y + \boldsymbol{k}z = \boldsymbol{i}t + \boldsymbol{j}t^2 + \boldsymbol{k}t^3 \tag{1}$$

$$\therefore \quad \frac{d\boldsymbol{r}}{dt} = \boldsymbol{i} + \boldsymbol{j}2t + \boldsymbol{k}3t^2 \tag{2}$$

Here t varies from 0 to 2.
Hence,

$$\boldsymbol{F} = \boldsymbol{i}yz + \boldsymbol{j}zx^2 - \boldsymbol{k}xy = \boldsymbol{i}t^5 + \boldsymbol{j}t^5 - \boldsymbol{k}t^3 \tag{3}$$

Then

$$\begin{aligned} \boldsymbol{F} \cdot d\boldsymbol{r} &= \boldsymbol{F} \cdot \frac{d\boldsymbol{r}}{dt} dt \\ &= (\boldsymbol{i}t^5 + \boldsymbol{j}t^5 - \boldsymbol{k}t^3) \cdot (\boldsymbol{i} + \boldsymbol{j}2t + \boldsymbol{k}3t^2)dt \\ &= (t^5 + 2t^6 - 3t^5)dt = 2(t^6 - t^5)dt \end{aligned} \tag{4}$$

Thus

$$\int_C \boldsymbol{F} \cdot d\boldsymbol{r} = \int_0^2 2(t^6 - t^5)dt = 2\left(\frac{t^7}{7} - \frac{t^6}{6}\right)\Bigg|_0^2 = 2 \cdot 2^6 \left(\frac{2}{7} - \frac{1}{6}\right) = 2^6 \cdot \frac{5}{21}$$

EXAMPLE L.8 *Find the work done in moving a particle in the xy-plane once around a circle of radius 2 units with centre at the origin, if the force field is given by*

$$\boldsymbol{F} = \boldsymbol{i}(2x - y + 2z) + \boldsymbol{j}(x + y - z^2) + \boldsymbol{k}(3x - 2y - 5z)$$

Solution. Equation of the given circle is

$$x^2 + y^2 = 4 \tag{1}$$

Its parametric form is

$$x = 2\cos\theta,\ y = 2\sin\theta \tag{2}$$

Then radius vector of any point on the curve

$$\boldsymbol{r} = \boldsymbol{i}x + \boldsymbol{j}y = 2(\boldsymbol{i}\cos\theta + \boldsymbol{j}\sin\theta) \tag{3}$$

$$\therefore\ d\boldsymbol{r} = 2(-\boldsymbol{i}\sin\theta + \boldsymbol{j}\cos\theta)d\theta \tag{4}$$

The force field is given as

$$\boldsymbol{F} = \boldsymbol{i}(2x - y + 2z) + \boldsymbol{j}(x + y - z^2) + \boldsymbol{k}(3x - 2y - 5z)$$

On the xy-plane, $z = 0$. Then the force field active on the xy-plane is

$$\begin{aligned}\boldsymbol{F} &= \boldsymbol{i}(2x - y) + \boldsymbol{j}(x + y) + \boldsymbol{k}(3x - 2y)\\ &= \boldsymbol{i}(4\cos\theta - 2\sin\theta) + \boldsymbol{j}(2\cos\theta + 2\sin\theta) + \boldsymbol{k}(6\cos\theta - 4\sin\theta)\end{aligned} \tag{5}$$

From Eqns (4) and (5),

$$\begin{aligned}\boldsymbol{F}\cdot d\boldsymbol{r} &= [-2\sin\theta(4\cos\theta - 2\sin\theta) + 2\cos\theta(2\cos\theta + 2\sin\theta)]d\theta\\ &= (4 - 2\sin 2\theta))d\theta\end{aligned} \tag{6}$$

The work done is

$$\int \boldsymbol{F}\cdot d\boldsymbol{r} = \int_0^{2\pi} (4 - 2\sin 2\theta)d\theta = \left[4\theta + 2\frac{\cos 2\theta}{2}\right]_0^{2\pi} = 8\pi$$

L.15 Normal Surface Integral

We consider a continuous vector point function $\boldsymbol{F}$ and a smooth surface S.

We subdivide the area S into m elements of area δS_p, where $p = 1, 2, \ldots, m$. We take a point P_p on the area δS_p whose coordinates are (x_p, y_p, z_p).

Defining

$$\boldsymbol{F}_p = \boldsymbol{F}(x_p, y_p, z_p),$$

we form the sum

$$\sum_{p=1}^{m} \boldsymbol{F}_p \cdot \boldsymbol{n}_p \delta S_p, \tag{L.72}$$

where $\boldsymbol{F}_p \cdot \boldsymbol{n}_p$ denotes the normal component of $\boldsymbol{F}_p$ at the point P_p.

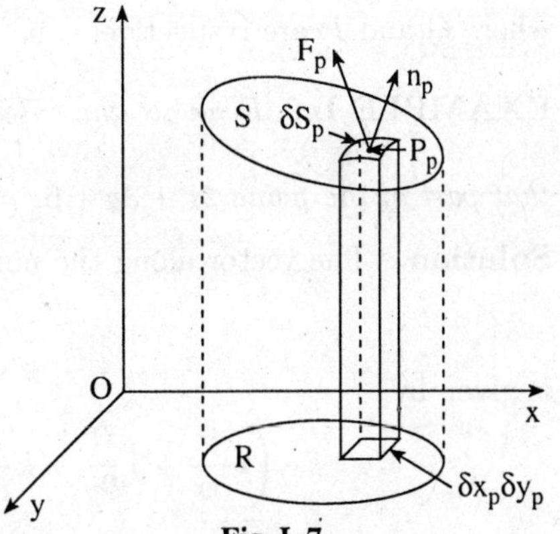

Fig. L.7

Now we take the limit of the sum (L.72) as $m \to \infty$ in such a way that the largest dimension of δS_p approaches zero. This limit, if it exists, is called as the normal surface integral of $\boldsymbol{F}$ over S and is denoted by

$$\iint_S \boldsymbol{F}\cdot\boldsymbol{n}dS \tag{L.73}$$

Now, $\boldsymbol{n}_p\delta S_p$ is the vector area of δS_p and hence its projection on xy-plane whose unit normal is $\boldsymbol{k}$, is

$$\boldsymbol{n}_p\delta S_p \cdot \boldsymbol{k} = (\boldsymbol{n}_p \cdot \boldsymbol{k})\delta S_p \tag{L.74}$$

But the projection of elementary area δS_p on xy-plane is $\delta x_p \delta y_p$. Hence,

$$(\boldsymbol{n}_p \cdot \boldsymbol{k})\delta S_p = \delta x_p \delta y_p, \text{ i.e., } \delta S_p = \frac{\delta x_p \delta y_p}{\boldsymbol{n}_p \cdot \boldsymbol{k}}. \tag{L.75}$$

Thus the sum (L.72) becomes

$$\sum_{p=1}^{m} \boldsymbol{F}_p \cdot \boldsymbol{n}_p \frac{\delta x_p \delta y_p}{\boldsymbol{n}_p \cdot \boldsymbol{k}} \tag{L.76}$$

The limit of this sum as $m \to \infty$ in such a manner that the largest $\delta x_p \delta y_p$ approach zero is

$$\iint_R \boldsymbol{F} \cdot \boldsymbol{n} \frac{dxdy}{|\boldsymbol{n} \cdot \boldsymbol{k}|} \tag{L.77}$$

Thus we have shown that

$$\iint_S \boldsymbol{F} \cdot \boldsymbol{n} dS = \iint_R \boldsymbol{F} \cdot \boldsymbol{n} \frac{dxdy}{|\boldsymbol{n} \cdot \boldsymbol{k}|}, \tag{L.78}$$

where R is the projection of the area S on xy-plane.

Similarly, we can prove that

$$\iint_S \boldsymbol{F} \cdot \boldsymbol{n} dS = \iint_Q \boldsymbol{F} \cdot \boldsymbol{n} \frac{dydz}{|\boldsymbol{n} \cdot \boldsymbol{i}|} \tag{L.79}$$

$$\text{and} \quad \iint_S \boldsymbol{F} \cdot \boldsymbol{n} dS = \iint_P \boldsymbol{F} \cdot \boldsymbol{n} \frac{dzdx}{|\boldsymbol{n} \cdot \boldsymbol{j}|}, \tag{L.80}$$

where Q and P are respectively the projections of the surface S on yz- and zx-planes.

EXAMPLE L.9 *Evaluate the integral* $\iint_S \boldsymbol{F} \cdot \boldsymbol{n} dS$, *where* $\boldsymbol{F} = 18z\boldsymbol{i} - 12\boldsymbol{j} + 3y\boldsymbol{k}$, *and* S *is that part of the plane* $2x + 3y + 6z = 12$ *which is located in the first octant.*

Solution. The vector along the normal to the surface

$$2x + 3y + 6z = 12 \tag{1}$$

is given by

$$\left(\boldsymbol{i}\frac{\partial}{\partial x} + \boldsymbol{j}\frac{\partial}{\partial y} + \boldsymbol{k}\frac{\partial}{\partial z}\right)(2x + 3y + 6z - 12) = 2\boldsymbol{i} + 3\boldsymbol{j} + 6\boldsymbol{k} \tag{2}$$

Hence unit normal vector to the surface (1)

$$\boldsymbol{n} = \frac{2\boldsymbol{i} + 3\boldsymbol{j} + 6\boldsymbol{k}}{\sqrt{4 + 9 + 36}} = \frac{2}{7}\boldsymbol{i} + \frac{3}{7}\boldsymbol{j} + \frac{6}{7}\boldsymbol{k} \tag{3}$$

Then

$$\boldsymbol{F} \cdot \boldsymbol{n} = (18z\boldsymbol{i} - 12\boldsymbol{j} + 3y\boldsymbol{k}) \cdot \frac{1}{7}(2\boldsymbol{i} + 3\boldsymbol{j} + 6\boldsymbol{k}) = \frac{1}{7}(36z - 36 + 18y) \tag{4}$$

Since $2x + 3y + 6z = 12$, we have

$$(3y + 6z) = 12 - 2x \tag{5}$$

Using this relation in Eqn (4), we get

$$\boldsymbol{F}\cdot\boldsymbol{n}=\frac{6}{7}(3y+6z-6)=\frac{6}{7}(12-2x-6)=\frac{6}{7}(6-2x) \qquad (6)$$

Also

$$\boldsymbol{n}\cdot\boldsymbol{k}=\frac{1}{7}(2\boldsymbol{i}+3\boldsymbol{j}+6\boldsymbol{k})\cdot\boldsymbol{k}=\frac{6}{7} \qquad (7)$$

Hence

$$\iint_S \boldsymbol{F}\cdot\boldsymbol{n}dS=\iint_R \boldsymbol{F}\cdot\boldsymbol{n}\frac{dxdy}{\boldsymbol{n}\cdot\boldsymbol{k}}=\iint_R (6-2x)dxdy, \qquad (8)$$

where R is the projection of S on xy-plane.

Putting $z=0$ in Eqn (1), we get

$$2x+3y=12, \qquad (9)$$

which shows that in the integral (8), y varies from 0 to $(12-2x)/3$ and x varies from 0 to 6.

Thus

$$\iint_S \boldsymbol{F}\cdot\boldsymbol{n}dS=\int_0^6\int_0^y (6-2x)dxdy=\int_0^6 (6-2x)\frac{(12-2x)}{3}dx$$

$$=\frac{1}{3}\int_0^6 (72-36x+4x^2)dx=\frac{1}{3}\left[72x-36\frac{x^2}{2}+4\frac{x^3}{3}\right]_0^6=24$$

L.16 Space Curves

Let C be a space curve defined by the function

$$\boldsymbol{r}=\boldsymbol{r}(s), \qquad (L.81)$$

where s denotes the arc length measured from some fixed point on C. Then the unit tangent vector to the curve C denoted by $\boldsymbol{T}$ is

$$\boldsymbol{T}=\frac{d\boldsymbol{r}}{ds} \qquad (L.82)$$

Since

$$\boldsymbol{T}\cdot\boldsymbol{T}=1,$$

we have

$$\boldsymbol{T}\cdot\frac{d\boldsymbol{T}}{ds}=0 \qquad (L.83)$$

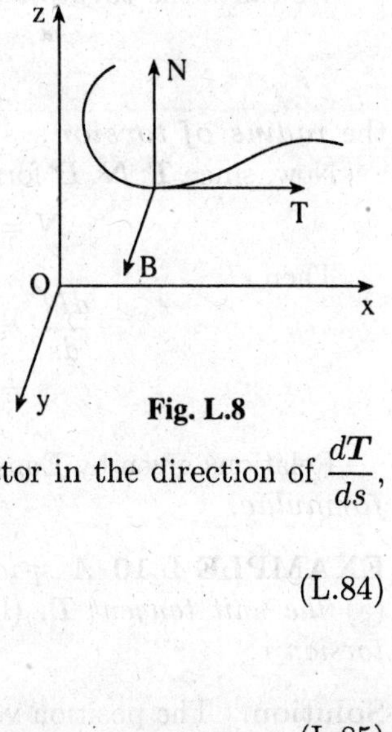

Fig. L.8

This implies that $\frac{d\boldsymbol{T}}{ds}$ is perpendicular to $\boldsymbol{T}$. If $\boldsymbol{N}$ is a unit vector in the direction of $\frac{d\boldsymbol{T}}{ds}$, then we can write,

$$\frac{d\boldsymbol{T}}{ds}=k\boldsymbol{N} \qquad (L.84)$$

We call $\boldsymbol{N}$ the principal normal, k the curvature and

$$\rho=\frac{1}{k}, \qquad (L.85)$$

the radius of curvature.

A unit vector $\boldsymbol{B}$ perpendicular to the plane containing $\boldsymbol{T}$ and $\boldsymbol{N}$ such that

$$\boldsymbol{B} = \boldsymbol{T} \times \boldsymbol{N} \tag{L.86}$$

is called the ***binormal to the curve***. It is clear that the directions of $\boldsymbol{T}, \boldsymbol{N}$ and $\boldsymbol{B}$ form a local right-handed rectangular coordinate system (FIG. L.8).

Since $\boldsymbol{B} \cdot \boldsymbol{B} = 1$, it follows that

$$\boldsymbol{B} \cdot \frac{d\boldsymbol{B}}{ds} = 0 \tag{L.87}$$

Thus $\dfrac{d\boldsymbol{B}}{ds}$ is perpendicular to $\boldsymbol{B}$ and hence lies on the plane containing $\boldsymbol{T}$ and $\boldsymbol{N}$.

Again, we have from Eqn (L.86)

$$\begin{aligned} \frac{d\boldsymbol{B}}{ds} &= \boldsymbol{T} \times \frac{d\boldsymbol{N}}{ds} + \frac{d\boldsymbol{T}}{ds} \times \boldsymbol{N} \\ &= \boldsymbol{T} \times \frac{d\boldsymbol{N}}{ds} + k\boldsymbol{N} \times \boldsymbol{N} \text{ [using (L.84)]} \\ &= \boldsymbol{T} \times \frac{d\boldsymbol{N}}{ds} \end{aligned} \tag{L.88}$$

Hence,

$$\boldsymbol{T} \cdot \frac{d\boldsymbol{B}}{ds} = \boldsymbol{T} \cdot \boldsymbol{T} \times \frac{d\boldsymbol{N}}{ds} = 0, \tag{L.89}$$

which implies that $\boldsymbol{T}$ is perpendicular to $\dfrac{d\boldsymbol{B}}{ds}$ lying on the plane containing $\boldsymbol{T}$ and $\boldsymbol{N}$.

Hence we can write,

$$\frac{d\boldsymbol{B}}{ds} = -\tau \boldsymbol{N} \tag{L.90}$$

We call τ the ***torsion of the curve*** and

$$\delta = \frac{1}{\tau}, \tag{L.91}$$

the ***radius of torsion***.

Now, since $\boldsymbol{T}, \boldsymbol{N}, \boldsymbol{B}$ form a right-handed system we have

$$\boldsymbol{N} = \boldsymbol{B} \times \boldsymbol{T} \text{ and } \boldsymbol{T} = \boldsymbol{N} \times \boldsymbol{B} \tag{L.92}$$

Then

$$\begin{aligned} \frac{d\boldsymbol{N}}{ds} &= \boldsymbol{B} \times \frac{d\boldsymbol{T}}{ds} + \frac{d\boldsymbol{B}}{ds} \times T \\ &= \boldsymbol{B} \times k\boldsymbol{N} - \tau \boldsymbol{N} \times \boldsymbol{T} = -k\boldsymbol{T} + \tau \boldsymbol{B} = \tau \boldsymbol{B} - k\boldsymbol{T} \end{aligned} \tag{L.93}$$

Relations given by Eqns (L.84), (L.90) and (L.93) are collectively known as ***Frenet-Serret formulae***.

EXAMPLE L.10 *A space curve is defined by* $x = 3\cos t, y = 3\sin t, z = 4t$. *Determine* (a) *the unit tangent* $\boldsymbol{T}$, (b) *the principal normal* $\boldsymbol{N}$, *curvature* k, (c) *the binormal* $\boldsymbol{B}$ *and torsion* τ.

Solution. The position vector of any point on the curve is

$$\boldsymbol{r} = \boldsymbol{i}(3\cos t) + \boldsymbol{j}(3\sin t) + \boldsymbol{k}(4t)$$

$$\therefore \frac{d\boldsymbol{r}}{dt} = \boldsymbol{i}(-3\sin t) + \boldsymbol{j}(3\cos t) + 4\boldsymbol{k} \tag{1}$$

Hence,

$$\frac{ds}{dt} = \left|\frac{d\boldsymbol{r}}{dt}\right| = \sqrt{\frac{d\boldsymbol{r}}{dt} \cdot \frac{d\boldsymbol{r}}{dt}}$$

$$\text{i.e.,} \quad \frac{ds}{dt} = \sqrt{(-3\sin t)^2 + (3\cos t)^2 + 4^2} = 5 \tag{2}$$

Then

$$\boldsymbol{T} = \frac{d\boldsymbol{r}}{ds} = \frac{d\boldsymbol{r}}{dt} \Big/ \frac{ds}{dt} = -\boldsymbol{i}\left(\frac{3}{5}\sin t\right) + \boldsymbol{j}\left(\frac{3}{5}\cos t\right) + \frac{4}{5}\boldsymbol{k} \tag{3}$$

Hence,

$$\frac{d\boldsymbol{T}}{dt} = -\boldsymbol{i}\left(\frac{3}{5}\cos t\right) - \boldsymbol{j}\left(\frac{3}{5}\sin t\right) \tag{4}$$

Then

$$\frac{d\boldsymbol{T}}{ds} = \frac{d\boldsymbol{T}}{dt} \Big/ \frac{ds}{dt} = -\boldsymbol{i}\left(\frac{3}{25}\cos t\right) - \boldsymbol{j}\left(\frac{3}{25}\sin t\right) \tag{5}$$

Hence,

$$\left|\frac{d\boldsymbol{T}}{ds}\right| = \sqrt{\left(-\frac{3}{25}\cos t\right)^2 + \left(-\frac{3}{25}\sin t\right)^2} = \frac{3}{25} \tag{6}$$

Since $\frac{d\boldsymbol{T}}{ds} = k\boldsymbol{N}$, we have

$$\left|\frac{d\boldsymbol{T}}{ds}\right| = |k||\boldsymbol{N}| = k \tag{7}$$

Thus, from Eqns (6) and (7), we get

$$k = \frac{3}{25}, \quad \therefore \rho = \frac{25}{3} \tag{8}$$

Again, the principal normal $\boldsymbol{N}$ is given by

$$\boldsymbol{N} = \frac{1}{k}\frac{d\boldsymbol{T}}{ds} = \frac{25}{3}\left[-\boldsymbol{i}\left(\frac{3}{25}\cos t\right) - \boldsymbol{j}\left(\frac{3}{25}\sin t\right)\right] = -\boldsymbol{i}\cos t - \boldsymbol{j}\sin t \tag{9}$$

The binormal $\boldsymbol{B}$ is given by

$$\boldsymbol{B} = \boldsymbol{T} \times \boldsymbol{N} = \begin{vmatrix} \boldsymbol{i} & \boldsymbol{j} & \boldsymbol{k} \\ -\frac{3}{5}\sin t & \frac{3}{5}\cos t & \frac{4}{5} \\ -\cos t & -\sin t & 0 \end{vmatrix}$$

$$= \boldsymbol{i}\left(\frac{4}{5}\sin t\right) - \boldsymbol{j}\left(\frac{4}{5}\cos t\right) + \frac{3}{5}\boldsymbol{k} \tag{10}$$

Hence,

$$\frac{d\boldsymbol{B}}{dt} = \boldsymbol{i}\left(\frac{4}{5}\cos t\right) + \boldsymbol{j}\left(\frac{4}{5}\sin t\right)$$

Then

$$\frac{d\boldsymbol{B}}{ds} = \frac{d\boldsymbol{B}}{dt} \Big/ \frac{ds}{dt} = \boldsymbol{i}\left(\frac{4}{25}\cos t\right) + \boldsymbol{j}\left(\frac{4}{25}\sin t\right) \tag{11}$$

Using Eqn (9),

$$-\tau \boldsymbol{N} = \tau(\boldsymbol{i} \cos t + \boldsymbol{j} \sin t) \tag{12}$$

Since $\dfrac{d\boldsymbol{B}}{ds} = -\tau \boldsymbol{N}$, we get from Eqns (11) and (12)

$$\tau(\boldsymbol{i} \cos t + \boldsymbol{j} \sin t) = \frac{4}{25}(\boldsymbol{i} \cos t + \boldsymbol{j} \sin t)$$

This yields

$$\tau = \frac{4}{25}, \ \therefore \delta = \frac{1}{\tau} = \frac{25}{4}.$$

Bibliography

[1] Besant, W.H. and Ramsey, A.S., *A Treatise on Hydro-mechanics*, G. Bell and Sons, Ltd., London, 1962.

[2] Routh, E.J., *An Analytical Statics* (Vol. II), Cambridge University Press, 1928.

[3] Ramsey, A.S., *An Introduction to the Theory of Newtonian Attraction*, Cambridge University Press, 1956.

[4] Mostkov, M.A., *Hydraulics, Gocydarstbennoe Transportnoe Jheleznodarojhnoe Izdatelsbo* (in Russian), Moscow, 1958.

[5] Bhattacharyya, B., *Fluid Dynamics*, New Central Book Agency (P) Ltd, Kolkata, 2009.

[6] Kar, J.M., *Hydrostatics*, K.P. Basu Publishing Co., Kolkata, 2007.

[7] Lewitt, E.H., *Hydraulics and Fluid Mechanics*, Sir Isaac Pitman & Sons Ltd., London, 1962.

Index

Some exciting gifts are up for grabs with this book!!

HYDROSTATICS

Dr Binoy Bhattacharyya

FREE Gifts Rs. 1000/-

Dear Readers

By now you must have gone through this book in the entirety. So we invite your assessment on this title with absolute impartiality and candid observations. After all, we publishers blindly depend on your benchmarking. Friends, please fill up the open spaces below with CAPITAL LETTERS. You may add some extra sheets of paper if the space provided is not sufficient for you.

i. What are the important topics of your syllabus which are not adequately covered in this book?

ii. In which chapters and on what topics the treatment of the subject-matter by the author is not satisfactory and why?

iii. Have you come across the typos, factual inaccuracies, inconsistencies, and other forms of mistakes in the book? If yes, then please specify the mistakes and page numbers.

iv. Name top three books on the same subject, which, in your consideration, are better than this book. Could you provide some reasons?

Your 1st Preference: ....

Your 2nd Preference: ....

Your 3rd Preference: ....

v. Any suggestion to improve the quality of illustrations, diagrams, etc.?

vi. How did you like the cover design? Any suggestion on colours?

vii. Suggestions for further improvement of production-standard:

viii. Who recommended you this book:

☐ Teacher ☐ Friend ☐ Parents ☐ Bookseller ☐ Librarian

ix. Name of the person who has recommended:

Designation: ....

Address: ....

Phone No.: ....

e-mail: ....

x. Name and address of the bookseller from whom you have purchased this book:

....

....

....

xi. Name and address of the library where you would like to find the book:

....

....

xii. Name and address of your institution (Please mention the University or Board):

....

....

xiii. Your name, e-mail, and complete postal address:

....

....

....

xiv. A few other titles published by us that you would like to read:

....

....

The best assessment will be rewarded half-yearly. The award will be in the form of our publications, as decided by the Editorial and Marketing Operations Division, amounting to Rs. 1000/- (Rupees one thousand) only.

....

Please mail the filled up coupon at your earliest to :

Editorial & Marketing Operations Division

NEW CENTRAL BOOK AGENCY (P) Ltd

8/1 Chintamoni Das Lane, Kolkata 700 009

West Bengal, INDIA

Phone + Fax: 033 + 2241 8411, 2241 8573, 2241 9094

e-mail: ncbapvtltd@eth.net, ncba@cal2.vsnl.net.in

Website: www.centralgroup.com